BANDICOOTS and BILBIES

BANDICOOTS
and
BILBIES

Edited by

J. H. Seebeck, P. R. Brown, R. L. Wallis
and C. M. Kemper

Published and Printed by

Surrey Beatty & Sons Pty Limited

In association with

The Australian Mammal Society

The National Library of Australia
Cataloguing-in-Publication entry:

Bandicoots and Bilbies.

Bibliography.
Includes index.
ISBN 0 949324 33 7.

1. Peramelidae. I. Seebeck, J. H. (John H.). II. Australian Mammal Society.

599.2

Published December, 1990

PRINTED AND PUBLISHED IN AUSTRALIA BY
SURREY BEATTY & SONS PTY LIMITED
43 Rickard Road, Chipping Norton, NSW 2170

Peroryctes raffrayanus, Papua New Guinea, June 1986. Photo: T. F. Flannery.

Foreword

The true or original bandicoots are large, rat-like rodents from the southern Indian region. That perameloids are still referred to as bandicoots is indicative of a lack of public interest in, or affection for, these marsupials; indeed, "bandicoot", remains a derogatory term in the Australian vernacular. Bandicoots have not fared much better in terms of scientific attention; in comparison with kangaroos, possums and dasyurids, little research has been conducted on them, possibly because we are not sure what sort of marsupials they are and also, perhaps, because of a feeling that all bandicoots are much the same — and therefore rather boring. I shall comment briefly upon these two points.

With knowledge of all of the currently recognized genera and of seven of the 11 Australian species, Waterhouse erected the family Peramelidae in 1841. In 1846 he made a systematic revision of the family, which included the following observations:

> In the structure of the hindfeet, the Peramelidae greatly resembles the Kangaroos . . . however the *Perameles'* foot . . . is proportionately shorter and the fourth . . . toe is less developed . . . I strongly suspect that they do not progress by leaps . . . [In *Chaeropus*] the forefeet differ from those of the Kangaroos in having the outer toes rudimentary . . . The Peramelidae may readily be distinguished by the number of their incisor teeth, of which there are ten in the upper and six in the lower jaw . . . The true molar teeth . . . are adapted to insect diet, though, according to the accounts of good observers residing in Australia, they will likewise eat vegetable substances.

Waterhouse recognized that the polyprotodont condition of didelphoids, dasyuroids and perameloids was very different from the diprotodont condition of the phalangeroids, vombatoids and macropodoids. However, his classification of the marsupials was in terms of the structure of the hindfoot. He separated the Dasyuridae, Didelphidae and *Myrmecobius* from the bandicoots, possums and kangaroos, in which the second and third toes are "slender and united in a common ligament" [syndactyly]. On this basis, he regarded the perameloids as having been derived from a *Phascogale*-like ancestor and themselves being ancestral to the macropodoids.

Some 40 years later, Oldfield Thomas (1888) reached the conclusion that a suite of dental and cranial characters linked the dasyuroids and perameloids so intimately that syndactyly must have evolved independently in the polyprotodont and diprotodont marsupials.

> This family [Peramelidae] is a very distinct and sharply-defined one, its only ally being *Dasyuridae,* from which it differs in many important respects. Owing to the syndactyly of the hind feet of its members, it has been commonly considered as forming a connecting link between the Polyprotodonts and Diprotodonts; but, judging from the wholly Polyprotodont character of the rest of their organization, even down to the structure of their carpal and tarsal bones, all of which show far greater resemblances to the *Dasyuridae* than to the *Phalangeridae,* it seems probable that this presumption is wrong, and that their syndactyly has arisen entirely independently of that of the Diprotodonts, a view under which we should look upon the *Peramelidae* as an offshoot of the *Dasyuridae,* with no special genetic relationship to the Diprotodonts whatever.

> Difficult of acceptance as this view may appear, it is probably the only one which will explain the extraordinary combination of characters presented by the *Peramelidae,* since it seems equally improbable either that the earliest Marsupials could have been syndactylous, or that the *Peramelidae* could have been originally Diprotodont and have independently assumed a Polyprotodont dentition, in default of one or other of which alternatives the view above put forward must be adopted.

In 1903, the pendulum swung back with Bensley's monograph on the evolution of the Australian marsupials. On the basis of an extensive study of tooth morphology and (external) characters of the hind feet, he concluded that:

> The difficulty presented by the Peramelidae, which while essentially polyprotodont possess a syndactylous type of pes otherwise characteristic of the diprotodont series, has already been commented upon by Flower, but the relations of this family have been explained by Thomas on the assumption that their syndactylism has been independently acquired.
>
> The diprotodont modification has obviously been derived from a polyprotodont one; it is traceable to an insectivorous specialization of the median lower incisors. The members of the diprotodont section may be shown to have passed through an omnivorous stage, and to have possessed omnivorous molar characters similar to those of the primitive Peramelidae. The primitive Phalangeridae are to be connected with the Peramelidae not only as syndactylous, but also as omnivorous forms, rather than separated as Diprotodontia, while the Dasyuridae are to be separated as eleutherodactylous ["free-toed"] and progressively carnivorous forms.
>
> In selecting a differential character on which a classification may be based we are obliged to turn to other structures than the dentition . . . Turning to the foot-structures, we find in the respectively eleutherodactylous and syndactylous condition of the pes a differential character of sufficient importance and applicability.

Bensley classified the Australian marsupials into the Diadactyla (dasyuroids) and the Syndactyla (all the others). He shared the view of most workers who have considered the origin of marsupial syndactyly, that the second and third digits of the hindfoot probably became reduced (as in some prosimians) to facilitate a grip between the first digit and the fourth and fifth. As has long been recognized, the opposability of the first pedal digit is plesiomorphic and indicative of the origin of marsupials from (or as) arboreal animals.

What is less clear is why the suppressed digits should have retained strong claws and become syndactylous. This question was tackled by Jones (1924), who argued that, if a primitive didactylous stock had divided into polyprotodont and diprotodont evolutionary lines, we should expect to find (as living or fossil forms) some diprotodont marsupials with simple feet: this is not the case. If, on the other hand, a primitive polyprotodont stock had separated into syndactylous and didactylous lines, we should expect to find polyprotodont didactylous forms (e.g., dasyuroids) and polyprotodont syndactylous forms (e.g., perameloids). Like Bensley, Jones envisaged the perameloids as the sister-group of the diprotodont marsupials, but he rejected the idea that the syndactylous toes are degenerate:

> Plenty of toes have degenerated, and become lost, in the story of the various marsupial groups; but this is the only occasion in which they seem, so to speak, to have become united in death.
>
> Maybe we shall do better to regard these very wonderful toes as forming a highly organized anatomical mechanism specialized as a hair comb . . .
>
> Were the condition of syndactyly to be one of simple degeneration, then it is perhaps not difficult to picture it as affecting the pedal digits of certain members of both series [polyprotodonts and diprotodonts] independently. But if the condition be the result of a very specialized development, begot

> in response to the demand of a very definite function . . . it is sufficiently unlikely to warrant the simple explanation of its only having been developed once . . .

Jones' further suggestion, that a syndactylous grooming comb was developed as a replacement for the combing function of numerous needle-like incisors in ancestral marsupials, carries less conviction but this does not detract from the strength of his argument that, far from being degenerate, the syndactylous toes comprise *a unique functional structure,* maintained in the hindfeet of all bandicoots and macropods despite the reduction or loss of the first digit and an extreme reduction of the fifth.

It is, of course, not difficult to combine the views of Bensley and Jones by postulating that the second and third pedal digits become reduced as an arboreal adaptation in a polyprotodont syndactylous common ancestor of the perameloids and diprotodont marsupials but that, at some stage they were "salvaged" to create a new, functional structure. The former event could well have occurred several times; the latter would seem to have a much lower probability.

In a study of the tarsal region of various Australian marsupials, Marshall (1972) proposed that the perameloids could have evolved from phalangeroids but this possibility raises more problems than it resolves. Szalay (1982) made a much more detailed study of the tarsal region and concluded that:

> The syndactylous condition of the foot . . . appears to be a monophyletic, uniquely shared and derived character complex. There is no known or documented evidence for the hypothesis that it has evolved more than once. Syndactyly probably postdates an adaptive modification in a dromiciopsian species in which the second and third toes became reduced in order to avoid injurious contact with the greatly enlarged and highly mobile big toes when grasping and encircling small terminal branches. The hypothesis of Jones (1924) that syndactyly is a grooming adaptation is probably correct, but it is probably based on a preadapted condition which is the result of a specific grasping adaptation early in an australidelphian lineage leading to the common ancestor of the Syndactyla.

Szalay's classification divides the Marsupialia into the cohorts Ameridelphia (all American taxa except *Dromiciops*) and the Australidelphia (all Australian and Melanesian taxa plus *Dromiciops*). His order Syndactyla includes the notoryctoids, perameloids and all the diprotodont groups. At the end of a long consideration of possible marsupial phylogenics, Archer (1984) summed up his opinions at the time in a "Grand Marsupial Tree" in which the Syndactyla is shown as the sister-group of the Dasyuroidea, and the Perameloidea as the sister-group of the diprotodont marsupials.

Szalay's concept of a monophyletic Australidelphia remains attractive but Reig, Kirsch and Marshall (1987) strongly contest his view that *Dromiciops* is a member of that group: they maintain that it is a didelphoid ameridelphian. They also offer detailed arguments against his concept of the Syndactyla, claiming that this cannot be supported if, in addition to tarsal morphology, one takes account of cranial and dental characters.

Reasoning from pedal morphology has led, on the whole, to the conclusion that dasyuroids and syndactylous marsupials separated from an ancestral australidelphian stock (perhaps about 50 mya). Two lines then diverged from the early (polyprotodont) syndactyl marsupials: the early perameloids and the early diprotodont marsupials. Previous strong arguments for a sister-group relationship between the perameloids and dasyuroids are now seen to have been based largely on symplesiomorphic or convergent characters which shed little light on relationships. We do not have much to work with.

It would be reassuring if the concept of the Syndactyla were supported by evidence from areas outside gross morphology, but little has been forthcoming. Hughes (1965) showed that the sperm heads of dasyuroids and perameloids show a greater resemblance to each other than to those of other marsupials. On the basis of number and shape of chromosomes, Sharman (1982) suggested that peramelids might share an immediate common ancestor with burramyids, vombatids and *Dromiciops* but the weight of such evidence is not great.

The most telling evidence *against* the Syndactyla as a natural group comes from serology. Kirsch's (1977) findings from cross reactions of polyvalent antisera to a wide variety of marsupials support the monophyly of the Australidelphia (minus the Dromiciopsia) and locate the perameloids as a sister-group of the dasyuroids. Baverstock and his colleagues (this volume) reach much the same conclusion from studies on microcomplement fixation of marsupial albumins. Until the latter work, it was possible to envisage that some quirk of surface chemistry had led Kirsch to an incorrect interpretation of the position of the perameloids but Baverstock's independent confirmation would seem to signal the death of the Syndactyla.

Archer abandoned the Syndactyla in a "new syncretic classification" (Aplin and Archer 1987) which takes account of evidence from morphology and serology and divides the cohort Australidelphia into five orders: Microbiotheria, Dasyuromorphia, Peramelomorphia, Notoryctemorphia and Diprotodonta. In respect of the Peramelomorphia, Aplin and Archer state that:

> While we suspect that on balance the evidence favours a special bandicoot/dasyuromorphian phylogenetic relationship, we have not followed Kirsch . . . in combining these taxa in a single order. The primary reason is that we remain cautious because not all of the data which suggests alliance of bandicoots with another order indicates that this alliance is necessarily with the Dasyuromorphia. Secondly, with Ride . . ., we consider that even were the monophyly eventually confirmed beyond reasonable doubt, the two groups would nevertheless warrant recognition at full ordinal rank. Relevant here is the growing palaeontological awareness of the former adaptive breadth of the bandicoot radiation and awareness of the considerable molecular distance of bandicoots from dasyuromorphians (e.g., Baverstock *et al.* in press) which indicates that bandicoots and dasyuromorphians may well have had an independent history at least as long as that which distinguishes the Diprotodontia.

Over the years, Archer's studies have tended to place the origins of the perameloids closer and closer to the base of the Australidelphian radiation. His remarks on the resemblance of some (as yet undescribed) fossil bandicoots from Riversleigh to dasyuroids *and* to didelphoids suggests the possibility that a perameloid (or proto-perameloid) may have been the common ancestor of the dasyuroids and the diprotodonts but, since the fossil evidence does not include hindfeet, we have no idea whether the stem-perameloid was, or was not, syndactylous.

The most recent commentary on the problem of syndactyly has been made by Hall (1987) in amplification of an earlier brief paper (Hall 1981). He strongly supports Jones' view that syndactyly is a grooming adaptation and suggests that the condition arose as a mutation "followed by intense selection pressure for its retention, possibly for grooming or climbing". In Hall's opinion, syndactyly arose only once in marsupials, probably in South America, but he concludes that "the problem remains unresolved". I would have thought that the possibility of a mutation which greatly reduced the size of the second and third pedal digits (while retaining their intrinsic musculature) provided a credible basis for the dual, or multiple, origin of syndactyly. Indeed, the partial fusion of the second and third toes of my own feet (and those of one of my daughters) are examples of a common mutation for syndactyly in humans. However, the human condition does not involve reduction in size of the digits, nor can a grooming function be attributed to a structure that ends with two nails. Hall has shed little light on the central evolutionary problem and we remain where Thomas left us a century ago, with a logical necessity to postulate at least a dual origin of the syndactylous condition.

Thomas remarked that the Peramelidae (present Perameloidea or Peramelomorphia) is "very distinct and sharply-defined". Probably the most extreme morphological variation among recent forms is seen in *Chaeropus*, which appears to have been almost digitigrade and probably a grazer (Dixon 1988). Yet there has been no difficulty in accommodating this genus with the typical bandicoots in the family Peramelidae. *Macrotis* is also distinctive in appearance and behaviour: on the basis of molar morphology, Bensley (1903) placed the genus in the subfamily Thylacomyinae and, in 1977, Archer and Kirsch raised this taxon to family rank. However, Groves and Flannery (this volume) return all the Australian genera except *Echymipera* to the family Peramelidae, and erect a new family for the predominantly New Guinean genera *Peroryctes, Microperoryctes, Rhynchomeles* and *Echymipera.* If their interpretation is correct, there has been remarkably little evolutionary radiation among the (described) bandicoots.

Cockburn (this volume) asks why this is so — why, for example, there appear to have been no large, omnivorous bandicoots, equivalent to eutherian bears. His interesting answer is based on the postulate that most major evolutionary transformation is the result of heterochrony, changes in timing of the development of organ systems in embryological development. He suggests that, in comparison with eutherians, marsupials have a high level of synchrony in their embryological development and that this has restricted the range of morphological novelty in the latter group. He further proposes that, if this be the case, evolutionary options would have been even more restricted in the bandicoots, which have the most truncated gestation of all mammals. In opting for high fecundity and rapid development, they may have severely restricted their evolutionary opportunities. His hypothesis is highly speculative and will carry little conviction until parallel instances can be cited in other mammalian groups (the murids, perhaps?). It may also be upset when we know more about Archer's discoveries of early perameloids: perhaps the extant bandicoots are a relic group representing only one line of a peramelomorph radiation.

Bandicoots continue to pose fundamental problems. They may, indeed, be descendants of a group close to the base of the Australidelphia but we need far more fossil evidence before this can be established. If we are to gain any evolutionary insights from the relatively similar living species, we must make much more detailed and diverse enquiries than are usually called for in the resolution of the relationships of other marsupial groups. This volume is welcome evidence that such research is being actively pursued.

Since 1982, the Australian Mammal Society and the Royal Zoological Society of New South Wales, working singly or together, have produced, in the same format, *Carnivorous marsupials, Possums and gliders, Possums and opossums* (which is the proceedings of a joint symposium of the mammal societies of Australia and America) and *Kangaroos, wallabies & rat-kangaroos.* With the publication of this volume a grand survey of the biology of the marsupials will have been accomplished in less than a decade. All responsible for this enterprise are to be congratulated but particular praise must go to Michael Archer, who set the project in motion and has been a notable contributor to the series.

Ronald Strahan
The Australian Museum

REFERENCES

APLIN, K. P. AND ARCHER, M., 1987. Recent advances in marsupial systematics with a new syncretic classification. Pp. xv–lxxii in *Possums and opossums* ed by M. Archer. Surrey Beatty & Sons: Sydney.

ARCHER, M., 1984. The Australian marsupial radiation. Pp. 633–808 in *Vertebrate zoogeography and evolution in Australasia* ed by M. Archer and G. Clayton. Hesperian: Perth.

ARCHER, M. AND KIRSCH, J. A. W., 1977. The case for the Thylacomyidae and Myrmecobiidae, Gill 1872, or why are marsupial families so extended? *Proceedings of the Linnean Society of New South Wales* **102:** 18–25.

BENSLEY, A., 1903. On the evolution of the Australian Marsupialia; with remarks on the relationships of the marsupials in general. *Transactions of the Linnean Society of London,* second series (Zool) **9:** 13–217.

DIXON, J. M., 1988. Notes on the diet of three mammals presumed to be extinct: the Pig-footed Bandicoot, the Lesser Bilby and the Desert Rat Kangaroo. *Victorian Naturalist* **105:** 208–11.

HALL, L. S., 1981. The problem with syndactyly in mammals. *Bulletin of the Australian Mammal Society* **7:** 36–37.

HALL, L. S., 1987. Syndactyly in marsupials—problems and prophecies. Pp. 245–55 in *Possums and opossums* ed by M. Archer. Surrey Beatty & Sons: Sydney.

HUGHES, R. L., 1965. Comparative morphology of spermatozoa from five marsupial families. *Australian Journal of Zoology* **13:** 533–43.

JONES, F. W., 1924. *The mammals of South Australia. Part II, The bandicoots and the herbivorous marsupials.* Government Printer: Adelaide.

KIRSCH, J. A. W., 1977. The comparative serology of marsupials and a classification of marsupials. *Australian Journal of Zoology Supplementary Series* **52:** 1–152.

MARSHALL, L. G., 1972. Evolution of the peramelid tarsus. *Proceedings of the Royal Society of Victoria* **85:** 51–60.

REIG, O. A., KIRSCH, J. A. W. AND MARSHALL, L. G., 1987. Systematic relationships of the living and Neocenozoic American "Opossum-like" marsupials. Pp. 1–89 in *Possums and opossums* ed by M. Archer. Surrey Beatty & Sons: Sydney.

SHARMAN, G. B., 1982. Karyotypic similarities between *Dromiciops australis* (Microbiotheriidae) and some Australian marsupials. Pp. 711–14 in *Carnivorous marsupials* ed by M. Archer. Royal Zoological Society of New South Wales: Sydney.

SZALAY, F. S., 1982. A new appraisal of marsupial phylogeny and classification. Pp. 621–40 in *Carnivorous marsupials* ed by M. Archer. Royal Zoological Society of New South Wales: Sydney.

THOMAS, O., 1888. *Catalogue of the Marsupialia and Monotremata in the collection of the British Museum (Natural History).* British Museum: London.

WATERHOUSE, G. T., 1846. *A natural history of the Mammalia. Vol. 1, Marsupialia or pouched mammals.* Baillière: London.

Contents

BEHAVIOUR

ECOLOGY, MANAGEMENT AND CONSERVATION

List of Colour Plates

Introduction and Acknowledgements

Bandicoots not only have the shortest gestation period known among mammals, they are also among the quickest to reach independence and maturity. Regrettably, publication of this book has not emulated the accellerated growth of its subject matter. Conceived in October 1983, parturition did not occur until May 1985, and independance not until now.

The Proceedings of the Symposium on which this volume is based were to be edited jointly by Peter Brown and Gordon Lyne, but after Gordon's untimely death there were some months of delay until John Seebeck and Robert Wallis were appointed co-editors with Peter. Catherine Kemper, as editor of the Society, joined the team for final production.

The 41 papers here presented are arranged in six sections, covering taxonomy and relationships, distribution and status, anatomy and physiology, reproductive biology, behaviour, and ecology, management and conservation. Generally, authors have been free to follow their own writing style, papers have been examined by two referees, and editorial high-handedness has been avoided. We have, however, imposed a consistency in bibliographic style and punctuation.

We have chosen to adopt the familial dichotomy proposed by Groves and Flannery (this volume), and thus we recognize two families within the Perameloidea, the Peramelidae and the Peroryctidae. We have also accepted Groves and Flannery's assignment of some *Peroryctes* to *Microperoryctes.* Australian genera of bandicoots are fairly clearly defined, but at species level the distinctions are often poorly recognized. This has resulted in a multitude of named species and subspecies, many of which are clearly not valid. The decline, and sometimes extinction of many arid and semi-arid zone forms has resulted in a confused taxonomy in most genera, much of which will remain forever unresolved. The taxonomy we have used is, we believe, the best synthesis available at this time. No doubt this statement itself will be a sufficient stimulus for rapid revision!

The Symposium, and the preparation of these proceedings, could not have taken place without the help and co-operation of many people and organizations. Convenor of the Symposium was Peter Brown, and the organizing committee was Graeme Coulson, Graeme George, John Seebeck and Robert Wallis. The University of Melbourne made conference facilities available, the Zoological Board of Victoria provided a fine venue for the Bandicoot Banquet, Rusden Campus of Victoria College supplied willing helpers (and biscuit bakers) and the Arthur Rylah Institute for Environmental Research and its then Director, Dr W. Chamley, provided Peter and John with support and encouragement.

Authors of the many papers deserve sincere thanks, especially because of the delay in publication. So, too, do the following referees for their assistance in getting the final manuscripts to their high standard: John Barnett, John Biggins, Peter Brown, Andrew Cockburn, Pete Copley, David Croft, Joan Dixon, Chris Driver, Tim Flannery, Barry Fox, Tony Friend, Greg Gordon, Michael Graydon, Bob Green, John Haight, Les Hall, Peter Hartman, George Heinsohn, Leon Hughes, Ken Johnson, Daz Kitchener, Tony Lee, Dan Lunney, Ian McDonald, Peter Menkhorst, John Menzies, Neil Murray, John Nelson, Marilyn Renfree,

David Ride, Tony Robinson, Randy Rose, John Seebeck, Lyn Selwood, Geoff Sharman, Bill Sherwin, Andrew Smith, Mike Stoddart, Ray Tedman, Peter Temple-Smith, Andrew Tribe, Steve Van Dyck, Rob Wallis, Chris Watts and Pam Whitely.

J. H. Seebeck
P. R. Brown
R. L. Wallis
C. M. Kemper

Melbourne, July 1989

Gordon Lyne, 1919–1986

This volume is dedicated to Gordon Lyne, Foundation (and Founding) Member of the Australian Mammal Society, for many years the foremost student of Australian bandicoots. He was honorary chairman for the Symposium on which this book is based and was to be co-editor of this volume but was prevented from doing this by the onset of his fatal illness.

ARTHUR Gordon Lyne was born in Launceston on 5 September, 1919 and died at Longford, Tasmania on 4 March, 1986. Gordon, as he was known to his family and friends, was the third of five sons of William James Walch Lyne of "York House", York Plains and later "Eskdale", Powranna. The Lynes have a long association with Tasmania, being decended from William Lyne of Gloucestershire who settled on the east coast with his wife, three sons and two daughters in 1826. There was then no road from Hobart Town to the east coast so "Apsley" (near Bicheno) was settled by chartering a boat which was so overladen that four men and four dogs had to be transferred to a rowing boat (presumably the life boat) during the outward voyage. Other hazards in the early days at "Apsley" included sheep spearing by aborigines and sheep predation by *Thylacinus.*

At "York House" Gordon and his brothers received their preliminary education from a governess but at about the time they moved to "Eskdale", in the early 1930s, Gordon went for secondary schooling to Launceston Church Grammar School. From the time he was a small boy he had a consuming interest in gardening and making things, so it is not surprising that his first job on leaving school (at 15), was at Launceston's City Park. Later he moved to Hobart and worked at the Botanic Garden.

Gordon's horticultural career was cut short when he joined the army in June 1940. He entered on active service with the 2/8th Field Regiment, A.I.F. in the Middle East in December 1940, and when the A.I.F. was brought home to defend the Pacific theatre, he transferred to the 2/2nd Malaria Control Unit and served in New Britain. This was his first zoological job. He married Betty (Parker) in 1943 while still in the army.

Gordon did his matriculation as a Commonwealth Reconstruction Training Scheme student in 1946 and entered the University of Tasmania, in a class swelled by ex-servicemen in 1947. As an undergraduate he displayed a great interest in marsupials and other mammals and his fellow students remember him in zoology classes for his meticulous practical books illustrated with beautiful drawings of skulls and external features, with the vibrissae always correctly placed and present in the right number. As an undergraduate he organized and took part in collecting expeditions during which he gathered together a great deal of the material subsequently described in his classic early papers on the external features of marsupials and monotremes. Immediately after graduation in 1950 he sold his home to go to England and take up a research studentship in the Zoology Department at Cambridge, accompanied by Betty and their first two children, William and Rosemary (David was born after their return to Australia). He took with him much of the material he had collected in Tasmania. It was also at Cambridge that he was introduced to tissue culture, an interest which he pursued on his return to Australia. Lack of funds prevented the family staying in Cambridge long enough for Gordon to complete the minimum residential requirement to submit a Ph.D. thesis there. In 1952 they returned to Tasmania and Gordon did research for his Ph.D. in the Zoology Department, University of Tasmania while looking for a permanent job. In England he had become acquainted with Frederic Wood Jones, then at the Hunterian Museum of the Royal College of Surgeons, who encouraged him to continue his Australian mammal work and corresponded with him after his return.

Gordon joined CSIRO in 1953 and worked firstly at the McMaster Laboratory (University of Sydney). It was from there the first of a series of papers, in collaboration with Margaret Hardy, on the development of wool follicles and the structure of skin in sheep were published. In 1957 he transferred to the Sheep Biology Laboratory (from March, 1960 known as the Ian Clunies Ross Laboratory) at Prospect and continued and extended the investigations that had been started at the McMaster Laboratory. At the same time, he began his series of great papers on reproduction, embryology, morphology and taxonomy of bandicoots which continued to appear until his death. In 1964 he organized, with the late B. F. Short, an international symposium "Biology of the Skin and Hair

Pages xix–xxii *in* BANDICOOTS AND BILBIES, ed by J. H. Seebeck, P. R. Brown, R. L. Wallis and C. M. Kemper. Surrey Beatty & Sons: Sydney, 1990.

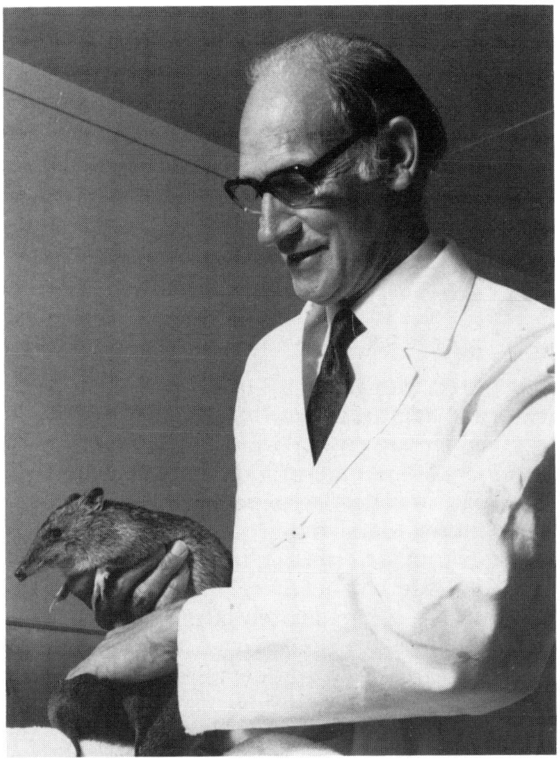

Fig. 1. A. Gordon Lyne (1919–1986), to whom this volume is dedicated, with some of his many bandicoot friends at CSIRO in Prospect, New South Wales.

Growth". Its published proceedings are standard references for present-day investigators. In July, 1974 a reorganization of the Division of Animal Production resulted in his transfer to the staff of the Division of Wildlife Research — a move for which Gordon had little enthusiasm. He remained and continued his researches at the Prospect laboratory until his retirement.

Gordon was present at the conception of the Australian Mammal Society at Portee Station, South Australia, on 23-24 August 1958. John Calaby, in his foreword to *Possums and gliders*, has described the future members of the society on that occasion as being suitably lubricated with Adelaide's excellent fermented liquors — a true enough statement in relation to some who were present. Gordon — the prime mover in the formation of the society — enjoyed fine wines and kept an excellent cellar but was a very moderate drinker in the field and at home. The Mammal Society was born, after a gestation period even shorter than that of bandicoots, during the ANZAAS meeting the following week.

The inaugural meeting of the Mammal Society, at which Gordon gave one of the eight presented papers, was clouded because of his heart attack during the ensuing ANZAAS meeting. Fortunately his recovery was rapid and a perusal of society publications shows that his enthusiasm for presenting papers at subsequent meetings was not dampened. He read papers at general meetings 3, 5, 10, 11, 12, 13, 14, 15, 16, 17 and 20, and at scientific meetings 23 and 24 and at other meetings. He was President in 1964-66 and in 1975-76, Ordinary member of Council 1961, 1971-72 and 1979-80 and Member of the Editorial Committee, *Australian Mammalogy* 1979-80. The Society is deeply indebted to Gordon, its most enthusiastic foundation member and its ardent supporter from its inception.

When Gordon retired from CSIRO in 1981 he and Betty bought "The Laurels", a gracious house in Evandale built in the first half of the nineteenth century. Here, in a very active retirement, he put his carpentry, restoration and horticultural skills to good use. With a keen eye for a bargain he attended the Evandale Sunday Market and various auctions and gathered together fine Tasmanian timbers and brass and other fittings for carpet, stairs and furniture. He made beautifully constructed tables and cabinets, repainted the house (employing painters only for the roof), repaired the fixtures, installed a damp course along one wall and maintained the lawns, trees and vegetable and flower gardens. He also continued his scientific works including preparing a paper which appears in this volume and writing a book about antelopes and other horned animals which remains unfinished.

Gordon will be remembered for many qualities. His colleagues will recall the breadth and thoroughness of every aspect of his scientific work and the emphasis he placed on research on Australian mammals. His special affection for bandicoots was well known; indeed he never wasted an opportunity to sing their praises. His film on bandicoots is a classic, and a particular segment, edited out be retained for special showings (e.g., the Bandicoot Symposium) is legendary. In this segment, Gordon is shown handling a sleek bandicoot and commenting on its affable nature. Whereupon the beast bites him firmly on the finger causing Gordon to curse lightly and drop it. Gordon's enjoyment of these special showings was typical of his rather mischevious sense of humour which so endeared him to friends. His cheerfulness was another of his features. He was seldom not so, except perhaps when attending a poorly presented talk.

To the general public he leaves several samples of his artistic talent. His book *Marsupials and monotremes of Australia* features his detailed sketches and can be found in most public libraries and his drawings of the marsupials illustrate the Mammals article in the latest editions of Encyclopaedia Britannica. Young investigators (and some no longer young) will remember Gordon's willingness to assist anyone who asked for help or advice in the way of materials or comments on research programmes or papers. His degrees were B.Sc. (1950), Ph.D. (1958), D.Sc. (1974), all from the University of Tasmania and his awards and distinctions included a Fulbright Research Scholarship at Brown University, Rhode Island (1961-62), Associate Member of the Australasian College of Dermatologists (1970), CSIRO Honorary Research Fellow (1981), Honorary Research Associate, Queen Victoria Museum (1983), Honorary Life Member (1982) and Ellis Troughton Memorial Medal (1985) of the Australian Mammal Society.

PUBLICATIONS 1951–1983

LYNE, A. G., 1951. Notes on external characters of the barred bandicoot (*Perameles gunnii* Gray), with special reference to the pouch young. *Proceedings of the Zoological Society London* **121:** 587–98.

LYNE, A. G. AND MCMAHON, T. S., 1951. Observations on the surface structure of the hairs of Tasmanian monotremes and marsupials. *Papers and Proceedings of the Royal Society of Tasmania* **1950:** 71–84.

LYNE, A. G., 1952. Notes on external characters of the pouch young of four species of bandicoot. *Proceedings of the Zoological Society London* **122:** 625–49.

LYNE, A. G., 1952. Syndactyly in the manus of a marsupial. *Papers and Proceedings of the Royal Society of Tasmania* **87:** 31–32.

HARDY, M. H., AND LYNE, A. G., 1956. Proposed terminology for wool follicles in sheep. *Nature* **177:** 705–06.

HARDY, M. H. AND LYNE, A. G., 1956. The pre-natal development of wool follicles in Merino sheep. *Australian Journal of Biological Sciences* **9:** 423–41.

HARDY, M. H. AND LYNE, A. G., 1956. Studies on the development of wool follicles in tissue culture. *Australian Journal of Biological Sciences* **9:** 559–74.

HARDY, M. H. AND LYNE, A. G., 1956. The histological development of the skin and wool in the Merino foetus. *Proceedings International Wool and Textile Research Conference of Australia* 1955 Vol. **F:** F26–31.

HARDY, M. H. AND LYNE, A. G., 1956. The development of wool follicles from foetal sheep skin in tissue culture. *Proceedings International Wool and Textile Research Conference of Australia* 1955 Vol. **F:** F32–35.

LYNE, A. G., 1956. Australian mammals. *Australian Museum Magazine* **12:** 121–25.

LYNE, A. G., 1957. Bundles of primary wool follicles in sheep. *Nature* **179:** 825–26.

LYNE, A. G., 1957. The development and replacement of pelage hairs in the bandicoot *Perameles nasuta* Geoffroy (Marsupialia: Peramelidae). *Australian Journal of Biological Sciences* **10:** 197–216.

LYNE, A. G., 1957. The development of wool in Merino sheep. *Wool Technology* **4:** 53–56.

LYNE, A. G., 1957. The development of the epidermis and hair canals in the Merino sheep foetus. *Australian Journal of Biological Sciences* **10:** 390–97.

LYNE, A. G. AND VERHAGEN, A. M. W., 1957. Growth of the marsupial *Trichosurus vulpecula* and a comparison with some higher mammals. *Growth* **21:** 167–95.

LYNE, A. G. AND HEIDEMAN, M. J., 1959. The pre-natal development of skin and hair in cattle (*Bos taurus* L.). *Australian Journal of Biological Sciences* **12:** 72–95.

LYNE, A. G., PILTON, P. E. AND SHARMAN, G. B., 1959. Oestrous cycle, gestation period and parturition in the marsupial. *Trichosurus vulpecula. Nature* **183:** 622–23.

LYNE, A. G., 1959. The systematic and adaptive significance of the vibrissae in the Marsupialia. *Proceedings of the Zoological Society London* **133:** 79–133.

DOWNES, A. M. AND LYNE, A. G., 1959. Measurement of the rate of growth of wool using cystine labelled with sulphur -35. *Nature* **184:** 1884–85.

BROOK, A. H., SHORT, B. F. AND LYNE, A. G., 1960. Formation of new wool follicles in the adult sheep. *Nature* **185:** 51.

LYNE, A. G. AND HEIDEMAN, M. J., 1960. The pre-natal development of skin and hair in cattle. II. *Bos indicus* L.×*B. taurus* L. *Australian Journal of Biological Sciences* **13:** 584–99.

LYNE, A. G., 1960. Pre-natal growth of cattle. *Proceedings of the Animal Production Society* **3:** 153–61.

DOWNES, A. M. AND LYNE, A. G., 1961. Studies on the rate of wool growth using [35S] cystine. *Australian Journal of Biological Sciences* **14:** 120–30.

MOLYNEUX, G. S. AND LYNE, A. G., 1961. Studies on experimental dermal cysts in sheep. *Australian Journal of Biological Sciences* **14:** 131–40.

LYNE, A. G., 1961. The post-natal development of wool follicles, shedding, and skin thickness in inbred Merino and Southdown-Merino crossbred sheep. *Australian Journal of Biological Sciences* **14:** 141–56.

DOWNES, A. M., LYNE, A. G. AND CLARKE, W. H., 1962. Radioautographic studies of the incorporation of [35S] cystine into wool. *Australian Journal of Biological Sciences* **15:** 713–19.

LYNE, A. G., 1964. Australian bandicoots. *Australian Natural History* Vol. xiv, No. 9, March 1964, 281–85.

LYNE, A. G. AND BROOK, A. H., 1964. Neogenesis of wool follicles. *Australian Journal of Biological Sciences* **17:** 514–20.

LYNE, A. G., 1964. Effect of adverse nutrition on the skin and wool follicles in Merino sheep. *Australian Journal of Agricultural Science* **15:** 788–801.

LYNE, A. G., 1964. Observations on the breeding and growth of the marsupial *Perameles nasuta* Geoffroy, with notes on other bandicoots. *Australian Journal of Zoology* **12:** 322–39.

LYNE, A. G., MOLYNEUX, G. S., MYKYTOWYCZ, R. AND PARAKKAL, P. F., 1964. The development, structure and function of the submandibular cutaneous (chin) glands in the rabbit. *Australian Journal of Zoology* **12:** 340–48.

LYNE, A. G. AND SHORT, B. F. (eds), 1965. *Biology of the skin and hair growth.* Angus & Robertson: Sydney.

LYNE, A. G., 1965. The hair cycle in the chinchilla. Pp. 467–89 in *Biology of the skin and hair growth.* Angus & Robertson: Sydney.

LYNE, A. G. AND SOMMERVILLE, R. I., 1965. A nematode parasite *Capillaria* sp., in the skin of the marsupial bandicoot, *Perameles nasuta. Australian Journal of Science* **28:** 205–06.

LYNE, A. G. AND CHASE, H. B., 1966. Branched cells in the epidermis of the sheep. *Nature, London* **209:** 1357–58.

LYNE, A. G., 1966. The development of hair follicles. *Australian Journal of Science* **28:** 370–77.

LYNE, A. G. AND HOLLIS, D. E., 1967. Asymmetric distribution of alkaline phosphatase activity in the hair and wool follicles of sheep. *Journal of Investigative Dermatology* **48:** 197–99.

LYNE, A. G., 1967. *Marsupials and monotremes of Australia.* Angus & Robertson: Sydney.

LYNE, A. G., HOLLIS, D. E., CHASE, H. B. AND HAYMAN, R. H., 1967. Changes experimentally produced in the pigmentation of the skin and coat of sheep. *Australian Journal of Science* **30:** 30–32.

LYNE, A. G. AND HOLLIS, D. E., 1968. The skin of the sheep: a comparison of body regions. *Australian Journal of Biological Sciences* **21:** 499–527.

LYNE, A. G. AND HOLLIS, D. E., 1968. Effects of freezing the skin and plucking the fibres in sheep, with special reference to pigmentation. *Australian Journal of Biological Sciences* **21:** 981–1000.

LYNE, A. G. AND JOLLY, M. G., 1969. Effect of tension along the fibres on the growth and structure of wool. Pp. 247–64 in *Advances in biology of skin. Vol. 9. Hair Growth.* Pergamon Press: Oxford.

JOLLY, M. AND LYNE, A. G., 1969. Effect of subdermal pressure on the skin and its appendages in the sheep. *Australian Journal of Biological Science* **22:** 433–48.

LYNE, A. G., 1969. Research on Australian marsupials. *Proceedings of the Royal Society of New South Wales* **1967–8:** 7–14.

LYNE, A. G., JOLLY, M. AND HOLLIS, D. E., 1970. Effects of experimentally produced local subdermal temperature changes on skin temperature and wool growth in the sheep. *Journal of Agricultural Science, Cambridge* **74:** 83–90.

LYNE, A. G., 1970. The melancoyte population in the skin during development of the marsupial *Trichosurus vulpecula. Australian Journal of Biological Sciences* **23:** 697–708.

LYNE, A. G., HENRIKSON, R. C. AND HOLLIS, D. E., 1970. Development of the epidermis of the marsupial *Trichosurus vulpecula. Australian Journal of Biological Sciences* **23:** 1067–75.

LYNE, A. G., 1970. The development of hair follicles in the marsupial *Trichosurus vulpecula. Australian Journal of Biological Sciences* **23:** 1241–53.

JOLLY, M. AND LYNE, A. G., 1970. The response of skin and wool growth to local subdermal temperature changes in a sheep. *Journal of Agricultural Sciences, Cambridge* **75:** 501–15.

LYNE, A. G. AND HOLLIS, D. E., 1971. Ultrastructural changes in Merino sheep epidermis during foetal development. *Journal of Anatomy* **108:** 211.

LYNE, A. G. AND HOLLIS, D. E., 1971. Merkel cells in sheep epidermis during foetus development. *Journal of Ultrastructural Research* **34:** 464–72.

LYNE, A. G., 1971. Bandicoots in captivity. *International Zoo Yearbook* **11:** 41–43.

LYNE, A. G. AND HOLLIS, D. E., 1972. The structure and development of the epidermis in sheep foetuses. *Journal of Ultrastructual Research* **38:** 444–58.

HOLLIS, D. E. AND LYNE, A. G., 1972. Acetylcholinesterase-positive Langerhans cells in the epidermis and wool follicles of the sheep. *Journal of Investigative Dermatology* **58:** 211–17.

LYNE, A. G. AND HOLLIS, D. E., 1973. Development of horns in Merino sheep. *Australian Journal of Zoology* **21:** 153–69.

LYNE, A. G., DOWNES, A. M. AND CHASE, H. B., 1974. Growth of vibrissae in the marsupial *Trichosurus vulpecula. Australian Journal of Zoology* **22:** 117–29.

HOLLIS, D. E. AND LYNE, A. G., 1974. Innervation of vibrissa follicles in the marsupial *Trichosurus vulpecula. Australian Journal of Zoology* **22:** 263–76.

LYNE, A. G., 1974. Gestation period and birth in the marsupial *Isoodon macrourus. Australian Journal of Zoology* **22:** 303–09.

LYNE, A. G., 1974. Monotremata. *Encyclopedia Britannica.* **12:** 384–87.

HOLLIS, D. E. AND LYNE, A. G., 1975. Observations on the structure of vibrissa follicles in the marsupial *Trichosurus vulpecula,* with special reference to keratinization. *Australian Journal of Zoology* **23:** 9–28.

LYNE, A. G. AND HOLLIS, D. E., 1976. Early embryology of the marsupials *Isoodon macrourus* and *Perameles nasuta. Australian Journal of Zoology* **24:** 361–82.

LYNE, A. G., 1976. Observations on oestrus and the oestrous cycle in the marsupials *Isoodon macrourus* and *Perameles nasuta. Australian Journal of Zoology* **24:** 513–21.

HOLLIS, D. E. AND LYNE, A. G., 1977. Endoderm formation in blastocysts of the marsupials *Isoodon macrourus* and *Perameles nasuta. Australian Journal of Zoology* **25:** 207–23.

LYNE, A. G. AND HOLLIS, D. E., 1977. An unusual blastocyst stage of the marsupial *Isoodon macrourus. Australian Journal of Zoology* **25:** 255–31.

LYNE, A. G. AND HOLLIS, D. E., 1977. The early development of marsupials, with special reference to bandicoots. Pp. 293–302 in *Reproduction and evolution: Proceedings of the 4th International Symposium on Comparative Biology of Reproduction, Canberra, 1976* ed by J. H. Calaby and C. H. Tyndale-Biscoe. Australian Academy of Science: Canberra.

LYNE, A. G. AND HOLLIS, D. E., 1979. Observations on the corpus luteum during pregnancy and lactation in the marsupials *Isoodon macrourus* and *Perameles nasuta. Australian Journal of Zoology* **27:** 881–99.

HOLLIS, D. E. AND LYNE, A. G., 1980. Ultrastructure of luteal cells in fully formed and regressing corpora lutea during pregnancy and lactation in the marsupials *Isoodon macrourus* and *Perameles nasuta. Australian Journal of Zoology* **28:** 195–211.

LYNE, A. G. AND HOLLIS, D. E., 1980. Effects of grafting the skin in Merino sheep, with special reference to the growth of wool and hair. *Australian Journal of Biological Sciences* **33:** 309–28.

LYNE, A. G. AND MORT, P. A., 1981. A comparison of skull morphology in the marsupial bandicoot genus *Isoodon:* its taxonomic implications and notes on a new species, *Isoodon arnhemensis. Australian Mammalogy* **4:** 107-33.

LYNE, A. G., 1981. Activity rhythms in the marsupials *Isoodon macrourus* and *Perameles nasuta. Australian Journal of Zoology* **29:** 821–38.

LYNE, A. G., 1982. Observations on skull growth and eruption of teeth in the marsupial bandicoot *Perameles nasuta* (Marsupialia: Peramelidae). *Australian Mammalogy* **5:** 113–26.

LYNE, A. G. AND HOLLIS, D. E., 1982. Observations on the lateral vaginae and birth canals in the marsupials *Isoodon macrourus* and *Perameles nasuta* (Mammalia). *Journal of Zoology, London* **198:** 263–77.

LYNE, A. G., 1982. The bandicoots *Isoodon macrourus* and *Perameles nasuta:* their maintenance and breeding in captivity. Pp. 47–52 in *The management of Australian mammals in captivity. Proceedings of the Scientific Meeting of the Australian Mammal Society, Healesville, Victoria, February 1979* ed by D. D. Evans. Zoological Board of Victoria: Melbourne.

LYNE, A. G. AND HOLLIS, D. E., 1983. Observations on Graafian follicles and their oocytes during lactation and after the removal of pouch young in the marsupials *Isoodon macrourus* and *Perameles nasuta. American Journal of Anatomy* **166:** 44–61.

A brief review of bandicoot studies

A. G. LYNE[1]

INTRODUCTION

ALTHOUGH the number of research studies of bandicoots (family Peramelidae) is now considerable there are still many gaps in our knowledge. My own incomplete list of references to articles in scientific journals contains about 400 titles, published from 1797 to 1984. More than half of these have been published during the last 20 years.

Despite their wide range of habitats, bandicoots form a fairly uniform group. They are small to medium-sized animals, most species weighing one to two kilograms when mature, and are mainly nocturnal and usually solitary, hiding in a nest or burrow by day. Although the bandicoots are less varied in size than the other groups of marsupials, they display a number of unique characters. They are polyprotodont and syndactylous whereas the other polyprotodonts are non-syndactylous and the other syndactylous marsupials are diprotodonts. The bandicoots thus occupy an intermediate position in respect of these two characters.

Reproduction in bandicoots is also unique. They are the only marsupials with a placenta similar to that found in eutherian mammals and, in at least two species — *Perameles nasuta* and *Isoodon macrourus,* the gestation period is only 12.5 days, the shortest gestation period recorded for any mammal (Lyne 1974).

Many investigators have contributed to our knowledge of bandicoots but only a small number of topics and contributors will be referred to in this review.

DISCOVERY AND TAXONOMY

Bandicoots are found only in Australia, Papua New Guinea and nearby islands. The first description of a bandicoot was based on a specimen sent to John Hunter in London from John White, the Chief Surgeon to the Settlement at Botany Bay. It was called *Didelphis obesula* by Shaw and Nodder (1797) and, in the state of exploration at that time, it is most likely that the type locality was near Sydney, New South Wales. The type specimen, a juvenile, is apparently no longer in existence, but the description and Shaw and Nodder's coloured figure indicate that it was a short-nosed bandicoot. The authors said that it showed some resemblance to a pig in miniature and because of this they gave it the common name — porculine opossum. The animal is now referred to as *Isoodon obesulus.*

The Long-nosed Bandicoot *P. nasuta* has the distinction of being the second bandicoot to be described (Geoffroy 1804). With the exception of two or three species, there was a lull in the description of new bandicoots until about the 1840s. Most of the New Guinea species were not described before 1875. The description of new Australian and New Guinean forms has continued until recent times.

The bandicoots are considered by some authors (e.g., Kirsch 1977) to be a troublesome group but this is mainly because of the anatomical features that link them with the two major groups of marsupials. As yet, the relative significance of dentition, foot structure, serology, chromosomes, and other features has not been determined. Consequently I consider that the bandicoots all belong to one family — the Peramelidae.

Unfortunately, the taxonomy of the bandicoots at the species level is very confused as are the scientific and common names used in some recent publications. With few exceptions, the authors of the popular books appear to have ignored the findings of investigators (Jones 1923; Troughton 1932; Tate 1948; Freedman 1967; Freedman and Joffe 1967a, 1967b; Freedman and Rightmire 1971; Lyne and Mort 1981) who have made detailed comparative studies of bandicoot material.

Some examples in support of this claim that the taxonomy is in a disorganized state are found in the short-nosed bandicoots of the genus *Isoodon,* reviewed by Lyne and Mort (1981).

The *Isoodon* forms *auratus* (Ramsay 1887), *barrowensis* (Thomas 1901), *nauticus* (Thomas 1922) and *peninsulae* (Thomas 1922) were originally described as separate species and Lyne and Mort (1981) confirmed this separation. In the original description of *I. barrowensis,* Thomas (1901) said that it was a small insular representative of mainland *I. obesulus* and that the skull, although

[1]Queen Victoria Museum and Art Gallery, Launceston, Tasmania, Australia 7250.

Pages xxiii-xxix *in* BANDICOOTS AND BILBIES, ed by J. H. Seebeck, P. R. Brown, R. L. Wallis and C. M. Kemper. Surrey Beatty & Sons, Sydney, 1990.

conspicuously smaller, agreed in most details with that of *I. obesulus*. The observations of Freedman and Rightmire (1971) were in agreement with those of Thomas (1901) in that *I. barrowensis* is very similar to *I. obesulus* in skull morphology. As the adult skull of *I. barrowensis* is much smaller than that of *I. obesulus,* the recognition of the former as a separate species seems to be justified and this is in agreement with various authors (Jones 1922; Glauert 1933; Iredale and Troughton 1934; Tate 1948; Troughton 1948; Marlow 1965). For some unexplained reason most recent authors (e.g., Ride 1970; Collins 1973; Kirsch and Calaby 1977) have included *I. barrowensis* with *I. auratus*.

The original description of *I. auratus* (Ramsay 1887) referred to the dental formula but no skull details were given. On the basis of measurements of the type specimen, Thomas (1904) said that *I. auratus* was a small form allied to *I. obesulus* and that it had larger tympanic bullae than in *I. barrowensis,* a feature which was confirmed by Lyne and Mort (1981). On this feature alone *I. auratus* and *I. barrowensis* can be separated.

Recent authors have included *I. nauticus* and *I. peninsulae* with *I. obesulus* despite the fact that they were originally described as separate species. The skull of *I. nauticus* was said by Thomas (1922) to be smaller than that of *I. obesulus*. He also said that the tympanic bullae in *I. nauticus* were shorter and that there was probably less sexual difference in size than in *I. obesulus*. Lyne and Mort (1981) confirmed Thomas's findings but only a small number of skulls was examined.

The skull of *I. peninsulae* was said by Thomas (1922) to be similar to that of *I. auratus* except that the muzzle was comparatively longer, and that it could be distinguished from that of *I. obesulus* by the almost complete absence of any sexual difference in size. This close similarity in size was confirmed by Lyne and Mort (1981).

The taxonomy of the rabbit-eared bandicoots or bilbies *(Macrotis)* is also in need of revision. To my knowledge, detailed comparative studies of the genus *Macrotis* have been published by only four investigators — Jones 1923; Troughton 1932; Finlayson 1935; Tate 1948. Jones (1923) admitted five species *(M. lagotis, M. sagitta, M. minor, M. nigripes* and *M. leucura)* and Troughton (1932) and Tate (1948) recognized three species *(M. lagotis, M. leucura* and *M. minor)*. However, Troughton (1948) recognized only *M. lagotis* and *M. leucura*. These two species are recognized by most modern authors.

There is a need for more detailed comparative studies of all the Australian bandicoots as well as the bandicoots which occur in New Guinea and nearby islands. The taxonomists need to look at more of the facts and explore them more fully before embarking on new classifications.

STATUS

Although at least 12 species of Australian bandicoots have been described, about half of them are now extinct, rare or threatened with extinction. They have suffered more than any other group of Australian marsupials.

Aitken (1979), in a review of the status of the Australian bandicoots, said that, in general, the bandicoots that live in the coastal forests and woodlands of northern, eastern, southern and southwestern Australia and in Tasmania have survived well. By contrast, he said, those that live on the inland plains and deserts have become endangered, critically rare or extinct. For example, the extraordinary Pig-footed Bandicoot *Chaeropus ecaudatus* once had an extremely wide distribution in southern Australia but it now appears to be extinct in all former areas. Our knowledge of *Chaeropus* as a living animal comes mainly from Krefft's (1866) paper on the vertebrates of the lower Murray and Darling Rivers.

There is an urgent need for more studies of the rare bandicoots before it is too late. The Greater Bilby *M. lagotis* is an example. The range of this species, which appears to be the only remaining species of the genus *Macrotis,* has decreased considerably since European settlement.

EXTERNAL CHARACTERS

Although the general features of the hairy coat, such as the colour, colour pattern and texture have been included among features of taxonomic value in the classification of the bandicoots, very little attention has been given to the detailed structure of the skin and hair.

Our knowledge of the microscopic structure of the hair is restricted to the studies of Lyne and McMahon (1951) and Brunner and Coman (1974). To date, only three species *(I. obesulus, P. nasuta* and *P. gunnii)* have been examined. The fact that these species can be separated on the structure of the grooved guard hairs suggests that the microscopic structure of the hair may prove to be of considerable taxonomic value.

The distribution of the vibrissae in a number of different Australian bandicoots has been studied by Jones (1922, 1924), Boardman (1943), Hill (1951) and Lyne (1951, 1952, 1959). These studies have revealed that all the typical facial groups of vibrissae are usually represented and the number of vibrissae is remarkably constant in most species. Of the bandicoots so far examined, it is only in the highly specialized

C. ecaudatus that one finds a great reduction in the number of vibrissae. *C. ecaudatus* has so many unique features, including its two-toed forefeet, that it may warrant a separate family. To my knowledge, the vibrissae of the bandicoots in the New Guinea region have not been examined.

Very little information is available on the structure of the skin in bandicoots. However, Lyne (1957) described the development and replacement of pelage hairs in *P. nasuta* and Green (1963) described the cutaneous glands in *I. obesulus*. Stoddart's (1980) comparative study of the skin gland complexes in bandicoots is the only paper in this field.

REPRODUCTION, GROWTH AND DEVELOPMENT

Considerable research has been done in the fields of reproduction, growth and development of bandicoots but only a few species have been studied. Outstanding contributions to our knowledge of marsupial embryology and reproduction were made by J. P. Hill. His description (Hill 1895, 1897) of the unique chorio-allantoic placenta in bandicoots was one of the most important early discoveries in the field of marsupial reproduction. Flynn (1923) also made a major contribution to our knowledge of the bandicoot placenta.

Although the female reproductive system of all marsupials is based upon a common plan, there are considerable species differences. In *Isoodon* and *Perameles* the urogenital sinus is extremely short, the lateral vaginae are straight and relatively long (about 50 mm in *I. macrourus* and *P. nasuta* and there are two large anterior expansions of the vaginal system — the vaginal caeca or seminal receptacles (Lyne and Hollis 1977, 1982). The only reference to the vaginal caeca in *M. lagotis* and *C. ecaudatus* is in a paper by Hill (1899). The sperm travel up the lateral vaginae and into the vaginal caeca before some of them pass through the uteri to the fallopian tubes, where it is presumed that fertilization occurs. During parturition, a birth canal forms in the centre of the urogenital strand, which contains the two lateral vaginae (Lyne and Hollis 1982).

It is, perhaps, significant that the bandicoot placenta, as shown by Hill (1895, 1897), is retained in the uterus and it may exert an endocrine effect preventing oestrus during the early part of lactation. In the kangaroo the placenta is expelled from the uterus at the time of giving birth (Sharman and Calaby 1964).

The bandicoots *P. nasuta* and *I. macrourus* are hardy animals during operative procedures. Since they resemble the Virginian Opossum *Didelphis virginiana* and some other marsupials in that they usually become pregnant in both uteri, embryos at two stages of development may be obtained by removing one pregnant uterus and allowing the embryos in the other uterus to develop further before removal (Lyne and Hollis 1977). These bandicoots are therefore ideal for securing accurately timed developmental stages.

Some bandicoots, such as *P. nasuta* in the Sydney region (Lyne 1964, 1982) and *I. macrourus* in the vicinity of Brisbane (Hall 1983), breed throughout the year although the latter have a peak of breeding in spring which Gemmell, Johnston and Barnes (1985) found to be associated with a peak in testosterone levels in males. Other species, such as *I. obesulus* in Tasmania and Victoria (Heinsohn 1966; Stoddart and Braithwaite 1979) usually have a more restricted breeding season. Our knowledge of the breeding in some species, such as *C. ecaudatus* is very sparse indeed. Spencer (1896) received a *C. ecaudatus* from Charlotte Water, Northern Territory, in August with two well-developed young in the pouch and he thought that they were born near the end of June. Krefft (1866) captured several females with pouch young and he concluded that the breeding season is in May or June. This appears to be the extent of our knowledge of breeding in this species.

The oestrous cycle length varies considerably in *P. nasuta* and *I. macrourus,* the two species which have received most attention, but the average oestrous cycle length (about 20 days) as well as the length of the gestation period (12.5 days) is shorter than it is in other marsupials (Lyne 1974, 1976). In most kangaroos (Calaby and Poole 1971) the oestrous cycle length is almost equal to the gestation length. This makes the bandicoots an excellent contrast to kangaroos for basic comparisons of reproduction in marsupials.

The effect of removal of pouch young on the time taken for breeding to recur has been studied in only two species — *P. nasuta* and *I. macrourus* (Mackerras and Smith 1960; Stodart 1966; Close 1977; Gemmell 1981; Lyne and Hollis 1983). In both species the return to breeding takes longer when the young are removed early in lactation than when they are removed later. For example, when young are removed when less than five days old, recurrence of breeding usually takes about 20 days, whereas when young are removed at 30 to 40 days, breeding recurs in only five to eight days. Oestrus also occurs normally in some animals of both species during lactation when the young reach about 50 days of the 60-day lactation period.

Gemmell (1979, 1981) showed that progesterone levels in *I. macrourus,* which were higher than those recorded for other marsupials, diminished sharply before day 11 of lactation. Luteal cells showed a corresponding decline in activity and were not secretory after day 19, but did not break down till day 45. He concluded that suckling appears to be the main inhibitor of ovulation in the latter stages of lactation and the continued maintenance of the corpora lutea remains a mystery.

In all the marsupials which have been studied, the rate of development is very slow at the beginning and exceedingly rapid during the final stages of pregnancy. At three days of pregnancy, the embryo is a tiny sphere 0.5 mm in diameter and is composed of no more the 32 cells, and the blastocyst cavity is not fully formed (Lyne and Hollis 1976). At about six days, that is about half-way through gestation, the embryo has formed the germ layers, ectoderm, endoderm and mesoderm (Hollis and Lyne 1977; Lyne and Hollis 1977), and it has a diameter of about 2 mm. It is not until 9.5 days that the allantoic placenta appears as a small vesicle and implantation begins. Even with this placenta the bandicoot embryo does not tarry long in the uterus as it is born three days later. The fullterm embryo is about 13 mm long and weighs about 200 mg.

Bandicoots are typical marsupials in that the young are born in a rudimentary condition and they are sheltered in a pouch during their later development. The pouch, which opens downwards and backwards, contains eight teats arranged in two curved rows of four.

The post-natal development of bandicoots (Lyne 1964; Hulbert 1972, 1982), like the pre-natal development (Lyne and Hollis 1977), is very rapid. In fact, the growth rate of bandicoot pouch young appears to be faster than it is in other marsupials so far investigated. The first hair emerges on the body at about 40 days after birth (Lyne 1957), eyes open at 45–50 days and weaning occurs at about two months after birth (Lyne 1964).

At least two species of *Isoodon (macrourus* and *obesulus)* and two species of *Perameles (nasuta* and *gunnii)* have a high reproductive rate (Lyne 1964; Heinsohn 1966; Stoddart and Braithwaite 1979; Hall 1983). In all of these species, some of the females start to breed before they are fully grown and they can produce several litters per year or during a breeding season.

ECOLOGY AND BEHAVIOUR

Some information is available on the preferred habitats of most species of bandicoots but very few ecological studies have been published. Fortunately, several good naturalists, such as Krefft in the 1850s and 1860s and Finlayson in the 1930s, spent considerable time in the field and they recorded a large amount of interesting and valuable information. Without their observations we would know next to nothing about several of the species which now appear to be extinct.

Finlayson (1934) also made some observations on the habitat preferences of several species which are still common in parts of eastern Australia. Referring to the bandicoots in Tasmania he said that *P. gunnii* is largely a grass or open country form whereas *I. obesulus,* though ubiquitous, prefers scrubby habitats. Finlayson (1934) also said that there is a great deal of overlapping, and it may be doubted whether either of the genera is decidedly adapted to any one type of habitat.

To date, the most complete published account of the ecology of bandicoots is the study by Heinsohn (1966) of *P. gunnii* and *I. obesulus* in Tasmania. More recently, valuable additional information on various aspects of bandicoot ecology has been published by Gordon (1974) and Stoddart and Braithwaite (1979). Despite these studies, there is a special need for long-term field work. A detailed knowledge of the ecology of all species, including the interaction between different species, must be acquired if intelligent conservation measures are to be applied. As the natural habitats of some species are disappearing rapidly there is an urgent need for these ecological studies.

Many fascinating ecological problems remain to be investigated. For example — why are some species common in certain areas and rare or absent in other areas which appear to be similar? What determines the distribution of the different forms of the genus *Isoodon* and how did they get there? Why doesn't *I. macrourus* occur south of the Hawkesbury River in New South Wales?

There have been very few behavioural studies of bandicoots. Being small and nocturnal they are difficult to observe in the field. However, Stodart (1966) kept *P. nasuta* in outdoor enclosures and she made some useful behavioural observations with the aid of a spotlight and binoculars. My behavioural studies (Lyne 1981), using exercise wheels and closed circuit television, clearly demonstrated the complexity of the rhythms of activity in *I. macrourus* and *P. nasuta* exposed to different lighting conditions. These experiments also showed that a large number of animals must be tested in order to show the variety of response patterns within a species, and that the test periods must not be too short, particularly when animals are exposed to abnormal lighting conditions. The bandicoots may be unique experimental animals for investigating circadian rhythms. There is a need for more behavioural studies, particularly of animals in the field and in outdoor enclosures.

GENETICS

Most of our knowledge of the chromosomes of bandicoots stems from the studies of Sharman (1961). Since then Martin and Hayman (1967), Hayman, Martin and Walker (1969), Hayman and Martin (1974), Hayman (1977), Close (1979, 1984), and a few others have contributed important cytological information.

About half of the species of bandicoots have now been studied cytologically and in all of them, except *M. lagotis,* the chromosome number is exclusively 2n = 14, the same as that of the Caenolestidae, Dasyuridae, and most of the Didelphidae. However, the bandicoots differ from these families in possessing larger X and Y chromosomes and in the elimination of one X chromosomes of females and the Y chromosomes of males in most somatic tissues (Sharman 1973). *Macrotis lagotis,* the only rabbit-eared bandicoot yet examined, has the same number of autosome arms as have the other bandicoots but a higher diploid number: 2n = 18♀, 19♂. Sex chromosome elimination from somatic tissue does not occur in *M. lagotis.*

As the development of "banding" techniques has greatly increased the capacity of cytologists to determine relationships between different species, a comparative study of bandicoot material using these techniques could be most rewarding.

CONCLUDING REMARKS

There are numerous gaps in our knowledge of Australian bandicoots and many aspects and problems remain to be investigated. With regard to the species in the New Guinea region, much basic work remains to be done.

In addition to long-term field studies and survey work, there is an urgent need for more studies of captive animals. We need to know much more about bandicoots in the broad sense. Many of the mysteries of bandicoot biology e.g., in reproduction, physiology, behaviour and nutrition) can only be conveniently studied in the laboratory. Until about 20 to 30 years ago, techniques for keeping bandicoots in captivity lagged behind those for some other marsupials, mainly because early attempts to breed them often resulted in fighting among adults and destruction of the pouch young by their mothers (Jones 1924; Mackerras and Smith 1960). Techniques for maintaining and breeding *I. macrourus* and *P. nasuta* in captivity are now fairly well established (Lyne 1982).

Isoodon macrourus is easier to breed in captivity than *P. nasuta* and is a particularly promising animal for many experimental studies. Since there are several young in most litters (more in *I. macrourus* than in *P. nasuta*) sampling can be sequential and control animals can be included. To this is the added advantage of fast-growing young and oestrus and pregnancy near the termination of lactation.

In view of the unique zoological position occupied by the bandicoots and the comparative rarity of many species, it would be most unfortunate if more species were added to the growing list of extinct animals.

REFERENCES

AITKEN, P., 1979. The status of endangered Australian wombats, bandicoots and the marsupial mole. Pp. 61–65 in *The status of endangered Australasian wildlife* ed by M. J. Tyler. Royal Zoological Society of South Australia: Adelaide.

BOARDMAN, W., 1943. On the external characters of the pouch young of some Australian marsupials. *Australian Zoologist* **10:** 138–60.

BRUNNER, H. AND COMAN, B. J., 1974. *The identification of mammalian hair.* Inkata Press: Melbourne.

CALABY, J. H. AND POOLE, W. E., 1971. Keeping kangaroos in captivity. *International Zoo Yearbook* **11:** 5–12.

CLOSE, R. L., 1977. Recurrence of breeding after cessation of suckling in the marsupial *Perameles nasuta. Australian Journal of Zoology* **25:** 641–45.

CLOSE, R. L., 1979. Sex chromosome mosaicism in liver, thymus, spleen and regenerating liver of *Perameles nasuta* and *Isoodon macrourus. Australian Journal of Biological Sciences* **32:** 615–24.

CLOSE, R. L., 1984. Rates of sex chromosome loss during development in different tissues of the bandicoots *Perameles nasuta* and *Isoodon macrourus* (Marsupialia: Peramelidae). *Australian Journal of Biological Sciences* **37:** 53–61.

COLLINS, L. R., 1973. *Monotremes and marsupials: a reference for zoological institutions.* Smithsonian Institution Press: Washington.

FINLAYSON, H. H., 1934. On mammals from the Dawson and Fitzroy Valleys; Central coastal Queensland — Part II. *Transactions of the Royal Society of South Australia* **58:** 218–31.

FINLAYSON, H. H., 1935. On mammals from the Lake Eyre Basin. Part II — The Peramelidae. *Transactions of the Royal Society of South Australia* **59:** 227–36.

FLYNN, T. T., 1923. The yolk-sac and allantoic placenta in *Perameles. Quarterly Journal of Microscopic Science* **67:** 123–82.

FREEDMAN, L., 1967. Skull and tooth variation in the genus *Perameles.* Part 1: Anatomical features. *Records of the Australian Museum* **27:** 147–65.

FREEDMAN, L. AND JOFFE, A. D., 1967a. Skull and tooth variation in the genus *Perameles.* Part 2: Metrical features of *P. nasuta. Records of the Australian Museum* **27:** 183–95.

FREEDMAN, L. AND JOFFE, A. D. 1967b. Skull and tooth variation in the genus *Perameles.* Part 3: Metrical features of *P. gunnii* and *P. bougainville. Records of the Australian Museum* **27:** 197–212.

FREEDMAN, L. AND RIGHTMIRE, G. P., 1971. Skull and tooth variation in Australian bandicoots (Peramelidae, Marsupialia): The genus *Isoodon* and multivariate comparisons with *Perameles. Journal of the Royal Society of Western Australia* **54:** 21–31.

GEMMELL, R. T., 1979. The fine structure of the luteal cells in relation to the concentration of progesterone in the plasma of the lactating bandicoot *Isoodon macrourus. Australian Journal of Zoology* **27:** 501–10.

GEMMELL, R. T., 1981. The role of the corpus luteum of lactation in the bandicoot *Isoodon macrourus* (Marsupialia: Peramelidae). *General and Comparative Endocrinology* **44:** 13–19.

GEMMELL, R. T., JOHNSTON, G. AND BARNES, A., 1985. Seasonal variations in plasma testosterone concentrations in the male marsupial bandicoot *Isoodon macrourus* in captivity. *General and Comparative Endocrinology* **59:** 184–91.

GEOFFROY, E., 1804. Memoire sur un nouveau genre de Mammiferes a bourse, nommé *Perameles. Annales de la Musée Histoire natureul de Paris* **4:** 56–65.

GLAUERT, L., 1933. The distribution of the marsupials in Western Australia. *Journal of the Royal Society of Western Australia* **19:** 17–32.

GORDON, G., 1974. Movements and activity of the shortnosed bandicoot *Isoodon macrourus* Gould (Marsupialia). *Mammalia* **38:** 405–31.

GREEN, L. M. A., 1963. Distribution and comparative histology of cutaneous glands in certain marsupials. *Australian Journal of Zoology* **11:** 250–72.

HALL, L. S., 1983. Observations on body weights and breeding of the northern brown bandicoot, *Isoodon macrourus*, trapped in south-east Queensland. *Australian Wildlife Research* **10:** 467–76.

HAYMAN, D. L., 1977. Chromosome number — constancy and variation. Pp. 27–48 in *The biology of marsupials* ed by B. Stonehouse and D. Gilmore. Macmillan Press: London.

HAYMAN, D. L. AND MARTIN, P. G., 1974. Monotremata and Marsupialia. In *Animal cytogenetics Vol. 4* ed by B. John. Gebrüder Borntraeger: Berlin.

HAYMAN, D. L., MARTIN, P. G. AND WALLER, P. F., 1969. Parallel mosaicism of supernumerary chromosomes and sex chromosomes in *Echymipera kalabu* (Marsupialia). *Chromosoma (Berlin)* **27:** 371–80.

HEINSOHN, G. E., 1966. Ecology and reproduction of the Tasmanian bandicoots (*Perameles gunni* and *Isoodon obesulus*). *University of California Publications in Zoology* **80:** 1–96.

HILL, J. P., 1895. Preliminary note on the occurrence of a placental connection in *Perameles obesula* and on the foetal membranes of certain macropods. *Proceedings of the Linnean Society of New South Wales* **10:** 578–81.

HILL, J. P., 1897. The placentation of *Perameles.* (Contributions to the embryology of the Marsupialia — I.) *Quarterly Journal of Microscopic Science* **40:** 385–446.

HILL, J. P., 1899. Contributions to the morphology and development of the female urogenital organs in the Marsupialia. *Proceedings of the Linnean Society of New South Wales* **24:** 42–82.

HILL, W. C. O., 1951. Observations on marsupials in the Royal Scottish Museum, with special reference to the foetal material. *Transactions of the Royal Society of Edinburgh* **62:** 145–67.

HOLLIS, D. E. AND LYNE, A. G., 1977. Endoderm formation in blastocysts of the marsupials *Isoodon macrourus* and *Perameles nasuta. Australian Journal of Zoology* **25:** 207–23.

HULBERT, A. J., 1972. Growth and development of pouch young in the rabbit-eared bandicoot, *Macrotis lagotis* (Peramelidae). *Australian Mammalogy* **1:** 38–39.

HULBERT, A. J., 1982. Notes on the management of a captive breeding colony of the greater bilby *Macrotis lagotis.* P.53 in *The management of Australian mammals in captivity* ed by D. D. Evans. The Zoological Board of Victoria: Melbourne.

IREDALE, T. AND TROUGHTON, E. LEG., 1934. A check-list of the mammals recorded from Australia. *Memoirs of the Australian Museum* **6:** xi + 1–122.

JONES, F. W., 1922. The external characters of pouch embryos of marsupials. No. 3 — *Isoodon barrowensis. Transactions of the Royal Society of South Australia* **61:** 39–45.

JONES, F. W., 1923. The marsupial genus *Thylacomys.* Review of the rabbit-bandicoots; with the description of a new species. *Records of the South Australian Museum* **2:** 333–52.

JONES, F. W., 1924. *The mammals of South Australia. Part 2. The bandicoots and the herbivorous marsupials.* Government Printer: Adelaide.

KIRSCH, J. A. W., 1977. The classification of marsupials. Pp. 1–50 in *The biology of marsupials* ed by D. Hunsaker II. Academic Press: New York.

KIRSCH, J. A. W. AND CALABY, J. H., 1977. The species of living marsupials: an annotated list. Pp. 9–26 in *The biology of marsupials* ed by B. Stonehouse and D. Gilmore. Macmillan Press: London.

KREFFT, G., 1866. On the vertebrated animals of the lower Murray and Darling, their habits, economy, and geographical distribution. *Transactions of the Philosophical Society of New South Wales* **1862–1865:** 1–38.

LYNE, A. G., 1951. Notes on external characters of the barred bandicoot (*Perameles gunnii* Gray), with special reference to the pouch young. *Proceedings of the Zoological Society of London* **121:** 587–98.

LYNE, A. G., 1952. Notes on external characters of the pouch young of four species of bandicoot. *Proceedings of the Zoological Society of London* **122:** 625–49.

LYNE, A. G., 1957. The development and replacement of pelage hairs in the bandicoot *Perameles nasuta* Geoffroy (Marsupialia: Peramelidae). *Australian Journal of Biological Sciences* **10:** 197–216.

LYNE, A. G., 1959. The systematic and adaptive significance of the vibrissae in the Marsupialia. *Proceedings of the Zoological Society of London* **133:** 79–133.

LYNE, A. G., 1964. Observations on the breeding and growth of the marsupial *Perameles nasuta* Geoffroy, with notes on other bandicoots. *Australian Journal of Zoology* **12:** 322–39.

LYNE, A. G., 1974. Gestation period and birth in the marsupial *Isoodon macrourus. Australian Journal of Zoology* **22:** 303–09.

LYNE, A. G., 1976. Observations on oestrus and the oestrous cycle in the marsupials *Isoodon macrourus* and *Perameles nasuta. Australian Journal of Zoology* **24:** 513–21.

LYNE, A. G., 1981. Activity rhythms in the marsupials *Isoodon macrourus* and *Perameles nasuta. Australian Journal of Zoology* **29:** 821–38.

LYNE, A. G., 1982. The bandicoots *Isoodon macrourus* and *Perameles nasuta:* their maintenance and breeding in captivity. Pp. 47–52 in *The management of Australian mammals in captivity* ed by D. D. Evans. The Zoological Board of Victoria: Melbourne.

LYNE, A. G. AND HOLLIS, D. E., 1976. Early embryology of the marsupials *Isoodon macrourus* and *Perameles nasuta. Australian Journal of Zoology* **24:** 361–82.

LYNE, A. G. AND HOLLIS, D. E., 1977. The early development of marsupials, with special reference to bandicoots. Pp. 293–302 in *Reproduction and evolution: proceedings of 4th Int. symposium on comparative biology of reproduction, Canberra* ed by J. H. Calaby and C. H. Tyndale-Biscoe. Australan Academy of Science: Canberra.

LYNE, A. G. AND HOLLIS, D. E., 1982. Observations on the lateral vaginae and birth canals in the marsupials *Isoodon macrourus* and *Perameles nasuta* (Mammalia). *Journal of Zoology, London* **198:** 263–77.

LYNE, A. G. AND HOLLIS, D. E., 1983. Observations of Graafian follicles and their oocytes during lactation and after the removal of pouch young in the marsupials *Isoodon macrourus* and *Perameles nasuta. American Journal of Anatomy* **166:** 41–61.

LYNE, A. G. AND MCMAHON, T. S., 1951. Observations on the surface structure of the hairs of Tasmanian monotremes and marsupials. *Papers and Proceedings of the Royal Society of Tasmania* **1950:** 71–84.

LYNE, A. G. AND MORT, P. A., 1981. A comparison of skull morphology in the marsupial bandicoot genus *Isoodon:* its taxonomic implications and notes on a new species, *Isoodon arnhemensis. Australian Mammalogy* **4:** 107–33.

MACKERRAS, M. J. AND SMITH, R. H., 1960. Breeding the short-nosed marsupial bandicoot, *Isoodon macrourus* (Gould), in captivity. *Australian Journal of Zoology* **8:** 371–382.

MARLOW, B. J., 1965. *Marsupials of Australia.* Jacaranda Press: Brisbane.

MARTIN, P. G. AND HAYMAN, D. L., 1967. Quantitative comparisons between the karyotypes of Australian marsupials from three different superfamilies. *Chromosoma (Berlin)* **20:** 290–310.

RAMSAY, E. P., 1887. Description of two new species of marsupials (*Perameles* and *Antechinus*), and a new species of *Mus* (*M. burtoni*), from the neighbourhood of Derby, N.W.A. *Proceedings of the Linnean Society of New South Wales* **11:** 551–53.

RIDE, W. D. L., 1970. *A guide to the native mammals of Australia.* Oxford University Press: Melbourne.

SHARMAN, G. B., 1961. The mitotic chromosomes of marsupials and their bearing on taxonomy and phylogeny. *Australian Journal of Zoology* **9:** 38–60.

SHARMAN, G. B., 1973. The chromosomes of non-eutherian mammals. Pp. 484–530 in *Cytotaxonomy and vertebrate evolution* ed by A. B. Chiarelli and E. Capanna. Academic Press: London.

SHARMAN, G. B. AND CALABY, J. H., 1964. Reproductive behaviour in the red kangaroo, *Megaleia rufa,* in captivity. *CSIRO Wildlife Research* **9:** 58–85.

SHAW, G. AND NODDER, F. P., 1797. *Didelphis obesula. Naturalist's Miscellany,* vol. 8, pl. 298.

SPENCER, W. B., 1896. Mammalia. Pp. 1–52 in *Report on the work of the Horn Scientific Expedition to Central Australia. Part II. — Zoology* ed by W. B. Spencer. Melville, Mullen and Slade: Melbourne.

STODART, E., 1966. Management and behaviour of breeding groups of the marsupial *Perameles nasuta* Geoffroy in captivity. *Australian Journal of Zoology* **14:** 611–23.

STODDART, D. M., 1980. Observations on the structure and function of cephalic skin glands in bandicoots (Marsupialia: Peramelidae). *Australian Journal of Zoology* **28:** 33–41.

STODDART, D. M. AND BRAITHWAITE, R. W., 1979. A strategy for utilization of regenerating heathland habitat by the brown bandicoot (*Isoodon obesulus*; Marsupialia, Peramelidae) *Journal of Animal Ecology* **48:** 165–79.

STRAHAN, R. ed, 1983. *Complete book of Australian mammals.* Angus & Robertson: Sydney.

TATE, G. H. H., 1948. Results of the Archbold Expeditions. No. 60. Studies in the Peramelidae (Marsupialia). *Bulletin of the American Museum of Natural History* **92:** 317–46.

THOMAS, O., 1901. On some kangaroos and bandicoots from Barrow Island, N.W. Australia, and the adjoining mainland. *Novitates Zoologicae* **8:** 394–96.

THOMAS, O., 1904. On a collection of mammals made by Mr J. T. Tunney in Arnhem Land, Northern Territory of South Australia. *Novitates Zoologicae* **11:** 222–90.

THOMAS, O., 1922. The bandicoots of Nuyts Archipelago, S. Australia, and of Cape York, N. Queensland. *Annals and Magazine of Natural History* **9:** 677–79.

TROUGHTON, E. LEG., 1932. A revision of the rabbit-bandicoots. *Australian Zoologist* **7:** 219–36.

TROUGHTON, E. LEG, 1948. *Furred animals of Australia.* Third edition. Angus & Robertson: Sydney.

Revision of the families and genera of bandicoots

COLIN P. GROVES[1] and TIM FLANNERY[2]

Skull characters clearly assort the bandicoots into two groups: a tropical rainforest group and a more open country group. We propose to recognize this division at the family level: Peroryctidae (n. fam.) for the former and Peramelidae for the latter. The Peramelidae are defined by a number of synapomorphies; the Peroryctidae have rather fewer synapomorphies, but the evidence supports their monophyly in addition. The family Thylacomyidae is rejected: its basis, the genus *Macrotis,* is not only a member of the family Peramelidae but even shares numerous synapomorphous states with another genus of that family, *Chaeropus.* Other relationships within the Peramelidae are ambiguous; within the Peroryctidae, relationships are more clear-cut. Certain species of New Guinea rainforest bandicoots previously assigned to *Peroryctes* are now reallocated to *Microperoryctes.*

Key words: Taxonomy, bandicoots, Peramelidae, Peroryctidae.

INTRODUCTION

STUDIES on the rainforest bandicoots, begun independently by the two of us, suggested that there might be more differentiation within each of the recognized genera than has previously been acknowledged, and this led naturally to a reconsideration of the relationships of the rainforest genera to each other, and to the dry-country bandicoots. We examined the skulls in the Australian Museum and elsewhere, and came to conclusions which so completely change the traditionally accepted arrangement, that we wish to present them in some detail here.

LITERATURE REVIEW

Most systematists of the last century placed the majority of bandicoot species into a single inclusive genus, *Perameles,* with only the two obviously divergent types, bilbies and the Pig-footed Bandicoot, in different genera (*Macrotis* or *Thylacomys,* and *Chaeropus,* respectively). Thomas (1888), who adopted just this scheme, did, however, recognize strong divisions among the species groups of his *Perameles.*

Bensley (1903) followed Thomas and others in this way, but on the grounds of certain striking dental differences, separated off the bilbies into a subfamily, Thylacomyinae.

In his later papers, Thomas began to split up the all inclusive *Perameles,* culminating in 1920 with his description of a new genus, *Rhynchomeles,* for a new species from Ceram (Seram). His recognition of the genera *Isoodon, Echymipera* and *Peroryctes* as separate from *Perameles* has been followed by all subsequent authors.

In 1932, Stein added a further genus, *Microperoryctes,* based on a diminutive new species from the New Guinea highlands, and this too has been accepted by all subsequent commentators.

Tate and Archbold (1937) recognized all previously described genera in the New Guinea region, and in addition divided *Peroryctes* into two subgenera: the nominate subgenus for *P. raffrayana* and related forms, and a new subgenus *Ornoryctes* to contain *P. longicauda, P. ornata* and relatives.

Later however, Tate (1948) suppressed *Ornoryctes* on the grounds that the different species described for both subgenera were probably in each case reducible to a single species, and "nothing is gained by maintaining two monotypic subgenera".

In the same paper, Tate gave a proposed phylogeny. In this, *Macrotis* is indicated as the most divergent genus, followed by *Chaeropus;* then a line leading to *Echymipera* (and *Rhynchomeles*) diverging from a line from which *Perameles* diverges, leaving *Peroryctes* and *Microperoryctes* as terminal twigs. Inspection of his phylogeny, and careful reading of the accompanying text, reveals that he regarded what today would be called autapomorphic characters as all important: thus *Macrotis* and *Chaeropus* are considered to have separated early because each is so distinctive, and the line leading to the cluster of the remaining five genera is characterized in the phylogeny merely as "little modified". In a similar vein, the *Echymipera/Rhynchomeles* pair have "I^5 lost" written against their common stem, contrasting with the stem leading to the other three genera which is characterized by "I^5 retained".

Since Tate's 1948 revision, the main alteration has been the proposal (Archer and Kirsch 1977) to revive Bensley's Thylacomyinae but at family level, on the grounds that the dental morphology as elucidated by Bensley requires that the posterolingual molar cusp has been evolved in a different way in the Thyacomyidae and Peramelidae. Subsequently however, Archer (1984)

[1]Colin P. Groves, Department of Prehistory and Anthropology, Australian National University, G.P.O. Box 4, Canberra, Australian Capital Territory, Australia 2601.
[2]Tim Flannery, Australian Museum, 6–8 College Street, Sydney, New South Wales, Australia 2000.
Pages 1–11 *in* BANDICOOTS AND BILBIES, ed by J. H. Seebeck, R. L. Wallis, P. R. Brown and C. M. Kemper. Surrey Beatty & Sons, Sydney, 1990.

has acknowledged that such a step could leave the Peramelidae a paraphyletic taxon, and has anticipated the possibility that some future study may split up the Peramelidae and/or add further genera to the Thylacomyidae. Kirsch (1968) and Baverstock, Flannery, Aplin, Birrell and Krieg (this volume) have also made observations on the phyletic relations of some bandicoots.

MATERIALS AND METHODS

We have studied all the skulls of bandicoots in the Australian Museum (AM), noting down all the consistent differences between taxa that we could find. One of us (CPG) had previously studied specimens in the British Museum (Natural History) (BM), and photographs taken at the time have been examined for the characters discovered in the Australian Museum sample. One of us (TF) has examined material in the Museum of Victoria; CPG has examined specimens in the Queensland Museum. Finally, we have borrowed a specimen of *Microperoryctes murinus*, a species not represented in the London or Sydney collections, from the Bernice P. Bishop Museum, Honolulu, Hawaii (BBM).

At first, we kept species separate as far as possible, but it soon emerged that the polytypic genera *Perameles, Isoodon, Macrotis* and *Echymipera* are each definable by a suite of character states unique to the group: hence, these genera are all probably monophyletic groups. It was on the contrary necessary to keep the species groups within *Peroryctes* — the *raffrayana* and *longicauda* groups (Tate and Archbold's subgenera *Peroryctes* and *Ornoryctes* respectively) — separate throughout. We also found that *Peroryctes, sensu* Tate (1948), is a paraphyletic group (see below).

Our analysis is cladistic: we attempt to sort the character states into primitive and derived. The polarity we assign to genus-level characters depends of course on how we view those characteristic of suprageneric groups. Within each family, we can use members of the other family as an outgroup when sorting out the characters of the genera. As the bandicoots are not clearly closely related to any other marsupials — they have been shuffled back and forth by recent authors between the diprotodontan and dasyuroid clusters — the choice of a suitable outgroup for the group as a whole is a very difficult one.

In the end, and given the prevailing political climate, we opted for a consensus approach: what is the usual state of a given character among the Australian marsupials as a whole? If a feature is widespread, but not in the bandicoots, we considered its absence as probably derived for bandicoots. In a few cases, we resorted to a kind of "usual evolutionary polarity" approach: digital reduction is most probably derived over a state of complete digital development, hypocone enlargement can be demonstrated to have occurred in other groups of mammals, and so on. This "predominant-states" method has been criticized as probabilistic yielding globally parsimonious cladograms (Maddison, Donoghue and Maddison 1984), but we do not feel that outgroup relationships are yet well enough understood to allow us to use the simultaneous outgroup-ingroup method.

THE FAMILIES OF BANDICOOTS

The skulls of bandicoots fall into two distinct phenetic groups, differing in a large number of characters. Assessing the polarities of the character states differentiating the two, as described below, each seems to have a suite of synapormorphous states, so that each is plausibly monophyletic (Table 1). One group contains species of *Perameles, Isoodon, Chaeropus* and *Macrotis;* the other contains species of *Echymipera, Rhynchomeles, Peroryctes* and *Microperoryctes.* If, following Szalay (1982), the bandicoots as a whole are awarded subordinal status as Perameliformes, this basic division would seem best recognized at family level. The

Table 1a,b. Distinguishing characters of bandicoot families (a) derived characters of Peramelidae (b) derived characters of Peroryctidae.

(a)

1. Cranium dorsally flattened in cross-section
2. Anteromedial orbital margins slope diagonally outward, forming a ridge
3. Facial extension of lacrimal bone very narrow, its suture with maxilla parallel to antorbital ridge
4. Molars squared
5. Auditory bulla complete
6. Inferior mandibular margin convex
7. Stylar cusp *c* enlarged

(b)

1. Foramen rotundum prolonged into a tube facing anteriad
2. Anterior zygomatic root low, at alveolar level
3. Infraorbital fossa low, compressed
4. Main palatal vacuities long, narrow, with no tendency to fuse across midline
5. Mesopterygoid fossa broad, parallel-sided
6. Antorbital surface flattened
7. Gonial hooks directed posteriad
8. Superior temporal lines nearly parallel throughout, not converging further back

first group is therefore the Peramelidae (Gray 1825). As none of the second group has hitherto been made the type of a suprageneric grouping, we hereby derive the most euphonious family name for them, Peroryctidae n. fam.

The Peramelidae are aptly characterized (Table 1) as flat-skulled (Fig. 1). The Peroryctidae are cylindrical-skulled, the crania being much higher and subcylindrical in cross-section in both rostral and neurocranial portions (Fig. 2). As the excessively flattened peramelid-type

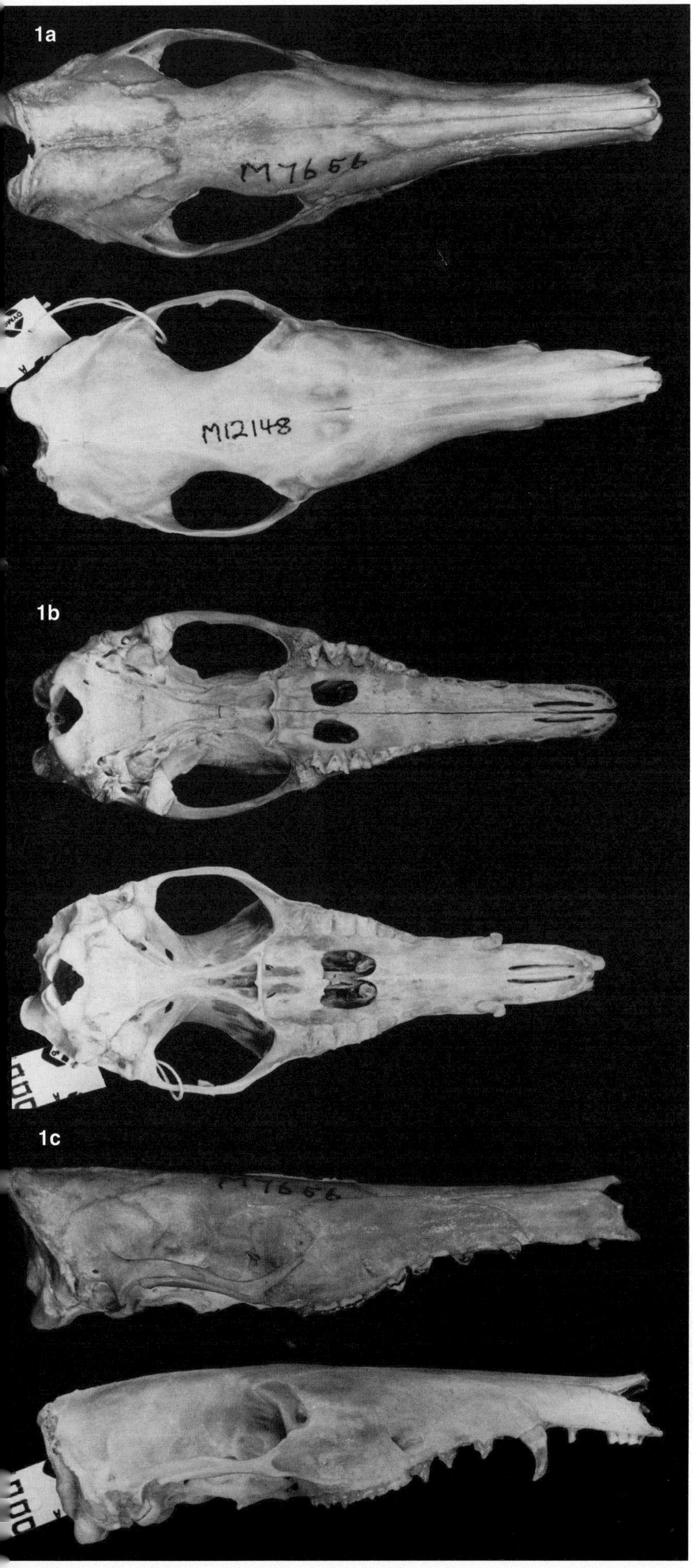

Fig. 1a,b,c. Crania of *Echymipera rufescens* (AM M7656) and *Perameles nasuta,* (AM M12148) to illustrate differences between Peroryctidae and Peramelidae. (a) dorsal view (b) ventral view (c) lateral view.

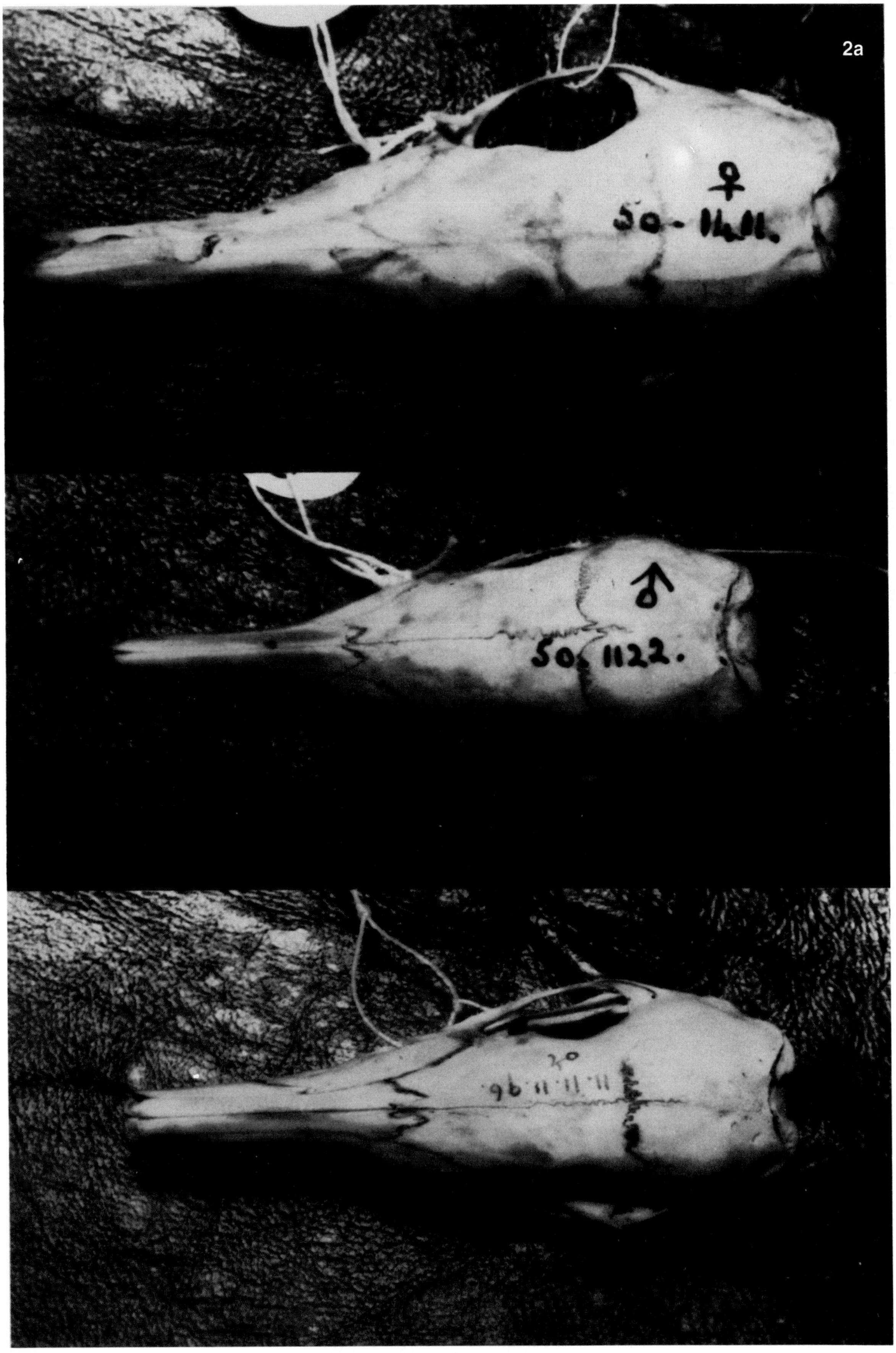
2a
50. 1122.

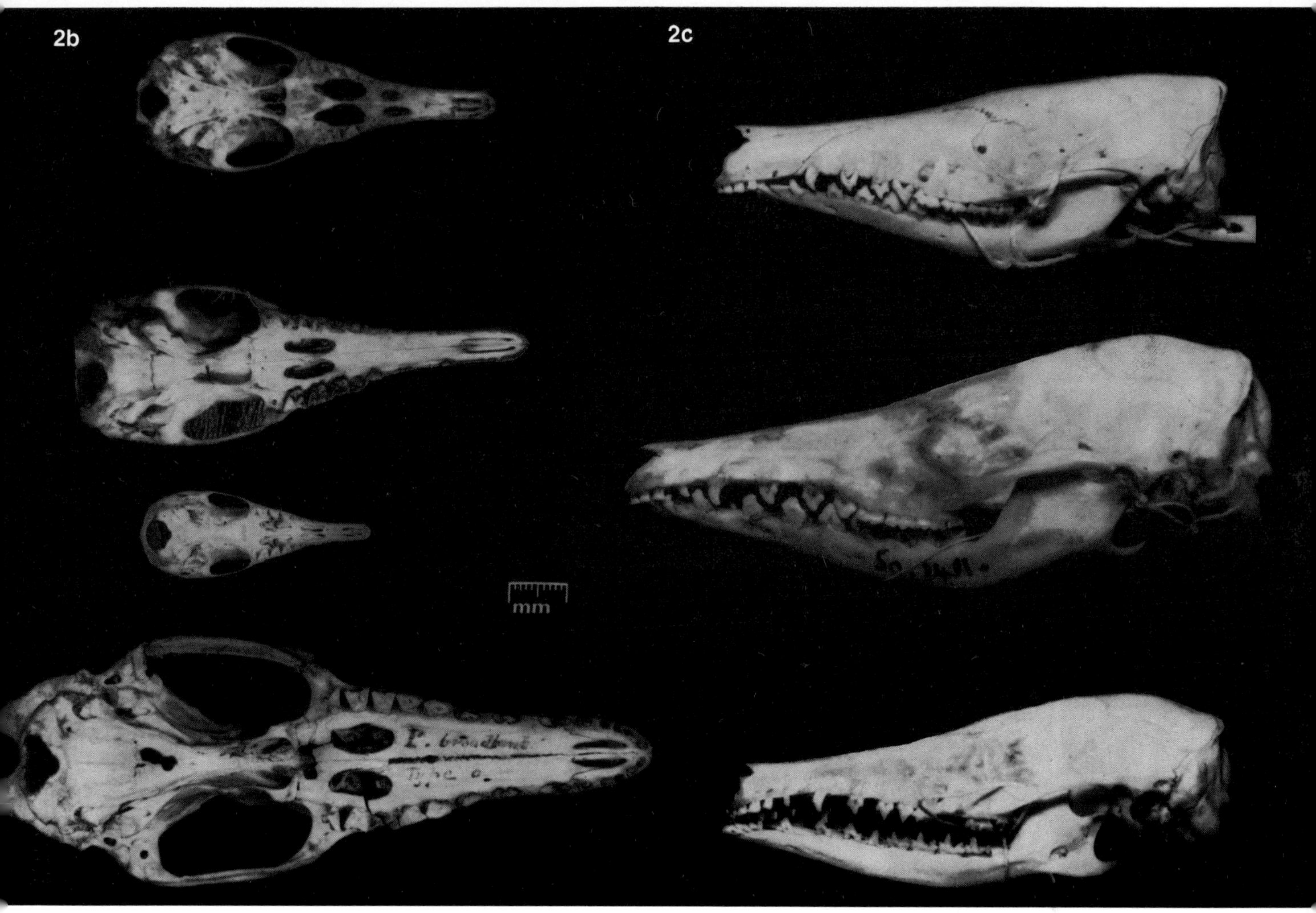

Fig. 2a,b,c (opposite and above). Cranial characters of genera of Peroryctidae. (a) *Echymipera kalubu* (BM 11.11.11.96), *Microperoryctes ornata* (BM 50.1122), *Peroryctes raffrayana* (BM 50.1411): dorsal view; (b) *Microperoryctes longicauda* (AM M9544), *Rhynchomeles prattorum* (BM NG 22457), *M. murina* (BBM NG 22457), *Peroryctes broadbenti* (AM A3238, the holotype): ventral view; (c) same specimens as (a): lateral view.

of skull is unusual in marsupials, we judge the Peramelidae to be derived in this feature. Similarly, we believe the lateral slope of the anteromedial orbital margin, with the remarkable ridge or crest along its edge, the reduction of the facial plate of the lacrimal, the convexity of the lower margin of the mandible, and all the features of Table 1(a), to be derived for the Peramelidae.

The most clearly-derived feature of the Peroryctidae is the curious tube-like foramen rotundum. (One of us, CPG, has observed a less pronounced, and differently formed, but in some ways superficially similar formation in some *Macrotis lagotis* skulls in the Queensland Museum. We must also admit that we are not certain of the identity of this foramen: it is at present under study by Dr Ray Tedman). The reduction of the alveolar plate in the molar region (so that the zygomatic arch swings forward low over the posterior molars), and the probably associated dorsoventral compression of the infraorbital fossa also seem likely to be derived. The characteristically incomplete state of the auditory bullae which we interpret as probably primitive in this family is a similar condition as that seen in didelphids. The characters listed in Table 1 in effect serve as the diagnosis of this new family.

We note with interest that the two family groups are, in effect, the rainforest (Peroryctidae) and the dry country (Peramelidae) forms respectively and draw the attention of zoogeographers and paleogeographers to the inferred primacy of this split. It seems unlikely that the differences between them relate directly to "wet" and "dry" adaptations, and so discount the possibility that either is polyphyletic though adaptive convergence.

Table 2a,b,c,d. Distinguishing characters of genera of Peroryctidae (a) synapomorphies shared between species of *Echymipera* and *Rhynchomeles* (b) synapomorphies shared between the species of *Microperoryctes* and the *Echymipera/Rhynchomeles* clade (c) synapomorph conditions of species here referred to *Microperoryctes* (d) synapomorph conditions of *Peroryctes*, shared with the *Echymipera/Rhynchomeles* clade and here interpreted as convergent.

(a)
1. Squamosal higher than in other Perameliformes
2. Rostrum projects forward, hardly at all downsloping
3. Palatal vacuities short, not extending forward of M^2
4. Paroccipital processes absent
5. Suprameatal foramen small *(Rhynchomeles)* to tiny *(Echymipera)*, lower, just above external auditory meatus
6. Occiput high, parallel-sided in rear view
7. Ventral margin of foramen magnum not deeply emarginate
8. Foramen magnum large, broadened
9. Posterior nasal sutures saw-edged
10. Nasals not so wedged back into frontals
11. Ascending ramus of mandible broad, nearly upright *(Echymipera)* to very broad but somewhat less upright *(Rhynchomeles)*
12. Precanine diastema of upper jaw very long
13. Mandibular diastema very elongated
14. Upper incisors reduced to four
15. Upper incisor edges straight, not pointed

(b)
1. Snout very narrow
2. Precanine diastema longer than in other Perameliformes
3. Anteromedial orbital margin quite flat, with no ridge
4. Nasal tips slightly upturned
5. Paroccipital process reduced in size, or absent
6. Premaxillae with lateral "wings" at nares
7. Zygomatic root moved forward, above $M^{3/4}$
8. Posterior nasal sutures together form a right angle, not an acute angle
9. Bulla lacking anterior spurs
10. Molars equal in size to premolars, or slightly larger
11. P^3 simple, pointed, not mesiodistally elongated
12. Canine entirely in maxilla, with alveolus not crossed by suture
13. Mandibular diastema elongated
14. Mammae reduced to three pairs

(c)
1. Braincase broadened
2. Coronal suture deeply wavy
3. Two pairs of palatal vacuities
4. Palatal vacuities shifted forward, extend from P^3 to M^3 or M^4
5. Posterior palatal margins straight, not bracket-shaped

(d)
1. Frontals send a deep wedge back between parietal and squamosal
2. Metatarsal elongated, 68–79% of foot length
3. Tail shortened

GENERIC RELATIONSHIPS IN THE PERORYCTIDAE

Tate (1948) aligned *Microperoryctes* with *Peroryctes* and *Rhynchomeles* with *Echymipera*, and we at first did likewise. It quickly became clear, however, that the features which unite a *Peroryctes/Microperoryctes* combination are wholly symplesiomorphic: not a single demonstrably derived state unites them. On the other hand, *Echymipera* and *Rhynchomeles* share many derived states (Table 2a), and they differ in only a few features. Species of *Echymipera* have a couple of more highly derived states (characters 5 and 11), as well as being derived in the intensely spiny fur, while *Rhynchomeles* has some more derived conditions of the skull (those pointed out by Thomas in his type description: the extreme length and slenderness of the muzzle, very small last molar and reduced, well-spaced cheekteeth in general) (Fig. 3).

When the species of *Peroryctes*, as conventionally conceived, are sorted into Tate's two subgenera, two surprising relationships become obvious. The first is that the characters of Tate and Archbold's *Ornoryctes* (the *longicauda/ornata* group) are also seen in *Microperoryctes murinus*: a fact missed by previous workers though perhaps hinted at by Tate (1948). The differences between them in fact concern mainly size; we have no alternative but to place the species of these groups in a single genus: *Microperoryctes* Stein, 1932 takes precedence over *Ornoryctes* Tate and Archbold, 1937. (It should be noted that Ziegler (1977:127) records that *M. murina* from the Bird's Head of Irian Jaya possess a dorsal stripe as do most taxa of so-called *Ornoryctes*.)

The second surprising result is recorded in Table 2(b): the genus *Microperoryctes*, as newly reconstituted, overwhelmingly shares derived character states with the *Echymipera/Rhynchomeles* clade. There are a few evident synapomorphies shared by all species of *Microperoryctes*, Table 2(c) confirming that the genus is, as constituted by us, monophyletic. There are also three features (Table 2(d)) in which *Peroryctes* resembles the *Echymipera* clade, but these are so outweighed by the list in Table 2(b) that they are most likely the result of convergence.

To conclude on the Peroryctidae: the genera *Echymipera* and *Rhynchomeles* form a clade, to which the genus *Microperoryctes* is the sister group; *Microperoryctes* contains the following species:

M. murina Stein, 1932
M. papuensis (Laurie, 1952)
M. longicauda (Peters and Doria, 1876)
M. ornata (Thomas, 1904) — if distinct from *longicauda* (Flannery, unpubl. data).

The genus *Peroryctes* is now left with two species:

P. raffrayana (Milne-Edwards, 1878)
P. broadbenti (Ramsay, 1879)

and it is the most divergent of the four genera in phyletic terms.

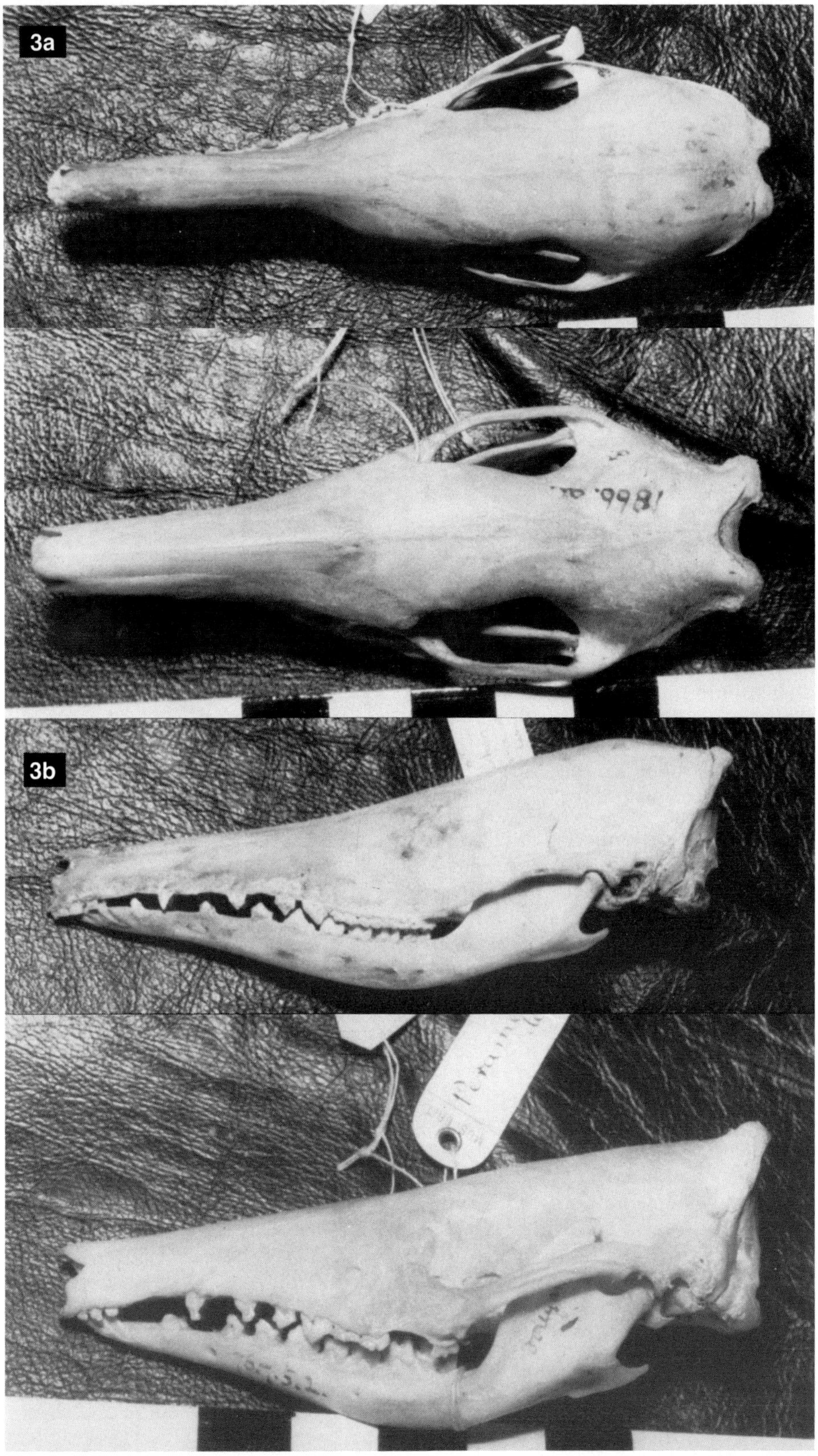

Fig. 3a,b. Echymipera kalubu (BM 1866a and BM 76.7.5.2 respectively), and *Rhynchomeles prattorum* (BM 20.7.26.36) crania: (a) dorsal view (b) lateral view.

GENERIC RELATIONSHIPS IN THE PERAMELIDAE

Interrelationships within the Peramelidae are not as clear cut as those within the Peroryctidae, with one exception: the genera *Macrotis* and *Chaeropus* share a very large number of derived character states (Table 3(a)) and are clearly monophyletic. The existence of all these common features, and their derived nature, has been lost amid the discussion, from Bensley (1903) via Tate (1948) to Archer and Kirsch (1977), of the undoubted wide phenetic separation of *Macrotis* from the other genera. The species of *Macrotis* have, for example, an extremely broadened braincase and narrowed snout, the most excessively flattened cranium of all, a directly forward-pointing rostrum, a long, thin paroccipital process, an enormous, pear-shaped bulla, a very elongated suprameatal foramen, lack hypocones on M^{2-4}, and lack paraconids on M_{3-5}: features which we identify as autapomorphous (Fig. 4).

Also, Kirsch (1968) found that *Macrotis* is well separated from a cluster containing *Perameles, Isoodon* and (reputedly) *Echymipera;* but of course, he did not (and could not) examine sera of *Chaeropus.* (It should be noted that he himself was skeptical of the validity of his findings on *Echymipera,* expressing dissatisfaction with the mode of preservation of the sample, and even querying its correct identification.)

Bensley (1903) drew attention to certain features of the molar teeth of *Macrotis,* especially the mode of formation of the distolingual cusp, which he showed is a metacone, not a true hypocone unlike other bandicoots. Archer and Kirsch (1977) interpreted this to mean that the two modes of development of this cusp had been entirely independent from an original tritubercular condition. We would draw attention, however, to presence of a tiny hypocone in one of Bensley's specimens; he illustrated this specimen, noting that it is rather unusual. It would seem more likely, therefore, that for whatever reason the *Macrotis* lineage has undergone an enormous expansion of the metacone, crowding out the hypocone in an already quadritubercular tooth. Moreover, as Figure 6 shows, *Chaeropus* does in fact show a less-modified version of the same condition: on M^4, the hypocone is very reduced, is not hypsodont as are the other cusps, and there is already an enlargement of the stylar cusps whose hypertrophy in *Macrotis* effectively displaces the main cusps lingually.

Isoodon and *Perameles* appear to share few or no derived states with each other. In both, the gonial hooks turn medially, and the suture on the paroccipital process runs down it, to (or anterior to) the tip; these are both different from the *Chaeropus/Macrotis* conditions, but we cannot assess their polarity and would not like to propose a clade on such evidence.

Table 3a,b,c. Characters of the genera of Peramelidae (a) synapomorphies of *Chaeropus* and *Macrotis* (b) derived conditions in common between *Chaeropus/Macrotis* clade and *Perameles* (c) conditions derived in common between the *Chaeropus/Macrotis* clade and *Isoodon.*

(a)

1. Cranial roof excessively flattened
2. Palatal vacuities enlarged, left and right vacuities fuse from an early age
3. Molar selenes separated by large gap
4. Suture between mastoid and exoccipital runs down paroccipital process and behind tip
5. Suprameatal foramen very big, elongated
6. Foramen magnum enlarged (convergent to *Echymipera/Rhynchomeles*), its anterior margin strongly emarginate
7. Posterior nasal sutures straight, not saw-edged
8. Glenoid fossa flattened
9. Inferior mandibular margin very convex
10. Rostrum projects forward with no downward slope
11. P_3 reduced, smaller than P_2
12. M_5 partly hidden behind ascending ramus
13. Stylar cusp *b* much enlarged on M_1
14. Hypocone much reduced or absent, much lower than other cusps
15. First pedal digit (hallux) absent
16. Tail crested
17. Ears greatly enlarged
18. Fur soft, no tendency to be spiny
19. Stylar cusp d on M^{1-3} large
20. Loss of the I^4/I^5 diastema (convergent with *Isoodon macrourus*)

(b)

1. First pedal digit reduced or absent
2. Foot elongated, its length more than 6.5 times its breadth
3. Sole hairy
4. Second and third pedal digits (syndactylous digits) at least somewhat reduced

(c)

1. Molars hypsodont
2. Bulla greatly enlarged, "pear-shaped" (less so in *Chaeropus*)
3. Precanine diastema short or absent, postcanine diastema reduced
4. Molars dominant to premolars
5. Sole wrinkled, with distinct pads

On the other hand, it is at present impossible to decide whether one of them is the sister group of the *Macrotis/Chaeropus* clade. Archer and Kirsch (1977) note that the presence of large, pear-shaped bullae in both *Macrotis* and *Isoodon* is probably the result of convergence, and cite the different position of the foramen ovale

Fig. 4a,b,c,d (opposite). Cranial characters of genera of Peramelidae. (a) *Macrotis lagotis* (AM S420), *Perameles nasuta* (AM M12148); *Isoodon macrourus* (AM M6561), *Chaeropus ecaudatus* (AM M422) dorsal view; (b) *Chaeropus, Isoodon, Macrotis, Perameles gunnii* (AM A6842), ventral view; (c) *Chaeropus, Isoodon, Macrotis,* lateral view; (d) Mandibles in lateral view of *Perameles nasuta* (AM M12148), *Echymipera rufescens* (AM M7656), *Macrotis* (AM M11205), *Chaeropus* (AM M422).

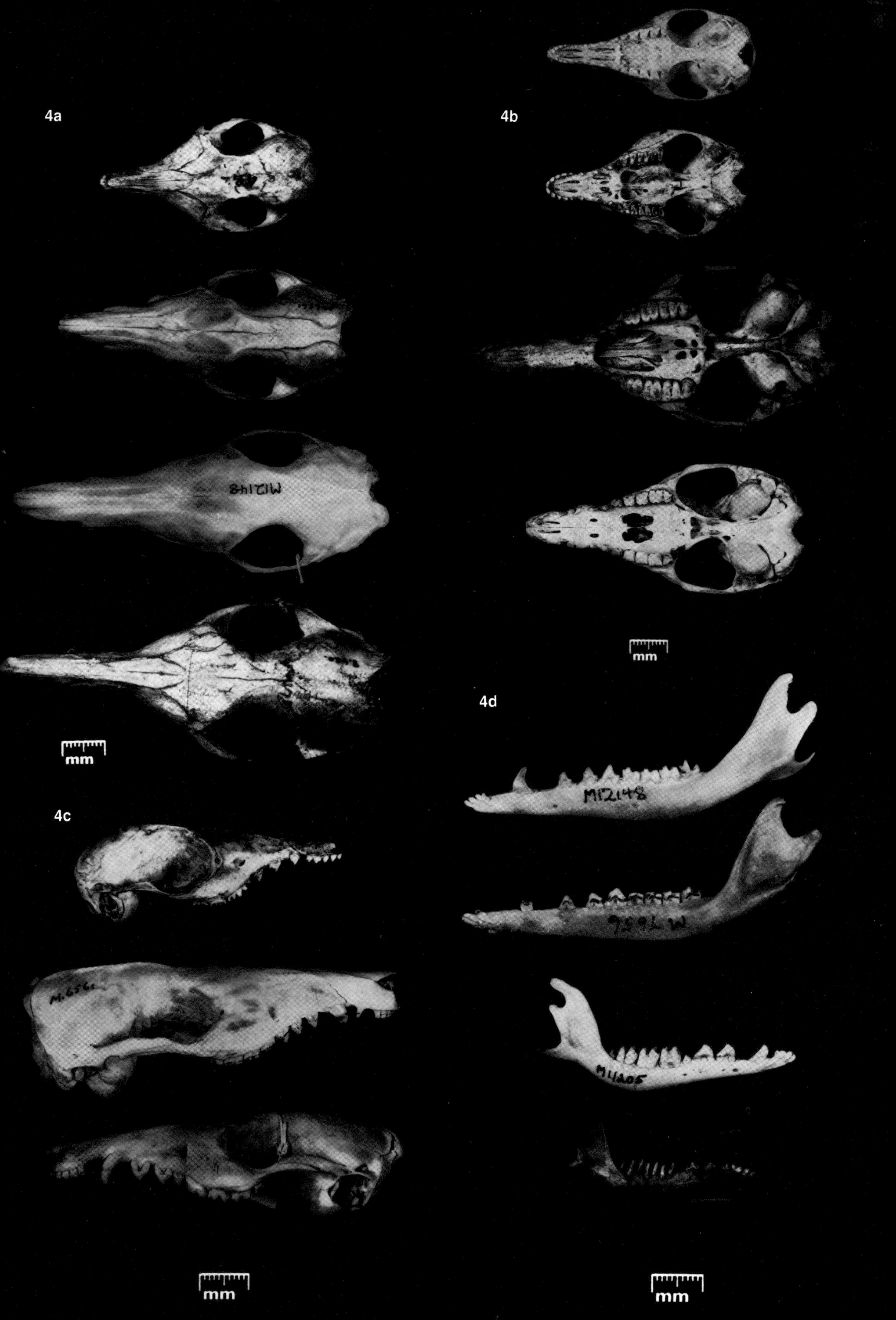
4a
4b
4c
4d
M12148
mm
mm
mm
mm

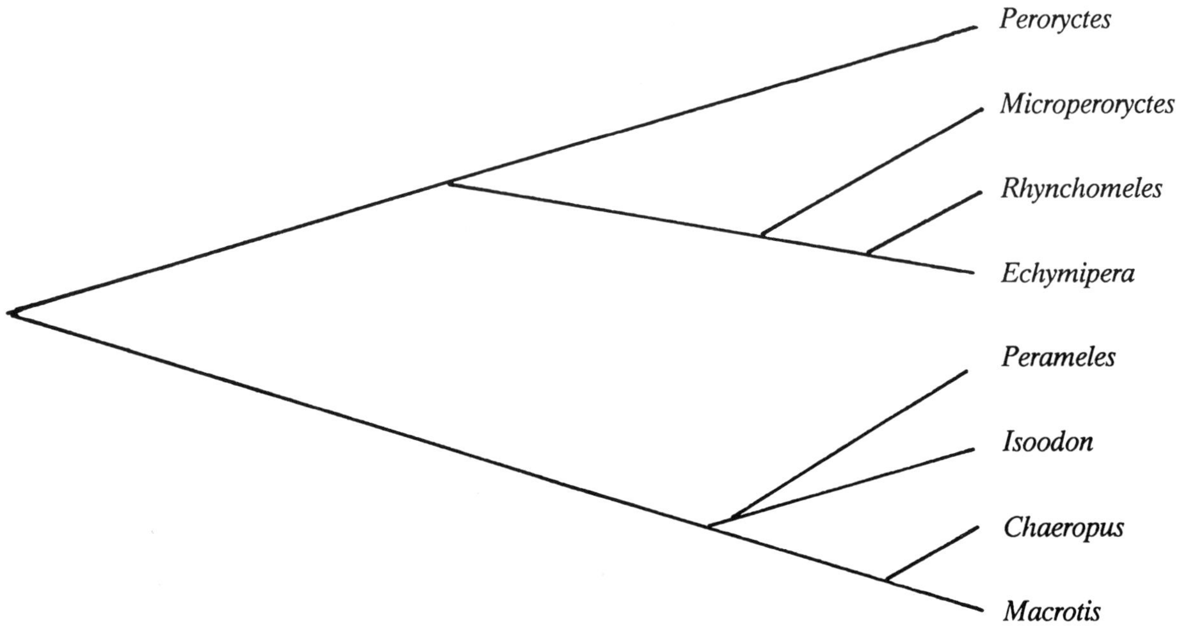

Fig. 5. A suggested cladogram of the Perameliformes.

relative to the bulla in these two forms as evidence of this. We concur with this assessment, and note in addition that the dental features listed in Table 3(c) are probably also convergent because the hypocone is hypsodont in the species of *Isoodon.* As the hypocone is not hyposodont in members of the *Chaeropus/ Macrotis* clade, it seems likely that hypsodonty has been independently derived in the two lineages. Four derived character states unite *Perameles* with the *Macrotis* and *Chaeropus* clade; five unite *Isoodon* with them. We find it impossible at present to decide between these alternatives; it is even possible that it is a genuine trifurcation, deriving from approximately synchronous divisions within a species showing clinal variation. (We use the term trifurcation here in preference to trichotomy.

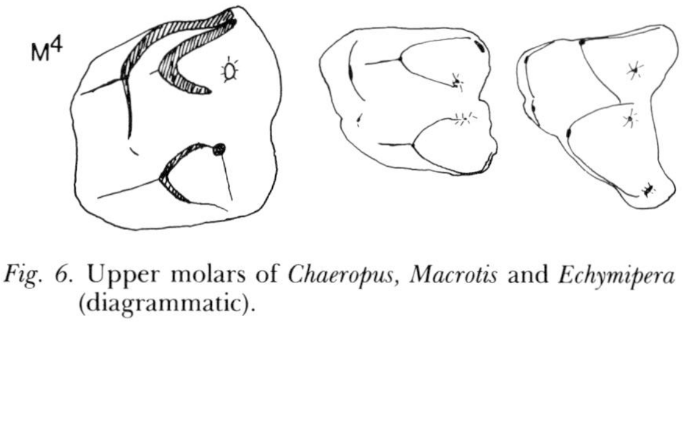

Fig. 6. Upper molars of *Chaeropus, Macrotis* and *Echymipera* (diagrammatic).

The Greek τομος means "a cutting", but the tricho- part could be derived equally well from "in three parts", or θριξ stem τριχ- "a hair". We do not wish to be labelled as hair splitters.)

That *Macrotis* and *Chaeropus* share so many derived states, yet each has in addition a long suite of uniquely derived features, is not very surprising. Lineages living in highly stressful environments — arid conditions, in the present instances — must of necessity have evolved very rapidly, piling on the autapomorphous states.

TAXONOMY OF THE PERAMELIFORMES

Within the Peroryctidae, where relationships are clear cut, there might be a case for erecting two subfamilies: one for *Peroryctes* alone, the other for the remaining three genera; within the latter, we might have two tribes, one for *Microperoryctes* and the other for *Echymipera* and *Rhynchomeles.* Where there are so few genera, however, it may not be especially helpful to do this.

This is even more the case in the Peramelidae. We could erect three subfamilies, but two would be monotypic. It has been suggested to us by an anonymous reviewer that a subfamily Thylacomyinae be retained to include *Macrotis* and *Chaeropus;* while we have much sympathy with this point of view, the uncertainty surrounding the relationships of *Perameles* and *Isoodon* makes us hesitate to do this, so at present we will leave the two families unsplit above the genus level.

SUMMARY AND CONCLUSIONS

1. The rainforest bandicoots form a family, Peroryctidae, separate from the dry-country/temperate forest one, Peramelidae. The family Thylacomyidae cannot be supported for the species of *Macrotis*.
2. The striped bandicoot, *"Peroryctes" longicauda* and its relatives, is separated from *P. raffrayana* and *broadbenti*, and referred to the genus *Microperoryctes*.
3. *Microperoryctes* is the sister group to the *Echymipera* clade rather than the *Peroryctes* clade.
4. *Macrotis* and *Chaeropus* are sister groups.

ACKNOWLEDGEMENTS

We gratefully acknowledge the assistance in various capacities of Ken Aplin, Mike Archer, Linda Gibson, John Edwards Hill, Paula Jenkins, Ralph Molnar, Steve van Dyck and Alan Ziegler.

REFERENCES

ARCHER, M., 1984. The Australian marsupial radiation. Pp. 633–808 in *Vertebrate zoogeography and evolution in Australia* ed by M. Archer and G. Clayton. Hesperian Press: Carlisle, Western Australia.

ARCHER, M. AND KIRSCH, J. W., 1977. The case for Thylacomyidae and Myrmecobiidae, Gill, 1872, or why are marsupial families so extended? *Proceedings of the Linnean Society of New South Wales* **102:** 18–25.

BENSLEY, B. A., 1903. On the evolution of the Australian Marsupialia: with remarks on the relationships of the Marsupials in general. *Transactions of the Linnean Society of London, Zoology (2)* **9:** 83–217.

GRAY, J. E., 1825. Outline of an attempt at the disposition of the Mammalia into tribes and families with a list of the genera apparently appertaining to each tribe. *Annals of the Philosophical Society* **10:** 336–44.

KIRSCH, J. A. W., 1968. Prodromus of the comparative serology of Marsupialia. *Nature* **217:** 418–20.

MADDISON, W. P., DONOGHUE, M. J. AND MADDISON, D. R., 1984. Outgroup analysis and parsimony. *Systematic Zoology* **33:** 83–103.

STEIN, G., 1932. Einige neue beuteltiere aus Neuguinea. *Zeitschrift für Saeugetierkunde* **7:** 254–57.

SZALAY, F. S., 1982. Phylogenetic relationships of the marsupials. *Geobios, Memoire speciale* **6:** 177–90.

TATE, G. H. H., 1948. Results of the Archbold Expeditions. No. 60. Studies in the Peramelidae (Marsupialia) *Bulletin of the American Museum of Natural History* **92:** 317–46.

TATE, G. H. H. AND ARCHBOLD, R., 1937. Results of the Archbold Expeditions. No. 16. Some marsupials of New Guinea and Celebes. *Bulletin of the American Museum of Natural History* **73:** 331–476.

THOMAS, O., 1888. *Catalogue of Marsupialia and Monotremata.* British Museum (Natural History): London.

THOMAS, O., 1920. On mammals from Ceram. *Annals and Magazine of Natural History* (9) **6:** 422–31.

ZIEGLER, A., 1977. Evolution of New Guinea's marsupial fauna in response to a forested environment. Pp. 117–138 in *The biology of marsupials* ed by B. Stonehouse and D. Gilmore. Macmillan: London.

Albumin immunologic relationships of the bandicoots (Perameloidea: Marsupialia) — a preliminary report

P. R. BAVERSTOCK[1], T. FLANNERY[2], K. APLIN[3], J. BIRRELL[1] and M. KRIEG[1]

Microcomplement fixation of albumin was used to examine the phylogenetic affinities of the bandicoots to other marsupials, and to assess relationships among the bandicoots. The results show that the dasyurids are the sister group to the bandicoots. Among the bandicoots available to us, *Perameles* and *Isoodon* are closely related and possibly paraphyletic with each other. *Macrotis, Echymipera* and the *Isoodon/Perameles* group form an unresolved trichotomy i.e., there is no support for a family Thylacomyidae (including *Macrotis*) separate from Peramelidae (including *Echymipera, Isoodon* and *Perameles*).

Key words: Bandicoot, phylogeny, microcomplement fixation, serology.

INTRODUCTION

MICROCOMPLEMENT fixation (MC'F) is an immunological technique that measures the extent of evolutionary divergence between homologous proteins of various taxa. Because proteins are a direct reflection of the DNA, MC'F data indicate the extent of genetic divergence for a particular length of DNA.

Briefly, a protein (usually plasma albumin) is purified and injected into several (usually three) rabbits, which raise antisera to that protein. The antisera are then cross-reacted with the homologous protein in other taxa, and the strength of cross-reaction measured by complement fixation to yield an Immunologic Distance (ID). It has been shown that the ID is directly proportional to the extent of amino acid sequence divergence between taxa, 1 ID being roughly equivalent to one amino acid substitution (Wilson, Carlson and White 1977). A more detailed exposé of the technical details is given in Champion, Prager, Wachter and Wilson (1974), and in Baverstock (1984).

It is important to appreciate the limitations of the technique of MC'F in phylogenetic analysis. Firstly, there are the purely technical limitations. With a single antibody (consisting of a pool from three rabbits), the reproducibility of the estimate of immunological distance is excellent being ± 1 or 2. However, reproducibility between antisera (i.e., between pools of three different rabbits) is poorer, of the order of ± 8 units (unpub. data).

Secondly, there is the problem of *reciprocity* (i.e., anti-A to B *versus* anti-B to A). This problem can be partly overcome by the use of a correction factor or CF (Sarich and Cronin 1976); that is, some antisera give consistently high or consistently low IDs compared to the antigen against which they were raised, and a CF is used to correct for this. However, even after correction, non-reciprocity usually remains. The non-reciprocity is estimated by the per cent standard deviation. (Maxson and Wilson 1975) and is usually of the order of 10–20% (Maxson, Tyler and Maxon 1982).

Finally, there is the problem of phylogenetic interpretation of the data i.e., tree-building. Numerous phylogenetic tree-building methods have been suggested, all of which attempt to produce a tree that minimizes or maximizes some quantity. This is known as *parsimony*. For reasons that will be outlined in more detail in a future paper, we prefer the Fitch-Margoliash criterion of parsimony over other criteria for MC'F data (Fitch and Margoliash 1967). However, any tree that is built is at best an approximation of the phylogenetic information supplied by the original data.

In view of the above limitations of MC'F data, how should they be interpreted phylogenetically? Our approach to this problem has been elaborated upon in more detail elsewhere (Richardson, Baverstock and Adams 1986). Briefly, we take the current classification as a null hypothesis of the phylogenetic relationship of the group concerned. The phylogenetic tree derived from the MC'F data is then compared to the classical picture. Where there is concordance, the MC'F data are taken as supporting the null hypothesis. Where there are discrepancies, the MC'F data are considered to suggest alternative and competing hypotheses. Additional independent data are then required to distinguish between the two hypotheses.

[1]Evolutionary Biology Unit, South Australian Museum, North Terrace, Adelaide, South Australia, Australia 5000.
[2]Australian Museum, 6–8 College Street, Sydney, New South Wales, Australia 2000.
[3]School of Zoology, University of New South Wales, P.O. Box 1, Kensington, New South Wales, Australia 2033.
Pages 13–18 *in* BANDICOOTS AND BILBIES, ed by J. H. Seebeck, P. R. Brown, R. L. Wallis and C. M. Kemper. Surrey Beatty & Sons, Sydney, 1990.

In order to fully resolve the branching relationships among taxa, it is necessary to raise antiserum to albumin from each of those taxa, and to cross-react each antiserum to albumin of each of the other taxa. Moreover, for the purposes of phylogenetic analysis, it is necessary to assess the extent of differential rates of albumin evolution in the various taxa. This can be done by using the outgroup method i.e., a taxon outside the group in question is included in the analysis (Sarich 1969). Ideally, the outgroup should be as close to the group in question as possible.

Our study of the Perameloidea is as yet far from complete and this should therefore be viewed as a preliminary report only. Firstly, we do not have material from any species of *Peroryctes,* and from only one species of *Echymipera.* Secondly, of the material we do have, the cross-reactions are as yet incomplete. Nevertheless the data do suggest that some of the more commonly held views about bandicoot evolution may be erroneous, and therefore indicate areas requiring more detailed analyses.

HISTORICAL REVIEW

The phylogenetic relationships of bandicoots to other marsupials have long been controversial, as, indeed, have the relationships of the bandicoots to each other. By the late nineteenth century most bandicoot species had become known to science, and the first comprehensive revision of the bandicoots, that of Thomas (1888) could be undertaken. Thomas argued that bandicoots are close relatives of dasyurids, and that syndactyly must have been independently derived in perameloids and diprotodontans. He recognized three bandicoot genera only, *Peragale* (for the species of *Macrotis*), *Chaeropus* (for *Chaeropus ecaudatus*) and *Perameles* (containing species recognized at present as belonging within the genera *Perameles, Isoodon, Peroryctes,* and *Echymipera*). In the 30 years following his 1888 review, Thomas split the species of the genus *Perameles* into the genera (listed above) recognized at present.

Bensley (1903) offered a different view of bandicoot relationships. He suggested that they are monophyletic with diprotodontans, and not close relatives of dasyurids. Bensley was also the first to suggest that the species of *Macrotis* should be placed within their own subfamily. He based this decision on the molar morphology of *Macrotis* species, emphasizing the absence of a hypocone in these forms. Despite the subfamilial status given to *Macrotis* species, Bensley's (1903:195) diagram illustrating the relationship of the bandicoots shows the species of *Macrotis* on a branch leading from *Perameles bougainville,* which itself is placed within the Peramelinae.

Thus Bensley's Peramelinae is a paraphyletic grouping, and *Macrotis* is envisaged by him as having evolved from *Perameles*-like ancestors.

Tate's (1948) study, on the other hand, is basically a phenetic division of the Peramelidae in which the most highly derived species (*Macrotis* and *Chaeropus*) branch first, and the more plesiomorphic species (*Perameles, Peroryctes* and *Microperoryctes*) are shown as close relatives.

Archer and Kirsch (1977) integrated much of Bensley's (1903) morphological data and Martin and Hayman's (1967) chromosome data along with a basicranial and immunological study and suggested that the species of *Macrotis* and the fossil *Ischnodon* should be placed within their own family, Thylacomyidae. Archer and Kirsch (1977) argue also that the Thylacomyidae are not derived from within the Peramelidae, but that both families are derived from an unknown ancestor.

Kirsch (1977) explored the relationships of the bandicoots by using antisera to whole plasma (so-called polyvalent antisera). He concluded that the bandicoots and the dasyurids were sister taxa.

Groves and Flannery (this volume) suggest that the bandicoots can be divided into two monophyletic groups. One group contains the predominantly New Guinean species of the genera *Peroryctes, Microperoryctes, Echymipera* and *Rhynchomeles*; the other the predominantly Australian species of the genera *Macrotis, Chaeropus, Isoodon* and *Perameles.* They further suggest that *Macrotis* and *Chaeropus* are monophyletic.

In the present study, we have used MC'F to address two general questions:

1) What are the relationships between the bandicoots and the other families of Australasian marsupials?
2) What are the relationships among the Australasian bandicoots?

MATERIALS AND METHODS

In order to test the albumin immunologic relationships between the bandicoots (superfamily Perameloidea) and the other superfamilies of Australasian marsupials, we raised antisera (three rabbits per antisera) to albumins of the following taxa:

Isoodon macrourus)	superfamily
Macrotis lagotis)	Perameloidea
Dasyurus viverrinus	— superfamily Dasyuroidea
Cercartetus nanus	— superfamily Phalangeroidea
Phascolarctos cinereus	— superfamily Vomatoidea
Tarsipes rostratus	— superfamily Tarsipedoidea

Caluromys lanatus — superfamily Didelphoidea — (antigen only) was included as an outgroup. Antigen of *Notoryctes typhlops* (Notoryctoidea) was also included in this set of reactions.

To explore relationships among the bandicoots, antisera were raised to albumins of the following species — *Isoodon macrourus, Perameles nasuta, Echymipera kalubu, Macrotis lagotis, Dasyurus viverrinus.*

Dasyurus viverrinus was included in these reactions as an outgroup. Cross-reactions were also performed involving antigens of *Perameles gunnii, Isoodon obesulus* and *I. auratus.*

RESULTS

Reactions among superfamilies

Table 1 shows the Albumin Immunological Distances (AIDs) among representatives of the six recognized superfamilies of extant Australian marsupials. Because we lack antisera to either *Caluromys* or *Notoryctes,* it was not possible to build trees simultaneously including both these taxa on the same tree. Therefore trees were first constructed excluding *Notoryctes.*

Table 1. Corrected IDs among albumins of representative taxa of each of the superfamilies of Australasian marsupials. *Caluromys lanatus* was included as an outgroup. CF is the correction factor for reciprocity (Sarich and Cronin 1976). % SD for reciprocity = 5.7%

		Antibody					
		A	B	C	D	E	F
Isoodon macrourus	A	0	54	61	76	87	90
Macrotis lagotis	B	59	0	65	80	91	110
Dasyurus viverrinus	C	60	72	0	64	102	78
Cercartetus nanus	D	76	74	65	0	92	54
Phascolarctos cinereus	E	76	105	106	78	0	68
Tarsipes rostratus	F	95	97	86	69	76	0
Notoryctes typhlops	G	79	109	99	70	116	112
Caluromys lanatus	H	80	81	63	62	90	90
	CF	0.94	1.19	1.10	0.94	0.86	1.08

Trees were built and assessed for goodness of fit by the Fitch-Margoliash criterion (Fitch and Margoliash 1967) using the PHYLIP package of J. Felsenstein. The order of input of the taxa was varied using the SHUFFLE option. The best-fit tree of over 200 tried was rooted using *Caluromys* as the outgroup, and the rooted tree is shown in Fig. 1.

The best-fit tree suggests that Dasyuroidea is the sister group of Perameloidea, and that the Dasyuroidea/Perameloidea cluster forms a sister group to the Tarsipedoidea/Vombatoidea/ Phalangeroidea cluster.

Next, unrooted trees were constructed from the data in Table 1, but this time including *Notoryctes* and excluding *Caluromys.* The resulting best-fit tree showed *Notoryctes* branching from the line leading directly to *Cercartetus.* Therefore, using the information in Fig. 1, the unrooted tree could be rooted on the branch between the Dasyuroidea/Perameloidea cluster and the Tarsipedoidea/Vombatoidea/Phalangeroidea cluster. The resulting rooted tree is shown in Fig. 2, which indicates no special affinity of Notoryctoidea to Perameloidea.

Reactions among genera and species

The next step in the analysis was to explore relationships among the extant genera of bandicoots. Table 2 shows the results of reciprocal cross-reactions among *M. lagotis, I. macrourus, P. nasuta* and *E. kalubu.* A best-fit unrooted tree for these data was constructed by the methods described previously. Unfortunately, this tree could not be rooted, because reciprocal reactions involving *Dasyurus* as an outgroup have not been performed for all of these taxa. However, we do have reciprocal cross-reactions involving *D. viverrinus* with *I. macrourus* and *M. lagotis* (Table 3). These data show that

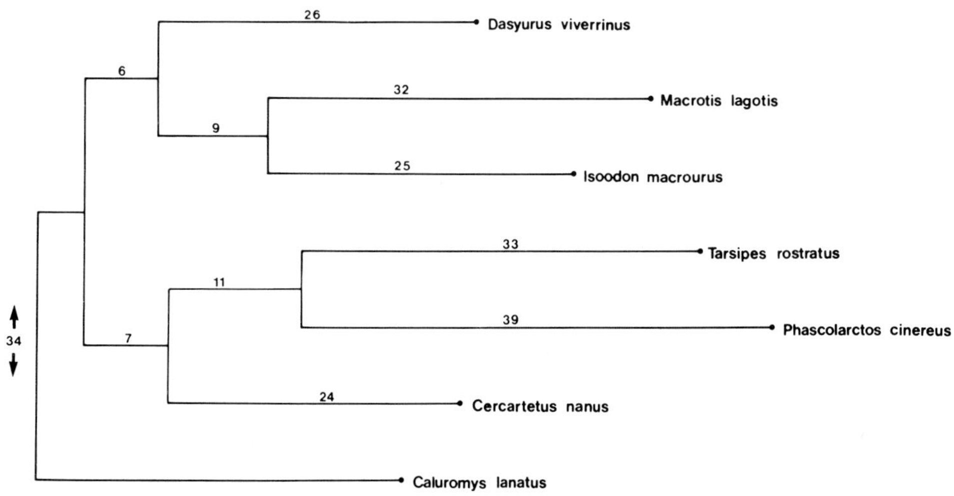

Fig. 1. Best-fit tree as assessed by the Fitch-Margoliash criterion for the albumin immunologic relationships among representatives of the superfamilies of Australasian marsupials including Notoryctoidea and using *Caluromys* as an outgroup. The tree shown is the best fit of over 200 tried. Per cent standard deviation for this tree 7.2%. The values indicate the amount of albumin evolution (in immunologic units) that has occurred along each branch.

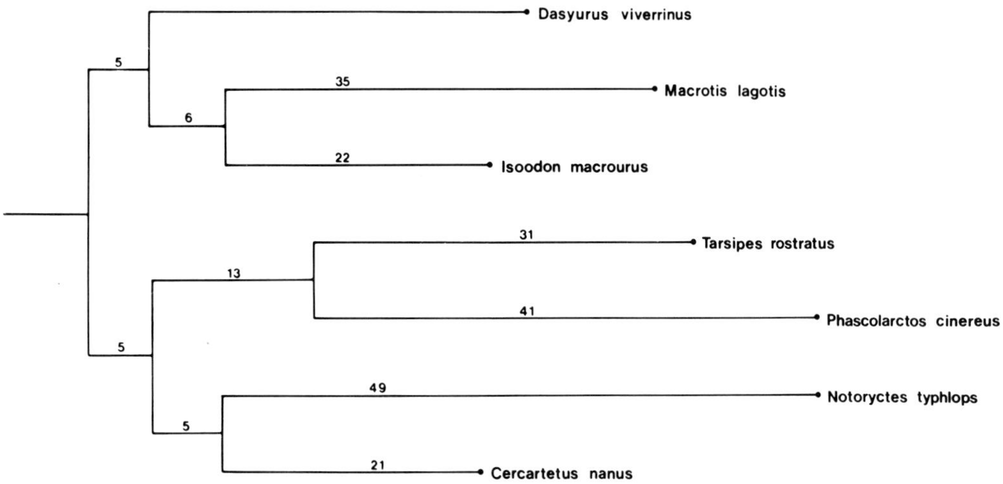

Fig. 2. Best-fit tree as assessed by the Fitch-Margoliash criterion for the albumin immunologic relationships among representatives of the superfamilies of Australian marsupials including Notoryctoidea. The tree shown is the best fit of over 200 tried. Per cent standard deviation for this tree is 8.5%. The values indicate the amount of albumin evolution (in immunologic units) that has occurred along each branch.

Table 2. Corrected AIDs between *Macrotis lagotis, Isoodon macrourus, Perameles nasuta* and *Echymipera kalubu.* % SD for reciprocity = 12.8%.

Antigen		Antibody			
		A	B	C	D
Macrotis lagotis	A	0	53	49	52
Isoodon macrourus	B	53	0	10	40
Perameles nasuta	C	49	5	0	41
Echymipera kalubu	D	50	46	36	0
	CF	1.19	0.83	1.08	0.97

the average ID of *D. viverrinus* to *I. macrourus* is 57 units, and of *D. viverrinus* to *M. lagotis* is 65 units. The unrooted tree involving *Isoodon, Macrotis, Echymipera* and *Perameles* was therefore rooted in such a way that the ratio of these distances was maintained (Fig. 3). This tree shows that *I. macrourus* and *P. nasuta* form a very tight grouping. However, the relationships among *Macrotis, Echymipera* and the *Isoodon/Perameles* cluster remain unresolved by these data.

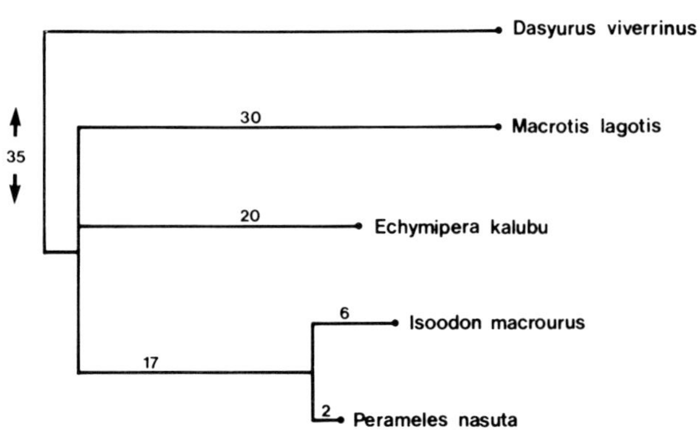

Fig. 3. A composite tree drawn from the data in Tables 2 and 3 showing the albumin immunologic relationships among the genera of bandicoots. The values indicate the amount of albumin evolution (in immunologic units) that has occurred along each branch.

The third step in the analysis involved cross-reactions of *I. macrourus* and *P. nasuta* to the various species of *Isoodon* and *Perameles* available to us, namely *I. obesulus, I. auratus,* and *P. gunnii.* The raw (uncorrected) data are shown in Table 4. The surprising feature of this table is that in neither genus do the species form a tight grouping.

Table 3. Corrected AIDs between *Dasyurus viverrinus, Macrotis lagotis,* and *Isoodon macrourus.* % SD for reciprocity = 4.2%.

Antigen		Antibody		
		A	B	C
Dasyurus viverrinus	A	0	55	66
Isoodon macrourus	B	60	0	49
Macrotis lagotis	C	64	54	0
	CF	1.08	0.86	1.08

Table 4. Uncorrected AIDs between albumins of various *Isoodon* and *Perameles* species to antisera of *Isoodon macrourus (Im)* and *Perameles nasuta (Pn).*

Antigen	Antibody	
	Im	*Pn*
Isoodon macrourus	0	11
Isoodon obesulus	9	12
Isoodon auratus	10	10
Perameles nasuta	7	0
Perameles gunnii	15	9

DISCUSSION

Our MC'F data on albumin suggest that the Dasyuroidea is the sister group to the Perameloidea. The evidence for such a relationship is not strong, with a total of only six units of albumin evolution being common to both these groups (Fig. 1). Maxson, Sarich and Wilson (1975) reached a similar conclusion from their albumin MC'F studies of marsupials. Moreover, Kirsch (1977), using polyvalent antisera, also concluded independently that the Dasyuroidea and Perameloidea were sister groups. All three studies agree in showing the branch common to these taxa as very short. Thus if the concept of a molecular clock (Maxson *et al.* 1975) is correct, then these data suggest that the Dasyuroidea/Perameloidea divergence occurred soon after the origin of the Australian marsupials.

If indeed the primary dichotomy of the Australian marsupials is Dasyuroidea plus Perameloidea versus the rest, then it implies that the common possession of syndactyly by Perameloidea and Diprotodonta is due to convergence (Thomas 1888), or that syndactyly was the character state possessed by the common ancestor of the Australian marsupials, and was secondarily lost in the Dasyuroidea. Further studies of foot structure of Australian and American marsupials will be needed to determine which of these possibilities might apply.

At the family level, the albumin MC'F data do not show the clear separation that might be expected based on Archer and Kirsch (1977) between *Macrotis* and the other genera of bandicoots. As indicated in the results, this assessment of albumin evolution in the Perameloidea is only tentative at this stage, since all possible pair-wise cross-reactions have yet to be completed. If the tree shown in Fig. 2 holds up under closer scrutiny, it would conflict with Kirsch's (1977) assessment, which showed *Macrotis* standing well apart from the other genera of bandicoots, and which in part led Archer and Kirsch (1977) to propose two families for bandicoots: Thylacomyidae for *Macrotis*, and Peramelidae for the remaining genera. However, he (like us) did not have *Peroryctes*, and his reactions involving *Echymipera* gave equivocal results. As Kirsch (1977) points out, his sample of *Echymipera* may not have been trustworthy.

A cladistic analysis of skull characters (Groves and Flannery, this volume) also concluded that *Macrotis* is monophyletic with a *Perameles/Isoodon* group to the exclusion of *Echymipera* and the other New Guinean genera, in full accord with the molecular data presented here.

Karyotypic data do not help resolve the conflict. Although *M. lagotis* has a highly derived karyotype, *Echymipera, Isoodon* and *Perameles* all possess the proposed ancestral karyotype for marsupials (Hayman and Martin 1974; Rofe and Hayman 1985). Thus the karyotypic similarities of *Echymipera, Isoodon* and *Perameles* are symplesiomorphic while the unusual karyotype of *Macrotis* is autapomorphic. Therefore karyotypic data yield no useful phylogenetic information in this case.

If we assume an average rate of albumin evolution for Australian marsupials of about 0.8 units per million years (Maxson *et al.* 1975; Baverstock 1984), then approximate dates of divergence can be calculated from Figs 1 to 3. The primary split between Dasyuroidea plus Perameloidea and the rest of the Australian marsupials would have occurred about 48 million years (MY) ago, about the time of the splitting of Gondwanaland. The Dasyuroidea/Perameloidea split would have occurred soon after. The trichotomy shown in Fig. 3 between *Macrotis, Echymipera* and the *Isoodon/Perameles* group dates at about 32 MY i.e., the Oligocene. The *Isoodon/Perameles* split is very recent, perhaps Pliocene.

The cross-reactions involving the species of *Isoodon* and *Perameles* are especially interesting. If these genera were clearly defined we would expect all *Isoodon* to be closer to *I. macrourus* than *Perameles* species, and all *Perameles* to be closer to *P. nasuta* than *Isoodon* species. As Table 4 shows, no such clear separation emerges. Therefore the data raise the possibility that one or other of the genera *Perameles* and *Isoodon* may be paraphyletic. The morphologic data defining these genera should therefore be reassessed cladistically. Unfortunately, albumin MC'F at this close level will not be able to resolve the pattern (see introduction), and other data such as morphology, allozyme electrophoresis, or MC'F of a faster evolving protein such as transferrin will be required.

We stress once again that the present analysis should be treated as a preliminary report only. However, it has indicated those areas requiring further study, namely i) the possibility that *Macrotis* may be more closely related to other bandicoots than was previously thought, ii) that *Echymipera* may represent a lineage distinct from *Isoodon* and *Perameles* (a possibility that emphasizes the need to include *Peroryctes* in the analysis), and iii) the possibility that the genera *Perameles* and *Isoodon* may be paraphyletic.

ACKNOWLEDGEMENTS

We are grateful to the many persons who contributed the material that made this study possible (see Appendix). Special thanks are due to Dr G. Maynes for his generous assistance with the collection of Papua New Guinea material. B. Young, S. Merrett and P. Altman assisted with the care of rabbits; P. Kidd typed the manuscript and R. Andrews gave critical comments on it. We are especially grateful to A. Gunjko for assistance with computer analyses.

REFERENCES

ARCHER, M. AND KIRSCH, J. A. W., 1977. The case for Thylacomyidae and Myrmecobiidae, Gill, 1872, or why are marsupial families so extended? *Proceedings of the Linnean Society of New South Wales* **102:** 18–25.

BAVERSTOCK, P. R., 1984. The molecular relationships of Australasian possums and gliders. Pp. 1–8 in *Possums and gliders* ed by A. P. Smith and I. D. Hume. Australian Mammal Society: Sydney.

BENSLEY, B. A., 1903. On the evolution of the Australian Marsupialia; with remarks on the relationships of the marsupials in general. *Transactions of the Linnean Society of London, Zoology* **9:** 83–217.

CHAMPION, A. B., PRAGER, E. M., WACHTER, D. AND WILSON, A. C., 1974. Microcomplement fixation. Pp. 397–416 in *Biochemical and immunological taxonomy of animals* ed by C. A. Wright. Academic Press: London.

FITCH, W. M. AND MARGOLIASH, E., 1967. Construction of phylogenetic trees. *Science* **155:** 279–84.

HAYMAN, D. L. AND MARTIN, P. G., 1974. Mammalia I: Monotremata and Marsupialia, Vol. 4: Chordata 4 in *Animal cytogenetics* ed by B. John. Gebruder Borntraeger: Berlin-Stuttgart.

KIRSCH, J. A. W., 1977. The comparative serology of Marsupialia, and a classification of marsupials. *Australian Journal of Zoology* Suppl. Ser. **52:** 1–152.

MARTIN, P. G., AND HAYMAN, D. L., 1967. Quantative comparisons between the karyotypes of Australian marsupials from three different superfamilies. *Chromosoma (Berlin)* **20:** 290–310.

MAXSON, L. R. AND WILSON, A. C., 1975. Albumin evolution and organismal evolution in tree frogs (Hylidae). *Systematic Zoology* **24:** 1–15.

MAXSON, L. R., SARICH, V. M. AND WILSON, A. C., 1975. Continental drift and the use of albumin as an evolutionary clock. *Nature* **255:** 397–400.

MAXSON, L. R., TYLER, M. J. AND MAXSON, R. D., 1982. Phylogenetic relationships of *Cyclorana* and the *Litoria aurea* species-group (Anura:Hylidae): a molecular AND perspective. *Australian Journal of Zoology* **30:** 643–51.

RICHARDSON, B. J., BAVERSTOCK, P. R. AND ADAMS, M., 1986. *Allozyme electrophoresis; a handbook for animal systematics and population studies.* Academic Press: Sydney.

ROFE, R. AND HAYMAN, D. L., 1985. G-Banding evidence for a conserved complement in the Marsupialia. *Cytogenetics and Cell Genetics* **39:** 40–50.

SARICH, V. M., 1969. Pinniped origins and the role of evolution of carnivore albumins. *Systematic Zoology* **18:** 286–95.

SARICH, V. M. AND CRONIN, J. E., 1976. Molecular systematics of the primates. Pp. 141–70 in *Molecular anthropology* ed by M. Goodman and R. E. Tashian. Plenum Press: New York.

TATE, G. H. H., 1948. Results of the Archbold Expeditions No. 60. Studies in the Peramelidae (Marsupialia). *Bulletin of the American Museum of Natural History* **92:** 317–46.

THOMAS, O., 1888. *Catalogue of the Marsupialia and Monotremata in the collection of the British Museum (Natural History).* British Museum: London.

WILSON, A. C., CARLSON, S. S. AND WHITE, T. J., 1977. Biochemical evolution. *Annual Review of Biochemistry* **46:** 573–639.

APPENDIX

Sources of Material

Species	Source
Caluromys lanatus	D. Hayman; San Diego Zoo
Cercartetus nanus	South Australian National Parks and Wildlife Service Bray Junction, South Australia
Dasyurus viverrinus	Adelaide Zoo, South Australia
Echymipera kalubu	J. Nelson; Papua New Guinea
Isoodon auratus	D. King; Barrow Island, Western Australia
I. macrourus	P. Baverstock; Nourlangie, Northern Territory
I. obesulus	P. Sharp; Uraidla, South Australia
Macrotis lagotis	C. Watts; Papunya, Northern Territory
Notoryctes typhlops	K. Johnson; Kintore Range, Northern Territory
Perameles gunnii	D. Spratt
P. nasuta	D. Spratt
Phascolarctos cinereus	T. Schwaner; Adelaide Zoo, South Australia
Tarsipes rostratus	D. King; captive colony, Murdoch University, Western Australia

Electrophoretic and chromosome surveys of the taxa of short-nosed bandicoots within the genus *Isoodon*

R. L. CLOSE[1], J. D. MURRAY[1,2] and D. A. BRISCOE[1]

Chromosomal and electrophoretic techniques were used to study the genetic affinities of eight of the 10 Australian taxa of *Isoodon* to better understand their evolutionary relationships. Only *I. auratus auratus* and *I. arnhemensis* were not examined. *Macrotis lagotis* and *Perameles nasuta* were studied for use as out-groups and while *M. lagotis* was the most genetically distinct taxon, *P. nasuta* was no more distant from *I. m. macrourus* than was *I. o. obesulus.*

No chromosome markers were found for distinguishing taxa of *Isoodon,* but an X chromosome was found in northern *I. macrourus macrourus* and *I. m. torosus* that was morphologically distinct from that found in *I. m. torosus* south of Cape York. The electrophoretic data for both subspecies were similar.

I. obesulus obesulus from New South Wales and Victoria were electrophoretically similar to *I. o. affinis* from Tasmania. *I. o. obesulus* from South Australia, however, aligned more closely with *I. o. fusciventer* from Western Australia, and with *I. o. nauticus. I. auratus barrowensis* aligned with southeastern *I. o. obesulus,* while *I. o. peninsulae* was distinct from all other taxa.

Key words: Bandicoot, taxonomy, chromosome cytology, protein electrophoresis.

INTRODUCTION

THE short-nosed bandicoots (*Isoodon*) are small (0.35–2.5 kg), omnivorous marsupials found in scattered coastal areas of all Australian states, the Northern Territory and New Guinea. Eleven species, which differ most obviously in size, have been described (Table 1), the most recent being *I. arnhemensis* by Lyne and Mort (1981). Fig. 1 shows Strahan's (1983) estimates of former distributions of these taxa.

All recent classifications since Iredale and Troughton (1934) have followed them in classifying *I. affinis* and *I. fusciventer* as subspecies of *I. obesulus,* but most have differed by including *I. torosus* in *I. macrourus.* While Iredale and Troughton (1934) recognized seven Australian species, Tate (1948) allowed only *I. barrowensis, I. macrourus* and *I. obesulus* (which included the remaining six forms); he stressed the division of the described forms into two distinct species groups, *I. obesulus* (including *I. barrowensis*), and *I. macrourus.* Ride (1970). Kirsch and Calaby (1977) and Strahan (1983) all listed three species, *I. auratus* (including *barrowensis*), *I. macrourus* and *I. obesulus,* which is the nomenclature followed in this paper. Lyne and Mort (1981), however, recognized seven species, *I. macrourus, I. obesulus, I. nauticus, I. peninsulae, I. auratus, I. barrowensis,* and *I. arnhemensis.*

Tate (1948) distinguished members of the *obesulus* group from those of the *macrourus* group by their overall smaller size, more fenestrated palate and additional palatal foramina, reduced metacone of M^4, smaller size of I^4 and molariform rather than blade-like deciduous DP^4 in juveniles. Thomas (1888, 1922) pointed out the greater angle (>105°) of the ramus of the lower jaw in *I.obesulus* and the disparity in size of the sexes in the southern forms of *I. obesulus, I. fusciventer* and *I. affinis* and the lack of such disparity in *I. auratus, I. nauticus* and *I. peninsulae.*

Freedman and Rightmire (1971) conducted multivariate comparisons of cranial and dental features within the genus and showed separation of *I. obesulus* from *I. macrourus.* Because only 60% of *I. torosus* individuals were excluded from the ranges of measurements for *I. macrourus,* these authors could not separate *I. macrourus* and *I. torosus* on morphological characters alone. In a detailed study of skull morphometrics, Lyne and Mort (1981) were unable to verify the former authors' claims that sexual dimorphism and ramus angle were less in *I. macrourus* than *I. torosus.* They found, however, that skull size, but not the width/length ratio, was larger for skulls from Queensland and New South Wales than those from Northern Territory and Western Australia, and that ramus angle of skulls from New South Wales exceeded that of northern specimens.

Continental forms of *Isoodon* occur widely and individuals from within a single taxon may be found in habitats as diverse as coastal heath and wet sclerophyll forest. The animals are mobile and apparently able to swim across small stretches of water (Gordon 1971), raising the

[1]School of Biological Sciences, Macquarie University, North Ryde, New South Wales, Australia 2109.
[2]CSIRO Division of Animal Production, Ian Clunies-Ross Laboratory, Prospect, New South Wales, Australia 2149.
Pages 19–27 *in* BANDICOOTS AND BILBIES, ed by J. H. Seebeck, P. R. Brown, R. L. Wallis and C. M. Kemper. Surrey Beatty & Sons, Sydney, 1990.

Table 1. Described species of *Isoodon*, with type localities and references to distributions shown in Figure 1.

Species	Type locality	Distribution
I. obesulus (Shaw, 1797)	Sydney, NSW	A
I. affinis (Gray, 1841)	Tasmania	B
I. fusciventer (Gray, 1841)	King George's Sound, WA	C
I. nauticus Thomas, 1922	Franklin Island, SA	D
I. peninsulae Thomas, 1922	Utingu, Cape York, Qld	E
I. auratus (Ramsay, 1887)	Derby, WA	F
I. barrowensis (Thomas, 1901)	Barrow Island, WA	G
I. macrourus (Gould, 1842)	Port Essington, NT	H
I. torosus (Ramsay, 1877)	Cooktown, Qld	I
I. arnhemensis Lyne and Mort, 1981	Melville Bay, NT	J

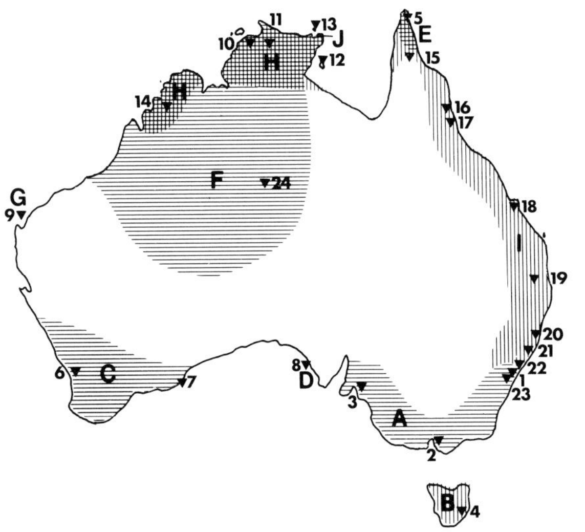

Fig. 1. Distributions of *Isoodon* taxa and collecting localities. Letters denote distributions (key in Table 1). Numbers denote collecting localities (key in Table 2).

question of how so many mainland forms could have become morphologically distinct. The present and recent distributions of *Isoodon* are poorly known; the recent rediscovery of *I. obesulus* near Sydney (Dixon 1981) highlights the possibility that distributions may be more extensive than is currently recognized.

Both *Isoodon* spp. and *Perameles* spp. have a diploid chromosome number 2n = 14, with similar autosome morphology for all taxa (Hayman and Martin 1974). The X chromosome shows greatest morphological variation; that of *I. o. fusciventer* from Perth is a submetacentric, that of *I. m. torosus* from Townsville a metacentric, and that of *P. nasuta* from Sydney a larger metacentric (see Hayman and Martin (1974), for measurements). *Macrotis lagotis* is markedly different with a chromosome number of 2n = 18,19 including an XY_1Y_2 sex chromosome system (Hayman and Martin 1974).

Haemopoietic cells of *Perameles* spp. and *Isoodon* spp. both lose an X chromosome from females and a Y from males early in development, yet the pattern of loss between the two genera is different from intestinal and corneal epithelia (Hayman and Martin 1974; Close 1979, 1984). Moreover, Close (1984) reported that not only are there sex differences in the time of sex chromosome loss from corneal cells of *I. macrourus* (New South Wales) but that the loss may occur several months later in development than was reported by Hayman and Martin (1974) for Tasmanian *I. obesulus.*

In this paper we report a survey utilizing chromosomal and electrophoretic studies on several taxa of Australian *Isoodon* to investigate their genetic relationships and to clarify the taxonomy but were unable to obtain the continental forms *I. auratus* and *I. arnhemensis.* We have included some data obtained from other bandicoots *P. nasuta* and *M. lagotis,* as comparisons.

Although our sample sizes were small, 29 loci were examined for most *Isoodon* taxa. Thus Nei's (1978) proviso that, for small sample sizes, a large number of loci should be considered in determining genetic distances between taxa is satisfied. We therefore believe that the conclusions we reach about genetic relationships among *Isoodon* taxa, although limited, are significant and will stimulate further investigation of the genus.

MATERIALS AND METHODS

Collection localities of animals used for chromosome and electrophoretic analysis are listed in Table 2 and shown in Fig. 1. Wherever possible, liver, kidney, muscle and blood samples were obtained, but for some specimens only blood samples were available.

Chromosomes

Cells were collected from bone marrow, peripheral blood cultures or fibroblast cultures following 1.5 h of exposure to colchicine (1.5 μg/g body weight or 1μg/ml culture medium). Cells were swollen for 12 min with KCl (0.53%) and fixed in 3:1 methanol-acetic acid. *Isoodon* leucocyte cultures grew poorly, and bone marrow was used in preference.

Table 2. Collection localities and number of animals examined cytologically and electrophoretically.

Species	Locality	Fig. 1 ref.	No. specimens	Examined cytologically	C-banded	Examined electrophoretically
I. o. obesulus (Shaw, 1797)	Sydney, NSW	1	3	3	3	3
	Cranbourne, Vic.	2	1	1	1	1
	Carey Gully, SA	3	2	2	2	2
I. o. affinis (Waterhouse, 1846)	Koonya, Tas.	4	2	2	2	2
I. o. peninsulae Thomas, 1922	Shelburne Bay, Cape York, Qld	5	1	1	1	1
I. o. fusciventer (Gray, 1841)	Perth, WA	6	2	1	0	2
	Daw Island, WA	7	1	0	0	2
I. o. nauticus Thomas, 1922	Franklin Island, SA	8	5	0	0	5
I. a. barrowensis (Thomas, 1901)	Barrow Island, WA	9	6	2	2	6
I. m. macrourus (Gould, 1842)	Batchelor, NT	10	1*	1	1	1
	Waterfall Creek, NT	11	2*	2	0	0
	Groote Eylandt, NT	12	5**	5	1	2
	Elcho Island, NT	13	1*	1	0	0
	Kimberley Range, WA	14	1*	1	1	1
I. m. torosus (Ramsay, 1877)	Coen-Lake Boronto, Cape York, Qld	15	5*	5	3	4
	Atherton, Qld	16	3	3	3	2
	Herbert River, Qld	17	3	3	0	3
	Rockhampton Qld	18	1	1	0	1
	Emu Vale, Qld	19	1	1	1	2
	Taree, NSW	20	2	2	2	2
	Myall Lakes, NSW	21	2	1	0	2
	Terrigal, NSW	22	5	5	3	2
Perameles nasuta (Geoffroy, 1803)	Sydney, NSW	23	6	2	0	6
Macrotis lagotis (Reid, 1837)	Tanami Desert, NT	24	1	1	0	1

**I. macrourus* Type II X chromosome; **three with Type II X chromosome.

Concentrated cell suspensions were dropped onto cold, wet (60% acetic acid) slides, air-dried, and stained with 1% lacto-aceto-orcein (CSL). C-bands were identified by the method of McKenzie and Lubs (1973) with the addition of a 2.5 min treatment in a 1.6 (v/v) 0.7 M NaOH — 2 X SSC (saline-sodium citrate) solution following treatment with 0.2 N HCl.

Electrophoresis

Erythrocytes in storage solution (400 ml ethylene glycol, 60 g Na_3 citrate in 600 ml distilled water), plasma, kidney, and liver were maintained at −20°C until used for electrophoresis. Tissues and erythrocytes were homogenized in 2–3 volumes of lysing solution (5 mg NADP, 0.2 ml 2-mercapto-ethanol in 10 ml distilled water).

Twenty-three protein-coding loci were examined on Cellogel (Chemetron, Milan) and a further six on 12% starch (Electrostarch, Otto Hiller, Madison, Wisconsin). Buffers, running conditions and staining reagents were minor modifications of accepted procedures (Harris and Hopkins 1976). The enzymes examined, their EC recommended nomenclature, their abbreviations used throughout this paper, and the electrophoretic media used for each are listed in Table 3. Because of the scarcity and scattered distribution of *Isoodon* spp., not all taxa were sampled during the main collecting period in 1977. As additional taxa subsequently became available, they were compared electrophoretically with the original samples. For some enzymes, long-term freezing reduced the activity such that comparisons could not be made.

Analysis

A matrix of genetic distances $\hat{D}$ for small sample sizes (Nei 1978) was determined for all taxa by using the GENDIST computer programme supplied by Dr G. McKay (unpublished 1984). This matrix was then analysed with the BMDP1M cluster analysis of variables programme (Hartigan 1983) using the maximum distance method to cluster the taxa and produce a tree of genetic relationships.

RESULTS

Chromosome number and morphology

In all 45 *Isoodon* spp. and *Perameles* spp. examined (Table 2), 13 chromosomes were observed, comprising six autosomal pairs plus

Table 3. Name, EC number and abbreviation of proteins surveyed, and the tissues and gel matrix used for each.

Abbr.	Name and EC No.	Source	Gel	Abbr.	Name and EC No.	Source	Gel
ADA	Adenosine deaminase (3.5.4.4)	M	S	IDH	Isocitrate dehydrogenase (1.1.1.41)	K	C
AK	Adenylate kinase (2.7.4.3)	E	C	LDH*	Lactate dehydrogenase (1.1.1.37)	E,K	C
ALB	Albumin	P	C	ME	Malic enzyme (1.1.1.40)	E,K	C
CA	Carbonic anhydrase(4.2.1.1)	E	C	MPI	Mannose phosphate (5.3.1.8) isomerase	K	S
EST	Esterase (3.1.1.1)	K	S				
αGAL*	αGalactosidase (3.2.1.22)	K	C	NP	Nucleoside phosphorylase (2.4.2.1)	K,L	S
GDH	Glutamate dehydrogenase (1.4.1.3)	K	C	6PGD	6-Phosphogluconate (1.1.1.44) dehydrogenase	E,K	C
GOT*	Glutamate oxaloacetate (2.6.1.1) transaminase	K	C,S				
				PGK*	Phosphoglycerate kinase (2.7.2.3)	E,K	C
G6PD	Glucose-6-phosphate (1.1.1.49) dehydrogenase	E,K	C	PGM*	Phosphoglucomutase (2.7.5.1)	K	C
				SOD	Superoxide dismutase (1.15.1.1)	K	S
GPI	Glucose phosphate (5.3.1.9) isomerase	E.K.	C	TRF	Transferrin	P	C
				PEP A	Peptidase (3.4.11 or 13)	E,K	C
HB	Haemoglobin	E	C	PEP D	Peptidase (3.4.13.9)	E,K	C

*2 loci scored; M, muscle; E, erythrocytes; P, plasma; K, kidney; L, liver; S, starch; C, cellogel.

one sex chromosome as described by Hayman and Martin (1965). The autosomes were similar in all short-nosed bandicoots examined (Fig. 2). C-banding analysis of 26 animals revealed a prominent block of centromeric heterochromatin on all autosomes and an absence of terminal or interstitial blocks (Fig. 3). In two individuals (Fig. 3: (a) pair 4; (c) pair 5) single autosome pairs displayed polymorphism for the amount of centromeric C-band material. In all other individuals the C-band pattern was constant.

The 12 *I. obesulus* group animals displayed the X chromosome morphology described by Hayman and Martin (1965) for *I. obesulus* from Western Australia (Fig. 2a). However, only 18 of 34 *I. macrourus* had a near-metacentric X the size of the fifth or sixth autosome as described by Hayman and Martin (1965, 1974) for *I. macrourus* from Townsville. The remaining 16 *I. macrourus*-group individuals possessed a metacentric X chromosome which was usually larger than that present in *obesulus* or the other *macrourus* animals (being between the number 4 and 5 autosomes in size) (Fig. 2c). This larger X chromosome is referred to as the *macrourus* Type II X, as opposed to the Type I X of the other *I. macrourus* and was found in all animals from Western Australia, mainland Northern Territory, Elcho Island and Coen, Cape York. On Groote Eylandt, three of the five animals examined carried the Type II X.

The three X chromosome morphologies observed in *Isoodon* spp. are characterized by distinct C-banding patterns (Fig. 3). The submetacentric *I. obesulus* X is distinguished by a small C-band in the short arm, a centromeric band, and two bands in the distal half of the long arm (Fig. 3a). The *I. macrourus* Type I X has a centromeric C-band and a large band in the shorter arm (Fig. 3b). In some preparations

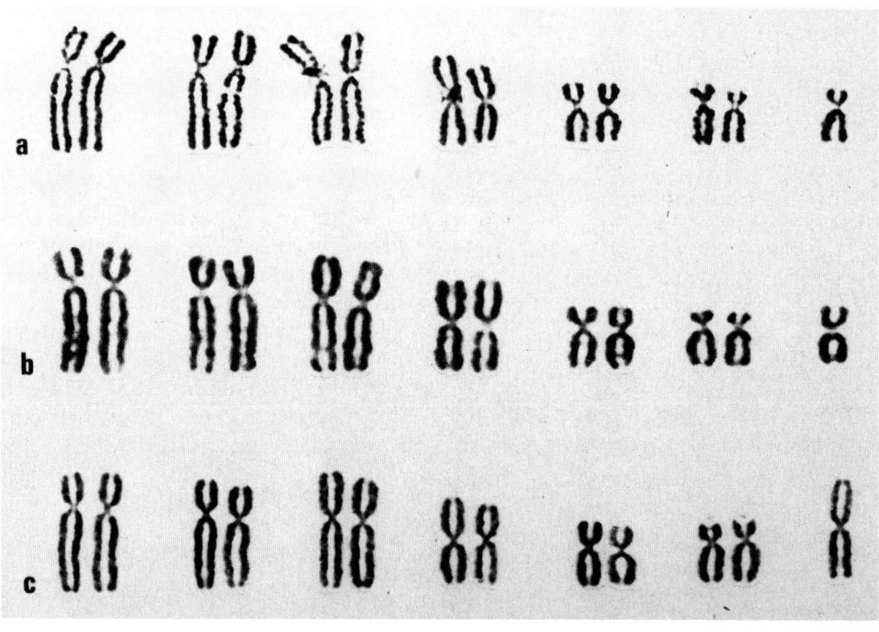

Fig. 2. Karyotypes of (a) *Isoodon obesulus;* (b) *I. macrourus* Type I X (c) *I. macrourus* Type II X.

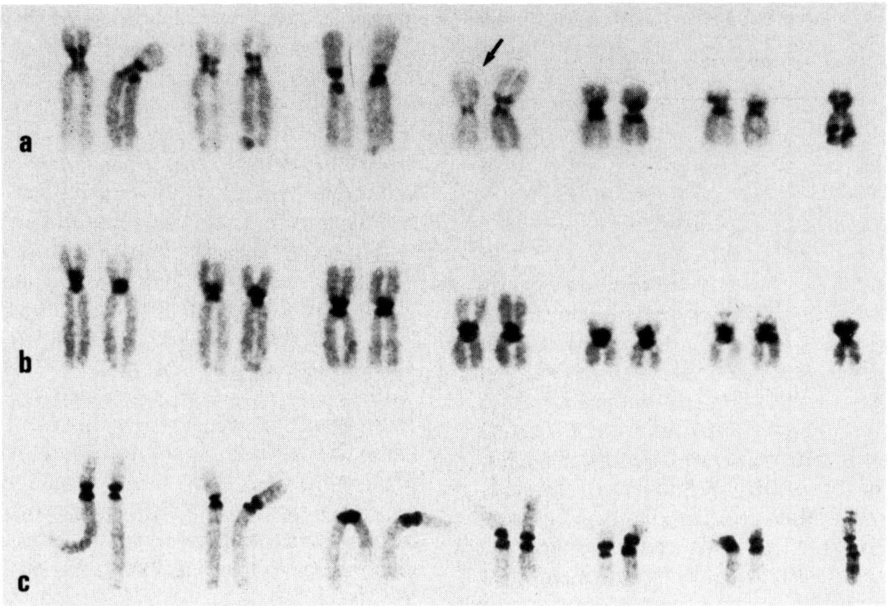

Fig. 3. C-banded karyotypes of (a) *Isoodon o. peninsulae*. Note size polymorphism for C-bands in chromosome pair 4 (arrow) (b) *I. macrourus* Type I X (c) *I. macrourus* Type II X.

nearly all of the shorter arm was C-band positive and, occasionally, a small faint band was observed in the distal half of the longer arm. The larger Type II X had a large centromeric C-band and two interstitial bands in each chromosome arm (Fig. 3c).

Electrophoresis

Of 53 animals examined electrophoretically, only 30 were screened for most loci. Of the 29 genetic loci surveyed, variation was detected for 16 in *Isoodon* spp. (Table 4). The invariant loci for *Isoodon* spp. were as follows: PEP A and D, G6PD, GPI, MDH, αGAL-2, IDH, ALB, LDH A and B, 6PGD, PGK A and NP. *Macrotis lagotis*, however, had different morphs for the latter five loci and also for HB, TRF and MPI from a total of 14 loci scored. *Perameles* spp. shared the slow NP electromorph with *M. lagotis* and had a unique AK electromorph, but for all other 14 loci had alleles that were also common to one or other of the *Isoodon* taxa.

Of the 16 variable loci in *Isoodon* spp. none could distinguish all members of the *I. obesulus* group of Tate (1948) from all of the *I. macrourus* group, yet several subgroups shared apparently unique alleles (Table 4). Because of their geographic proximity and similar electrophoretic profiles some populations were pooled before their data were analysed. Thus *I. o. obesulus* from Victoria and New South Wales were treated as one group but separated from the population from South Australia. Animals from Perth and Daw Island in Western Australia were pooled as *I. o. fusciventer*. Despite the chromosome polymorphism present in *I. macrourus*, all members of that genus from Queensland and New South Wales were treated as *I. m. torosus* while those from Western Australia and the Northern Territory were regarded as *I. m. macrourus*.

Table 5 shows the matrix of genetic distances ($\hat{D}$ for small sample sizes, Nei 1978), while Fig. 4 shows the tree produced by cluster analysis (Hartigan 1983) of the maximum genetic distances between taxa. *Macrotis lagotis* was separated from *Perameles* and *Isoodon* by distances greater than 1.0. *Perameles nasuta*, however, was scarcely more distant from the other taxa in *Isoodon* than was *I. m. macrourus*. According to Nei (1978), genetic distances between taxa are only reliable, when computed from small samples, when the number of genetic loci used is large (30–50), when heterozygosity within taxa is low, and when there is a large genetic distance between taxa (>0.15). Heterozygosities were low; only six heterozygotes were found in the *I. obesulus* group and only seven in the *I. macrourus* group for any loci, and genetic distances between most taxa were relatively large. Hence Nei's requirements are met for most comparisons among *Isoodon* taxa. Comparisons with *M. lagotis* and *Perameles* spp. are less reliable because of the smaller number of loci and animals examined.

The genetic distance separating *I. m. macrourus* and *I. m. torosus* was 0.102 when the genus was partitioned on a geographical basis. When the genus was divided according to chromosome type, and any population carrying the Type II X regarded as belonging to *I. m. macrourus*, the genetic distance became 0.103. Thus no matter whether eastern and northwestern populations

Table 4. Distribution of alleles detected for 16 variable loci in *Isoodon* spp., *Perameles nasuta* and *Macrotis lagotis*; Each column records which alleles from a particular locus were found in each taxon.

Taxon	Protein loci															
	AK	HB	CA	SOD	PGK-B	GAL-1	PGM-1	EST	GDH	TRF	GOT-C	PGM-2	GOT-A	ME	MPI	ADA
Isoodon auratus barrowensis Barrow Is.	A	A	B	B	B	B	A	A/B	A	B	A	B	C	B	B/D	C
Isoodon obesulus affinis Vic.	A	A	B	B	B	B	A	A	A	B	A	B	B	B	D	B
Isoodon o. obesulus NSW	A	A	B/C	B	B	B	A	A	A	B	A	A/B	B	B	D	B
Isoodon o. obesulus Tas.	A	A	B	B	B	B	A	A	A	B	A	B	—	—	D	A
Isoodon o. obesulus SA	A	A	B	A	A	C	A	B	A	A/B	A	B	B	B	D	A
Isoodon o. fusciventer WA	A	A	A/B	A	—	C	—	B	—	B	A	B	B	B	B/C/D	A
Isoodon o. nauticus Franklin Is.	A	A	A	A	—	—	—	B	—	B	A	B	B	B	D	—
Isoodon o. peninsulae Cape York	A	B	B	A	A	C	A	A/B	A	A/B	A	B	A	B	B	C
Isoodon macrourus macrourus NT, WA	C	B	C/E	B	B	B	A/B	B	A	B/C	A	A/B	—	A/B	B	—
Isoodon m. torosus Cape York	A	B	E	B	B	A	B	B	A/B	A/C	A/B	B	—	—	B	—
Isoodon m. torosus Cent. Qld	C	B	B/C/E	B	B	A	A/B	B	A	C	A/B	B	B	B	B	D
Isoodon m. torosus NSW	C	B	C	A/B	B	A	A/B	B	A	C	A	B	—	A	B	D
Perameles nasuta NSW	B	A	—	B	—	—	B	B	—	A	—	—	—	B	C	E
Macrotis lagotis NT	—	D	—	B	—	—	—	—	—	D	—	—	—	—	A	—

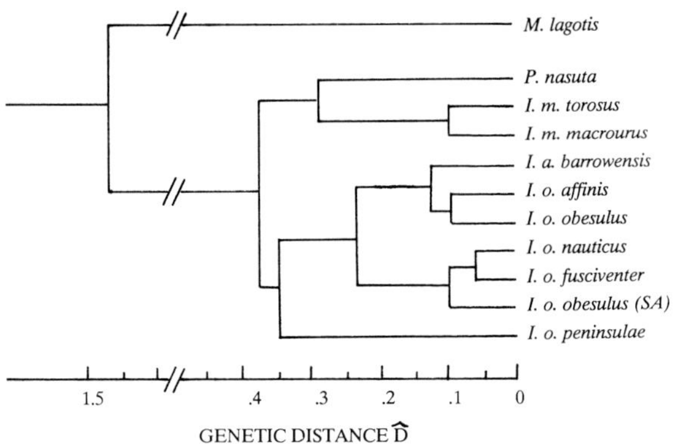

Fig. 4. Relationships of bandicoot taxa produced by cluster analysis of genetic distances (D̂ for small sample sizes, Nei (1978)).

of *I. macrourus* were separated on geographical or chromosomal grounds, no major electrophoretic distinction was discerned.

Within the *I. obesulus* group, *I. a. barrowensis* from Barrow Island, Western Australia forms a cluster with the populations from southeastern Australia, namely, *I. o. obesulus* from New South Wales and Victoria and *I. o. affinis* from Tasmania. The population of *I. o. obesulus* from South Australia has its affinities not with the southeastern forms but with *I. o. fusciventer* from Western Australia and *I. o. nauticus* from Franklin Island. The most genetically distinct form is *I. o. peninsulae* from Cape York which is almost equally distant from all other taxa in the genus, yet shares three unique alleles with the southwestern taxa (Table 5, Fig. 4).

DISCUSSION

The genetic distances separating *M. lagotis* from *Perameles* spp. and *Isoodon* spp. reflect the chromosomal differences between the two evolutionary lines. On the other hand, the similarity of the electrophoretic data of *Perameles* spp. and *Isoodon* spp. is mirrored by that of the chromosomal data. Baverstock *et al.* (this volume), however, while agreeing with our general findings on divergence of the *M. lagotis* and *Isoodon/Perameles* lineages, showed that the degree of chromosomal rearrangement of *Macrotis* spp. cannot be used as a measure of its separation from other bandicoots. They found that *Echymipera* spp. diverged from the *Isoodon/Perameles* line at the same time as *M. lagotis* yet has retained the 2n=14 karyotype.

Avise and Aquadro (1982) showed that great heterogeneity in genetic distances occurs among 313 congeneric species comparisons of mammals. The mean value of 0.3 is similar to those by which the *I. macrourus* and *I. obesulus* groups are separated from each other and from *P. nasuta* and, surprisingly, by which *I. o. peninsulae* is separated from all others.

Although Fig. 4 shows *P. nasuta* and the *I. macrourus* group on the same branch of the tree, this is certainly not due to their evolutionary relatedness but to their retention of the same ancestral alleles. By contrast, there is more protein variation apparent in the *I. obesulus* group (Table 4), and this is reflected in the greater geographic distribution of this group than of the *I. macrourus* group and the greater number of morphologically distinguishable taxa.

While electrophoretic data are valuable in detecting genetic similarity of taxa, they must be used with great care in predicting reproductive isolation. Genetic change, as determined by protein divergence, may be unaccompanied by reproductive isolation; the converse is also true. The decision to confer specific rank, therefore, in cases of allopatry is an arbitrary one, and the only method of verifying the decision is to conduct breeding trials. On the other hand, where taxa in sympatric or parapatric contact are genetically distinct, they can be considered good biological species (see discussion by Baverstock *et al.* 1977). Fig. 4 shows that results for *Isoodon* taxa are consistent with the subdivision of the genus into an *I. obesulus* group and a *I. macrourus* group (Tate 1948; Freedman and Rightmire

Table 5. Matrix of 11 populations of bandicoots showing genetic distances between each.

Population name/No.		1	2	3	4	5	6	7	8	9	10	11
I. o. obesulus (NSW, Vic)	1	0										
I. o. obesulus (SA)	2	0.206	0									
I. o. affinus	3	0.094	0.230	0								
I. o. peninsulae	4	0.350	0.294	0.345	0							
I. a. barrowensis	5	0.108	0.232	0.128	0.301	0						
I. o. fusciventer	6	0.210	0.087	0.228	0.307	0.224	0					
I. o. nauticus	7	0.141	0.101	0.189	0.314	0.194	0.061	0				
I. m. macrourus	8	0.240	0.370	0.233	0.352	0.220	0.349	0.344	0			
I. m. torosus	9	0.234	0.284	0.264	0.299	0.223	0.225	0.193	0.102	0		
P. nasuta	10	0.301	0.294	0.340	0.381	0.293	0.301	0.233	0.297	0.264	0	
M. lagotis	11	1.113	1.247	1.129	1.472	1.061	1.961	1.205	1.219	1.209	1.111	0

1971). The reproductive isolation of the *I. obesulus* and *I. macrourus* groups was verified by the genetic distance between *I. m. torosus* and *I. o. obesulus* despite their parapatric contact in New South Wales, and between *I. m. torosus* of Cape York and *I. o. peninsulae* despite their sympatry (Tate 1952).

Isoodon macrourus includes *I. m. macrourus* from Western Australia and Northern Territory and *I. m. torosus* from New South Wales and Queensland (Strahan 1983). The species can also be subdivided on the basis of size of the X-chromosome, the distributions of which correspond to those of *I. m. macrourus* and *I. m. torosus* except for the presence of populations with the Type II X on Cape York, and a polymorphic population on Groote Eylandt.

There is little electrophoretic evidence for subdividing *I. macrourus*. Only the CA^D allele was unique to Type II X animals while the CA^B allele was found in both groups including all animals from Cape York (Table 4). The $\alpha GAL\text{-}1^B$ allele was found in all specimens from Western Australia and Northern Territory but in none from New South Wales and Queensland. The differences between populations could be explained as the result of clinal differentiation over the range of *I. macrourus*. Preliminary breeding trials, however, indicate that northern and southern populations of *I. macrourus* may not interbreed. After seven weeks caged with a male *I. m. macrourus* (Type I X) from Groote Eylandt, four female *I. m. torosus* (Type I X) from New South Wales, had not produced young. When the male was replaced with a male *I. m. torosus* (Type I X) from New South Wales, two of the females produced young within four weeks (A. G. Lyne, pers. comm.). In other trials, Cape York animals (Type II X) were inter-fertile in captivity, yet failed to reproduce when caged with Type I X animals from Atherton, Queensland. In contrast, Type I X animals, from localities as far apart as Brisbane and Townsville, have bred successfully (G. Gordon, pers. comm.).

The chromosome and breeding data point to a division of *I. macrourus* on Cape York. It may be that a precise border between taxa referable to *I. m. macrourus* and *I. m. torosus* will be found there. Besides the presence of a chromosome marker with which to detect such a border, there is also an enzyme marker; AK^C, found only on Cape York (Table 4). Should there be a reproductive barrier between two populations on Cape York, then the allozyme and chromosome markers should enable it to be found.

Although no chromosomal markers could be used to identify members of the *I. obesulus* group, the electrophoretic data proved useful in distinguishing affinities. The Tasmanian form, *I. o. affinis*, differed from individuals from Victoria and New South Wales only at the ADA (muscle) locus and appears to have changed little since its separation from the mainland population.

Animals from South Australia, on the other hand, which were considered to be *I. o. obesulus* (Strahan 1983), clearly aligned with *I. o. fusciventer* from Western Australia (Tables 4, 5, Fig. 4). An area of contact between *I. o. fusciventer* and *I. o. obesulus* is predicted in an apparently continuous distribution of bandicoots between Adelaide and Melbourne. Should the genetic differences discovered during this study be maintained at the point of contact, then *I. o. fusciventer* would deserve specific ranking.

Isoodon o. nauticus from Franklin Island, was also genetically similar to *I. o. fusciventer*. Although the island and mainland taxa have diverged morphologicallly (Lyne and Mort 1981) since their separation by rising seas within the last 18 000 years (Chappell 1981), there has been little apparent associated divergence of proteins.

The genetic similarity of *I. a. barrowensis* and *I. o. obesulus* in southeastern Australia, despite their geographical separation, is striking. There has either been gene flow between the two taxa or both have independently retained a similar suite of ancestral alleles. Recent gene flow could only be explained by predicting a close relationship of *I. a. barrowensis* and *I. a. auratus* and relatively recent contact of the latter with *I. o. obesulus*. Such contact is feasible, as *I. auratus* once extended throughout the Northern Territory at least as far as the border with South Australia (Finlayson 1961; Parker 1973). Should *I. a. auratus* and *I. a. barrowensis* be genetically distinct, then the observed similarity of the latter to *I. o. obesulus* would be due to retention of similar ancestral alleles. Therefore, before the taxonomic status of *I. a. barrowensis* can be decided, *I. auratus*, preferably from central Australia, must be obtained and compared electrophoretically with both *I. a. barrowensis* and *I. o. obesulus*.

The genetic identity of *I. auratus* must also be determined before the status of *I. o. peninsulae* can be decided. The latter taxon is the most genetically distinct in the genus having genetic distances of 0.301–0.350 from others in the *I. obesulus* group and 0.299–0.352 from the *I. macrourus* group (Tables 4 and 5, Fig. 4). It carries five alleles not also found among *I. macrourus* and *Perameles* and its closest affinities are with *I. obesulus* from South Australia and Western Australia with which it shares three unique alleles. Should it prove genetically distinct from *I. auratus* and *I. arnhemensis*, we propose that *I. o. peninsulae* receive specific rank.

ACKNOWLEDGEMENTS

We acknowledge the co-operation of the wildlife authorities from all states and territories of Australia, and the many land owners for allowing the collection of animals. Professor G. B. Sharman obtained funding for the project and provided advice and assistance. Drs J. Calaby, A. G. Lyne, R. Braithwaite, C. Kemper, D. King, P. Baverstock, K. Johnson, G. Gordon, E. Thornber and Mr B. Prince and Mr P. Copley provided us with animals or material they had collected. Dr G. McKay provided the computer programme and advice on its use. Ms C. Martin and Mr A. Pavlov assisted with the electrophoresis.

This work was supported by Macquarie University, the Australian Research Grants Committee and the Australian Department of Environment, Housing and Community Development to Professor G. B. Sharman. J. D. Murray received support from the Australian-American Educational Foundation.

REFERENCES

AVISE, J. C. AND AQUADRO, C. F., 1982. A comparative summary of genetic distances in the vertebrates: patterns and correlations. *Evolutionary Biology* **15:** 151–86.

BAVERSTOCK, P. R., WATTS, C. H. S. AND COLE, S. R., 1977. Electrophoretic comparison of five Australian Pseudomyine rodents (Muridae). *Australian Journal of Biological Sciences* **30:** 471–85.

CHAPPELL, J., 1981. Sea level changes, 0 to 40 KA. *Proceedings of the first CLIMANZ conference of Victoria, Australia* 120–22.

CLOSE, R. L., 1979. Sex chromosome mosaicism in liver, thymus, spleen, and regenerating liver of *Perameles nasuta* and *Isoodon macrourus*. *Australian Journal of Biological Sciences* **32:** 615–25.

CLOSE, R. L., 1984. Rates of sex chromosome loss during development in different tissues of the bandicoots *Perameles nasuta* and *Isoodon macrourus* (Marsupialia: Peramelidae). *Australian Journal of Biological Sciences* **37:** 53–61.

DIXON, J. M., 1981. Selection of the neotype for the southern short-nosed (brown) bandicoot, *Isoodon obesulus* (Shaw and Nodder, 1797). *Victorian Naturalist* **98:** 130–35.

FINLAYSON, H. H., 1961. On Central Australian mammals. Part IV — The distribution and status of Central Australian species. *Records of the South Australian Museum* **14:** 141–91.

FREEDMAN, L. AND RIGHTMIRE, G. P., 1971. Skull and tooth variation in Australian bandicoots (Peramelidae, Marsupialia): the genus *Isoodon* and multivariate comparisons with *Perameles*. *Journal of the Royal Society of Western Australia* **54:** 21–31.

GORDON, G., 1971. *A study of island populations of the short-nosed bandicoot,* Isoodon macrourus *Gould.* Ph.D. Thesis: University of New South Wales: Sydney.

HARRIS, H. AND HOPKINS, D. A., 1976. *Handbook of enzyme electrophoresis in human genetics.* North Holland: Oxford.

HARTIGAN, J., 1983. Cluster analysis of variables. Pp. 447–55 in *BMDP statistical software* ed by W. J. Dixon. University of California Press: Berkeley.

HAYMAN, D. L. AND MARTIN, P. G., 1965. Sex chromosome mosaicism in the marsupial genera *Isoodon* and *Perameles*. *Genetics* **52:** 1201–06.

HAYMAN, D. L. AND MARTIN, P. G., 1974. Mammalia I: Monotremata and Marsupialia. In *Animal cytogenetics,* Vol. 1. Chordata 4 ed by B. John. Gebrüder Borntraeger: Berlin/Stuttgart.

IREDALE, T. AND TROUGHTON, E. LeG., 1934. A checklist of the mammals recorded from Australia. *Memoirs of the Australian Museum.* **6:** 1–122.

KIRSCH, J. A. W. AND CALABY, J. H., 1977. The species of living marsupials: an annotated list. Pp. 9–26 in *The biology of marsupials* ed by B. Stonehouse and D. Gilmore. Macmillan: London.

LYNE, A. G. AND MORT, P. A., 1981. A comparison of skull morphology in the marsupial bandicoot genus *Isoodon:* its taxonomic implications and notes on a new species, *Isoodon arnhemensis*. *Australian Mammalogy.* **4:** 107–33.

MCKENZIE, W. H. AND LUBS, H. A., 1973. An analysis of the technical variables in the production of C bands. *Chromosoma (Berlin)* **41:** 175–82.

NEI, M., 1978. Estimation of average heterozygosity and genetic distance from a small number of individuals. *Genetics* **89:** 583–90.

PARKER, S. A., 1973. An annotated checklist of the native land animals of the Northern Territory. *Records of the South Australian Museum* **16:** 1–57.

RIDE, W. D. L., 1970. *A guide to the native mammals of Australia.* Oxford University Press: Melbourne.

STRAHAN, R. (ed), 1983. *The Australian Museum complete book of Australian mammals.* Angus & Robertson: Sydney.

TATE, G. H. H., 1948. Results of the Archbold Expeditions. No. 60. Studies on the Peramelidae (Marsupialia). *Bulletin of the American Museum of Natural History* **92:** 313–46.

TATE, G. H. H., 1952. Notes on the occurrence of rain forest in Queensland. Results of the Archbold Expeditions. No. 66. *Bulletin of the American Museum of Natural History.* **98:** 563–616.

THOMAS, O., 1888. *Catalogue of the Marsupialia and Monotremata in the collection of the British Museum of Natural History.* British Museum: London.

THOMAS, O., 1922. The bandicoots of Nuyts Archipelago, S. Australia and of Cape York, northern Queensland. *Annals and Magazine of Natural History* **9:** 677–79.

Echymipera davidi, a new species of Perameliformes (Marsupialia) from Kiriwina Island, Papua New Guinea, with notes on the systematics of the genus *Echymipera*

TIM FLANNERY[1]

Echymipera davidi n. sp. is a plesiomorphic species of *Echymipera* thus far known only from Kiriwina Island, Trobriands Group, Papua New Guinea. It can be distinguished from other species of *Echymipera* by its relatively uniform colouration, sparse coat, relatively elongate anterior palatal foramina and short snout, weak mesial inflection of the angle relative to the rest of the dentary, small hypocones and large P3. Its relationships with mainland species of *Echymipera* remain obscure. However, it is clearly not closely related to any other species, and has presumably been isolated on Kiriwina for a considerable time.

Echymipera clara, while highly autapomorphic, appears to be a close relative of *E. rufescens*. The *E. clara*/*E. rufescens* group, *E. kalubu*, and *E. davidi* form an unresolved trichotomy in the preferred phylogeny.

Key words: *Echymipera,* bandicoot, Perameloidea, Kiriwina, Papua New Guinea, systematics.

INTRODUCTION

THE genus *Echymipera* as currently constituted consists of three species: *E. kalubu* (Lesson, 1828), *E. rufescens* (Peters and Doria, 1875) and *E. clara* Stein, 1932. The only speculation regarding relationships within the genus is that of Tate (1948) who concluded that *E. rufescens* and *E. kalubu* were close relatives, and that *E. clara* was more distantly related to that group.

Echymipera kalubu is widespread throughout mainland New Guinea below about 1 600 m (Ziegler 1977) and is found on islands off the Vogelkop, and on islands to the north of New Guinea. *Echymipera rufescens* is also widespread on mainland New Guinea below about 800 m (Ziegler 1977), and in addition is found on Cape York Peninsula, and islands to the south of New Guinea as well as possibly the D'Entrecasteaux Group. *Echymipera clara* is found below about 1 200 m (Zeigler 1977) north of the central New Guinean cordillera between the Huon Peninsula and Japan Island.

The peramelids of the islands off north-eastern Papua have remained poorly-studied. Thomas (1896) recorded a single juvenile specimen similar to *E. kalubu* from Woodlark Island. Unfortunately this specimen can no longer be located in the collections of the British Museum (P. Jenkins, pers. comm.). Brass (1959) recorded *E. rufescens* from both Kiriwina Island in the Trobriands and Fergusson Island in the D'Entrecasteaux Group. The material reported on by Brass (1959) from Kiriwina Island, as well as a specimen from Kiriwina held by the Australian Museum, are here described as a new species of *Echymipera, E. davidi.* The systematics of the genus *Echymipera* are also examined, and two alternative cladograms for the species are discussed. Colours, where capitalized, follow Ridgeway (1886), and cheektooth ennumeration follows Archer (1976). AM M is a prefix for Australian Museum mammal specimens, and AMNH for American Museum of Natural History specimens. The measuremens and notes of the holotype stated as being taken while the animal was freshly dead have been extracted from Ellis Troughton's notebook from his 1945 New Guinea expedition, now held in the Fisher Library, Sydney University.

SYSTEMATICS

Peroryctidae Groves and Flannery 1990
Echymipera Lesson, 1842
Echymipera davidi n. sp.
(Figs 1–3, Table 1)

Holotype. AM M7154 (field no. 100), adult male, puppet skin, skull extracted. Collected on 28 April 1945 by E. LeG. Troughton, and lodged in the mammal collection of the Australian Museum, Sydney.

Type locality. Kiriwina Island (08°30′ S, 151° 00′ E), Trobriand Group, Papua New Guinea.

Paratypes. All from Liluta, altitude 10 m, Kiriwina Island, Papua New Guinea. Collected 15 December 1956 by R. F. Peterson. AMNH 190765, adult male, puppet skin, skull

[1]Australian Museum, 6–8 College Street, Sydney, New South Wales, Australia 2000.
Pages 29–35 *in* BANDICOOTS AND BILBIES, ed by J. H. Seebeck, P. R. Brown, R. L. Wallis and C. M. Kemper. Surrey Beatty & Sons, Sydney, 1990.

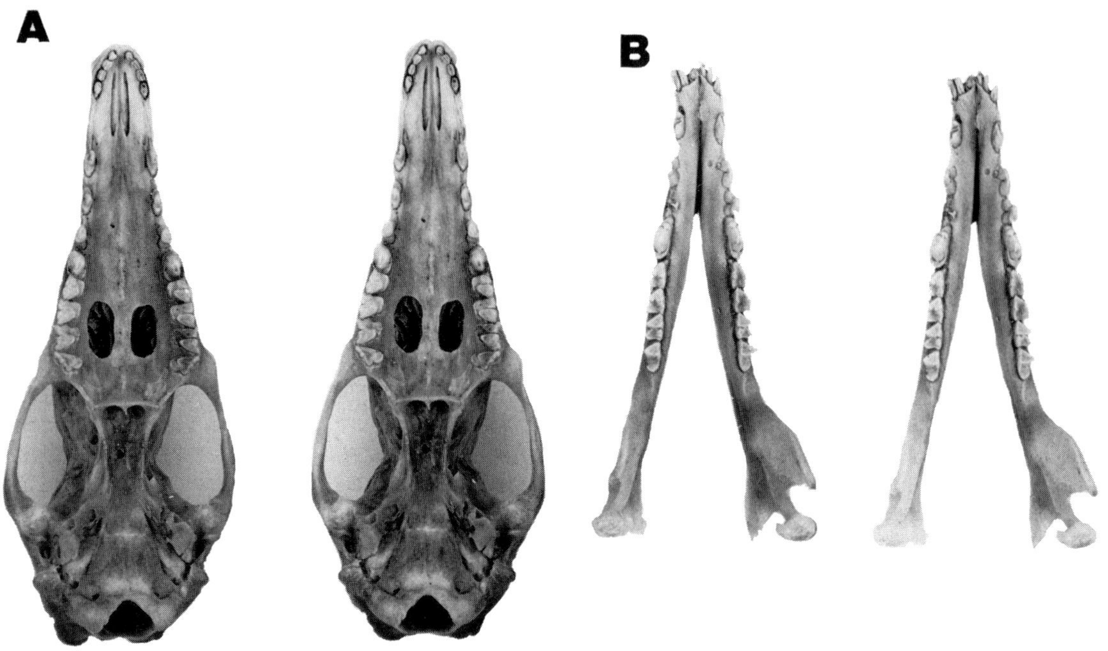

Fig. 1. A, stereopair of occlusal view of the skull, and B, stereopair of occlusal view of the dentary of AMNH 159472, paratype of *Echymipera davidi* n. sp., adult male.

extracted; AMNH 159478, adult male, puppet skin, skull extracted; AMNH 159472 adult male, puppet skin, skull extracted.

Etymology. For my son, David, whose birth, and my ejection from the neonatal ward due to termination of visiting hours, allowed me to wander into the mammal collections of the Australian Museum. Here I first recognized the holotype of *E. davidi* as a different kind of bandicoot.

Suggested common name. Kiriwina Bandicoot.

Diagnosis. The following features are unique within *Echymipera* to *E. davidi.* It is thinly-haired, the skin being visible through the spines and hair on the ventral surface, flanks and lower back. The overall "brindled" colouration present in the other species of *Echymipera* is much muted. The crown of the head is the same colour as the dorsal surface of the body and the face lacks a contrasting border between the light ventral surface and the dark crown. The anterior palatal foramina extend posteriorly to the anterior edge of the C^1 root (Figs 2, 3). The angle of the dentary is not as sharply deflected medially relative to the body of the dentary.

In addition to these features, *E. davidi* differs from *E. kalubu* as follows: The hypocone of M^{2-4} is less well-developed; the anterior cingulum of M^{2-4} extends only a small way lingual of the parastyle, and does not contact the protocone (in *E. kalubu* the anterior cingulum extends from the parastyle to the protocone, forming a continuous shelf); the stylar cusps are less well-developed; the P3s are larger and more bulbous. The posterior palatal foramina are smaller.

Echymipera davidi further differs from *E. rufescens* and *E. clara* as follows: the light tips of the hairs are shorter (about 1/5 of the total hair length versus 1/3); the hindfoot lacks a raised black pad; a hair whorl is present on either side of the rostrum between the eyes and nostrils; the ears are shorter and more rounded (noted when freshly dead).

It further differs from *E. rufescens* as follows: it possesses more bulbous, larger P3s; the rostrum is less elongate and the skull relatively broader; on M_2 the paraconid is larger and placed closer to the metaconid; the saggital crest is better-developed; the ascending ramus of the dentary is generally broader and taller; the dentary condyle is mediolaterally broader.

It further differs from *E. clara* as follows: the P3 and C1 of the males are smaller; the spines of the coat are broader, the C1–P1 diastema is shorter; the postparacrista of M^{2-4} is better-developed. The condyle of the dentary is mediolaterally narrower; dentaries of males are far less robust. The posterior palatal vacuities are larger.

Description. The following description of the pelage is based on the holotype only. The holotype skin is a puppet and is well-preserved. The anterior-most 9 mm of the nose is naked. Two hair whorls are located approximately 31 mm posterior to the nose tip, on the dorsolateral surface of the rostrum. The eyes appear to have been small. The ears are clothed externally in very fine hairs that are barely visible to the naked eye. The ears are dark, short and rounded. They are most similar to those of

Table 1. Some cranial and dental measurements (in mm) for *Echymipera davidi*. * = holotype.

Measurements	AM M7154*	AMNH 190765	AMNH 159478	AMNH 159472
Basicranial length	83.4	75.6	74.2	78.2
Bizygomatic width	32.9	29.7	30.0	32.4
Anterior of P^3 to premaxilla tip length	29.9	26.4	26.7	28.4
P^3–M^5 length	17.0	16.7	16.0	17.4
P^3 maximum width	2.5	2.6	2.4	2.8
M^2 maximum width	2.9	3.0	2.9	3.2
Length of the anterior palatal foramen	9.3	8.2	8.4	8.1
Length of the posterior palatal foramen	7.7	6.9	7.3	7.0
Minimum interorbital width	11.7	13.0	12.5	12.7
Height of ascending ramus above condyle	7.6	8.8	8.8	7.0
Length of condyle to angular process tip	9.5	8.4	8.0	9.6
M_4 maximum width	2.3	2.3	2.2	2.3
$M_{2\text{-}4}$ length	15.1	14.7	14.2	14.0
P_3 length	4.7	4.5	4.3	4.9
P_3 maximum width	2.1	1.9	2.1	2.1
M^2–M^5 length	13.5	13.2	12.9	13.1

E. kalubu, the ears of *E. clara* and *E. rufescens* being longer and more pointed. When freshly dead the tragus formed a simple, fleshy triangle 5 mm tall and 4 mm broad. The ear was 29 mm long and 19 mm broad. The hair of the crown of the head is short, but the hairs lengthen posterior to the shoulders. There are two distinct kinds of hair on the dorsal surface; spines and finer hairs. The spines are flattened, and are grooved longitudinally. They are narrowed at the base and tip, and are a uniform Seal Brown throughout. More abundant thinner hairs are subovate in cross section, and are Seal Brown for the basal four-fifths of their length, and are nearly always tipped with Burnt Umber on their distal fifth. On the ventral surface the spines are very reduced or absent. Only sparse, thinner, light-tipped hairs remain (Fig. 3). The body was 397 mm long when freshly dead. The dorsal surface of the fore and hindfeet is sparsely covered with short, Chocolate fur. The feet were 72 mm long and 21 mm broad when freshly dead. The tail is short and is covered in prominent scales interspersed with occasional, short, Chocolate hairs — it is undamaged and was 108 mm long when freshly dead.

On the forefeet digits I and V are reduced and clawless. Digit II is shorter than III and IV which possess larger, more robust claws. There is no indication of striate or tesserate pads on the forefeet.

The hallux is short and clawless. Digits II and III are syndactylous, and the ventral surfaces of their claws differ from those of the other digits in being sharply convex. Digit IV is longer than digit V, and both possess sharp, robust claws. The plantar surface of the foot, from the heel to the base of the digits, is covered in irregularly-shaped (but rougly square) tubercles. There is no area where the tubercles coalesce to form a raised pedal pad as in *E. rufescens* and *E. clara*. The skin as preserved is orange-brown.

The cranium of *E. davidi* exhibits fewer specializations than those of the other species of *Echymipera*. In common with *E. kalubu* (but distinct from *E. rufescens*), the rostrum is not as relatively elongate nor narrowed (Figs 1, 2). The well-developed saggital crest of *E. davidi* is similar to that seen in male *E. clara*, and is better developed than that seen in the remaining species. Overall the skull shape shows the greatest similarity to that of *E. kalubu*. The bullae appear to be slightly less domed than those of the other species of *Echymipera*.

Dentition. There are four upper incisors on each side. I^1 is smaller than I^{2-4}, and on the holotype I^4 is anteroposteriorly longer than I^{2-3}. The crown of C^1 is very small (smaller than P^1). However, the roots of the canines are large and in the holotype only the roots remain. They extend will beyond the alveoli and give the misleading impression that C^1 was large and robust.

P^1 is smaller and narrower then P^2. Both are 3-cusped teeth. There is a major central cusp between a smaller anterior and posterior cusp. These smaller cusps are better-developed on P^2 than P^1. P^3 is relatively large and bulbous (Figs 1, 2). It consists of a principal cusp and a broad, although variable, posterolingual cingulum.

All of the upper molars examined are somewhat worn so that details of crests and stylar cusp positions are obscured. A prominent parastyle is present on M^2. It is not connected with the protocone. The hypocone is small on M^{2-4}. A posthypocrista runs steeply towards the crown base, and terminates just posterior to and basal to the apex of the metacone on M^{2-4}. M^5 is variable in robustness. In AM M7154, AMNH 159472 and AMNH 190765 it is much broader (anteroposteriorly) than in AMNH 159478.

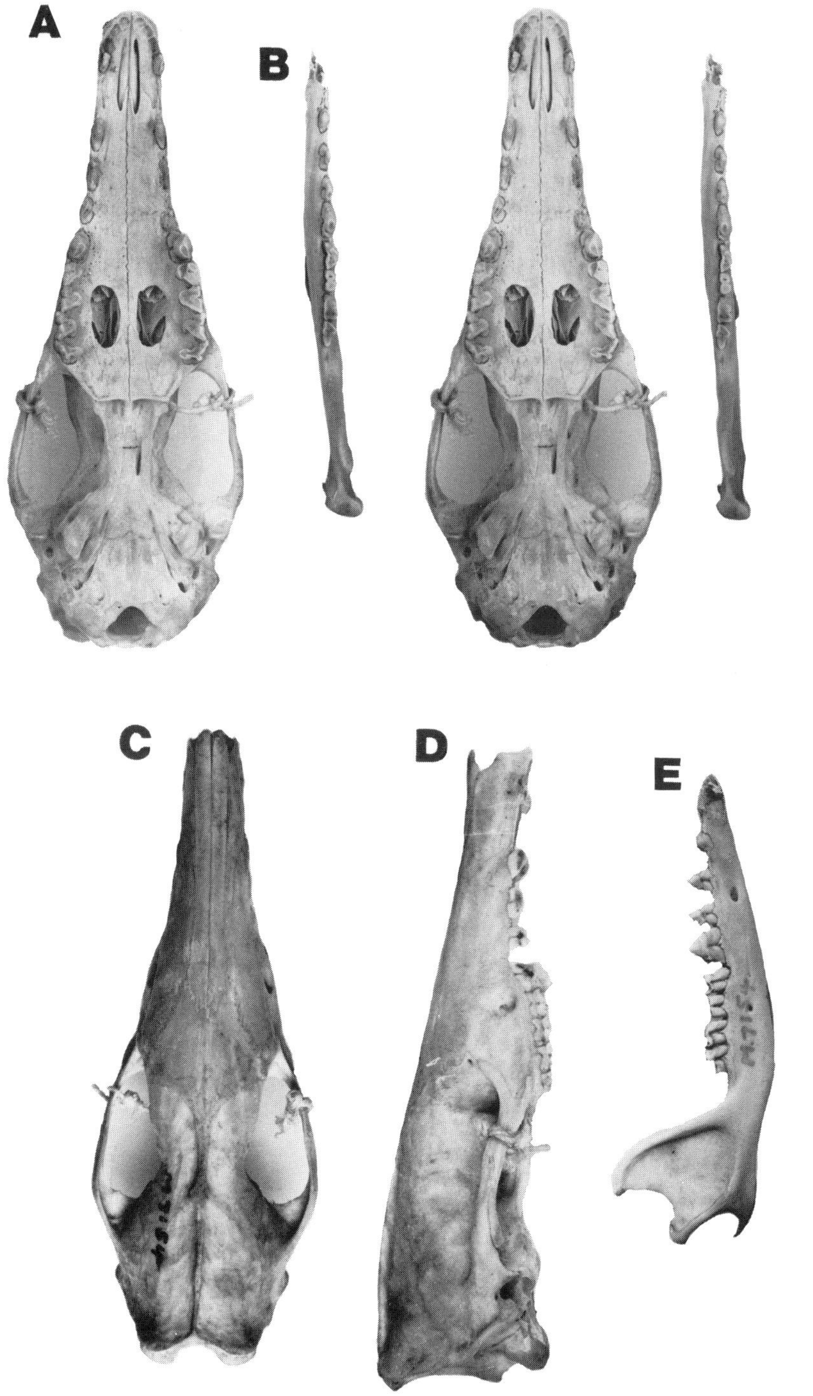

Fig. 2. A, stereopair of occlusal view of the skull; B, stereopair of occlusal view of the dentary; C, dorsal view of the skull; D, buccal view of the skull; and E, buccal view of the dentary of AM M7154, the holotype of *Echymipera davidi* n. sp., adult male.

There are three lower incisors on each side. I_1 is smaller than I_2, which in turn is smaller than I_3. I_3 is slightly bilobed, being ornamented by a groove in the middle of its buccal surface. C_1 is similar to C^1. P_1 and P_2 are less bulbous than P^1 and P^2. These teeth consist of a principal central cusp between a smaller anterior and posterior cusp. P_2 is slightly larger than P_1. P_3 is large and bulbous. It consists of a principal cusp, a small anteriorly-placed cusp and a posterolingual cingulum that terminates buccally in a small posterior cusp.

There is a small diastema between P_3 and M_2. Wear obscures many features of the lower molars. M_2 lacks an anterobuccal cingulum, a feature that is well-developed on M_{3-5}. The paraconid is placed anterior to and slightly lingual of the protoconid. The trigonids of M_{2-4} are all heavily worn, and it is difficult to

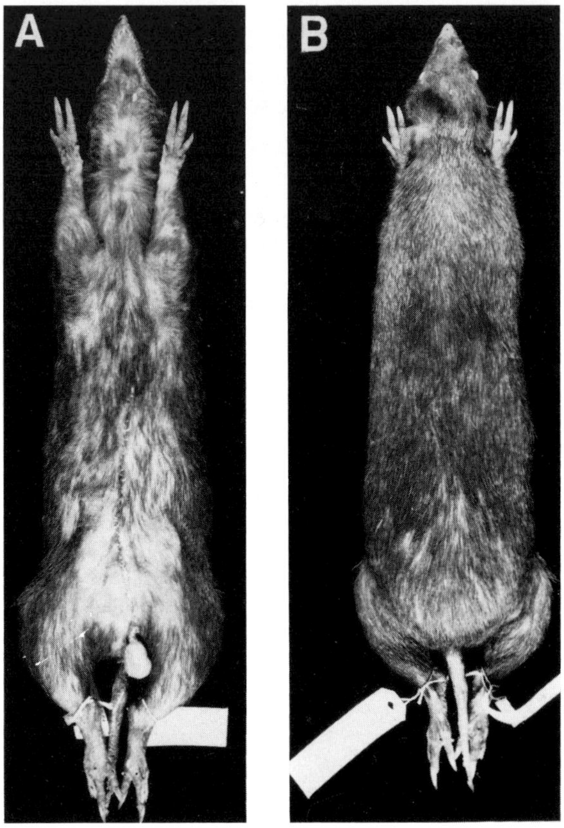

Fig. 3. A, ventral view, and B, dorsal view of the puppet skin of the holotype of *Echymipera davidi* n. sp. AM M7154, adult male.

determine the morphology of the trigonid cusps. However, it is clear that the paraconid is larger, and further posterolingually placed than that in *E. rufescens*. The cristid obliqua terminates slightly lingual and basal to the apex of the protoconid on M_{2-4}. The talonids are also all worn, but it is clear that a hypoconid, entoconid and hypoconulid (placed posterior to the entoconid) are all present. The talonid and trigonid are subequal in height. M_5 differs from more anterior molars in having a much lower, narrower talonid with a much reduced hypoconulid. The cristid obliqua on M_5 terminates posterior to and just buccal of the apex of the metaconid.

DISCUSSION

The harshly spinous fur, short thinly-haired tail, and short broad pes confirm that *E. davidi* is a species of *Echymipera*. Within *Echymipera*, *E. davidi* does not appear to be closely related to any previously known species, and exhibits many plesiomorphic features, some of which are retained in other lineages. These include the presence of a small hypocone on M^{2-4} (seen also in *E. clara* and *E. rufescens*), the lack of a raised pad on the foot, and the presence of a less elongate rostrum (also seen in *E. kalubu*). *E. davidi* also exhibits some features that are autapomophic. These include the uniform dorsal and ventral colouration, sparse covering of spines and hairs, the less mesially inflected angle of the dentary, and the enlarged premolars.

Echymipera clara and *E. rufescens* possess a number of features which indicate that these forms are monophyletic. Clear synapomorphies include the presence of a distinct raised black pad on the hindfoot, and loss of the hair whorls from the rostrum. Additional features shared between these species but not seen in other *Echymipera* include the presence of elongate, triangular-shaped ears, and the presence of a longer light tip on the hairs than is seen in other species. The polarity of these latter features is, however, unclear.

Several hypotheses regarding the interrelationships of the three groups of *Echymipera* (*E. kalubu*, *E. davidi* and *E. clara/rufescens*) recognized here need to be examined. It is possible that *E. davidi* is the plesiomophic sister-group to the other species, or that *E. davidi* is more closely related to either *E. kalubu* or the *E. rufescens/E. clara* clade. Data to test these hypotheses however are few. The former hypothesis hinges on the interpretation of the unique features of *E. davidi*. If the uniform colour pattern and the relatively long anterior palatal foramena of *E. davidi* are plesiomorphic, then the contrasting colour pattern and shorter anterior palatal foramena seen in other species could be interpreted as being derived, and an indication of monophyly for those species. However, outgroup comparisons do not support this interpretation, as most peramelids have a light venter and darker dorsal surface. Also, some have long (i.e., *Peroryctes broadbenti*) and others shorter (i.e., *P. longicauda*) anterior palatal foramena.

There are a number of features shared between *E. davidi* and other *Echymipera* species. Some of these could be interpreted as synapomorphies, including short light tips on the hairs (shared with *E. kalubu*), a well-developed saggital crest, and large, bulbous P3s (shared with *E. clara*). The case for an *E. davidi/E. kalubu* relationship is weak, being based solely upon hair colour, a notoriously homoplasious feature. The case for an *E. davidi/E. clara/rufescens* link is also weak. If the similarities between *E. clara* and *E. davidi* are indeed synapomorphies, then they must have been lost in *E. rufescens* which lacks them. On the basis of parsimony this is considered to be unlikely.

Thus it appears that the relationships of the three lineages of *Echymipera* are best described at present as an unresolved trichotomy. However, *E. davidi* appears to retain a greater number of symplesiomorphies than the other lineages.

Several specimens of a species of *Echymipera* were collected by members of the Fifth Archbold Expedition on Goodenough and Fergusson Islands in the D'Entrecasteaux Group. As the D'Entrecasteaux Group lies between Kiriwina Island and mainland New Guinea, this population is of considerable interest. Surprisingly, individuals from the D'Entrecasteaux Group show similarities with *E. rufescens* from mainland New Guinea in cranial and dental proportions (Figs 4, 5, 6). In addition, these specimens show the typical brindled colour pattern of *E. rufescens,* and have a dark crown. They also lack rostral hair whorls. However, they are similar to *E. davidi* and *E. kalubu* in possessing short light tips to the hair, in apparently lacking raised footpads (this is difficult to interpret because of the way the feet have been skinned on the specimens available), and in having the paraconid on M_2 placed close to the metaconid. A single specimen of *Echymipera* from Fergusson Island (AMNH 159469) shows rostral proportions similar to *E. davidi* (Fig. 6). However, in this specimen the rostrum is markedly twisted to the left, and it is possible that this deformity has retarded normal growth of the rostrum.

The status of the D'Entrecasteaux Group *Echymipera* can only be regarded as indeterminate pending further investigation. They may simply represent a subspecies of *E. rufescens* which retains a few plesiomorphic features, but it is also possible that they represent a distinct species allied to *E. davidi.*

The distinctive morphology of *E. davidi* (Fig. 6) suggests that this species has been isolated on Kiriwina Island for considerable time. Diamond (1972) suggested that Kiriwina Island would have been joined to the D'Entrecasteaux Group as little as 10 000 years ago because of lowered sea levels during glacial maxima. Ruling out an extrememly rapid evolution of *E. davidi,* other factors must be responsible for the presence of *E. davidi* on Kiriwina Island and an *E. rufescens*-like form on the D'Entrecasteaux Group. It is possible that even though these islands were joined, the intervening land was unsuitable for peramelids, or even that *E. rufescens* has recently invaded the D'Entrecasteaux Group, displacing *E. davidi* there.

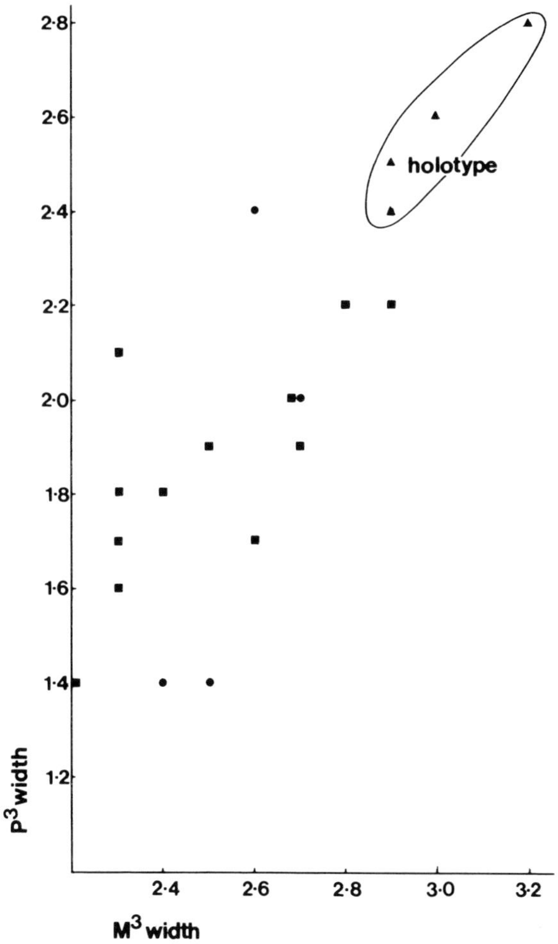

Fig. 4. Bivariate plot of P^3 width over M^3 width for: *Echymipera rufescens* (mainland New Guinea), squares; *Echymipera rufescens* (D'Entrecasteaux Group), circles; and *Echymipera davidi,* triangles.

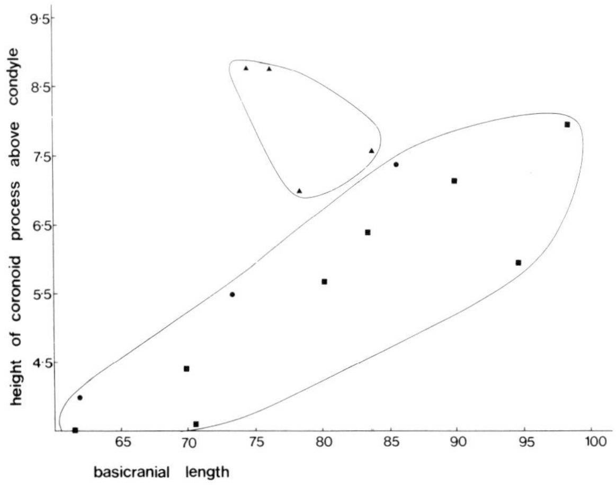

Fig. 5. Bivariate plot of the height of the coronoid process above the condyle over basicranial length for: *Echymipera rufescens* (mainland New Guinea), squares; *Echymipera rufescens* (D'Entrecasteaux Group), circles; and *Echymipera davidi,* triangles.

SUMMARY

Echymipera davidi is thus far known only from Kiriwina Island in the Trobriand Group. It retains many plesiomorphic features, and is not closely related to any mainland species. The species of *Echymipera* can be divided into three lineages; *E. kalubu, E. davidi* and *E. clara*/*E. rufescens*. The relationships of these lineages to each other cannot at present be resolved.

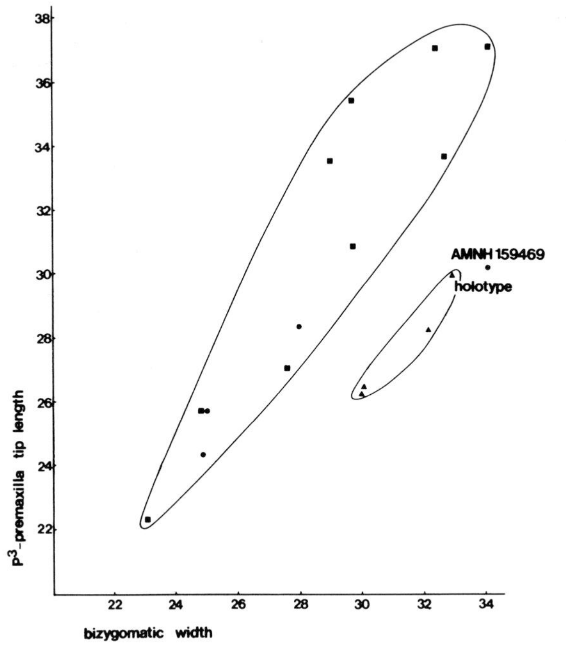

Fig. 6. Bivariate plot of P^3 anterior edge — premaxilla tip length over bizygomatic width for: *Echymipera rufescens* (mainland New Guinea), squares; *Echymipera rufescens* (D'Entrecasteaux Group), circles; and *Echymipera davidi,* triangles.

ACKNOWLEDGEMENTS

Drs M. Archer and C. Groves are thanked for their fruitful discussions of this paper. Material contributing to this study was collected during fieldwork in Papua New Guinea that was financed by a National Geographic Society Grant to M. Archer, and a Joyce Vickery Research award from the Linnean Society of New South Wales to the author. The Division of Wildlife, Papua New Guinea, is particularly thanked for help in obtaining the necessary permits and contacts to undertake this fieldwork. Dr Guy Musser of the American Museum of Natural History is also thanked for his loan of specimens. Kate Lowe of the Australian Museum produced the photographs.

REFERENCES

ARCHER, M., 1976. Phascolarctid origins and the potential of the selenodont molar in the evolution of diprotodont marsupials. Memoirs of the Queensland Museum **17:** 367–71.

BRASS, L. J., 1959. Results of the Archbold Expeditions No. 79. Summary of the fifth Archbold Expedition to New Guinea (1956–1957). *Bulletin of the American Museum of Natural History* **118:** 1–70.

DIAMOND, J. M., 1972. Biogeographic kinetics: estimation of relaxation times for avifaunas of southwest Pacific Islands. *Proceedings of the National Academy of Sciences* **69:** 3199–203.

RIDGWAY, R., 1886. *A nomenclature of colors for naturalists.* Little, Brown and Co: Boston.

TATE, G. H. H., 1948. Results of the Archbold Expeditions No. 60. Studies in the Peramelidae (Marsupialia). *Bulletin of the American Museum of Natural History* **92:** 313–46.

THOMAS, O., 1896. On mammals collected by Mr Albert Meek on Woodlark Island, and on Kiriwina in the Trobriand Group. *Novitates Zoologicae* **3:** 526–29.

ZIEGLER, A. C., 1977. Evolution of New Guinea's marsupial fauna in response to a forested environment. Pp. 117–38 in *The biology of marsupials* ed by B. Stonehouse and D. Gilmore. Macmillan Press: London.

Status of bandicoots in Queensland

G. GORDON[1], L. S. HALL[2] and R. G. ATHERTON[1]

Extant bandicoots are represented in Queensland by two species of *Isoodon*, one of *Perameles*, one of *Echymipera*, and one of *Macrotis*. Four species occur in higher rainfall zones and one in the arid interior. It is possible that other arid or semi-arid species have extended into Queensland in recent times. *Perameles nasuta* occurs commonly in closed forest and occasionally in open forest, woodland, and many disturbed habitats. *Perameles nasuta* and *I. macrourus* appear to be free of major threats. *Isoodon obesulus peninsulae* is of unknown status and restricted in distribution. Its habitats include eucalypt woodland, grassland and heath. *Echymipera rufescens* is locally common and is restricted to northern Cape York Peninsula, largely in closed forest. The range of the arid zone species, *Macrotis lagotis*, has contracted to far southwestern Queensland. It is regarded as an endangered species.

Key words: Bandicoots, status, conservation, distribution, Queensland.

THE SPECIES

Echymipera rufescens australis

THIS bandicoot was first discovered in Australia in 1932 by P. J. Darlington. It had been known previously from New Guinea (Tate 1948). The Archbold Expedition of 1948, to Cape York Peninsula, failed to collect any "despite intensive efforts" (Tate 1952), and it was then considered rare in Australia (Ride 1970). Tate (1948) gave the type locality of the Australian subspecies as the Rocky River, but later (Tate 1952) corrected this to the Nesbit River, farther north. The species was rediscovered at the Rocky River and on the McIlwraith Range in 1970 (Hulbert, Gordon and Dawson 1971). Subsequent workers have collected them at a number of localities on the eastern side of Cape York Peninsula north of the latitude of Coen (Gordon and Lawrie 1977). Fig. 1 shows the distribution of *E. rufescens* in Australia.

Despite its relatively restricted range in Australia, its status appears to be satisfactory, as the species is locally common where it occurs. Some of the earlier concern about its status possibly arose from difficulties that are usually encountered when trapping bandicoots in tropical areas, which seem to abound in other foods more attractive than those offered as bait in bandicoot traps.

There is no indication that its status has changed since European settlement of Australia and there are no obvious threats to its present survival. Cape York Peninsula is likely to remain relatively undeveloped for some time, and much of its rainforest habitat is, or will be, conserved in national parks. A potential and major threat that could arise at some time would be intensive development of the cattle industry on Cape York, which at present consists mostly of extensive operations. Due to transport difficulties, Cape York Peninsula is marginal cattle country and further development of the cattle industry is not economically feasible at present.

Echymipera rufescens has a broad habitat usage, occurring in closed forest, layered open forest, grassy woodland and closed heath (windswept coastal scrub). All of the collecting localities are close to the large rainforest tracts and *E. rufescens* seems to be dependent on these areas. However, in view of the fact that bandicoots are markedly sensitive to grazing pressure (Newsome 1971) it is possible that this pattern of habitat preference and distribution is an artefact of settlement and development of Cape York and its present distribution may have contracted from a much broader one. It is possible that the distribution of *E. rufescens* once extended into the riverine rainforests of central and western Cape York. The southern limit of its present distribution is set by a dry corridor extending from Coen to Cooktown.

In New Guinea, *E. rufescens* is found in lowland rainforest below 800 m (Ziegler 1977) and this factor possibly allowed it to cross the land bridge between Australia and New Guinea (unlike the New Guinea montane bandicoots). *Echymipera rufescens* forms part of the faunal transition between northern Australia and southern New Guinea.

One of the most interesting aspects of its occurrence on Cape York Peninsula relates to its ecological relationships with the other rainforest bandicoot of Cape York, *Perameles nasuta.* This is the only part of Australia where two predominantly rainforest species are sympatric. *Perameles nasuta* has not been collected on Cape York since the Archbold Expeditions and may be rare, so a comparative study would be

[1]Southern Regional Office, Queensland National Parks and Wildlife Service, P.O. Box 42, Kenmore Queensland, Australia 4069.
[2]Department of Anatomy, University of Queensland, St. Lucia, Queensland, Australia 4067.
Pages 37–42 *in* BANDICOOTS AND BILBIES, ed by J. H. Seebeck, R. L. Wallis, P. R. Brown and C. M. Kemper. Surrey Beatty & Sons, Sydney, 1990.

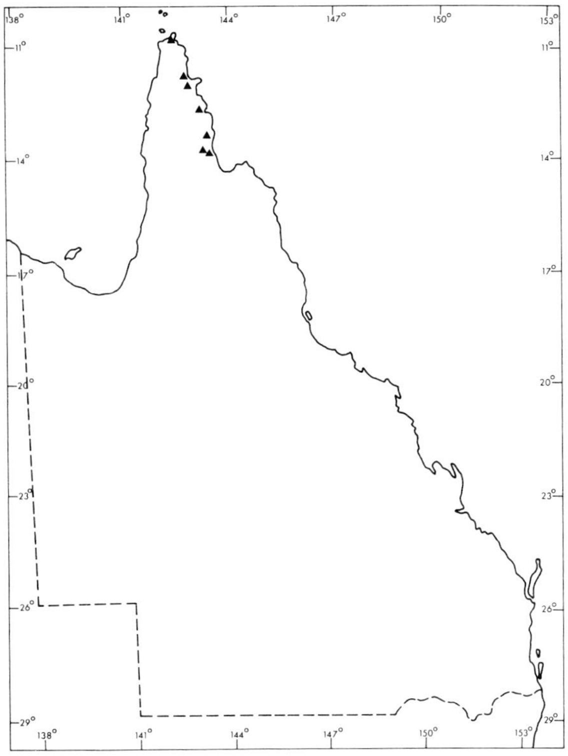

Fig. 1. Distribution of *Echymipera rufescens australis* in Queensland.

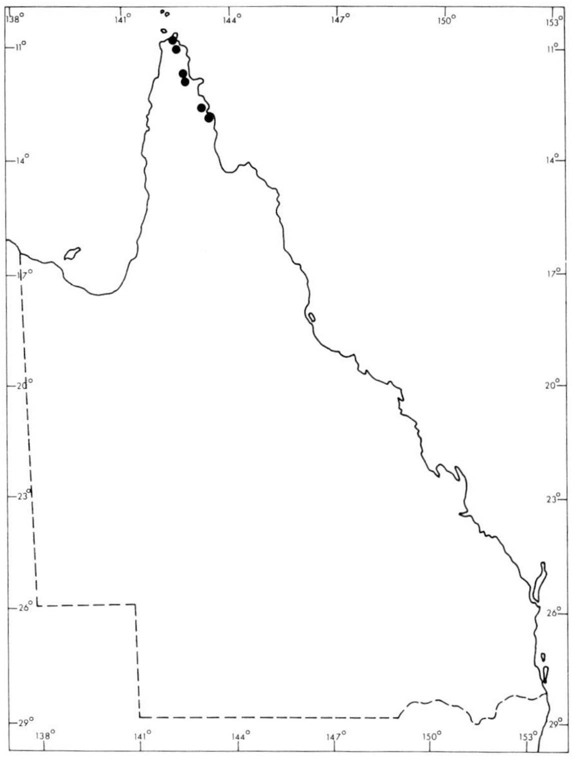

Fig. 2. Distribution of *Isoodon obesulus peninsulae* in Queensland.

difficult. *Echymipera rufescens* shows some distinct differences from *P. nasuta* that are suggestive of a differing ecological niche; these include a greater development of a slightly mobile proboscis-like nose, smaller eyes, and (in cages) a greater climbing ability.

Isoodon obesulus peninsulae

This is another species found on Cape York Peninsula but is of uncertain taxonomic status. It has been regarded as a subspecies of *I. obesulus,* but may eventually come to be regarded as a form of *I. auratus.* It was first collected in 1912 by R. Kemp and subsequently by Thomson (in the Iron Range area) and during the Archbold Expeditions of 1948 (Tate 1948, 1952). The only recent records have been two animals collected during the 1970s from heathlands (J. Winter and R. Close, pers. comm.) (Fig. 2).

Its current status appears to be uncommon or even rare and it is difficult to capture. The greater capture success of earlier workers could indicate that it was more common then, but it may also reflect the fact that aborigines were used to collect their specimens.

The extant populations of *I. o. peninsulae* occur on land that is zoned as a future national park. However, it is difficult to predict any threats with our current knowledge of the species on Cape York.

Habitats recorded for the Archbold Expeditions specimens were listed as savannah forest (i.e., eucalypt woodland) and grassland. The specimens collected by J. Winter and R. Close were from a 2 m high closed heath with emergent *Grevillea pteridifolia.* The woodland site occurred on deep sands and the heath site on hard reddish soil with termite mounds. The habitat of the Thomson specimens was recorded as "open grassy lightly timbered country" (Dixon and Huxley 1985). This area of Cape York is one of the few places in Australia where two species of *Isoodon* are sympatric and thus is of some interest.

Isoodon macrourus

I. macrourus is common to abundant in eastern Australia in a wide variety of situations. It is the common outer urban bandicoot of many coastal Queensland towns (Lavery and Johnson 1968, 1974; Gordon 1974b) except in a few areas where *P. nasuta* is more common.

Its distribution covers the humid sub-coastal regions, and in southern and central Queensland it formerly extended 300 km inland (Fig. 3). In central inland Queensland it now occurs mainly in riverine forest, perhaps extending further inland in this habitat than would otherwise be the case, with an occasional occurrence in other nearby habitats.

Although *I. macrourus* is common to abundant in coastal areas of Queensland, its status in central inland areas is rare to extinct. It appears that this has occurred because of its sensitivity to grazing animals. In higher rainfall areas, there are no apparent threats to this species' survival.

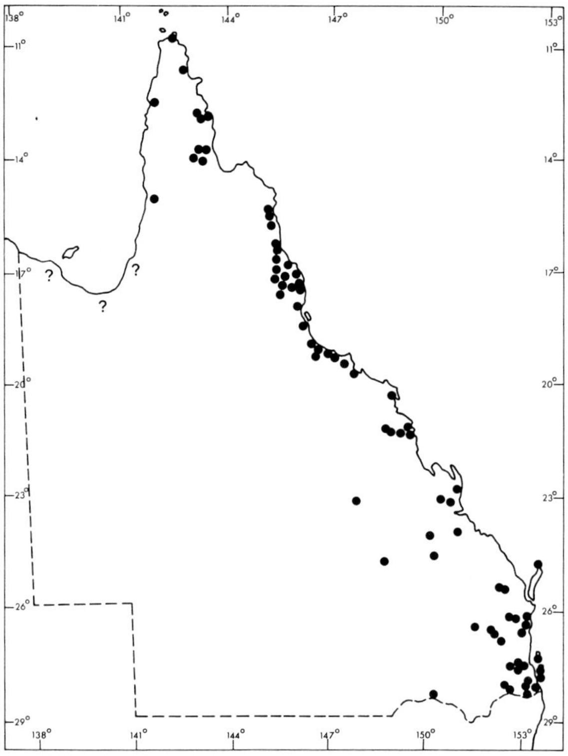

Fig. 3. Distribution of *Isoodon macrourus* in Queensland. Question marks indicate possible range extension.

Isoodon macrourus shows a preference for areas with low ground cover (Gordon 1974a; Lavery and Grimes 1974), and is found in a wide range of vegetation types which occasionally include rainforest (Harrison 1962). It also occurs in suitable types of eucalypt open forest, woodland and grassland. The habitat preference of *I. macrourus* in the dry season in the Northern Territory is strongly associated with floristic and structural attributes but lacks any definite relationships with vegetation in the wet season, when food is more abundant and widespread (Friend and Taylor 1985). It could be reasonably expected that the same situation occurred in northern Queensland. Recent studies (Gordon, unpublished) suggest that *I. macrourus* consists of two morphologically similar species, with a break between the two groups north of Cooktown.

An interesting aspect of the distribution of this bandicoot, and the following species, *P. nasuta,* is the inward bulge of its distribution in central and southern Queensland. This correlates with a zone of overlapping summer and winter rainfall. It would appear that climatic factors could have limited the inland distribution of *I. macrourus* prior to European settlement.

Perameles nasuta

Although not as extensive, the distribution of *P. nasuta* is somewhat similar to that of *I. macrourus,* suggesting a climatic barrier to its extension inland.

Its status is generally secure, being common to abundant in its preferred habitat through most of its distribution. However, there are two exceptions:

(i) On Cape York Peninsula *P. nasuta* has been collected very rarely and from only two sites, both during the Archbold Expeditions (Tate 1952). This seems to indicate a genuine scarcity in the region, although, as mentioned for *I. o. peninsulae,* there are difficulties involved in trapping bandicoots in the tropics. Subsequent efforts to capture *P. nasuta* on Cape York have failed.
(ii) In inland Queensland the distribution of *P. nasuta* has contracted greatly, possibly to the extent of extinction. The most recent record for that area was in the mid-1970s near Emerald in a bonewood (softwood) scrub (Gordon, unpublished data).

Conservation prospects for *P. nasuta* appear good except in the inland pastoral zones. Like *I. macrourus,* it appears to be very sensitive to grazing and cannot coexist with broad scale stocking. In the last two decades extensive clearing of the softwood scrubs (vine thickets) has caused further reduction of habitat for *P. nasuta* in inland central Queensland.

Along the eastern coastal zones of Queensland *P. nasuta* occurs principally in rainforests (Harrison 1962; Lavery and Johnson 1968; Lavery and Grimes 1974). Other habitats used are softwood scrub, occasionally open eucalypt forest, that is, dry and wet sclerophyll forest, and short-grassed areas, such as on dairy and other small farms. *Perameles nasuta* appears to select areas which have an open ground stratum, a feature shared by the above habitats.

Its habitat utilization in southern Queensland includes open forest, but is much narrower than in central and southern New South Wales where it extends into shrubby open forests, vacant overgrown urban land, shrubland, and coastal heath. Habitat utilization narrows from central New South Wales to northern Queensland. In the latter zone it becomes far more restricted to rainforest or farmlands on former rainforest lands. The disturbance and destruction of rainforests is an obvious threat to *P. nasuta* in northern Queensland, and further studies on its ability to utilize degraded land are necessary.

Softwood scrubs are, or were, used by *P. nasuta* in inland Queensland pastoral zones, permitting it to extend well inland (Fig. 4). It is possible that other vegetation types were also important in this area, such as the better kinds of brigalow forest, which are now rare. *Perameles nasuta* is known from Pleistocene deposits in Russenden Cave in south-east Queensland (Archer 1978) which is on the western limits of its present distribution.

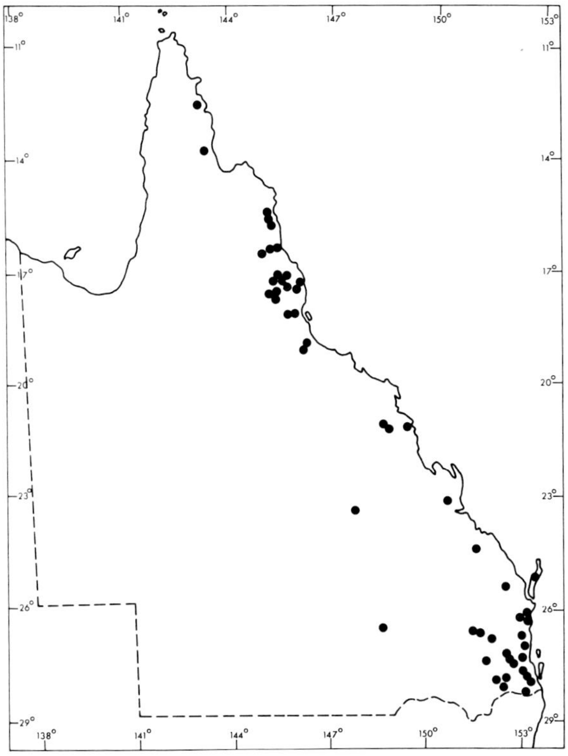

Fig. 4. Distribution of *Perameles nasuta* in Queensland.

Macrotis lagotis

Macrotis lagotis once occurred in much of southern Queensland, at least as far east as Surat. It appears to have vanished from a large part of this area and is now only found in south-western Queensland (Fig. 5). The extent of its previous range in Queensland is uncertain, possibly extending further east to the Great Dividing Range, as it did in New South Wales (Ashby, Lunney, Robertshaw and Harden, this volume). Its distribution may have also extended further north into central Queensland. Two old reports given to one of us (G. G.) in 1975 by property owners indicated that the species occurred near Rolleston and Alpha (see Fig. 5). Although these reports could not be verified, they appear plausible.

Macrotis lagotis has changed in status greatly, contracting towards the arid areas. It appears to survive in two regions:

(i) Near Cunnamulla in southwestern Queensland. Recently (1984) two live animals were reported to have been captured in a 1.5 m deep oil pipeline trench by a digging crew. Further investigations are necessary to assess the status of this population.

(ii) East of a line between Birdsville and Boulia in far southwestern Queensland. Several scattered but localized populations occur in this region. These have been surveyed by the Queensland National Parks and Wildlife Service, and by a project sponsored by the World Wildlife Fund (Southgate, this volume). Six properties are known to have *M. lagotis* colonies and there are a number of sites from which *M. lagotis* has disappeared relatively recently (Fig. 5). Only one of the colonies appears to have a satisfactory number for a viable population. The remainder appear to be low density populations, and are isolated from other *M. lagotis* colonies.

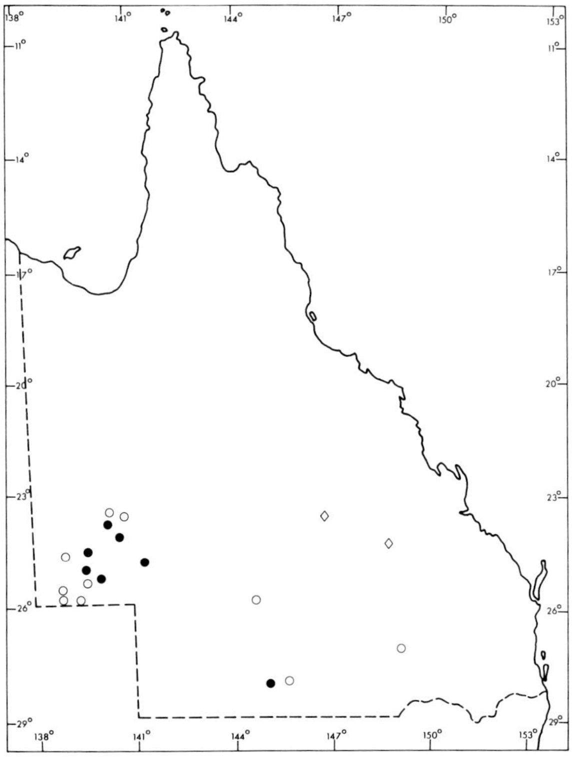

Fig. 5. Distribution of *Macrotis lagotis* in Queensland. Solid circle = recent sightings; open circle = old records or specimens; open diamond = unverified records.

There is evidence that *M. lagotis* has recently declined in the Birdsville region and is still declining. This is indicated by:

(a) the sparse, low-density distribution of most remaining colonies. The low number of *M. lagotis* may result in a lack of reproductive and genetic viability.

(b) *Macrotis lagotis* survives in country that mostly represents poorer cattle country (e.g., gibber plains). Many abandoned burrow sites are present in better quality cattle country, suggesting a withdrawal of *M. lagotis* from those areas in the face of competition from stock.

(c) Survival of *M. lagotis* on some properties (or paddocks), and not on others seems to be correlated with cattle management.

(d) Land use in the form of improved pastoral management is changing slowly. These developments mean that more land is used more often, e.g., trucking of stock rather than droving; provision of extra watering points; subdivision of larger paddocks; helicopter mustering, etc. It follows that higher cattle capacity must mean a lower capacity for the native ground fauna.

Macrotis lagotis populations in Queensland seem to be highly endangered, mainly due to stocking, as few other environmental changes have occurred in far southwestern Queensland since European settlement. Monitoring of the known colonies is urgently requried. Introduced predators (foxes and cats) and rabbits are present but not abundant in areas where *M. lagotis* is presently found, and broad-scale clearing is not practised.

Habitats used by extant populations of *M. lagotis* are Mitchell Grass *(Astrebla pectinata)* plains on clay soils, as well as gravelly plains and gibber plains which mostly occur on loamy soils. Clearly, *M. lagotis* had a much wider range of habitat utilization in its original Queensland range which extended much farther to the east.

DISCUSSION

Although five species of bandicoots are extant in Queensland, there is a possibility that other species were present prior to European settlement. Arid and semi-arid bandicoots have undergone a marked decline in Australia since European settlement and five such species, now probably extinct, occurred in adjacent states to Queensland (Newsome 1971; Frith 1973).

Macrotis leucura and *P. eremiana* occurred in arid central Australia. *Isoodon auratus* was widespread in arid and semi-arid areas of the Northern Territory and *P. bougainville* and *Chaeropus ecaudatus* occurred in arid and semi-arid areas in New South Wales and South Australia. All of these species could potentially have occurred in Queensland, and ecological gaps are present in the Queensland bandicoot fauna which they may have occupied. Thus, there is no known extant bandicoot in northern inland Queensland.

There are over six unnamed species of bandicoots representing at least two undescribed genera in the hard middle Miocene limestone deposits (approx. 15 myo) at Riversleigh in northeastern Queensland (Archer and Clayton 1984) (Fig. 6). This deposit represents the richest concentration of bandicoots (both extinct and extant) recorded for the one locality and further analysis of the material could be rewarding in interpreting modern distributions.

A fossil bandicoot, *Perameles allinghamensis*, and several unnamed forms of bandicoots have been recorded in early Pliocene deposits at Bluff Downs, northeastern Queensland by Archer and Wade (1976) (Fig. 6).

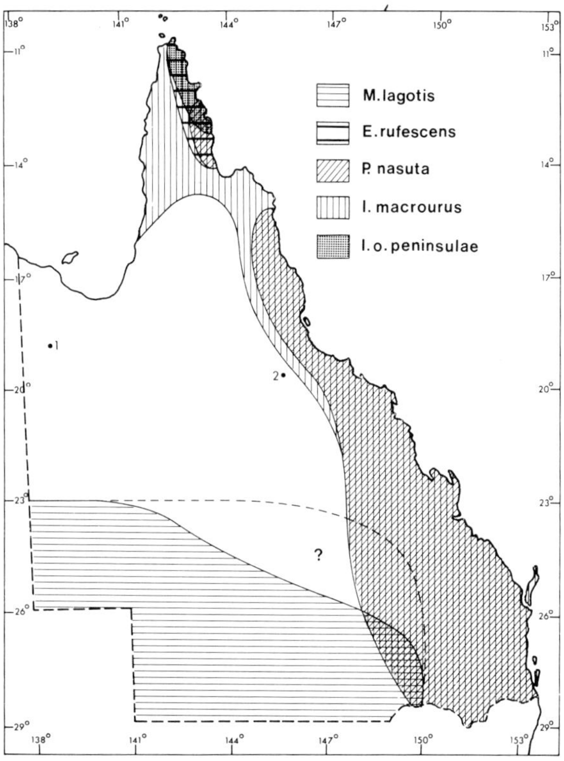

Fig. 6. Distribution of bandicoots in Queensland. 1 = Riversleigh, 2 = Bluff Downs, and ? = unverified occurrence of *Macrotis lagotis*.

All bandicoots are now rare or locally extinct in southern inland Queensland in the pastoral zones. It is conceivable that arid and semi-arid Queensland once had a richer bandicoot fauna which became extinct rapidly after European settlement, as happened in most other parts of Australia in the arid and semi-arid zones. At present, eastern Cape York Peninsula has Australia's richest bandicoot fauna. Four species occur there in a small area near Iron Range (Fig. 6). This is comparable to the bandicoot species diversity occurring in parts of New Guinea and is partly due to the presence on Cape York Peninsula of a New Guinean species, *E. rufescens*. Regional faunas elsewhere in Australia usually consist of only one or two species of bandicoots. Highest species diversities are reached in the high-rainfall tropics possibly reflecting a greater abundance and diversity of insect food in these regions.

Queensland bandicoots have survived very well in the coastal and subcoastal areas and declined in the semi-arid and arid inland. This pattern of decline and survival correlates best with two environmental factors; the climatic zonation between the humid coastal areas and the semi-arid inland, and the presence/absence of broad-scale grazing by introduced herbivores, cattle, sheep and rabbits and associated land use changes. It does not correlate nearly as well with the presence of introduced predators, foxes and cats, at least when compared on a regional basis. Thus bandicoots appear to have declined in central inland Queensland, an area where foxes are generally uncommon, and they have survived well in the more coastal regions and in outer urban areas where feral cats are often common. They have also coped well with the dingo, another introduced predator. Although the effect of introduced predators should not be excluded, there is no conclusive evidence from the Queensland scene to show that they have sufficient impact on bandicoot populations to cause local extinction.

The favourable status of bandicoots in the higher rainfall subcoastal zone of Queensland is ultimately a consequence of the topographic diversity of this region which ensures that a broad range of land use types is present including those that are compatible with survival of bandicoots. Bandicoots coexist well with introduced predators (cats and possibly foxes) in parts of this zone and are interspersed among grazing areas, tending to occur in ungrazed or more lightly disturbed areas. Queensland bandicoots have disappeared or become rare in regions where topographic diversity is such that broad-scale intensive grazing is possible. In such areas grazing or other disturbance to the understorey is pervasive in that it can occur with little interruption over very large areas, whereas in coastal zones there are many refuge areas free from disturbance to the understorey by grazing and other agents.

It is likely that the Queensland bandicoot declines have been a direct result of grazing disturbance to the understorey or associated disruption of their food supply. However, it is possible that associated changes, such as changed fire regimes, and use of pesticides and herbicides, have also been important.

REFERENCES

ARCHER, M., 1978. Quaternary vertebrate faunas from the Texas Caves of southeastern Queensland, Australia. *Memoirs of the Queensland Museum* **19:** 61–109.

ARCHER, M. AND CLAYTON, G., 1984. *Vertebrate zoogeography and evolution in Australasia.* Hesperian Press: Perth.

ARCHER, M. AND WADE, M., 1976. Results of the Ray E. Lemley Expeditions, Part 1. The Allingham Formation and a new Pliocene vertebrate fauna from northern Queensland. *Memoirs of the Queensland Museum* **17:** 379–97.

DIXON, J. M. AND HUXLEY, L. (eds), 1985. *Donald Thomson's mammals and fishes of northern Australia.* Thomas Nelson: Melbourne.

FRIEND, G. R. AND TAYLOR, J. A., 1985. Habitat preferences of small mammals in tropical open-forest of the Northern Territory. *Australian Journal of Ecology* **10:** 173–85.

FRITH, H. J., 1973. *Wildlife conservation.* Angus and Robertson: Sydney.

GORDON, G., 1974a. Movements and activity of the short-nosed bandicoot *Isoodon macrourus* Gould (Marsupialia). *Mammalia* **38:** 405–31.

GORDON, G., 1974b. Queensland bandicoots. *Queensland Agricultural Journal* **100:** 16–20.

GORDON, G. AND LAWRIE, B. C., 1977. The rufescent bandicoot, *Echymipera rufescens* (Peters and Doria) on Cape York Peninsula. *Australian Wildlife Research* **5:** 41–45.

HARRISON, J. L., 1962. Mammals of Innisfail. I. Species and distribution. *Australian Journal of Zoology* **10:** 45–83.

HULBERT, A. J., GORDON, G. AND DAWSON, T. J., 1971. Rediscovery of the marsupial *Echymipera rufescens* in Australia. *Nature* **231:** 330–31.

LAVERY, H. J. AND GRIMES, R. J., 1974. Mammals and birds of the Ingham district, north Queensland. I. Introduction and Mammals. *Queensland Journal of Agricultural and Animal Sciences* **31:** 383–90.

LAVERY, H. J. AND JOHNSON, P. M., 1968. Mammals and birds of the Townsville district, north Queensland. I. Introduction and Mammals. *Queensland Journal of Agricultural and Animal Sciences* **25:** 29–37.

LAVERY, H. J. AND JOHNSON, P. M., 1974. Mammals and birds of the lower Burdekin River district, north Queensland. I. Introduction and Mammals. *Queensland Journal of Agricultural and Animal Sciences* **31:** 97–104.

NEWSOME, A. E., 1971. Competition between wildlife and domestic livestock. *Australian Veterinary Journal* **47:** 577–86.

RIDE, W. D. L., 1970. *A guide to the native mammals of Australia.* Oxford University Press: Melbourne.

TATE, G. H. H., 1948. Results of the Archbold Expeditions, No. 60. Studies in the Peramelidae (Marsupialia). *Bulletin of the American Museum of Natural History* **92:** 313–46.

TATE, G. H. H., 1952. Results of the Archbold Expeditions, No. 66. Mammals of Cape York Peninsula, with notes on the occurrence of rain forest in Queensland. *Bulletin of the American Museum of Natural History* **98:** 563–616.

ZIEGLER, A. C., 1977. Evolution of New Guinea's marsupial fauna in response to a forested environment. Pp. 117–38 in *The biology of marsupials* ed by B. Stonehouse and D. Gilmore. University Park Press: Baltimore.

Distribution and status of bandicoots in New South Wales

ELIZABETH ASHBY[1], DANIEL LUNNEY[1], JOHN ROBERTSHAW[2] and ROBERT HARDEN[1]

Six species of bandicoots (family Peramelidae) have been recorded from New South Wales. The species of the inland plains and woodlands are now probably extinct in New South Wales — the Pig-footed Bandicoot *Chaeropus ecaudatus* Ogilby, the Western Barred Bandicoot *Perameles bougainville* Quoy and Gaimard and the Greater Bilby *Macrotis lagotis* (Reid). The ranges of the three extant species have contracted since European settlement and continuing habitat change without specific conservation measures is a threat to all species. The Long-nosed Bandicoot *P. nasuta* Geoffroy and the Northern Brown Bandicoot *Isoodon macrourus* (Gould) are common, and the Southern Brown Bandicoot *I. obesulus* (Shaw) is rare.

Key words: Bandicoot, status, conservation, distribution, *Perameles nasuta, Perameles bougainville, Isoodon macrourus, Isoodon obesulus, Macrotis lagotis, Chaeropus ecaudatus.*

INTRODUCTION

THE popular concept of bandicoots has little improved since Geoffroy's dismissive remarks in 1805 when describing *Perameles nasuta:* "their muzzle, which is much too long, gives them an air exceedingly stupid" (Geoffroy 1805). Indeed the very name bandicoot reflects the public perception. Bandicoot is an Indian word for large rodents of India and Sri Lanka. This was brought to the colony and first applied to *Isoodon* or *Perameles* by George Bass (Strahan 1981). The aim of this paper is to assess the impact of this attitude on the distribution and status of this group of marsupials.

Bandicoots are unique among Australian marsupials in possessing polyprotodont dentition together with syndactylous toes (Jones 1924). Furthermore, bandicoots possess eutherian-like features such as a chorioallantoic placenta (Hill 1895) and persistent corpora lutea through pregnancy (Close 1977). Thus, there has been much active scientific interest in taxonomic relationships (e.g., Mackerras and Mackerras 1960; Freedman 1967; Freedman and Rightmire 1971; Fosse and Risnes 1972; Marshall 1972; Archer 1976; Clemens 1977; Kirsch and Calaby 1977; Lyne and Mort 1981; Lyne 1982) and reproductive biology (e.g., Hill 1895, 1899, 1900; Flynn 1923; Mackerras and Smith 1960; Lyne 1964, 1974, 1976; Stodart 1966, 1977; Lyne and Radford 1974; Lyne and Hollis 1979; Gemmell 1981; Hall 1983; Barnes and Gemmell 1984; Gemmell, Walker, Johnson and Cepon 1984). Other aspects of bandicoot physiology have been studied, including thermoregulation and water metabolism (Hulbert and Dawson 1974a and b; Hulbert and Gordon 1972) as well as activity rhythms (Day, Kirkby and Stenhouse 1974; Lyne 1981) and anatomy (Grant 1828; Poulton 1883; MacKenzie 1918; Stoddart 1980).

There has been relatively little ecological work undertaken, and far less than is considered necessary to prepare a plan of management for any species. Heinsohn (1966) compared the ecology and reproduction of *Perameles gunnii* with *Isoodon obesulus* in farmland and associated woodland of northern Tasmania. He found *I. obesulus* to be trap-shy and to prefer areas with "good" cover and avoid open spaces.

Braithwaite and Gullan (1978) studied the habitat preferences of *I. obesulus* within a patch of isolated heathland east of Melbourne. Here they found that this bandicoot's habitat could be resolved by measuring structural and floristic parameters of the vegetation. However, they concluded that the underlying factors regulating the distribution and abundance of *I. obesulus* are "enigmatic".

The population dynamics of *I. obesulus* were investigated by Stoddart and Braithwaite (1979) in the same heathland. They found a change in habitat preference, pattern of dispersion and reproductive performance with age. Stoddart and Braithwaite (1979) concluded that these features, coupled with a high reproductive rate and a low rate of juvenile recruitment, served to maximize the fitness of *I. obesulus* to the unstable habitat of heathland regenerating after bulldozing.

The natural history of *Macrotis lagotis* has been reviewed by Smyth and Philpott (1968) and Watts (1969). Both concluded that the Greater Bilby has suffered a severe contraction in range which Watts (1969) attributed principally to predation by foxes and competition with rabbits.

[1]New South Wales National Parks and Wildlife Service, P.O. Box 1967, Hurstville, New South Wales, Australia 2220.
[2]Department of Zoology, University of New England, Armidale, New South Wales, Australia 2350.
Pages 43–50 *in* BANDICOOTS AND BILBIES, ed by J. H. Seebeck, P. R. Brown, R. L. Wallis and C. M. Kemper. Surrey Beatty & Sons, Sydney, 1990.

The location of *M. lagotis* burrows seemed to be determined by the qualities of the soil. Within arid and semi-arid habitats, *M. lagotis* was found wherever the ground was soft enough to burrow in and was not subject to waterlogging.

The population dynamics of *I. macrourus* have been investigated by Gordon (1971) on several small islands along the New South Wales central coast and also in a semi-urban district of Brisbane (Gordon 1974a). In his New South Wales study he found this species in open forests with a grassy and/or shrubby understorey. However, in Brisbane, the preferred habitat of *I. macrourus* was a medium dense shrub layer which was provided by disturbed areas where the natural forest cover had been removed.

METHODS

The past and present distributions of bandicoots in New South Wales (Figs 1 to 6) were compiled using the registers of the Australian Museum, the Museum of Victoria, the Queensland Museum, the New South Wales National Parks and Wildlife Service inventory programme, and published records in: Mitchell (1839), Gould (1845), Krefft (1866), Marlow (1958), Troughton (1962), Calaby (1966), Wakefield (1966), Watts (1969), Dixon (1978), Aitken (1979) and Braithwaite, Clayton, Maclean and Parker (1984). Doubtful records were omitted. On this basis we have assigned a conservation status to each bandicoot species. We recognize that there has been more effort in mammal surveys in some parts of the state than in others, thus introducing a possible bias in the distribution records. However, it is our view that the previous work has been extensive enough to determine the ranges and to assign a conservation status to each species.

Ride and Wilson (1982) discussed the confusion and problems of assigning a conservation status to species, and we follow their adaptations of the International Union for the Conservation of Nature and Natural Resources classifications of "probably extinct" and "rare". "Common" is used in this paper to apply to a species that is regularly caught throughout its range with little contraction over time.

RESULTS

Western species

The distribution of the Pig-footed Bandicoot *Chaeropus ecaudatus* (Fig. 1) once encompased the woodland, shrubland and grassland plains of semi-arid Australia, including southwestern New South Wales (Aitken 1983). The type specimen was collected by Surveyor-General Mitchell in 1836 from the grassy plains of the southern banks of the Murray River (Mitchell 1839) The only specimens from New South Wales were collected by Krefft in 1857 (Krefft 1866) at the junction of the Murray and the Darling Rivers (Wakefield 1966). Krefft (1866) reported that "it is exceedingly rare, and is disappearing as fast as the native population . . . I had the greatest difficulty in obtaining a few specimens". He also considered that the "large flocks of sheep and herds of cattle occupying the country" were responsible for the increasing rarity of this species. This bandicoot was last recorded at Lake Eyre North, South Australia in 1907 (Aitken 1979). *Chaeropus ecaudatus* must therefore be classified as probably extinct.

The Western Barred Bandicoot *Perameles bougainville* (Fig. 2), in its many subspecific forms, once extended over much of southern Australia. Gould (1845) nominated the stony ridges near the Namoi and Darling Rivers as "localities in which it may always be found". Krefft collected many specimens on the Blandowski expedition and found it to be "common on all parts of the Murray River and is also found in . . . the immediate neighbourhood of Sydney" (Krefft 1866). This is the only record of this species near Sydney, and since no specimens were lodged with the Australian Museum, one can never be certain of the identity of this specimen. The specimens from the Blandowski expedition were the last collected from New South Wales. *Perameles bougainville* is now probably extinct on mainland Australia. It now occurs only on Bernier and Dorre Islands off the coast of Western Australia (Burbidge 1983).

The Greater Bilby *M. lagotis* (Fig. 3) was formerly widespread and common west of the Great Dividing Range and south of 18° latitude (Watts 1969). It is now probably extinct in New South Wales, the last recorded pair being shot for sport at Wagga in 1912 (Troughton 1962). Krefft foreshadowed the contraction of its range by his remark that "this beautiful animal . . . has long ago retreated to the north of the Murray" (Krefft 1866). Small and scattered populations of *M. lagotis* still exist in central Australia (Smyth and Philpott 1968; Watts 1969; Johnson 1983).

Eastern species

The Northern Brown Bandicoot *Isoodon macrourus* (Fig. 4) is considered to be common throughout most of its former range along the coast of New South Wales north of the Hawkesbury River although fragmentation of suitable habitat has caused some range contraction (Gordon 1983). However, given its ability to live in the semi-urban environment (Gordon 1974a,b), it is not a priority for conservation action. Whether its population size has changed since development is unkown. This question is exacerbated by the confusion of *I. obesulus* and *I. macrourus* by early workers, a confusion that persisted until Mackerras and Mackerras (1960)

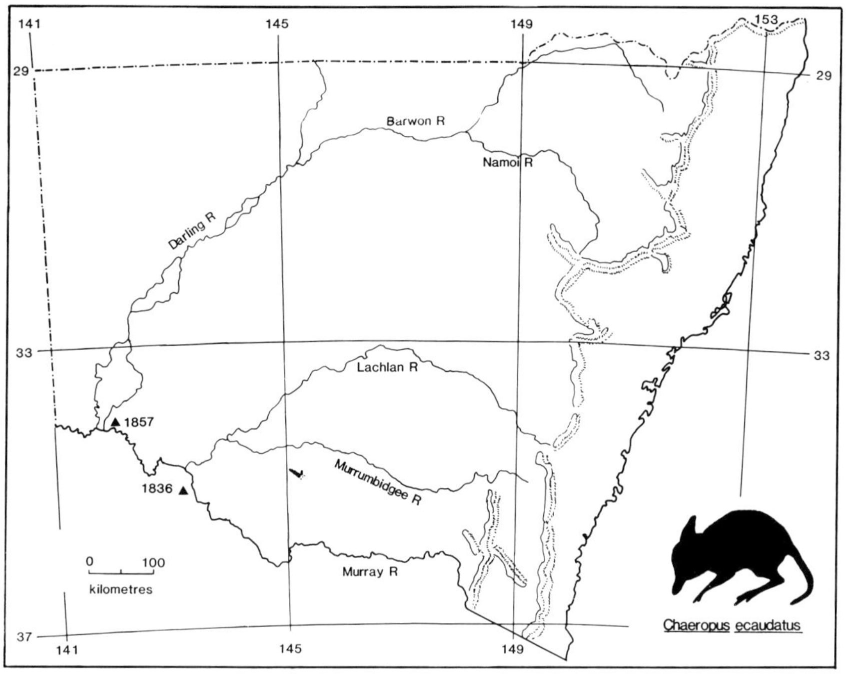

Fig. 1. Locality records of *Chaeropus ecaudatus* in New South Wales. Specimens held in museums are represented by triangles, other reliable records by circles.

traced the specific identities of short-nosed bandicoots in the literature and Lyne and Mort (1981) separated the two species using skull morphology. They identified specimens from the north of Sydney as *I. macrourus* (Fig. 4) and those from the south of Sydney as *I. obesulus* (Fig. 5) with an overlap zone in Ku-ring-gai Chase National Park on the southern side of the Hawkesbury River.

The Southern Brown Bandicoot *Isoodon obesulus* now has a patchy distribution over a reduced range (Aitken 1979; Braithwaite 1983), and is now rare in New South Wales. This species was once "one of the very commonest of Australian mammals" (Gould 1845) and, more particularly, "the most common of the Peramelidae" (Krefft 1866). This is evidenced also by its very early description by Shaw in 1797

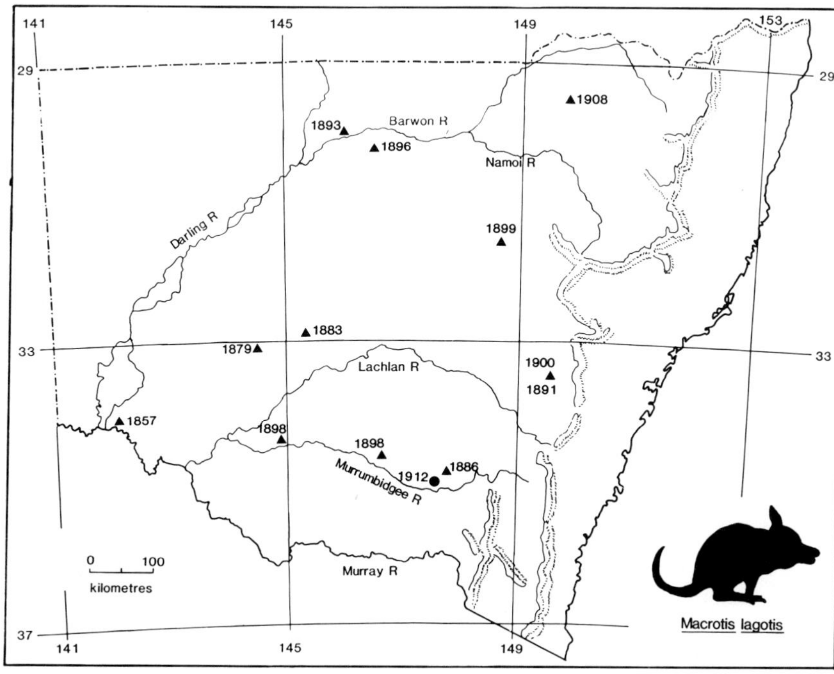

Fig. 2. Locality records of *Perameles bougainville.* Symbols same as Fig. 1.

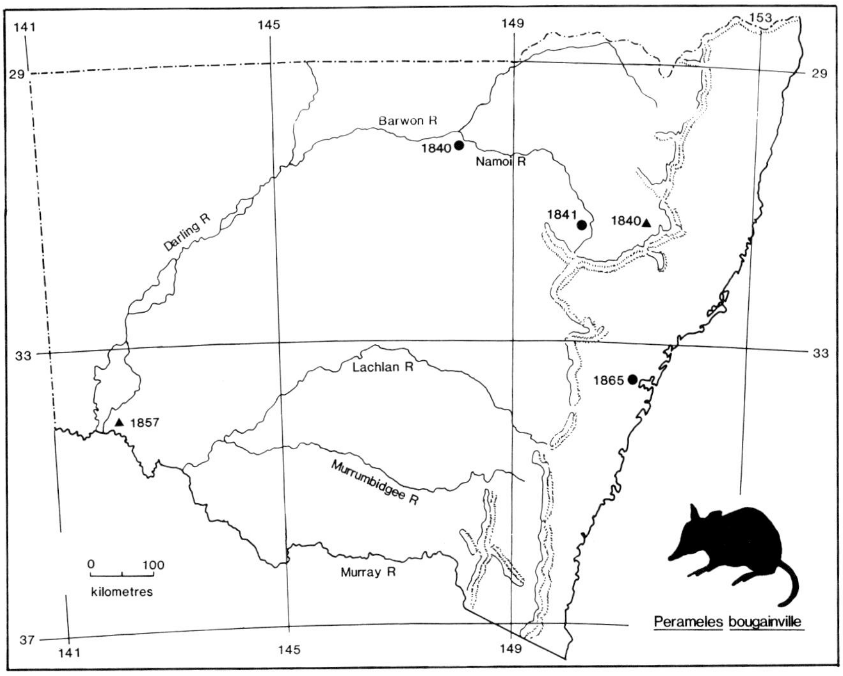

Fig. 3. Locality records of *Macrotis lagotis*. Symbols same as Fig. 1.

(Shaw 1789–1813). *Isoodon obesulus* once occupied the Murray-Darling basin (Krefft 1866; Wakefield 1966) and was abundant in the Bega District on the New South Wales south coast at the turn of the century (Lunney and Leary 1988). The inland population no longer exists (Frith 1974; Braithwaite 1983), and there are few recent records of *I. obesulus* between the Ku-ring-gai population at Sydney and the population in and near Nadgee Nature Reserve on the New South Wales/Victorian border. Dixon (1978) carried out a survey on the New South Wales south coast specifically looking for *I. obesulus,* but caught none. A survey by the National Parks and Wildlife Service of New South Wales of the forests on the coast near Bega found that *I. obesulus* was rare, the only evidence of its presence being hair in one of 2 249 predator scats (Lunney and Barker 1986).

Although Heinsohn (1966) regarded the species as trap-shy, we attribute the paucity of

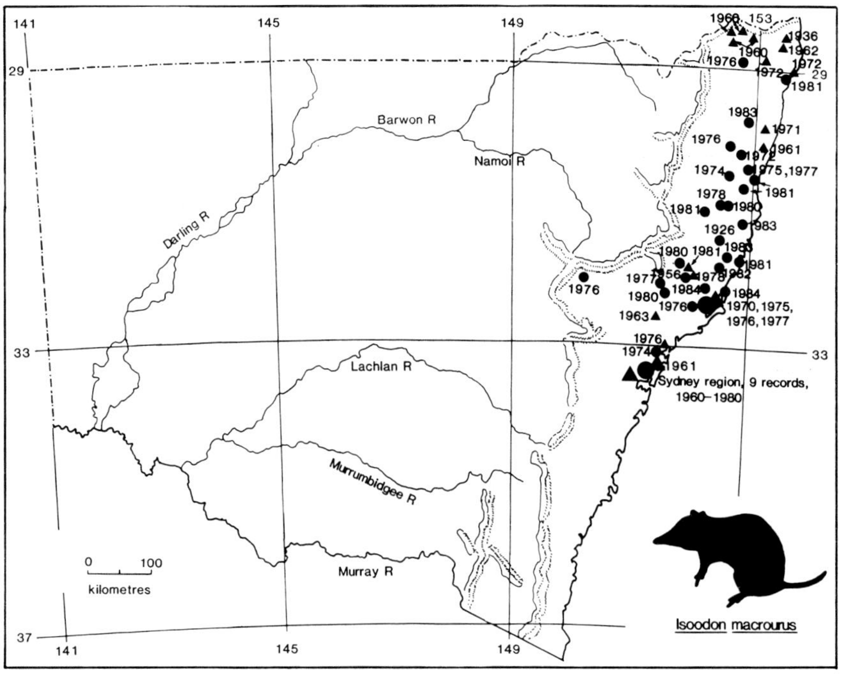

Fig. 4. Locality records of *Isoodon macrourus*. Symbols same as Fig. 1.

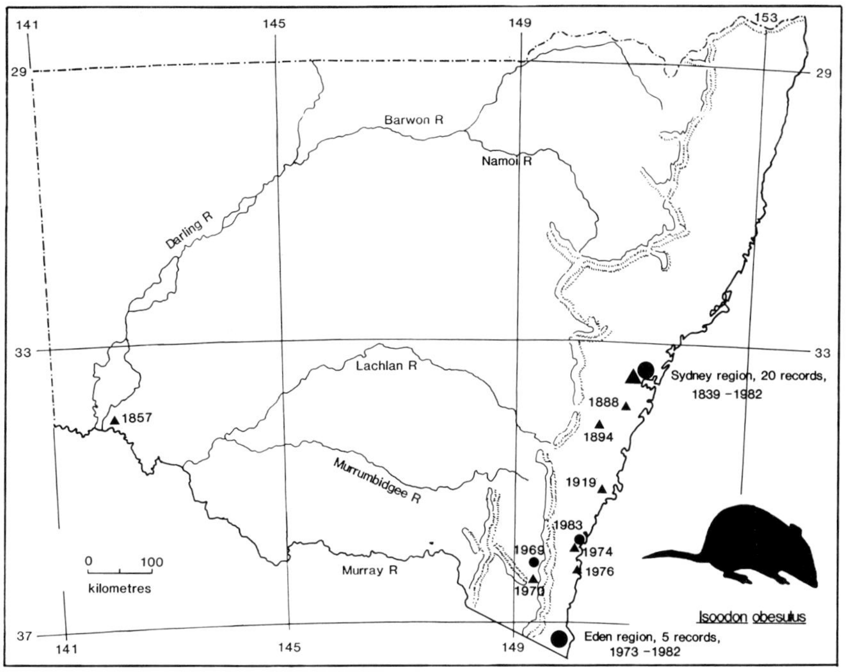

Fig. 5. Locality records of *Isoodon obesulus*. Symbols same as Fig. 1.

records of *I. obesulus* in New South Wales to its rarity. Heinsohn's observation of this bandicoot's aversion to open areas could partly explain the drop in numbers with the loss of habitat due to development and clearing for agriculture. Similarly, the study by Lunney and Leary (1988) of the historical changes in the numbers of the mammals in the Bega District identified the conversion of valley forest to farmland as being the principal cause of the decline of many species including *I. obesulus*. Stoddart and Braithwaite (1979) suggested that the frequency of disturbance (e.g., fire) may be important for *I. obesulus* conservation.

The Long-nosed Bandicoot *Perameles nasuta* is regarded as common throughout most of its range and is found in a variety of forested habitats (Stodart 1983). Comments by Gould (1845), however, indicate that this may not

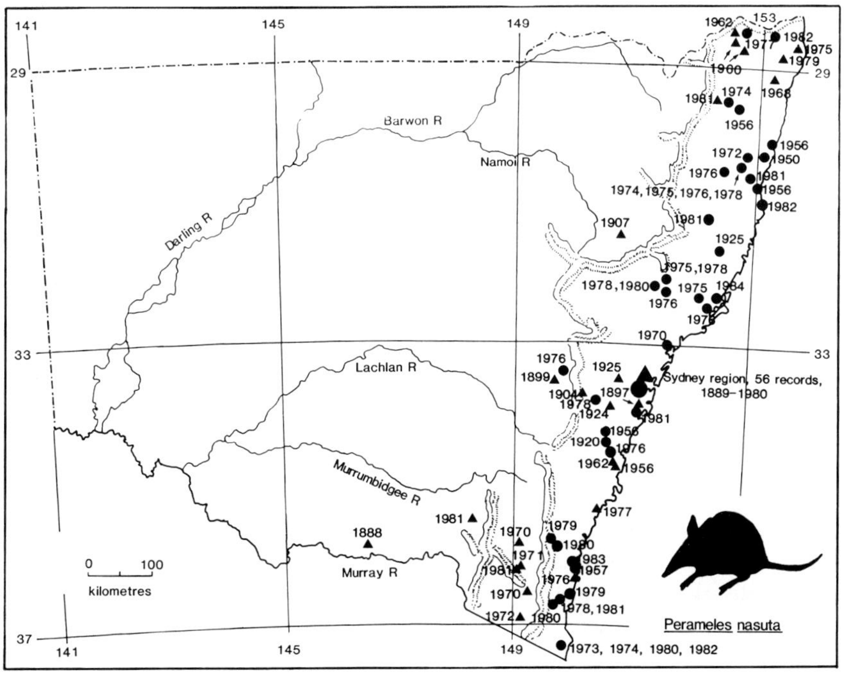

Fig. 6. Locality records of *Perameles nasuta*. Symbols same as Fig. 1.

always have been the case. Gould "ascertained that it is sparingly dispersed over the districts lying between the mountain ranges and the sea" and cited that as the reason for so little being known about the animal at that time. This raises the possibility, although without corroborative evidence, that this bandicoot may have become more common after European settlement. In contrast, there is clear evidence that this species has suffered a contraction of its range. Specimens were collected west of the Great Dividing Range last century and early this century (Marlow 1958) (Fig. 6). Recent specimens, however, are largely restricted to the coastal strip. The ecology of this species is unkown.

McIlroy (1983) investigated the sensitivity of *I. obesulus, P. nasuta* and *P. gunnii* to 1080 poison. This poison is commonly laid throughout Australia for rabbits, pigs and dingoes, and the effect on non-target native species is an important consideration in wildlife management and conservation. McIlroy concluded that only the meat baits intended for dingoes did not appear to pose a serious hazard to these bandicoots. Thus, populations of the rare *I. obesulus* and the common *P. nasuta* could be under threat by pest eradication programmes of pigs and rabbits. McIlroy pointed out that monitoring of field populations needs to be carried out for an accurate assessment of the importance of these laboratory results.

CONSERVATION MEASURES

Bandicoots were not afforded protection in New South Wales until the *Fauna Protection Act* of 1948. By that time *C. ecaudatus, P. bougainville, M. lagotis* and the inland populations of *P. nasuta* and *I. obesulus* had disappeared. As protected fauna under the *National Parks and Wildlife Act* (1974) and the *National Parks and Wildlife Amendment Act* (1983), scientific licences must be issued before any bandicoot may be taken or killed.

A goal of the National Parks and Wildlife Service of New South Wales is to preserve, preferably in free-living and self-perpetuating populations, all species of plants and animals native to New South Wales. The principal means employed to this end has been conservation of species by reservation of habitat in national parks and nature reserves. However, bandicoot preservation has never been the primary aim for reservation of any area, nor have management procedures been instituted solely to benefit bandicoots. The reservation of land will not reinstate the populations now extinct from inland New South Wales, and may not be enough to save the New South Wales populations of *I. obesulus* or secure the status of *P. nasuta* and *I. macrourus*.

The formulation of effective management strategies is not possible until much more is known about bandicoot habitat requirements and population dynamics throughout their range. For example, the effects of drought, development, road building, changing fire frequency, introduced predators and forestry practices on free-living populations are largely unkown. Until studies which specifically address these questions are undertaken, the retention of the remaining New South Wales bandicoot fauna cannot be assured and it can be reasonably concluded that there will be further losses as land-use changes continue.

ACKNOWLEDGEMENTS

We wish to acknowledge Gordon Lyne, Chris Watts and Tony Rose for access to their original data on bandicoot distribution; Tim Flannery and Linda Gibson for access to records of the Australian Museum; Linda Huxley and Joan Dixon for the records of the National Museum of Victoria; Greg Gordon for access to his records from the Queensland Museum; John Calaby for his clarification of the Tilba Tilba specimen; Liz Dovey for providing information from the New South Wales National Parks and Wildlife Service inventory; and John Barker for his comments on the manuscript.

REFERENCES

AITKEN, P., 1979. The status of endangered Australian wombats, bandicoots and the marsupial mole. Pp. 61–66 in *The status of endangered Australian wildlife* ed by M. Tyler. Royal Zoological Society of South Australia: Adelaide.

AITKEN, P., 1983. Pig-footed Bandicoot *Chaeropus ecaudatus*. Pp. 104 in *The Australian Museum complete book of Australian mammals* ed by R. Strahan. Angus & Robertson: Sydney.

ARCHER, M., 1976. Dasyurid dentition and its relationship to that of didelphids, thylacinids, borhyaenids (Marsupicarnivora) and peramelids (Peramelina: Marsupialia). *Australian Journal of Zoology, Supplementary Series* **39:** 1–34

BARNES, A. AND GEMMELL, R. T., 1984. Correlations between breeding activity in the marsupial bandicoots, and some environmental variables. *Australian Journal of Zoology* **32:** 219–26.

BRAITHWAITE, R. W., 1983. Southern Brown Bandicoot *Isoodon obesulus*. Pp. 94 in *The Australian Museum complete book of Australian mammals* ed by R. Strahan. Angus & Robertson: Sydney.

BRAITHWAITE, L. W., CLAYTON, M., MACLEAN, L. AND PARKER, B. S., 1984. Vertebrate fauna of a 144–ha water catchment within eucalypt forest being harvested for woodpulp at Eden, New South Wales. *CSIRO Division of Wildlife and Rangelands Research Technical Paper* **35.**

BRAITHWAITE, R. W. AND GULLAN, P. K., 1978. Habitat selection by small mammals in a Victorian heathland. *Australian Journal of Ecology* **3:** 109–27.

BURBIDGE, A. A., 1983. Western barred bandicoot *Perameles bougainville*. Pp. 101 in *The Australian Museum complete book of Australian mammals* ed by R. Strahan. Angus & Robertson: Sydney.

CALABY, J. H., 1966. Mammals of the upper Richmond and Clarence Rivers, New South Wales. *CSIRO Division of Wildlife Technical Paper* **10**.

CLEMENS, W. A., 1977. Phylogeny of the marsupials. Pp. 51–68 in *The biology of the marsupials* ed by B. Stonehouse and D. Gilmore. Macmillan: London.

CLOSE, R. L., 1977. Recurrence of breeding after cessation of suckling in the marsupial, *Perameles nasuta*. *Australian Journal of Zoology* **25:** 641–45.

DAY, B., KIRKBY, R. AND STENHOUSE, D., 1974. The behaviour of marsupials III. The short-nosed bandicoot, *Isoodon macrourus* (Peramelidae), in the open wild. *Australian Mammalogy* **1:** 255–59.

DIXON, J. M., 1978. Long-nosed bandicoots elude searchers. *Forest and Timber* **14:** 13–17.

DIXON, J. M., 1981. Selection of a neotype for the southern short-nosed (brown) bandicoot, *Isoodon obesulus* (Shaw and Nodder, 1797). *Victorian Naturalist* **98:** 130–35.

FLYNN, T. T., 1923. The yolk-sac and allantoic placenta in *Perameles*. *Quarterly Journal of Microscopical Science* **67:** 123–82.

FOSSE, G. AND RISNES, S., 1972. Development of the teeth in a pouch young specimen of *Isoodon obesulus* and one of *Perameles gunnii* (Peramelidae: Marsupialia). *Archives of Oral Biology* **17:** 829–38.

FREEDMAN, L., 1967. Skull and tooth variations in the genus *Perameles* Part I. Anatomical features. *Records of the Australian Museum* **27:** 147–66.

FREEDMAN, L. AND RIGHTMIRE, G. P., 1971. 4. Skull and tooth variations in Australian bandicoots (Peramelidae: Marsupialia), the genus *Isoodon* and multivariate comparisons with *Perameles*. *Journal of the Royal Society of Western Australia* **54:** 21–31.

FRITH, H. J., 1974. Wildlife. Pp. 78–90 in *The Murray waters — man, nature and a river system* ed by H. J. Frith and G. Sawer. Angus & Robertson: Sydney.

GEMMELL, R. T., 1981. The role of the corpus luteum of lactation in the bandicoot, *Isoodon macrourus*, in captivity. *General and Comparative Endocrinology* **44:** 13–19.

GEMMELL, R. T., WALKER, M. T., JOHNSON, G. AND CEPON, G., 1984. The number of young present in sequential litters of marsupial bandicoot, *Isoodon macrourus*, in captivity. *Australian Journal of Zoology* **32:** 623–29.

GEOFFROY, E., 1805. Memoir on a new genus of mammalia with pouches named *Perameles*. *Philosphical Magazine of London* **21:** 28–34.

GORDON, G., 1971. *A study of island populations of the short-nosed bandicoot* Isoodon macrourus *Gould*. Ph.D. Thesis: University of New South Wales, Sydney.

GORDON, G., 1974a. Movements and activity of the short-nosed bandicoot *Isoodon macrourus* Gould (Marsupialia). *Mammalia* **38:** 405–31.

GORDON, G., 1974b. Queensland bandicoots. *Queensland Agricultural Journal* **100:** 416–20.

GORDON, G., 1983. Northern Brown Bandicoot *Isoodon macrourus*. Pp. 96 in *The Australian Museum complete book of Australian mammals* ed by R. Strahan. Angus & Robertson: Sydney.

GOULD, J., 1845. *The mammals of Australia. Vol 1.* The author: London.

GRANT, R. E., 1828. The anatomy of *Perameles nasuta*. *Memoirs of the Wernerian Natural History Society* **6:** 187–202.

HALL, L. S., 1983. Observations on body weight and breeding of the northern brown bandicoot, *Isoodon macrourus*, trapped in south-eastern Queensland. *Australian Wildlife Research* **10:** 467–76.

HEINSOHN, G. E., 1966. Ecology and reproduction of the Tasmanian bandicoots (*Perameles gunni* and *Isoodon obesulus*). *University of California Publications in Zoology* **80:** 1–107.

HILL, J. P., 1895. Preliminary note on the occurrence of a placental connexion in *Perameles obesula*, and on the foetal membranes of certain macropods. *Proceedings of the Linnean Society of New South Wales* **20:** 579–81.

HILL, J. P., 1899. Contributions to the morphology and development of the female urogenital organs in the Marsupialia. I. On the female urogenital organs of *Perameles* with an account of the phenomenon of parturition. *Proceedings of the Linnean Society of New South Wales* **24:** 42–82.

HILL, J. P., 1900. Contributions to the embryology of the Marsupialia. II. On a further stage in the placentation of *Perameles*. *Quarterly Journal of Microscopical Science* **43:** 1–22.

HULBERT, A. J. AND DAWSON, T. J., 1974a. Thermoregulation in perameloid marsupials from different environments. *Comparative Biochemistry and Physiology* **47:** 591–616.

HULBERT, A. J. AND DAWSON, T. J., 1974b. Water metabolism in perameloid marsupials from different environments. *Comparative Biochemistry and Physiology* **47:** 617–33.

HULBERT, A. J. AND GORDON, G., 1972. Water metabolism of the bandicoot, *Isoodon macrourus* Gould in the wild. *Comparative Biochemistry and Physiology* **41:** 27–34.

JOHNSON, K. A., 1983. Bilby *Macrotis lagotis*. Pp. 107–08 in *The Australian Museum complete book of Australian mammals* ed by R. Strahan. Angus & Robertson: Sydney.

JONES, F. W., 1924. *The mammals of South Australia. Part II.* Government Printer: Adelaide.

KIRSCH, J. AND CALABY, J., 1977. The species of living marsupials: an annotated list. Pp. 9–26 in *The biology of marsupials* ed by B. Stonehouse and D. Gilmore. Macmillan: London.

KREFFT, G., 1866. On the vertebrated animals of the Lower Murray and Darling, their habits, economy, and geographical distribution. *Transactions of the Philosophical Society of New South Wales* (for 1862–1865) **1866:** 1–33.

LUNNEY, D. AND BARKER, J., 1986. Mammals of the coastal forests near Bega, New South Wales. I Survey. *Australian Zoologist* **23:** 19–28.

LUNNEY, D. AND LEARY, T., 1988. The impact on native mammals of land-use changes and exotic species in the Bega district, New South Wales, since settlement. *Australian Journal of Ecology* **13:** 67–92.

LYNE, A. G., 1964. Observations on the breeding and growth of the marsupial *Perameles nasuta* Geoffroy, with notes on other bandicoots. *Australian Journal of Zoology* **12:** 322–39.

LYNE, A. G., 1974. Gestation period and birth in the marsupial *Isoodon macrourus*. *Australian Journal of Zoology* **22:** 303–09.

LYNE, A. G., 1976. Observations on oestrus and the oestrous cycle in the marsupials *Isoodon macrourus* and *Perameles nasuta*. *Australian Journal of Zoology* **24:** 513–21.

LYNE, A. G., 1981. Activity rhythms in the marsupials *Isoodon macrourus* and *Perameles nasuta* in captivity. *Australian Journal of Zoology* **29:** 821–38.

LYNE, A. G., 1982. Observations on skull growth and eruption of teeth in the marsupial bandicoot *Perameles nasuta* (Marsupialia: Peramelidae). *Australian Mammalogy* **5:** 113–26.

LYNE, A. G. AND HOLLIS, D. E., 1979. Observations on the corpus luteum during pregnancy and lactation in the marsupials *Isoodon macrourus* and *Perameles nasuta*. *Australian Journal of Zoology* **27:** 881–99.

LYNE, A. G. AND MORT, P. A., 1981. A comparison of skull morphology in the marsupial bandicoot genus *Isoodon:* its taxonomic implication and notes on a new species, *Isoodon arnhemensis*. *Australian Mammalogy* **4:** 107–34.

LYNE, A. G. AND RADFORD, H. M., 1974. Some observations on reproduction in bandicoots. *Australian Mammalogy* **1:** 293–94.

MACKENZIE, W. C., 1918. *The gastro-intestinal tract in monotremes and marsupials*. Parker Press: Melbourne.

MACKERRAS, I. M. AND MACKERRAS, M. J., 1960. Taxonomy of the common short-nosed marsupial bandicoot of eastern Queensland. *Australian Journal of Science* **23:** 51–53.

MACKERRAS, M. J. AND SMITH, R. H., 1960. Breeding the short-nosed bandicoot, *Isoodon macrourus* (Gould), in captivity. *Australian Journal of Zoology* **8:** 371–82.

MARLOW, B. J., 1958. A survey of marsupials of New South Wales. *CSIRO Wildlife Research* **3:** 71–114.

MARSHALL, L. G., 1972. Evolution of the peramelid tarsus. *Proceedings of the Royal Society of Victoria* **85:** 51–60.

McILROY, J. C., 1983. The sensitivity of Australian animals to 1080 poison VI. Bandicoots. *Australian Wildlife Reasearch* **10:** 507–12.

MITCHELL, T. L., 1839. *Three expeditions into the interior of eastern Australia, with descriptions of the recently explored region of Australia Felix, and of the present colony of New South Wales. Vol II. (2nd ed).* T. & W. Boone: London.

POULTON, E. B., 1883. The tongue of *Perameles nasuta*, with some suggestions as to the origins of the taste bulbs, *Quarterly Journal of Microscopical Science* **23:** 69–86.

RIDE, W. D. L. AND WILSON, G. R., 1982. The conservation status of Australian animals. Pp. 27–44 in *Species at risk: research in Australia* ed by R. H Groves and W. D. L. Ride. Australian Academy of Science: Canberra.

SHAW, G., 1789–1813. *The naturalist's miscellany: or, coloured figures of natural objects drawn and described immediately from nature*. Nodder: London.

SMYTH, D. R. AND PHILPOTT, C. M., 1968. Field notes on rabbit bandicoots, *Macrotis lagotis* Reid (Marsupialia), from central Western Australia. *Transactions of the Royal Society of South Australia* **92:** 3–14.

STODART, E., 1966. Management and behaviour of breeding groups of the marsupial *Perameles nasuta* Geoffroy in captivity. *Australian Journal of Zoology* **14:** 611–23.

STODART, E., 1977. Breeding and behaviour of Australian bandicoots. Pp. 179–191 in *The biology of marsupials* ed by B. Stonehouse and D. Gilmore. Macmillan: London.

STODART, E., 1983. Long-nosed Bandicoot *Perameles nasuta*. Pp. 99 in *The Australian Museum complete book of Australian mammals* ed by R. Strahan. Angus & Robertson: Sydney.

STODDART, D. M., 1980. Observations on the structure and function of the cephalic skin gland in bandicoots (Marsupialia: Peramelidae). *Australian Journal of Zoology* **28:** 33–41.

STODDART, D. M. AND BRAITHWAITE, R. W., 1979. A strategy for the utilization of regenerating heathland habitat by the brown bandicoot (*Isoodon obesulus;* Marsupialia: Peramelidae). *Journal of Animal Ecology* **48:** 165–79.

STRAHAN, R., 1981. *A dictionary of Australian mammal names*. Angus & Robertson: Sydney.

TROUGHTON, E., 1962. *Furred animals of Australia*. (9th ed). Angus & Robertson: Sydney.

WAKEFIELD, N. A., 1966. Mammals of the Blandowski expedition to north-western Victoria, 1856–1857. *Proceedings of the Royal Society of Victoria* **79:** 371–91.

WATTS, C. H. S., 1969. Distribution and habits of the rabbit bandicoot. *Transactions of the Royal Society of South Australia* **93:** 135–41.

Distribution and conservation status of bandicoots in Victoria

PETER W. MENKHORST[1] and J. H. SEEBECK[1]

In Victoria, the survival of bandicoots has been more adversely affected by European settlement than has the survival of any other mammalian family except the Potoroidae. Of the seven bandicoot taxa which are considered to have been present before European settlement, only three remain. *Perameles nasuta* and *Isoodon obesulus* are widespread and are not under immediate threat of a decline in conservation status. *Perameles gunnii* has been reduced to a small population in semi-natural habitat on the outskirts of Hamilton and is classified as endangered. *Chaeropus ecaudatus, P. bougainville, Macrotis lagotis* and a small form of *Isoodon* (cf. *auratus*) all inhabited the semi-arid Murray-Darling region and all disappeared from Victoria remarkably soon after the first European settlers and flocks of sheep penetrated the region. Reasons for the decline of these four taxa are discussed and distribution maps are presented for all taxa.

Key words: Peramelidae, *Perameles bougainville fasciata, P. gunnii, P. n. nasuta, Isoodon o. obesulus, Chaeropus ecaudatus, Macrotis lagotis,* distribution, abundance, conservation status and threats.

INTRODUCTION

IN this review we summarize information pertaining to the historical occurrence and present distribution and abundance of bandicoots in the state of Victoria. Special attention is given to an assessment of the present conservation status of each species and to threats which may adversely affect them in future. Information used in this review was derived from the following sources: the results of fauna surveys conducted over much of Victoria since 1972 by the Fisheries and Wildlife Service (FWS) (now National Parks and Wildlife Division (NPWD); the data base of the Atlas of Victorian Wildlife (NPWD); the collections of the Museum of Victoria (MV); and a synthesis of ecological research conducted on Victorian bandicoots (e.g., Seebeck 1979; Stoddart and Braithwaite 1979; Opie 1980; Brown 1989; Lobert (this volume) and Lobert and Lee (this volume).

In Victoria bandicoots have suffered more from European settlement than all other mammalian families except the Potoroidae; half the bandicoot taxa known to have been present at European settlement are no longer present in the state (Table 1). Five species from three genera have been formally recorded from Victoria and a sixth species was possibly present until about the mid-1800s. Three of the six species, and a small form of another, are no longer present in Victoria, one species is endangered and two are common and widespread (Table 2). The taxa which have disappeared all inhabited the semi-arid northwest of the state whereas the two common and widespread species are inhabitants of forest and moist heathland. Each taxon is discussed separately and the past and present distribution, as defined by records held in the Atlas of Victorian Wildlife data base, are shown in Figs 1 to 4. The mapping units are grid cells of five minutes of latitude by five minutes of longitude. The time periods of pre-1970 and post-1969 were chosen because extensive fauna surveys were started in Victoria during the early 1970s (Emison, Menkhorst, Mansergh, Norman, Robertson, Corrick and Alderson 1984). Therefore, post-1969 records are likely to provide a reasonable approximation of present-day distributions.

SPECIES ACCOUNTS

Western Barred Bandicoot
Perameles bougainville fasciata

In 1857, Krefft found this species to be "common on all parts of the Murray River, and is also found in Victoria . . . "(Krefft 1866). He collected numerous individuals from near the junction of the Murray and Darling Rivers but only four specimens are known to have survived (MV C5921, C5922, C5923 and C5884). These are the only confirmed records of the species in Victoria in historical times (Fig. 1). The species is now presumed to be extinct on mainland Australia and is known to survive only on Bernier and Dorre Islands in Shark Bay, Western Australia.

Perameles bougainville formerly occurred across a large portion of semi-arid to arid southern Australia on plains and sand ridges supporting woodlands, saltbush or heath (Aitken 1979). The species was probably widespread in north-western Victoria until the mid to late-1800s when pastoral and subsequently agricultural

[1]Arthur Rylah Institute for Environmental Research, National Parks and Wildlife Division, Department of Conservation and Environment, P.O. Box 137, Heidelberg, Victoria, Australia 3084.

Pages 51–60 *in* BANDICOOTS AND BILBIES, ed by J. H. Seebeck, P. R. Brown, R. L. Wallis and C. M. Kemper. Surrey Beatty & Sons, Sydney, 1990.

Table 1. Number of native species of each mammalian family which have disappeared from Victoria since European settlement began in the 1830s. (Source, Menkhorst 1987.) Percentages are for number of species in family. * 57% if *Macrotis lagotis* is included as a former Victorian species.

Family	No. of Victorian species	Number of species extinct in Victoria
Tachyglossidae	1	0
Ornithorhynchidae	1	0
Dasyuridae	15	4 (27%)
Peramelidae	6	3 (50%)*
Phalangeridae	2	0
Burramyidae	5	0
Petauridae	6	0
Potoroidae	5	3 (60%)
Macropodidae	11	4 (36%)
Phascolarctidae	1	0
Vombatidae	1	0
Pteropodidae	2	0
Rhinolophidae	1	0
Emballonuridae	1	0
Molossidae	2	0
Vespertilionidae	14	0
Muridae	16	7 (44%)
Canidae	1	0
Total	91	21 (23%)

settlement was rapidly spreading (Kenyon 1914). In much of northwestern Victoria, heavy-soiled plains supporting tall mallee or grassy woodlands of box eucalypts or *Casuarina* spp. and *Callitris preissii,* were interspersed amongst sand dunes carrying *Triodia* hummock grassland and heathland dominated by *Callitris* spp. or *Leptospermum laevigatum.* Such areas may have provided suitable habitat for this species because they would have offered a range of substrates in which to forage as well as dense cover amongst the grass tussocks, *Triodia* hummocks and heath. The heavy- soiled plains were also the most productive pastoral land and were the first areas to be utilized by pastoralists who arrived from the 1840s onwards with large flocks of sheep and, later, cattle (Kenyon 1914).

Apparently *P. bougainville* was one of 15 mammalian species from the Murray-Darling area (including the bandicoots *Chaeropus ecaudatus, Macrotis lagotis* and an inland form of *Isoodon*)

Table 2. Present conservation status of bandicoots in Victoria.

Species	Conservation status
Western Barred Bandicoot	
Perameles bougainville fasciata	Presumed extinct
Eastern Barred Bandicoot	
Perameles gunnii	Endangered
Long-nosed Bandicoot	
Perameles nasuta nasuta	Widespread and secure
Southern Brown Bandicoot	
Isoodon obesulus obesulus	Widespread and secure
Isoodon sp. (cf. *auratus*)	Extinct
Pig-footed Bandicoot	
Chaeropus ecaudatus	Extinct
Greater Bilby	
Macrotis lagotis	Extinct

which were unable to survive the ecological disruption caused by European settlement (Bennett, Lumsden and Menkhorst 1989). Causes of the declines of these species probably included: decreases in vegetation cover and diversity which resulted from grazing by European rabbits, sheep and cattle and from altered fire regimes; predation by introduced foxes and cats; and, later, the clearing of vast areas for agriculture.

Recent unconfirmed reports of bandicoot-like animals from the Big Desert give reason for a faint hope that the species still survives in this state.

Eastern Barred Bandicoot
Perameles gunnii

The past and present distribution was described by Seebeck (1979) and Seebeck, Bennett and Dufty (1990) and is illustrated in Fig. 1. Records of the Museum of Victoria show that before the 1930s *P. gunnii* was widely distributed across the native tussock grasslands of the Western Volcanic Plains from near Melbourne in the east, to slightly west of Hamilton in the west, north to near Beaufort and south to the coast near Port Fairy. Other localities which yielded early specimens include Bacchus Marsh, Werribee, Warncoort, Birregurra, Mortlake, Trawalla and Penshurst. Since about 1930 its range has decreased to the extent that it is now considered an endangered species (NPWD unpublished). The species was present at Cressy until 1932, at Noorat near Terang as late as 1951 (Fisheries and Wildlife Service now National Parks and Wildlife Division files) and at Woolsthorpe until the mid-1940s (Bennett 1982). However, since 1930 most of the records are from around three localities; Mount Gellibrand near Colac (where C. W. Brazenor collected specimens in 1949), the Toolong–Orford area near Port Fairy, and the Hamilton–Penshurst area.

The fluctuating fortunes of the colony at Mount Gellibrand during the first half of this century were documented by Seebeck (1979) from the diary entries of a local resident. Those entries indicated a population steadily declining from abundant to locally extinct, but with numerous minor fluctuations which were claimed by the diarist to correspond to the cycle of wet and dry periods, and to fluctuations in the population of "grass grubs" which the bandicoots ate. Bandicoots were last seen at Mount Gellibrand during the early 1950s and subsequent searches by J. M. Dixon (Museum of Victoria) and J. H. Seebeck have failed to locate the species (Seebeck 1979).

The colonies in the Port Fairy district were a few kilometres north of the town in the Toolong–Orford area. It is not known whether they were

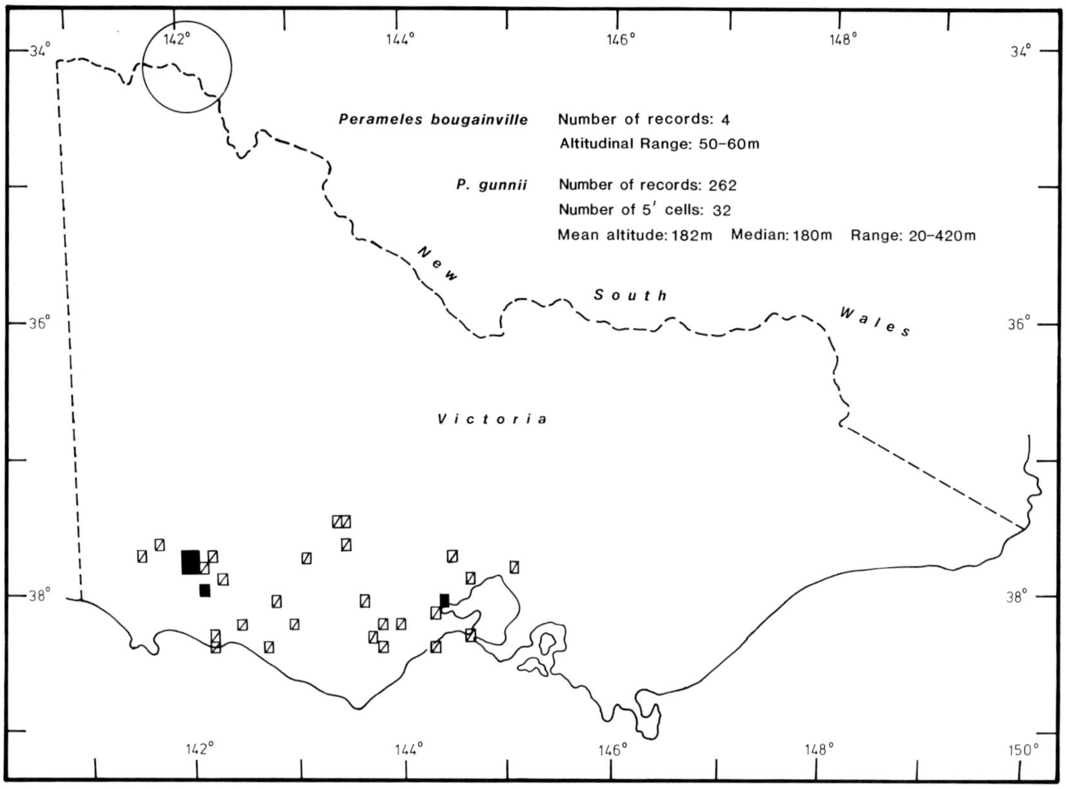

Fig. 1. Victorian distributions of *Perameles gunnii* and *Perameles bougainville.* ■ post-1969 record of *P. gunnii* ▨ pre-1970 record of *P. gunnii.* The eastern post-1969 record (38°00′S 144°20′E) is a feral population derived from escapees from a captive colony held at the Serendip Wildlife Research Station, Lara. The large open circle represents the general area from which specimens of *P. bougainville* were obtained during the Blandowski expedition of 1857.

continuous with the Hamilton populations, but as the intervening geological formations, soil types and original vegetation were similar they probably were. However, the last specimen collected near Port Fairy was in 1971 (National Parks and Wildlife Division files).

Early records of bandicoots in the Hamilton district are few although bandicoots were apparently common. Their continued existence there was confirmed in 1960 and since then reports by local naturalists and studies by Seebeck (1979, 1984), Moon (1984) and Brown (1989) have resulted in a large number of records in and around Hamilton, particularly from near the Grange Burn. The population was estimated at about 650 in 1985 (Sherwin and Brown, this volume), between 200 and 300 in 1968 (Dufty 1988) and is now believed to number less than 200 (Lacy and Clark 1990).

The reasons for the species' decline are complex. They have been detailed by Seebeck (1979) and Seebeck *et al.* (1990) and include the clearing of vegetation, cropping, grazing, pasture improvement and physical consolidation of soils by stock. Predation and disturbance by introduced carnivores is also presumed to have played a role in reducing the population. Foxes, dogs and cats have been responsible for the deaths of hundreds of bandicoots (Seebeck 1979; Brown 1989).

The precise effect of the introduction of European rabbits on the bandicoot's ecology is not known, but the rabbit plagues that formerly occurred in western Victoria must have contributed to a reduction in habitat, and the widespread use of gin traps to catch rabbits has caused the death of many bandicoots. Others may have been killed after eating 1080 poison in baits laid during rabbit control programmes. It is theoretically possible for *P. gunnii* to consume enough rabbit baits to obtain a lethal dose of 1080 (McIlroy 1983).

The original vegetation of the basalt plains was essentially perennial tussock grassland dominated by Wallaby Grasses *Danthonia* spp., Speargrasses *Stipa* spp., Tussock Grass *Poa labillardieri,* and Kangaroo Grass *Themeda triandra* (Willis 1974). There were also areas of savannah woodland mostly composed of River Red Gum *Eucalyptus camaldulensis,* but with some Yellow Box *E. melliodora,* Swamp Gum *E. ovata* and Drooping Sheoak *Casuarina stricta* and a Tussock Grass understorey. With increasing rainfall from east to west the grassy understorey changed from dry tussock near Geelong to wet tussock at Hamilton (Willis 1974).

Farming practice on these grasslands during the past 140 years has changed the original species composition and structure in most areas and very little of the original habitat now exists. Species such as Kangaroo Grass and Wallaby Grass have largely disappeared or have been reduced and replaced by exotic grasses, clovers and weeds. In the city of Hamilton *P. gunnii* occurs in semi-natural situations, for example along the Grange Burn as well as in golf courses, public parks and gardens, cemeteries and private suburban gardens. Near Grange Burn the vegetation consists of introduced pasture grasses, Monterey Cypress, thickets of White Poplar and domestic gardens. On the outskirts of Hamilton and further afield bandicoots are found in farmland, especially in ungrazed areas and in or near swampy depressions, and in road reserves. At Toolong–Orford the habitat was relatively unimproved farmland, particularly in poorly drained areas where the ground vegetation was dominated by tussocks of *Poa* sp. On improved farmland the pasture grasses were closely grazed, as were swampy depressions, and the remaining rushes and sedges were scattered and offered little cover.

In Tasmania Heinsohn (1966) found that *P. gunnii* used open areas extensively for foraging, but selected nest sites in more sheltered situations. The same is evident in and around Hamilton and the additional cover available in suburban gardens, Gorse *Ulex europaeus* thickets and rubbish dumps probably contributes to the bandicoot's relative abundance in such situations.

No specific reserves have been created for this species. However, in 1983 the World Wildlife Fund and a private donor provided financial support to enable a two-year, full-time study of *P. gunnii* to start. The research included studies of the population dynamics and reproductive biology of the species, its dietary habits, the effects of introduced predators and the utilization of natural and artificial habitat. This study, together with more recent work (e.g., Dufty 1988; Brown 1989) has resulted in the preparation of a recovery strategy (Seebeck 1990) aimed at stabilizing the population at Hamilton, and the successful re-introduction of the species into suitable areas within their former range.

Long-nosed Bandicoot
Perameles nasuta nasuta

Perameles nasuta is widely distributed in Gippsland, northeastern Victoria and the periphery and foothills of the Great Dividing Range west to the Dandenong and Mount Disappointment Ranges near Melbourne. An isolated population occurs in the Otway Ranges and westwards to the Naringal district (38°20′S 142°45′E). The species is not known to occur at altitudes above about 800 m and is absent from the plains of the La Trobe Valley and from the Western Victorian Uplands (Fig. 2). There is little evidence of a reduction in the range of this species since European settlement although it

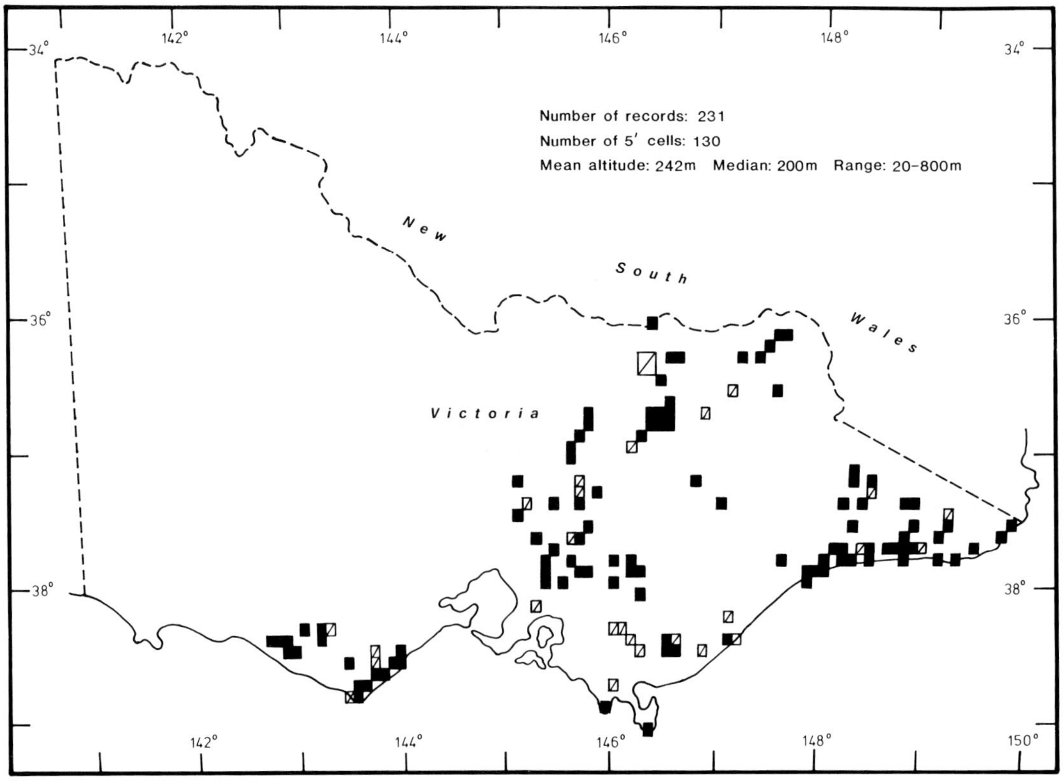

Fig. 2. Victorian distribution of *Perameles nasuta*. ■ post-1969 record, ▨ pre-1970 record at 5′ resolution, ▨ pre-1970 record at 10′ resolution.

has presumably been excluded from much of the land within its range that is now used for agriculture. There are no confirmed records from the Melbourne metropolitan area (Seebeck 1977).

In much of the Great Dividing Range, *Perameles nasuta* is uncommon and restricted to wetter and more fertile sites such as those along streams. However, the species is common in high rainfall areas of northeastern Victoria, Gippsland and the Otway Ranges. In north-eastern Victoria, the Dandenong Ranges and parts of coastal Gippsland, *Perameles nasuta* inhabits rank grassland in paddocks and gardens where it is occasionally considered a pest (Norris, Mansergh, Ahern, Belcher, Temby and Walsh 1983; Menkhorst, Emison and Marcius 1984).

Habitats utilized by *P. nasuta* include damp areas with dense ground or shrub layers within vegetation formations ranging from closed forest through open forest to heath and grass-land. It is common in riparian vegetation such as fern or *Rubus* sp. thickets on river and creek flats in the ranges, and in open forests or wood-lands where the understorey is dense heath (Emison, Porter, Norris and Apps 1975; Norris *et al.* 1983; Menkhorst *et al.* 1984). On coastal plains and foothills *P. nasuta* occasionally occurs in sympatry with *Isoodon obesulus* in moist situa-tions in heath or heathy woodlands (e.g., Emison *et al.* 1975; Norris, Gilmore and Menkhorst 1979).

Perameles nasuta is difficult to capture in con-ventional small mammal traps and is doubtless more common than trapping results suggest. For example, at Naringal East, near the western limits of the species' distribution, only seven individuals were captured a total of eight times during 7 200 trapnights between October 1972 and February 1975. During the same period 130 *Potorous tridactylus* were captured a total of 1 600 times at the site (Seebeck, unpublished data). Yet the 11 reports of *P. nasuta* killed on nearby roads between 1972 and 1983, plus another 11 incidental observations or specimen records from the Naringal area (Atlas of Victorian Wildlife data base) suggest that the species is not uncommon there. Similarly, no *P. nasuta* were trapped during 4 126 trapnights in Boola Boola State Forest between 1973 and 1977 yet remains of at least 10 were identified in predator scats collected there (Loyn, MacFarlane, Chesterfield and Harris 1980).

At present this species is not considered threatened and it occurs in many conservation reserves including most of those in Gippsland, the north-east and the Otway Ranges. However, the effects of 1080 poison, which is widely used in Victoria to control European rabbits, require further investigation and monitoring. *Perameles nasuta* could theoretically consume enough baits to obtain a lethal dose (McIlroy 1983).

Southern Brown Bandicoot
Isoodon obesulus obesulus

In Victoria, *I. obesulus* is an animal of the coastal or fluviatile plains although populations exist in the Grampians and Dandenong Ranges. Its range extends along the entire coastal region except for apparent gaps corresponding to the Western Volcanic Plains and the La Trobe Valley (Fig. 3). The species is generally absent from the Eastern and Western Victorian Uplands although it occurs at altitudes of up to 1 000 m in the Grampians. In southwestern Victoria its range extends inland for almost 100 km (to about 37°10'S) along a band of sandy coastal plains but elsewhere, except for the Grampians, it rarely occurs more than 50 km from the coast. The population in the Grampians apparently extends eastwards to the Great Western area (37°10'S 142°50'E) where a road-killed specimen was found in grassy River Red Gum woodland beside Concongella Creek. Surprisingly, the species is absent from Victorian islands even though some, such as French, Phillip, Snake and Sunday Islands support large tracts of suitable-looking habitat (Norris *et al.* 1979; Seebeck 1981).

On the coastal plains of southwestern Victoria, the species appears to be more abund-ant than in eastern Victoria and, during two similar fauna surveys, was captured eight times more frequently in the south-west than in East Gippsland (44 captures from 5 280 trapnights (Menkhorst and Beardsell 1982) compared with five captures in 4 000 trapnights (NPWD, un-published data). It was described as common in the Grampians (Emison, Porter, Norris and Apps 1978).

In recent decades the *I. obesulus* has disappeared from some areas of intensive agricultural or urban development. For example, it was common in pockets of heath-land, rough pasture and market gardens in some outer southeastern suburbs of Melbourne until the late 1960s (Dixon 1966; Seebeck 1977), but records since then have been few (Table 3), probably because of the development of extensive housing estates. In these areas the species is now confined to a few golf courses and undeveloped bush remnants. Unlike the Northern Brown Bandicoot *Isoodon macrourus* in northern New South Wales and Queensland (Gordon 1983), *I. obesulus* has not prospered in suburban environments.

Isoodon obesulus inhabits heathland, shrubland and heathy open forest and woodland. It is usually associated with well-drained soils and "dry" heath communities and does not occur in tall open forest (e.g., Emison *et al.* 1975). In

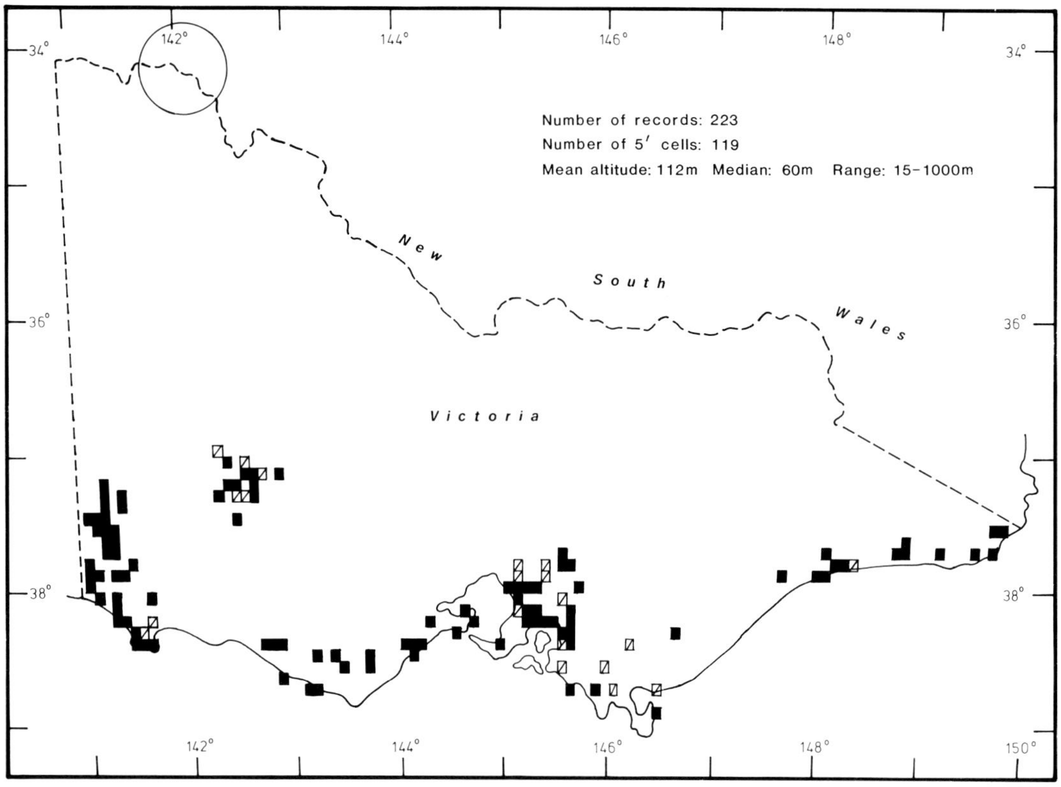

Fig. 3. Victorian distribution of *Isoodon obesulus*. ■ post-1969 record, ▨ pre-1970 record. The large open circle represents the general area from which specimens of a smaller inland form (possibly *Isoodon auratus*) were collected during the Blandowski expedition of 1857.

southwestern Victoria, trapping success in Brown Stringybark open forest (3.7%) was similar to that in Silver Banksia closed heath (3.3%) but was lower in the wetter Messmate open forest and Scented Paper-bark closed scrub communities (Menkhorst and Beardsell 1982). At Cranbourne in central southern Victoria, Braithwaite and Gullan (1978) found that *I. obesulus* preferred regenerating heath about 18 years old, but Opie (1980), working at the same site, found that captures were highest in 12 year old heath which supported a greater abundance of invertebrate prey than did older heath with a different floristic composition and structure. Lobert (1985) concluded that vegetation of an early successional stage is generally preferred to later successional stages, unless the early stages provide insufficient cover, in which case they may still be used for feeding.

Table 3. Numbers of records of *Isoodon obesulus* in 5′ latitude/longitude cells covering Melbourne's eastern suburbs. Data from Atlas of Victorian Wildlife.

5′ cell centred on	pre–1955	1955–69	1970–84
Kew	1		
Doncaster	1		
Mt Evelyn		4	
Mt Waverley		3	
Kallista		1	
Black Rock	2	1	1
Dingley	2	16	1
Dandenong			1
Lysterfield			1
Chelsea		1	2
Frankston	1		
Skye		13	
Totals	7	39	6

Isoodon obesulus is well catered for by the system of conservation reserves in Victoria and is not considered under threat. Its extraordinary fecundity (Lobert and Lee, this volume) and behavioural adaptations, including high juvenile dispersal, enable the species to quickly colonize patches of suitable habitat as these develop (Stoddart and Braithwaite 1979). This makes the species amenable to ecological management, especially the use of controlled fires to produce a mosaic of heathlands of different ages. As for *P. nasuta* and *P. gunnii*, *I. obesulus* may be susceptible to 1080 poison laid to control European rabbits (McIlroy 1983). Monitoring of population numbers before and after poisoning campaigns is necessary to establish the effects on bandicoot populations.

A form of *Isoodon* that formerly occurred in the Mildura area must have been common because the specimen catalogue of the Blandowski expedition includes at least 62 numbers referred to this form (Wakefield 1966a). Only five specimens have survived (MV C5789–C5793). Wakefield (1966b) considered this form to be a distinct small form of *I. obesulus*, perhaps more closely related to *Isoodon auratus* of Central Australia. An examination of cranial characters of the Blandowski specimens led Dixon (1978) to assign them to *I. auratus*.

Whatever its taxonomic status, this form quickly declined after European settlement, presumably for similar reasons to the other bandicoots which formerly inhabited that region (see text on *P. bougainville* for further details) and no further specimens were obtained.

Pig-footed Bandicoot
Chaeropus ecaudatus

The type specimen of this now extinct species was collected in Victoria by Sir Thomas Mitchell a few miles south-east of the confluence of the Murray and Murrumbidgee Rivers on 16 June 1836 (Mitchell 1839; Wakefield 1966b). The habitat was described as "grassy plains bounded by sandhills on which grew pines (callitris); and open forests of goborro (or box tree), prevailed very generally nearer the river" (Mitchell 1839).

In 1857, Gerard Krefft camped for six months "not far from the spot where Sir Thomas Mitchell secured his tail-less animal" (Krefft 1866) [This camp was at Mondellimin near where Mildura is now (Wakefield 1966a) and about 110 km west-north-west of the type locality]. Krefft obtained about eight specimens (Krefft 1866) and according to a list of those specimens retained for the museum, which was compiled by Krefft (Museum of Victoria archives), all were procured on the New South Wales side of the Murray River (Wakefield 1966a). Therefore, the type specimen is the only valid record of this species from Victoria (Fig. 4).

Krefft (1866) accurately predicted that —

"This singular animal . . . is still found on the plains of the Murray; though it is exceedingly rare, and is disappearing as fast as the native population. The large flocks of sheep and herds of cattle occupying the country will soon disperse those individuals which are still to be found in the so-called settled districts, and it will become more and more difficult to procure specimens for our national collection".

The notes of Mitchell (1839) and Krefft (1866) are the only sources of information about the habitat of *C. ecaudatus* in Victoria or New South Wales. The species was last positively recorded in 1907 on the western bank of Lake Eyre North and is now presumed extinct. Aitken (1979) reasoned that the paucity of specimen records suggests that the species was already rare at the time of European settlement and that the species rapidly declined to extinction in the face of pastoral development. However, we consider the meagre evidence available from 19th century collections insufficient to allow conclusions about the species' status before European settlement. Reasons for the decline of *C. ecaudatus* in the lower Murray Valley do not include competition with European rabbits or predation by foxes because the decline preceded the arrival of these two alien species in the lower Murray Valley (Rolls 1969; Aitken 1979).

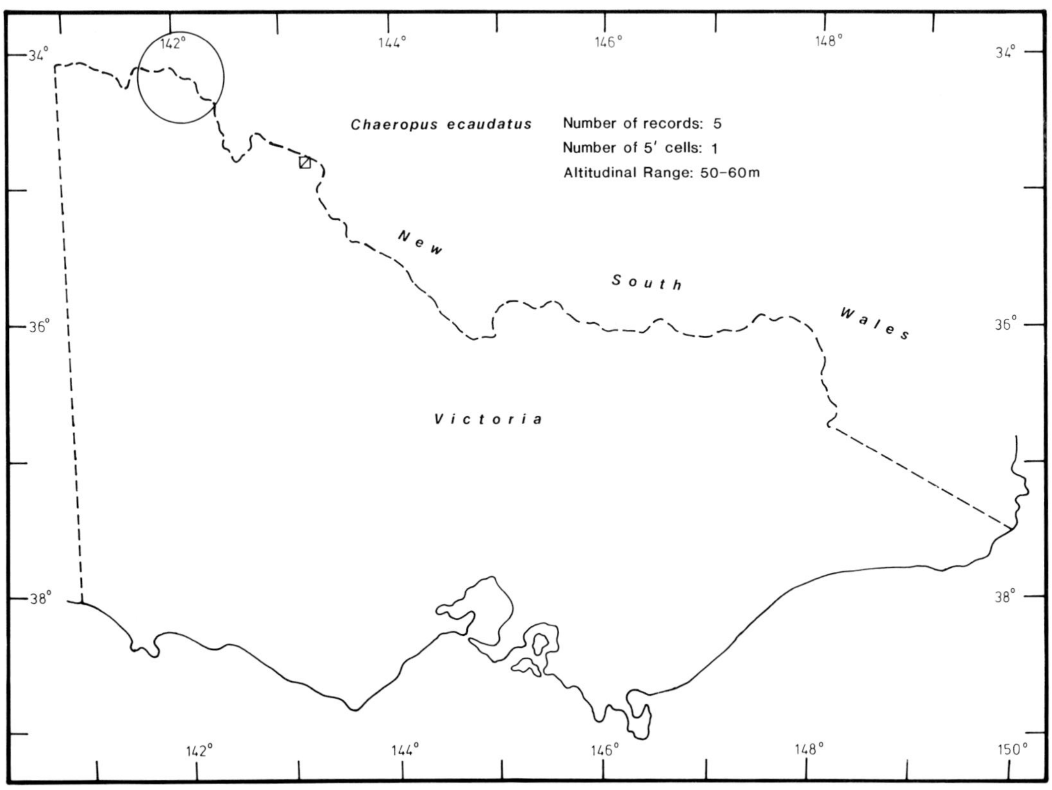

Fig. 4. Distribution of records of the *Chaeropus ecaudatus* and *Macrotis lagotis* in Victoria. ☒ — type locality of *C. ecaudatus*, the large open circle represents the general area from which specimens of *C. ecaudatus* and *M. lagotis* were collected during the Blandowski expedition of 1857.

Rather, the intensive grazing of sheep and cattle in the optimum habitat (open woodland with an understorey of shrubs and grasses) may have rapidly altered the floristic composition and structure of the ground vegetation to the detriment of this species whose diet was apparently more vegetarian than that of other bandicoots (Krefft 1866; Dixon 1988) and which required dense ground vegetation for shelter from climate and predators (Newsome 1971).

Greater Bilby
Macrotis lagotis

Northwestern Victoria is often included in descriptions of the former range of this species (e.g., Frith 1973; Ovington 1978; Aitken 1979; Johnson 1983), apparently in uncritical acceptance of the comment by Brazenor (1950) that the species "... originally recorded from the north-west area of the state, was reported extinct in 1866". This comment was presumably based upon Krefft's (1866) statement (reporting on observations made in 1857) that: "this beautiful animal, like so many other species, has long ago retreated to the north of the Murray". Subsequently, Krefft's comment has been misrepresented in the literature by statements such as: "some remained in Victoria until at least 1866" (Frith 1973) and "last reported in Victoria in 1866" (Ovington 1978). The provenance of Krefft's specimens is not clear. Wakefield (1966b) considered that they were all obtained in New South Wales but provided no evidence for this assertion. Krefft himself appears to have been uncertain about the possible occurrence of *M. lagotis* in Victoria; in his catalogue of specimens in the Australian Museum (Krefft 1864), Victoria was not included in the description of the species' distribution but in his subsequent book (Krefft 1871) he said it occurred in northern Victoria. In the Museum of Victoria are two other specimens labelled "banks of the Murray" which were donated by Brodribb. One was presented in 1896 (C5896), the other is undated (C5883). Because of the vague locality details the provenance of these specimens remains equivocal and we maintain that there are no tangible records of this species from Victoria although it quite possibly occurred in the north-west until the late 1800s (C. H. S. Watts, pers. comm., 1985).

Like *C. ecaudatus, M. lagotis* quickly succumbed to ecological changes associated with European settlement of the Murray-Darling region. *Macrotis lagotis* is more dependent on plant material for food than most bandicoots (Lee and Cockburn 1985). Thus, they may have been particularly susceptible to floristic and structural changes to the vegetation induced by the introduction of sheep and cattle (Newsome 1971). If the species persisted in Victoria until the late 1870s they may have suffered from the arrival of European rabbits. Elsewhere, rabbits usurped the burrows of *M. lagotis* thus leaving them open to exposure and predation, and many may have been unintentionally killed by rabbit-trappers (Wood 1984).

The suggestion that *M. lagotis* may have occurred on the volcanic plains of southwestern Victoria (Scarlett 1969) was based upon Aboriginal vocabularies documented by Dawson in 1881. However, as Scarlett points out, the evidence is inadequate. The words could just as readily be interpreted as referring to *Bettongia lesueur,* but, if so, Dawson must have confused these with a dasyurid, perhaps *Dasyurus viverrinus,* as neither bandicoots nor potoroids have open pouches or produce four or five young.

ACKNOWLEDGEMENTS

We thank J. M. Dixon for allowing access to the catalogue of the Museum of Victoria mammal collection. Dr C. H. S. Watts and Helen McCracken kindly provided information on the provenance of specimens of *M. lagotis* from southeastern Australia. Our colleagues Peter Brown and Robert Warneke were always available with helpful discussions about bandicoots in Victoria. Lance Williams and Karl Gleeson have collated distribution data for the Atlas of Victorian Wildlife with admirable thoroughness and good humour. Helpful comments on drafts of this paper were kindly provided by Darwin Evans, Lindy Lumsden and Linda Huxley.

REFERENCES

Aitken, P., 1979. The status of endangered Australian wombats, bandicoots and the marsupial mole. Pp. 61–65 in *The status of endangered Australasian wildlife* ed by M. J. Tyler. Royal Zoological Society of South Australia: Adelaide.

Bennett, A. F., 1982. The mammals of the Woolsthorpe area, western Victoria: and the changes that have occurred since European settlement. *Victorian Naturalist* **99:** 229–40.

Bennett, A. F., Lumsden, L. F. and Menkhorst, P. W., 1989. Mammals of the mallee region of southeastern Australia. Pp. 191–220 in *Mediterranean Landscapes in Australia: Mallee ecosystems and their management* ed by J. C. Noble and R. A. Bradstock. CSIRO: Melbourne.

Braithwaite, R. W. and Gullan, P. K., 1978. Habitat selection by small mammals in a Victorian heathland. *Australian Journal of Ecology* **3:** 109–27.

Brazenor, C. W., 1950. *The mammals of Victoria and the dental characteristics of monotremes and Australian marsupials.* National Museum of Victoria. Handbook No. 1, Melbourne.

Brown, P. R., 1989. Management Plan for the conservation of the Eastern Barred Bandicoot, *Perameles gunnii,* in Victoria. *Arthur Rylah Institute Technical Report Series* No. **62.** National Parks and Wildlife Division: Victoria.

DIXON, J. M., 1966. Bandicoots — partial survival in times of possible extinction. *Victoria's Resources* **8:** 62–63.

DIXON, J. M., 1978. Taxonomy of Short-nosed Bandicoots of the genus *Isoodon* (Marsupialia: Peramelidae), based on cranial characteristics. *2nd International Theriological Congress Abstracts of papers:* **79.**

DIXON, J. M., 1988. Notes on the diet of three mammals presumed to be extinct: the Pig-footed Bandicoot, the Lesser Bilby and the Desert Rat Kangaroo. *Victorian Naturalist* **105:** 208–11.

DUFTY, A. C., 1988. *The distribution, population abundance, status, movement and activity of the Eastern Barred Bandicoot,* Perameles gunnii, *at Hamilton.* B.Sc.(Hons) Thesis. La Trobe University, Melbourne.

EMISON, W. B., PORTER, J. W., NORRIS, K. C., AND APPS, G. J., 1975. Ecological distribution of the vertebrate animals of the volcanic plains-Otway Range areas of Victoria. *Fisheries and Wildlife Papers, Victoria* No. 6.

EMISON, W. B., PORTER, J. W., NORRIS, K. C. AND APPS, G. J., 1978. Survey of the vertebrate fauna in The Grampians-Edenhope area of southwestern Victoria. *Memoirs of the National Museum of Victoria* **39:** 281–363.

EMISON, W. B., MENKHORST, P. W., MANSERGH, I. M., NORMAN, F. I., ROBERTSON, P., CORRICK, A. H. AND ALDERSON, J. A., 1984. Wildlife surveys conducted by the Fisheries and Wildlife Service in Victoria, 1972–1983. Pp. 166–94 in *Survey methods for nature conservation* Vol. 2 ed by K. Myers, C. R. Margules and I. Musto. Australian Government Printing Service: Canberra.

FRITH, H. J., 1973. *Wildlife conservation.* Angus & Robertson: Sydney.

GORDON, G., 1983. Northern Brown Bandicoot *Isoodon macrourus.* P. 96 in *Complete book of Australian mammals* ed by R. Strahan. Angus & Robertson: Sydney.

HEINSOHN, G. E., 1966. Ecology and reproduction of the Tasmanian bandicoots *(Perameles gunni* and *Isoodon obesulus). University of California Publications in Zoology* **80:** 1–107.

JOHNSON, K. A., 1983. Bilby *Macrotis lagotis.* Pp. 107–08 in *Complete book of Australian mammals* ed by R. Strahan. Angus & Robertson: Sydney.

KENYON, A. S., 1914. The story of the mallee. Part II — Settlement — The Squatters. *Victorian Historical Magazine* **4:** 121–50.

KREFFT, G., 1864. *Catalogue of Mammalia in the collection of the Australian Museum.* Government Printer: Sydney.

KREFFT, G., 1866. On the vertebrated animals of the lower Murray and Darling, their habits, economy, and geographical distribution. *Transactions of the Philosophical Society of New South Wales,* **1862-65:** 1–33.

KREFFT, G., 1871. *The mammals of Australia; with a short account of all the species hitherto described.* Government Printer: Sydney.

LACY, R. C. AND CLARK, T. W., 1990. Population viability assessment of the Eastern Barred Bandicoot in Victoria. Pp. 131–46 in *Management and conservation of small populations* ed by Tim W. Clark and John H. Seebeck. Chicago Zoological Society: Chicago

LEE, A. K. AND COCKBURN, A., 1985. *Evolutionary ecology of marsupials.* Cambridge University Press: Cambridge.

LOBERT, B., 1985. *The ecology of the Southern Brown Bandicoot in south-east Australian heathland.* M.Sc Thesis. Monash University, Melbourne.

LOYN, R. H., MACFARLANE, M. A., CHESTERFIELD, E. A. AND HARRIS, J. A., 1980. Forest utilization and the flora and fauna in Boola Boola State Forest in south-eastern Victoria. *Forests Commission, Victoria. Bulletin* No. **28.**

MCILROY, J. C., 1983. The sensitivity of Australian animals to 1080 poison VI. Bandicoots. *Australian Wildlife Research* **10:** 507–12.

MENKHORST, P. W., 1987. Working list of Victorian mammals. Second edition. *Arthur Rylah Institute Technical Report Series* No. **50**. National Parks and Wildlife Division: Victoria.

MENKHORST, P. W. AND BEARDSELL, C. M., 1982. Mammals of southwestern Victoria from the Little Desert to the coast. *Proceedings of the Royal Society of Victoria* **94:** 221–47.

MENKHORST, P. W., EMISON, W. B. AND MARCIUS, J. M., 1984. A survey of mammals and birds in the Whitfield area of north-eastern Victoria. *Arthur Rylah Institute Technical Report Series* No. **14**. Fisheries and Wildlife Service: Victoria.

MITCHELL, MAJOR T. L., 1839. *Three expeditions into the interior of Eastern Australia; with descriptions of the recently explored region of Australia Felix, and of the present colony of New South Wales, Vol. 2., 2nd ed.* T. & W. Boone: London.

MOON, B. R., 1984. Current distribution of the Eastern Barred Bandicoot, *Perameles gunnii,* in Victoria. *Arthur Rylah Institute Technical Report Series* No. **5**. Fisheries and Wildlife Service: Victoria.

NEWSOME, A. E., 1971. Competition between wildlife and domestic stock. *Australian Veterinary Journal* **47:** 577–86.

NORRIS, K. C., GILMORE, A. M. AND MENKHORST, P. W., 1979. Vertebrate fauna of south Gippsland, Victoria. *Memoirs of the National Museum of Victoria* **40:** 105–99.

NORRIS, K. C., MANSERGH, I. M., AHERN, L. D., BELCHER, C. A., TEMBY, I. D. AND WALSH, N. G., 1983. Vertebrate fauna of the Gippsland Lakes Catchment, Victoria. *Fisheries and Wildlife Victoria. Occasional Papers Series* No. **1**.

OPIE, A. M., 1980. Habitat selection and the diet of *Isoodon obesulus. Australian Mammal Society Bulletin* **6**(2): 56 [abstract].

OVINGTON, D., 1978. *Australian endangered species: mammals, birds and reptiles.* Cassell Australia: Sydney.

ROLLS, E. C., 1969. *They all ran wild: the story of pests on the land in Australia.* Angus & Robertson: Sydney.

SCARLETT, N., 1969. The Bilby, *Thylacomys lagotis* in Victoria. *Victorian Naturalist* **86:** 292–94.

SEEBECK, J. H., 1977. Mammals in the Melbourne metropolitan area. *Victorian Naturalist* **94:** 165–71.

SEEBECK, J. H., 1979. Status of the Barred Bandicoot, *Perameles gunnii,* in Victoria: with a note on husbandry of a captive colony. *Australian Wildlife Research* **6:** 255–64.

SEEBECK, J. H., 1981. *Potorous tridactylus* (Kerr) (Marsupialia: Macropodidae) its distribution, status and habitat preferences in Victoria. *Australian Wildlife Research* **8:** 285–306.

SEEBECK, J. H., 1984. Mammals of the plains or, where have all the wombats gone? Pp. 39–47 in *The Western Plains — a natural and social history* ed by David Conley and Claire Dennis. Australian Institute of Agricultural Science: Melbourne.

SEEBECK, J. H., 1990. Recovery management of the Eastern Barred Bandicoot in Victoria: Statewide strategy. Pp. 165–78 in *Management and conservation of small populations* ed by Tim W. Clark and John H. Seebeck. Chicago Zoological Society: Chicago.

SEEBECK, J. H., BENNETT, A. F. AND DUFTY, A. C., 1990. Status, distribution and biogeography of the Eastern Barred Bandicoot, *Perameles gunnii* in Victoria. Pp. 21–32 in *Management and conservation of small populations* ed by Tim W. Clark and John H. Seebeck. Chicago Zoological Society: Chicago.

STODDART, D. M. AND BRAITHWAITE, R. W., 1979. A strategy for utilization of regenerating heathland habitat by the Brown Bandicoot (*Isoodon obesulus;* Marsupialia, Peramelidae). *Journal of Animal Ecology* **48:** 165–79.

WAKEFIELD, N. A., 1966a. Mammals of the Blandowski expedition to north-western Victoria, 1856–57. *Proceedings of the Royal Society of Victoria* **79:** 371–91.

WAKEFIELD, N. A., 1966b. Mammals recorded for the mallee, Victoria. *Proceedings of the Royal Society of Victoria* **79:** 627–36.

WILLIS, J. H., 1974. Vegetation of western Victoria. Pp. 24–33 in *The natural history of western Victoria* ed by M. H. Douglas and L. O'Brien. Australian Institute of Agricultural Science: Horsham.

WOOD, D. H., 1984. The Rabbit (*Oryctolagus cuniculus* L.) as an element in the arid biome of Australia. Pp. 273–87 in *Arid Australia* ed by H. G. Cogger and E. E. Cameron. Australian Museum: Sydney.

Status of bandicoots in Tasmania

G. J. HOCKING[1]

Two species of bandicoot, *Isoodon obesulus* and *Perameles gunnii* are found in Tasmania. There, in contrast to the situation on the Australian mainland, both species remain widespread and abundant. Habitat changes brought about by settlement have, to date, mostly favoured these species. However, long-term changes in agricultural practices could change this situation. This is of concern in view of the poor representation of bandicoot habitats, particularly that of *P. gunnii* in the reserve system.

Key words: *Isoodon obesulus, Perameles gunnii,* Tasmania, distribution, habitat preferences, status.

INTRODUCTION

Two species of bandicoot occur in Tasmania, the Southern Brown Bandicoot, *Isoodon obesulus*, and the Eastern Barred Bandicoot, *Perameles gunnii.* Both species are widespread, occurring sympatrically over a large part of the state. They are common and form a conspicuous part of the Tasmanian mammal fauna.

This situation contrasts with that elsewhere in Australia where these species have much reduced and restricted ranges. *Perameles gunnii* in particular has undergone a massive decline in numbers and range in Victoria in historic times, and its future there is highly uncertain (Seebeck 1979).

This paper discusses the distribution of bandicoot species in Tasmania, and describes their habitat and conservation status.

DISTRIBUTION

Data on the distribution of bandicoots in Tasmania were obtained from the following sources:

(i) The TASPAWS Biological Records Scheme maintained by the Tasmanian Department of Parks, Wildlife and Heritage. This records scheme contains details of the locations of sightings of bandicoots made by service personnel and the public since 1967 Between 1967 and 1981 over 1 000 sightings of bandicoots were recorded. Of these 86% were sightings of roadkilled animals.

(ii) The results of systematic surveys of roadkilled animals. These surveys were commenced in 1979 in an effort to monitor long-term trends in the abundance of the Red-bellied Pademelon *Thylogale billardierii* and the Brush-tailed Possum *Trichosurus vulpecula;* however, all species of roadkilled mammals observed have been noted (Johnson 1977). Surveys have been carried out three times each year, with each survey covering 12 standard routes with a total length of 1 563 km covering all but the southwestern part of the state.

(iii) The published literature. This provided information on the prehistoric and historic ranges of the species (Bowdler 1974, 1984; Hope 1973; Murray and Goede 1977), as well as some additional information on their present ranges, particularly in the west and south-west of the state (Green 1979, 1984; Hocking and Guiler 1983; Taylor, Bryant, Pemberton and Norton 1985).

Isoodon obesulus

This species is widespread in Tasmania (Fig. 1). Over the past 18 years it has frequently been recorded from the high rainfall regions of the south-east and north. Scattered records also come from East Coast, Midlands and Western regions. There are no records from the Central Plateau.

Isoodon obesulus is known from a number of islands including Bruny Island, and Maria Island to which it was successfully introduced following the release of 44 individuals in 1969–70 (Tasmanian Department of Parks, Wildlife and Heritage records). It has frequently been recorded on the essentially insular Tasman Peninsula.

On the islands of Bass Strait, *I. obesulus* is currently known to exist only on West Sister Island in the Furneaux Group (Green 1967; Whinray 1971). However, elsewhere in the Furneaux Group it is known from sub-fossil deposits on Passage, Cape Barren, and Flinders Islands and persisted on Flinders Island until as recently as the early part of this century (Hope 1973). Apparently suitable habitat for *I. obesulus* remains on both Flinders and Cape Barren Islands. At the western side of Bass Strait the species is recorded in fossil deposits on Hunter

[1]Tasmanian Department of Parks, Wildlife and Heritage, 134 Macquarie Street, Hobart, Tasmania, Australia 7001.

Pages 61–66 *in* BANDICOOTS AND BILBIES, ed by J. H. Seebeck, P. R. Brown, R. L. Wallis and C. M. Kemper. Surrey Beatty & Sons, Sydney, 1990.

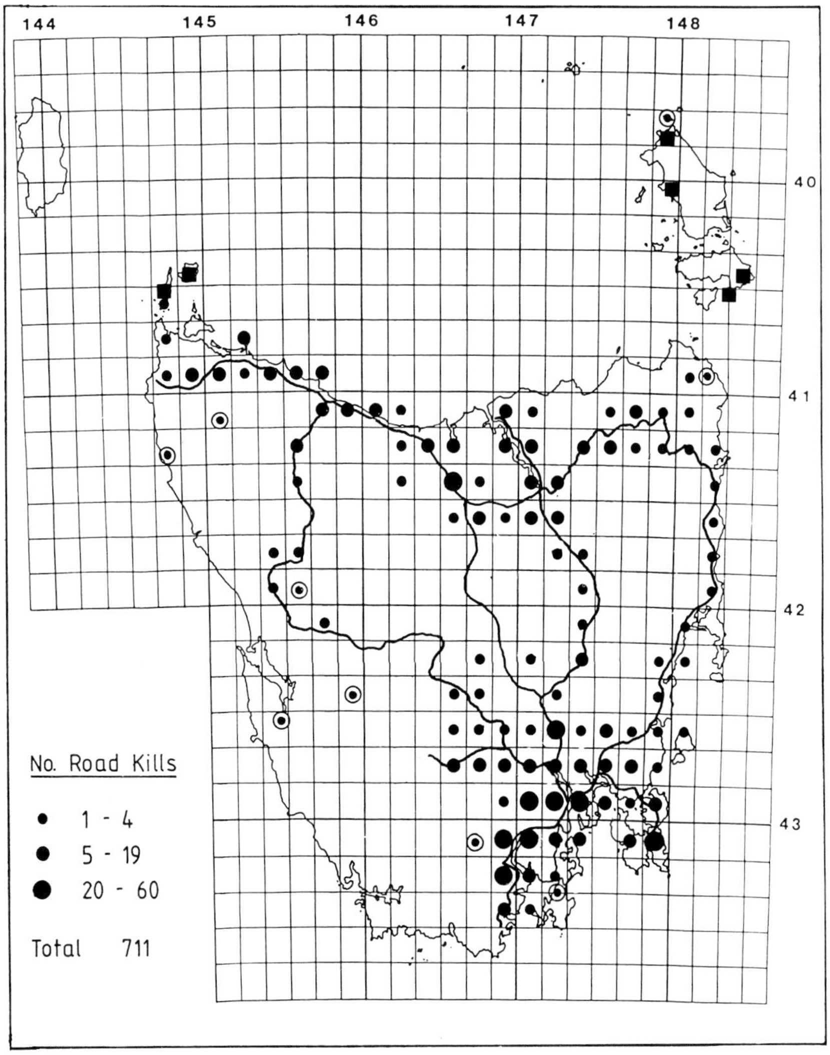

Fig. 1. Distribution of *Isoodon obesulus* in Tasmania. Records include sightings of road-killed animals (●), fossil deposits (■), and trapping results (⦿). Major roads are marked.

and Three Hummock Islands (Bowdler 1984; Hope 1973) although it is unknown there during historic times.

Perameles gunnii

This species is widespread in Tasmania, having a distribution that broadly parallels that of *I. obesulus* (Fig. 2). Over the past 18 years *P. gunnii* has frequently been recorded in the higher rainfall regions of the southeastern, northwestern and northeastern parts of the state. There are no records from the alpine areas of the Central Plateau. Similarly, with the exception of one record from Strahan, there are no records of this species from the southwestern part of Tasmania.

Perameles gunnii has not been recorded in historic times from the islands of Bass Strait, although it occurs in fossil deposits on both Hunter (Bowdler 1974, 1984) and Flinders Island (Hope 1973). These deposits are, however, of Pleistocene age, at which time these islands were linked by land to Tasmania (Hope 1973). Similarly, this species is unknown from Bruny Island and, until recently, Maria Island, where it was successfully introduced following the transfer of 55 animals in 1969–70 (Tasmanian Department of Parks, Wildlife and Heritage records). This population has persisted for 15 years. The absence of *P. gunnii* from islands is notable, particularly in its implications for maintaining populations of this species in those reserves consisting of isolated areas of habitat.

HABITAT

In broad terms the distribution of bandicoot species in Tasmania follows that of eucalypt-dominated forest communities, and includes extensive areas of cleared land (Fig. 3). Both species are, however, absent from alpine

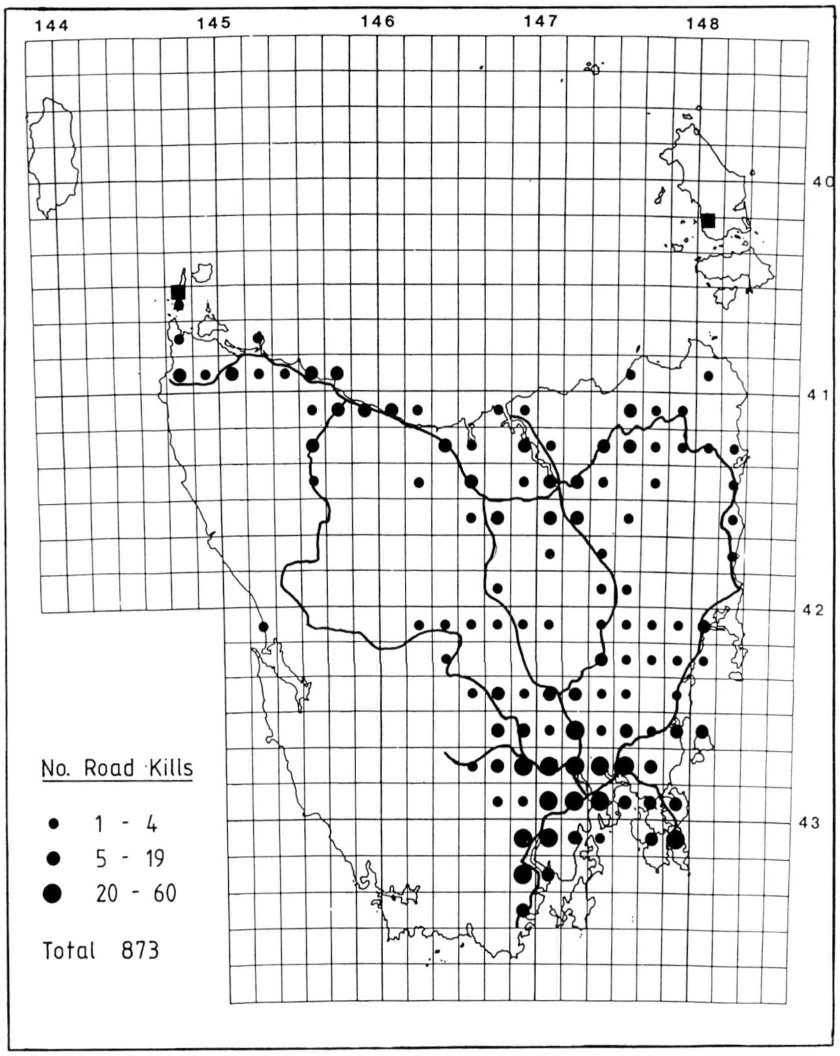

Fig. 2. Distribution of *Perameles gunnii* in Tasmania. Records include sightings of roadkilled animals (●) and fossil deposits (■). Major roads are marked.

habitats above approximately 950 m and are uncommon in the rainforests and sedgelands of the western half of the island (Hocking and Guiler 1983; Taylor *et al.* 1985). Within their overlapping ranges the two species show a considerable overlap in habitat preferences and they frequently occur together in the same area.

Isoodon obesulus occurs primarily in dry sclerophyll forest, scrub, and heathland communities. It has also been recorded infrequently from temperate rainforest dominated by Myrtle Beech, *Nothofagus cunninghamii* (Green 1979). This species is able to rapidly recolonize burnt habitat (Green 1982; Stoddart and Braithwaite 1979). It is also frequently found feeding in areas of pasture of crop-land lying close to dense cover (Heinsohn 1966; Maloney 1982). Indeed, such areas can support high densities of *I. obesulus* (Heinsohn 1966).

Generally *I. obesulus* favours those areas providing dense ground cover. This cover is often provided by shrub species such as *Melaleuca* spp. and *Leptospermum* spp., or tussock-forming species such as Long Mat-rush *Lomandra longifolia*, and other sedges or grasses (*Poa* sp.). In agricultural areas suitable cover is available from exotic species including brambles (*Rubus* sp.), Gorse (*Ulex europaeus*) and Boxthorn (*Lycium ferrocissimum*).

By comparison, *P. gunnii* occurs in more open habitats including both native and exotic grasslands, and woodland or open forest with a grassy understorey. This species is recorded widely in agricultural areas where it feeds extensively in pastures and crops and where, in contrast to *I. obesulus*, it often forages well away from cover (Heinsohn 1966; Maloney 1982). As was the case for *I. obesulus*, Heinsohn (1966) found that a mosaic of pasture and scrub supported the highest densities of *P. gunnii*, with densities of up to 1.5 animals/ha being recorded. He attributed this to the increased food supply of soil

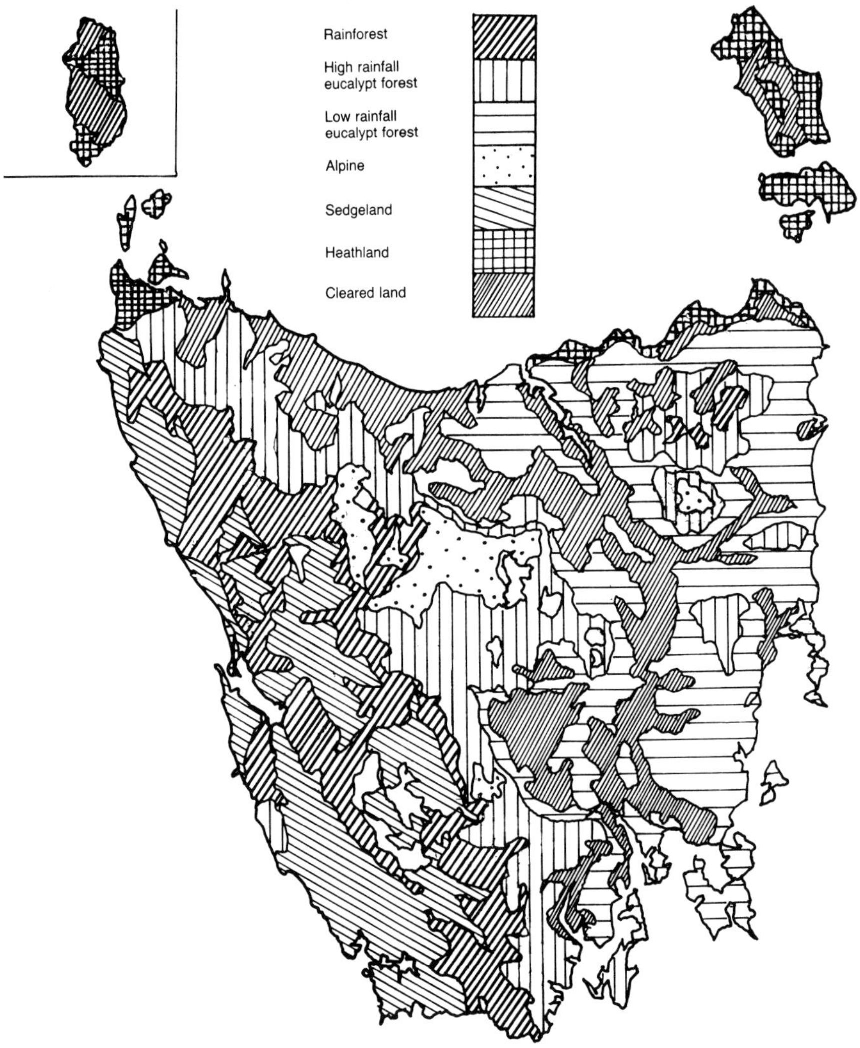

Fig. 3. Distribution of vegetation types in Tasmania (following Kirkpatrick and Dickinson 1984).

invertebrates resulting from cultivation and the application of fertilizer. More generally, clearing for agriculture has considerably extended the range of this species into much of northern, and parts of southeastern Tasmania that previously supported forest habitat unsuitable for *P. gunnii*. Cover in these areas is provided by remnant patches of scrub comprising native species such as *Melaleuca* spp., or hedgerows and waste-ground supporting exotics such as brambles, gorse and boxthorn. On the other hand the widespread development for pasture of native grasslands in areas such as the Midlands during the early phase of settlement undoubtedly destroyed much of what was originally the prime habitat of this species. Today the species remains in these areas only by virtue of shelterbelts and pockets of uncleared land being available.

CONSERVATION STATUS

Both *P. gunnii* and *I. obesulus* remain widespread and abundant in Tasmania, having undergone no major decline in range since European settlement. Elsewhere in Australia these species have suffered major declines in range during this time. This is particularly true of *P. gunnii* which is regarded as critically endangered outside of Tasmania (Seebeck 1979; Moon 1984). The conservation of the Tasmanian populations of *P. gunnii* and *I. obesulus* is therefore of considerable importance in the overall conservation of these species.

Several factors have been identified which account for recent declines in bandicoot abundance on the Australian mainland. Introduced predators were considered by Seebeck (1979) to be a significant factor in the decline of *P. gunnii* in Victoria. In Tasmania cats, both feral and domestic, and dogs are present and are known to kill large numbers of bandicoots of both species. The fox, *Vulpes vulpes,* is however, absent and this may in part explain the continued abundance of bandicoots in Tasmania.

Change in habitat brought about by land management practices is however, viewed as the major factor in the decline of bandicoot populations on the Australian mainland (Heinsohn 1966; Seebeck 1979). It is seen as constituting the greatest threat to bandicoot species in Tasmania.

Suitable habitat can, of course, be maintained through the provision and appropriate management of reserves. In this regard over 20% of Tasmania is presently reserved for nature conservation purposes. However, despite this seemingly high figure bandicoots are but poorly protected by the reserve system. This is particularly true for *P. gunnii*. Thus, of those reserves having an area greater than 1 000 ha, most contain habitat of only low to medium suitability for either *P. gunnii* or *I. obesulus* (Table 1). Moreover, those records of *P. gunnii* from reserves typically come from areas on the periphery of the reserve, adjacent to agricultural land.

In view of this poor representation in reserves it is clear that the long-term conservation of bandicoot species in Tasmania depends upon land management practices outside reserves. Here two forms of land-use predominate, namely forestry and agriculture.

Approximately 30% of Tasmania supports forest of commercial quality which is utilized for timber production. This includes large areas of *I. obesulus* habitat. The practice of clear-felling and burning of native forest has the greatest potential impact, with over 20 000 ha of forest being affected annually. Nevertheless, evidence suggests that *I. obesulus* is able to recolonize such areas rapidly when they are allowed to regenerate to native forest (Green 1982).

Land used for agriculture and grazing comprises 30% of the area of Tasmania and includes most of the range of *P. gunnii* and much of that of *I. obesulus*. Approximately 40% of agricultural land is cleared for pasture or crops and this area is currently increasing at a rate of approximately 12% per decade. Clearing for agriculture has in the past resulted in some increase in the range and abundance of bandicoots. This has happened where land clearance produces the mosaic of pasture and forest or scrub which Heinsohn (1966) found leads to an increase in bandicoot numbers through an increase in their food supply. This is particularly so for *P. gunnii* which has increased its range in the higher rainfall and previously forested areas, in the north-east, north-west and southern parts of the state. On the other hand, extensive land clearance, including the removal of weed species such as brambles and gorse, is expected to lead to some loss of habitat through the loss of cover for nesting sites. This has occurred over much of the range of *P. gunnii* in Victoria (Seebeck 1979). In

Table 1. Conservation of *Isoodon obesulus* and *Perameles gunnii* in Tasmanian reserve system. *=low, **=medium, ***=high, P=species known to be present.

Name of reserve	Area (ha)	Suitability of habitat	
		I. obesulus	*P. gunnii*
STATE RESERVES			
South-west	442 240	**(P)	*
Franklin-Gordon River	195 200	**(P)	*
Cradle Mtn-Lake St Clair	131 915	**	*
Ben Lomond	16 526	**	*(P)
Mt Field	16 257	**(P)	*(P)
Mt William	13 816	***(P)	***(P)
Walls of Jerusalem	11 510	*	*
Freycinet	10 010	**	*
Maria Island	9 672	***(P)	***(P)
Three Hummock Island	7 284	—	—
Hartz Mountain	6 470	*	*
Lavinia	4 622	—	—
Asbestos Range	4 281	**	**
Strzelecki	4 215	—	—
Pieman River	3 328	**	*
Cape Pillar	3 200	**	*
Rocky Cape	3 050	**	*
Labillardiere	2 332	**	*
Cape Raoul	2 066	**(P)	*
Alum Cliffs	1 540	**	*
Lime Bay	1 310	**(P)	*(P)
CONSERVATION AREAS			
South-west	791 351	**(P)	*
Central Plateau	22 780	*	*
Tooms Lake	22 663	**	**
Ben Lomond	2 665	*	*
Southport Lagoon	3 616	**	*

Tasmania, this process has taken place in parts of the Midlands and is widely occurring elsewhere. Similarly, other changes, such as the increased use of insecticides in agriculture, have the potential to reduce bandicoot numbers through a decline in their food supply of soil invertebrates.

In summary, both *P. gunnii* and *I. obesulus* are common in Tasmania where they have survived and in many instances benefited from the changes in habitat wrought by man. Nevertheless, since most of the habitat of *P. gunnii* occurs in agricultural areas subject to changing management practices, the long-term security of that species cannot be assumed. The dramatic decline of *P. gunnii* on the mainland of Australia provides ample warning of this.

ACKNOWLEDGEMENTS

I thank all those members of the Tasmanian Department of Parks, Wildlife and Heritage who have, over the years, provided records of bandicoot distribution. I particularly thank David Rounsevell for providing access to the TASPAWS record system, and for criticizing this manuscript.

REFERENCES

BOWDLER, S., 1974. An account of an archeological reconnaissance of Hunters Isles, North-west Tasmania, 1973/74. *Records of the Queen Victoria Musuem* **54:** 1–25.

BOWDLER, S., 1984. Hunter Hill, Hunter Island: Archaeological investigations of a prehistoric Tasmanian site. *Terra Australis* **8:** 1–148.

GREEN, R. H., 1967. Birds of Flinders Island. *Records of the Queen Victoria Museum* **34:** 1–32.

GREEN, R. H., 1979. A survey of the vertebrate fauna of the Sumac Forest, and Dempster Plains, North-West Tasmania. *Records of the Queen Victoria Museum* **65:** 1–9.

GREEN, R. H., 1982. The activity and movement of fauna in Compartment 2, Maggs Mountain, Tasmania, in the first five years of forest regeneration. *Records of the Queen Victoria Museum* **75:** 1–24.

GREEN, R. H., 1984. The fauna of Ordnance Point, Northwestern Tasmania. *Records of the Queen Victoria Museum* **84:** 32–52.

HEINSOHN, G. E., 1966. Ecology and reproduction of the Tasmanian bandicoots (*Perameles gunni* and *Isoodon obesulus*). *University of California Publications in Zoology* **80:** 1–96.

HOCKING, G. J. AND GUILER, E. R., 1983. The mammals of the Lower Gordon River Region, South-west Tasmania. *Australian Wildlife Research* **10:** 1–23.

HOPE, J. H., 1973. Mammals of Bass Strait Islands. *Proceedings of the Royal Society of Victoria* **85:** 163–96.

JOHNSON, K. A., 1977. Methods for the census of wallaby and possum in Tasmania. *Tasmanian National Parks and Wildlife Service Technical Report No.* **5.**

KIRKPATRICK, J. B. AND DICKINSON, K. J. M., 1984. *Vegetation of Tasmania.* Forestry Commission of Tasmania: Hobart.

MALONEY, D. J., 1982. A comparison of the behaviour and ecology of the Tasmanian bandicoots, *Perameles gunnii* (Gray 1838), and *Isoodon obesulus* (Shaw and Nodder 1797). Hons. Thesis. University of Tasmania: Hobart.

MOON, B. R., 1984. Current distribution of the Eastern Barred Bandicoot, *Perameles gunnii,* in Victoria. *Arthur Rylah Institute of Environmental Research Technical Report No.* **5.**

MURRAY, P. AND GOEDE, A., 1977. Pleistocene vertebrate remains from a cave near Montagu, N.W. Tasmania. *Records of the Queen Victoria Museum* **60:** 1–30.

SEEBECK, J. H., 1979. Status of the barred bandicoot, *Perameles gunnii,* in Victoria with a note on husbandry of a captive colony. *Australian Wildlife Research* **6:** 255–64.

STODDARD, D. M. AND BRAITHWAITE, R. W., 1979. A strategy for utilization of regenerating heathland habitat by the Brown Bandicoot (*Isoodon obesulus:* Marsupialia, Peramelidae). *Journal of Animal Ecology* **48:** 165–79.

TAYLOR, R. J., BRYANT, S. L., PEMBERTON, D. AND NORTON, T. W., 1985. Mammals of the Upper Henty River Region, Western Tasmania. *Papers and Proceedings of the Royal Society of Tasmania* **119:** 7–14.

WHINRAY, J. S., 1971. The present distribution of some mammals in the Furneaux Group, Bass Strait, Tasmania. *Victorian Naturalist* **88:** 279–85.

Status of bandicoots in South Australia

CATHERINE KEMPER[1]

At the time of European man's arrival in South Australia, eight species of bandicoot (*Perameles bougainville, P. eremiana, P. gunnii, Isoodon auratus, I. obesulus, Chaeropus ecaudatus, Macrotis lagotis* and *M. leucura*) were present. One species *(I. obesulus)* remains. Last recorded specimens of extinct species were generally dated the 1920s and 1930s while *P. gunnii* and possibly *C. ecaudatus* disappeared about the turn of the century. Actual geographic distributions of species were generally greater than the specimen record indicates. Subfossil deposits and literature records were included here to augment specimen records and the approximate number of museum specimens collected in South Australia was given to aid in determining conservation status in this state.

Isoodon obesulus is now found in forest and woodland habitat on the Eyre Peninsula, Kangaroo Island, Fleurieu Peninsula, in the Mt Lofty Ranges and lower southeastern part of the state. *Isoodon obesulus nauticus*, the insular form, is extant on St Francis and Franklin Islands in Nuyts Archipelago Conservation Park. The status of *I. obesulus* is considered vulnerable due to the patchiness of its habitat and the effects of severe disturbance on already low population levels.

Key words: Status, distribution, bandicoots, bilbies, South Australia.

INTRODUCTION

THE extinction of mammal species in South Australia is the highest of any Australian state — 22% have disappeared since the arrival of European man (Aslin 1985). Medium-sized species, including the bandicoots and bilbies, seem to have fared the worst. As early as the 1920s, Jones (1924) noted the decline in several species of bandicoots and bilbies in South Australia. He gave hunting pressure (from Europeans) and foxes as two possible causes. Since then, many other reasons, largely conjectural, have been put forward including: other exotic predators, introduced grazing stock and rabbits, changes in the fire regime, Aboriginal hunting, increased aridity, exotic diseases and poisoning (Aitken 1978; Burbidge and Fuller 1979; Finlayson 1961). Whatever the reasons for the decline in numbers, they are certainly not restricted to South Australia. Why, then, has this state lost such a large proportion of its mammal fauna? The answer appears to lie in the fact that South Australia is the most arid state and arid-zone species have been most affected by the changes that have occurred since settlement. Coupled with this is the large reduction (75%) in natural vegetation of open forest and mallee scrub in the more mesic coastal regions (South Australian Department of the Environment 1976).

Eight species of bandicoot were known to have existed in South Australia at the time of European settlement (Aslin 1985). Three species *(Perameles eremiana, Isoodon auratus, Macrotis leucura)* were restricted to the arid-zone of the far north, three *(Perameles bougainville, Chaeropus ecaudatus, Macrotis lagotis)* were widely distributed and found in both the arid and semi-arid regions and two *(Perameles gunnii, Isoodon obesulus)* occurred in the coastal woodlands and forests.

This paper collates information on the status, in both the recent past and present, of bandicoots in South Australia. Species distributions are discussed as well as estimates of abundance and date of extinction for those no longer inhabiting the state.

SOURCES OF INFORMATION

Museum records of South Australian specimens were obtained from the following institutions: Western Australian Museum, South Australian Museum, Northern Territory Museum of Arts and Sciences, Museum of Victoria and Australian Museum. Publications which provided many records were Thomas (1888), Jones (1924) and Tate (1948) as well as those referred to specifically in the text. Subfossil information was obtained from G. Medlin (pers. comm. 1985), Tidemann (1967), Smith (1977), Pledge (1979), Watts and Ling (1985) and South Australian Museum records. Several localities for *I. obesulus* were provided by Field Naturalists Society of South Australia — Mammal Club and B. Gepp (pers. comm. 1985).

RESULTS AND DISCUSSION

Perameles bougainville

This species had a wide distribution in southern and western South Australia (Fig. 1) and there are at least 30 specimens from this state in museums. Although only one subspecies, *P. bougainville notina*, has been recorded from South Australia, *P. b. fasciata* occurred in western New South Wales and Victoria and was therefore likely to have been in South Australia as well.

[1]South Australian Museum, North Terrace, Adelaide, South Australia 5000.

Pages 67–72 *in* BANDICOOTS AND BILBIES, ed by J. H. Seebeck, P. R. Brown, R. L. Wallis and C. M. Kemper. Surrey Beatty & Sons, Sydney, 1990.

The type specimen of *P. b. notina* is from the head of Gulf St Vincent and is held by the British Museum (Natural History) (BMNH). Other BMNH specimens date from the 1800s and were collected in the Adelaide/River Murray region. Krefft (1865) recorded *P. bougainville* as common along the Murray River and presumably this included the South Australian portion of that river. By the 1920s it had long disappeared from the southeastern part of its range (Jones 1924). There is no evidence of it having occurred on the Yorke Peninsula (Aitken 1973). In the early 1920s good numbers were obtained in the Ooldea/Fisher area. The last specimen was collected before 1931 (Table 1).

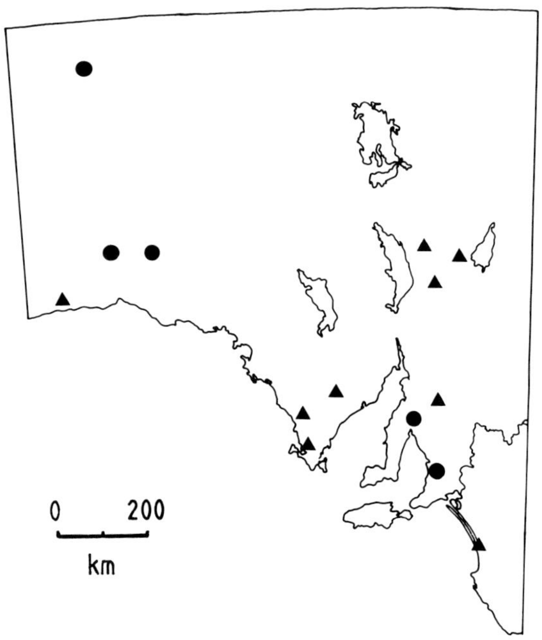

Fig. 1. Distribution of *Perameles bougainville* in South Australia. ● = specimen; ■ = literature record; ▲ = subfossil material.

The South Australian Museum (SAM) has three specimens of *P. bougainville* from "south of Mt Crombie" (in the far north-west of the state) registered in 1931. Finlayson (1961), however, did not mention this species but rather *P. eremiana* from that region. The taxonomy of these two species is being studied at present (G. Gordon, pers. comm., 1984) and his study may resolve the issue.

Perameles eremiana

Jones (1924) knew of no South Australian specimens of this species. The only two individuals in the SAM were collected at about the turn of the century by R. T. Maurice but they were not registered until 1936. Both were previously identified as *P. bougainville* but G. Gordon (pers. comm. 1984) has tentatively assigned them to *P. eremiana* pending a resolution of the taxonomic status of the inland barred bandicoots.

Finlayson (1961) noted that *P. eremiana* was well known and plentiful in the north-west of South Australia in the early 1930s but that by the time of writing his article it was rare or absent. He also believed that some material collected by C. G. A. Winnecke, east of Stuart's Line, could have been this species. Extensive field surveys by Philpott and Smyth (1967) in 1966 failed to find any evidence that *P. eremiana* or *P. bougainville* were extant in the northern part of the state.

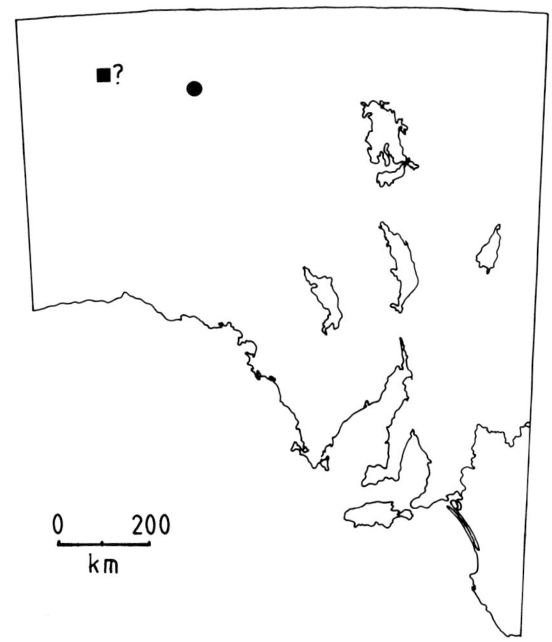

Fig. 2. Distribution of *Perameles eremiana* in South Australia. Symbols as in Fig. 1. ? = imprecise locality.

Perameles gunnii

There are four SAM specimens of *P. gunnii*, all collected in the 1890s. Three of these have precise locality data and one is labelled "South Australia". Specimens from Mt Monster and Kongal — the label for this specimen reads "The Rocks, Kougal" but as no such place exists we interpret this to be Kongal, west of Bordertown — are of interest. They are from an area of lower rainfall (500 mm pa) than the Victorian populations and on sandy soils with a clay subsoil as opposed to basalt plains in Victoria (Seebeck 1983). The original vegetation of the Kongal area was described by a government surveyor as "Open Forest", "Honeysuckle etc." and "Mallee" (Survey Map 1880, South Australian Department of Lands).

Freedman (1967) referred to a single skull specimen of *P. gunnii* from South Australia, but he did not list the registration number or museum of origin. Presumably it was one of the SAM specimens referred to above, either M3956 or M1607. M3956, from Kongal, is a spirit specimen with the skull separately preserved.

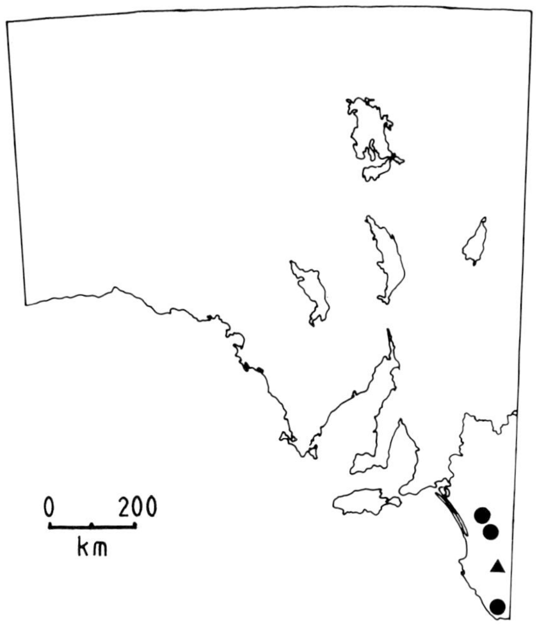

Fig. 3. Distribution of *Perameles gunnii* in South Australia. Symbols as in Figs 1 and 2.

External characteristics described by Lyne (1951) to distinguish between *P. bougainville fasciata* and *P. gunnii* are, in this specimen, consistent with *P. gunnii*. Subfossil material has been found at Naracoorte by Tidemann (1967). There seems little doubt, therefore, that *P. gunnii* did occur in South Australia but it is now considered extinct.

Isoodon auratus

The first South Australian specimen of this species was collected by R. T. Maurice in 1903 in the Musgrave Ranges area. There are only three South Australian individuals in museum collections but several without precise locality data may also have originated in the state. In the early 1930s Finlayson (1961) recorded *I. auratus* from the Everard Ranges, from south of the Musgrave Ranges and from the Barcoo River. There are no museum specimens of *I. auratus* from the north-east of the state to substantiate the latter locality but Smith (1977) recorded subfossil material from the northern Flinders Ranges. When Finlayson (1961) wrote on the status of central Australian mammals he believed that numbers of *I. auratus* had been greatly reduced in the preceeding 25 years and therefore considered it rare or absent. Although the region it inhabited is remote and rarely visited, there is no evidence to support that it is extant in South Australia.

Isoodon obesulus

Two subspecies of *I. obesulus* occur in South Australia — *I. o. nauticus* and *I. o. obesulus.*

Museum specimens of *I. o. nauticus* have been collected on Franklin and St Francis Islands (both in Nuyts Archipelago Conservation Park) and subfossil material has been found on Flinders Island (South Australia). Fifty specimens are currently held by SAM of which most were collected during the past 25 years. The type specimen of *I. o. nauticus,* from Franklin Island, is in the BMNH collection. Jones (1924) considered this insular form to be rare on St Francis Island but abundant on Franklin Island and this status remains unchanged today. Its overall status is considered vulnerable due to the limited nature of its distribution and the susceptability of such island populations to detrimental changes in their environment (e.g., the possible introduction of feral predators to Franklin Island).

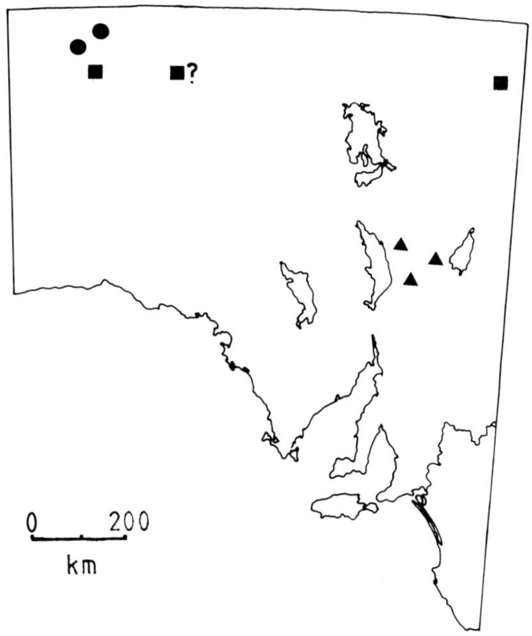

Fig. 4. Distribution of *Isoodon auratus* in South Australia. Symbols as in Figs 1 and 2.

Isoodon o. obesulus was, according to Krefft (1865), the most common bandicoot inhabiting southern Australia. By the 1920s its numbers had dwindled to the extent that Jones (1924) considered it rare in South Australia, except on Kangaroo Island. There are about 100 specimens at SAM, of which one half have been obtained since 1960. This is probably a result of increased awareness of the value of museum collections rather than an increase in abundance of the species. There appears to be no evidence of its existence on Yorke Peninsula (Aitken 1973) nor have any specimens, with locality information specifying South Australia, been recorded from the River Murray. BMNH specimens collected by Grey may have originated from the South Australian portion of that river (Thomas 1888).

Isoodon o. obesulus is now found in forest and woodland habitat on Eyre Peninsula, Kangaroo Island, in the Mt Lofty Ranges, on the Fleurieu

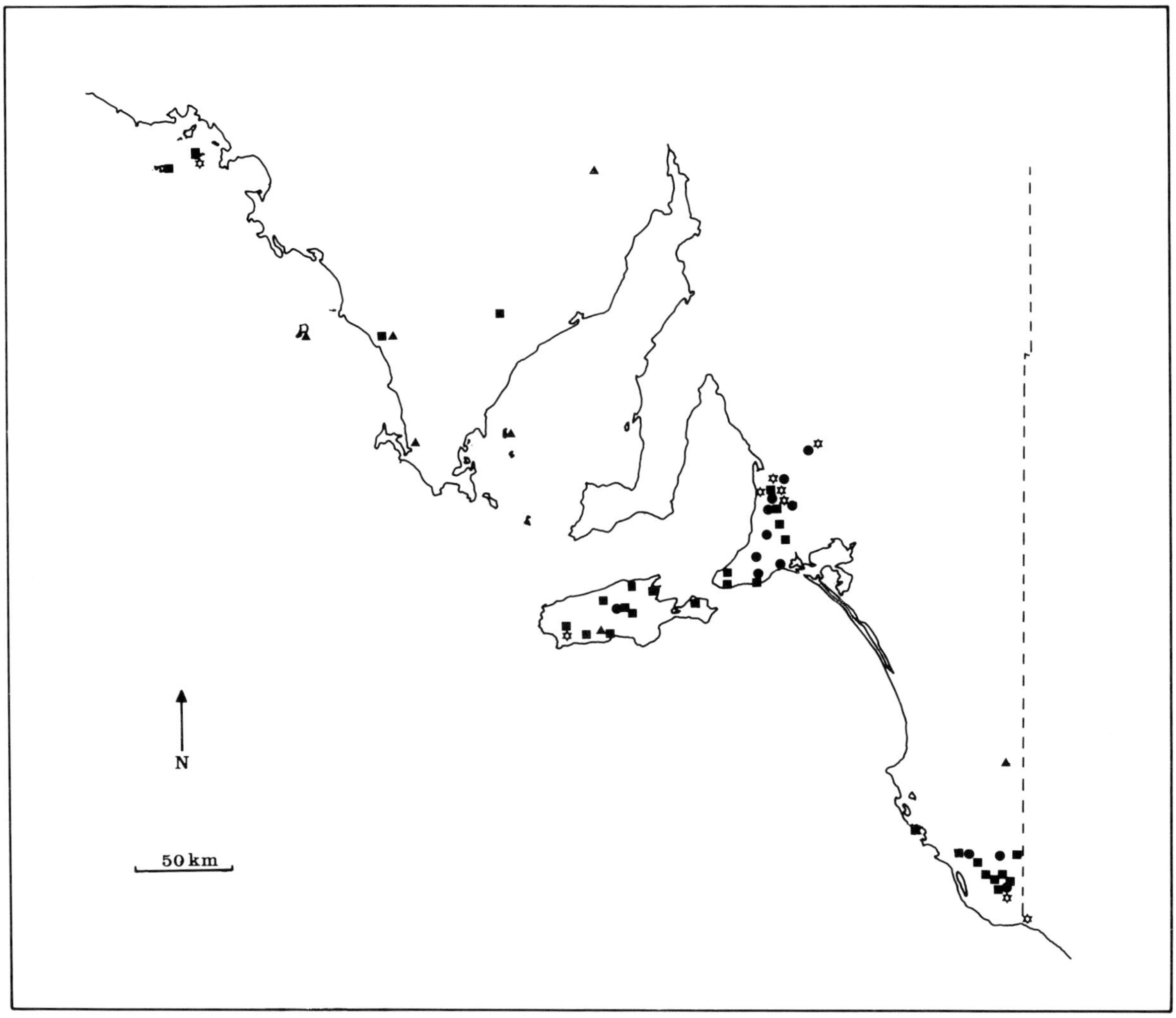

Fig. 5. Distribution of *Isoodon obesulus* in South Australia. ▲ = subfossil; ☆ = pre 1930; ● = 1931–1960; ■ = 1961–1985; – – – = state border.

Peninsula and in the lower southeastern portion of the state. Nowhere is it abundant. Aitken (1983) reported that its future in the south-east was dismal: it was not known to inhabit any conservation park and the forest reserves in which it occurred were being degraded by too-frequent burning. Since Aitken's article the Ash Wednesday fires in February 1983 may have caused more local extinctions. *Isoodon o. obesulus* is known from several state parks, the largest of which is Flinders Chase National Park on Kangaroo Island. There it is also protected from predation by the fox, which does not occur on the island.

The status of *I. o. obesulus* is considered vulnerable in South Australia due to the patchy nature of its distribution and overall low abundance within those isolates of native vegetation.

Chaeropus ecaudatus

This species is known from several localities in South Australia but only two modern specimens with locality data are found in museums. BMNH

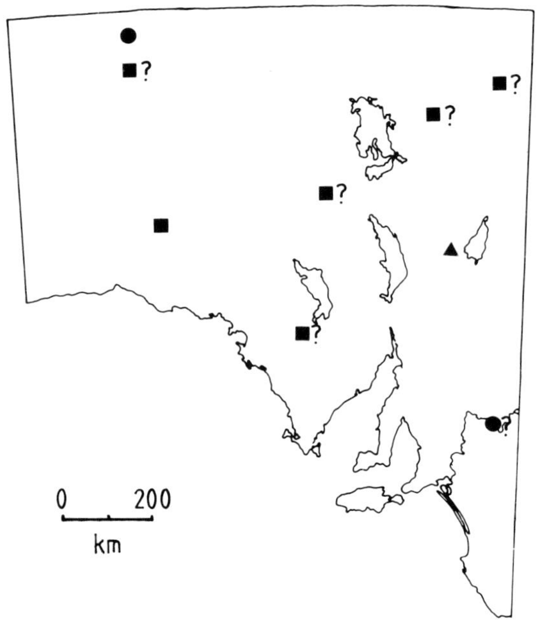

Fig. 6. Distribution of *Chaeropus ecaudatus* in South Australia. Symbols as in Figs 1 and 2.

Table 1. Species of bandicoot known from South Australia, their status and details of last collected specimens. *=registration date.

Species	Status (SA)	Last collected	Where	Collector
Perameles bougainville	Extinct	1921	Ooldea	Le Souef
		*1931	South of Mt Crombie	Brumby
Perameles eremiana	Extinct	1930s	Musgrave/Mann Ranges	Finlayson
Perameles gunnii	Extinct	1893	Mt Gambier	
Isoodon auratus	Extinct	1933	Mt Crombie	Tindale and Hackett
Isoodon obesulus obesulus	Vulnerable	1985		
Isoodon o. nauticus	Vulnerable	1985		
Chaeropus ecaudatus	Extinct	1901	Musgrave Ranges	Maurice
Macrotis lagotis	Extinct	1933	Musgrave/Everard Ranges	Finlayson and Brumby
Macrotis leucura	Extinct	1931	Diamantina River	Finlayson

has one collected by Grey at "Murray River, South Australia" (Thomas 1888) and SAM has one from the Musgrave Ranges collected in 1901 (Table 1). Jones (1924) reported a specimen, unfortunately not retained, from southwest of Lake Eyre in 1920. At the time of writing he believed that *C. ecaudatus* may still have been extant in the Ooldea area. Finlayson (1961) recorded it from south of the Musgrave Ranges (last seen by Pitjanjarra men in 1926) and the lower Barcoo River in the north-east of the state. It is now extinct throughout Australia.

Macrotis lagotis

Macrotis lagotis was the most widely distributed of the South Australian bandicoots, being apparently absent only from the south-east and Kangaroo Island. Specimens were not obtained from all regions but supplementary information such as subfossil deposits helps to complete the picture. Evidence of its occurrence can also be gleaned from Aboriginal languages in different regions. Using this method, Aitken (1973) concluded it had been present on Yorke Peninsula. It was also recorded from Aboriginal dialects in the Flinders Ranges and Eyre Peninsula (D. Tunbridge, pers. comm., 1985).

There have been more then 40 *M. lagotis* specimens collected in South Australia. Three subspecies have been described — *M. l. grandis* (type in SAM), from the Adelaide/River Murray area; *M. l. sagitta* (type in BMNH) from the Lake Eyre Basin; *M. l. nigripes* (type in BMNH, Aitken 1976) from the Ooldea area.

In the southeastern portion of its range *M. lagotis* was apprently abundant in the 1890s but it had disappeared from there by the early 1900s (Jones 1924). Jones (1924) believed it to have been extant in the Lake Eyre Basin and around Ooldea. Finlayson (1935, 1961), who had collected the species in the far north-west and north-east of the state, concluded that it was gone from the former but still present in the Lake Eyre Basin. The last dated specimens were obtained in the early 1930s in both the above regions. Surveys conducted in the 1960s by Philpott and Smyth (1967) and Watts (1969) failed to secure any evidence of *M. lagotis* in northern South Australia.

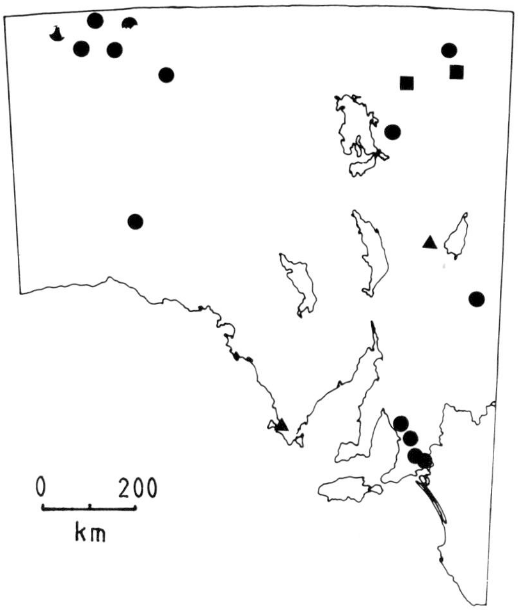

Fig. 7. Distribution of *Macrotis lagotis* in South Australia. Symbols as in Figs 1 and 2.

Macrotis leucura

Finlayson (1932, 1961) recorded this species from three localities in the far north-east of South Australia. Museum material collected in 1931 is available only from Cooncheri, on the Diamantina River. Finlayson (1961) believed it to have been plentiful in the early 1930s but no specimens have been collected since that time. There are only two South Australian specimens in SAM and one in the BMNH (Thomas 1888) which probably came from the northern part of the state. *Macrotis leucura* is considered extinct.

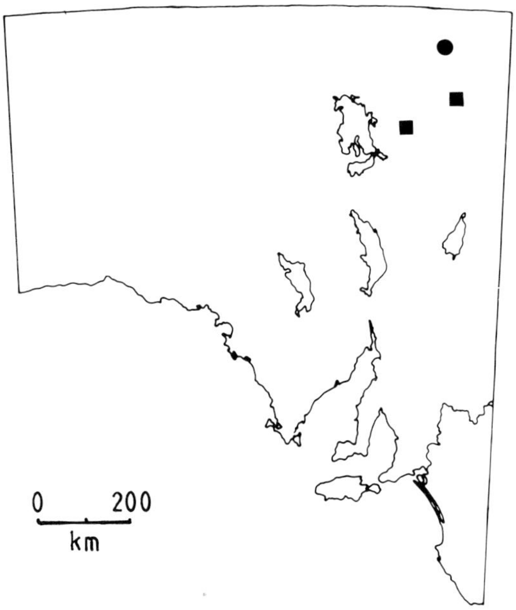

Fig. 8. Distribution of *Macrotis leucura* in South Australia. Symbols as in Figs 1 and 2.

ACKNOWLEDGEMENTS

I would like to thank Graeme Browning for his help with collating information on South Australian Museum specimens, Graham Medlin for information on subfossil deposits, Katherine Ashby for preparing some of the figures and Debbie Brunker for typing the manuscript.

REFERENCES

AITKEN, P. F., 1973. Mammals. Pp. 156-66 in *Yorke Peninsula a natural history* ed by D. Corbett. Department of Adult Education, University of Adelaide: Adelaide.

AITKEN, P. F., 1976. Vertebrate type-specimens in the South Australian Museum V. Mammals. *Records of the South Australian Museum* **17:** 197–203.

AITKEN, P. F., 1978. The status of endangered Australian wombats, bandicoots and the marsupial mole. Pp. 61–65 in *The status of endangered Australian wildlife.* Royal Zoological Society of South Australia: Adelaide.

AITKEN, P. F., 1983. Mammals. Pp. 127–33 in *Natural history of the south east* ed by M. J. Tyler, C. R. Twidale, J. K. Ling and J. W. Holmes. Royal Society of South Australia: Adelaide.

ASLIN, H. (ed), 1985. *A list of vertebrates of South Australia.* Biological Survey Co-ordinating Committee and Department of Environment and Planning: South Australia.

BURBIDGE, A. A. AND FULLER, P. J., 1979. Mammals of the Warburton Region, Western Australia. *Records of the Western Australian Museum* **8:** 57–73.

FINLAYSON, H. H., 1932. Preliminary descriptions of two new animals from South Australia. *Transactions of the Royal Society of South Australia* **56:** 168–71.

FINLAYSON, H. H., 1935. On Mammals from the Lake Eyre Basin. Part II — the Peramelidae. *Transactions of the Royal Society of South Australia* **59:** 227–36.

FINLAYSON, H. H., 1961. On central Australian mammals. Part IV — The distribution and status of central Australian species. *Records of the South Australian Museum.* **14:** 141–91.

FREEDMAN, L., 1967. Skull and tooth variation in the genus *Perameles.* Part 1 — Anatomical features. *Records of the Australian Museum* **27:** 147–66.

JONES, F., 1924. *The mammals of South Australia. Part II.* Government Printer: Adelaide.

KREFFT, G., 1865. On the Vertebrata of the Lower Murray and Darling; their habits, economy, and geographical distribution. Reading and Wellbank Printers: Sydney. Pp. 1–33.

LYNE, A. G., 1951. Notes on the external characters of the barred bandicoot (*Perameles gunnii* Gray), with special reference to the pouch young. *Proceedings of the Royal Society of London* **121:** 587–98.

PHILPOTT, C. M. AND SMYTH, D. R., 1967. A contribution to our knowledge of some mammals from inland Australia. *Transactions of the Royal Society of South Australia* **91:** 115–29.

PLEDGE, N. S., 1979. Fossil Vetebrates. Pp. 123–27 in *Natural history of Kangaroo Island* ed by M. J. Tyler, C. R. Twidale and J. K. Ling. Royal Society of South Australia: Adelaide.

SEEBECK, J. H., 1983. Eastern Barred Bandicoot. P. 100 in *Complete book of Australian mammals* ed by R. Strahan. Angus & Robertson: Sydney.

SMITH, M. J., 1977. Remains of mammals, including *Notomys longicaudatus* (Gould) (Rodentia: Muridae), in owl pellets from the Flinders Ranges, South Australia. *Australian Wildlife Research* **4:** 159–70.

SOUTH AUSTRALIAN DEPARTMENT FOR THE ENVIRONMENT, 1976. *Vegetation clearance in South Australia. Report of the interdepartmental committee on vegetation clearance.* South Australian Government: Adelaide.

TATE, G. H. H., 1948. Results of the Archbold Expeditions. No. 60. Studies in the Peramelidae (Marsupialia). *Bulletin of the American Museum of Natural History* **92:** 317–46.

THOMAS, O., 1888. *Catalogue of the Marsupialia and Monotremata in the collection of the British Museum (Natural History).* British Museum (Natural History): London.

TIDEMANN, C. R., 1967. Some mammal remains from cave deposits in the south-east of South Australia. *South Australian Naturalist* **42:** 21–27.

WATTS, C. H. S., 1969. Distribution and habitats of the rabbit bandicoot. *Transactions of the Royal Society of South Australia* **93:** 135–41.

WATTS, C. H. S. AND LING, J. K., 1985. Marine and terrestrial mammals. Pp. 139–47 in *Natural history of Eyre Peninsula* ed by C. R. Twidale, M. J. Tyler and M. Davies. Royal Society of South Australia: Adelaide.

Haddleton (1952), whose records of the fauna of the Katanning district in the southern wheatbelt cover the same period, made no mention of the species. Between 1904 and 1907, Shortridge (1910) learned that an animal known to the colonists as "Camel-foot" which he thought to be *C. ecaudatus* was very rare, if not extinct, around Beverley and York.

Chaeropus ecaudatus is thought to be extinct (Aitken 1983). Its persistence in the desert regions well into the twentieth century is indicated by the verbal evidence recorded by Burbidge and Fuller (1979) and Burbidge, Johnson, Fuller and Southgate (1988). It is unlikely that the sightings reported to these workers occurred much earlier than 1920. This and other evidence (Finlayson 1961) indicates that the final decline of the species occurred between 1920 and 1950.

Macrotis lagotis

The relatively large number of museum specimens of the Greater Bilby (or Dalgyte, as this species was known in the south-west) testify to its previous abundance and wide distribution in Western Australia (Fig. 7). The regions in which it was not recorded are those that experience annual rainfall well in excess of 500 mm; that is, the extreme south-west of the state and most of the Kimberley. There are also no records from the western parts of the pastoral district (western Murchison and Gascoyne) but, given the former wide occurrence of the species in similar terrain, this is probably due to the lack of collecting in these areas. *Macrotis* sp. bone material was found in surface deposits from rock shelters now submerged beneath Lake Argyle (Kitchener 1978). These deposits are thought to postdate Aboriginal contact with Europeans, suggesting that *Macrotis* (presumably *M. lagotis*) also occurred in this part of the Kimberley until recently.

A large proportion of Western Australian specimens were collected in the central and southern wheatbelt (Fig. 7) before the mid-1930s. The sudden extinction of *M. lagotis* there has been attributed to the massive proliferation of European rabbits and attempts to control them (Leake 1962; Haddleton 1952; Jenkins 1974). Foxes were building up in numbers by this stage also (Long 1970).

The Western Australian distribution of *M. lagotis* now consists of several apparently discrete areas of occurrence, in the Gibson Desert, in the inland Pilbara and along the southern edge of the Kimberley and the northern Great Sandy Desert (Fig. 7: post-1970 specimens WAM; other references: McKenzie *et al.* 1979; Youngson, Henry and McKenzie 1981; McKenzie 1983b; McKenzie and Youngson 1983). Within these areas, colonies are scattered, and numbers are far below those described in earlier accounts (e.g., Jones 1924).

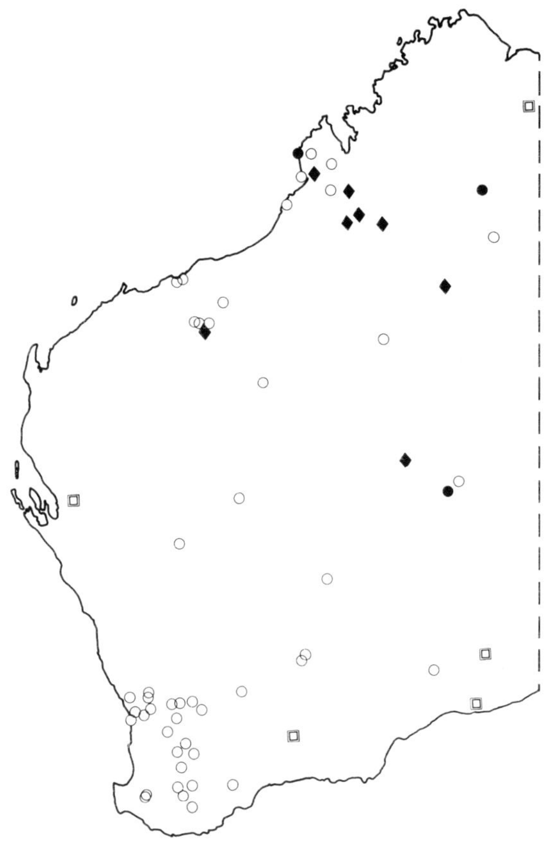

Fig. 7. Recorded occurence of *Macrotis lagotis* in Western Australia. ● museum specimens collected since 1969; ○ museum specimens collected before 1970; ◆ other published records since 1969; □ subfossil material.

Macrotis leucura

The Lesser Bilby *M. leucura* is known to have occurred in the arid zone in South Australia and the Northern Territory and was last recorded alive in 1931 (Johnson 1983a). The similarity of the mammal faunas of the western and central parts of the arid zone (e.g., Finlayson 1961) and the paucity of early records from the Western Australian desert indicate that *M. leucura* may well have occurred in this state before the recent demise of so many of the medium-sized species. In fact Burbidge *et al.* (1988) have collected much data through interviews of older Aboriginal people which confirm that *M. leucura* has occurred in Western Australia in living memory.

DISCUSSION

Western Australia has the largest area of any of the states of Australia and encompasses parts of the three major Australian biogeographic provinces of Spencer (1896: Bassian, Eyrean and Torresian). It is not surprising, then, that of the 11 Australian species of bandicoots recognized here, eight occurred within present state boundaries at the time of white settlement. In common with other groups of medium-sized mammals in Western Australia, however, most species of bandicoot have suffered massive declines in both number and range, and several species have become extinct (Table 1).

Table 1. Present status of bandicoots in Western Australia.

Perameles bougainville	*Very restricted, locally abundant
Perameles eremiana	*Extinct (1943)
Isoodon auratus	*Locally abundant
Isoodon macrourus	Locally abundant
Isoodon obesulus	Locally abundant
Chaeropus ecaudatus	*Extinct (1843)
Macrotis lagotis	*Rare, scattered
Macrotis leucura	*Extinct (date?)

*Species which have experienced massive declines since white settlement. Collection date of last museum specimen from Western Australia in brackets (extinct species).

At the same time, however, several bandicoot species which have declined or disappeared from their former range in the rest of Australia persist in special situations in the west. Thus while the state's bandicoot fauna has become severely depleted since 1827, Western Australia is an important refuge for the group as a whole.

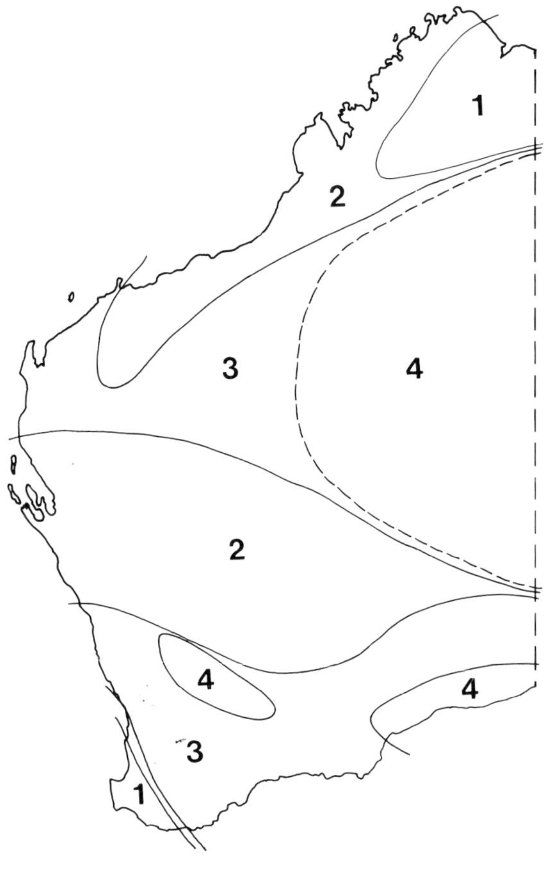

Fig. 8. Isocoots (lines indicating areas with equal numbers of bandicoot species) for Western Australia in 1827. *Macrotis leucura* is omitted.

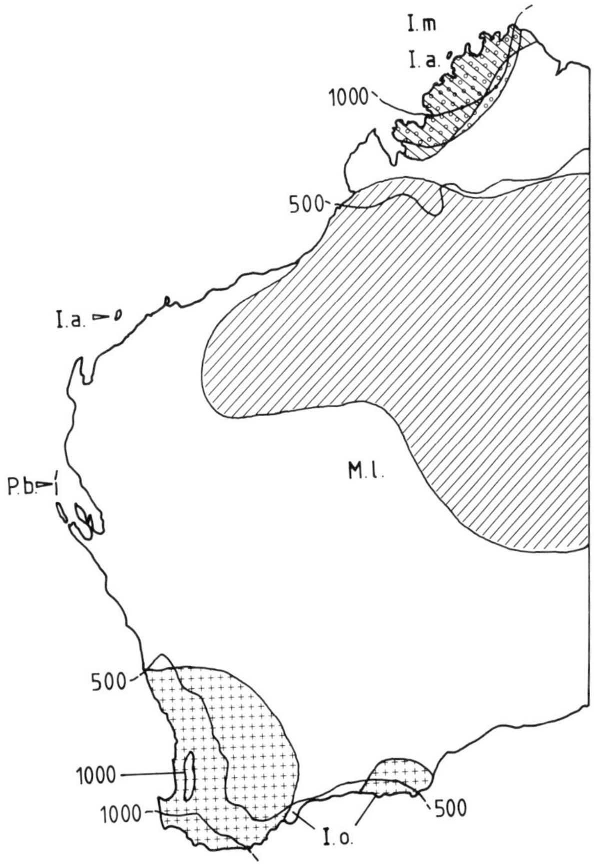

Fig. 9. Present distribution of bandicoots in Western Australia, showing average annual rainfall in millimetres. I.m. = *Isoodon macrourus;* I.a. = *I. auratus* (circles); P.b. = *Perameles bougainville;* M. l. = *Macrotis lagotis;* I. o. = *I. obesulus.*

PREVIOUS DISTRIBUTIONS AND PATTERNS OF DISAPPEARANCE

An approximate picture of the distribution of bandicoot species richness in Western Australia immediately before white settlement is shown in Fig. 8 (omitting *M. leucura*). The lines (or "isocoots") delimit areas with similar numbers of bandicoot species. This shows that the state's richest bandicoot faunas were found in the Gibson and Great Sandy deserts and in the central wheatbelt. These areas are virtually devoid of bandicoots today. By contrast, the areas in which bandicoots survive, that is, the extreme south-west, the north-west Kimberley, and some coastal islands, supported depauperate bandicoot faunas at the time of white settlement.

The distribution of the remnant bandicoot fauna of Western Australia (Fig. 9) can be summarized as follows:

(1) *Isoodon* species have survived in those areas of the state experiencing the highest rainfall.

(2) *Macrotis lagotis* has survived in the northern part of its former range in the arid zone, in the areas probably least drastically altered by European agriculture and pastoral practices.

(3) The bandicoot faunas of islands are intact except where cats have become established (Table 2).

It has been pointed out (Kitchener 1978) that in Western Australia, bandicoots are amongst the first mammals to disappear in the face of European settlement and farming. It may be useful to examine the group closely in an attempt to detect the attributes that make particular bandicoots vulnerable to change. We might look at interspecific differences in diet, reproduction, nesting habits, territorial behaviour and habitat. Unfortunately there is least information about the species which disappeared most rapidly. However, from the little that is known (various accounts in Strahan 1983), there appear to be no major differences in dietary specialization or in average litter size between Western Australian bandicoots. Breeding frequency is known with certainty only for the *Isoodon* species. In nesting habits and territorial behaviour, however, differences are apparent.

With the exception of the bilbies, all bandicoots treated in this account built nests in shallow dug-out scrapes concealed in the vegetation. The *Macrotis* species dug true burrows, however, and this habit may have given them some protection against exotic predators. While the smaller *M. leucura* has probably become extinct throughout its range, *M. lagotis* is unusual amongst medium-sized mammals of the arid zone in persisting there (Burbidge and McKenzie 1983), albeit in scattered colonies and only in northern areas. Bolton and Latz (1978) have pointed out that changes in frequency and pattern of burning in the arid zone have followed the movement into settlements by the previously nomadic Aboriginal inhabitants which commenced in the 1930s. They postulated that consequently the mosaic of fire-ages previously imposed on the vegetation by Aboriginal burning regimes is no longer present, and wildfires now periodically devastate huge areas. The timing of disappearance of a number of medium-sized desert mammals coincided with the change of lifestyle of the Aboriginal people (Burbidge 1985). These mammals include *P. eremiana, I. auratus, C. ecaudatus* and *M. leucura* (see species accounts above). Under the new fire regimes, populations of animals which show strong fidelity to their home ranges even in the event of a catastrophic fire, as does the Brush-tailed Bettong (or Woylie), *Bettongia penicillata,* in the south-west forests (Christensen 1980), would appear to be vulnerable to extinction without the refuge of nearby unburnt areas. In general, studies of bandicoots have revealed that animals do not readily move from an established home range (Gordon 1974; Heinsohn 1966; Stoddart and Braithwaite 1979). *Macrotis lagotis,* however, will move its home range in response to changes in food availability (Johnson 1983b). This behavioural feature may allow individuals of this species to respond to the lack of nearby food after one of the extensive fires of the post-traditional Aboriginal era (Bolton and Latz 1978) by moving large distances to suitable habitat. This hypothesis does not explain the survival of *M. lagotis* only in the northern arid zone. It is likely that the failure of the fox *(Vulpes vulpes)* to become strongly established in the north of Western Australia (King and Smith 1985) has lessened predator pressure on *M. lagotis* in this region.

The persistence of *I. obesulus* and *I. macrourus* in most of their previous range might indicate simply that these two species are inherently tolerant to environmental change. However, taken with the disappearance of *I. auratus* in all of its continental range except where *I. macrourus* has also survived (the north-west Kimberley), this suggests that less disastrous changes have occurred in the high-rainfall areas of the state than those which have caused extinctions within the desert bandicoot fauna.

The distributions of a number of previously widespread mammals have contracted to small remnants in the south-west at the high rainfall end of their previous ranges. Examples are the Numbat *(Myrmecobius fasciatus), B. penicillata,* the Western Quoll *(Dasyurus geoffroii)* and the Red-tailed Phascogale *(Phascogale calura).* This pattern apparently corresponds to the survival of mainland *I. auratus* only in the north-west Kimberley, where indeed no mammal species is known to have been lost since the arrival of Europeans (King and Smith 1985). The persistence of the mammal fauna in these areas is probably due to a number of factors. The greater availability of cover provided by the more dense and structurally diverse vegetation in the high rainfall areas may afford more secure refuge from exotic predators than is found in the semi-arid and arid zones. The dependable seasonal rainfall may also be important, as the combined effect of large fluctuations in food availability in regions of irregular rainfall, changes in fire regimes and increased rates of predation by introduced species may have pushed some native mammals into population decline. In the north-west Kimberley, the intact nature of the mammal fauna may also be due to the failure of the fox to become established there (King and Smith 1985).

Bandicoots on islands

While contractions of range and extinctions have been common amongst Western Australian bandicoots, most island populations (Table 2) have survived and appear to be flourishing. The majority of islands off the coast of Western

Table 2. Bandicoots on Western Australian islands. Records of live specimens only.

Island	Species	Area (ha)
Daw	*Isoodon obesulus*	330
Dorre	*Perameles bougainville*	4 420
Bernier	*Perameles bougainville*	4 640
Barrow	*Isoodon auratus*	20 700
Middle	*Isoodon auratus*	450
Hermite	*Isoodon auratus* (extinct)	939
Augustus	*Isoodon auratus*	17 950
St Andrew	*Isoodon macrourus*	1 410

Australia have escaped extensive agricultural or pastoral activity. Dirk Hartog Island is under pastoral lease, and mammal species have been lost; however, cats have also been introduced there (Burbidge and George 1978). Separation of the effects of grazing and of the introduction of cats on island mammal faunas is possible as goats, but not cats, have been present on Bernier Island for much of this century, without the loss of any mammal species (Ride and Tyndale-Biscoe 1962). The introduction of cats onto Hermite Island apparently led to the extinction of both *I. auratus* and *Lagorchestes conspicillatus* (Burbidge 1971). This is the only instance of the establishment of cats on an island where bandicoots were known to be present, except for a period when cats were reportedly present on Bernier Island (Shortridge 1910) but apparently did not survive (Ride and Tyndale-Biscoe 1962). The abundance of mammals on some of these islands compared with the adjacent mainland appears to be a consequence of higher rainfall (e.g., Dirk Hartog Island versus Carnarvon, Ride 1962) and perhaps low fire frequency rather than being solely due to the absence of exotic predators. Indeed, Peron, who visited Shark Bay in 1804, remarked on the far greater abundance of hare-wallabies on the islands than on the mainland (Alexander 1915), while the high density of Quokkas *(Setonix brachyurus)* on Rottnest Island has prompted comments from even the earliest visitors (de Vlamingh).

Conditions prevailing on these islands make their mammal populations extremely vulnerable if new predators are introduced. The ready availability of prey should lead to a rapid rate of increase in the predator population. On a small island, dispersal of the predator's young, which would occur on the mainland, can take place only to a limited degree. This will lead to an unnaturally high density of the predator during the early stages, so that immense pressure is placed on the prey species. If the predator has a non-specialized diet, as do cats and foxes, scarcity of one type of prey will not have a drastic effect on the island's carrying capacity for that species of predator, and the remnant prey population will be at great risk.

CONCLUSIONS

(1) Bandicoots have suffered drastic declines in Western Australia. Three species have been lost entirely from the fauna, and three others have been reduced to a small part of their previous range. The other two species have undergone a much lesser change in status.

(2) *Isoodon* species have survived well in regions of higher rainfall.

(3) The range of *M. lagotis* has contracted towards the northern part of the arid zone. Its survival there, where four other bandicoot species have become extinct, may be due to its burrowing habit and its propensity to shift home range when conditions are unsuitable, as well as to lower densities of foxes in the north.

(4) Island populations have shown the least tendency to become extinct if undisturbed. Because of their nature, however, they are extremely vulnerable to the effects of introduced predators.

ACKNOWLEDGEMENTS

I would like to thank a number of people for their assistance in the preparation of this paper: Ms Barbara Jones and Ms Jan Henry (Western Ausralian Museum), Ms Linda Huxley (Museum of Victoria), Ms Linda Gibson (Australian Museum) and Dr Catherine Kemper (South Australian Museum), who kindly extracted bandicoot records from museum registers, Ms Ruth Clark and Mr Bruce Turner who prepared the distribution maps, and Ms Diana Morgan who provided information on the history of the Monte Bello Islands. British Airways provided an airfare to London, enabling me to examine the bandicoot collection of the British Museum (Natural History). I am particularly grateful to Dr Andrew Burbidge of the Western Australian Wildlife Research Centre for critically reviewing this manuscript and allowing me access to unpublished data. This study was supported by the Western Australian Department of Conservation and Land Management.

REFERENCES

AITKEN, P. F., 1983. Pig-footed Bandicoot *Chaeropus ecaudatus*. Pp. 104–05 in *The complete book of Australian mammals* ed by R. Strahan. Angus & Robertson: Sydney.

ALEXANDER, W. B., 1915. History of zoology in Western Australia. Part II. 1791–1829. *Journal of the Royal Society of Western Australia* **1:** 83–149.

BAYNES, A., 1984. Native mammal remains from Wilgie Mia ochre mine: evidence on the pre-european fauna of the western arid zone. *Records of the Western Australia Museum* **11:** 297–310.

BOLTON, B. L. AND LATZ, P. K., 1978. The Western Hare-wallaby, *Lagorchestes hirsutus* (Gould) (Macropodidae), in the Tanami Desert. *Australian Wildlife Research* **5:** 285–93.

BURBIDGE, A. A., 1971. The fauna and flora of the Monte Bello Islands. *Department of Fisheries and Fauna, Western Australia* Report No. **9**.

BURBIDGE, A. A., 1983. Western Barred Bandicoot *Perameles bougainville*. P. 101 in *The complete book of Australian mammals* ed by R. Strahan. Angus & Robertson: Sydney.

BURBIDGE, A. A., 1985. Fire and mammals in hummock grasslands of the arid zone. Pp. 91–94 in *Fire ecology and management in Western Australian ecosystems* ed by J. R. Ford. *WAIT Environmental Studies Group Report* **14**.

BURBIDGE, A. A. AND FULLER, P. J., 1979. Mammals of the Warburton region, Western Australia. *Records of the Western Australian Museum* **8:** 57–73.

BURBIDGE, A. A. AND FULLER, P. J., 1984. Finding out about desert mammals. *SWANS* **14:** 9–13.

BURBIDGE, A. A. AND GEORGE, A. S., 1978. The flora and fauna of Dirk Hartog Island, Western Australia. *Journal of the Royal Society of Western Australia* **60:** 71–90.

BURBIDGE, A. A., JOHNSON, K. A., FULLER, P. J. AND SOUTHGATE, R. I., 1988. Aboriginal knowledge of the mammals of the central deserts of Australia. *Australian Wildlife Research* **15:** 9–39.

BURBIDGE, A. A. AND MCKENZIE, N. M., 1983. Patterns of decline in Western Australian vertebrates. Zoology abstracts, ANZAAS Congress, Perth.

CALABY, J. H., 1954. Comments on Gilbert's note-book on marsupials. *Western Australian Naturalist* **4:** 147–48.

CHRISTENSEN, P. E. S., 1980. The biology of *Bettongia penicillata* Gray, 1837, and *Macropus eugenii* (Desmarest, 1817) in relation to fire. *Forests Department, Western Australia, Bulletin* **91**.

DAHL, K., 1897. Biological notes on North-Australian Mammalia. *Zoologist* series 4, **1:** 189–216.

FINLAYSON, H. H., 1961. On central Australian mammals. Part IV. The distribution and status of central Australian species. *Records of the South Australian Museum* **14:** 141–91.

FREEDMAN, L., 1967. Skull and tooth variation in the genus *Perameles*. Part I: Anatomical features. *Records of the Australian Museum* **27:** 147–166, pls 16–23.

FRIEND, J. A., FULLER, P. J. AND DAVIS, J. A., 1982. The Numbat in central Australia. *SWANS* **12:** 20–26.

GLAUERT, L., 1948. The development of our knowledge of the marsupials of Western Australia. *Journal of the Royal Society of Western Australia* **34:** 115–34.

GORDON, G., 1974. Movements and activity of the Shortnosed Bandicoot *Isoodon macrourus* Gould (Marsupialia). *Mammalia* **38:** 405–31.

GORDON, G., 1983a. Desert Bandicoot *Perameles eremiana*. P. 102 in *The complete book of Australian mammals* ed by R. Strahan. Angus & Robertson: Sydney.

GORDON, G., 1983b. Northern Brown Bandicoot *Isoodon macrourus*. Pp. 96–97 in *The complete book of Australian mammals* ed by R. Strahan. Angus & Robertson: Sydney.

GOULD, J., 1844. *Proceedings of the Zoological Society of London* **12:** 104.

GOULD, J., 1863. *The mammals of Australia* Part 1, Volume 1. The Author: London.

HADDLETON, J. F., 1952. *Katanning pioneer*. Paterson Brokensha: Perth.

HEINSOHN, G. E., 1966. Ecology and reproduction of the Tasmanian bandicoots (*Perameles gunni* and *Isoodon obesulus*). *University of California Publications in Zoology* **80:** 1–107.

HILL, F. L., 1955. Notes on the natural history of the Monte Bello Islands. *Proceedings of the Linnean Society of London* **165:** 113–24.

IREDALE, T. AND TROUGHTON, E. LEG., 1934. A checklist of the mammals of Australia. *Memoirs of the Australian Museum* **6:** 1–120.

JENKINS, C. F. H., 1974. The decline of the Dalgite *(Macrotis lagotis)* and other wildlife in the Avon valley. *Western Australian Naturalist* **12:** 169–72.

JOHNSON, K. A., 1983a. Lesser Bilby *Macrotis leucura*. P. 109 in *The complete book of Australian mammals* ed by R. Strahan. Angus & Robertson: Sydney.

JOHNSON, K. A., 1983b. Bilby *Macrotis lagotis*. Pp. 107–08 in *The complete book of Australian mammals* ed by R. Strahan. Angus & Robertson: Sydney.

JOHNSON, K. A. AND ROFF, A. D., 1981. The Western Quoll *Dasyurus geoffroii* (Dasyuridae, Marsupialia) in the Northern Territory: Historical records from venerable sources. Pp. 221–26 in *Carnivorous marsupials* ed by M. Archer. Royal Zoological Society of New South Wales: Sydney.

JONES, F. W., 1924. *The mammals of South Australia*. Part 2. Government Printer: Adelaide.

KENDRICK, G. W. AND PORTER, J. K., 1973. Remains of a thylacine (Marsupialia: Dasyuroidea) and other fauna from caves in the Cape Range, Western Australia. *Journal of the Royal Society of Western Australia* **56:** 116–22.

KING, D. R. AND SMITH, L. A., 1985. The distribution of the European Red Fox *(Vulpes vulpes)* in Western Australia. *Records of the Western Australian Museum* **12:** 197–205.

KITCHENER, D. J., 1978. Mammals of the Ord River area, Kimberley, Western Australia. *Records of the Western Australian Museum* **6:** 189–219.

KITCHENER, D. J., CHAPMAN, A. AND BARRON, G., 1978. Mammals of the northern Swan coastal plain. Pp. 54–92 in *Faunal studies of the northern Swan Coastal Plain, a consideration of past and future changes.* Western Australian Museum unpublished report.

KITCHENER, D. J., KELLER, L. E., CHAPMAN, A., MCKENZIE, N. L., START, A. N. AND KENNEALLY, K. F., 1981. Observations on mammals of the Mitchell Plateau area, Kimberley, Western Australia. Pp. 123–68 in *Biological survey of Mitchell Plateau and Admiralty Gulf, Western Australia.* Western Australian Museum: Perth.

LEAKE, B. W., 1962. *Eastern wheatbelt wildlife*. B. W. Leake: Perth.

LIPFERT, O., 1912. A visit to Bernier and Dorre Islands during August and September, 1910. *Records of the Western Australian Museum* **1:** 98–101.

LONG, J. L., 1970. Introduced birds and mammals in Western Australia. *Agriculture Protection Board of Western Australia Technical Series* 1.

LUNDELIUS, E. jr., 1957. Additions to knowledge of the ranges of Western Australian mammals. *Western Australian Naturalist* **5:** 173–82.

McKenzie, N. L., 1981. Mammals of the Phanerozoic South-west Kimberley, Western Australia: biogeography and recent changes. *Journal of Biogeography* **8:** 263–80.

McKenzie, N. L., 1983a. Golden Bandicoot *Isoodon auratus*. P. 98 in *The complete book of Australian mammals* ed by R. Strahan. Angus & Robertson: Sydney.

McKenzie, N. L., 1983b. Mammals. Pp. 40–53 in *Wildlife of the Dampier Peninsula, South-west Kimberley, Western Australia* ed by N. L. McKenzie. *Wildlife Research Bulletin, Western Australia* **11**.

McKenzie, N. L., Chapman, A. and Youngson, W. K., 1975. Mammals. Pp. 69–74 in *A biological survey of the Prince Regent River Reserve, North-west Kimberley, Western Australia* ed by J. M. Miles and A. A. Burbidge. *Wildlife Research Bulletin, Western Australia* **3**.

McKenzie, N. L., Burbidge, A. A., Chapman, A. and Youngson, W. K., 1978. Mammals. Pp. 22–28 in *The islands of the North-west Kimberley, Western Australia* ed by A. A. Burbidge and N. L. McKenzie. *Wildlife Research Bulletin, Western Australia* **7**.

McKenzie, N. L., Youngson, W. K., Burbidge, A. A. and Chapman, A., 1979. Mammals. Pp. 16–21 in *The wildlife of some existing and proposed nature reserves in the Gibson, Little Sandy and Great Victoria Deserts, Western Australia* ed by N. L. McKenzie and A. A. Burbidge. *Wildlife Research Bulletin, Western Australia* **8**.

McKenzie, N. L. and Youngson, W. K., 1983. Mammals. Pp. 63–93 in *Wildlife of the Great Sandy Desert, Western Australia* ed by A. A. Burbidge and N. L. McKenzie. *Wildlife Research Bulletin, Western Australia* **12**.

Merrilees, D., 1968. Remains of the Pig-footed Bandicoot in Nullarbor caves. *Western Australian Naturalist* **11:** 19.

Montague, P. D., 1914. A report on the fauna of the Monte Bello Islands. *Proceedings of the Zoological Society of London* **1914:** 625–52.

Morris, K. D., 1985. Feral animal control on NW island nature reserves. Department of Conservation and Land Management, Western Australia. Wildlife Research Branch Research Programmes 1985 (unpublished).

Parker, S. A., 1973. An annotated checklist of the native land mammals of the Northern Territory. *Records of the South Australian Museum* **16:** 1–57.

Quoy, J. R. C. and Gaimard, P., 1824. *Voyage autour du monde . . . sur les corvettes l'Uranie et la Physicienne. Zoologie*. Paris.

Ride, W. D. L., 1962. The physical environment. Pp. 19–30 in *The results of an expedition to Bernier and Dorre Islands, Shark Bay, Western Australia* ed by A. J. Fraser. *Fisheries Department of Western Australia, Fauna Bulletin* **2.**

Ride, W. D. L. and Tyndale-Biscoe, C. H., 1962. Mammals. Pp. 54–97 in *The results of an expedition to Bernier and Dorre Islands, Shark Bay, Western Australia* ed by A. J. Fraser. *Fisheries Department of Western Australia, Fauna Bulletin* **2**.

Schreber, J. C. D. von, 1855. *Die Saugethiere* (Suppl.) **5:** 217. (Not seen).

Serventy, D. L. and Marshall, A. J., 1964. A natural history reconnaissance of Barrow Island and Monte Bello Islands, 1958. *CSIRO Division of Wildlife Research Technical Paper* **6**.

Sheard, K., 1950. A visit to the Monte Bello Islands. *Western Australian Naturalist* **2:**150–51.

Shortridge, G. C., 1910. An account of the geographical distribution of the marsupials and monotremes of south-west Australia, having special reference to the specimens collected during the Balston Expedition of 1901–1907. *Proceedings of the Zoological Society of London* **1909:** 803–48.

Smith, E., 1966. *The beckoning west. The story of H. S. Trotman and the Canning Stock Route*. Angus & Robertson: Sydney.

Spencer, W. B., 1896. Narrative. Pp. 1–136 in *Report on the work of the Horn Scientific Expedition to Central Australia* ed by W. B. Spencer. Part 1. Dulau: London.

Stoddart, D. M. and Braithwaite, R. W., 1979. A strategy for utilization of regenerating heathland habitat by the brown bandicoot (*Isoodon obesulus;* Marsupialia, Peramelidae). *Journal of Animal Ecology* **48:** 165–79.

Strahan, R., 1982. *Mammal Index* (3rd ed.). The Australian Museum: Sydney.

Strahan, R. (ed), 1983. *The complete book of Australian mammals*. Angus & Robertson: Sydney.

Thomas, N. D., 1985. Breeding biology and behaviour of the Quenda *(Isoodon obesulus)* in captive and natural environments. Honours Thesis: Murdoch University: Perth.

Thomas, O., 1888. *Catalogue of the Marsupialia and Monotremata in the collection of the British Museum (Natural History). British Museum Trustees: London.*

Thomas, O., 1901. On some kangaroos and bandicoots from Barrow Island, N.W. Australia, and the adjoining mainland. Novitates Zoologicae **8:** 394–96.

Thomas, O., 1907. List of further collections of mammals from Western Australia, including a series obtained for Mr W. E. Balston; with field-notes by the collector, Mr G. C. Shortridge. *Proceedings of the Zoological Society of London* **1906:** 763–77.

Thomas, O., 1922. *Annals and Magazine of Natural History* Series 9, **10:** 144.

Wagner, J. A., 1841. *Archiv für Naturgeschichte* **7:** 289–97. (Not seen).

Whittell, H. M., 1938. Notes on field-trips of J. T. Tunney. *Emu* **38:** 322–26.

Whittell, H. M., 1941. A review of the work of John Gilbert in Western Australia. *Emu* **41:** 112–242.

Whittell, H. M., 1951. A review of the work of John Gilbert in Western Australia. Part IV. *Emu* **51:** 17–29.

Whittell, H. M., 1954. John Gilbert's notebook on marsupials. *Western Australian Naturalist* **4:** 104–14.

Youngson, W. K., Henry, J. and McKenzie, N. L., 1981. Mammals. Pp. 39–53 in *Wildlife of the Edgar Ranges area, South-west Kimberley, Western Australia* ed by N. L. McKenzie. *Wildlife Research Bulletin, Western Australia* **10**.

Present and former status of bandicoots in the Northern Territory

K. A. JOHNSON[1] and R. I. SOUTHGATE[1]

European settlement of the Northern Territory began in 1824 but it was not until the 1890s that any substantial collections were made of the mammal fauna. The existence of six species of bandicoots had been established by 1900. Each species appears to have occupied a broad range and most were apparently relatively common.

A wave of extinction swept over the southern border in the first few decades of this century and by the early 1960s had rolled north to take *Chaeropus ecaudatus, Perameles eremiana, Isoodon auratus* and *Macrotis leucura* from the desert region. *Macrotis lagotis* remains as a seriously depleted and probably declining species. *Isoodon macrourus* remains in the northern wet-dry tropics and it is possible that isolated relict populations of *I. auratus* remain in the same region.

The impact of exotic mammals, drought, and the pastoral industry was probably implicated in the decline of the bandicoots. A changing fire regime associated with the movement of Aborigines from their country is more coincident with, and universally applicable to, the change in bandicoot fauna.

Key words: Bandicoots, Northern Territory, range, extinction, fire.

INTRODUCTION

THE Northern Territory has an early history of attempts to establish a European outpost along the northern coast. The first attempt, in 1824, was at Melville Island, 10 years or more before the settlements at Melbourne, Adelaide or Perth. Three more spectacularly unsuccessful attempts were made: Raffles Bay (1828–29), Port Essington (1838–49) and Adelaide River (1864–67). When Port Essington was abandoned in 1849, the inhabitants demonstrated their despair with the northern environment when the departing ship was turned broadside, to shell what buildings were left after 11 years of termite activity (Lockwood 1974).

Occupation of the central region occurred sometime later, following Stuart's explorations of 1858–62 and the completion of the overland telegraph line in 1872.

Spasmodic occupation and the remoteness of the region made travel a hazardous undertaking and scientific collecting a rare occurrence. Significant collections were delayed until towards the end of the 1800s when more permanent settlements were established and the pastoral industry had begun to develop.

The first scientific specimens obtained were probably the Dugongs acquired by R. Tytler from Raffles Bay in 1827 (Johnson 1964: 429). Gilbert collected in the vicinity of Port Essington in 1840 and 1841 and provided type material used by Gould (1842, in Tate 1848) for the description of *Isoodon macrourus*. Elsey collected near the mouth of the Victoria River during Gregory's 1855–56 expedition across northern Australia. Knut Dahl travelled through country west from the South Alligator River and made important collections and observations during 1894–95 (Collett 1897; Dahl 1897).

The first major collecting expedition to central Australia did not take place until 1894 when the outstandingly successful Horn Scientific Expedition visited the central ranges and country to the south-west as far as Ayers Rock (Spencer 1896). Baldwin Spencer enlisted the help of Aborigines and established a network of white Australian contacts who continued to supply him with specimens for many years. A particularly industrious officer at Charlotte Waters Telegraph Station at the edge of the Simpson Desert, P. M. Bryne, provided Spencer with specimens for many years including the type material of *Perameles eremiana* and valuable specimens of *Macrotis leucura* (Spencer 1897).

Collecting proceeded intermittently until major expeditions were made by H. H. Finlayson in the south (variously during 1931–35) (Finlayson 1943, 1958, 1961) and Donald Thomson in the north (variously during 1935–43) (Dixon and Huxley 1985).

It is regrettable that the deserts were not subject to any systematic collecting in the period between Finlayson's 1931–35 and 1950–56 expeditions. In this short period there were massive changes to the faunal assemblage. About

[1]Conservation Commission, P.O. Box 1046, Alice Springs, Northern Territory, Australia 0871.

Pages 85–92 *in* BANDICOOTS AND BILBIES, ed by J. H. Seebeck, P. R. Brown, R. L. Wallis and C. M. Kemper. Surrey Beatty & Sons, Sydney, 1990.

35% of Northern Territory desert mammal species disappeared and a further 9% underwent considerable reductions in range.

The bandicoots were among the most severely affected groups and from an original six species known from the Northern Territory only two remain with any certainty.

In the absence of properly documented scientific field work during this period one can only piece together the fragments of published information and sift through the anecdotal material available from Aboriginal and white Australian observers.

This paper collates what distributional information was available to the authors and makes an assessment of the factors potentially involved in the species' decline.

THE STATUS OF NORTHERN TERRITORY BANDICOOTS

Pig-footed Bandicoot Chaeropus ecaudatus

Specimens from Ryans Well (1891; 22°43′S 133°23′E), Alice Springs (donated by Spencer in 1916; 23°42′S 133°51′E) and "40 miles north-east of Charlotte Waters" (1895; 25°55′S 134°54′E) are recorded by Parker (1973). Our discussions with Aborigines have shown it to be well known to many of the older people living north and west of Alice Springs and it appears to have had a distribution extending further west into the spinifex (*Triodia* spp. and *Plectranchne* spp.) deserts than indicated on Aitken's (1983) map (Fig. 1). Some Aborigines said that it used to occupy the Mulga *(Acacia aneura)* shrubland.

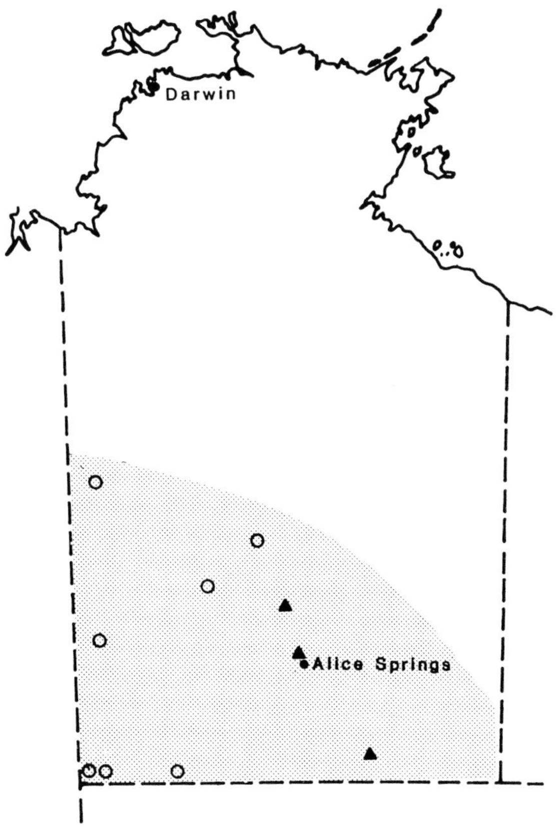

Fig. 1. Former range of *Chaeropus ecaudatus* ▲ = specimen records; ○ = localities referred to by Aborigines.

Finlayson (1961) was unable to locate a specimen during his expeditions in the early 1930s and 1950s and considered the species to be exceedingly rare or extinct.

Warlpiri Aborigines younger than 60 years have little or no first hand knowledge of the species. Gibson's (1986) extensive survey of the Tanami Desert with assistance from skilled Aboriginal trackers found no sign of it. The species must be considered extinct in the Northern Territory.

Golden Bandicoot Isoodon auratus

The Golden Bandicoot in the Northern Territory had a former distribution which extended from the border with South Australia north to at least the South Alligator River (12°12′S 132°23′E) where it appears to have been marginally sympatric with *I. macrourus* (Parker 1973). Information gathered by us from Aborigines indicates that the species had a distribution extending well to the north-east of Alice Springs.

The last specimen was collected in 1952 at The Granites (20°34′S 130°21′E) in the Tanami Desert (Parker 1973). Newsome (pers. comm.) saw tracks and dung identified as belonging to the "pakaru" by an Aboriginal informant in the Tanami Desert at about 20°52′S 130°34′E during a field trip in December, 1959. The name is now known by us to apply widely and exclusively to *I. auratus*.

Finlayson (1961) reported a steep decline in numbers in the southwestern portion of the Northern Territory in the period between his 1931–35 and 1950–56 series of expeditions. He believed, however, that it survived in considerable numbers in the area north-west of the latitude of Alice Springs at the time of his 1950–56 expeditions.

Extensive surveys undertaken since 1964 in the Tanami Desert and south to the region of Ayers Rock (Gibson 1986; Johnson, unpublished; Conservation Commission files) have failed to locate the species. Aborigines living in this extensive desert region consider the species to be extinct in their country and all available evidence supports this view.

Donald Thomson (Dixon and Huxley 1985) noted during his travels in the Northern Territory during 1942 that Aborigines in the eastern Arnhem Land region used three names for bandicoots. He considered the name "wangura" to be applicable to *I. macrourus* whereas "moiny moiny", a smaller bandicoot, he considered to be probably *I. auratus*. Thomson nevertheless obtained only specimens of *I. macrourus*.

Discussions with a man from Robinson River station (16°45′S 136°58′E) in the Gulf of Carpentaria region provided two names for bandicoots recognized as being distinct (Johnson and Latz, unpublished data). "Wakanya" was given for one species and "Mutari" for a smaller animal. Neither had been seen in the area for about 20 years but the informant believed them to persist in association with the coastal vine thickets.

Johnson (1964) considered *I. macrourus* to be the only bandicoot inhabiting the Arnhem Land area investigated by the American-Australian Scientific Expedition in 1948. Collections were gathered near Darwin, Groote Eylandt, Yirrkalla, Cape Don and Oenpelli. Nineteen specimens of *I. macrourus* were collected but no specimens or anecdotal accounts of *I. auratus* were recorded. Museum specimens are known however from the South Alligator River and Mount McMinn (on the Roper River) at the western and southern edges respectively of Arnhem Land (Parker 1973).

Dahl (1897) was undoubtably aware of the two northern species of *Isoodon* as he observed *I. auratus* (referred to by him as *Perameles obesula*) to be numerous in the coastal country in the vicinity of where Broome now stands in Western Australia but Collett (1897), reporting on Dahl's activities, noted that the species was not seen by him in Arnhem Land.

On the basis of these observations it is considered probable that *I. auratus* has not occupied Arnhem Land in historic times, as shown by McKenzie (1983). It possibly occurs in the coastal region of the Gulf country (Fig. 2) and survey effort is required in the vine thicket habitat associated with springs and coastal dunes.

Testimony from Aboriginals suggests that *I. auratus* was a common species and a favoured food item. In the southwestern Northern Territory it appears to have suddenly declined in the period from the mid-1930s to the mid-1950s. In the more northerly Tanami Desert this decline was delayed until the mid-1940s to late 1950s. It is now considered extinct in the desert region but may still persist in the Gulf region.

Northern Brown Bandicoot
Isoodon macrourus

A locally common species, confined mainly to the coastal and subcoastal areas from the Daly River east to Cape Arnhem Peninsula and south

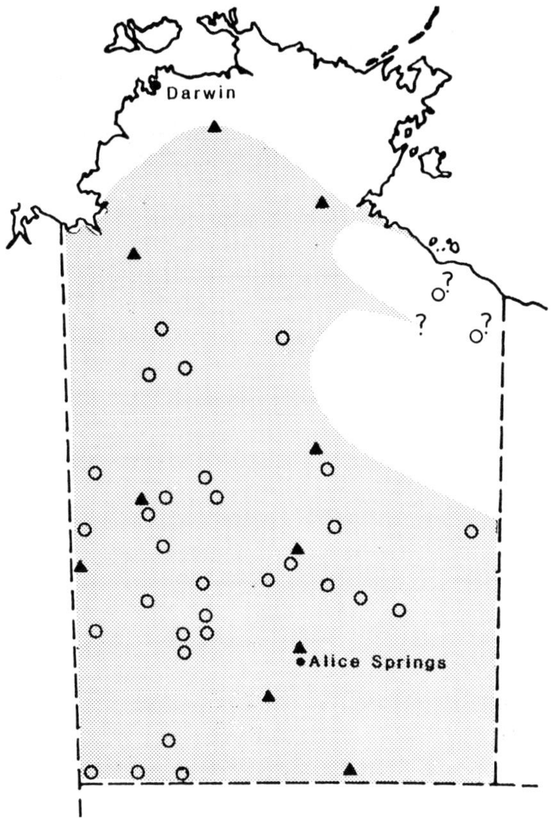

Fig. 2. Former range of *Isoodon auratus* showing the region of the Gulf of Carpentaria requiring further survey to check for the species presence. Symbols same as Figure 1.

to a latitude of about 15°30′S (Fig. 3). It occurs on many of the large offshore islands such as Groote Eylandt, Elcho Island (Dixon and Huxley 1985) and Centre Island (Calaby 1976). McKean and Howe (1972) found *I. macrourus* to be abundant on Maria Island and up to 20 could be seen on occasions, foraging and scavenging around their camp. Aboriginals from the Robinson River region of the Gulf (16°35′S 136°59′E) know of bandicoots inhabiting vine thickets growing in moist spring-fed gorges near Redbank Mine (17°11′S 137°45′E) and on coastal dune systems (Johnson and Latz, unpublished). Tracks corresponding to a bandicoot the size of *I. macrourus* were found at the mouth of the Calvert River (16°15′S 137°44′E), at the base of coastal dunes in 1987 (Johnson, unpublished). This information contrasts with that obtained by the former CSIRO Division of Wildlife Research from local residents familiar with the species in other parts of the Northern Territory (e.g., Darwin) who stated that they had never seen bandicoots on the mainland in the Gulf region (Calaby 1976).

Biological surveys in the catchments on the Nicholson and Cox Rivers (Conservation Commission files) and the lower McArthur River region have not revealed the presence of any bandicoot species. However, the presence of

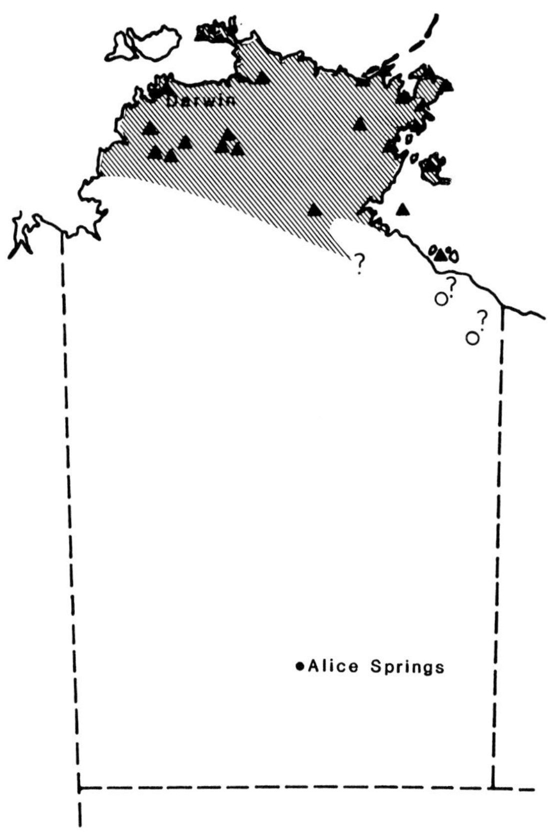

Fig. 3. The present range of *Isoodon macrourus* showing the Gulf of Carpentaria region where the species may have formerly or may still exist. Symbols same as Figure 1.

I. macrourus on nearby Centre Island and the likelihood that Aborigines are thoroughly familiar with the medium-sized mammals of their country indicates that the species is, or was until recently, an inhabitant of the Gulf hinterland. Gordon's (1983) map should be tentatively extended on these bases to include the area to at least 60 km inland from the coast along the Northern Territory portion of the Gulf of Carpentaria.

Dahl (1897) reported that *I. macrourus* "abounded" in Arnhem Land especially in the vicinity of large rivers and creeks. Johnson (1964) noted that all bandicoots observed during the American-Australian Scientific Expedition to Arnhem Land were "in forested areas" except for two which were trapped in a large patch of coarse grass on dry soil adjacent to a billabong. Calaby (1976) observed *I. macrourus* to be uncommon in open eucalypt forest on Centre Island and Calaby and Keith (1974) took one specimen from eucalypt forest at Port Essington. Intensive survey in the Keep River area of the Victoria River region (15°24′S 129°03′E) failed to record any bandicoots (Conservation Commission files). The absence of bandicoots from surveys in the Gulf region of the Northern Territory where Aborigines maintain they formerly existed raises the question of whether the species is coming under pressure at the southern limits of its range.

Desert Bandicoot Perameles eremiana

Parker (1973) listed museum specimens from the Burt Plain about 25 km north of Alice Springs, and the type locality of "40 miles (64.3 km) north-east of Charlotte Waters". He questions the claim by Finlayson (1961) of records from the Tanami Desert.

Finlayson (1961) examined material collected by Winnecke on the eastern side of the Northern Territory and provisionally attributed it to *P. eremiana.* No specific collection locality was given. Finlayson (1961) considered it to be common in the southwestern region of the Northern Territory in 1932–35, but less numerous than *I. auratus,* with which it was sympatric. Like *I. auratus* he considered it to have declined markedly in the period from 1932–35 to 1950–56 to a point where it was rare or absent from the southwestern areas but persisting in the Tanami Desert.

Information obtained by us from Aborigines corroborates Finlayson's (1961) views of abundance relative to *I. auratus*, its sympatric existence with this species, and range extending well into the Tanami Desert (Fig. 4).

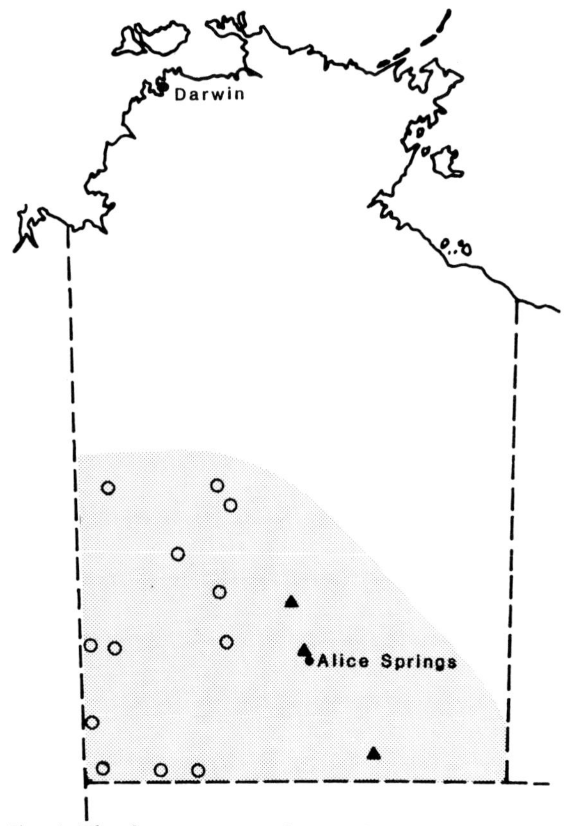

Fig. 4. The former range of *Perameles eremiana.* Symbols same as Figure 1.

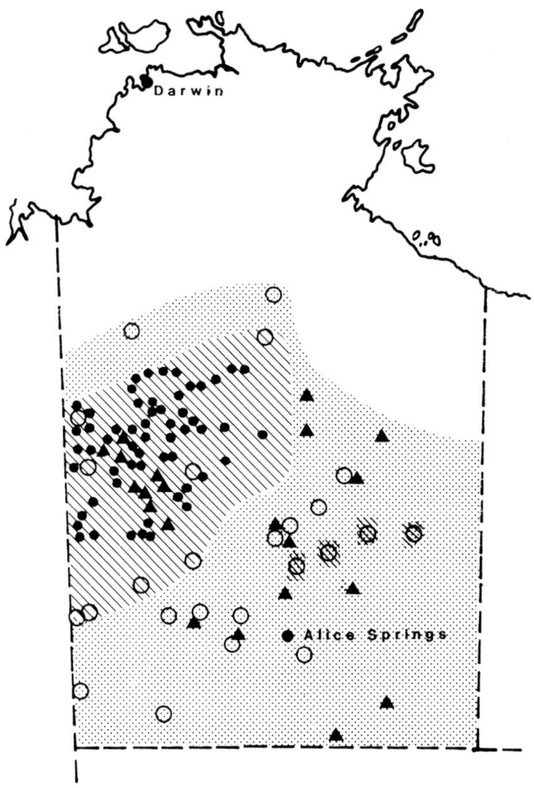

Fig. 5. The present (hatched) and former (stippled) range of *Macrotis lagotis*. ● = Gibson (1986) records, other symbols same as Figure 1.

There have been no museum specimens from the Northern Territory since those collected by Spencer (1897) and no specimen records since those of Finlayson (1961). Extensive field surveys (Gibson 1986) have failed to locate any sign of the species in the Tanami Desert and Aborigines living within its former range consider it to have vanished. Aborigines familiar with the Tanami Desert maintain that this decline took place concurrently with *I. auratus* after the Second World War during the late 1940s and 1950s. It should be regarded as extinct in the Northern Territory.

Greater Bilby Macrotis lagotis

The Greater Bilby in the Northern Territory once occupied a range extending north from the South Australian border to a latitude of about 16°40'S, with records 80 km to the north of Newcastle Waters (Finlayson 1961), and near Kalkaringi (Aboriginal records). It was formerly abundant. Spencer and Gillen (1912) commented on the extensive use of tail tips for decoration by Aborigines and concluded that the species must have remarkable powers of procreation to survive such extraordinary depredations. Finlayson (1961) considered it to have been one of the most plentiful and universally distributed mammals of central Australia. He remarked that it had been eliminated from much of the south-west Northern Territory in the 25 years since his 1931–35 expeditions and that elsewhere it was rapidly being reduced to the status of a rare form.

It is now confined to the Tanami Desert north-west of Alice Springs and to a few pockets in pastoral country to the north-east. Museum specimen and other reliable records are shown in Figure 5.

Numbers seem to fluctuate considerably in relation to seasonal conditions. Gibson (1986) recorded the species as being common and widespread during his 1981–84 surveys. Our observations show that the saline drainage systems winding through the Tanami Desert were particularly favoured habitats during the recent wet years which persisted until 1984. The subsequent two-year dry spell caused its status in these systems to change from one of almost universal occupation to one of sporadic occurrence.

Testimony of a local pastoralist, Mr E. Hayes, indicated that *M. lagotis* was reasonably common in pastoral country to the north of Alice Springs before the catastrophic 1958 to 1965 drought. Watts (1969) was able to locate the species in pastoral country to the north-west of Alice Springs in 1968. It is now extremely rare or

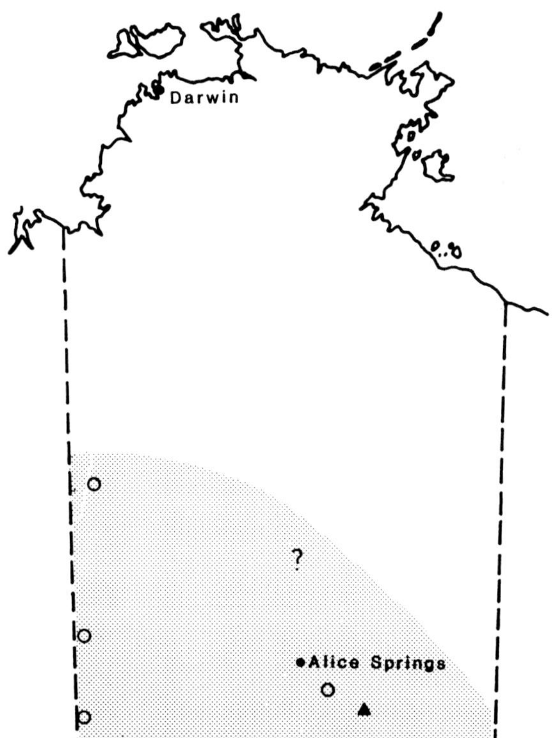

Fig. 6. The former range of *Macrotis leucura*. ? = dubious Barrow Creek records of Spencer, other symbols same as Figure 1.

absent from this area. Surveys by one of us (RIS) indicate that its distribution is continuing to contract and fragment despite the period of exceptionally favourable seasons from 1974 to 1984.

Lesser Bilby **Macrotis leucura**

Parker (1973) listed three localities from which specimens have been taken. These are Barrow Creek, 1901 (280 km north of Alice Springs), "40 (64.3 km) miles north-east of Charlotte Water", 1895, and Steele Gap, 1967 at the northern edge of the Simpson Desert.

We agree with Parker's (1973) uncertainty of the Barrow Creek locality on the basis that Spencer and Gillen (1912) remarked that *M. lagotis* was the only interesting mammal collected during their expedition of 1901 and although *M. leucura* was discussed, its presence (surely a remarkable find) was not mentioned. Three specimens obtained by Spencer and labelled Barrow Creek, nevertheless exist in the Museum of Victoria spirit collection (C7093–C7095). The Steele Gap specimen was a skull of indeterminate age collected from a Wedge-tailed Eagle *(Aquila audax)* nest. Testimony from Aboriginals indicate that the species once extended across the southern part of the Northern Territory into the Great Sandy and Gibson deserts (Burbidge *et al.*, in press) thus greatly extending Johnson's (1983) distribution map (Fig. 6).

DISCUSSION

The tragedy of the bandicoots of central Australia is that they were once a reasonably common component of the fauna yet most species disappeared before any detailed understanding of their biology or place in the ecosystem could be determined. When Finlayson was busy collecting *Caloprymnus* in northeastern South Australia in 1931 he was rightly concerned with obtaining as much information as possible on this little-known, re-discovered species. With assistance from an Aboriginal he collected 12 specimens of *M. leucura* and fortunately published our only first-hand account of the species. However, he regarded them as "plentiful" and "widespread". *Macrotis lagotis,* with which it occurred, was considered to be "by no means uncommon" at the time (Finlayson 1935) and he did not pursue either species with great effort (H. Finlayson, pers. comm.). It was the last time a scientific specimen of *M. leucura* was collected alive.

A wave of extinction appears to have swept up from the south and removed, or severely reduced, the range of all the desert bandicoots. For the most part, this change took place in less than 25 years, which, on an ecological scale, is little more than an instant.

A dubious exception is *C. ecaudatus* for which there are few data. Finlayson (1961), on advice from Aborigines, considered it to be rare by 1926 in the south-west and to be a southern form. Aborigines of 60 to 70 years of age in the more northerly Tanami Desert region nevertheless still remember the animals well and in their country it appears it survived until the 1930s at least. Pintupi people from the western edge of the Northern Territory recall seeing the species until about 30 years ago (Burbidge *et al.* 1988).

Few data are available for *M. leucura* but a skull (NTM 5053) found in a Wedge-tailed Eagle nest at the northern edge of the Simpson Desert in 1967 was still in resaonable condition. As exposed bone material readily deteriorates in central Australia we would guess that the skull was less then 10–15 years old indicating that the species probably persisted in this region until at least the 1950s.

Finlayson (1961) considered *M. lagotis, P. eremiana* and *I. auratus* to have vanished from the south-west in the 25 year period separating his 1931–35 and 1950–56 expeditions. Aboriginal testimony together with museum records and other anecdotal reports indicate that *P. eremiana* and *I. auratus* persisted in the Tanami Desert region until the late 1950s. *Macrotis lagotis* persists there still.

The desert region of the Northern Territory adjoining the Western Australian border extends from the South Australian border north approximately 900 km. It is an interesting study area in that until after the decline in species occurred it had never been used for pastoral purposes and had only sporadic, mainly short-lived mining installations. It was examined for pastoral potential and prospected over for minerals by white Australians and then largely ignored (Gibson 1986).

The grazing by livestock cannot therefore by implicated in the demise of the bandicoots from this area. The associated land management effects of erosion, use of poisons to control pests, introduction of weed species, and installation of artificial waters causing an increase in predators can be similarly eliminated.

Feral cats, *Felis catus,* have occupied the entire region for a very long time. Aborigines we spoke to, no matter how old, regarded them as being always present (i.e., before they were born and probably early in their parent's lives). Corroboration is provided by early explorers and prospectors such as Carnegie (1898) who . . . "was astonished to see a common domestic black cat, evidently just killed". It was in the possession of two Aboriginal women who had been disturbed from hunting by the explorer and his team. The observation was made more than 380 km west of the nearest

European outpost of the time, the tiny mission of Hermannsburg. The now-extinct bandicoots must have survived the depredations of cats for at least 50 years and probably longer before swiftly disappearing from the desert. It seems improbable that cats, having lived with the bandicoots for such a long time, should suddenly and singularly be responsible for such a dramatic and catastrophic decline.

Rabbits have been implicated in the demise of *M. lagotis* (Rolls 1969) and through their devastation of the vegetation during plagues and following droughts might be expected to compete directly or indirectly with bandicoots for food. Rabbits eat the bulbs of a sedge *(Cyperus bulbosus)* which is of great importance to *M. lagotis* and no doubt eat other plants that produce the seeds and fruits that these bandicoots eat (Johnson, unpublished). Rabbits had entered the Northern Territory from South Australia by 1894 (Strong 1983). They spread rapidly and had been recorded 112 km south of the Tanami goldfields, some 550 km north of the Northern Territory–South Australian border, by 1910 (Worgan 1911; Gibson 1986). Rabbits appear to have never been common in the Tanami Desert, especially in the northern areas (Sweeney 1947). Finlayson (1958) considered that the rabbit probably had an adverse effect on "small native herbivores" but noted that . . . "in the 60 years they have been together in the southwestern areas of central Australia, it has not proved a fatal (factor) . . .".

Foxes *(Vulpes vulpes)* were considered by Finlayson (1961) to have been a crucial factor in the demise of many native mammals. They were uncommon at the time of his 1931–35 expeditions but had become plentiful in the south-west when he travelled the area in the 1950s. The disappearance of bandicoots from the south-west therefore, coincides broadly with the firm establishment of the fox and it is tempting to label it, as did Finlayson (1961), as the single responsible factor.

Foxes, however, did not penetrate the Tanami Desert until the 1970s; well after the decline of the bandicoots (Dick Kimber, pers. comm.). Foxes were so rare as to be noteworthy to Aboriginals at Yuendumu (at the southern edge of the Tanami) in the late 1970s (Karl Benz, pers. comm.). *Perameles eremiana, I. auratus* and *C. ecaudatus* have nevertheless disappeared from this country which was never densely occupied by European rabbits and not inhabited by foxes until after they disappeared.

Other factors must be involved and fire seems to be implicated. Aborigines relentlessly burned their country before European settlement and this produced a tight mosaic of patches in different stages of recovery (Latz and Griffin 1978). This contrasts with the present situation in the remote desert country where fires are mostly initiated by lightning strike during the spring and summer months and burn hotly over large areas. The effect of such large burns is to homogenize the landscape in a way similar to intensive grazing by cattle and sheep in pastoral country. Structural and floristic diversity is critical to the survival of *Lagorchestes hirsutus* in the one remaining colony in the Tanami Desert (Bolton and Latz 1978).

It is similarly important in producing a flush of highly productive "fire weeds" in the first 1–2 years after burning. These native species produce a flush of seed, fruits and insect larvae which form the majority of food for *M. lagotis* (Johnson, unpublished data) and were most probably important foods for the other bandicoots.

Aborigines had begun moving to cattle stations and missions in the southwestern region when they were established in the 1930s (e.g., Ernabella in 1932). Movement probably quickened as others became established (e.g., Everard Park Station 1932, Areyonga 1942, Haasts Bluff 1947).

A similar pattern emerges with the Warlpiri of the Tanami Desert and the timing coincides with the disappearance of most native mammals.

The harshest drought on record in central Australia occurred between 1958 and 1965 and this may be implicated with the final demise of many species. Animals living in a desert already debilitated by the decline in floristic and structural patchiness created via Aboriginal burning practices would be more vulnerable to the effects of a devastating drought.

Whatever factor or factors caused the decline in bandicoots in the western spinifex deserts of the Northern Territory, white Australian land-use practices are not responsible. Foxes may have been important in the southern parts but in the Tanami Desert region they arrived after the effect and cannot be implicated. Rabbits might likewise have been significant in the southern region but to the north their effect must be regarded as marginal only.

In the pastoral lands of the Northern Territory it is likely that the grazing by cattle exacerbated the situation for native mammals.

ACKNOWLEDGEMENTS

We would like to thank the Aborigines from communities at Kaltukajara (Docker River), Mutijtula (Ayers Rock), Ntaria (Hermannsburg), Yuendumu, Willowra, Lajamanu, Papunya, Walangkura (Kintore), Nyirrpi, Utopia, Neutral Junction, Stirling, Murray Downs, Tea Tree,

Wave Hill, Ali Curung, Tennant Creek and Elliott for their enthusiastic contribution of information used in this paper.

We are grateful to Mary Laughren and David Nash for their readiness to assist in interpretation, and to Dick Kimber for his encouragement in tapping the considerable knowledge of Aborigines.

REFERENCES

AITKEN, P. F., 1983. Pig-footed Bandicoot *Chaeropus ecaudatus*. Pp. 104–05 in *The Australian Museum complete book of Australian mammals* ed by R. Strahan. Angus & Robertson: Sydney.

BOLTON, B. L. AND LATZ, P. K., 1978. The Western Hare-wallaby *Lagorchestes hirsutus* (Gould) (Macropodidae) in the Tanami Desert. *Australian Wildlife Research* **5:** 285–93.

BURBIDGE, A. A., JOHNSON, K. A., FULLER, P. J. AND SOUTHGATE, R. I., 1988. Aboriginal knowledge of the mammals of the central deserts of Australia. *Australian Wildlife Research* **15:** 9–39.

CALABY, J. H., 1976. VI Mammals. Pp. 1–17 in *A survey of the fauna of the lower McArthur River region, Northern Territory* ed by J. H. Calaby. CSIRO Division of Wildlife Research (unpublished).

CALABY, J. H. AND KEITH, K., 1974. Mammals. Pp. 1–208 in *Fauna survey of the Port Essington District, Cobourg Peninsula, Northern Territory of Australia* ed by H. J. Frith and J. H. Calaby. *CSIRO Division of Wildlife Research Technical Paper* No. **28**.

CARNEGIE, D. W., 1898. *Spinifex and sand*. Reprint (1982) Hesperian Press: Carlisle, Western Australia.

COLLETT, R., 1897. On a collection of mammals from North and North-west Australia. *Proceedings of the Royal Zoological Society of London* **1897:** 317–36.

DAHL, K., 1897. Biological notes on North Australian mammalia. *The Zoologist* 4th series, Volume 1. May 189–216.

DIXON, J. AND HUXLEY, L., 1985. *Donald Thomson's mammals and fishes of northern Australia*. Nelson: Melbourne.

FINLAYSON, H. H., 1935. On mammals from the Lake Eyre basin. Part III — The Peramelidae. *Transactions of the Royal Society of South Australia* **59:** 227–36.

FINLAYSON, H. H., 1943. *The Red Centre: man and beast in the heart of Australia*. Angus & Robertson: Sydney.

FINLAYSON, H. H., 1958. On central Australian mammals (with notice of related species from adjacent tracts). Part III. The Potoroinae. *Records of the South Australian Museum* **13:** 235–302.

FINLAYSON, H. H., 1961. On Central Australian Mammals. Part IV. The distribution and status of central Australian species. *Records of the South Australian Museum* **14:** 141–91.

GIBSON, D. F., 1986. A biological survey of the Tanami Desert in the Northern Territory. *Conservation Commission of the Northern Territory Technical Report* No. **30**.

GORDON, G., 1983. Desert Bandicoot *Perameles eremiana*. P. 102 in *The Australian Museum complete book of Australian mammals* ed by R. Strahan. Angus & Robertson: Sydney.

JOHNSON, D. H., 1964. Mammals of the Arnhem Land expedition. Pp. 427–513 in *Records of the American–Australian Scientific Expedition to Arnhem Land*. Zoology ed by R. L. Specht. Melbourne University Press: Melbourne.

JOHNSON, K., 1983. Lesser Bilby *Macrotis leucura*. P. 109 in *The Australian Museum complete book of Australian mammals* ed by R. Strahan. Angus & Robertson: Sydney.

LATZ, P. K. AND GRIFFIN, G. F., 1978. Changes in Aboriginal land management in relation to fire and to food plants in central Australia. Pp. 77–85 in *The nutrition of Aborigines in relation to the ecosystem of Central Australia* ed by B. S. Hotzel and H. J. Frith. CSIRO: Melbourne.

LOCKWOOD, D., 1974. *The front door. Darwin 1869–1969*. Rigby: Adelaide.

MCKEAN, J. AND HOWE, D., 1972. *Report on the mammals of Maria Island* (unpublished). Conservation Commission of the Northern Territory.

MCKENZIE, N. L., 1983. Golden Bandicoot *Isoodon auratus*. P. 98 in *The Australian Museum complete book of Australian mammals* ed by R. Strahan. Angus & Robertson: Sydney.

PARKER, S. A., 1973. An annotated checklist of the native land mammals of the Northern Territory. *Records of the South Australian Museum* **16**(11)**:** 1–57.

ROLLS, E., 1969. They all ran wild: the story of pests on the land in Australia. Angus & Robertson: Sydney.

SPENCER, W. B. (ed), 1896. *Report on the work of the Horn Scientific Expedition to Central Australia* (4 volumes). Melville, Mullin and Slade: Melbourne.

SPENCER, W. B., 1897. Description of two new species of marsupials from Central Australia. *Proceedings of the Royal Society of Victoria* **9:** 5–11.

SPENCER, W. B. AND GILLEN, F. J., 1912. *Across Australia* (2 volumes). MacMillan: London.

STRONG, B. W., 1983. The invasion of the Northern Territory by the wild European Rabbit *Oryctolagus cuniculus*. *Conservation Commission of the Northern Territory Technical Report* No. **3**.

SWEENEY, G., 1947. Food supplies of a desert tribe. *Oceania* **17**(4)**:** 289–99.

TATE, G. H. H., 1948. Results of the Archbold Expeditions. No. 60. Studies in the Peramelidae (Marsupialia). *Bulletin of the American Museum of Natural History* **92**(6)**:** 314–46.

WATTS, C. H. S., 1969. Distribution and habits of the rabbit bandicoot. *Transactions of the Royal Society of South Australia* **93:** 135–41.

Status of New Guinea bandicoots

GRAEME G. GEORGE[1] and GERALD M. MAYNES[2]

The New Guinea sub-region has five genera and 11 species of bandicoot. Three genera and nine species are endemic, and two genera and two species are shared with Australia. The majority of species are confined to New Guinea itself.

Peroryctes raffrayanus, P. longicauda, Echymipera rufescens and *E. kalubu* are widespread, common, and widely hunted for food. *Peroryctes papuensis* and *P. broadbenti* are restricted to the south-east, *Isoodon macrourus* to the southern grasslands and the rare *Echymipera clara* to primary rainforests of the north-central lowlands and Japen Island. *Microperyctes murina* is a small cryptic species of the north-west, rarely encountered. *Rhynchomeles prattorum* is known only from the type specimens collected on Seram Island. A fourth species of *Echymipera* has recently been identified from Kiriwina Island in the south-east.

Peroryctes has a wide altitudinal range over the New Guinea mainland whereas *Echymipera* is restricted to lowlands. *Echymipera rufescens* extends south from New Guinea to the Kei and Aru islands, the D'Entrecasteaux group and Cape York. *Echymipera kalubu* extends north to Waigeo Island and the Bismarck Archipelago.

Key words: New Guinea, bandicoots, distribution, status.

INTRODUCTION

THE New Guinea sub-region of the Australasian Faunal Region includes the main island of New Guinea and the surrounding island groups of the Moluccas, Lesser Sunda Islands, Admiralty Islands and Bismarck Archipelago. The western half of the island of New Guinea is included in the Indonesian Province of Irian Jaya. Papua New Guinea comprises the eastern half of New Guinea along with island groups to the north-east and east. The faunal affinities of these islands to the north of the Australian continent, are in general, with Australia, due to these land masses being part of the Australian tectonic plate. Many of the island chains and mountain ranges were formed as a result of tectonic forces associated with contact between the Australian and other plates during the Miocene and Pliocene (Doutch 1972).

New Guinea itself is a composite of land which was once the northern edge of the Australian continent (the present central cordillera of the island) and land which began as archipelagoes during the Pliocene (now forming mountain ranges along the north coast) (Doutch 1972; Archer 1984). New Guinea is rich in marsupials, whilst other islands in the sub-region carry a depauperate marsupial fauna which reflects limited past opportunities for dispersal from the north Australian/New Guinea land mass during and since the Miocene (George 1987). Bandicoots have dispersed further from New Guinea than any other marsupial group, with the exception of the cuscuses, having reached Seram Island in the Moluccas and the Bismarck Archipelago.

The New Guinea sub-region has five genera and at least 11 species of bandicoots. Two of these genera (*Echymipera* and *Isoodon*) are shared with Australia, two (*Peroryctes* and *Microperyctes*) and endemic to New Guinea and one (*Rhynchomeles*) is endemic to Seram Island. *Echymipera* is a New Guinea centred genus with four species, one of which extends to Australia. The other shared genus, *Isoodon,* is Australia centred, with one species extending to New Guinea.

Several species are common in New Guinea and the family forms a more significant part of the terrestrial mammal fauna in the New Guinea rainforests than it does on the Australian continent. In this context, the absence in New Guinea of Potoroids such as *Hypsiprymnodon* may be of some significance.

Bandicoots are widely hunted as game animals throughout New Guinea where they are known collectively by the Melanesian Pidgin term "Mumut".

The only thorough review of the family in recent years is that of Tate (1948). He recognized eight species in five genera in the New Guinea area. Since that review a new species of *Peroryctes (P. papuensis)* has been described (Laurie 1952), *P. broadbenti* has been reinstated as a full species (Van Deusen 1972) and a fourth species of *Echymipera* has been distinguished (Flannery, this volume).

This review of the status of bandicoots in the New Guinea sub-region is based on literature records, the personal observations and records

[1]Zoological Board of Victoria, P.O. Box 74, Parkville, Victoria, Australia 3052.
Present address: AZDANZ Species Management Program, c/- 156 Union Road, Surrey Hills, Victoria, Australia 3127.
[2]Division of Wildlife, Department of Lands, Survey and Environment, P.O. Box 6601, Boroka, Papua New Guinea.
Present address: Australian National Parks and Wildlife Service G.P.O. Box 636, Canberra 2601.
Pages 93–105 *in* BANDICOOTS AND BILBIES, ed by J. H. Seebeck, P. R. Brown, R. L. Wallis and C. M. Kemper. Surrey Beatty & Sons, Sydney, 1990.

of George during 13 years residence in the Central Highlands of Papua New Guinea, the results of field surveys carried out by the Division of Wildlife in Papua New Guinea, and unpublished observations of other field workers.

SPECIES ACCOUNTS

New Guinea Long-nosed Bandicoots — Genus Peroryctes

The genus *Peroryctes,* usually referred to as the New Guinea Long-nosed Bandicoots, may be more appropriately known as the New Guinea mountain bandicoots as the genus is centred in the New Guinea mountains and most genera of bandicoot have an elongated rostrum. The species referred to the genus *Peroryctes* are mostly mountain dwelling, and are distinguished from the other common genus, *Echymipera,* of the New Guinea lowlands, by their coarse but not spiny pelage, relatively longer tails and limbs, the retention of I^5 and a premolariform canine. As with *Echymipera* the rostrum in *Peroryctes* is elongate and the ears short.

Broadbent or Giant Bandicoot —
Peroryctes broadbenti *(Ramsay)*

Peroryctes broadbenti is the largest species of bandicoot known. Thomas (1888) doubted its specific distinction from *P. raffrayanus* and later authors (Tate and Archbold 1937; Tate 1948; Laurie and Hill 1954) treated it as a subspecies of *P. raffrayanus.* It is now generally accepted as a full species on the basis of morphological distinctness and sympatry (Van Deusen 1972; Ziegler 1977) although Ziegler (1982) qualifies this recognition.

Description: *Peroryctes raffrayanus* and *P. broadbenti* together constitute the sub-genus *Peroryctes* in the sense of Tate and Archbold (1937). Both have an unpatterned pelage in contrast to the sub-genus *Ornoryctes* in which the pelage is variably ornamented.

Peroryctes broadbenti attains a greater maximum size than any other known bandicoot. One specimen, collected on the Kwagira River by the Fourth Archbold Expedition in 1953 measured 790 mm from nose to tail tip and weighed 4.8 kg (Brass 1956; Van Deusen 1972). Another specimen in the collection of the University of Papua New Guinea weighed 4.9 kg (J. Menzies, pers. comm.). The pelage is a dark red-brown grading into a light rufous-buff ventrally (Ziegler 1977). Thomas (1888) commented that the tail, covered with scales above and flat scaly tubercles below, appeared to be prehensile. However, there is no other evidence that this might be so.

Distribution (Fig. 1): Literature records of this species from a number of localities in the Owen Stanley Ranges of south-east Papua New Guinea

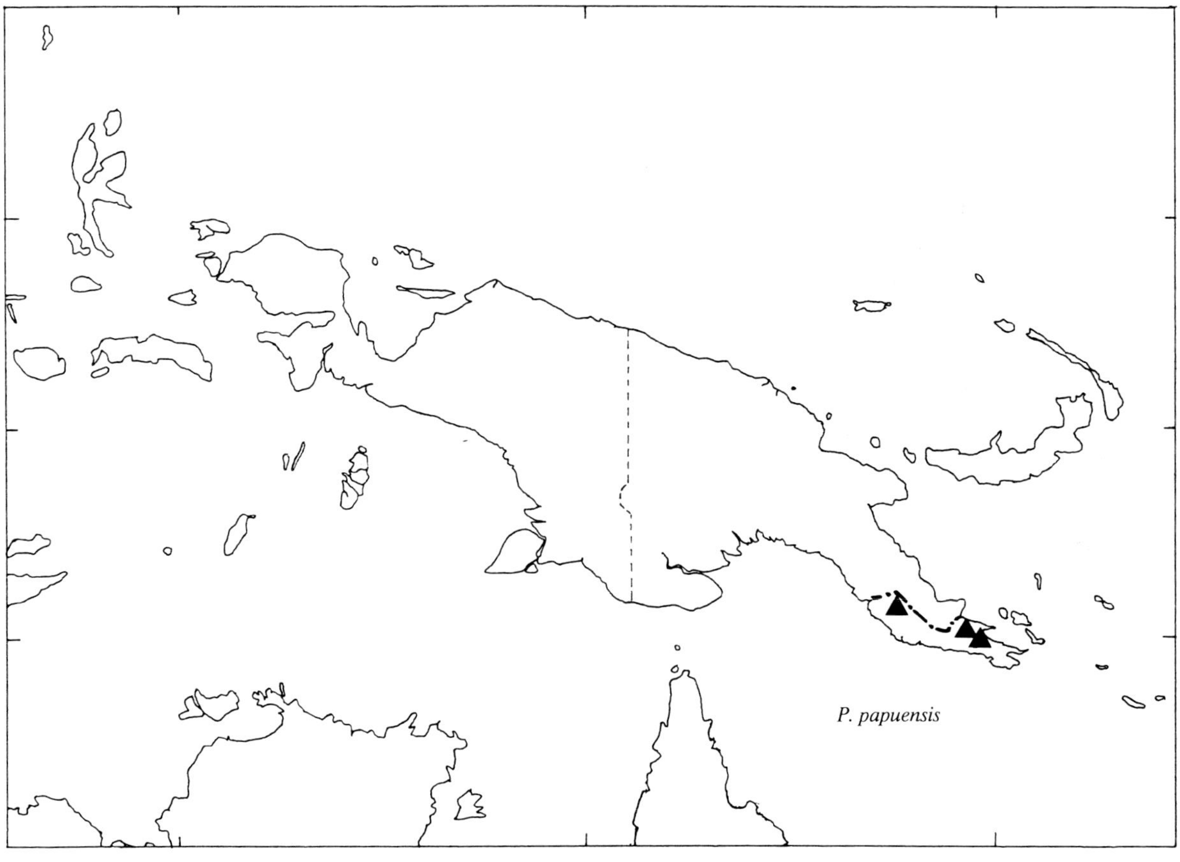

Fig. 1. Known localities (▲) and presumed range (— · —) of *Peroryctes broadbenti.*

suggest a wide altitudinal range from sea level to 2 860 m (Tate 1948; Laurie and Hill 1954). However, until recently it was customary to refer all large *Peroryctes* from south-east New Guinea to the taxon *P. raffrayanus broadbenti* before it became evident that the giant form was specifically distinct and sympatric with the more widespread form. Specimens referred to *P. r. broadbenti* by Tate and Archbold (1937) and Tate (1948) may therefore include examples of *P. r. raffrayanus*. The published measurements of the specimens collected at Mafulu, 1 250 m and at Murray Pass, 2 860 m in the Owen Stanley Ranges (Tate and Archbold 1937) do, in fact, fall within the size range of *P. r. raffrayanus*. Specimens subsequently collected by Tate at Kagi and Sogeri in the foothills of the Owen Stanleys were compared by him with *P. raffrayanus* from the Idenburg River in Irian Jaya and only minor differences detected (Tate 1948). It seems likely that Tate did not have true *P. broadbenti* for comparison.

The holotype of *P. broadbenti* came from dense mountain scrub on the Goldie River, southern slopes of the Owen Stanleys. The Kwagira River specimen was collected at 50 m at the foot of the northern slopes inland from Collingwood Bay. The habitat at this locality was predominantly tall primary lowland rainforest (Brass 1956). A third specimen, obtained at Brown River, a lowland rainforest locality inland from Port Moresby, is in the Papua New Guinea National Museum and Art Gallery. These low altitude records suggest that *P. broadbenti* is a species of the lowland rainforest, separated altitudinally from *P. raffrayanus* of the middle altitude forests. Ziegler (1982) suggests that *P. broadbenti* is the ecological replacement of *P. raffrayanus* at lower altitudes in the south-east. Because collectors in New Guinea usually rely on local hunters bringing specimens in to base camps, the actual site of capture is not always precisely known. *Peroryctes broadbenti* may extend into or be more typical of the lowland hill forests.

Status: The species is poorly known, presumably quite rare and with an apparently restricted range. Its large size renders it vulnerable to hunting pressure, to which it may already have succumbed, although extensive tracts of suitable habitat and a sparse human population through its presumed range offer some hope that this is not so (J. Menzies, pers. comm.).

The probability that the species is much rarer than published records suggest was not appreciated when George prepared a survey of threatened New Guinea mammals in 1978 (George 1979).

The southeastern records suggest an origin for the species through differentiation in the south-east mountains in a manner similar to that postulated for the bowerbird *Amblyornis subalaris* and the bird of paradise *Parotia helenae* (Schodde and McKean 1973a, 1973b). The rare wallaby *Dorcopsis (Dorcopsulus) macleayi* has a similar distribution pattern to *A. subularis* (George 1979) and a search for this wallaby and the rare *P. broadbenti* in primary forests of the southeastern foothills is warranted.

Raffray's Bandicoot —
Peroryctes raffrayanus *(Milne-Edwards)*

Also known as the Long-legged Bandicoot. *Peroryctes raffrayanus* is common and widespread through the New Guinea mountains.

Description: *Peroryctes raffrayanus* is a medium to large species up to 600 mm in total length, similar to but smaller than *P. broadbenti*. The pelage is a dark grey brown with white on the chin and chest (Thomas 1888). The juvenile is a rich reddish-tan which gradually changes to the adult colouration as it matures (Tate 1948; Lindgren 1975). The Huon Peninsula race, *P. r. rothschildi*, is more rufescent than the nominotypical form (Tate 1948). No subspecific variation has been described along the central cordillera.

Distribution (Fig. 2): Found throughout the central cordillera of New Guinea from the mountains of the eastern Vogelkop to the southeast mountains of Papua; also in the north-coastal ranges and on the Huon Peninsula.

Laurie and Hill (1954) give the altitudinal range along the central cordillera as sea level to 2 700 m, however records known to us from Papua New Guinea are all above 500 m and the species appears to be commonest in mid-mountain forests between 1 200 and 2 000 m. Along the central cordillera *P. raffrayanus* is less common than *P. longicauda* at the higher altitudes suggesting competitive replacement. In the absence of *P. longicauda* on the Huon Peninsula, the race *P. r. rothschildi* occurs up to 3 900 m (Ziegler 1977, 1982).

The Sixth Archbold Expedition obtained this species at a collecting camp at 500 m on the Umi River, Markham Valley, at the foot of the Finisterre Ranges (Brass 1964). The only other low altitude record we know of for eastern New Guinea is a specimen in the British Museum from the Goldie River (type locality for *P. broadbenti*) collected by A. Goldie. In both cases the actual altitude at which the animals were taken could be higher than the camp locality.

The Third Archbold Expedition, working in the Idenburg Valley of the Snow Mountains of Irian Jaya in 1939 collected a good series at a succession of collecting stations between 50 m and 2 150 m (Tate 1948). Descriptions of the plant communities at each site given by Archbold,

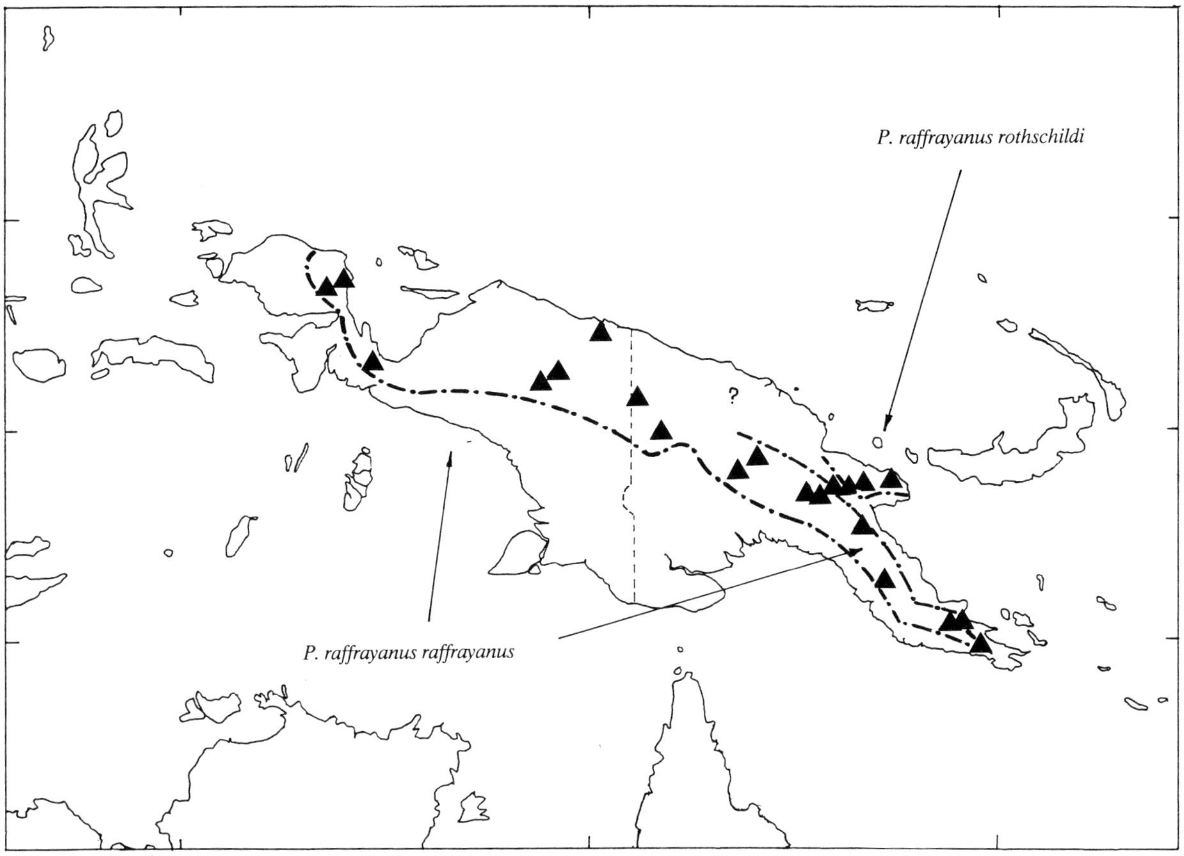

Fig. 2. Known localities and presumed range of *Peroryctes raffrayanus*.

Rand and Brass (1942) indicate varied habitats occurring at lower than usual altitudes compared to other parts of New Guinea, because of local topographic and climatic factors. Bearing in mind that specimens would have been collected at various altitudes within range of each camp, a correlation between abundance of *P. raffrayanus* and the presence of lowland hill forest and mid-mountain oak forests, as follows, seems likely:

(1) Bernhard Camp, 50 m, lowland rainforest, with a subsidiary camp at the base of the nearby mountains. One specimen.

(2) Araucaria Creek, 850 m, lowland hill forests with some oaks and emergent *Agathis*, well mossed; beech *(Nothofagus)* forest on the ridges 50–100 m above the camp. 18 specimens.

(3) 1 200 m Camp. Mixed forest on mountain slopes with some patches of mid-mountain oak *(Castanopsis, Lithocarpus)* forest on narrow ridges down to 1 100 m and generally above 1 500 m. Nine specimens.

(4) 1 800 m Camp. *Nothofagus* forest on ridges, mixed forest on lower slopes. One specimen.

(5) 2 150 m Camp. *Nothofagus* and sub-alpine shrubs with mixed forest in the valleys. Two specimens.

Based on his experience on several post-war Archbold Expeditions, Van Deusen (1972) regarded *P. raffrayanus* as a species of the mid-mountain oak forests. The species has one of the widest altitudinal ranges of any mammal in Papua New Guinea and cannot be considered a species "of the foothills and lowlands" as Tate (1948) regarded it.

Status: Widespread and apparently common in primary forest but less common in areas subjected to hunting pressure.

The Third Archbold Expedition did not obtain *P. raffrayanus* at collecting stations in the Balim or Bele valleys or from the slopes of Mount Wilhelmina, despite good results from the Idenburg River sites, and the collection of good series of the smaller *P. longicauda* from the Bele Valley and slopes of Mount Wilhelmina (Tate 1948). The Balim and Bele River valleys are densely populated and only remnants of the original mid-mountain oak forests remain due to the gardening activities of the human population up to the frost-line (Archbold, Rand and Brass 1942). Populations of *P. raffrayanus* which may have occurred in the original oak forests and in the *Nothofagus* forests of the mountain slopes may have been reduced by hunting pressure as well as loss of habitat.

Bulmer and Menzies (1973) did not obtain any direct evidence of this species in the upper Kaironk Valley of the Schrader Ranges (1 500–2 600 m) of Papua New Guinea but the animal known locally as "Pakam" was presumed to be *P. raffrayanus*. This was said to be "found both in local mountain forests (though rarely) and in more open country at lower altitudes".

George obtained trophy material of this bandicoot taken from oak forest at 2 100 m in the upper Mendi Valley of the Southern Highlands Province of Papua New Guinea in 1963. A large collection of mammal trophy material purchased at the Government Rest House at Winja in the Lai River valley further south in 1964, consisted of skulls and mandibles of large rodents and medium to large-sized marsupials. It contained a minimum of 365 individuals, hunted from surrounding oak and *Nothofagus* forests above 2 000 m. This collection contained 21 *P. raffrayanus* (5.7%) and 39 *P. longicauda* (10.7%).

Striped Bandicoot —
Peroryctes longicauda *(Peters and Doria)*

Also known as the Long-tailed Bandicoot.

Description: This medium-sized species is smaller than *P. raffrayanus*, measuring up to 500 mm, but with a proportionately longer tail. On the basis of a number of cranial differences, Tate and Archbold (1937) created a sub-genus *Ornoryctes* with *P. longicauda* as type species, to emphasize the difference between it and *P. raffrayanus*.

Four subspecies of *P. longicauda* are recognized. These form a stepped cline from unornamented animals in the north-west to strikingly marked animals in the south-east.

Peroryctes longicauda — Vogelkop Peninsula and Weyland Mountains. Pelage dull grey-brown dorsally, yellowish to whitish ventrally, usually without any dorsal ornamentation. Tate (1948) mentions two specimens collected by F. Shaw Mayer in the Gebroeders, Weyland Ranges, which show traces of a dorsal stripe.

Peroryctes dorsalis — Central highlands of Irian Jaya. A well-marked median dorsal stripe almost always present. A few individuals also show a trace of lateral rump stripes (Tate 1948).

Peroryctes ornatus — Central highlands of Papua New Guinea and the Owen Stanley Ranges. In addition to the black mid-dorsal stripe, this race shows a black streak through the eye and a pair of lateral rump stripes.

Peroryctes magnus — Mountains of south-eastern Papua (the south-east extension of the main Owen Stanley Range separated by the low altitude Keveri Hills). Distinguished by Laurie (1952) on the basis of larger size, longer tail and darker venter than *P. l. ornatus* with smaller, sometimes minute, anterior palatal foramina.

Distribution (Fig. 3): Mountains of the central cordillera of New Guinea, from the Vogelkop Peninsula in the north-west, to the mountains of south-east Papua. It may be found as low as 1 000 m but is most common above 1 500 m and occurs up to the alpine grasslands at 4 000 m. It is usually regarded as a forest species (e.g., Ziegler 1977). Bulmer and Menzies (1973) record it as "not uncommon at all altitudes in forest and bush fallow" in the upper Kaironk Valley of the Schrader Ranges. Dwyer (1984) records the trapping of one in primary forest at 1 450 m on the southern slopes of Mount Haliago (=Mt Sisa) in the Southern Highlands of Papua New Guinea.

At Wapenamanda in the Western Highlands of Papua New Guinea, George was told that the species favoured tall cane-grass (*Saccharum* sp.) thickets.

Status: This species appears to be more common than *P. raffrayanus* along the central cordillera, particularly at higher altitudes. In the collection of trophy material purchased by George in the Upper Lai Valley of the Southern Highlands, *P. longicauda* was twice as common as *P. raffrayanus*. The Third Archbold Expedition to New Guinea, working in the Irian Jaya highlands, collected a good series at eight collecting stations between the Idenberg River at 1 200 m and Mr Wilhelmina at 3 950 m. Fifty specimens were obtained in the Bele Valley at 2 200 m (Tate 1948). Bulmer and Menzies (1973) obtained 10 specimens of this species in the Upper Kaironk Valley, but did not see any examples of *P. raffrayanus* though local descriptions indicated that it was also present in the area. No information is available on the status of *P. l. longicauda* in north-west Irian Jaya.

Papuan Bandicoot —
Peroryctes papuensis *Laurie*

This diminutive species is one of New Guinea's least known mammals.

Description: *Peroryctes papuensis* is a small replica of *P. longicauda ornatus* differing chiefly in size and recognized as a full species because of sympatry with *P. l. magnus* in the southeastern mountains of Papua. It has similar markings to *P. l. ornatus* and *P. l. magnus* but is generally darker. Adults show a rich orange-buff venter which is light grey in juveniles (Laurie 1952).

Distribution (Fig. 4): Limited to mid-mountain and lower montane forests (1 200 to 2 000 m) of southeastern New Guinea.

The species was described (Laurie 1952) from specimens collected on Mount Mura, near Mount Simpson, and at Enaena in the southeastern

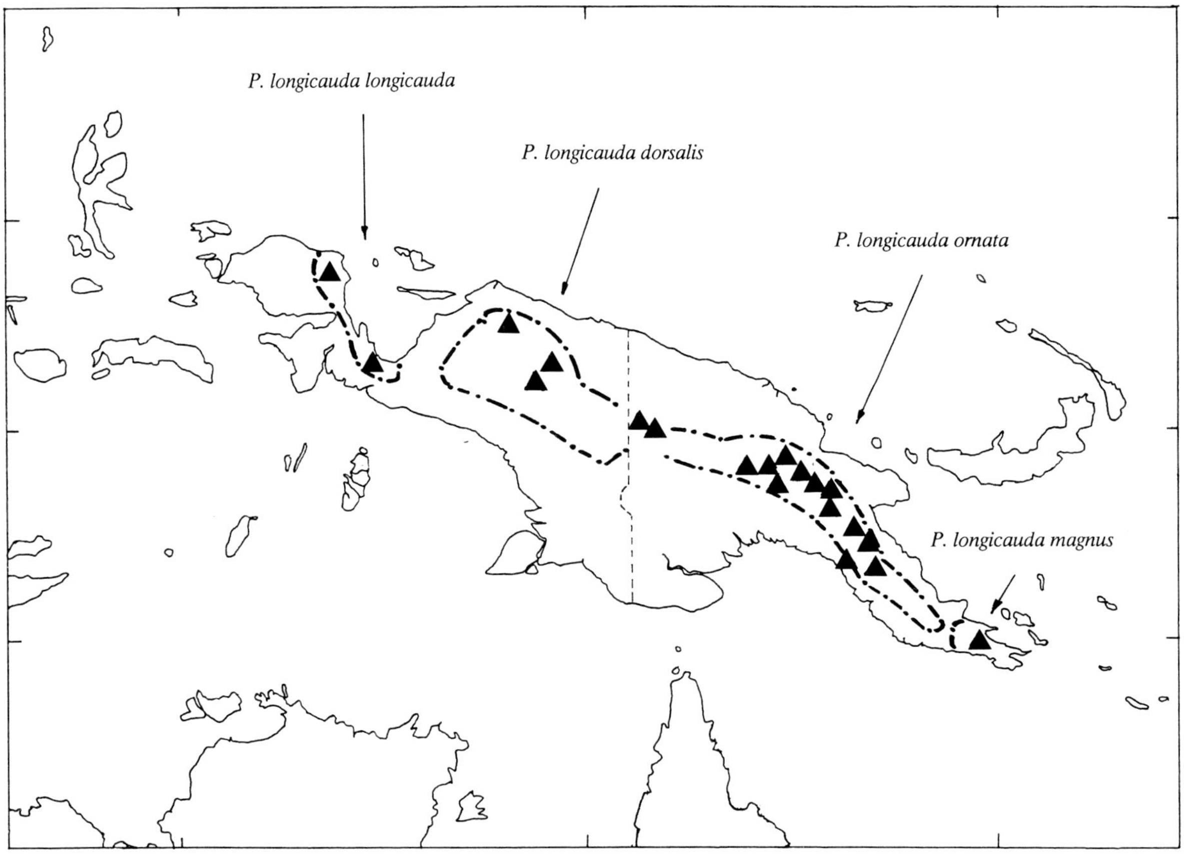

Fig. 3. Known localities and presumed range of *Peroryctes longicauda*.

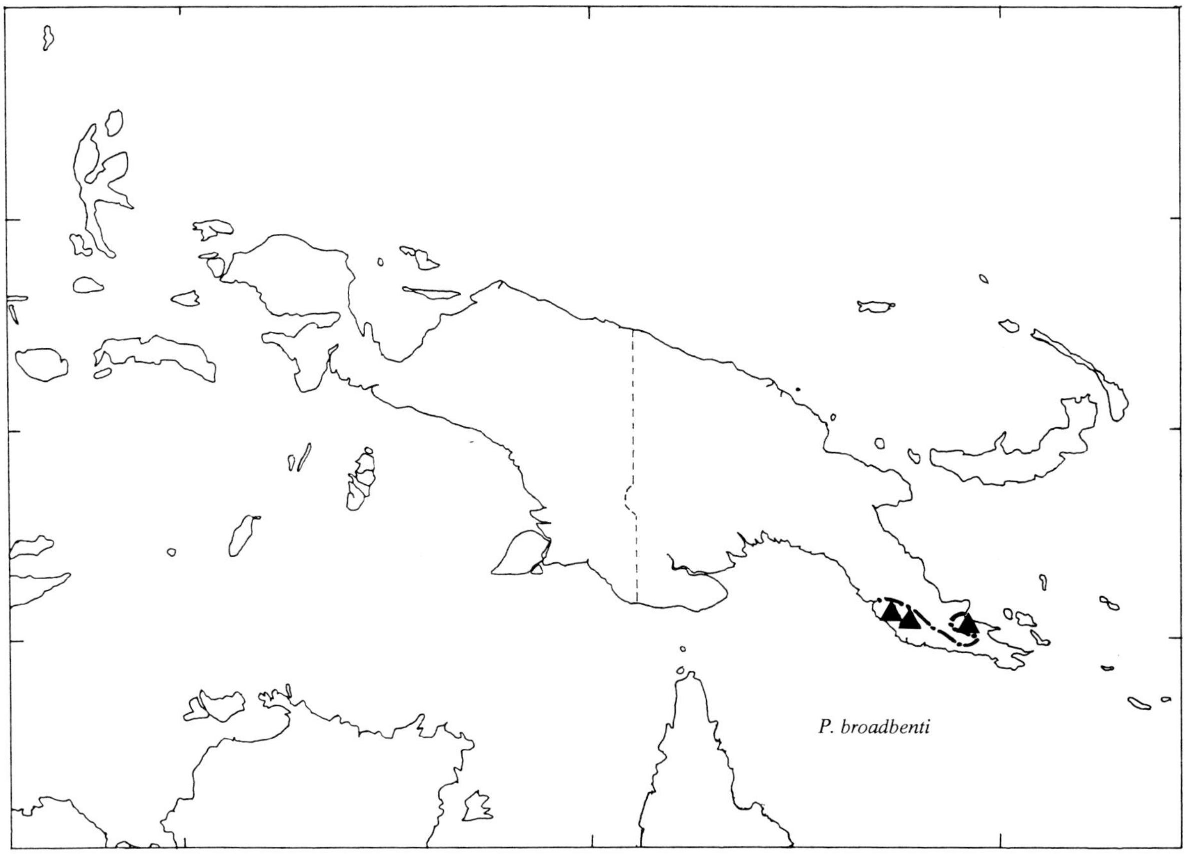

Fig. 4. Known localities and presumed range of *Peroryctes papuensis*.

mountains by F. W. Shaw Mayer. Shaw Mayer found the new species between 1 200 and 1 500 m and collected *P. l. magnus* at altitudes below 1 200 m in the same general area. Jim Menzies (pers. comm.) caught one specimen each of *P. papuensis* and *P. longicauda* in the same trapline on Mount Dayman, in forest at 2 000 m. Juvenile specimens collected at Kagi (1 200 m) in the southern foothills of the Owen Stanley Ranges, recorded as *P. longicauda* by Tate (1948), have also been identified as *P. papuensis* by Ken Aplin (pers. comm.), thus extending the previously known range considerably further west.

Status: Very little is known about this small ornate species. It must be considered rare due to a limited distribution and in the absence of any evidence to the contrary.

Mouse Bandicoot — Microperoryctes murina *Stein*

This cryptic animal is one of New Guinea's rarest and least-known species.

Description: Very small with a head and body length of only 150 mm. The pelage is a dark smoky grey and soft, compared by Tate and Archbold (1937) with that of moles and shrews. They also described the claws as being "highly fossorial". The skull is similar to that of *Peroryctes (Ornoryctes)*, generic distinction being based on the very small size and quality of the pelage (Tate 1948).

Distribution (Fig. 5): Known only from the three specimens (one male, two females) collected by Georg Stein at Sumuriberg in the Weyland Range of northwestern New Guinea at 2 500 m in 1931.

Another small specimen of bandicoot, tentatively referred to this species, has been collected in the eastern Vogelkop (Cenderawasih Peninsula) of Irian Jaya. This differs in having paler fur, not especially soft, with a dark grey mid-dorsal stripe and white tail tip (Ziegler 1977).

Status: Very rare, at least in collections, and poorly known.

Genus Echymipera — *Spiny Bandicoots*

The Spiny Bandicoots are characterized by the spinous flattened guard hairs of the pelage, extremely elongated rostrum, short ears, short tail, short limbs, and loss of I^5. Three species are reviewed below. The fourth species is discussed by Flannery (this volume).

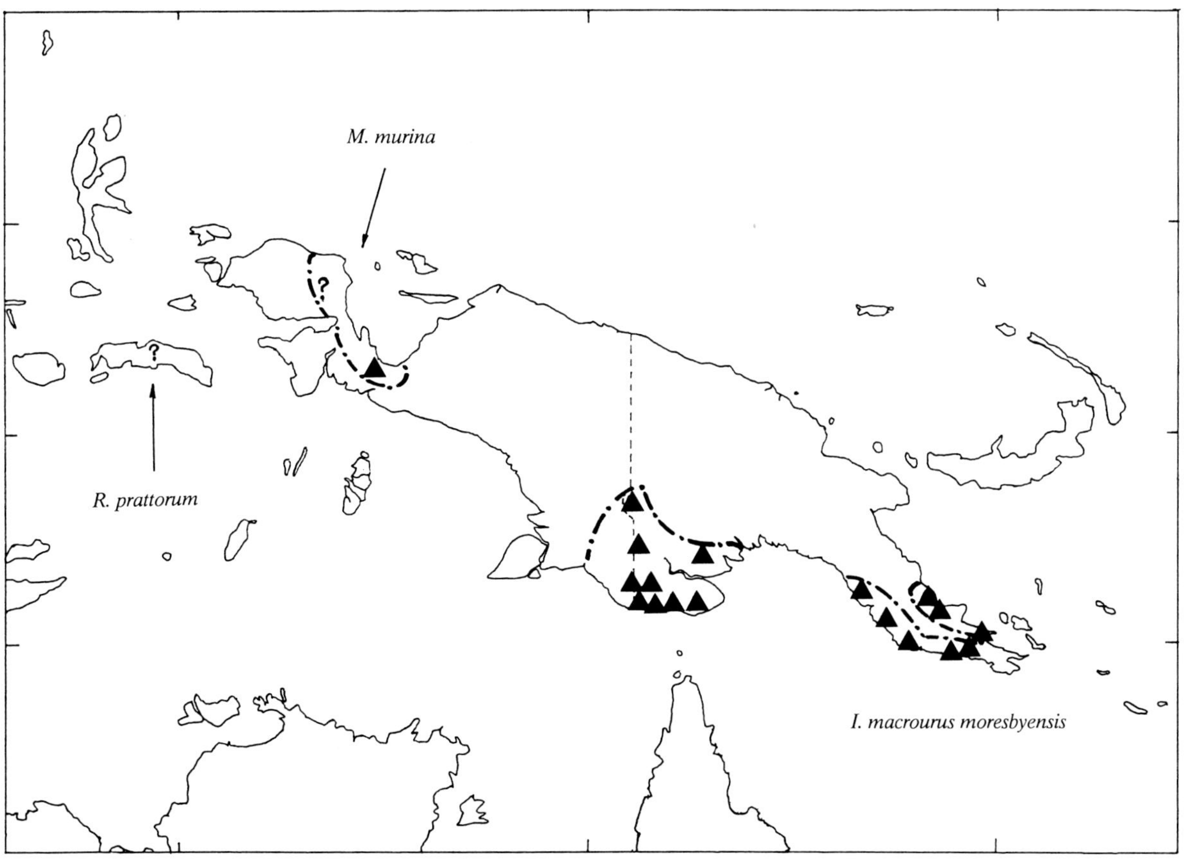

Fig. 5. Known localities and presumed range of *Microperoryctes murina, Rhynchomeles prattorum* and *Isoodon macrourus moresbyensis.*

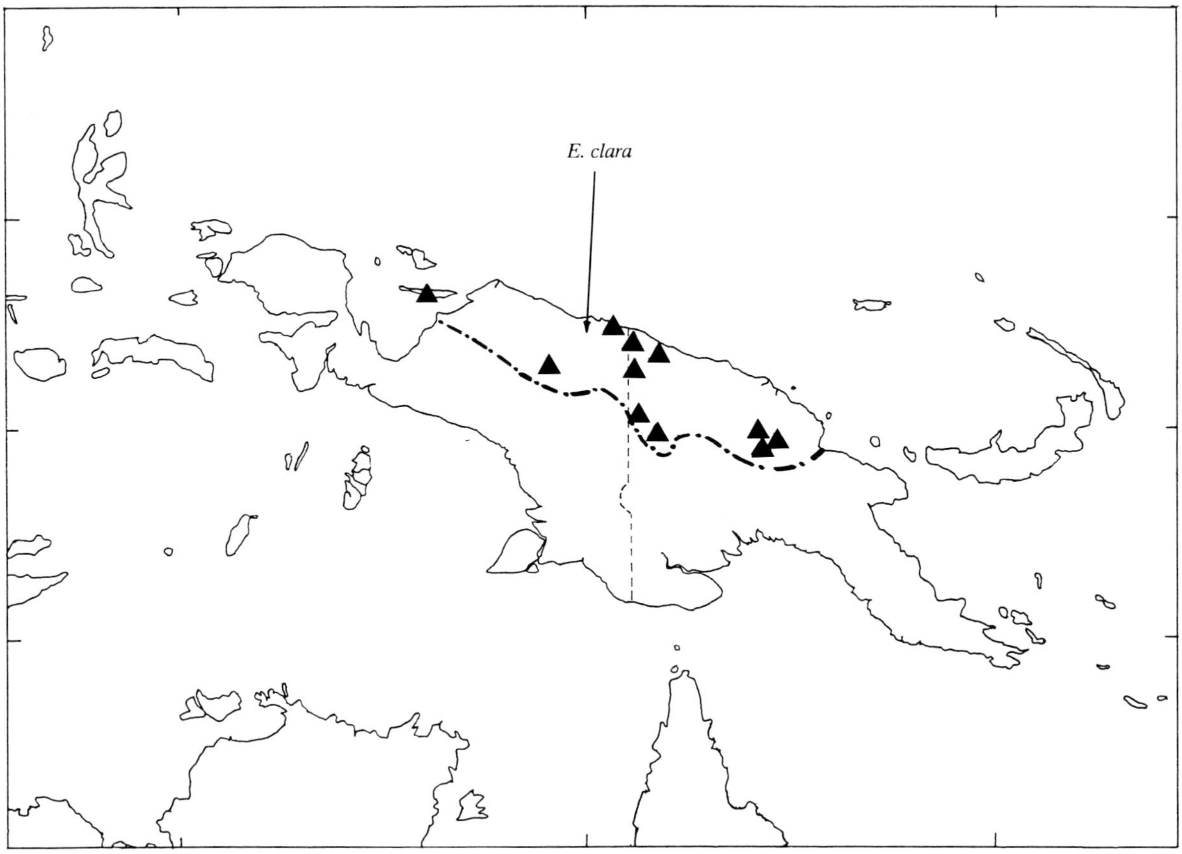

Fig. 6. Known localities and presumed range of *Echymipera clara.*

Clara or White-lipped Bandicoot —
Echymipera clara *Stein*

Echymipera clara is another of New Guinea's rarer species with a limited distribution in the northern lowlands.

Description: *Echymipera clara* is externally similar to its congenors. Tate (1948) mentioned white upper and lower lips (both hair and skin) but specimens examined by George in 1974 in the American Museum of Natural History collections differed from other *Echymipera* spp. in degree only. Stein (1932) and Tate (1948) both comment on a specialized pad on the hindfoot of *E. clara* which is not seen in *E. kalubu* and is less well developed in *E. rufescens.* The skull and mandibles are very distinctive. The canines and P^3_3 are large and heavy, the dentary is correspondingly much more robust than in the other two species and the zygomata are wider.

Distribution (Fig. 6): *Echymipera clara* is limited to altitudes below 1 200 m in the lowlands of northern New Guinea, from Japen Island in Geelvink Bay, where it was first collected by Georg Stein in 1931, through the Mamberano-Idenburg River basin in Irian Jaya to the Sepik-Ramu basin in Papua New Guinea (Van Deusen and Keith 1966). It appears to be restricted to, or dependent on, primary rainforest and may not occur above 1 000 m. Specimens collected by a Wildlife Branch survey team in the Torricelli Mountains of the Sepik area, on slopes below 1 000 m at the foot of Mount Somoro, included one taken in a garden adjoining forest. At Yapsiei, in the upper Sepik basin, Tim Flannery (pers. comm.) collected several in primary foothill forests between 200 and 400 m.

Status: Generally rare with a restricted range, but may be locally more common. The low frequency of collection of this species suggests that it should be classed as rare. George collected quantities of trophy mandibles of bandicoots and other mammals in the Jimi River valley (tributary of the Sepik River) in the Western Highlands in 1966. The material included six individuals of *E. clara* among several dozen of each of the other two species common in the area — *E. kalubu* and *E. rufescens.* The material was purchased from villagers around the Tabibuga Patrol Post, which is surrounded by steeply forested slopes and ridges with much secondary forest. Primary rainforest below 1 000 m occurred some distance north-west of the station. The rarity of the species in this trophy material may be due to the distance of *E. clara* habitat from inhabited areas. Flannery (pers. comm.) found *E. clara* to be not uncommon in the sparsely populated foothill forests at Yapsiei.

Kalubu Bandicoot —
Echymipera kalubu *(Lesson)*

Also known as the Black-headed Bandicoot. This common and widespread spiny bandicoot is sympatric with *E. rufescens* throughout much of the New Guinea lowlands. Tate (1948) and previous authors treated this species under the name *E. doreyana*. Laurie and Hill (1954) listed *E. doreyana* as a junior synonym of *E. kalubu*.

Description: Similar to *E. rufescens* externally, though not reaching the large size of old males of that species. A series of 41 animals trapped at Baiyer River in the Western Highlands of Papua New Guinea weighed up to 1 800 g, with males averaging 1 173 g and females 685 g (Anderson, Berry, Amos and Cook 1988).

The crown is blackish and sharply demarcated from the paler throat and cheeks. Specimens from the Sepik River plains collected by Kent Keith for the CSIRO Division of Wildlife Research were unusually melanistic, being blackish over the whole of the dorsum (Keith, pers. comm.).

The skull of *E. kalubu* differs from that of *E. rufescens* chiefly in the presence of hypocones on the molars giving these teeth a trapezoidal rather than a sub-triangular outline. Tate (1948) also noted smaller I^1 and I^4 space between first upper incisors, broader P^3, shorter incisive foramina, as well as three pairs of teats in females (four pairs in *E. rufescens*).

Specimens from the Fly River basin of southern New Guinea, which were distinctively smaller with smaller teeth, were described by Tate and Archbold (1936) as *Echymipera oriomo*. As this form shares other characteristics with *E. kalubu* and is allopatric with that species, Tate (1948) relegated the form to subspecific level (as *E. doreyana oriomo*).

Distribution (Fig. 7): The large-toothed race *E. k. kalubu* occurs over the whole of the northern lowlands of New Guinea up to 1 200 m, but only as far west as the Central Province of Papua New Guinea in the southern lowlands. It extends to Waigeo and Japen Islands in the north-west, and to Goodenough Island, New Britain and New Ireland in the east. The small-toothed race *E. k. oriomo* was collected by the Archbold Expedition throughout the Fly River basin of southern New Guinea and shows a disjunct distribution with the nominotypical form. The species is not recorded from the southwestern lowlands of Irian Jaya, nor from the lowlands of the Gulf of Papua. However, Dwyer (1984) found the species in grassland, secondary forest and primary forest between 1 000 and 1 550 m on the southern slopes of Mt Haliabu

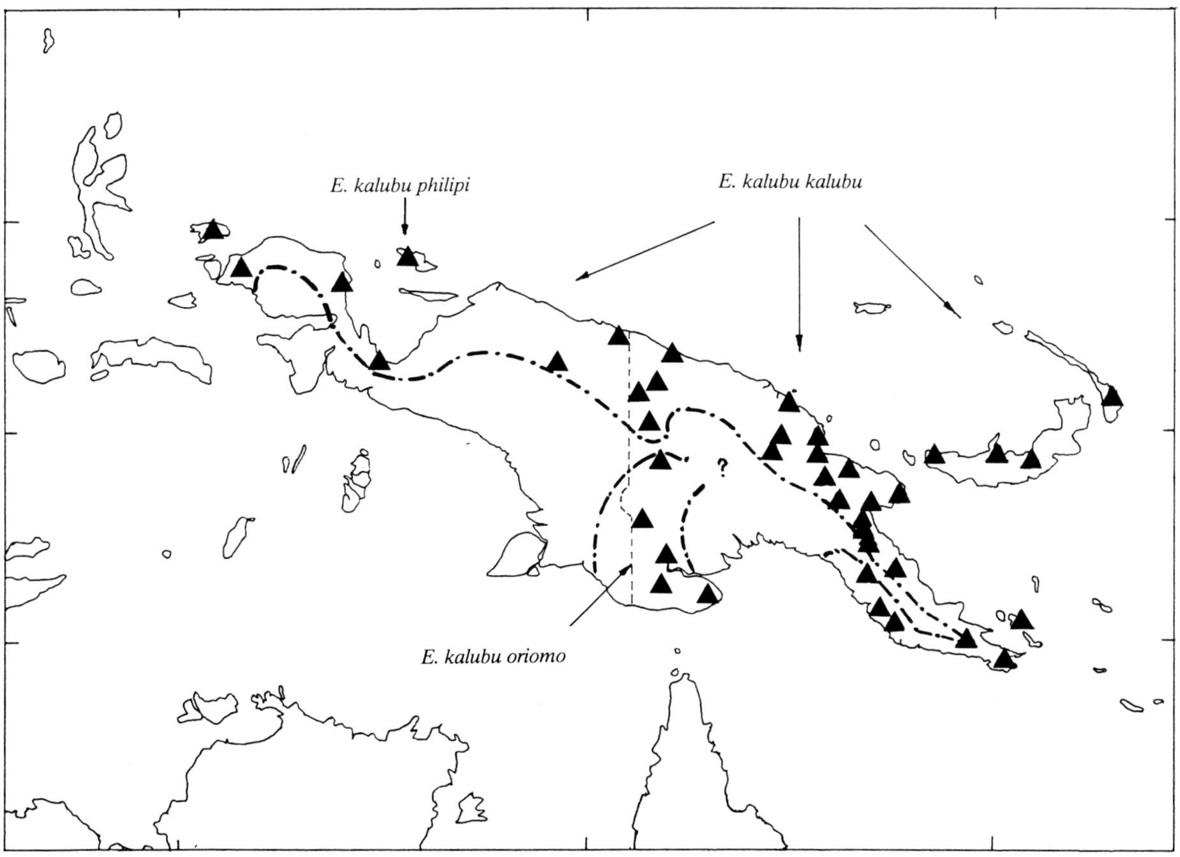

Fig. 7. Known localities and presumed range of *Echymipera kalubu.*

(= Mt Sisa) in the Southern Highlands of Papua New Guinea. The subspecific affinities of Dwyer's material is unknown to us.

The insular form *E. k. philipi*, described by Ellis Troughton as a new species in 1945, is known only from the type series collected by the United States Army during the Second World War. Troughton (1945) gives the provenance as Owi Island, Schouten Group, New Guinea. Laurie and Hill (1954) locate these islands in Geelvink Bay, Irian Jaya, but there is also a Schouten Group off the north coast of Papua New Guinea. Flannery (pers. comm.) advises that the Geelvink Bay location is correct.

The records of *E. kalubu* and *E. rufescens* from islands in the D'Entrecasteaux Archipelago of Papua New Guinea should be re-examined in the light of the discrimination of a fourth species of *Echymipera* from the Trobriand Islands (Flannery, this volume).

Echymipera kalubu kalubu is sympatric with *E. rufescens* in the rainforests and forest margins but is more frequently encountered in secondary grasslands and regrowth forests. Keith collected only *E. kalubu* on the grassy floodplains of the Sepik River, south of Maprik, where it was common (Van Deusen and Keith 1966). Flannery (pers. comm.) found the species in all habitats at Yapsiei in the upper Sepik basin, and sympatric with *E. clara* in foothill forests. At Baiyer River, Anderson *et al.* (1988) found *E. kalubu* to be common in the rainforest at 1 200 m, favouring areas of tree-fall and subsequent dense regrowth. Activity at this site centred on rotting logs and areas of fruit fall, but individuals also ventured into a garden and cane-grass (*Saccharum* sp.) thickets on the forest edge. Anderson does not record other species trapped during this Oxford University study in 1985, but George found both *E. kalubu* and *E. rufescens* to occur in the Baiyer River area during residence there 1968–1975. In south-east Papua New Guinea, where *Isoodon macrourus* occupies secondary grasslands, Wildlife Survey teams have found *E. kalubu* to be more restricted to forest. The southern subspecies *E. k. oriomo* is, however, typical of the open country of the Fly River plains (Tate 1948).

Status: This is the commonest and most widespread of the New Guinea bandicoots and is an important game species in the New Guinea lowlands.

The Archbold Expeditions of the American Museum of Natural History obtained large series of *E. kalubu* in the Sentani Lake–Cyclops Mountains area and on the Idenburg River in Irian Jaya.

George found that a high proportion of trophy material he purchased from traditional hunters in the Tabibuga area of the Jimi River valley in 1966 consisted of bandicoot mandibles. These were predominantly *E. kalubu* and *E. rufescens* with a few *E. clara*. The ratio of *E. rufescens* to *E. kalubu* was not recorded at the time of collection and much of this material, which was stored at the Baiyer River Sanctuary, seems to have been subsequently destroyed or discarded (P. Morris, pers. comm. 1985).

Rufous Spiny Bandicoot —
Echymipera rufescens *(Peters and Doria)*

Echymipera rufescens is one of the two species of bandicoot shared between Australia and New Guinea.

Description: This is a large species with males reaching up to 3 kg in weight. The dorsal pelage is blackish with varying amounts of tawny to rufous ticking. In this respect, all members of the genus are similar and the specific epithet of this species is not a distinguishing feature of the pelage. The crown is blackish, but is not sharply demarcated from the pale fur of the throat and jaws as it is in *E. kalubu.* The rostrum is narrowed. The molars lack a hypocone giving the tooth a sub-triangular outline.

Distribution (Fig. 8): Lowlands of northern, eastern and southern New Guinea, the Aru and Kei Islands and Cape York. Not recorded from the Vogelkop Peninsula of Irian Jaya or the Bismark Archipelago. Usually found in rainforest and rainforest margins and may venture into adjoining habitats.

Over much of its range it is sympatric with *E. kalubu.* Tim Flannery (pers. comm.) notes that *E. rufescens* appears to be absent from foothill forests on the northern side of New Guinea in contrast to wider habitat usage in other parts of its range. At Yapsiei, foothill forests were occupied by *E. clara* and *E. kalubu,* which suggests that *E. clara and E. rufescens* occupy mutually exclusive habitats. At this locality Flannery encountered only *E. rufescens* in short grass on the flood plains. However, Kent Keith did not find *E. rufescens* on the Sepik plains south of Maprik, where *E. kalubu* was common (Van Deusen and Keith 1966). George found both *E. rufescens* and *E. kalubu* to be common in trophy material collected north-west of Tabibuga in the middle Jimi River valley of the Western Highlands of Papua New Guinea.

Populations throughout New Guinea and the Aru and Kei islands are referred to the nominotypical subspecies *E. rufescens rufescens.* The Queensland populations are referred to *E. r. australis.*

Status: Common and frequently hunted as a game animal throughout the New Guinea lowlands but does not appear to be anywhere as locally common as *E. kalubu* is in parts of northern New

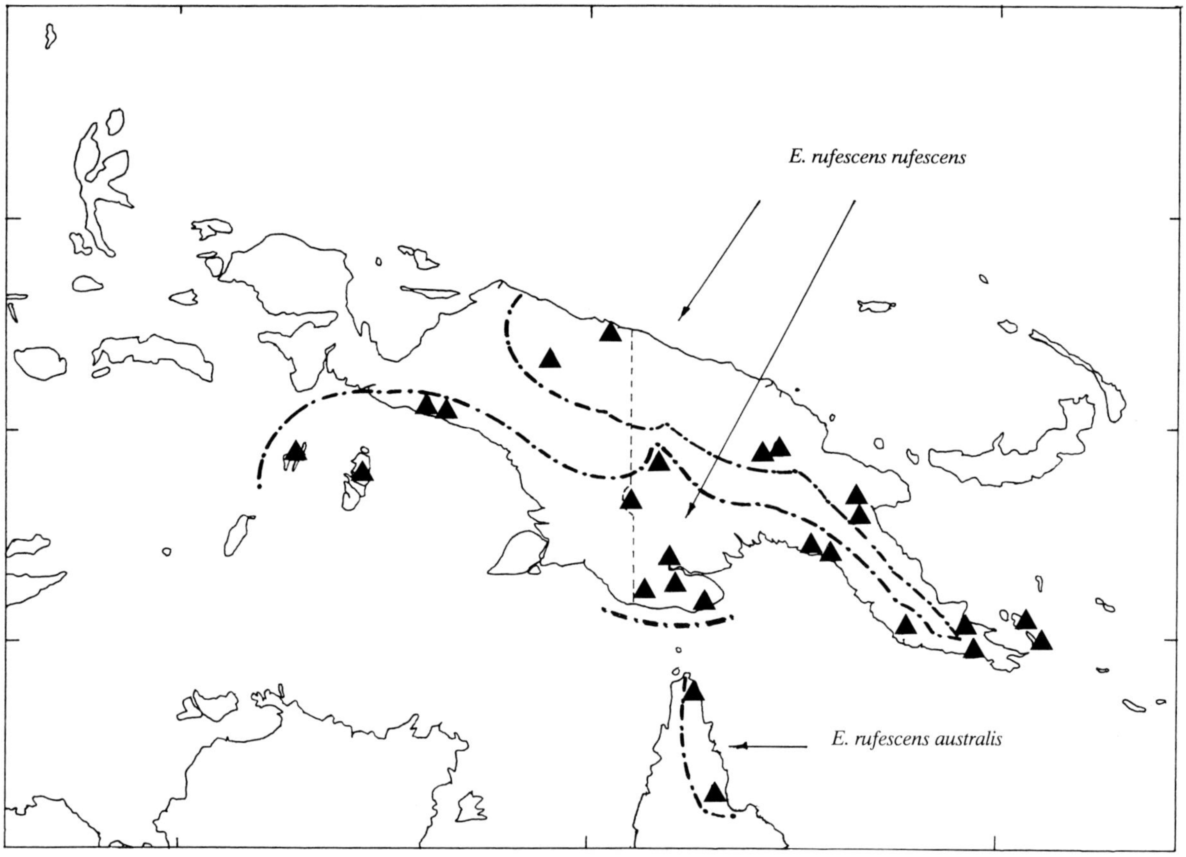

Fig. 8. Known localities and presumed range of *Echymipera rufescens.*

Guinea. Tate (1948) regarded *E. rufescens* of the Fly River forests as being rarer than *E. kalubu oriomo* of the Fly River savannahs.

Seram Island Bandicoot —
Rhynchomeles prattorum *Thomas*

This isolated relict form is known only from the type series of four specimens collected by the Pratt brothers on Mt Manusela, Seram, 1 800 m, in 1920.

Description: Deep glossy grey-brown with a white chest patch and relatively long tail. On cranial and dental characters this genus appears to be related to *Echymipera.* It lacks I^5 and has an extremely elongated rostrum. It has enlarged P^3_3 as in *E. clara,* slender zygomata and slit-like incisive foramina as in *E. rufescens.* The pelage is much less spinous than in *Echymipera* and the protocone on M^5 is greatly reduced (Tate 1948).

Distribution (Fig. 5): Known only from the type locality — Mt Manusela, Seram Island, Moluccas. This is one of the few Australasian marsupials to occur beyond New Guinea and its offshore islands. Its presence on Seram suggests that there may have been a low-level land connection during the Miocene/Pliocene that allowed the dispersal of the ancestral stock of *Rhynchomeles* along with other elements of the lowland New Guinea fauna (such as cassowaries and cuscuses) to Seram (George 1987).

Status: No information is available.

Northern Brown or Brindled Bandicoot —
Isoodon macrourus morebyensis *(Ramsay)*

The genus *Isoodon* is distinguishable from other New Guinea genera by its shortened rostrum and inflated auditory bullae.

This species is known as the Northern Brown Bandicoot in Australia, which is an inappropriate name in the New Guinea context. It is one of several mammals distributed through the drier savannah habitats on both sides of Torres Strait which have differentiated very little from each other since isolated by the rise in sea level over the Sahul Shelf some 8–10 000 years ago (Schodde and Calaby 1972).

Description: This is the New Guinea race of *I. macrourus* of northeastern Australia. According to Tate (1948) *I. m. moresbyensis* is much darker than Queensland specimens and old males from the Fly River savannahs may be strongly yellowish on the venter.

Distribution (Fig. 5): Savannah grasslands of the Fly River district, the central Papuan coast south of the Owen Stanley Ranges and secondary grasslands in higher rainfall areas of south-east Papua New Guinea as far as the Dobodura area of the Northern Province on the northern coast, at low altitudes.

Status: This is a very common species in the eucalypt savannahs around Port Moresby. It is moderately common in secondary grasslands in the south-east.

SUMMARY

For its size in comparison to Australia, New Guinea is relatively rich in bandicoots with two genera *(Perorcytes, Microperoryctes)* and five species endemic, and another genus *(Echymipera)* centred in New Guinea but with one species shared with north Queensland. *Rhynchomeles,* endemic to Seram, is related to *Echymipera.* The Australia-centred genus *Isoodon* has one species shared with southern New Guinea. As physical isolation of New Guinea from Australia is a relatively recent geological event, the apparent endemism in the family must be related more to habitat adaptations than to geographical isolation. The New Guinea-centred genera are rainforest animals. In comparison, Australia-centred genera are adapted to sclerophyl forests and more arid plant communities.

Four of the New Guinea species are widespread, common and secure. *Isoodon macrourus,* though restricted to grasslands of the south and east, is also common and secure. *Echymipera clara,* though rare in the sense of it having a limited distribution, may be locally common in parts of its range. *Peroryctes papuensis,* with restricted geographical range, is of doubtful status as little is known of its ecology and records are few. The largest species, *P. broadbenti* and the smallest, *M. murina* are known from few specimens and may be quite insecure. The endemic *R. prattorum* on Seram is of unknown status.

Peroryctes raffrayanus is widespread throughout the central and north-coastal mountains of New Guinea. It is more common in mid-mountain oak forests but also occurs in lowland hill forest and montane southern beech *(Nothofagus)* forests. The very rare *P. broadbenti* appears to occupy a restricted range in the south-east in lowland and possibly foothill forests below the range of *P. raffrayanus.* The ornamented forms of *P. longicauda* overlap the range of *P. raffrayanus* in the central mountains, but are more common at higher altitudes. The related *P. papuensis* of the southeastern mountains occurs sympatrically with *P. longicauda* but the extent of any altitudinal or ecological separation of the two species is unknown. In contrast to the genus *Peroryctes,* the spiny bandicoots of the genus *Echymipera* are characteristic of the mixed rainforests of the lowlands and foothills from sea level to 1 200 m, occasionally found as high as 1 500 m. The two common species, *E. kalubu* and *E. rufescens,* are broadly sympatric, but *E. kalubu* usually occupies a broader range of habitats, including grassy floodplains where it may be quite common. The rarer *E. clara,* known only from the lowlands of northern New Guinea, appears to exclude *E. rufescens,* but not *E. kalubu,* from foothill forests below 1 000 m. There appears to be a correlation between range and habitat usage between the two widespread spiny bandicoots. *Echymipera kalubu,* with its wider habitat tolerance in the northern parts of its range also has a more extensive distribution in the north and apparent higher population densities. *Echymipera rufescens,* with a narrower habitat usage in the north, has a more extensive southerly distribution. These distribution patterns suggest that the two species may have originated respectively in the northern and southern lowlands. However, the factors determining habitat use may be competiton between *E. clara* and *E. rufescens* in the north and between *E. kalubu* and *I. macrourus* in the south-east, rather than the geographical origins of the two common spiny bandicoots.

The New Guinea bandicoots appear to be unaffected as yet by the kind of large scale habitat changes that have affected many Australian species. The rarity of some forms may be due more to ecological specialization and/or vulnerability to hunting pressure than to habitat changes as it appears to be for other threatened mammal species in New Guinea (George 1979).

ACKNOWLEDGEMENTS

Our thanks go to many friends and colleagues who have assisted with field work or provided information: former residents Tas Hammersley of Baiyer River and William Peckover of Port Moresby; Eric Lindgren of Port Moresby; James Menzies of the National Museum and Art Gallery, Port Moresby; Lester Seri, Division of Wildlife, Port Moresby; Tim Flannery, Australian Museum; Ken Aplin, University of New South Wales.

REFERENCES

Anderson, T. J. C., Berry, A. J., Amos, J. N. and Cook, J. M., 1988. Spool and line tracking of the New Guinea spiny bandicoot, *Echymipera kalubu* (Marsupialia, Peramelidae). *Journal of Mammalogy* **69:** 114–20.

Archbold, R., Rand, A. L. and Brass, L. J., 1942. Results of the Archbold Expeditions No. 41. Summary of the 1938–1939 New Guinea Expedition. *Bulletin of the American Museum of Natural History* **79:** 197–288.

Archer, M., 1984. Earth-shattering concepts for historical zoogeography. Pp. 45–59 in *Vertebrate zoogeography and evolution in Australasia* ed by M. Archer and G. Clayton. Hesperian Press: Perth.

Brass, L. J., 1956. Results of the Archbold Expeditions No. 75. Summary of the Fourth Archbold Expedition to New Guinea (1953). *Bulletin of the American Museum of Natural History* **11:** 77–152.

BRASS, L. J., 1964. Results of the Archbold Expeditions No. 86. Summary of the Sixth Archbold Expedition to New Guinea (1959). *Bulletin of the American Museum of Natural History* **127:** 147–215.

BULMER, R. N. H. AND MENZIES, J. I., 1973. Karam classification of marsupials and rodents, Part 2. *Journal of the Polynesian Society* **82:** 87–88.

DOUTCH, H. F., 1972. The palaeogeography of northern Australia and New Guinea and its relevance to the Torres Strait area. Pp. 1–10 in *Bridge and barrier: the natural and cultural history of Torres Strait* ed by D. Walker. Australian National University: Canberra.

DWYER, P. D., 1984. From garden to forest: small rodents and plant succession in Papua New Guinea. *Australian Mammalogy* **7:** 29–36.

GEORGE, G. G., 1979. The status of endangered Papua New Guinea mammals. Pp. 93–100 in *The status of endangered Australasian wildlife* ed by M. Tyler. Royal Zoological Society of South Australia: Adelaide.

GEORGE, G. G., 1987. Characterization of the living species of Cuscus (Marsupialia: Phalangeridae) Pp. 507–26 in *Possums and oppossums: studies in evolution* ed by M. Archer. Royal Zoological Society of New South Wales: Sydney.

LAURIE, E. M. O., 1952. On mammals collected by Mr Shaw Mayer in New Guinea 1932–1949. *Bulletin of the British Museum (Natural History), Zoology* **1:** 291.

LAURIE, E. M. O. AND HILL, J. E., 1954. *List of land mammals of New Guinea, Celebes and Adjacent Islands 1758–1952.* British Museum (Natural History): London.

LINDGREN, E., 1975. *Wildlife in Papua New Guinea.* Golden Press: Sydney.

SCHODDE, R. AND CALABY, J. H., 1972. The biogeography of the Australo-Papuan bird and mammal faunas in relation to Torres Strait. Pp. 257–300 in *Bridge and barrier: the natural and cultural history of Torres Strait* ed by D. Walker. Australian National University: Canberra.

SCHODDE, R. AND MCKEAN, J. L., 1973a. Distribution, taxonomy and evolution of the gardener bowerbirds *Amblyornis* spp. in eastern New Guinea. *Emu* **73:** 51–60.

SCHODDE, R. AND MCKEAN, J. L., 1973b. The species of the genus *Parotia* (Paradisaeidae) and their relationships. *Emu* **73:** 145–56.

STEIN, G., 1932. Einige neue Beuteltiere aus Neuguinea. *Zeitschrift für Säugertierkunde.* **7:** 256.

TATE, G. H. H., 1948. Results of the Archbold Expeditions No. 60. Studies in the Peramelidae (Marsupialia). *Bulletin of the American Museum of Natural History* **92:** 313–46.

TATE, G. H. H. AND ARCHBOLD, R., 1936. Results of the Archbold Expeditions No. 8. Four apparently new polyprotodont marsupials from New Guinea. *American Museum Novitates* **823:** 1–4.

TATE, G. H. H. AND ARCHBOLD, R., 1937. Results of the Archbold Expeditions No. 16. Some marsupials of New Guinea and Celebes. *Bulletin of the American Museum of Natural History* **68:** 331–467.

THOMAS, O., 1888. *Catalogue of the Marsupialia and Monotremata in the collection of the British Museum (Natural History).* British Museum (Natural History): London.

TROUGHTON, E. LEG., 1945. Diagnoses of new mammals from the south-west Pacific. *Records of the Australian Museum* **21:** 373.

VAN DEUSEN, H. M., 1972. Marsupials. Pp. 711–14 in *Encyclopaedia of Papua and New Guinea* ed by P. Ryan. Melbourne University Press/University of Papua and New Guinea: Melbourne.

VAN DEUSEN, H. M. AND KEITH, K., 1966. Range and habitat of the bandicoot *Echymipera clara* in New Guinea. *Journal of Mammalogy* **47:** 721–23.

ZIEGLER, A. C., 1977. Evolution of New Guinea's marsupial fauna in response to a forested environment. Pp. 117–38 in *The biology of marsupials* ed by B. Stonehouse and D. Gilmore. Macmillan Press: London.

ZIEGLER, A. C., 1982. An ecological checklist of the New Guinea recent mammals. Pp. 867–93 in *Biogeography and ecology of Papua New Guinea* ed by J. L. Gressit. W. Junk: The Hague.

Some observations on the visceral anatomy of the bandicoot *Isoodon macrourus* (Marsupialia: Peramelidae)

R. A. TEDMAN[1]

The visceral anatomy of the peramelid marsupial *Isoodon macrourus* was examined. The alimentary system features well-developed submaxillary glands, a simple, oval stomach, short intestine with a tubular caecum, and diffuse pancreatic tissue. A pair of rectal glands are located adjacent to the distal part of the rectum. The large triangular spleen overlaps the stomach, proximal small intestine and left kidney. The liver possesses four lobes. The right adrenal is firmly attached to the spigelian lobe of the liver, adjacent to the posterior vena cava, while the left adrenal is located adjacent to the junction of the left renal vein and posterior vena cava. Four lobes are present in the right lung while the left lung is undivided. The male posseses a large heart-shaped prostate, two pairs of well-developed bulbourethral glands and a cleft glans penis. The vaginal apparatus consists of a median cul-de-sac into which open two uterine canals and two bulbous vaginal caeca. Each vaginal caecum opens into a long, narrow lateral vaginal canal which terminates in the extremely short urogenital sinus slightly posterior to the urethral opening. Comparisons with comparable data from other perameloids are discussed.

Key words: Anatomy, viscera, *Isoodon macrourus.*

INTRODUCTION

BECAUSE bandicoots are unique in being syndactylous polyprotodonts, they are of particular interest to taxonomists. The superfamily Perameloidea contains the family Thylacomyidae (bilbies) and the family Peramelidae (bandicoots) (Strahan 1983). The relationship of these two families and the origins of the different peramelid genera remain unclear (Archer 1984). In recent years there has been increasing interest in bandicoot ecology, behaviour, and physiology as well as taxonomy. However, there is still a paucity of information on detailed visceral anatomy.

The first general description of the visceral anatomy of a peramelid was that of *Perameles nasuta* by Grant in 1832. While a detailed account of the female urogenital systems of *P. nasuta* and *Perameles (Isoodon) obesulus* was given by Hill in 1899, other body systems have received little attention. Fragmentary and often incidental and comparative observations have been reported by a number of authors, for example, Owen (1868), Alix (1879), Johnstone (1898), Mitchell (1905), Fraser (1915), Mackenzie (1918), Sonntag (1921b) and Pearson (1940). Parsons (1903) contributed a general description of the visceral anatomy of the presumed extinct *Choeropus (Chaeropus) castanotis (ecaudatus).*

Several studies have described the microscopic appearance of various peramelid tissues, for example, intestinal mucosa (van Lennep 1962, 1964), cloacal glands (Green 1963), ovary (Ullmann 1979; Hollis and Lyne 1980; Lyne and Hollis 1979, 1983), lateral vaginae (Lyne and Hollis 1982) and male accessory glands (Rodger and Hughes 1973). In addition, between 1955 and 1970, Cleland, Sapsford and co-workers published detailed accounts of the ultrastructural morphology of sperm and spermiogenesis in *P. nasuta* (Sapsford, Rae and Cleland 1970).

The present study of visceral anatomy of *I. macrourus* was initiated to provide a general account of the viscera of this species and in turn to secure a basis for comparison with other perameloids and marsupials in general.

MATERIALS AND METHODS

Adult male (N=2) and female (N=3) specimens of *I. macrourus* with mean body weights of 2.1 kg and 1.4 kg respectively were obtained from a captive breeding colony maintained in open-air enclosures by the School of Anatomy, University of Queensland, Brisbane. All animals were killed by intraperitoneal injection of pentobarbital sodium. One of each sex was embalmed with 10% phosphate buffered formalin and the others were dissected fresh before being preserved in embalming fluid for subsequent examination. Photographs, sketches and measurements were recorded at regular intervals during each progressive dissection.

RESULTS

Alimentary system

The oral cavity receives secretions from well developed salivary glands. The submaxillary gland is located ventrocaudal to the masseter and consists of a rostral lobe ventral to the

[1]Department of Anatomy and Histology, University of Adelaide, Box 498, G.P.O. Adelaide, South Australia, Australia 5001.

Pages 107–16 *in* BANDICOOTS AND BILBIES, ed by J. H. Seebeck, P. R. Brown, R. L. Wallis and C. M. Kemper. Surrey Beatty & Sons, Sydney, 1990.

Table 1. Morphometrics of *Isoodon macrourus.* $\bar{x}^1$ = mean of two specimens; $\bar{x}^2$ = mean of three specimens; $\bar{x}^3$ = mean of five specimens.

	Total body length (TBL) (cm)	Total body weight (kg)	Length of stomach (cm)	Length small intestine (cm)	Length of caecum (cm)	Length of large intestine (cm)	Total intestine length (TIL) (cm)	TIL/TBL
Male	39	1.7	5.8	61.0	7.0	26.0	96.0	2.5
	41	2.5	4.1	62.0	5.6	22.5	94.1	2.3
$\bar{x}^1$	40	2.1	4.5	61.5	6.4	24.3	95.1	2.4
Female	32	1.1	2.4	49.5	5.0	19.5	74.0	2.3
	37	1.5	4.2	49.0	7.1	24.5	80.6	2.2
	38	1.7	5.2	55.0	6.5	27.4	88.9	2.3
$\bar{x}^2$	35.7	1.4	3.9	51.2	6.2	23.8	81.2	2.3
$\bar{x}^3$	37.9	1.8	4.2	56.4	6.3	24.1	88.2	2.3

geniohyoid and anterior belly of digastric plus a caudal lobe ventrolateral to the sternohyoid muscle (Figs 1, 5). Medially, a single duct exits from the deep surface of each lobe to pass rostrally deep to the mylohyoid, uniting to form a common duct at the level of the rostral border of the masseter. This duct then lies successively on hyoglossus and genioglossus adjacent to the ramus of the mandible before opening into the oral cavity at the base of the frenulum of the tongue. The sublingual gland persists as a small, flat, elongate structure deep to the mylohyoid muscle and attached to the duct leading from the rostral lobe of the submaxillary gland. The parotid gland is visible ventral to the external acoustic meatus and caudal to the masseter muscle. The parotid duct passes rostrally lateral to the masseter before entering the oral cavity adjacent to the upper jaw.

The oesophagus persists as a 4–5 mm wide tube passing from the oropharynx into the thorax dorsal to the trachea and the azygos lobe of the right lung. In the caudal part of the thorax, the oesophagus lies ventral to the descending aorta before passing through the oesophageal opening in the left side of the diaphragm to reach the stomach. The simple, oval stomach is covered ventrally by liver and spleen (Fig. 1).

The small intestine is not clearly demarcated into different regions although part of the proximal small intestine is attached to an extension of the gastrohepatic ligament (hepatoduodenal ligament), as well as a 5–6 cm long band of peritoneum from the dorsal abdominal wall (Figs 2, 3). The proximal part of the common mesentery of the small intestine is covered by this latter band of peritoneum.

A tubular caecum, varying in length from 5.0 to 7.1 cm, is attached to the colon by peritoneum and marks the beginning of the large intestine (Fig. 2). The large intestine has a long straight portion which lies in the midline and passes through the pelvis to terminate at the cloaca. Two large, oval rectal glands open into the cloaca about 5 mm from the external opening. The total length of the intestine (64% of which is small intestine) is about twice the total body length (Table 1).

The liver has left, right, middle and spigelian lobes (Figs 1, 4). The largest lobe, the middle or cystic lobe, is demarcated by a cleft to the left of the suspensory fissure and a cleft to the right of the cystic fissure. The gall bladder is located within the cystic fissure. The spigelian or caudate lobe, which is located on the posterior (visceral) surface of the liver, partially surrounds the posterior vena cava. The heaviest liver, weighing 37.4 g, was found in a female weighing 1.5 kg (Table 2).

Table 2. Organ weights of *Isoodon macrourus.* $\bar{x}^1$ = mean of two specimens; $\bar{x}^2$ = mean of three specimens; $\bar{x}^3$ = mean of five specimens.

	Total body weight (kg)	Heart (g)	Spleen (g)	Liver (g)	Kidneys (g)	Lungs (g)
Male	1.7	11.9	6.3	23.8	4.0	9.4
	2.5	12.3	15.9	25.8	5.6	11.7
$\bar{x}^1$	2.1	12.1	11.1	24.8	4.8	10.6
Female	1.1	5.5	4.7	13.7	3.6	—
	1.5	11.1	12.5	37.4	—	11.6
	1.7	9.8	17.4	34.7	4.4	10.0
$\bar{x}^2$	1.4	8.8	11.5	28.6	4.0	10.8
$\bar{x}^3$	1.8	10.5	11.3	26.7	4.4	10.7

The bile duct enters the duodenum 1.5–2.0 cm from the pyloric sphincter (Fig. 2). The distal end of the duct has an expanded portion 11–15 mm long and 4–6 mm wide into which a single pancreatic duct enters proximally. Diffuse pancreatic tissue consisting of subdivided and dendritic processes are present dorsal to the stomach and spleen and along the mesenteric border of the duodenum.

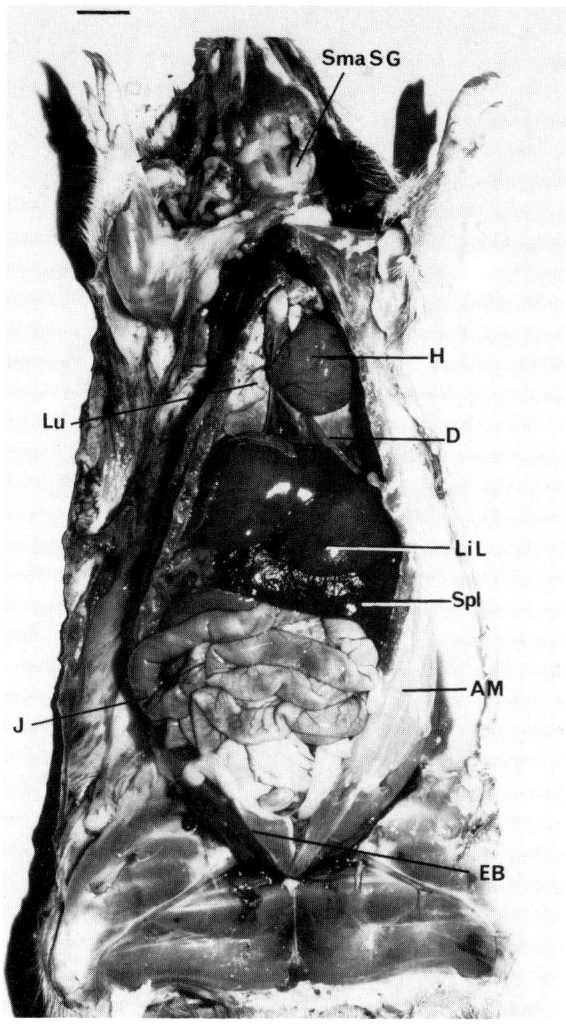

Fig. 1. Ventral view of body cavity of a female *Isoodon macrourus* with viscera *in situ*. Bar = 20 mm. Abbreviations: AcF, acetabular fossa; Ad, adrenal; AM, abdominal muscles; AnVC, anterior venal cava; Ao, aorta; AoA, arch of aorta; BD, bile duct; Bl, bladder; BuG, bulbouretheral gland; Ca, caecum; Cl, cloaca; C-mA, coeliacomesenteric artery; ComCA, common carotid artery; CrP, crus penis; CyF, cystic fissure; D, diaphragm; DuD, ductus deferens; Duo, duodenum; EB, epipubic bone; ExIA, external iliac artery; GB, gall bladder; GoA, gonadal artery; H, heart; HepL, hepatoduodenal ligament; I, ileum; InIA, internal iliac artery; J, jejunum; K, kidney; LI, large intestine; LaV, lateral vaginae; LiC, cystic lobe of liver; LiL, left lobe of liver; LiR, right lobe of liver; LiS, spigelian lobe of liver; Lu, lung; LuAz, azygos lobe of lung; M, masseter; Mes, mesentery of small intestine; MeSA, median sacral artery; O, oesophagus; OG, oesophageal groove; Ov, ovary; OvV, ovarian vessels; P, pancreas; Pe, pelvis; PerF, peritoneal fold; PoVC, posterior vena cava; Pr, prostate; PsMa, psoas major muscle; PsMi, psoas minor muscle; PyS pyloric sphincter; RG, rectal gland; Re, rectum; RenA, renal artery; RenV, renal vein; SclA, subclavian artery; SmaSG, submaxillary salivary gland; Spl, spleen; St, stomach; Sthy, sternohyoid muscle; Stma, sternomastoid muscle; SusF, suspensory fissure; T, testis; Th, thyroid; Tr, trachea; U, ureter; Ur, urethra; UrB, urethral bulb; UroS, urogenital sinus; Ut, uterus; Utn, uterine neck; VC, vaginal caeca; VS, vaginal cul-de-sac.

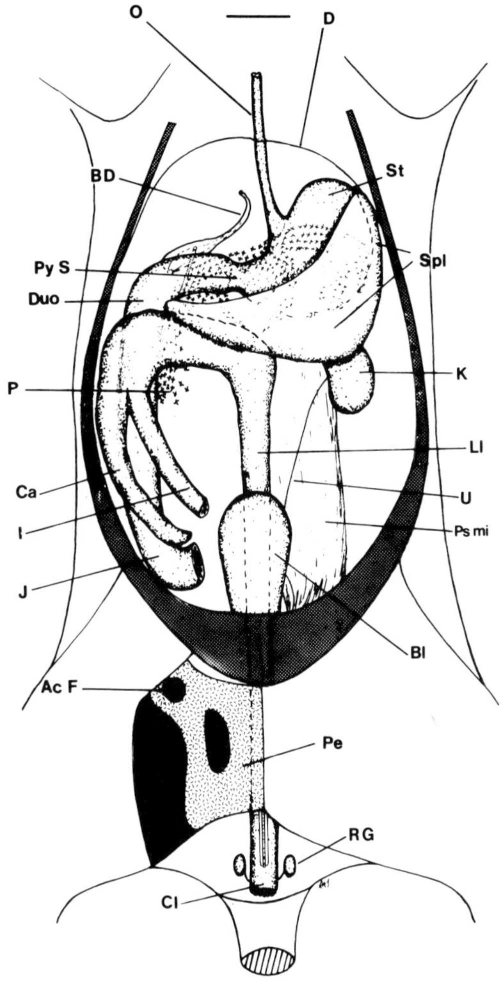

Fig. 2. Ventral view of the abdominal cavity of a female *Isoodon macrourus* with diaphragm, most of the small intestine and right medial thigh muscles removed. Bar = 20 mm. See Figure 1 for abbreviations.

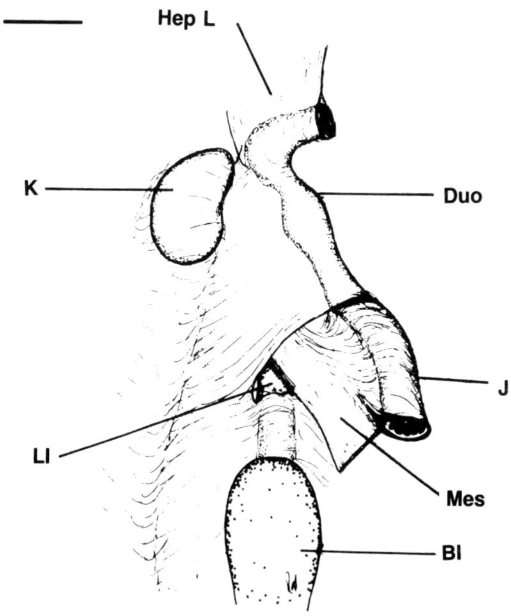

Fig. 3. Ventral view of peritoneal folds attaching to proximal small intestine of *Isoodon macrourus*. Duodenum has been pulled to the left. Bar = 20 mm. See Figure 1 for abbreviations.

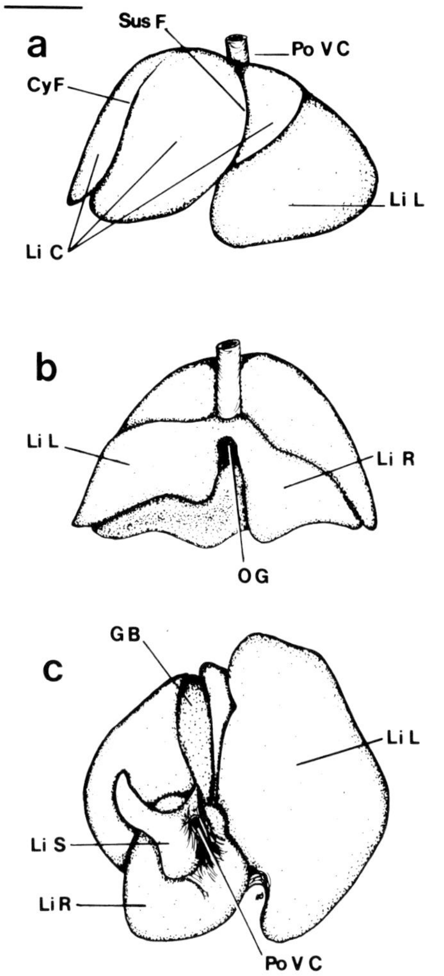

Fig. 4. Ventral (a), dorsal (b) and posterior (c) views of liver of *Isoodon macrourus*. Bar = 20 mm. See Figure 1 for abbreviations.

Cardiovascular system

The right atrium of the heart receives two anterior venae cavae and a single posterior vena cava while five pulmonary veins enter the left atrium (Fig. 5). Two pulmonary arteries branching from a common pulmonary trunk pass dorsal to the corresponding left and right anterior venae cavae to reach the lung roots. The arch of the aorta features (1) a large brachoicephalic (innominate) artery, which branches into right subclavian and right and left common carotid arteries, and (2) a left subclavian artery.

The abdominal aorta lies in a deep midline groove between the well-developed psoas muscles. It has a comon trunk (6–8 mm long) for coeliac and mesenteric branches which arises adjacent to the right crus of the diaphragm. This trunk supplies blood to the entire gastrointestinal tract as far caudally as the pelvic part of the large intestine. The right renal artery

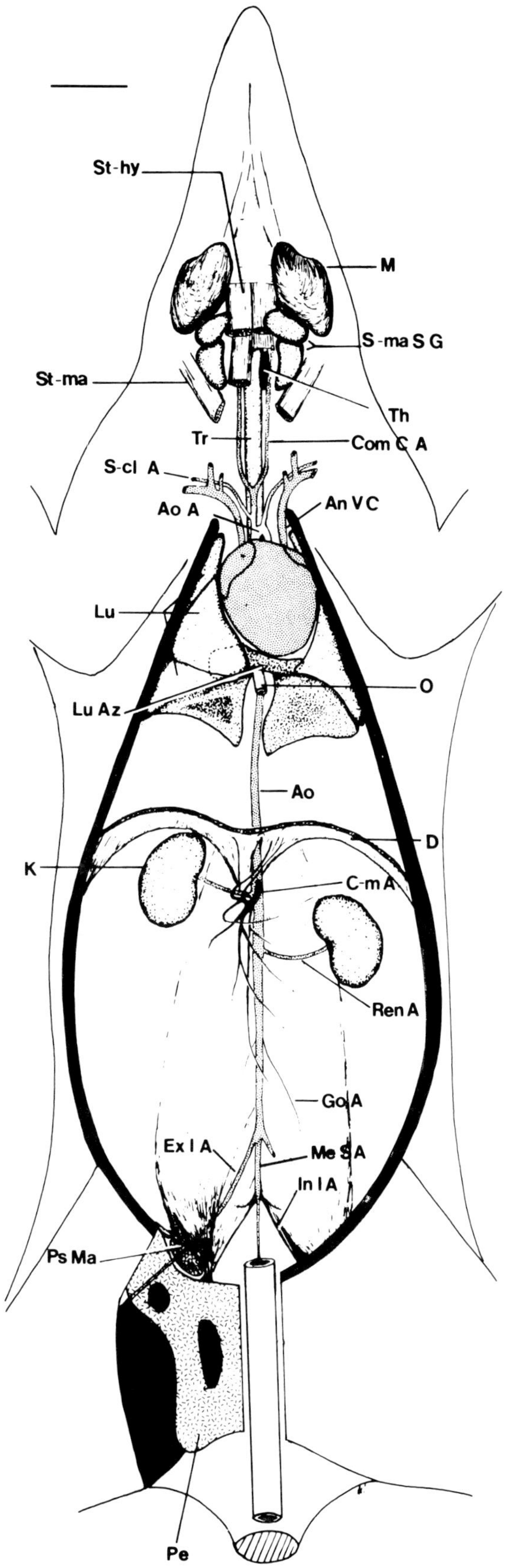

Fig. 5. Ventral view of a female *Isoodon macrourus* with gastrointestinal tract, liver, spleen, reproductive tract and right medial thigh muscles removed. Bar = 20 mm. See Figure 1 for abbreviations.

arises slightly caudal to the coeliacomesenteric trunk while the left renal branches from the aorta 1–2 cm further caudally. Left and right phrenic arteries arise from the coeliacomesenteric trunk and the right renal artery respectively. Paired gonadal arteries branch from the aorta further caudally. The aorta divides into left and right external iliac arteries before continuing into the pelvis to give rise to the paired internal iliac arteries and terminating as the median sacral artery. The external iliac crosses part of the psoas major muscle, passes between psoas minor and psoas major and then enters the thigh.

The posterior vena cava lies central and slightly to the right of the abdominal aorta throughout its course before piercing the liver adjacent to the spigelian lobe (Figs 4, 6). Left and right renal veins lie ventral and cranial to corresponding renal arteries.

Isoodon macrourus possesses a large triangular spleen with rostral and lateral processes which cover much of the greater curvature of the stomach and extend to the right to reach the duodenum (Fig. 2). The left lobe and part of the middle lobe of the liver cover part of the ventral surface of the spleen. While the average spleen weight was 11.3 g the largest spleen weighed 17.4 g (see Table 2).

A thoracic thymus gland lies dorsal to the cranial part of the sternum and adjacent to the ventral surface of the heart, especially the atria. No cervical thymus glands were found.

Urogenital system

In the female, each kidney is located slightly lateral to the respective psoas muscle. The right kidney lies adjacent to the dorsal attachment of the diaphragm to the thoracic skeleton while the left kidney lies further caudally (Figs 5, 6). The ureters pass caudally on the psoas muscles before passing dorsal to the bladder to enter the bladder a few millimetres cranial to the opening of the urethra. The ureters pass dorsal to the vaginal caeca and between the lateral vaginal canals before entering the bladder.

The female reproductive tract consists of paired uteri and a complex vaginal apparatus (Figs 6, 7). The paired uteri are firmly bound together by connective tissue. The caudal-most part of "neck" region lies adjacent and dorsal to the median vaginal cul-de-sac. Internally, the uterine canals remain separate and enter the undivided vaginal cul-de-sac. The vaginal cul-de-sac also receives openings from the two bulbous vaginal caeca which lie ventral to the uteri. The vaginal caeca, although separated by a median partition, appear undivided with a bilobed appearance externally. Each vaginal caecum is connected to a long narrow lateral vagina which passes to the urogenital sinus, lateral to the urethra. The extremely short urogenital sinus opens into the cloaca.

The kidneys and ureters in the male *I. macrourus* are identical in appearance and location to those found in the female. The caudal part of the bladder lies dorsal to the large, heart-shaped prostate (Fig. 8).

The scrotum lies cranial to the cloaca. The ductus deferens from each testis passes through the inguinal canal in the ventral abdominal wall before passing cranially adjacent to the dorsolateral surface of the prostate before bending sharply, medially, to enter the proximal part of the urethral canal without looping over the ureters (Fig. 8). The thick-walled urethra passes through the prostate and in the pelvis lies ventral to the colon before entering the penis between the bulbous crura of the penis. Immediately upon entering the penis the urethra receives the openings of the paired common ducts of the two pairs of bulbourethral glands (Fig. 9).

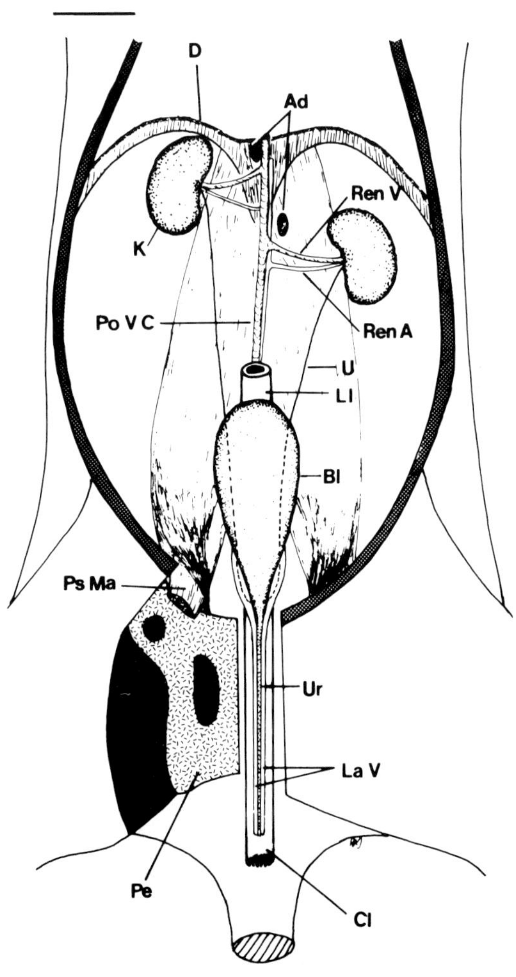

Fig. 6. Ventral view of abdominal cavity of a female *Isoodon macrourus* with gastrointestinal tract, liver, spleen and right medial thigh muscles removed. Bar = 20 mm. See Figure 1 for abbreviations.

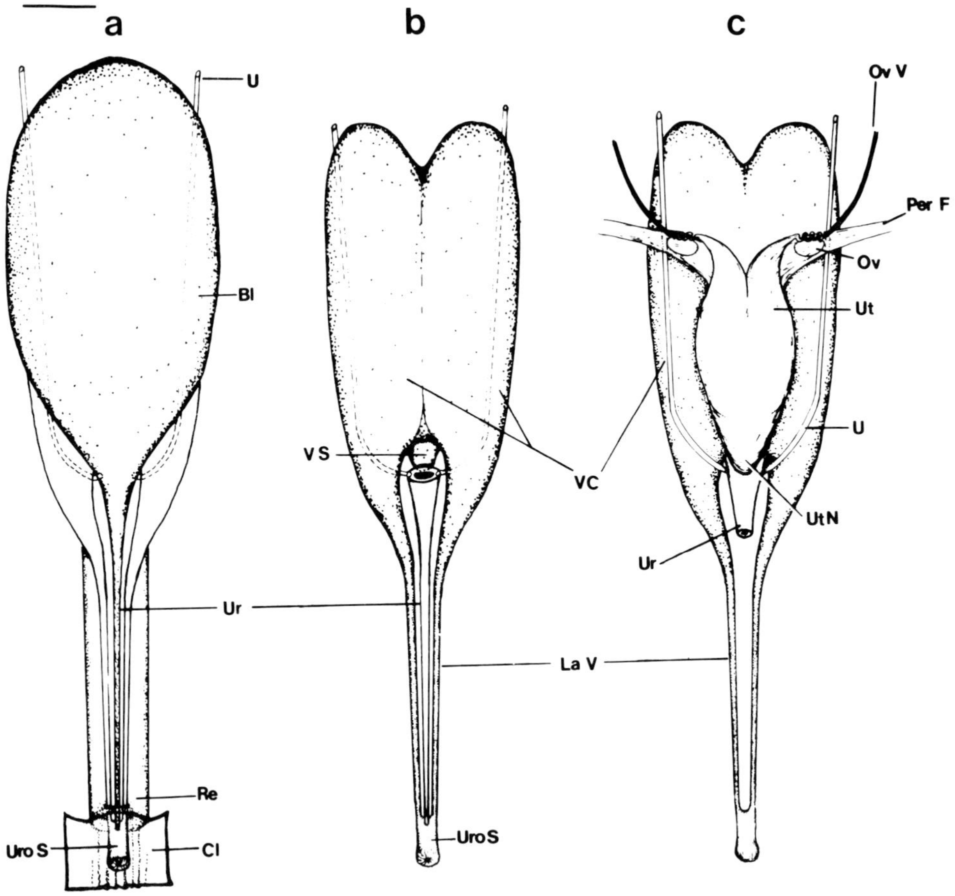

Fig. 7. Ventral (a,b) and dorsal (c) views of female reproductive tract of *Isoodon macrourus*. Bladder and part of urethra have been removed in (b) and (c) respectively. Bar = 10 mm. See Figure 1 for abbreviations.

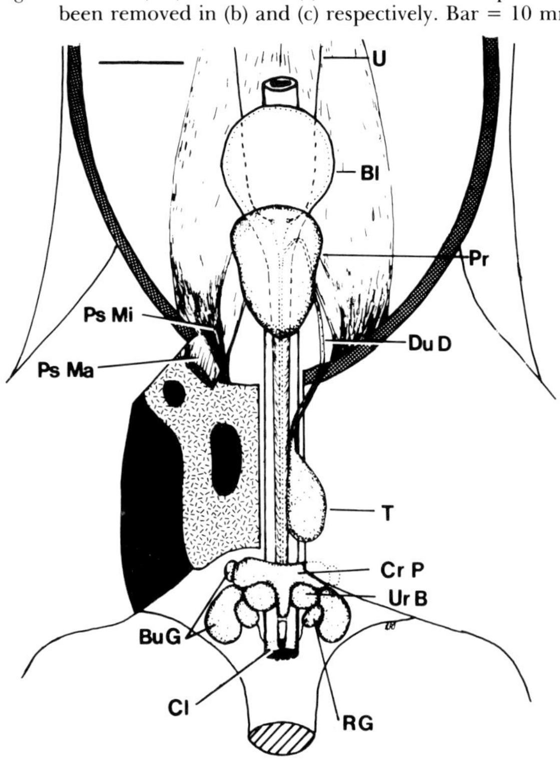

Fig. 8. Ventral view of abdominal and pelvic region of a male *Isoodon macrourus* with gastrointestinal tract, right medial thigh muscles, and perineal fat and deep fascia removed. Bar = 20 mm. See Figure 1 for abbreviations.

The urethra passes caudally into the 'S'-shaped curvature of the penis before dividing into two canals near the distal end of the penis. The two urethral canals are completely separate in the bifid glans penis. The urethral bulbs on each side unite to surround the urethra.

Respiratory system

The diaphragm arches well into the rostral part of the thorax so that the heart and ventral part of the lungs are restricted to the cranial half of the thoracic skeleton (Figs 1, 5). However, the dorsal parts of the lungs project further caudally. The left lung is undivided whereas the right lung has four lobes including an azygos lobe which projects across the midline caudal to the heart and ventral to the oesophagus.

Adrenal glands

Each adrenal is about 6.5 mm long and 5.0 mm wide, and is covered by a tough connective tissue capsule. The right adrenal lies medial to the cranial end of the right kidney and is firmly attached to the spigelian lobe of the liver and the posterior vena cava (Fig. 6). The left adrenal,

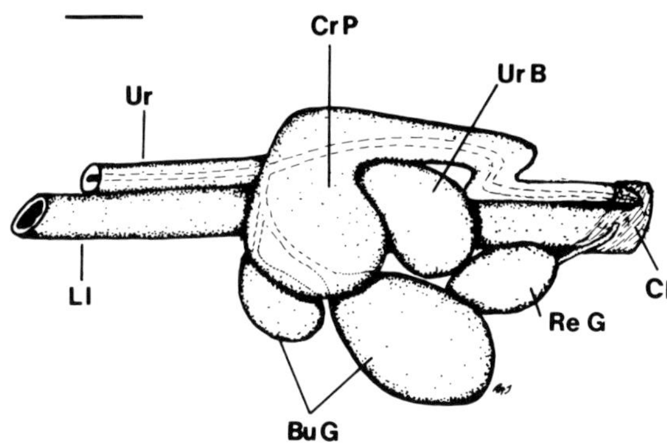

Fig. 9. Lateral view of the accessory organs of a male genital tract of *Isoodon macrourus.* Bar = 5 mm. See Figure 1 for abbreviations.

although medial to the cranial end of the left kidney, lies adjacent to, and is firmly attached to, the posterior vena cava near the junction with the left renal vein.

DISCUSSION

Because of the fragmentary and often incidental nature of most of the current knowledge of visceral anatomy of bandicoots, only superficial comparisons can be made between *I. macrourus* and other perameloids. Furthermore, there is some confusion in the nomenclature of animals described in early reports.

Alimentary system

Relatively small parotid glands and large submaxillary glands are characteristic of polyprotodonts including the Peramelidae (Owen 1868; Sonntag 1921b). Sonntag (1921b) suggested that the large size of the salivary glands in *I. obesulus* might be due to the need for abundant secretion to entangle and contain the insects trapped within the cheek pouches of this species. However, the submaxillary glands in *I. macrourus* are not as extensive as those shown for *I. obesulus* in Sonntag's work. Marsupial sublingual glands vary greatly in size and shape in different species (Sonntag 1921b). In *I. obesulus* these glands are small, oval bodies placed cranial to the submaxillary glands (Sonntag 1921b), whereas in *I. macrourus* the sublingual glands are small, flat, elongate structures.

The general appearance of the stomach and intestine of *I. macrourus* is similar to that of *P. nasuta* (Grant 1832; Owen 1868; Harrop and Hume 1980), *Perameles* sp. (Shultz 1976), *Perameles (Isoodon) obesula (obesulus)* (Sonntag 1921b), *Choeropus (Chaeropus) castanotis (ecaudatus)* (Parsons 1903) and *Peragale (Macrotis) lagotis* (Mitchell 1905). Sonntag (1921b) referred to a "duodenal loop" in *I. obesulus,* but there is no such delineation of duodenum in *I. macrourus.* Rather, the proximal part of the small intestine of *I. macrourus* (present study), which is attached to an extension of the gastrohepatic ligament as well as a 5–6 cm long band of peritoneum from the dorsal abdominal wall, may correspond to the duodenum.

The ratio of intestinal length to body length of about 2.3 in *I. macrourus* is similar to that of *P. nasuta* (2.1) (Grant 1832; Owen 1868) but slightly less than that in *C. ecaudatus* at 3.3 (Parsons 1903).

The caecum of *I. macrourus* is similar in relative and absolute size to that of *P. nasuta* (Grant 1832; Owen 1868), much longer than the 2 cm length reported by Schultz (1976) for *Perameles* sp. and shorter than the 10 cm organ found in *C. ecaudatus* (Parsons 1903). Hume (1982) believed that the relatively longer caecum in *C. ecaudatus* suggested that a significant amount of plant material was part of the normal diet of this species. Hill and Rewell (1954) recognized an evolutionary series in size and specialization of the caecum with decreasing complexity represented by *M. lagotis, P. nasuta, I. obesulus* and *Echymipera rufescens* respectively. However, except for *E. rufescens* the caeca in these perameloids generally appear to be similar in shape and orientation. The caecum reported by Hume (1982) for *M. lagotis* looks more like the flask-like organ found in *E. rufescens* than the caecum of *M. lagotis.* Although the caecum of *I. macrourus* (present study) is similar to that shown for *I. obesulus* by Hill and Rewell (1954), the peritoneal attachments resemble more those described for *M. lagotis* by those authors.

The dendritic and subdivided appearance of the pancreas in *I. macrourus* resembles that observed in *I. obesulus* and all other marsupials examined by Sonntag (1921b), except *Phascolarctos cinereus.* He considered this dendritic form of the pancreas to be more primitive than that found in *P. cinereus.* As observed by Owen (1868) and Sonntag (1921b) for many marsupials, the pancreatic duct in *I. macrourus* opens into the dilated terminal part of the bile duct.

In 1963, Green described the microscopic appearance of several glands found in the cloacal region of four species of marsupials. In *I. obesulus* she found three pairs of paracloacal glands that resembled the sebaceous glands of the skin and produced an oily secretion. Only one pair of rectal glands are found adjacent to the rectum of *I. macrourus.* Further histological work is required to determine whether these glands resemble the paracloacal glands described by Green.

With four lobes and a large gall bladder lying within a cleft in the middle or central lobe, the livers of *C. ecaudatus* (Parsons 1903) and *I. macrourus* are very similar. Parsons, however, did not report on the location of the liver nor its relationship to other viscera. As in most marsupials, the spigelian lobe is an appendage of the right lobe (Owen 1868).

Cardiovascular system

According to Owen (1868) the aortic arch of most marsupials gives off an innominate (brachiocephalic) artery which branches into a right subclavian and the two common carotids. However, Pearson (1940) found just as much variation in the arrangement of these vessels within the Marsupialia as there is in the whole of the Mammalia. In *I. obesulus* and *Perameles gunnii* (Pearson 1940) and *I. macrourus* (present study) the arrangement of these arteries is the same as that described by Owen (1868). Marsupials have a separate coeliac branch of the dorsal aorta (to liver, spleen, pancreas, stomach and duodenum) and a separate mesenteric branch to small and large intestine, but lack a posterior or inferior mesenteric branch (Owen 1868; Pearson 1940). In addition, like all other marsupials, the external iliac arteries arise directly from the aorta which continues as the median sacral (caudal) artery from which the internal iliac arteries arise. In possessing a common coeliacomesenteric trunk, *I. macrourus* differs significantly from this pattern, but a similar arrangement has been found in *P. cinereus* (Sonntag 1921a).

Except for *Petaurus breviceps*, which has a single anterior vena cava, all the marsupials studied by Pearson (1940) possessed two anterior venae cavae and a single posterior vena cava. The vessels in *I. macrourus* also conform to this pattern.

The large triangular spleen found in *I. macrourus* resembles that found in *I. obesulus* but is unlike the thin spleen which lacks prolongations that is found in *P. nasuta* (Sonntag 1921b). Parsons (1903) claimed that a spleen was absent in *C. ecaudatus*, however, this has not been confirmed since no other detailed studies of this presumed extinct bandicoot are available. Schultz (1976) reported a small, flattened triangular spleen in *Perameles* sp. Clearly there seems to be considerable variation in spleen morphology within the Peramelidae.

Foetal *P. gunnii* examined by Fraser (1915) confirmed the observations on adult material by Johnstone (1898) that this peramelid possesses a bilobed thoracic thymus but no cervical thymus. In 1973 Yadav reported similar observations for *P. gunnii* and *I. obesulus* and noted that the absence of a cervical thymus is a feature of all the polyprotodont marsupials he examined. The present study confirms this for *I. macrourus*.

Urogenital system

In 1868 Owen briefly commented on the general arrangement of the female reproductive tract of *I. obesulus* and in 1899 Hill published a detailed account of this system in several marsupials including *P. nasuta* and *I. obesulus*. Hill considered that, except for some "minute differences" which might have existed between these two species, the female reproductive tracts were identical. The present description of *I. macrourus* differs only slightly from those early descriptions. Differences in the shape of the two uteri and median vaginal cul-de-sac between *I. macrourus* and *I. obesulus* are minor in comparison to the remarkable features common to the Peramelidae such as the two large vaginal caeca, forward position of the bladder and therefore long urethra, long pseudovaginal passage, long lateral vaginae and very short urogenital sinus (Hill 1899). Pearson (1945) regarded these features as quite a departure from the primitive condition he envisaged for a prototypal marsupial. Another feature which is not only characteristic of the Peramelidae but also a distinctive feature of all marsupials is the passage of the ureters dorsal to the vaginal caeca and between the lateral vaginal canals before entering the bladder (Hill 1899).

There is no detailed report of the male reproductive tract of a peramelid apart from a short description of *C. ecaudatus* by Parsons (1903). Only brief reports of the structure of the penis (Owen 1868) and of the accessory glands (Rodger and Hughes 1973) of this group are available. The reproductive tract of male marsupials is basically the same with paired testes, epididymides, vasa deferentia which open into the anterior end of the large disseminate prostate, no seminal vesicles, large bulbourethral glands (Cowper's glands) which open into the urethra, and a penis posterior to the scrotum (Biggers 1966; Woolley and Webb 1977). In characteristic marsupial fashion, each ductus deferens enters the proximal part of the urethral canal without looping over the corresponding ureter (Owen 1868; Barbour 1977). Unlike eutherians, marsupials, including *I. macrourus*, have separate urethral bulbs which unite medially to surround the urethra (Woolley and Webb 1977). The penis in all marsupials is bent upon itself when passive and retracted. The glans is concealed just within the cloacal aperture from which it emerges when the penis is turgid and erect (Owen 1868). According to Owen (1868) a bifurcate glans penis is present in multiparous marsupials such as phalangerids, opossums and peramelids while a single glans is found in uniparous marsupials such as kangaroos; intermediate forms are found in the wombat and koala. As in *M. lagotis* described by Owen (1868) the bifurcate glans penis of

I. macrourus features two separate urethral canals. *Perameles nasuta* also possesses a bifurcate glans penis (Rodger and Hughes 1973).

Whereas the prostate in the Peramelidae, including *I. macrourus*, and the Phascolarctidae is heart-shaped, in all other marsupials studied it is carrot-shaped (Rodger and Hughes 1973). Marsupials possess from one to three pairs of bulbourethral glands with two pairs present in the Peramelidae (Rodger and Hughes 1973). Parsons (1903) reported three pairs of Cowpers' glands in *C. ecaudatus*, but one of these pairs may have been wrongly identified since they occupy roughly the same location as the rectal glands in *I. macrourus* (present study). The two pairs of bulbourethral glands found in *I. macrourus* differ in position and relative size from those glands described for *I. macrourus* and *P. nasuta* by Rodger and Hughes (1973).

Respiratory system

An undivided left lung has been reported for all peramelids examined. However, the right lung has been described as trilobate in *P. nasuta* (Grant 1832), and trilobate plus an azygos appendage in *I. obesulus* (Sonntag 1921b) and *Perameles* sp. (Owen 1868). Parsons (1903) referred to a well-marked azygos lobe in *C. ecaudatus*, and said that the right lung had a "slight fissure near the ventral margin". *Isoodon macrourus* resembles the other peramelids except for *C. ecaudatus* with the presence of four lobes, one of which is a well-marked azygos lobe.

Adrenal glands

Bourne (1949) divided the Peramelidae into two groups based principally on the position of the left adrenal. Accordingly, all peramelids except *P. gunnii* were supposed to share a dual vena cava and kidney attachment for the right adrenal. However, Bourne's illustrations clearly show the right adrenal of *P. nasuta* firmly attached not to the kidney but to a lobe of the liver and vena cava, remarkably similar to the situation with *I. macrourus* described in the present study. The left adrenal in *I. macrourus* shares a firm attachment to the junction of the vena cava and left renal vein like *Thalacomys (= Macrotis), I. obesulus* and *P. gunnii* (Bourne 1949).

Sexual dimorphism

The body weights of adult male (500 g–3 100 g) and female (500 g–1 700 g) *I. macrourus* given by Gordon (1983) indicate a marked sexual dimorphism by the largest adults. Not all the animals used in the present study displayed such a large size difference. Although a few of the variations in intestinal length and organ weights presented in Tables 1 and 2 seem to indicate some sexual dimorphism, not enough specimens were available to fully document individual variation in this species. There appear to be no consistent differences in length of stomach, caecum or large intestine between the two sexes although the small intestine is longer in the males. Of the two male *I. macrourus*, the heaviest specimen had the heaviest heart, lungs, spleen, liver and kidney; however, the heaviest female had a larger liver and spleen than those found in the males. Only the heart weights were consistently larger in the male animals.

CONCLUSIONS

Comparisons with other bandicoots and marsupials in general are difficult because of the fragmentary nature of the information available, but the following summary can be made: (a) the ratio of intestinal length to body length of about 2.3 in *I. macrourus* is similar to that of *P. nasuta* (2.1) but different from *C. ecaudatus* (3.3); (b) the caecum of *I. macrourus* (6.3 cm) is similar in length to that of *P. nasuta* and *I. obesulus* but different from *C. ecaudatus* (10 cm); (c) the spleen of *I. macrourus* is similar in size and shape to that in *I. obesulus* but different from that of *P. nasuta;* (d) the left adrenal in *I. macrourus* is located in a different position from that of *P. nasuta* and *C. ecaudatus* but in the same position as that in *I. obesulus,* while the right adrenal in *I. macrourus* is in a different position from that of *I. obesulus* and *C. ecaudatus* but the same as in *P. nasuta;* (e) the male accessory glands in *I. macrourus* (present study) are different in position and relative size from those in *I. macrourus* and *P. nasuta* described by Rodger and Hughes (1973); (f) the common origin of the coeliac and mesenteric arteries from the dorsal aorta is the same as in *P. cinereus* but unlike that found in other marsupials; (g) there is a large thoracic thymus but no cervical thymus, like that of other polyprotodont marsupials; (h) the heart-shaped prostate, although common in the Peramelidae and Phalangeridae is different from other marsupials studied; and (i) the female urogenital tract of *I. macrourus,* like *I. obesulus,* differs from other marupials in the anterior position of the bladder, the long urethra, expanded vaginal caeca, small vaginal cul-de-sac and short urogenital sinus.

Although many characteristics of the viscera appear to be similar in all perameloids so far examined, *I. macrourus* shares more features in common with *I. obesulus* and *P. nasuta* than with *C. ecaudatus.* Unfortunately not enough is known about detailed visceral anatomy in the Perameloidea to make any detailed comparison with *I. macrourus* or to comment on the taxonomic affinities of different members within this group.

ACKNOWLEDGEMENTS

I am grateful to Dr L. Hall, School of Anatomy, University of Queensland for drawing my attention to the need for the present work and for providing the specimens. Some of this work was done at the School of Anatomy, University of Queensland and I thank the staff concerned for their assistance. I thank R. Barbour, L. Hughes and G. Lyne for their comments on an early draft of the manuscript.

REFERENCES

ALIX, P. H. E., 1879. Sur les organes de la parturition chez les Marsupiaux. *Bulletin de la Societé Zoologique de France* **1879:** 118.

ARCHER, M., 1984. The Australian marsupial radiation. Pp. 633–808 in *Vertebrate zoogeography and evolution in Australasia* ed by M. Archer and G. Clayton. Hesperian Press: Carlisle, Western Australia.

BARBOUR, R. A., 1977. Anatomy of marsupials. Pp. 237–72 in *The biology of marsupials* ed by B. Stonehouse and D. Gilmore. Macmillan: London.

BIGGERS, J. D., 1966. Reproduction of male marsupials. *Symposia of the Zoological Society of London* **15:** 251–80.

BOURNE, G. H., 1949. *The mammalian adrenal gland.* Oxford University Press: Oxford.

FRASER, E. A., 1915. The development of the thymus, epithelial bodies, and thyroid in the Marsupialia. Part II. *Phascolarctos, Phascolomys* and *Perameles. Philosophical Transactions of the Royal Society of London* **B 207:** 87–112.

GORDON, G., 1983. Northern Brown Bandicoot. Pp. 96–97 in *The Australian Museum complete book of Australian mammals* ed by R. Strahan. Angus & Robertson: Sydney.

GRANT, R. E., 1832. Observations on the anatomy of *Perameles nasuta* from New Holland. *Memoirs of the Wernerian Natural History Society* **6:** 184–202.

GREEN, L. M. A., 1963. Distribution and comparative histology of cutaneous glands in certain marsupials. *Australian Journal of Zoology* **11:** 250–72.

HARROP, C. J. F. AND HUME, I. D., 1980. The digestive tract and digestive function in monotremes and non-macropod marsupials. Pp. 63–77 in *Comparative physiology: primitive mammals* ed by K. Schmidt-Nielson, L. Bolis and C. R. Taylor. Cambridge University Press: London.

HILL, J. P., 1899. Contributions to the morphology and development of the female urogenital organs in the Marsupialia. 1. On the female urogenital organs of *Perameles,* with an account of the phenomena of parturition. *Proceedings of the Linnean Society of New South Wales* **24:** 42–82.

HILL, W. C. O. AND REWELL, R. E., 1954. The caecum of monotremes and marsupials. *Transactions of the Zoological Society of London* **28:** 185–240.

HOLLIS, D. E. AND LYNE, A. G., 1980. Ultrastructure of luteal cells in fully formed and regressing corpora lutea during pregnancy and lactation in the marsupials *Isoodon macrourus* and *Perameles nasuta. Australian Journal of Zoology* **28:** 195–211.

HUME, I. D., 1982. *Digestive physiology and nutrition of marsupials.* Cambridge University Press: London.

JOHNSTONE, J., 1898. The thymus in the marsupials. *Journal of the Linnean Society of London* **26:** 537–57.

LYNE, A. G. AND HOLLIS, D. E., 1979. Observations on the corpus luteum during pregnancy and lactation in the marsupials *Isoodon macrourus* and *Perameles nasuta. Australian Journal of Zoology* **27:** 881–99.

LYNE, A. G. AND HOLLIS, D. E., 1982. Observations on the lateral vaginae and birth canals in the marsupials *Isoodon macrourus* and *Perameles nasuta* (Mammalia). *Journal of Zoology, London* **198:** 263–77.

LYNE, A. G. AND HOLLIS, D. E., 1983. Observations on graffian follicles and their oocytes during lactation and after removal of pouch young in the marsupials *Isoodon macrourus* and *Perameles nasuta. American Journal of Anatomy* **166:** 41–61.

MACKENZIE, W. C., 1918. *The gastrointestinal tract in monotremes and marsupials.* Critchley Packer Pty Ltd: Australia.

MITCHELL, P. C., 1905. On the intestinal tract of mammals. *Transactions of the Zoological Society of London* **17:** 437–537.

OWEN, R., 1868. *On the anatomy of vertebrates.* Vol. 3. Longmans, Green and Co.: London.

PARSONS, F. G., 1903. On the anatomy of the pig-footed bandicoot *(Choeropus castanotis). Journal of the Linnean Society of London, Zoology* **29:** 64–80.

PEARSON, J., 1940. Notes on the blood system of the Marsupialia. *Papers and Proceedings of the Royal Society of Tasmania* **1939:** 77–94.

PEARSON, J., 1945. The female urogenital system of the Marsupialia with special reference to the vaginal complex. *Papers and Proceedings of the Royal Society of Tasmania* **1944:** 71–98.

RODGER, J. C. AND HUGHES, R. L., 1973. Studies of the accessory glands of male marsupials. *Australian Journal of Zoology* **21:** 303–20.

SAPSFORD, C. S., RAE, C. A. AND CLELAND, K. W., 1970. Ultrastructural studies on the development and form of the principal piece sheath of the bandicoot spermatozoon. *Australian Journal of Zoology* **18:** 21–48.

SCHULTZ, W., 1976. Der Magen-Darm-Kanal der Monotremen und Marsupialier. *Handbuch der Zoologie* **8:** 1–177.

SONNTAG, C. F., 1921a. The comparative anatomy of the koala *(Phascolarctos cinereus)* and vulpine phalanger *(Trichosurus vulpecula). Proceedings of the Zoological Society of London* **1921:** 547–77.

SONNTAG, C. F., 1921b. Contributions to the visceral anatomy and myology of the Marsupialia. *Proceedings of the Zoological Society of London* **1921:** 851–82.

STRAHAN, R. (ed), 1983. *The Australian Museum complete book of Australian mammals.* Angus & Robertson: Sydney.

ULLMANN, S. L., 1979. Observations on the primordial oocyte of the bandicoot *Isoodon macrourus* (Peramelidae, Marsupialia). *Journal of Anatomy* **128:** 619–31.

VAN LENNEP, E. W., 1962. The histology of the mucosa of the small intestine of the long-nosed bandicoot (Marsupialia: *Perameles nasuta* Geoffroi) with special reference to intestinal secretion. *Acta Anatomica* **50:** 73–89.

VAN LENNEP, E. W., 1964. Ultrastructural differentiation of the intestinal epithelium in the bandicoot *Perameles nasuta. Zietschrift fuer Zellforschung* **62:** 485–94.

WOOLLEY, P. AND WEBB, S. J. V., 1977. The penis of dasyurid marsupials. Pp. 307–23 in *The biology of marsupials* ed by B. Stonehouse and D. Gilmore. Macmillan: London.

YADAV, M., 1973. The presence of the cervical and thoracic thymus lobes in marsupials. *Australian Journal of Zoology* **21:** 285–301.

The eye of the Northern Brown Bandicoot, *Isoodon macrourus*

JULIA CHASE[1] and MAL L. GRAYDON[2]

The Northern Brown Bandicoot *(Isoodon macrourus)* has small, laterally positioned eyes, well adapted for nocturnal vision. The retina is thin and avascular but contains many glycogen-rich Müller cells. Rod photoreceptors predominate over a scattering of cones which are characterized by oil droplets. There is a high receptor to retinal ganglion cell convergence ratio of 120 to 1. Retinal ganglion cell topography shows a small degree of regional specialization, having a low centroperipheral cell density gradient indicating that the bandicoot eye is adapted for panoramic surveillance rather than acuity function. The retina is nourished primarily by a sinus-like choriocapillaris which is supplied by larger calibre deep vessels of the choroid. A system of capillary loops penetrates into the vitreous from the margin of the optic disc. Among marsupials this conus-like structure combined with an avascular retina are characteristic of diprotodonts. Neither feature has been noted in non-peramelid polyprotodonts to date.

Key words: Bandicoot, retina, photoreceptor, conus, choriocapillaris, Müller cell, marsupial, taxonomy.

INTRODUCTION

THE histological studies of Franz (1913, 1934) and the artistic illustrations of Johnson (1901) currently provide the most comprehensive information on the anatomy of the marsupial eye. Among 15 marsupials, Johnson included two bandicoots, *Macrotis lagotis* and *Isoodon obesulus*. Both were found to have avascular retinae and a vascular, conus-like structure in the vitreous. More recently, Tancred (1981) described the retinal ganglion cell topography of *I. obesulus* and other marsupials. Apart from these studies there appears to be little other information on the eye of the bandicoot or indeed, on most marsupial eyes (Johnson 1977). There is, however, quite a bit of information available on the central anatomical organization of marsupial visual systems (see particularly Sanderson, Haight and Pettigrew 1984).

The aim of the present study was to improve our knowledge of the marsupial visual system and in particular that of bandicoots, which is of special interest for two reasons. Firstly, bandicoots occupy a strategic phylogenetic position by being both polyprotodont and syndactylous (Archer 1984). Secondly, bandicoots occupy an interesting ecological niche which is similar to that of a rat. Both of these aspects make bandicoots an ideal experimental model for comparative vision research.

MATERIALS AND METHODS

The animals used for this work were taken from breeding stock kept at the University of Queensland, Department of Anatomy Native Animal Research Unit. We used four adult Northern Brown Bandicoots *(Isoodon macrourus)* weighing about 1.5–2 kg.

The eyes of *I. macrourus* were examined and photographed with a fundus camera while the animal was lightly anaesthetized with Nembutal and the pupil dilated by application of n-ethyl-n (4-pyridylmethyl) tropamide (Mydriacyl).

Light microscopic examination of retinal histology and histochemistry were done on 1 μm sagittal sections of formalin of glutaldehyde fixed eyes, embedded in glycol methacrylate (JD4, Polysciences). Alternate slides were stained with haematoxylin and eosin and periodic acid Schiff's reagent (PAS). PAS was used to localize glycogen in the retina. Control slides were treated with amylase prior to staining. The densities of the receptor, interneuron and ganglion cell layers were determined by counting all cells within an area defined by an eyepiece grid and correcting for differences in cell size (Abercrombie 1946; Chase 1972). For electron microscopy, eyes were fixed by perfusion with 4% glutaraldehyde and 2% paraformaldehyde in 0.1M-phosphate buffer (pH 7.2), washed with phosphate buffer, post-fixed in osmium tetroxide, stained with uranyl acetate, and embedded in Epon. Material was sectioned at 750 (Å) A for the examination of fine structure.

Two retinal wholemounts (Stone 1981) were prepared by perfusing an anaesthetized bandicoot with 10% phosphate buffered (0.1M, pH 7.2) formalin, removing each eye, and dissecting out the lens and vitreous. The other wholemounts

[1]Department of Biological Sciences, Barnard College, Columbia University, New York, New York, U.S.A. 10027.
[2]Department of Anatomy, University of Queensland, St Lucia, Queensland, Australia 4067.
Pages 117–22 *in* BANDICOOTS AND BILBIES, ed by J. H. Seebeck, P. R. Brown, R. L. Wallis and C. M. Kemper. Surrey Beatty & Sons, Sydney, 1990.

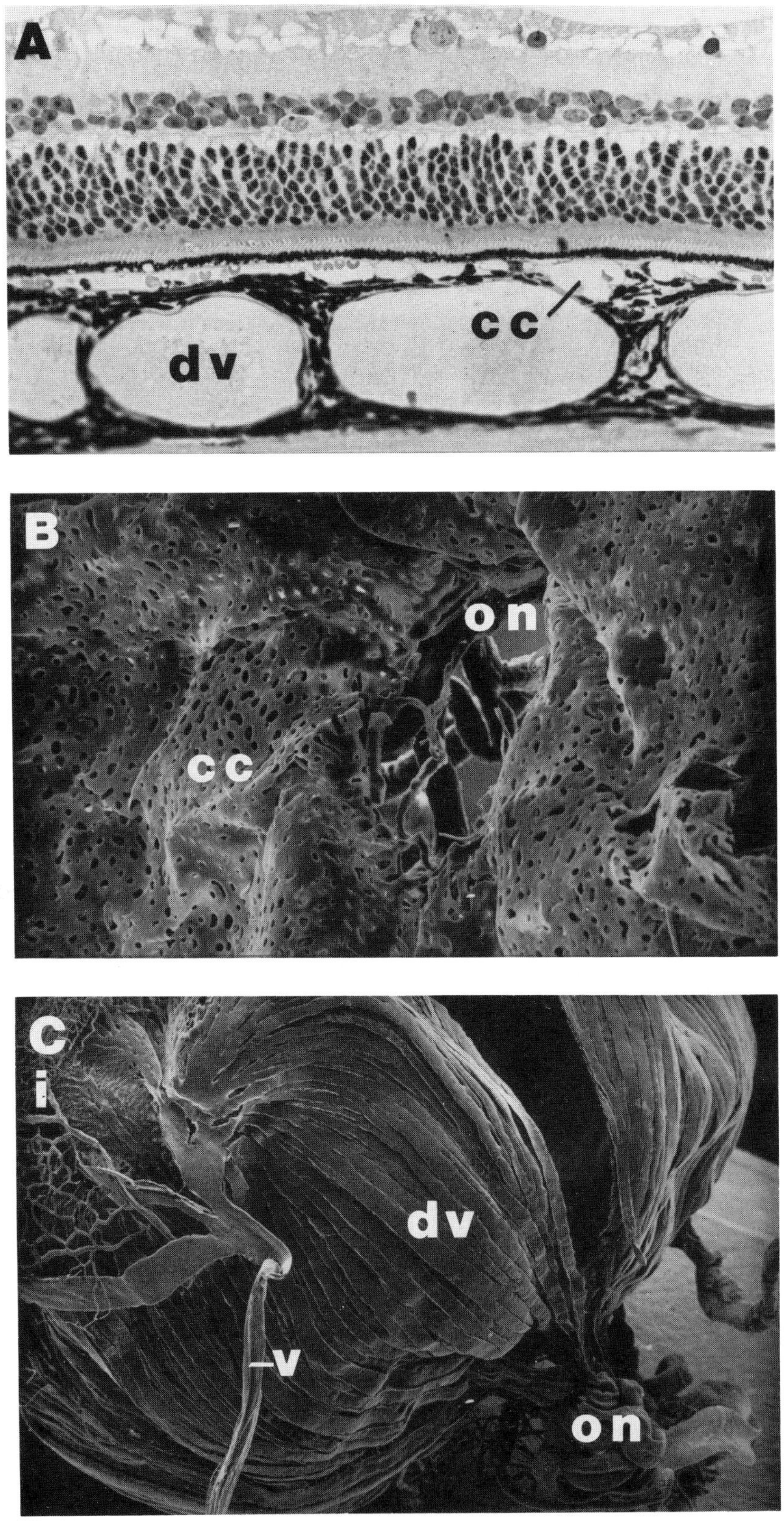

Fig. 1A, B, C. Choroidal vessels supplying the avascular retina of *Isoodon macrourus.* Note the extensive system of deep vessels (dv) which arise from the vessels of the optic nerve (on) and supply the sinus-like choriocapillaris (cc) subtending the retina. Anteriorly, the deep vessels connect with the circulation of the iris (i) and four large vessels (v) that appear to drain the four quadrants of the eye. (A): Epon section of choroid and retina, haematoxylin-eosin stain (264×). (B) and (C): SEM views of methylmethacrylate casts (16×); the view in (B) from within the eye and (C) posteromedial aspect of the eye.

were prepared from simple immersion fixed specimens obtained using a procedure to histochemically demonstrate blood vessels vitread to the retina (see below). Radial cuts were made in the cornea and sclera at eight locations to allow the retina to be pinned out flat. The cells of the ganglion cell layer were stained with cresyl violet and counted using a lens net graticle that superimposed a grid covering 0.016 mm^2 at 1 mm steps over the field of view (Graydon and Giorgi 1984; Giorgi and Graydon 1987). Appreciable shrinkage of the wholemount was avoided by leaving the iris attached to the retina and allowing the iris to dry and adhere to the glass slide in advance of the retina.

To examine the three dimensional structure of the choroidal vasculature, the vessels of the eye were perfused with 0.9% heparinized NaCl followed by methyl methacrylate solution (Batson's compounds A, B, C, Polysciences). The eye was then placed in 50°C water for 30 min to speed polymerization of the plastic. All tissues were digested with a 12% KOH solution, leaving a cast of arteries, capillaries and veins that could be examined microscopically or with SEM.

Conus-like central vitreal vessels were demonstrated histochemically with diaminobenzidene tetrahydrochloride (DAB), which reacted with the endogenous peroxidase of red blood cells. For this purpose the retinal wholemounts were prepared as above, but from eyes with the ophthalmic vessels clamped to reduce blood efflux after dissection.

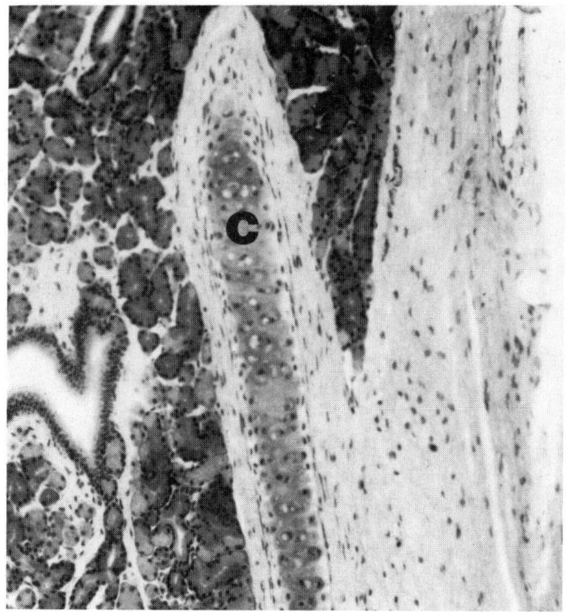

Fig. 2. Cartilage plate (c) in the sclera. Glycolmethacrylate section, haematoxylin-eosin stain. (170X).

RESULTS

The eyes of our specimens of *I. macrourus* were nearly spherical, with a diameter of approximately 7 mm and a white nictitating membrane which swept across the cornea in a temporal to nasal direction.

Ophthalmoscopic examination revealed a pink fundus with prominent bundles of axons converging on the circular white optic disc. The ventral half of the retina was more darkly pigmented than the dorsal half. Although the bandicoot eye returned a dull yellow eyeshine, there was no evidence ophthalmoscopically or histologically of a tapetum. The lens was biconvex with a 0.7 ratio of axial (2.5 mm) to radial (3.7 mm) diameters.

The retina of *I. macrourus* was completely avascular, i.e., there were no intraretinal vessels although both the choriocapillaris and the deep choroidal vessels were well-developed (Fig. 1A). The choriocapillaris was formed by many sinus-like vessels bathing the posterior surface of the retina (Fig. 1B). The deep vessels, which measure 70 μm in diameter, filled the entire space behind the choriocapillaris. Packed side by

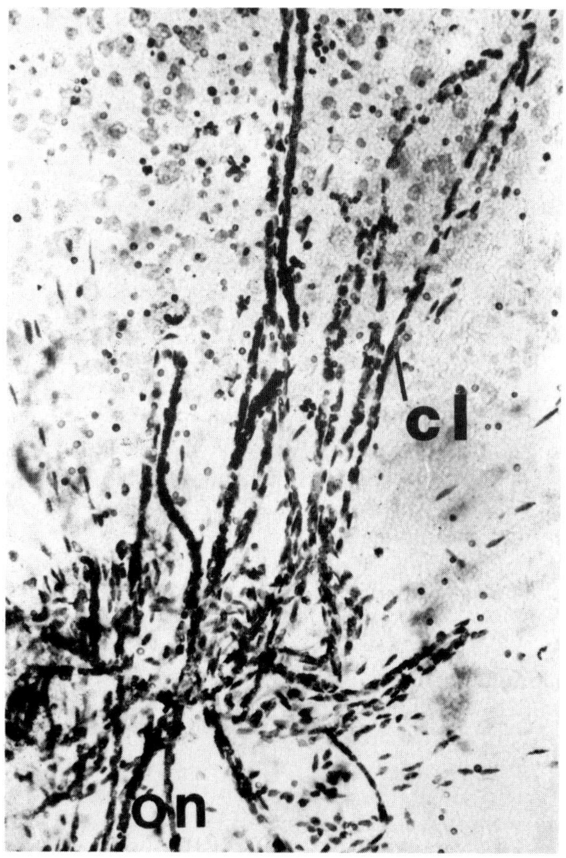

Fig. 3. Retinal wholemount showing capillary loops (cl) of the conus emerging from the optic nerve head (on). DAB-H_2O_2 reaction for endogenous peroxidase. (140X).

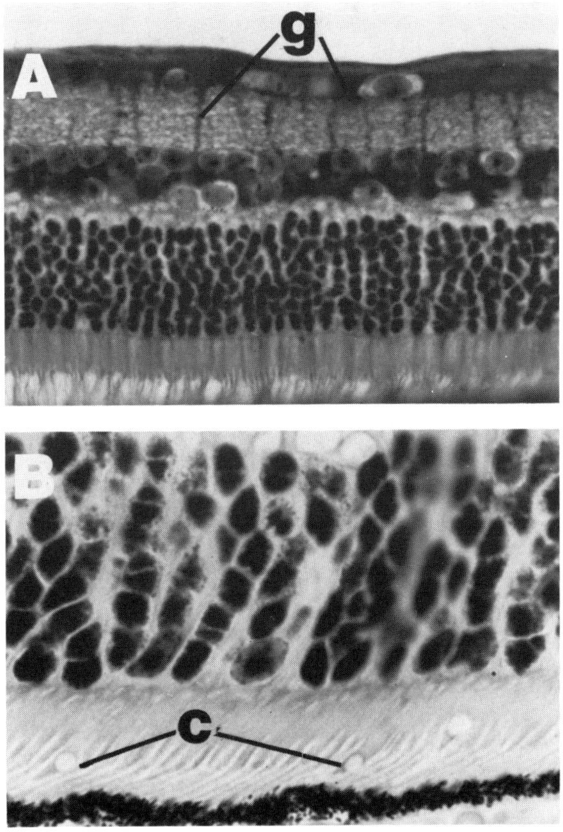

Fig. 4A,B. (A): Periodic acid and Schiff's reagent (PAS) used to demonstrate glycogen storage (g) in the inner plexiform and inner nuclear layers. Glycolmethacrylate section, PAS (325X). (B): Cones (c) with oil droplets interspersed among the rods. Glycomethacrylate section, haematoxylin-eosin stain. (540X).

side, these vessels run from the optic nerve to the ora serrata where they connect with the circulation of the iris and appear to empty into four large vessels that emerge from the four quadrants of the eye (Fig. 1C). The sclera was characterized by the presence of a cartilagenous plate (Fig. 2).

The most unusual feature we found in the bandicoot eye was the presence of a conus-like structure in the vitreous (Fig. 3). This was studied by ophthalmoscopic examination, histology and microdissection. In consisted of a bundle of capillary loops rising along the optic axis from the optic disc and protruding approximately 800 μm into the vitreous. This conus consisted of approximately 12 capillary loops each, by definition comprised only of endothelium and having diameters of 8 to 12 μm. These straight, thin-walled vessels emerged from larger vessels at the margin of the optic disc and eventually formed simple, hairpin turns with occasional interconnecting links. There was neither pigmentation nor an obvious connective tissue support structure associated with the conus. Its position at the optic nerve head suggests that it may occupy the tunnel left by the receding hyaloid vessels and be supported by the remaining connective tissue fibres that run between the

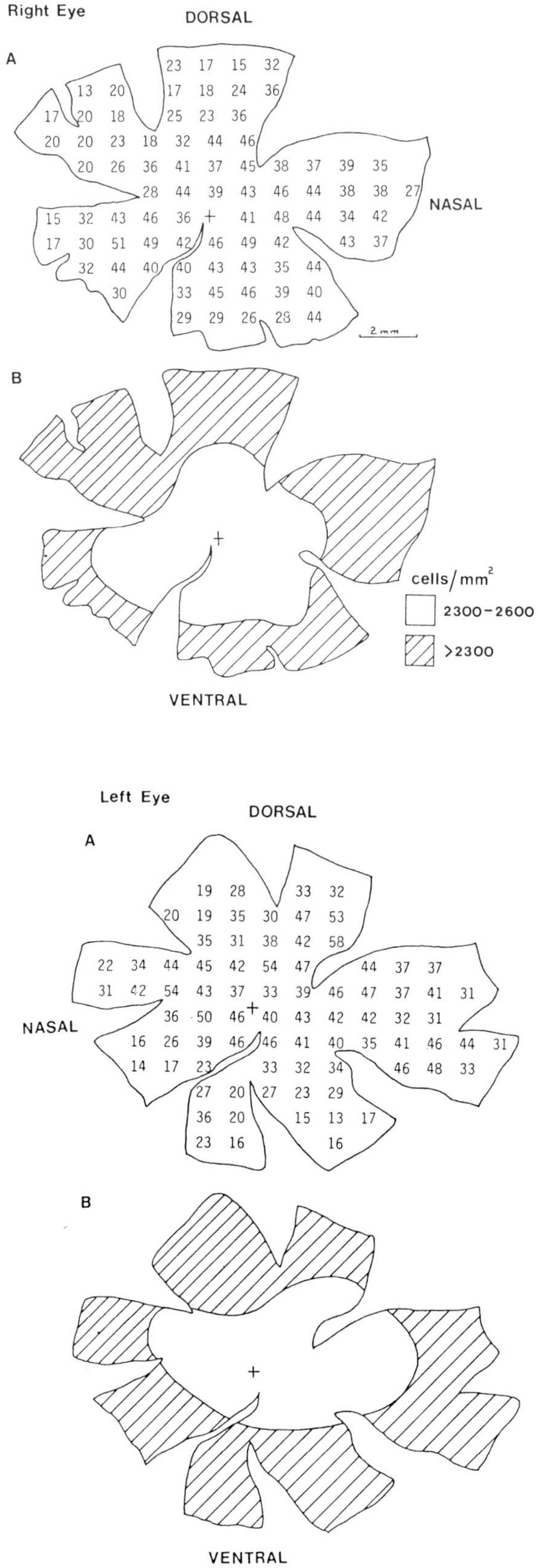

Fig. 5A,B. Right and left eyes retinal wholemount outlines for an individual specimen of *Isoodon macrourus*. The central cross represents the optic disc. (A): Sampling sites and retinal ganglion cell layer cell numbers per site. (B): Isodensity line profile of cell density (N/mm^2) in the retinal ganglion cell layer.

optic nerve head and the posterior face of the lens, but we have no evidence to support this suggestion.

The neural component of the retina of *I. macrourus* measured only 125 μm from choroid to vitreous. The rod outer segments were short (14 μm) for a nocturnal animal. The outer plexiform later was barely 6 μm thick. The Müller cells of the inner retina were well-supplied with glycogen as shown in Figure 4A. The receptors were predominantly rods with a scattering of cones throughout the retina. Cones could be distinguished by the presence of oil droplets in the inner segments, short conical outer segments, a tigroid nucleus, and under electron microscopy, by triangular multisynaptic pedicles. As expected for a nocturnal animal, the receptors were densely packed (432 000 receptors per mm^2) and there was a high degree of neural convergence with a ratio of 120 receptors:1 ganglion cell. The map of the ganglion cell distribution shows a fourfold decrease in the number of ganglion cells (2 300 vs $500/mm^2$) as one moves from the fundus to the periphery (Fig. 5). This is a relatively low centroperipheral cell density gradient.

DISCUSSION

The bandicoot eye is primarily designed for nocturnal vision. The lateral placement of the eyes provides an almost panoramic view of the animal's surroundings at the expense of binocular vision (Pettigrew 1972). Johnson (1901) found bandicoots to have an optic axis divergence of 75°, one of the most extreme noted in mammals and nearly equal to the divergence he measured in rabbits (85°). Rod densities are similar to those recorded in such placental mammals as bats, rats and voles. The predominance of rods is characteristic of nocturnal species as is the high convergence of receptors onto ganglion cells. Convergence increases sensitivity at the price of reduced acuity. In comparison to other nocturnal mammals, the bandicoot retina shows little regional specialization. There is no morphologically distinct fovea or area. The peak ganglion cell density is only half that noted in other nocturnal marsupials such as the Tasmanian Devil *(Sarcophilus harrisii)* and the Red-bellied Pademelon *(Thylogale billardierii)* (Tancred 1981) but is comparable with that in the Virginia opossum *Didelphis virginiana* (Rapaport, Wilson and Rowe 1981) and only one-fourth that noted in the cat (Hughes 1975). In the bandicoot, the ganglion cell density is only three to four times greater in the central fundus than in the periphery while in *S. harrisii* there is a 20 fold centroperipheral gradient and in the cat a 30 fold gradient, suggesting a reduced visual acuity in bandicoots when compared with other nocturnal species (Rodieck 1973; Frisen and Frisen 1976). Tancred (1981) reported the presence of a visual streak and a weak area centralis in the Southern Brown Bandicoot *(I. obesulus)* but these features were not defined in our preparations of *I. macrourus*.

The general histological features of the bandicoot retina are typical of an avascular retina. The thinness of the retina, the lack of a tapetum, the rich choroidal vasculature, the shortness of the photoreceptors, and the presence of rich glycogen deposits in the Müller cells of the inner retina are all characteristics shared with other mammalian avascular retinae (Chase 1982). In such eyes, metabolically dependent on diffusion from the choroid, retinal thickness theoretically should not exceed the oxygen diffusion maximum of 142 μm (Dollery, Bulpitt and Kohner 1969). This restriction in the overall thickness would appear to account for the abbreviated outer segments of the photoreceptors. Since the choriocapillaris is the primary source of nutrients, it would be expected to be particularly well-developed as it is in the bandicoot. The absence of a tapetum is a characteristic of mammalian avascular retinae. Presumably there is selective pressure against the evolution of fibrous or crystalline structures in the pigment epithelium or choroid which could impede diffusion exchange between retina and choroid. The prominent, stored glycogen of mammalian avascular retinae is thought to provide a reserve of glucose for the inner retina, which is distant from choroidal blood vessels (Kuwabara and Cogan 1963; Chase 1982).

To our knowledge, this is the first report of a cartilagenous plate in the sclera of a marsupial eye. Such plates, supposed to support the shape of the eyeball, have been described in the Platypus (Owen 1968) and lower vertebrates.

The discovery of a conus with retinal avascularity in all three species of bandicoots examined to date, *I. macrourus* (this study), *I. obesulus*, and *M. lagotis* (Johnson 1901), raises questions once again about the classification of bandicoots with the polyprotodonts. The retinae of all other polyprotodonts described to date, *D. virginiana* and *Marmosa mexicana* (Walls 1942), *S. harrisii* (Johnson 1901; Tancred 1981), and *Thylacinus cynocephalus* (Franz 1913, 1934; Walls 1942) have well-vascularized retinae and none is reported to have a conus. All but *M. mexicana* are reported to have a tapetum. With one exception, all diprotodonts examined to date, including the 10 species described by Johnson (1901) (Macropodoidea, Phalangeridae, Vombatidae) and the two possum species we have studied, *Pseudocheirus peregrinus* and *Trichosurus vulpecula* (see also Prince 1949), have an avascular retina, a conus, and no tapetum. Among diprotodonts only *Petaurus* sp. (Duke-Elder 1958) is said to have a retinal circulation, but it is unclear from Duke-Elder's description whether the vessels are true retinal vessels or vitreal vessels that run on the surface of the retina. Thus, the evidence to date suggests that retinal vascularity is characteristic of polyprotodonts and retinal avascularity is characteristic of diprotodonts.

The characteristics of the retina discussed above may be relevant to the taxonomic classification of bandicoots. In mammals, the evolution of retinal blood vessels appears to have been a necessary condition for the evolution of the two main adaptations for increased light sensitivity in nocturnal vision: the tapetum, and long photoreceptors (Chase 1982). Retinal circulation seems to be a "key character" in van Valen's sense of the term, that is, "an element of physiology that makes a taxon . . . appreciably more efficient than . . . its ancestors" (van Valen 1971). Moreover, the evolution of retinal blood vessels is a rare event that, according to Walls (1942), occurred only three times in vertebrates: in eels, in polyprotodont marsupials, and in placental mammals. Furthermore, the differences between retinae with and without blood vessels are not trivial: they differ anatomically, biochemically and probably functionally (Chase 1982). Finally, the presence of a conus is extremely rare in mammals. It has been found only in the diprotodont marsupials and the hystricomorph rodents.

Clearly, we need to know more about the visual systems of marsupials before we can infer phylogenetic relationships. Fundus patterns have been described in less than 10% of the living marsupial species. The apparent similarities between the eyes of bandicoots and diprotodonts should, however, contribute to the long-standing debate as to whether bandicoots should be classed with the polyprotodonts, with whom they share patterns of dentition and blood protein affinities, or with the diprotodonts, with whom they share the characteristics of syndactyly (Archer 1984) and retinal morphology.

ACKNOWLEDGEMENTS

We thank L. Hall for providing the bandicoots used in the study. J. Pettigrew loaned us the fundus camera and J. Massey assisted in the histology. We thank P. Giorgi, L. Hall and two anonymous referees for their comments and suggestions. Travel funds to J. Chase were provided by Barnard College. M. Graydon was supported by a Commonwealth Postgraduate Award.

REFERENCES

ABERCROMBIE, M., 1946. Estimation of nuclear population from microtome sections. *Anatomical Record* **94:** 239–47.

ARCHER, M., 1984. The Australian marsupial radiation. Pp. 633–808 in *Vertebrate zoogeography and evolution in Australasia* ed by M. Archer and G. Clayton. Hesperian Press: Carlisle, WA.

CHASE, J., 1982. The evolution of retinal vascularization in mammals: a comparison of vascular and avascular retinae. *Ophthalmology* **89:** 1518–25.

CHASE, J., 1972. The role of vision in echolocating bats. Ph.D. Thesis: Indiana University.

DOLLERY, C. T., BULPITT, C. J. AND KOHNER, E. M., 1969. Oxygen supply to the retina and choroidal circulation at normal and increased arterial oxygen tensions. *Investigative Ophthalmology* **8:** 588–94.

DUKE-ELDER, SIR W. S., 1958. *The eye in evolution.* Volume 1. Henry Kimpton: London.

FRANZ, V., 1913. *Lehrbuch der vergleichenden microskopiechen Anatomie der Wirbeltiere.* Schorpan Verlag: Jena.

FRANZ, V., 1934. Anatomie des Wirbeltierauges. In *Handbuch der vergleichenden Anatomie der Wirbeltiere,* Bd 2, Abt. 3, Kap. 1, Urban u. Schwarzenberg: Berlin u. Wien.

FRISEN, L. AND FRISEN, M., 1976. A simple relationship between the probability distribution of visual acuity and the density of retinal output channels. *Acta Opthalmologica* **54:** 437–44.

GIORGI, P. P. AND GRAYDON, M. L., 1987. Retinal wholemounts: a simple method for precise mapping. *Journal of Neuroscience Methods* **22:** 119–24.

GRAYDON, M. L. AND GIORGI, P. P., 1984. Topography of the retinal ganglion cell layer of *Xenopus. Journal of Anatomy, London* **139:** 145–57.

GRIFFITHS, M., 1968. *Echidnas.* Pergamon: Oxford.

HUGHES, A., 1975. A quantitative analysis of the cat retinal ganglion cell topography. *Journal of Comparative Neurology* **163:** 107–28.

JOHNSON, G. L., 1901. Contributions to the comparative anatomy of the mammalian eye, chiefly based on ophthalmoscopic examination. *Philosophical Transactions of the Royal Society of London* Series B., **194:** 1–82.

JOHNSON, J. I., 1977. Central nervous system of marsupials. In *The biology of marsupials* ed by D. Hunsaker. Academic Press: New York.

KUWABARA, T. AND COGAN, D., 1963. Glycogen in the retina. *Annals of Histochemistry* **8:** 223–28.

OWEN, R., 1868. *Anatomy of vertebrates,* Volume III, *Mammals.* Longmans, Green and Co.: London.

PETTIGREW, J. D., 1972. The neurophysiology of binocular vision. *Scientific American* **227:** 84–95.

PRINCE, J. H., 1949. *Visual development.* Livingstone: England.

RAPAPORT, D. H., WILSON, P. D. AND ROWE, M. H., 1981. The distribution of ganglion cells in the retina of the North American opossum *(Didelphis virginiana). Journal of Comparative Neurology* **199:** 465–80.

RODIECK, R. W., 1973. *The vertebrate retina: Principles of structure and function.* W. H. Freeman and Co.: San Francisco.

SANDERSON, K. J., HAIGHT, J. R. AND PETTIGREW, J. D., 1984. The dorsal lateral geniculate nucleus of macropodid marsupials: cytoarchitecture and retinal projections. *Journal of Comparative Neurology* **224:** 85–106.

STONE, J., 1981. *The wholemount handbook.* Maitland Publication Pty Ltd: Sydney.

TANCRED, E., 1981. The distribution and sizes of ganglion cells in the retinas of five Australian marsupials. *Journal of Comparative Neurology* **196:** 585–603.

VAN VALEN, L., 1971. Adaptive zones and the orders of mammals. *Evolution* **25:** 420–28.

WALLS, G. L., 1942. *The vertebrate eye and its adaptive radiation.* Cranbrook Institute of Science: Bloomfield Hills, Michigan.

Growth and a description of the development of external features of pouch young of captive *Isoodon macrourus*

LESLIE S. HALL[1]

Measurements of the length of hind foot, head, tail, body weight, and observations on external features were made at regular intervals on 63 known age pouch young *Isoodon macrourus*. Front feet digits are separate and clawed at birth but hind feet digits and claws do not appear until around seven days. The pouch is evident at seven days and the scrotum appears at nine days. Eyes commence to open around 45 days, mystacial vibrissae appear at 18 days, and there is a fine cover of body hair at 40 days. Ears become free from the investing epitrichium at 14 days.

Head length showed a linear relationship with age in days [head length (mm) = 4.429 + 0.894 day] and was found to be the best predictor of age. Hind foot and tail length also showed a linear relationship with age, but were less accurate in predicting age. Body weight showed a distinct curvilinear relationship with age but was the least reliable method of predicting age.

These observations closely resemble those from other members of the Peramelidae and represent one of the shortest periods of development and pouch life in marsupials.

Key words: *Isoodon*, pouch young, development, growth.

INTRODUCTION

THE rapidity of bandicoot growth and development has been recorded and commented on by Mackerras and Smith (1960) and Lyne (1964). Because of a very short gestation and the rapid development and growth of pouch young, Russell (1982) considered that bandicoots have followed a very different path to other marsupials, which are regarded as one of the slowest growing mammals (Case 1978). *Isoodon macrourus* has a gestation period of 12.5 days (Lyne 1974) and pouch life lasts approximately 50±7 days (Gordon 1974) after which only a few days are spent in weaning. The mean litter size for wild-caught *I. macrourus* around Brisbane is 2.9 (Gemmell 1982; Hall 1983) but variations can be caused by seasonal effects (Gordon 1971), stage of lactation (Gemmell, Walker, Johnston and Cepon 1984b), while additional trapping and handling in captivity subsequently reduces litter size (Mackerras and Smith 1960; Gemmell 1982).

Mackerras and Smith (1960) gave brief data on the growth and development of *I. macrourus* up to weaning and Gemmell *et al.* (1984b) looked at the post weaning growth rate from 65 to 105 days after birth.

As there is always a need for an accurate ageing technique for pouch young marsupials based on known age specimens, this study reports on the growth of captive *I. macrourus* from birth to the end of pouch life. Observations are also made on the development of external features during this period. These data are compared with observations on the growth and development of other species of bandicoots as well as for other marsupials.

MATERIALS AND METHODS

Bandicoots were trapped in the vicinity of Brisbane between September 1977 and May 1984 (see Hall 1983) and housed in outdoor yards at the University of Queensland's veterinary farm, Pinjarra Hills. The animals were fed dry dog cubes (Provincial Traders Pty Ltd, Brisbane) *ad libitum*, supplemented by invertebrate rich compost (2 kg/animal/week) and sweet potato (200 g/animal/week). Water was available at all times. The yards were of varying size (max. 565 m^2, min. 60 m^2) with a dense ground cover provided by a tall introduced pasture grass, *Panicum maximum*. Wooden nest boxes (300×200×1 000 mm) were placed in two of the yards to facilitate capture of animals. Bandicoots were also trapped in the yards using a similar technique to that described by Hall (1983).

The density of animals in the yards varied, but was generally of the ratio one male to three females per 150 m^2, The technique for obtaining known age pouch young *I. macrourus* has been described by Hall and Hughes (1985). Head, tail and hind foot length measurements of young were made with vernier calipers. Body weights of pouch young which had recently died or were removed from the pouch and teat for

[1]Department of Anatomy, University of Queensland, St Lucia, Queensland, Australia 4067.

Pages 123–33 *in* BANDICOOTS AND BILBIES, ed by J. H. Seebeck, P. R. Brown, R. L. Wallis and C. M. Kemper. Surrey Beatty & Sons, Sydney, 1990.

other studies (e.g., Hall and Hughes 1985), were made on a Mettler electric balance. Measurements of litters were taken at three to seven day intervals and captures of females were staggered to give an approximate continuous series of measurements of pouch young from birth up to 60 days.

RESULTS

Twenty-three litters of *I. macrourus* were born in the yards, producing 82 young (average 3.57 per litter), 63 of which (77%) survived to the end of pouch life. Of the 19 pouch young lost, 10 died when they were ejected from the pouch of females trapped overnight, but the fate of the remaining nine is unknown. There was no obvious relationship between mortality and the age, sex and position on teats of the young, nor with month of year or size of female.

External features

At birth *I. macrourus* had a highly transparent skin, giving the newborn a reddish pink colour. Numerous blood vessels could be seen. In particular the facial, brachial/axillary, lateral thoracic and femoral blood vessels were prominent (Fig. 1). The pale cartilage rib models and their intercostal blood vessels were visible. A superficial layer of epitrichium up to six cells thick covered the lower and upper jaws and extended caudally to cover the eye and ear bulge. The rest of the body was covered by a thin layer of epitrichium. The tail was short, curved and closely appressed to the lower abdomen. The immobile and unrotated hind feet showed early digital rays and clasped the tail from either side. In contrast, the forelimbs were pronated, had separate clawed digits, and were in constant motion. The permanent claws on the 2nd, 3rd and 4th digits were covered with additional long yellowish deciduous claws which were eventually shed between three and seven days. The head was disproportionately large and bent at a right angle to the body. The retinal pigmentation of the eye appeared as a dark ring and the nostrils were open and prominent. Following the ingestion of milk, the intestines were visible as a whitish bulge in the lower abdomen (Fig. 2). Twenty min. (±15 m) after birth the crown-rump length was 13.3 mm±0.5 SD (n=15), head lenth 6.16 mm±0.3 SD (n=15) and body weight 0.21 g±0.005 SD (n=7).

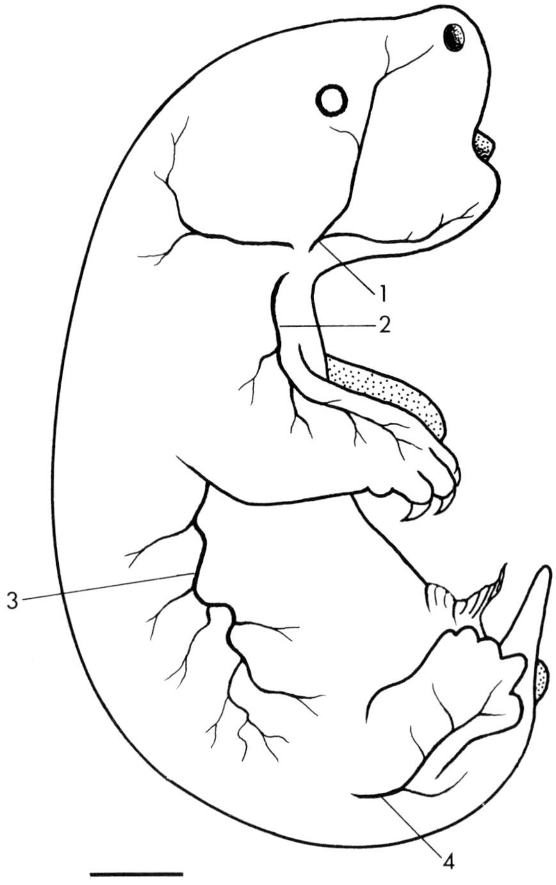

Fig. 1. Outline of *Isoodon macrourus* at birth showing the facial (1), brachial/axillary (2), lateral thoracic (3) and femoral (4) blood vessels. The pigment of the retina is visible through the epitrichium, forelimbs are pronated and clawed, hind limbs show early digital separation. Scale bar = 1 mm.

The subsequent development of external features is summarized in Table 1 and shown in Figures 3, 4 and 5. There was very little difference in the rate of development between litters, sex, teat locations and maternal body size. No litter runts were observed.

The sex of pouch young could be determined by unaided gross manual examination at seven days in females when a small pouch appeared, and at nine days, when the scrotum was present in males.

Table 1. The development of external features of *Isoodon macrourus* from birth to weaning. Small variations in timing occur and days cited represent the average.

Age	External development features
0	See text for detailed description and Lyne (1974) for description of the birth process
2	Umbilicus disappears, thick epitrichium covers most of head region
4	Deciduous claws on digits 2, 3 and 4 of forelimb lost
5	Prominent ear bulge
7	Pouch present in females
9	Scrotum present in males
10	Hindlimbs commence rotation
12	Teats present in female pouch
14	Ears free of epitrichium
15	Claws appearing on hindlimb digits
18	Mystacial vibrissae appear
23	Eyelids separating
25	Righting reflex present
30	All vibrissae present
35	Ears unplugged
37	Lips opening
39	Pelage pigment shows on dorsum
42	Short fur all over body
45	Eyes commence to open
48	Detaching from teat
50	Eyes fully open, teat swapping occurs
52	Young leave pouch but are not fully weaned
60	Mostly independent

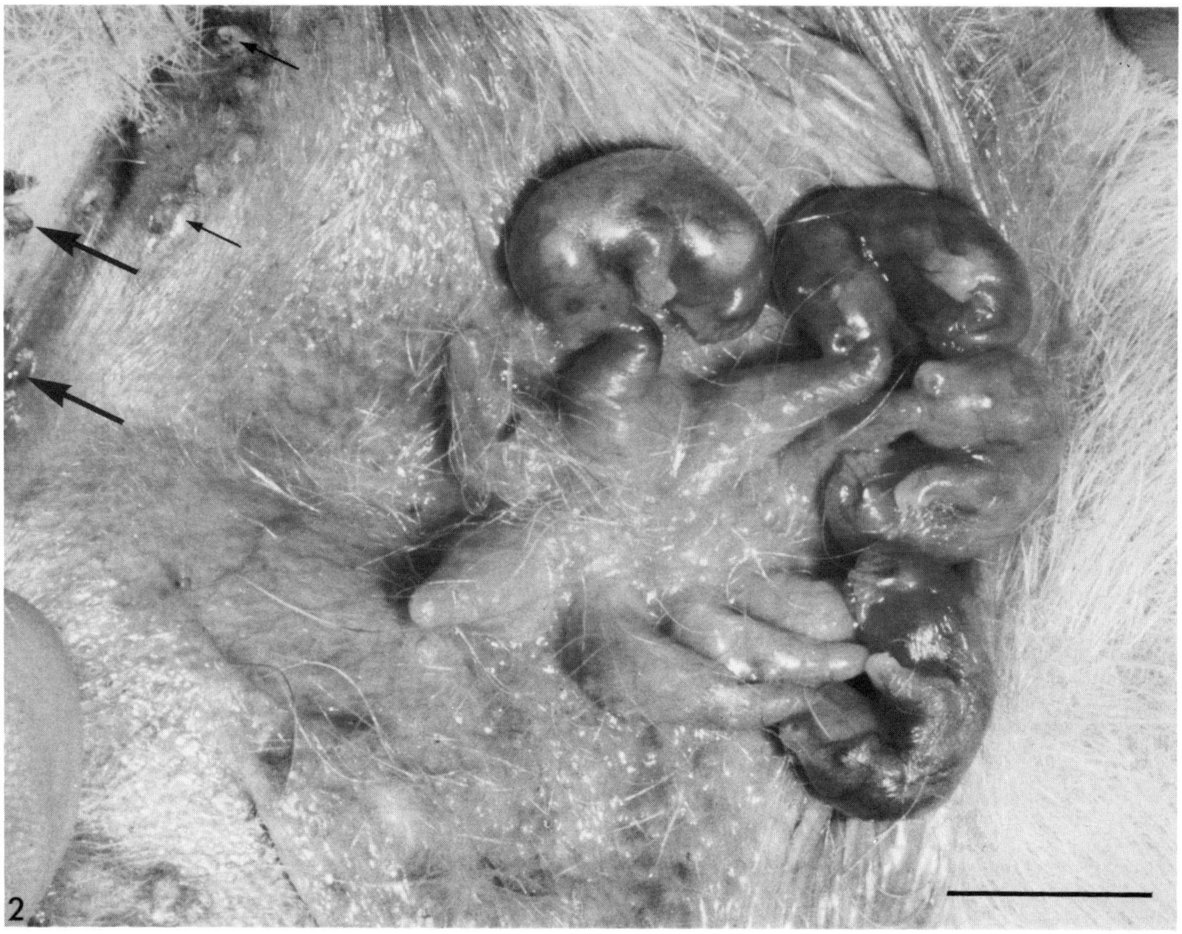

Fig. 2. Exposed pouch of *Isoodon macrourus* containing four young which are 5 h old. Note prominent blood vessels identified in Figure 1, and milk already visible in top specimen. There are ticks at the entrance of the pouch (large arrows) and scar tissue resulting from ticks and trombiculid mites (small arrows). Scale bar = 10 mm.

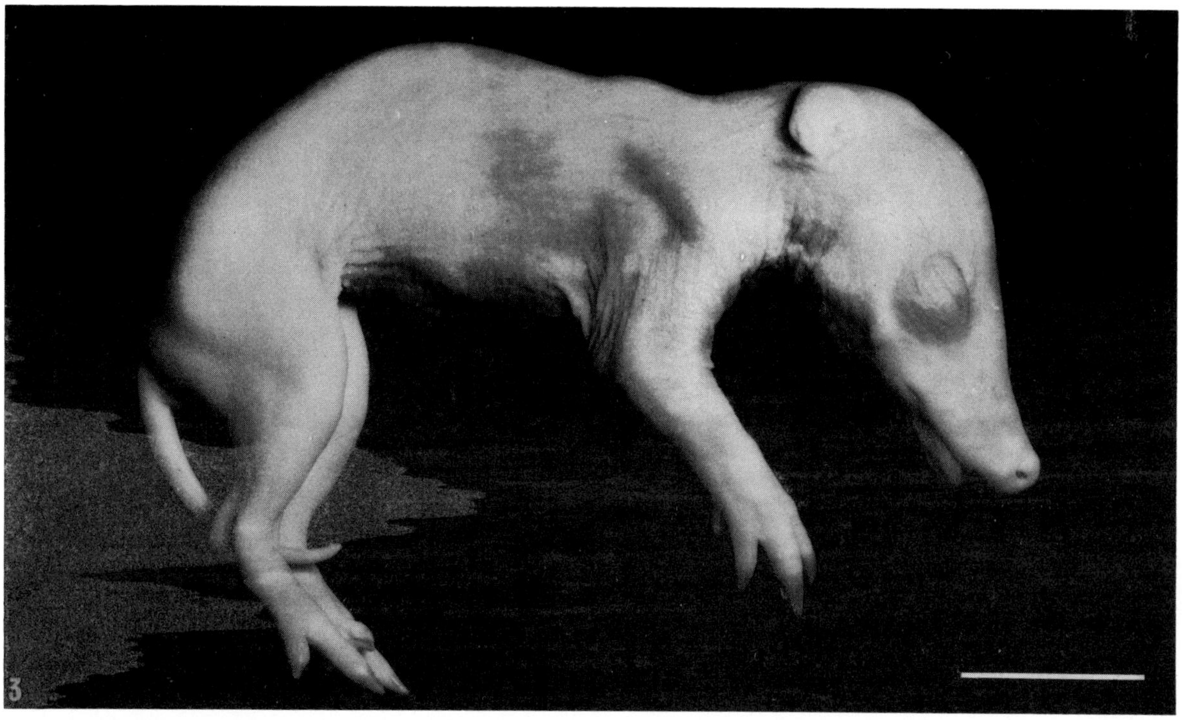

Fig. 3. Twenty day old pouch young *Isoodon macrourus.* Note claws on all digits, ear free of epitrichium and eyelids incompletely formed. Scale bar = 10 mm.

Fig. 4. Thirty-five day old pouch young *Isoodon macrourus*. Note prominent mystacial vibrissae but no body fur. Eyelids are formed but closed and the ear is no longer filled with epitrichium. Scale bar = 10 mm.

The ears became free of epitrichium at 14 days and were unplugged at 35 days (Fig. 4). However, a basic righting reflex in pouch young was evident around 25 days. By one day epitrichium had fused the lips tightly around the teat. The fused lips commenced to open at 37 days and teat swapping by young occurred after 48 days when their lips had become totally separated (Fig. 5). The first signs of skin pigmentation occurred on dorsal areas at 39 days and short fur covered the pouch young three or four days later.

The formation of eyelids occurred in the first week of pouch life, and the line of separation of the eyelids became apparent at 23 days. There was some variation in the timing of the eye opening with the earliest open at 45 days, and the latest at 50 days. In several litters (45 and 47 days), different young showed stages from fully closed, partly open, to fully open eyes at the same time.

By 55 days most young had permanently vacated the pouch, but teats often showed signs of continued use up to 60 days.

No ticks (mostly *Ixodes holocyclus* and *I. tasmani* on adults) were seen on pouch young, even during periods of heavy infestation of the adults (January to May) (Fig. 2). Trombiculid mites (family Trombiculidae) which are frequently found in large aggregations on the area of short hair surrounding the posterior margin of the pouch, were never observed on pouch young (Fig. 2).

Measurements from the 63 young were pooled to construct Figures 6–9 which show the growth of head, hind foot, tail and body weight of pouch young *I. macrourus* as a function of age from birth to 60 days. Due to the lack of differentiation, no foot length measurements prior to nine days were possible (Fig. 7), and the small size curvature of the tail made its measurement before nine days difficult (Fig. 8).

In order to use the data shown in Figures 6–9 for predictive purposes, they were subjected to regression analysis.

Based on 482 measurements, the relationship between head length and age could be expressed by:

$$\text{Head} = 4.429 + 0.894\ \text{day}.$$

The hind foot length equation from 119 measurements was:

$$\text{Hind foot length} = -4.858 + 0.789\ \text{day}.$$

The tail length equation from 119 measurements was:

$$\text{Tail length} = -12.33 + 1.415\ \text{day}.$$

The body weight equation from 141 measurements was:

$$\text{Body weight}^{1/3} = 0.3745 + 0.0746\ \text{day} \quad \text{(i)}$$

Body weight$^{1/3}$ was obviously not linear when expressed as weight$^{1/3}$/day and the following equation improved the fit of the data quite substantially.

$$\text{Body weight}^{1/3} = 0.6311 + 0.0455\ \text{day} + 0.000502\ \text{day}^2 \quad \text{(ii)}$$

Fig. 5. Forty-five day old pouch young *Isoodon macrourus*. At this stage the body has a full cover of short fur and the eyes commence to open. The lips are open at the front and angle of the mouth, but are still fused in the mid-section (between arrows).

Standard deviations for the above formulae are head/days = 0.975, foot/days = 2.241, tail/days = 3.913, weight$^{1/3}$/days = 0.18617 (ii)

An analysis of log weight did not provide a simple model. When analysed separately, there was very little difference between litters in the equations. There were no litter effects for head or hind foot length versus day, and only trivial effects of litter on tail length and body weight. Thus litter effects can reasonably be ignored in age prediction.

There was strong evidence of some curvature in all the relationships (Figs 6–9) but it was much less apparent in the head measurements.

To compare the measurements for efficiency of predicting age in *I. macrourus* between birth and 60 days, the following formula can be used.

The standard error of a predicted day is approximately (accurate to within one or two per cent) given by:

SE (days) $= \frac{s}{b}$ where s is the DS (Y) given day, and b is the coefficient in the fitted equation.

Confidence intervals for days are then given by:

$$\text{Days}_l = \text{Days}^* - t\frac{s}{b} \text{ and}$$

$$\text{Days}_u = \text{Days}^* + t\frac{s}{b}$$

Where Days* is the estimated days from the fitted regression equation, t is the Student's *t* value for $P = 0.05$, and Days_l and Days_u are the lower and upper confidence limits.

For comparative purposes it was necessary to look at the SE (Days) for the difference predictions of head, foot, tail and weight$^{1/3}$. It was found that only the linear equations could be easily used for predictive purposes.

SE (Days/Head) = 1.091
SE (Days/Hind foot) = 2.841
SE(Days/Tail) = 2.765
SE (Days/Weight) = 4.092

The values for hind foot and tail are relatively large because of greater curvature in the data (Figs 7 and 8). The body weight value is greater because the relationship between body weight$^{1/3}$/days is not fully linear (Fig. 9).

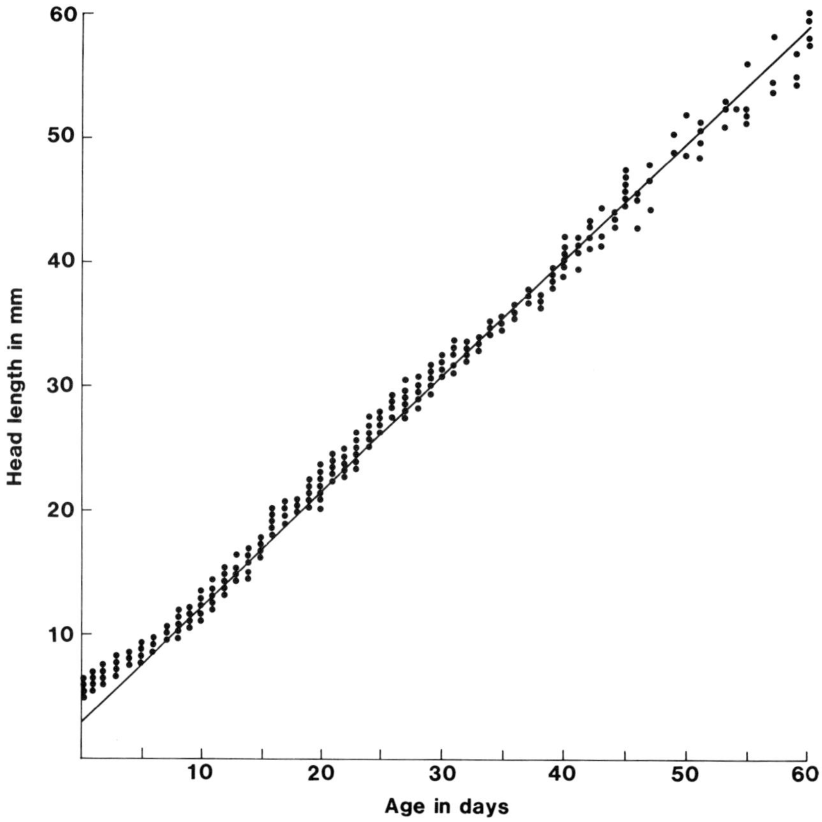

Fig. 6. Relationship between head length and age in pouch young *Isoodon macrourus.* Fitted linear regression line is — Head length = 4.429+0.894 days.

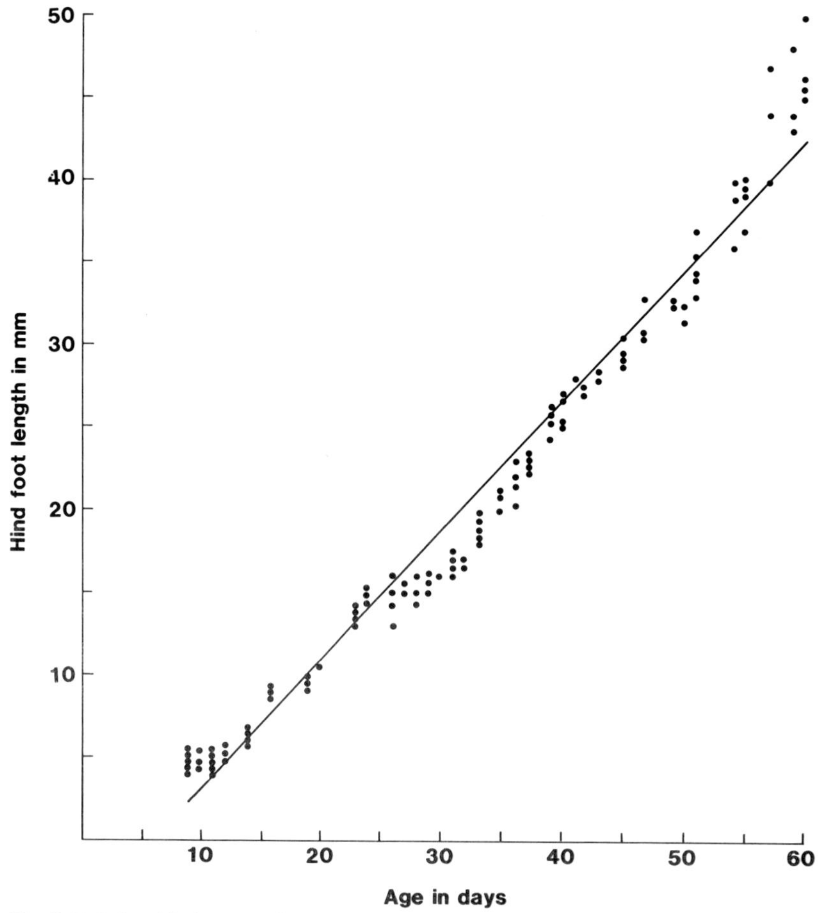

Fig. 7. Relationship between hind foot length and age in pouch young *Isoodon macrourus.* Fitted linear regression line is — Hind foot length = −4.858+0.789 days.

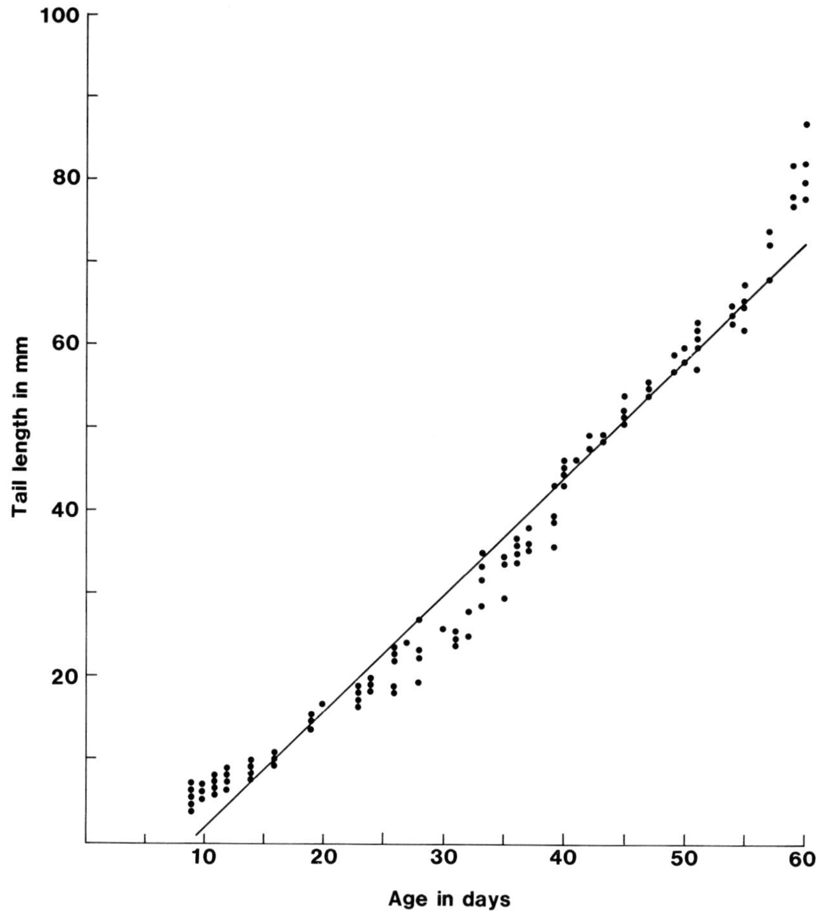

Fig. 8. Relationship between tail length and age in pouch young *Isoodon macrourus*. Fitted linear regression line is — Tail length = −12.33+1.415 days.

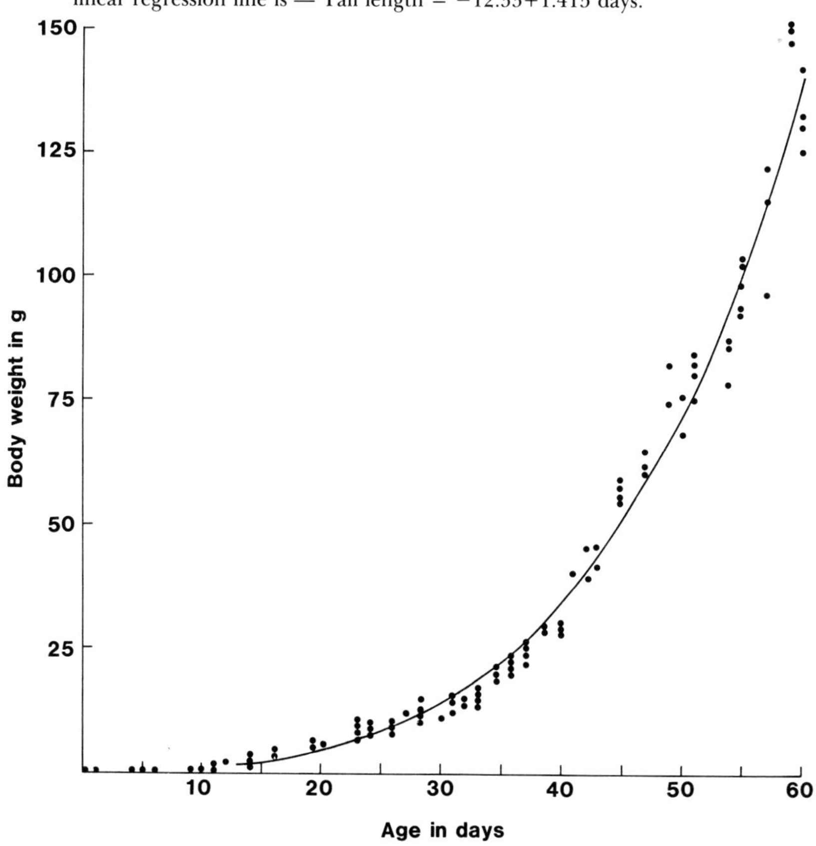

Fig. 9. Relationship between body weight and age in pouch young *Isoodon macrourus*. Fitted quadratic expression is — $\text{Weight}^{1/3} = 0.6311+0.0455\ \text{days}+0.000502\ \text{days}^2$.

From the above analysis it can be seen that head length is the best measurement for age prediction in pouch young *I. macrourus* between birth and 60 days.

DISCUSSION

The mean litter size of 3.57 for captive-bred *I. macrourus* observed in this study and 4.0 calculated from Mackerras and Smith (1960) is larger than the 2.9 recorded for wild-caught individuals by Gemmell (1982) and Hall (1983). Recently Gemmell *et al.* (1984b), investigated the litter size in captive *I. macrourus* and found that it decreased from 3.03 early in lactation to 2.31 at weaning. In a previous study Gemmell (1982) showed a decrease in average litter size from 2.9 early in lactation to 1.8 at weaning using recently wild-caught *I. macrourus* in captivity. Mean litter size in the present study decreased from 3.57 at birth to 2.74 at the end of weaning and pouch life.

The mortality of pouch young *I. macrourus* in captivity has been briefly reviewed by Gemmel *et al.* (1984b) who considered that a survival rate for young of 72.8% could be expected from a well managed captive population. A survival rate of 77% was achieved for young in this study, with the major cause of mortality (10 of 19, or 53%) being recorded when females were caught overnight in wire traps. Stress caused by excessive handling was considered by Gemmell *et al.* (1984b) to be the reason for the low (16.5%) survival of pouch young recorded by Gemmell (1982). Gemmell (1982) believed that females selectively cull their litter and that this was an adaptive mechanism for an unpredictable environment (Gemmell, Johnston and Baines 1984a). However, one needs to be wary when extrapolating data from captive experimental animals to natural life history strategies. Myers, Hale, Mykytowycz and Hughes (1971) have shown that animal density, food supply, and availability of shelter all have pronounced effect on reproductive success of European rabbits *(Oryctolagus cuniculus)* held in various sized outdoor enclosures, and that the effects of such stressful factors can be measured at the embryonic, nestling and young animal levels. In the present study there was a high survival rate of young from females kept at low densities in yards with dense ground cover and when females were caught instantly in the nest boxes. Stodart (1966) considered that a ⅛ acre (=0.05 ha) enclosure may be near the minimum size for successfully breeding from one male and two female *P. nasuta*. Mortality of pouch young bandicoots increases as a result of handling (Stodart 1966; Gemmell *et al.* 1984b) and perhaps the culling of pouch young in *I. macrourus* (Gemmell 1982) is a phenomenon of captivity. Litter size is positively correlated with female body size in wild-caught *I. macrourus* (Hall 1983) and in *I. obesulus* (Stoddart and Braithwaite 1979) but this was not apparent in captive-bred *I. macrourus* (Gemmell *et al.* 1984b; this study). It is clear that further work is required to establish a protocol for successfully breeding *I. macrourus* in captivity.

A summary of the development of external features of pouch young *I. macrourus* in shown in Table 1. The newborn *I. macrourus* has been previously described by Mackerras and Smith (1960) and Lyne (1974) who noted the rich vascular network visible through the transparent skin. Lyne (1974) considered the newborn *I. macrourus* was similar in most respects to the newborn *P. nasuta,* which had been described by Hill (1897), Lyne (1964) and Stodart (1966). Despite their small size and state of development at birth, most newborn marsupials have distinct external features, perhaps the most notable being the cervical swelling in dasyurids (Hill and Hill 1955; Hall 1984).

The external features of *I. macrourus* at birth are shown in Figures 1 and 2. One of the most obvious features is the pronated forelimbs with separate clawed digits. Digits 2, 3, and 4 also possess an additional deciduous claw which covers the dorsal aspect of the permanent claw. These deciduous claws are lost between three and seven days, which is similar to that recorded for *P. nasuta* by Lyne (1964).

Following teat attachment, *I. macrourus* is still anchored by the umbilical cord to the urogenital opening for a short period after birth. It was suspected that the umbilical cord at this stage functioned as a safety cord (Lyne 1974) and was present up to 50 min. after birth in *P. nasuta* (Stodart 1966). This is in contrast to the present study where there was no sign of an umbilical cord 5 h after birth in the specimens shown in Figure 2.

A layer of epitrichium covers the newborn *I. macrourus* and is thickest in the oral region and thinnest in dorso-caudal areas. The epitrichium is a transitory superficial layer of the skin that is found in the embryonic stages of most amniotes. It is usually only one cell layer thick, does not become keratinized, and gradually detaches and is lost prior to the appearance of hair follicles or scales (Sengel 1976). In *I. macrourus*, the epitrichium is up to five or six cell layers thick in the oral region where it seals the lips in the angle of the mouth closed, ensuring firm attachment of the newborn to the teat. The epitrichium also completely covers the eye, which has no eyelids at birth, and fills the external auditory canal, thus functioning as a protective layer.

The eyes open at approximately 45 days in *I. macrourus* and as Lyne (1964) recorded in *P. nasuta,* littermates can show stages from fully open to completely closed on the same day. All eyes were open by 50 days in *I. macrourus* which agrees with the observations of Mackerras and Smith (1960).

Although the ear bulge can be seen at birth, the pinna has not formed (Hall, unpublished) and the ear does not become free until 14 days in *I. macrourus*. The external auditory canal remains plugged with epitrichium until 35 days and although not thoroughly tested, there appeared to be a righting reflex in pouch young of *I. macrourus* at 25 days.

Sex could be determined visually in *I. macrourus* at seven days when a pouch was visible in females, and at nine days when the scrotum appeared in males.

Hair first appeared on the dorsal surface of *I. macrourus* at 39 days and by 42 days there was a short coat of fur all over the body. This was similar to the observations of Mackerras and Smith (1960). A detailed description of the hair patterns and hair development in a number of species of bandicoots can be found in Boardman (1943) and in Lyne (1951, 1952) who also described the histological development and replacement of pelage hairs in *P. nasuta* (Lyne 1957).

The lack of external parasites on all stages of pouch young *I. macrourus* is a result of the grooming and pouch cleaning activities of females. Even when females were heavily infested with ticks and trombiculid mites around the area of short fur close to the pouch, the pouch young were always free of ectoparasites (Fig. 2). It is also possible that the pouch itself may have some inhibitory effect on ectoparasites.

The firm attachment of the marsupial pouch young to its mother's teat is substantially aided by the growth of the epitrichium sealing the lips. The breakdown of this tissue is necessary prior to pouch evacuation. The lips of *I. macrourus* begin to open at 37 days, and by 48 days young are often detached from the teat and their eyes fully open. Teat swapping can occur at 50 days. No attempts were made to remove and reattach pouch young on teats as was done on macropodids by Merchant and Sharman (1966) and Ashman, Shield and Waring (1975). Mackerras and Smith (1960) successfully reattached to a teat a pouch young *I. macrourus* which was estimated to be 35 days old.

Descriptions of the development of external features of *P. gunnii* and *I. obesulus* are given by Heinsohn (1966) and for *P. nasuta* by Lyne (1964) and Stodart (1966). Descriptions of pouch young bandicoots of unknown ages are given by Hill (1897) *(P. nasuta, I. obesulus)*, Jones (1922) *(I. barrowensis)*, and Lyne (1951, 1952) *(P. gunnii, I. obesulus, P. eremiana, M. lagotis* and *Chaeropus ecaudatus)*. The development of the teeth in pouch young of unknown age of *I. macrourus, I. obesulus* and *P. gunnii* is discussed by Fosse and Risnes (1972a,b).

All the growth parameters of pouch young *I. macrourus* measured in the present study could be expressed by linear equations (Figs 6-9). However, there was evidence that some curvature existed in all the relationships, and a quadratic equation improved the fit for body weight.

Head length was found to be the best predictor of age in pouch young *I. macrourus* with a standard error of 1.091 days, compared with 2.765 days for tail length, 2.841 days for hind foot length, and 4.092 days for body weight. The formula for predicting the age of pouch young from a head length measurement is:

$$\text{Head length} = 4.429 + 0.894 \text{ days}$$

The standard error of a predicted day is:

SE (Days) $= \frac{s}{b}$ where s is the SD (Y) given days and b is the coefficient in the fitted equation.

Hind foot and tail measurements are less accurate because of a greater curvature and the inability to measure them prior to nine days (Figs 7 and 8). The SE for body weight$^{1/3}$ is greater because the relationship between body weight$^{1/3}$ and age in days is not fully linear (Fig. 9).

Due to the ease of measurement and the less likelihood of error, head length was also considered to be the best index of age in *P. nasuta* (Lyne 1964). Because of the lack of definition in form, hind foot length in *I. macrourus* was difficult to measure before nine days. After nine days hind foot length became one of the easiest features to measure, and provided a reasonably accurate method for predicting age (Fig. 7).

The birth weights recorded in the present study (0.21 g, n = 7) are similar to those recorded for *I. macrourus* by Mackerras and Smith (1960), who gave the weight of a new born (0.18 g) and a recently attached young (0.22 g). These authors also gave the weights of three unattached young obtained from Mr D. Fleay (0.148 g, 0.170 g and 0.1729 g).

Because detaching pouch young from the teat may decrease chances of survival, Russell (1982) considered that body weight was not a good criterion for measuring the growth of marsupial pouch young. In the present study the availability of known age pouch young *I. macrourus* removed for other studies (e.g., Hall and Hughes 1985) was utilized to plot body weight as a function of age (Fig. 9). The relationship of body weight with age can best be expressed by a quadratic equation, but was found to be the least reliable predictor of age in pouch young *I. macrourus*.

Without detailed continuous observations it is difficult to determine precisely when young leave the pouch and lactation ceases in bandicoots. Previous observations on the length of pouch life in *I. macrourus* have been eight to nine weeks for pouch life (Mackerras and Smith 1960) and approximately 60 days for weaning in captivity (Gemmell 1982), and 50 ± 7 days for pouch occupancy in the wild (Gordon 1974). In the present study *I. macrourus* commenced to leave the pouch at 45 days and no young were found in their mother's pouch after 55 days, although there were occasional indications that teat sucking still occurred up to 60 days.

Heinsohn (1966) found that the pouch young of *P. gunnii* were about 48 to 53 days old when they leave the pouch, but young could still be found in the same nest as their mother at 65 days. In *P. nasuta,* females carry their young for 50–54 days and milk can be expressed from the females at 63 days (Stodart 1966). Lyne (1964) recorded the young of *P. nasuta* first leaving the pouch at 51 days and last seen in the pouch at 69 days after birth. Observations on three litters of captive *M. lagotis* by Hulbert (1972) showed young first leaving the pouch at 67–68 days and last seen in the pouch at 72 days after birth.

Little is known about the behaviour of bandicoots from the time they leave the pouch until they are sub-adults (Hall 1983). Gordon (1974) has recorded that of 26 young *I. macrourus* known to survive pouch life, only three reached maturity. It appears that they lead a solitary life and do not remain as a litter group after weaning (Heinsohn 1966). This stage of the life cycle of bandicoots warrants further studies.

Bandicoots all share a characteristic rapid development during the shortest pouch life of all marsupials. Hulbert (1972) considered that, in view of the divergence between *M. lagotis* and other bandicoots (Kirsch 1968; Martin and Hayman 1967), it appeared that this rapid growth and development were features present early in the evolution of peramelid marsupials.

ACKNOWLEDGEMENTS

I would like to thank M. McKeon and G. Cepon for assistance in maintaining the bandicoots. Drs R. L. Hughes and A. G. Lyne critically read and made helpful comments on the manuscript and Tony Barnes provided statistical advice.

REFERENCES

ASHMAN, R. M., SHIELD, J. AND WARING, H., 1975. Nipple reattachment by dislodged pouch young of the quokka *Setonix brachyurus. Journal of Reproduction and Fertility* **42:** 179–81.

BOARDMAN, W., 1943. On the external characters of the pouch young of some Australian marsupials. *Australian Zoologist* **10:** 138–60.

CASE, T. J., 1978. On the evolution and adaptive significance of postnatal growth rates in the terrestrial vertebrates. *Quarterly Review of Biology* **53:** 243–82.

FOSSE, G. AND RISNES, S., 1972a. Development of the teeth in a pouch young specimen of *Isoodon obesulus* and one of *Perameles gunnii* (Peramelidae: Marsupialia). *Archives of Oral Biology* **17:** 829–38.

FOSSE, G. AND RISNES, S., 1972b. Development of the incisors in two pouch-young stages of *Isoodon macrourus. Archives of Oral Biology* **17:** 839–45.

GEMMELL, R. T., 1982. Breeding bandicoots in Brisbane (*Isoodon macrourus;* Marsupialia, Peramelidae). *Australian Mammalogy* **5:** 187–93.

GEMMELL, R. T., JOHNSTON, G. AND BARNES, A., 1984a. The uniformity of growth within the litter of the marsupial, *Isoodon macrourus. Growth* **48:** 221–33.

GEMMELL, R. T., WALKER, M. T., JOHNSTON, G. AND CEPON, G., 1984b. The number of young present in sequential litters of the marsupial bandicoot, *Isoodon macrourus,* in captivity. *Australian Journal of Zoology* **32:** 623–29.

GORDON, G., 1971. A study of island populations of the short-nosed bandicoot *Isoodon macrourus* Gould. Ph.D. Thesis, University of New South Wales, Sydney.

GORDON, G., 1974. Movements and activity of the short-nosed bandicoot *Isoodon macrourus* Gould (Marsupialia). *Mammalia* **38:** 405–31.

HALL, L. S., 1983. Observation on body weights and breeding of the Northern Brown Bandicoot, *Isoodon macrourus,* trapped in south-east Queensland. *Australian Wildlife Research* **10:** 467–76.

HALL, L. S., 1984. Some comparisons of developmental stages of different marsupials at birth. *Bulletin of the Australian Mammal Society* **8:** 82 (abstract).

HALL, L. S. AND HUGHES, R. L., 1985. The embryological development and cytodifferentiation of the anterior pituitary in the marsupial, *Isoodon macrourus. Anatomy and Embryology* **172:** 353–63.

HEINSOHN, G. E., 1966. Ecology and reproduction of the Tasmanian bandicoots (*Perameles gunni* and *Isoodon obesulus*). *University of California Publications in Zoology* **80:** 1–96.

HILL, J. P., 1897. The placentation of *Perameles.* (Contributions to the embryology of the Marsupialia 1). *Quarterly Journal of Microscopical Science* **40:** 385–446.

HILL, J. P. AND HILL, W. C. O., 1955. The growth stages of the native cat *(Dasyurus viverrinus)* together with observations on the anatomy of the newborn young. *Transaction of the Zoological Society of London* **28:** 349–452.

HULBERT, A. J., 1972. Growth and development of pouch young in the Rabbit-eared Bandicoot, *Macrotis lagotis* (Peramelidae). *Australian Mammalogy* **1:** 38–39.

JONES, F. W., 1922. The external characters of pouch embryos of marsupials. No. 3 *Isoodon barrowensis. Transactions of the Royal Society of South Australia* **46:** 39–45.

KIRSCH, J. A. W., 1968. Prodromus of the comparative serology of Marsupialia. *Nature (London)* **217:** 418–20.

LYNE, A. G., 1951. Notes on external characters of the Barred Bandicoot (*Perameles gunnii* Gray), with special reference to the pouch young. *Proceedings of the Zoological Society of London* **121:** 587–98.

LYNE, A. G., 1952. Notes on external characters of the pouch young of four species of bandicoot. *Proceedings of the Zoological Society of London* **122:** 625–49.

LYNE, A. G., 1957. The development and replacement of pelage hairs in the bandicoot *Perameles nasuta* Geoffroy (Marsupialia: Peramelidae). *Australian Journal of Biological Science* **10:** 197–216.

LYNE, A. G., 1964. Observations on the breeding and growth of the marsupial *Perameles nasuta* Geoffroy, with notes on other bandicoots. *Australian Journal of Zoology* **12:** 322–39.

LYNE, A. G., 1974. Gestation period and birth in the marsupial *Isoodon macrourus. Australian Journal of Zoology* **22:** 303–09.

LYNE, A. G. AND VERHAGEN, A. M. W., 1957. Growth of the marsupial *Trichosurus vulpecula* and a comparison with some higher mammals. *Growth* **21:** 167–95.

MACKERRAS, M. J. AND SMITH, R. H., 1960. Breeding the short-nosed marsupial Bandicoot, *Isoodon macrourus* (Gould), in captivity. *Australian Journal of Zoology* **8:** 371–82.

MARTIN, P. G. AND HAYMAN, D. L., 1967. Quantitative comparisons between the karyotypes of Australian marsupials from three different superfamilies. *Chromosoma* **20:** 290–310.

MERCHANT, J. C. AND SHARMAN, G. B., 1966. Observations on the attachment of marsupial pouch young to the teats and on the rearing of pouch young by foster-mothers of the same or different species. *Australian Journal of Zoology* **14:** 593–609.

MYERS, K., HALE, C. S., MYKYTOWYCZ, R. AND HUGHES, R. L., 1971. The effects of varying density and space on sociality and health in animals. Pp. 148–87 in *Behaviour and environment* ed by A. H. Esser. Plenum Press: New York.

RUSSELL, E. M., 1982. Patterns of parental care and parental investment in marsupials. *Biological Review* **57:** 423–86.

SENGEL, P., 1976. *Morphogenesis of skin.* Cambridge University Press: Cambridge.

STODART, E., 1966. Management and behaviour of breeding groups of the marsupial *Perameles nasuta* Geoffroy, in captivity. *Australian Journal of Zoology* **14:** 611–23.

STODDART, D. M. AND BRAITHWAITE, R. W., 1979. A strategy for utilization of regenerating heathland habitat by the brown bandicoot (*Isoodon obesulus;* Marsupialia: Peramelidae). *Journal of Animal Ecology* **48:** 165–79.

WILSON, J. T. AND HILL, J. P., 1896. Observations upon the development and succession of the teeth in *Perameles,* together with a contribution to the discussion of the homologies of the teeth in marsupial animals. *Quarterly Journal of Microscopic Science* **39:** 427–588.

Morphological development of the adrenal and pituitary glands in *Isoodon macrourus*

LESLIE S. HALL[1] and R. J. LESLIE[1]

Observations on the morphological development of the adrenal and pituitary glands were made on *Isoodon macrourus* from their first appearance in foetal life until the end of pouch life at 60 days.

The adrenal cortex was first observed at 9.5 days gestation as a thickening of the coelomic mesoderm. At birth the adrenal consists of irregular cords of acidophilic cells separated by capillaries. These cells contained granules and are known to produce cortisol. The cortical tissue was invaded by undifferentiated cells from the autonomic ganglion and four weeks after birth the adrenal consists of a thin cortex incompletely surrounding a large medulla. Growth and differentiation of the cortical zones and invasion by large sympathetic neurones occurred during the remaining period of pouch life.

The pituitary arises from an area of close attachment of the roof of the stomodeum and the floor of the forebrain and was clearly seen at 9.5 days gestation. At birth the pituitary was composed of undifferentiated uniform cells with only partial contact between the adeno- and neurohypophysis. A small amount of secretory activity was evident at birth in both the pars distalis and pars nervosa and five granulated cell types could be identified in the pars distalis seven days after birth. At 42 days the pituitary had a similar morphological appearance to that of the adult.

Considering the short gestation period in *I. macrourus* (12.5 days), the development of the pituitary and adrenal glands can be regarded as precocious and they may have some influence on parturition and the development of neonatal tissues. Most other aspects of the origin and development of the adrenal and pituitary glands parallel studies on other marsupials and eutherians.

Key words: *Isoodon,* pituitary, adrenal, development.

INTRODUCTION

ALTHOUGH there are a number of basic similarities in the reproductive hormonal relationships of eutherians and marsupials (Hearn 1977; Tyndale-Biscoe and Evans 1980), several characteristic features have emerged which typify marsupials. Pregnancy in marsupials appears to exert little endocrine influence on the mother and at present, there is little evidence for a foetal role in parturition (Stewart and Tyndale-Biscoe 1983).

During late pregnancy in Eutheria, the foetal pituitary-adrenocortical axis influences developmental activities both within and outside the foetus. Maturation of the small intestine (Moog 1953), lung (Buckingham, McNary, Sommers and Rothschild 1968) and hepatic glycogen synthesis (Jones and Rolph 1981) are influenced by corticosteroid hormones, while outside the foetus this axis has a primary role in the initiation of parturition in a number of species (Liggins 1968; Nathanielsz 1978).

As in the foetus of eutherians, the adrenal cortex of several marsupials has been shown to be actively secreting hormones several days prior to birth. In the Tammar Wallaby *(Macropus eugenii)*, Call, Catling and Janssens (1980) suspected that the adrenal may be capable of synthesizing hormones at 22 days gestation and cortisol and androgen can be detected in the plasma of embryos at 24 days (Catling and Vinson 1976). On the basis of both ultrastructure and radioimmunoassay, Gemmel, Singh-asa, Jenkin and Thorburn (1982) showed that the adrenal of the Northern Brown Bandicoot *(Isoodon macrourus)* had evidence of cortisol secretion at birth and suggested that the foetal adrenal axis in *I. macrourus* was possibly involved in parturition.

The pituitary of foetal eutherians is known to influence the time of parturition (Liggins, Fairclough, Grieves, Kendall and Knox 1973) and in mammals with a short gestation (e.g., the rat) granule formation and secretion of pituitary hormones commences well before birth (Svalander 1974). There appears to be some variation in the morphological development of the pituitary in marsupial neonates. Early postnatal cytodifferentiation of the pars distalis occurs in macropodids, which have a relatively long gestation (Leatherland and Renfree 1983; Walker and Gemmell 1983), however, cytodifferentiation occurs at a later stage in non-macropodid marsupials with a short gestation (Wheeler 1943; Shorey 1968). This difference could be related to the relatively longer post-primitive streak embryogenic period in the macropods. Despite the short period of 5.5 days from primitive streak stage to birth

[1]Department of Anatomy, University of Queensland, St Lucia, Queensland, Australia 4067.
Pages 135–51 *in* BANDICOOTS AND BILBIES, ed by J. H. Seebeck, P. R. Brown, R. L. Wallis and C. M. Kemper. Surrey Beatty & Sons, Sydney, 1990.

(Lyne and Hollis 1977) the pituitary gland at birth in *I. macrourus* contains secretory granules (Hall and Hughes 1980, 1985) thus exhibiting a state of development not typical of short gestation marsupials at birth.

In eutherians it is well established that the control of parturition and the stimulus to organ maturition in the foetus is associated with enhanced activity of the foetal pituitary-adrenal axis (Challis, Manchester, Mitchell and Patrick 1981). Such detail is lacking in marsupials. A brief description of the cellular development of the adrenal in *I. macrourus* during pouch life has been given by Leslie and Hughes (1978) and Leslie (1979) and its appearance at birth by Gemmell *et al.* (1982). The development of the adrena medulla during the first 12 post-natal days has been described for *Didelphis virginiana* by Carmichael, Spagnoli, Frederickson, Krause and Culberson (1987) and Spagnoli, Frederickson, Robinson and Carmichael (1987). Hall and Hughes (1985) have described the development of the anterior pituitary in *I. macrourus* from its first appearance in embryonic life until the end of pouch life. Brief descriptions of the developing pituitary of *I. macrourus* can also be found in Hall (1977) and Hall and Hughes (1980). Since a complete understanding of the morphological development of an organ is essential for a functional interpretation we describe the development of the adrenal and the anterior and posterior pituitary gland in *I. macrourus* from the neonate to the end of pouch life, with particular emphasis on their early origins and state of development at birth. These morphological data will be examined to see if they indicate the early establishment of a reciprocal interaction between these two developing endocrine glands.

MATERIAL AND METHODS

Pouch young were obtained from wild caught *I. macrourus* (Hall 1983) which were kept and bred in captivity (see Hall, this volume). Known age embryos were obtained by the method described by Hall and Hughes (1985) and the exact age of pouch young was known or calculated from the growth curves found in Hall (this volume).

Light microscopy

Adrenal and pituitary glands from pouch young older than three days were removed with the aid of a stereo-microscope and fixed in 10% buffered formalin or aqueous Bouin's fixative. Pouch young less than three days and embryos were fixed *in toto* in 10% buffered formalin. Specimens were stored for 2–50 weeks, then blocked in wax, serially sectioned at 5 μm and stained with Harris's haematoxylin and alcoholic eosin (H and E).

Electron microscopy

Samples of adrenal and pituitary glands were taken in an identical manner to the light microscope specimens and immediately immersed in glutaraldehyde fixative. The tissue was stained with 1% osmium tetroxide and uranyl acetate. In pouch young less than three days and in all embryos the adrenal and pituitary were located by cutting a block of tissue from the general area of the gland. After fixation in glutaraldehyde, the blocks were trimmed, post fixed and stained as above. Sections were cut with an LKB ultramicrotome and viewed with a Zeiss EM 10A electron microscope. Thick (1 μm) sections were stained with toluidine blue. Table 1 shows the age and treatment of the embryos and pouch young used in the study

Table 1. Age of specimens of *Isoodon macrourus* used in this study and their histological treatment. dp=day of pregnancy.

Age	Light microscopy Adrenal	Light microscopy Pituitary	Electron microscopy Adrenal	Electron microscopy Pituitary
8.5 dp	+	+	−	−
9.5 dp	+	+	−	−
10.5 dp	+	+	−	−
11.5 dp	+	+	−	+
Birth	+	+	−	+
1 day	+	+	+	+
2 days	−	+	+	+
5 days	−	−	+	−
7 days	+	+	−	+
11 days	−	+	+	−
13 days	+	+	−	+
17 days	−	−	+	+
28 days	+	−	+	−
34 days	+	+	−	+
42 days	+	+	−	−
55 days	+	+	−	+
60 days	+	−	−	+

RESULTS

Adrenal gland

Light microscopy

In the embryo of 9.5 days gestation, the adrenal appeared as a loose aggregation of undifferentiated cells in a region of mesoderm, medial to the mesonephros. By 10.5 days gestation the adrenal cortical primordium was a nodule of closely aggregated cells situated in the mesoderm lateral to the dorsal aorta and medial to the mesonephros. This cell mass had bulged into the coelomic cavity in the region between the gastric mesentery and the medial edge of the cranial pole of the mesonephros. The adrenal groove separated this mass from the genital ridge which was developing on the coelomic surface of the mesonephros. The sympathetic ganglia were increasing in size as a result of cell division and growth of nerve fibres. This nervous tissue was situated in the para-aortic region and strands of cells were seen connecting with the cortical primordium at this stage. The cells of the developing cortex were relatively undifferentiated with a thin rim of cytoplasm surrounding a pale, vesicular nucleus. A few cords of cells were still attached to the nodule from the proliferating columnar cells which lined the coelomic cavity. The cells of the nodule began to group around the neighbouring veins, on the left near a branch of the posterior cardinal vein, and around the inferior vena cava on the right. This nodule of cells was invaded by capillary channels which multiplied, dividing the cortical mass into loose, irregular cords of cells.

At birth (Fig. 1), the adrenals were prominent oval shaped bodies situated dorsomedially to the mesonephros and the developing metanephros. The right adrenal was closely associated with the right lateral lobe of the liver, the tissues being separated by only a few fine fibres. This adrenal was closely applied to the wall of the vena cava and its venous blood drained almost directly into the lumen of this major vein. The cortical tissue was composed of large acidophilic cells arranged in thin irregular cords which were separated by capillaries. The developing sympathetic and para-aortic complexes consisted of nerve fibres and small cells with densely staining nuclei. Aggregations of this tissue were visible on the dorsomedial aspect of the cortical tissue. Nerve fibres and cords of these small basophilic cells entered the gland in finger-like processes. The gland was not surrounded by a definite connective tissue capsule at birth.

During the first two weeks of pouch life, the structure changed from the thin cord-like arrangement to an irregular clumping of cells into wide cords, four to five cells thick separated by sinusoids. A thin capsule of collagen fibres now surrounded the gland. The cells in the associated sympathetic ganglia had undergone differentiation and two cell types could be distinguished; typical large sympathetic ganglion cells with large, pale staining nuclei and small densely staining cells. The latter cells were the ones observed in migration at this stage. The adrenal cords were composed of groups of acidophilic cortical cells intermingled with clusters of basophilic neural cells which were randomly distributed throughout the gland (Fig. 2). These groups of basophilic cells moved towards the central regions of the gland and

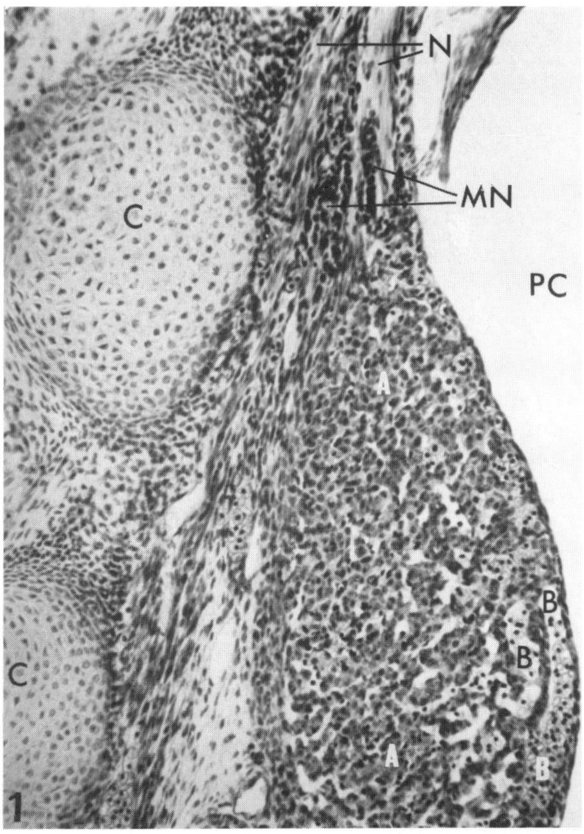

Fig. 1. At birth the adrenal gland of *Isoodon macrourus* lies on the dorsal wall of the peritoneal cavity (PC) and is composed of cords of cells (A) interspersed by a network of sinusoids (B). Cartilage of developing vertebrae (C), nerve fibres from sympathetic ganglion (N) and migrating neurons (MN). Magnification ×124.

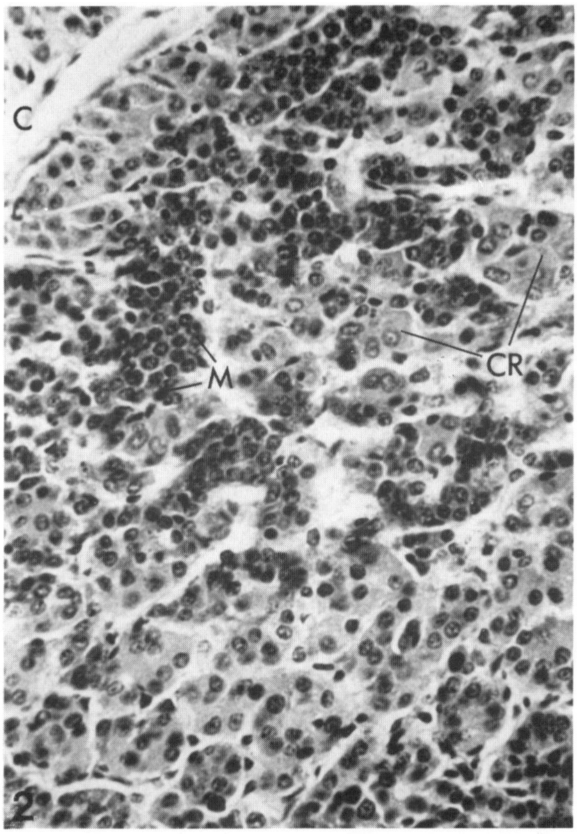

Fig. 2. The adrenal gland of *Isoodon macrourus* 14 days after birth is composed of dark migrating neural cells (presumptive medullary cells) (M) and light acidophilic cells (CR) which are presumptive cortical cells. A thin capsule of connective tissue (C) is present. Magnification ×318.

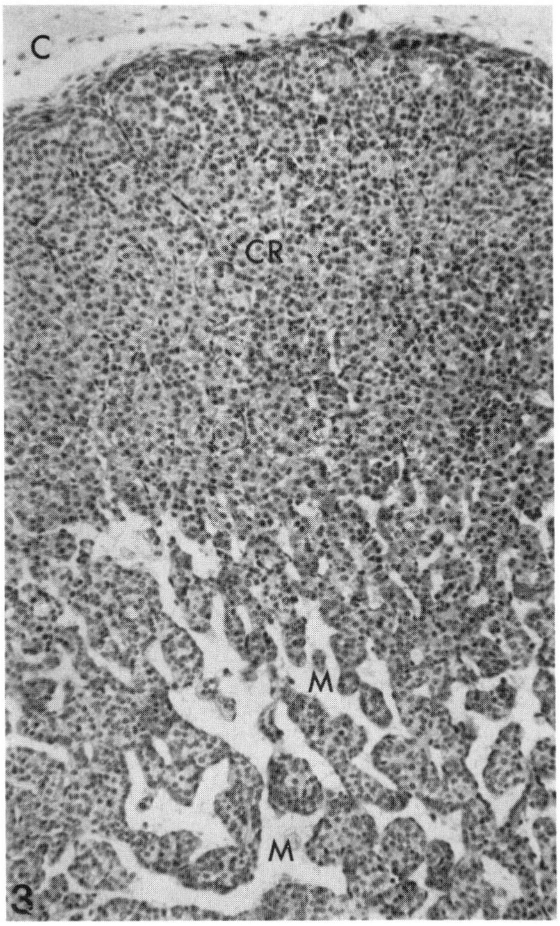

Fig. 3. Twenty-five days after birth the adrenal gland of *Isoodon macrourus* shows a zone of cortical tissue (CR) and an area of medullary sinusoids (M). Connective tissue capsule (C). Magnification ×124.

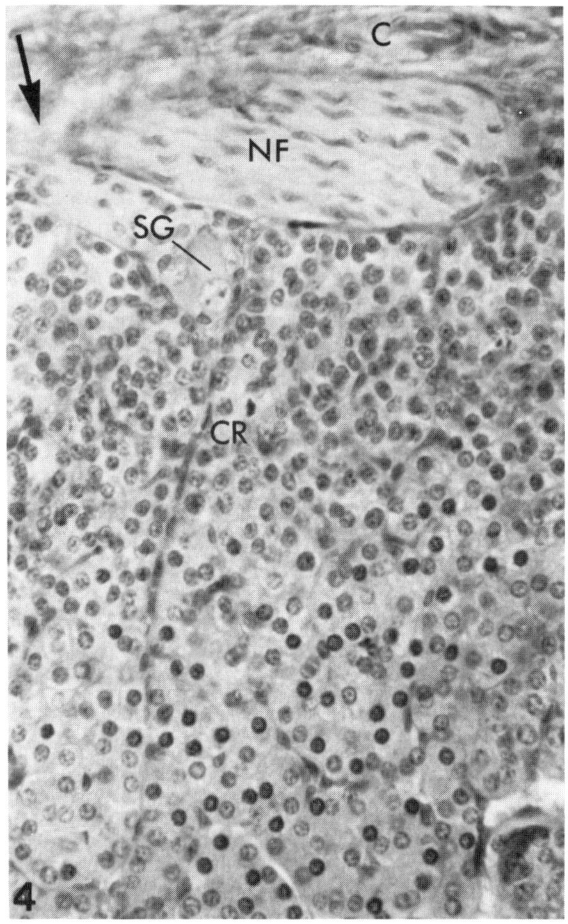

Fig. 4. The adrenal gland of *Isoodon macrourus* 34 days after birth shows a thick capsule (C) penetrated by nerve fibres (NF) via pores in the capsule (arrow). Large migrating neuroblasts (SG) from the sympathetic ganglion can be seen entering the outer cortex (CR). Magnification ×318.

early in the fourth week of pouch life there was a definite arrangement of the adrenal into irregular cortical and medullary areas (Fig. 3). The cortex was a thin rim of tissue incompletely surrounding a large medulla of basophilic cells. The medulla was a loose arrangement of cords of varying thicknesses separated by sinusoids. Two cell types could be distinguished, one staining more intensely basophilically than the other. Groups of cortical cells were continuous with those of the medulla and the cells in peripheral regions of the cortex were closely packed. The cortical tissue was interrupted in several areas where cords of medullary cells connected to the peripheral regions. This was seen particularly in regions where active migration of sympathetic ganglion elements was still evident.

A final stage of development occurred in pouch young from 34 days until 60 days when large sympathetic neurons entered the developing adrenal. These cells migrated from the ganglion along nerve fibre tracts which penetrated the gland through pores in the capsule (Fig. 4). Initially these neurons lay just beneath the capsule but later moved toward the central regions of the medulla where they were seen arranged singly, or in groups, in pouch young at 60 days. Active invasion by ganglion cells was still evident at 60 days. The associated sympathetic ganglion now consisted predominantly of large neurons with a few small dark cells.

Growth in width and differentiation of the cortical zones occurred during this final phase of development. In the adrenal of a 34-day pouch young, the cortex did not show typical zone or stratum. However, three general areas could be distinguished. The cells beneath the capsule were small and tightly packed, with pale staining cytoplasm. These merged to a narrow zone of cells with highly vacuolated cytoplasm. The cells in the area nearest the medulla stained intensely acidophilically and there was considerable interlocking of cortical and medullary tissue in junctional areas.

At 60 days the zonas glomerulosa and reticularis were not well defined. The zona glomerulosa was a narrow zone of small tightly-packed cells which did not show a typical glomerular arrangement.

The zona fasciculata was wide with inner and outer areas. In the outer fasciculata, large, polyhedral cells were arranged in irregular radial cords and the cell cytoplasm was filled with many vacuoles. Nuclei were spherical, large and pale. Cells comprising the inner fasciculata were less vacuolated with smaller and darker nuclei. This zone, together with the thin zona reticularis, increased in width with further growth as seen in the adult.

At 60 days, the adrenals of *I. macrourus* were well developed, oval shaped bodies, the right adrenal being closely attached to the vena cava and a medial lobe of the liver. The cortex was variable in thickness and usually surrounded the medulla except in areas where blood vessels entered or drained the gland. The cortex occupied about half the volume of the gland and the medulla was composed of groups of closely packed cells, separated by sinusoids. The adrenal was surrounded by a capsule of dense collagen fibres, orientated parallel to the surface of the gland.

Electron microscopy

The earliest adrenal tissue examined by electron microscopy was one day after birth. At this stage the adrenal was composed of cells of two types together with (or interspersed with) unmyelinated nerve fibres and sinusoids lined with endothelial cells. One cell type had a relatively pale nucleus, well-developed Golgi stacks composed of three or four lamellar vesicles, and tubular to spherical mitochondria. The RER and SER were extensive and frequently shared the same vesicle. The cytoplasm was dense and contained centrioles, many polyribosomes and lipid droplets (Fig. 5). The other cell type had only a small amount of pale cytoplasm which contained polyribosomes and scattered electron-dense bodies (200–400 nm diameter). In this cell the mitochondria were less prominent and were frequently in close association with RER. The clumps of these paler cells were interspersed by intercellular spaces (Fig. 5). Nucleoli were observed in both cell types. Also in the adrenal tissue one day after birth could be seen unmyelinated nerve fibres, and sinusoids lined with endothelial cells.

At two days most of the adrenal tissue was composed of two distinct cell types. One was a large pale cell containing electron dense granules (150–350 nm diameter) and scattered tubular mitochondria with loosely arranged cristae. The nucleus was large and spherical and contained mainly euchromatin. This cell was regarded as a presumptive catecholamine-secreting cell. The other cell was darker and its cytoplasm contained numerous large lipid droplets which frequently coalesced. These dark cells had irregularly shaped nuclei and their cytoplasm was densely packed with tubular to spherical mitochondria which contained numerous tubulo-vesicular cristae and a dark matrix, well-developed Golgi stacks, RER and extensive SER. The dark cells frequently formed small clumps and several contained electron dense bodies (300–350 nm diameter). They were regarded as presumptive steroid hormone synthesizing cells (Fig. 6). Ocassionally a second type of granular cell was found at two days. These cells were smaller, darker and less numerous than the other granular cell and their nucleus was usually irregular in outline.

By 11 days the light and dark granulated cells (presumptive catecholamine-secreting) were intermixed in equal numbers (Fig. 7). There was no apparent change in the density of granules within these cells.

Cortical and medullary areas could be distinguished at 17 days in the developing adrenal. In the large central medulla, nerve fibres were seen along with the two granular cell types (presumptive catecholamine-secreting cells) and there was a network of sinusoids lined with endothelial cells. In the cortex, two cell types could be identified. Adjacent to the capsule were the dark presumptive steroid synthesizing cells containing lipid droplets. In between these groups of cells were pale granulated cells (presumptive catecholamine-secreting cells) migrating through the cortex.

At 28 days there were still pale, granule-laden, presumptive catecholamine-secreting cells migrating through the cortex. The difference between light and dark presumptive catecholamine-secreting cells in the medulla was much less noticeable than recorded at 17 days. In the cortex two types of presumptive steroid synthesizing cells could be distinguished. One had a pale matrix, dark mitochondria, scattered lipid droplets and sparse RER. The other cell type was more numerous, had a dark matrix and mitochondria, abundant SER, and contained large numbers of lipid droplets (Fig. 8).

Pituitary

Light microscopy

At 9.5 days gestation the stomodeal epithelium had a thickened roof which was closely applied to the floor of the forebrain. This formed the beginning of Rathke's Pouch and was separated from the foregut by the bucco-pharyngeal membrane. A deep cranial invagination of the foregut marked Seessel's Pouch (Fig. 9). Rathke's and Seessel's pouches became closely attached at 10.5 days gestation and by 11.5 days gestation Rathke's Pouch had formed a large irregular vesicle (portio buccalis), the cranial surface of which lay in close contact with a slight depression in the forebrain floor. The

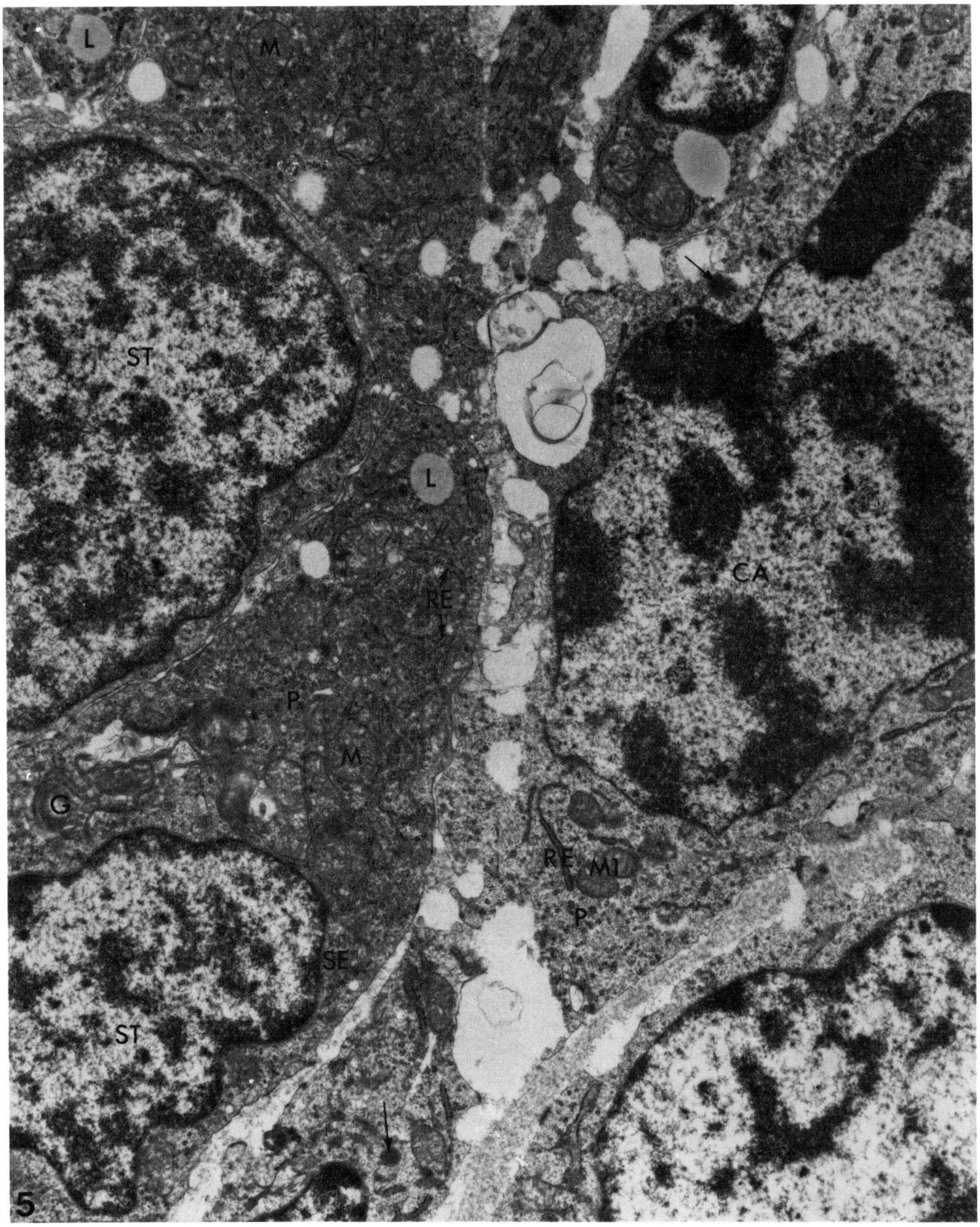

Fig. 5. One day after birth the adrenal gland of *Isoodon macrourus* exhibits two types of cells. One cell (ST) contains scattered lipid droplets (L), Golgi stacks (G), mitochondria (M), rough and smooth endoplasmic reticulum (RE and SE) and polyribosomes (P). The other cell (CA) has a paler cytoplasm containing small mitochondria (MI) which are often in close association with rough endoplasmic reticulum (RE), electron-dense bodies (arrows) and polyribosomes (P). Magnification ×11 472.

Fig. 6 (opposite). The adrenal gland of *Isoodon macrourus* two days after birth. A presumptive steroid synthesizing cell (ST) contains numerous coalescing lipid droplets (L), tubular to spherical mitochondria (M) which have tubulo-vesicular cristae, and a dark cytoplasm. The presumptive catecholamine-secreting cell has a pale cytoplasm (CA) containing numerous electron-dense (150–350 nm diameter), small tubular mitochondria (MI) and rough endoplasmic reticulum (RE). Intercellular space (IS). Magnification ×17 500.

Fig. 7 (opposite). Eleven days after birth the adrenal gland of *Isoodon macrourus* has two types of presumptive catecholamine-secreting cells. A large pale cell (PC) shows sparse heterochromatin in a large regular nucleus, while a smaller darker cell (DC) has an irregular dark nucleus. The size of the electron-dense granules is relatively constant in both cells (150–350 nm diameter). Magnification ×7 245.

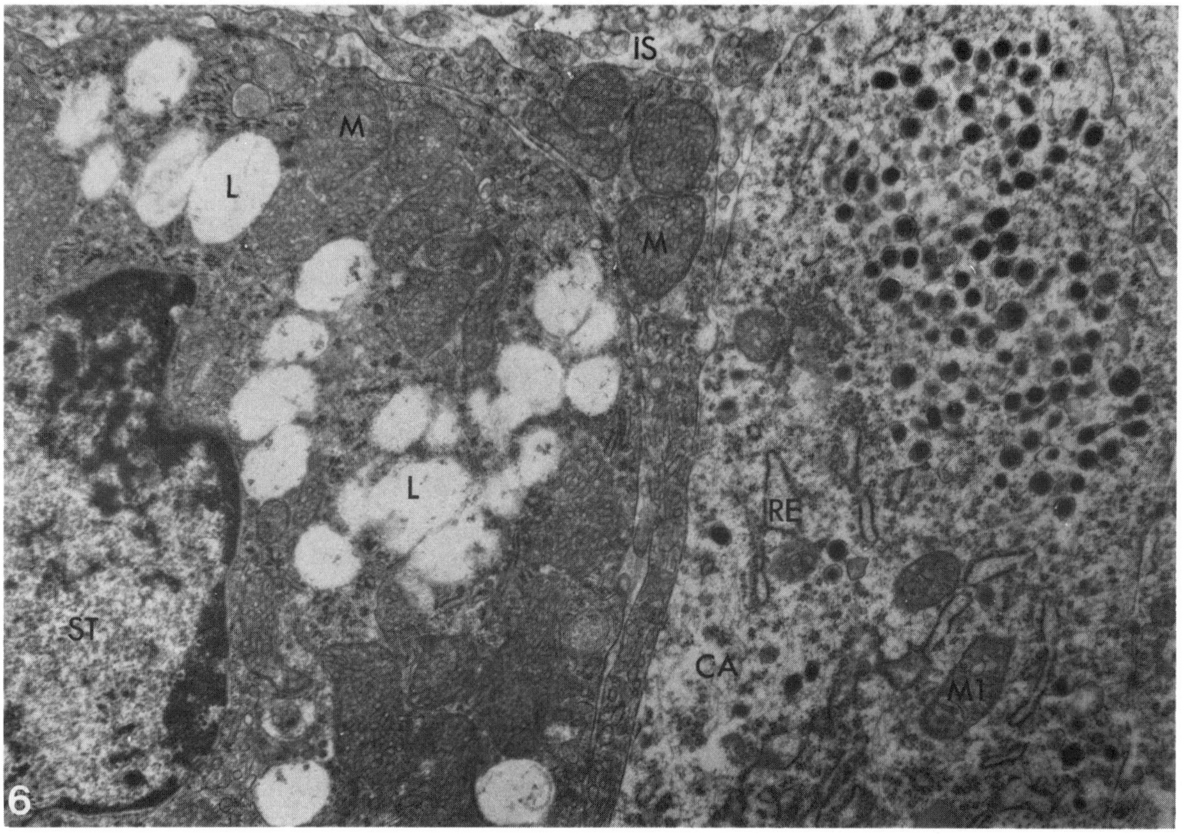
IS
M
L
M
L
RE
ST
CA
M1
6

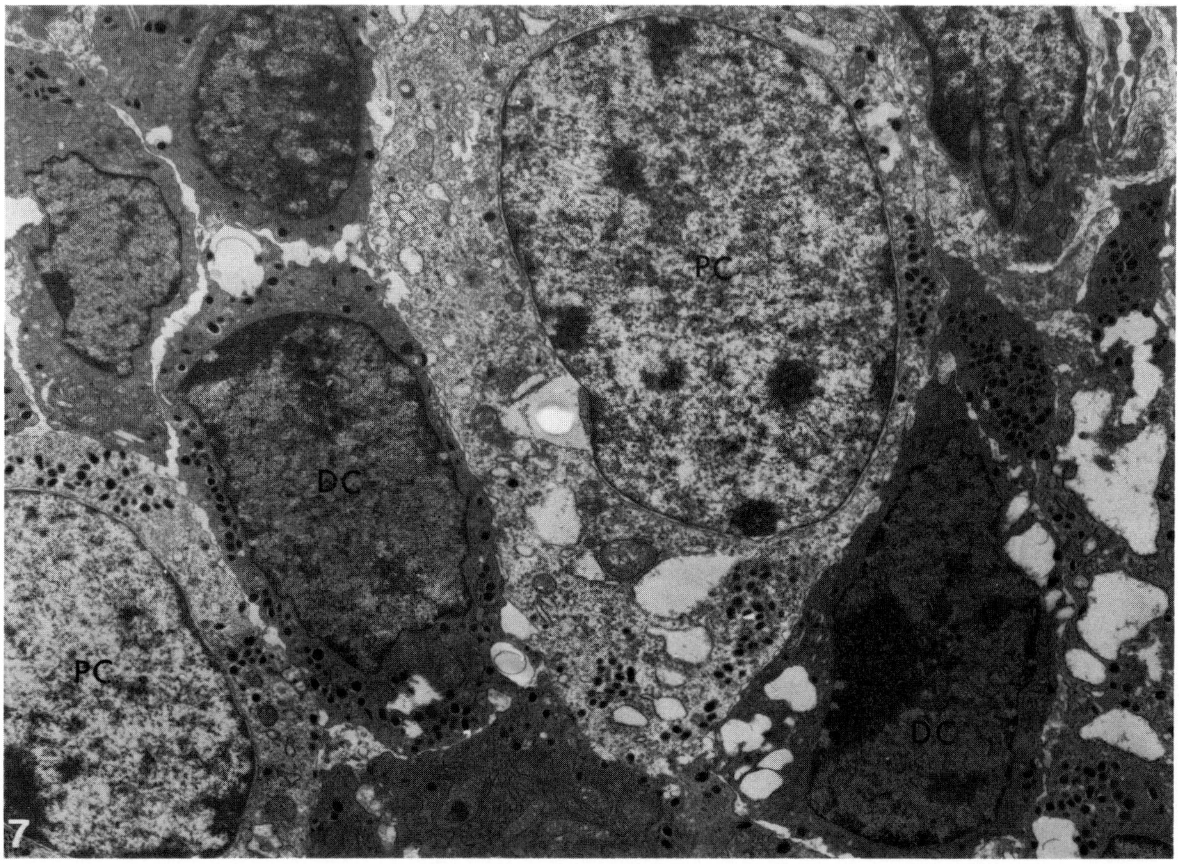
PC
DC
PC
DC
7

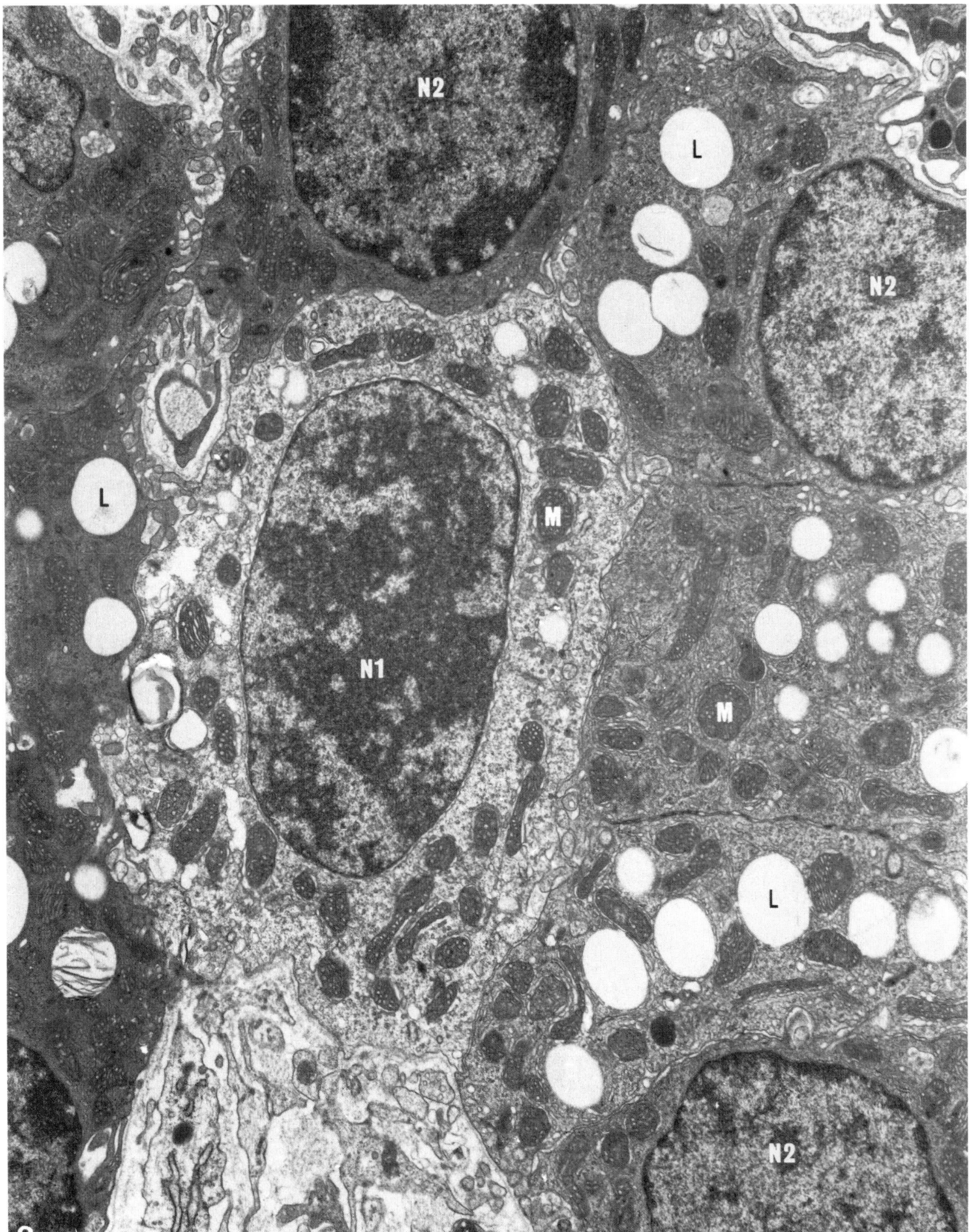

Fig. 8. In the adrenal cortex of *Isoodon macrourus* 28 days after birth, two types of presumptive steroid synthesizing cells are found in the adrenal cortex. One cell has pale cytoplasm and its nucleus contains abundant heterochromatin (NI) and the other cell has dark cytoplasm containing abundant endoplasmic reticulum and its nucleus contains mainly euchromatin (N2). Both cells contain dark tubular to spherical mitochondria (M) and lipid droplets (L). Magnification ×9 560.

portio buccalis was connected to the buccal cavity by patent ducts which were the remnants of Rathke's and Seessel's pouches. The future infundibulum was now a prominent pocket in the floor of the forebrain and contained dense nuclei.

At birth the multi-chambered vesicle resulting from Rathke's and Seessel's pouches was lined with a varying thickness of cells, being greatest in the most caudal zone (six to eight nuclei) and narrowest at its point of contact with the pars nervosa (two to four nuclei). The remnants of Rathke's and Seessel's pouches were still present as a solid cord of cells although some areas exhibited a distinct lumen (Fig. 10). The pars nervosa was densely packed with nuclei, numbers of which were seen undergoing mitosis. Blood vessels were prominent, particularly in the area between the diencephalon and the closely applied area of the vesicle which is the future pars intermedia (Fig. 10). The basisphenoid was cartilaginous and blood vessels were prominent in the tissue surrounding the pituitary. The pars distalis appeared avascular.

Two days after birth there was a thickening of the walls of the pars distalis. Connective tissue surrounded the developing pituitary and small amounts penetrated the pars distalis. The future pars intermedia area was a thin layer (two to five) of densely packed nuclei. The infundibulum was a deep pocket with an expanded terminal lumen. The pars nervosa contained dense nuclei and the surface of the infundibular lumen was ciliated. Blood vessels surrounded all parts of the developing pituitary and penetrated the pars distalis. The interface of the pars nervosa and pars intermedia was highly vascular. The contact of Rathke's and Seessel's pouches with the nasal cavity had almost been separated by the developing basisphenoid which was still cartilaginous.

From two to 11 days the pars distalis underwent further cellular expansion but the vesicle had now become a single chamber (future hypophyseal cleft). The pars intermedia was a distinct thin zone of two to four layers of basophilic (H and E) cells closely attached to and encircling the pars nervosa. The pars distalis contained strands of connective tissue and a distinct zone (H and E) of acidophils had developed at its anterior end. There were scattered basophils throughout the pars distalis.

There was a pronounced expansion of the pars nervosa around seven days after birth. The infundibular lumen was almost obliterated and the developing neuroblasts became scattered, except for the residual area of the infundibular pouch which was densely lined with nuclei. During this stage the basisphenoid commenced to ossify beneath the pituitary and the remnants of Rathke's and Seessel's pouches had disappeared. A dense layer of connective tissue surrounded the developing pituitary which was now compressed in a cranio/caudal direction. The pituitary extended forward to a point underlying the optic chiasma and was surrounded by large blood vessels (see Fig. 5 in Hall and Hughes 1985).

By 34 days after birth numerous acidophils were present throughout the pars distalis and colloid was present in the partes distalis, intermedia and nervosa. The pars nervosa was a dense homogeneous mixture of nerve fibres, pituicytes, areas of colloid and small blood vessels.

The anterior lobe of the pituitary had completely encircled the neural lobe 42 days after birth and acidophils were concentrated in the thin caudal part of the pars distalis.

At 60 days after birth the pituitary gland was similar in morphology to that of the adult.

Electron microscopy

At 11.5 days gestation the cells lining the vesicle formed from Rathke's Pouch were columnar and had a high nuclear to cytoplasmic ratio. The cells immediately adjacent to the vesicle were compressed, contained scattered spherical mitochondria and possessed occasional cilia. The cells of the infundibular process were large, contained few organelles and were electron lucid. The cells bordering the infundibular recess were densely ciliated. At birth the cells of the pars distalis lining the vesicle were columnar and uniform in appearance. They contained numerous Golgi stacks; areas of endoplasmic reticulum, and several contained small electron dense granules. Intercellular spaces were prominent. Both spherical and tubular mitochondria were present and there was a noticeable reduction of cilia on cells lining the vesicle. The pars nervosa consisted of densely packed elongated neuroblasts. These cells contained few organelles, but abundant ribosomes. Scattered between the neuroblasts were small vesicles containing electron dense neurosecretory granules (Fig. 11).

One week after birth a large percentage of the cells of the pars distalis and pars intermedia were granulated. Five cell types could be identified on the basis of granule size, cytoplasmic intensity and shape. Two had dark cytoplasm, one with large granules and the other with small granules. the other three were pale cells with large, medium and small granules (for a detailed description of these cells, see Hall and Hughes 1985). In addition to the five cell types, the pars distalis contained scattered fibroblasts and non-granulated cells, endothelial cells lining and pericytes surrounding capillaries.

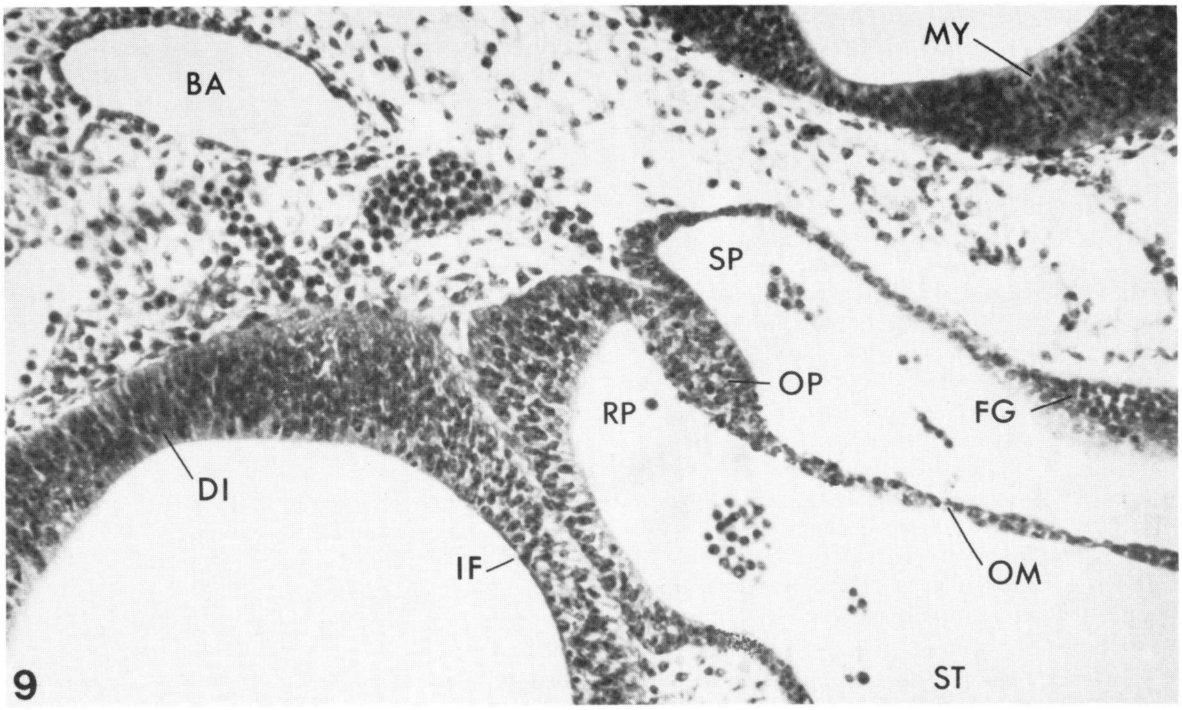

Fig. 9. Longitudinal section through the area of the developing pituitary gland of *Isoodon macrourus* at 9.5 days gestation. Basilar artery (BA), myencephalon (MY), diencephalon (DI), infundibular area (IF), Rathke's Pouch (RP), Seessel's Pouch (SP), oropharyngeal membrane (OM), oral plate (OP), foregut (FG) and stomodeum (ST). Magnification ×360.

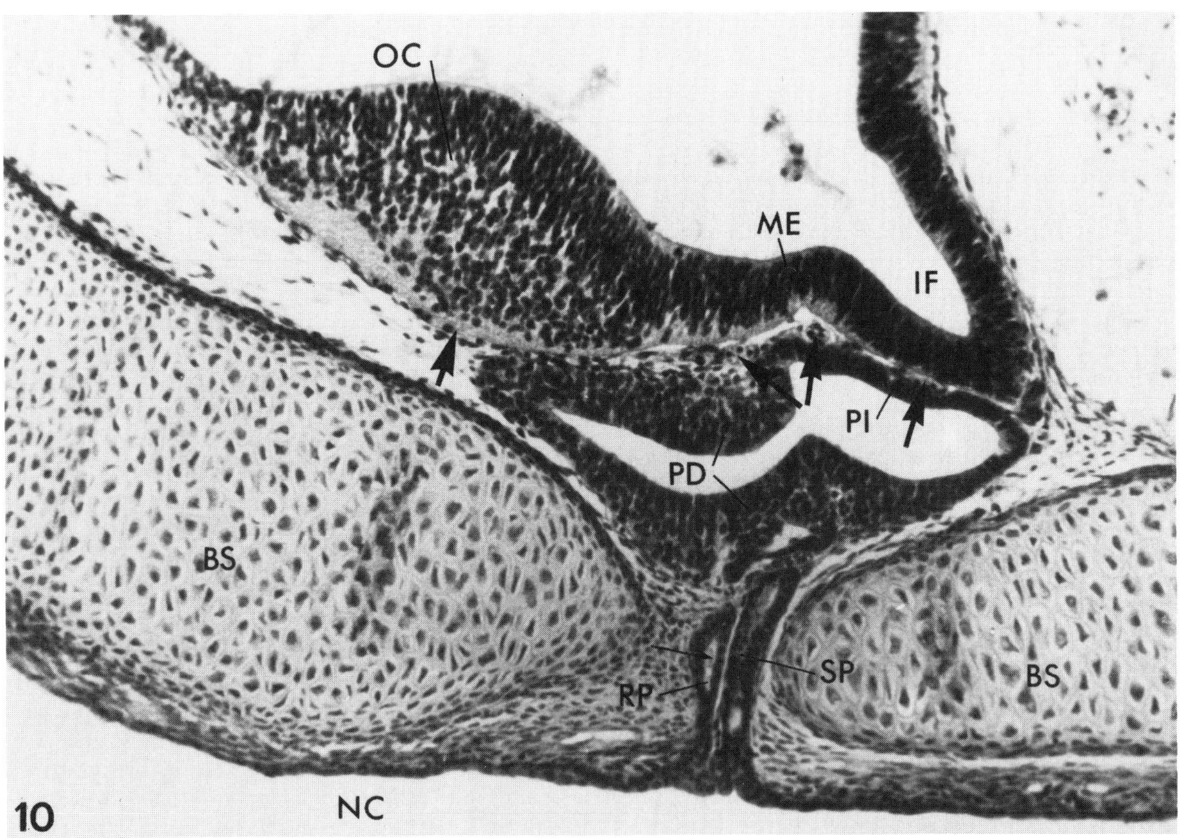

Fig. 10. Longitudinal section through the developing pituitary gland of *Isoodon macrourus* at birth. Remnants of Rathke's and Seessel's pouches (RP, SP), nasal cavity (NC), cartilaginous basisphenoid (BS), developing partes distalis and intermedia (PD, PI), infundibular process (IF), median eminance (ME) and optic chiasma (OC). There is a small lumen still visible in Rathke's Pouch (small arrow) and there are numerous small blood vessels in the zone of contact between the developing adenohypophysis and neural tissue (large arrows). Magnification ×238.

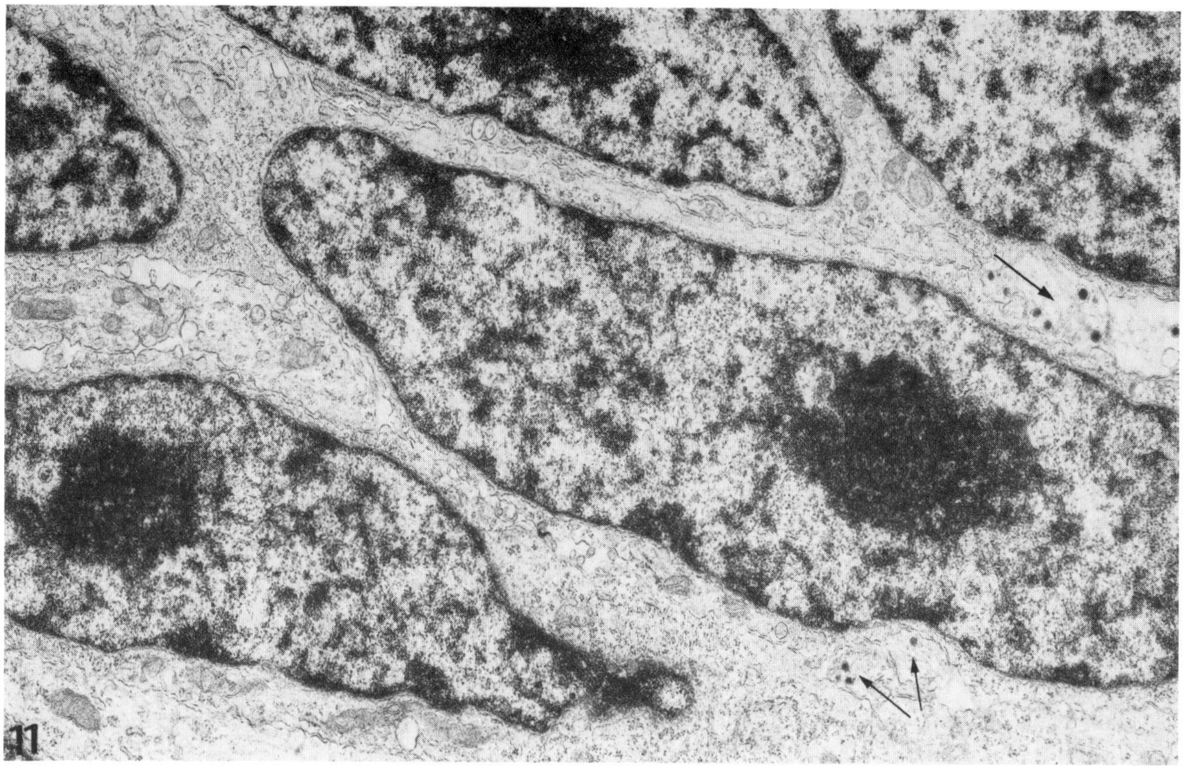

Fig. 11. The future pars nervosa of *Isoodon macrourus* at birth contains a dense concentration of cells which exhibit only a small amount of cytoplasm. Between these cells are found a small number of nerve fibres containing secretory granules (arrows). Magnification ×8 125.

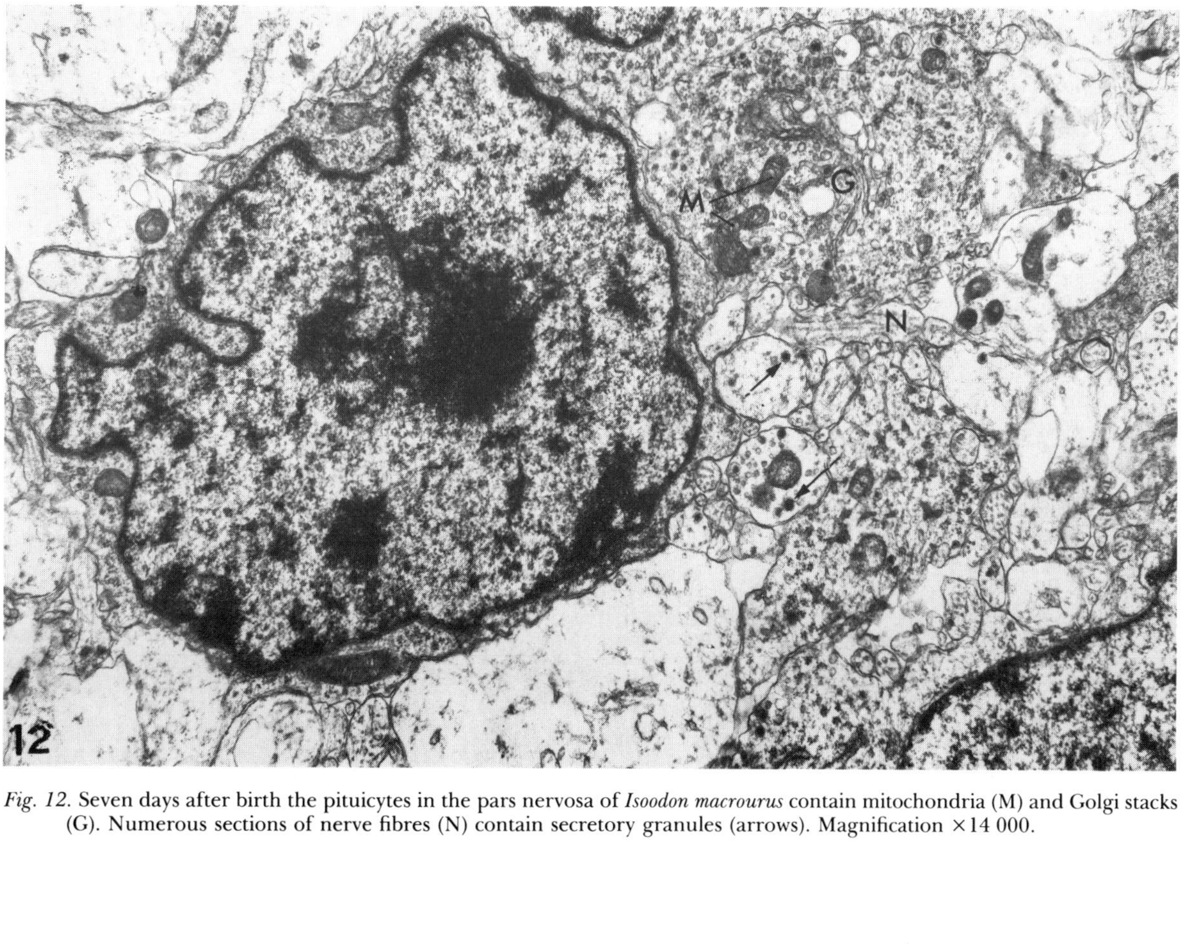

Fig. 12. Seven days after birth the pituicytes in the pars nervosa of *Isoodon macrourus* contain mitochondria (M) and Golgi stacks (G). Numerous sections of nerve fibres (N) contain secretory granules (arrows). Magnification ×14 000.

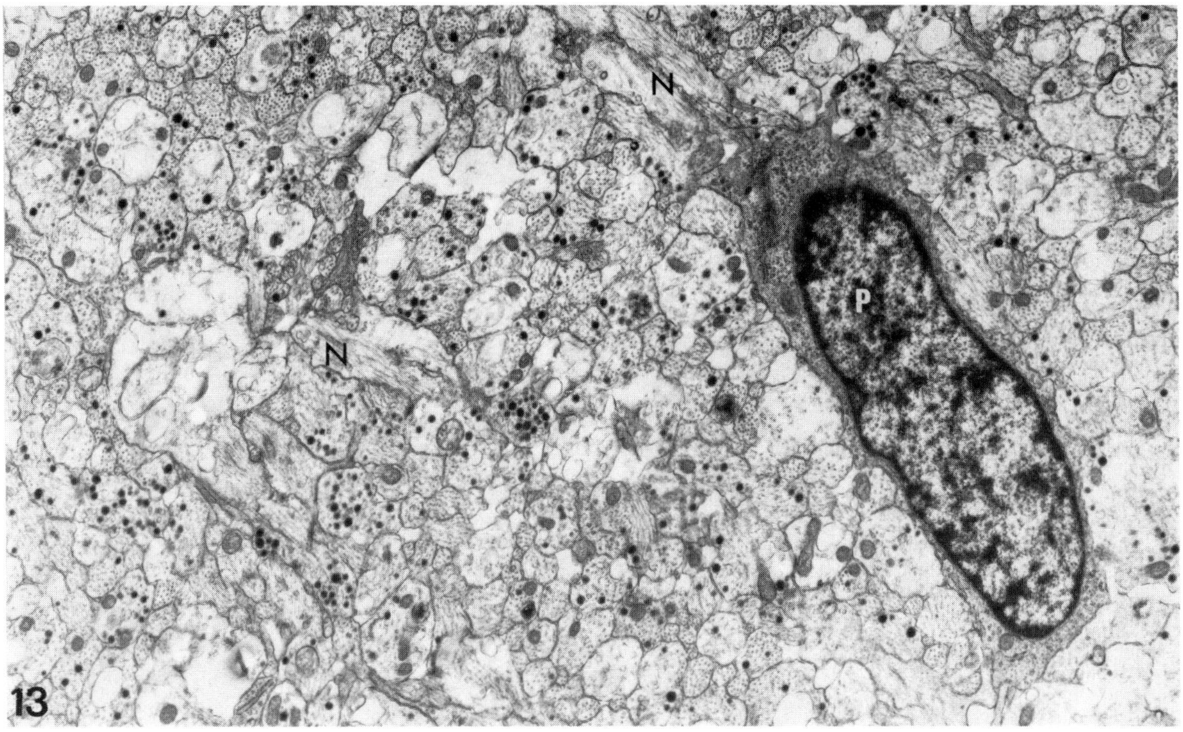

Fig. 13. The pars nervosa of *Isoodon macrourus* at 12 days contains scattered pituicytes, which have very little cytoplasm (P) and a dense array of nerve fibres (N), most of which contain secretory granules. Magnification × 7 875.

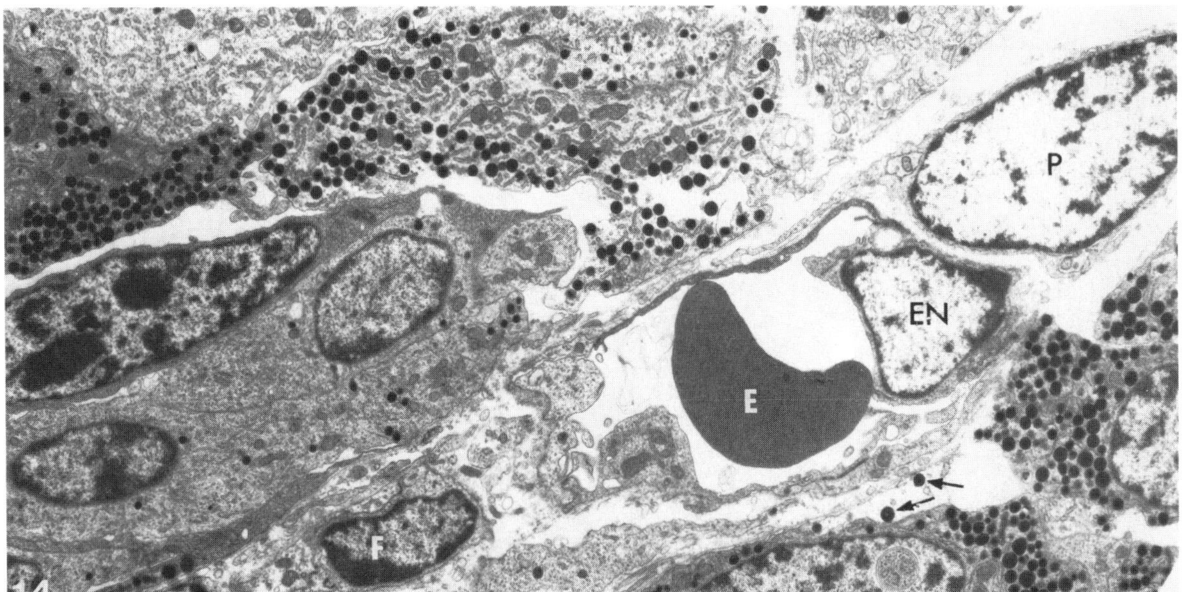

Fig. 14. The pars distalis of *Isoodon macrourus*, 12 days after birth is composed of five types of granulated cells (see text). This electromicrograph shows a granulated cell releasing granules into an intercellular space (arrows) adjacent to a capillary lined with an endothelial cell (EN) and containing an erythrocyte (E), fibroblast (F) and pericyte (P). Magnification ×4 400.

The pars nervosa seven days after birth consisted mainly of unmyelinated axons containing numerous neurosecretory granules. There were also scattered pituicytes, small capillaries, intercellular spaces and bundles of neurotubules (Fig. 12).

Thirteen days after birth the pars nervosa was densely packed with unmyelinated axons containing numerous neurosecretory granules and bundles of neurotubules (Fig. 13). The pituicytes contained only a small amount of dark cytoplasm with few organelles and some of the capillaries in the pars nervosa appeared fenestrated. In the pars distalis there was an increase in the granule content of cells and a reduction in the amount of intercellular space. Granules could be seen in the prominent perivascular spaces (Fig. 14). The hypophyseal portal system consisted of numerous fenestrated capillaries surrounded by perivascular spaces and small amounts of collagen fibres.

From 13 days to the end of pouch life at 60 days no new granulated cell types appeared in the pars distalis. The major changes during this period were an increase in fenestrated capillaries and a decrease in the number of non-granulated cells and intercellular spaces in the pars distalis. The pars nervosa became more vascular and the number of neurosecretory granules per vesicle increased.

DISCUSSION

In *I. macrourus* the adrenal was first seen as a loose aggregation of undifferentiated cells in a region of mesoderm medial to the mesonephros in the 9.5-day embryo. The presumptive adrenal cells had prominent nucleoli and mitotic figures were frequent. There was no apparent vascularization of the adrenal tissues at this stage. This tissue proliferated and at birth the adrenal consisted of cords of cells separated by sinusoids (Fig. 1). Nervous tissue from a nearby sympathetic ganglion invades the developing adrenal before birth. This method of formation and development of the adrenal in *I. macrourus* is similar to that described for another marsupial, *M. eugenii*, by Call *et al.* (1980) and for a number of eutherians (human, Keene and Hewer 1927; mouse, Waring 1935; rabbit, Roaf 1935; and rat, Pankratz 1931).

On the basis of ultrastructural features, Gemmell *et al.* (1982) suggested that the adrenal of *I. macrourus* is capable of producing steroid hormones at birth. This was supported by the measurement of low levels of cortisol from pooled adrenal tissue. Those authors described the presence of two cell types in the adrenal of *I. macrourus* at birth. One had the features of a presumptive steroid hormone synthesizing cell and the other was a presumptive catecholamine-secreting cell. In the present study the two types of cells described by Gemmell *et al.* (1982) could be identified by electron microscopy two days after birth (Fig. 3). Also at two days another granular cell with dark cytoplasm was occasionally seen. One day after birth the presumptive steroid synthesizing cell could be seen but the catecholamine-secreting cell appeared to be in a pre-granular state (Fig. 2).

It was shown by Carmichael and Blair (1973) that differences in the cytoplasmic density in cells of the adrenal of the day-old dog were caused by fixation procedures. Immersion fixation gave both light and dark staining catecholamine-secreting cells while perfused material gave only light cells with no cytoplasmic difference. In the present study all fixation was by immersion, but the differences between light and dark catecholamine-secreting cells and steroid producing cells could represent developmental stages of the same cells.

The anatomical relationship of the right adrenal lying in an indentation of the liver and emptying directly into the IVC, and the left adrenal emptying into the renal vein and then the IVC was first seen at 11.5 days gestation and persisted through pouch life in *I. macrourus*. This relationship has previously been commented on by Bourne (1949) who used the position of the left adrenal to divide the Peramelidae into two groups. All Peramelidae, with the exception of *P. gunnii*, show a typical dual IVC and kidney attachment for the right adrenal. *P. myosura (=P. bougainville)*, *P. nasuta* and *Chaeropus ecaudatus* have the left adrenal attached to the antero-mesial border of the left kidney, and *Thalacomys (=Macrotis)*, *Isoodon* and *P. gunnii* have the left adrenal attached firmly to the junction of the IVC and the left renal vein (Bourne 1949).

Bourne (1949) described clear demarcation for cortex and medulla in seven species of adult bandicoots. In *I. macrourus* separation of medulla and cortex is possible in the fourth week. The zones of the adrenal cortex in *I. macrourus* were difficult to distinguish at the end of pouch life. Bourne (1949) observed that the cortical zones were indistinct in the closely related *I. obesulus* and Call *et al.* (1980) found that zonation in the adrenal of *M. eugenii* was also difficult to distinguish. The adrenal cortex of the female Brush-tailed Possum *Trichosurus vulpecula* contains an additional hypertrophied area of cells (Bourne 1949) which is absent in males and immature animals (Vinson and Kenyon 1978). This zone responds to FSH and estradiol by a change in the pattern of steroidogenesis (Risbridger and Weiss 1985).

A foetal cortex zone is a feature of human adrenals (Lanman 1953; Ross 1960; Luse 1967;

Johannisson 1968). Luse (1967) described a foetal zone in the cortex of the hare foetal adrenal and a foetal zone which degenerates after birth has been recorded in the adrenal of the Nine-banded Armadillo by Enders, Schlafke and Warren (1966). In the present study there was no indication of a foetal zone in *I. macrourus*.

The appearance of the adrenal at birth in *I. macrourus* is quite different from that of an adult. At birth the adrenal is composed of cords of cells separated by sinusoids (Fig. 1) and shows no sign of the typical zonation exhibited by the adult.

The morphological stage of development of the adrenal at birth in *I. macrourus* is similar to that described for other marsupials (e.g., *Didelphis virginiana* (McCrady 1938; Spagnoli 1979; Carmichael *et al.* 1987), *M. eugenii* (Call *et al.* 1980), *M. rufogriseus* (Walker and Gemmell 1983) and *T. vulpecula* (Bourne 1949; Hughes and Hall 1984). Although there are a number of published observations on the morphology of the adrenal at birth in marsupials, only the study on *M. eugenii* by Call *et al.* (1980) followed the development of the adrenal through to the end of pouch life. In their study Call *et al.* (1980) found that from day 22 of embryogenesis until at least day 20 of pouch life a sympathetic nerve plexus is pressed to the medial side of the adrenal. Cells migrate from the plexus through pores in the capsule and by 60 days have formed a distinct medulla. This is similar to the present study on *I. macrourus* where invasion by nervous tissue also occurred before birth and a separate cortex and medulla were distinguishable early in the fourth week. In their study on the development of the adrenal in *D. virginiana* Carmichael *et al.* (1987) used specimens from birth to 12 days. They found that during the third to fifth days the cortical and medullary regions began to separate and primitive sympathetic cells were found in both intra and extra-adrenal locations in newborn.

Weiss (1980) reported that marsupial adrenal glands secrete the same glucocorticoids and mineralocorticoids as their eutherian counterparts, the major mineralocorticoid being aldosterone and the major glucocorticoid being cortisol. However, there are intriguing differences in the apparent metabolic role of glucocorticoids in different species of adult marsupials (McDonald 1980). There is very little information on the nature of the tropic hormones of the foetal adrenal gland in any mammal. Glucocorticoids accelerate the maturation of foetal lungs in a range of eutherians (Farrell 1977; Hitchcock 1980) and is also responsible for hepatic glycogen synthesis (Jones and Rolph 1981) and the development of the small intestine (Moog 1953). In sheep the control of parturition and the stimulus to organ maturation in the foetus is associated with enhanced activity of the foetal pituitary-adrenal axis (Challis *et al.* 1981). An increase in the concentration of cortisol in the plasma of the sheep foetus near term, results from an increase in the rate of secretion of cortisol by the foetal adrenal gland (Challis *et al.* 1981). Call *et al.* (1980) also showed that the adrenal of *M. eugenii* can synthesize steroids five days before birth and Catling and Vinson (1976) reported the presence of corticosteroids in the plasma of neonatal *M. eugenii*. Cortisol has been found in low levels in the adrenal tissue at birth in *I. macrourus* (Gemmell *et al.* 1982) and *M. rufogriseus* (Walker and Gemmell 1983). Several studies have examined the relationship between cell ultrastructure and steroidogenesis in both the foetal adrenal (Idelman 1970; Jost 1975; Coffigny and Dupouy 1978; Boshier, Holloway and Liggins 1980), and adult adrenal gland (Malamed 1975). These studies have a general agreement on the significance of the relationship between a well-developed SER, mitochondria with tubular or vesicular cristae, lipid droplets, and corticosteroid secretion by adrenocortical cells. In the present study and that of Gemmell *et al.* (1982) the adrenocortical cells examined at birth in *I. macrourus* demonstrated these features.

From the early morphological appearance of the developing pituitary gland first located at 9.5 days, it is apparent that the origin of the pituitary in *I. macrourus* is the same as that described for a large number of mammals (see Hall and Hughes 1985 for review). It is generally considered that the pituitary has tissue from two sources, the neurohypophysis, which is derived from ectoderm of the diencephalon, and the adenohypophysis from the ectodermal lining of the stomodeum. In this study the tissue of the adenohypophysis anterior to the oropharyngeal membrane (i.e., Rathke's Pouch) was ectodermal in origin, and the tissue just caudal to the membrane (i.e. Seessel's Pouch) was from foregut endoderm. This could be clearly seen at 9.5 days (Fig. 9). Subsequent combining of both these pouches to form the adenohypophysis (Fig. 10) means that this structure contains tissue derived from both ectoderm and endoderm. This combination of tissues in the formation of the adenohypophysis of marsupials has been commented on previously by Parker (1917) who showed that the endodermal component (Seessel's Pouch) forms a significant part of the adenohypophysis of the Koala *(Phascolarctos cinereus)* and Common Wombat *(Phascolomys mitchelli)* (=*Vombatus ursinus*). The fate of the two pouches in the developing adenohypophysis in *I. macrourus* could not be followed. However, Rathke's Pouch contributed to the bulk of the pars distalis, and

Seessel's Pouch appeared to form only a small posterior segment of the pars distalis.

The early state of development of the pars distalis at birth in *I. macrourus* has been recorded by Hall (1977) and Hall and Hughes (1980, 1985) who considered that, due to the presence of granules in cells, the pituitary could be functional. In the present study additional data on the pars nervosa and pars tuberalis of *I. macrourus* at birth is consistent with the morphologically incomplete developmental state of the pituitary. In the newborn *I. macrourus* the pars nervosa consisted of undifferentiated neuroblasts which contained few organelles. Scattered between these cells were small vesicles containing neurosecretory granules (Fig. 11). The absence of an identifiable pars tuberalis and the lack of vascularization in the pars nervosa and pars distalis indicated that the classical functional pathway of hormone secretion as seen in most mammals (Holmes and Ball 1974) was not operating at birth in *I. macrourus*. The highly vascular nature of the junctional area of the pars nervosa, pars distalis and median eminence (Fig. 10) suggests that some form of interaction between these tissues was possible. Glydon (1957) has shown that foetal rat pituitary can function in the absence of the distinctive vascular arrangement seen in the adult, and Pasteels, Gausset, Danguy and Ectois (1974) have shown that in humans, early differentiated gonadotrophs were capable of releasing hormones via mesenchymal tissue.

The pars nervosa grows as a result of the growth process occurring in the surrounding head region, principally by the pars distalis. The lack of mitosis in the infundibular area of the diencephalon supports the theory that the pituitary is formed from the original attachment of the stomodeum to the brain floor, which is then relocated by folding and bending of the area (Gilbert 1935). At birth the remnants of Rathke's and Seessel's pouches were still connected to the upper surface of the nasal cavity in *I. macrourus* (Fig. 10). This attachment persists for several days but has disappeared by seven days when the basisphenoid commences to ossify. By 42 days the pituitary of *I. macrourus* had a similar morphological appearance to that of the adult, however, its secretory ability at all stages still awaits physiological testing.

The presence of five types of granular cells and morphological evidence of their secretion in *I. macrourus* seven days after birth (Hall and Hughes 1985), combined with the present observations which showed neurosecretory granules in the pars nervosa and a developing hypophyseal/portal system at the same stage, suggests that the pituitary may be elaborating a number of hormones at seven days.

This study has further emphasized the precocious development of two endocrine glands in *I. macrourus*. After an exceptionally short period of organogenesis, which is restricted to the last five days of the 12.5 days gestation period (Lyne and Hollis 1977), *I. macrourus* is born with incompletely developed, but apparently functional adrenal and pituitary glands. Although their functional ability awaits critical testing, the morphological evidence suggests that the neonatal adrenal and pituitary could have a functional role in parturition and organ development in *I. macrourus*.

ACKNOWLEDGEMENTS

The authors would like to thank Drs R. T. Gemmell and R. L. Hughes for their helpful comments on the manuscript.

REFERENCES

Boshier, D. P., Holloway, H. and Liggins, G. C., 1980. Growth and cytodifferentiation of the fetal lamb adrenal cortex prior to parturition. *Journal of Anatomy* **130:** 97–111.

Bourne, G. H., 1949. *The mammalian adrenal gland.* Clarendon Press: Oxford.

Buckingham, S., McNary, W. F., Sommers, S. C. and Rothschild, J., 1968. Is lung an anolog of Moog's developing intestine? I. Phosphatases and pulmonary alveolar differentiation in fetal rabbits. *Federal Proceedings* **27:** 328.

Call, R. N., Catling, P. C. and Janssens, P. A., 1980. Development of the adrenal gland in the Tammar wallaby, *Macropus eugenii* (Desmarest) (Marsupialia: Macropodidae). *Australian Journal of Zoology* **28:** 249–59.

Carmichael, S. W. and Blair, B. G. H., 1973. Normal ultrastructure of the day old dog adrenal medulla. *Journal of Anatomy* **115:** 113–18.

Carmichael, S. W., Spagnoli, D. B., Frederickson, R. G., Krause, W. J. and Culberson, J. L., 1987. Opossum adrenal medulla: I. Postnatal development and normal anatomy. *American Journal of Anatomy* **179:** 211–19.

Catling, P. C. and Vinson, G. P., 1976. Adrenocortical hormones in the neonate and pouch young of the Tammar wallaby *Macropus eugenii. Journal of Endocrinology* **69:** 447–48.

Challis, J. R. G., Manchester, E. L., Mitchell, B. F. and Patrick, J. E., 1981. The development of fetal adrenal function. Pp. 43–65 in *The fetus and independent life* (Ciba Foundation Symposium 86). Pitman: London.

Coffigny, H. and Dupouy, J.-P., 1978. The fetal adrenals of the rat: correlations between growth, cytology, and hormonal activity, with and without ACTH deficiency. *General Compendium of Endocrinology* **34:** 312–22.

Enders, A. C., Schlafke, S. and Warren, R., 1966. Cytology of the fetal zone of the adrenal gland of the Armadillo. *Anatomical Record* **154:** 807–22.

Farrell, P. M., 1977. Fetal lung development and the influence of glucocorticoids on pulmonary surfactant. *Journal of Steroid Biochemistry* **8:** 463–70.

Gemmell, R. T., Singh-Asa, P., Jenkin, G. and Thorburn, G. D., 1982. Ultrastructural evidence for steriod-hormone production in the adrenal of the marsupial, *Isoodon macrourus,* at birth. *Anatomical Record* **203:** 505–12.

GILBERT, M. S., 1935. Some factors influencing the early development of the mammalian hypophysis. *Anatomical Record* **62:** 337–59.

GLYDON, J., 1957. The development of the blood supply of the pituitary in the albino rat, with special reference to the portal vessels. *Journal of Anatomy* **91:** 237–44.

HALL, L. S., 1977. Preliminary observations on the development of the pituitary gland in the marsupial, *Isoodon macrourus. Journal of Anatomy* **124:** 537 (Abstract).

HALL, L. S., 1983. Observations on body weights and breeding of the Northern Brown Bandicoot, *Isoodon macrourus,* trapped in south-east Queensland. *Australian Wildlife Research* **10:** 467–76.

HALL, L. S. AND HUGHES, R. L., 1980. Early cytodifferentiation of the pituitary gland in the Brindled Bandicoot *(Isoodon macrourus). Bulletin of the Australian Mammal Society* **6:** 36 (Abstract).

HALL, L. S. AND HUGHES, R. L., 1985. The embryological development and cytodifferentiation of the anterior pituitary in the marsupial *Isoodon macrourus. Anatomy and Embryology* **172:** 353–63.

HEARN, J. P., 1977. Pituitary function in marsupial reproduction. Pp. 337–44 in *The biology of marsupials* ed by B. Stonehouse and D. Gilmore. Macmillan: London.

HITCHCOCK, K. R., 1980. Lung development and the pulmonary surfactant system: hormonal influences. *Anatomical Record* **198:** 13–34.

HOLMES, R. L. AND BALL, J. N., 1974. *The pituitary gland — a comparative account.* Cambridge University Press: Cambridge.

HUGHES, R. L. AND HALL, L. S., 1984. Embryonic development in the Common Brushtail Possum *Trichosurus vulpecula.* Pp. 197–212 in *Possums and gliders* ed by A. P. Smith and I. D. Hume. Australian Mammal Society: Sydney.

IDELMAN, S., 1970. Ultrastructure of the mammalian adrenal cortex. *International Review of Cytology* **27:** 181–281.

JOHANNISSON, E., 1968. The fetal adrenal cortex in the human. *Acta Endocrinologica* (Supplement 130) **58:** 1–104.

JONES, C. T. AND ROLPH, T. P., 1981. Metabolic events associated with the preparation of the fetus for independent life. Pp. 214–33 in *The fetus and independent life.* (Ciba Foundation Symposium 86). Pitman: London.

JOST, A., 1975. The fetal adrenal cortex. In *Handbook of physiology, Vol. VI: Adrenal gland* ed by R. O. Greep and E. B. Astwood. American Physiological Society: Washington.

KEENE, M. F. L. AND HEWER, E. E., 1927. Observations on the development of the human suprarenal gland. *Journal of Anatomy* **61:** 302–24.

LANMAN, J. T., 1953. The fetal zone of the adrenal gland. *Medicine* **32:** 389–430.

LEATHERLAND, J. F. AND RENFREE, M. B., 1983. Structure of the pars distalis in pouch-young tammar wallabies *(Macropus eugenii). Cell Tissue Research* **230:** 587–603.

LESLIE, R. J., 1979. Aspects of adrenal development in the marsupial, *Isoodon macrourus. Journal of Anatomy* **128:** 657 (Abstract).

LESLIE, R. J. AND HUGHES, R. L., 1978. Preliminary observations on the development of the adrenal gland in the bandicoot, *Isoodon macrourus. Bulletin of the Australian Mammal Society* **5:** 42–43 (Abstract).

LIGGINS, G. C., 1968. Premature parturition after infusion of corticotrophin or cortisol into foetal lambs. *Journal of Endocrinology* **42:** 323–29.

LIGGINS, G. C., FAIRCLOUGH, R. J., GRIEVES, S. A., KENDALL, J. Z. AND KNOX, B. S., 1973. The mechanisms of initiation of parturition in the ewe. *Recent Progress in Hormone Research* **29:** 111–59.

LUSE, S. A., 1967. Electron microscopy and fine structure of adrenal cortex in rat, rabbit, dog, man and armadillo. Pp. 1–59 in *The adrenal cortex* ed by A. B. Eisenstein. Little Brown: Boston.

LYNE, A. G. AND HOLLIS, D. E., 1977. The early development of marsupials, with special reference to bandicoots. In *Reproduction and evolution* ed by J. H. Calaby and C. H. Tyndale-Biscoe. Australian Academy of Science: Canberra.

MCCRADY, E., 1938. The embryology of the opossum. *American Anatomical Memoirs* **16:** 1–233.

MCDONALD, I. R., 1980. Physiology of the adrenal cortex in marsupials. In *Comparative physiology: primitive mammals* ed by K. Schmidt-Nielsen, L. Bolis and C. R. Taylor. Cambridge University Press: London.

MALAMED, S., 1975. Ultrastructure of the mammalian adrenal cortex in relation to secretory function. In *Handbook of physiology, Vol. VI: Adrenal gland* ed by R. O. Greep and E. B. Astwood. American Physiological Society: Washington.

MOOG, F., 1953. The functional differentiation of the small intestine. III. The influence of the pituitary-adrenal system on the differentiation of phosphatases of the duodenum of the suckling mouse. *Journal of Experimental Zoology* **124:** 329–46.

NATHANIELSZ, P. W., 1978. Endocrine mechanisms of parturition. *Annual Review of Physiology* **40:** 411–45.

PANKRATZ, D. S., 1931. The development of the suprarenal gland in the albino rat with a consideration of its possible relation to the origin of foetal movements. *Anatomical Record* **49:** 31–49.

PARKER, K. M., 1917. The development of the hypophysis cerebri preoral gut and related structures in the Marsupialia. *Journal of Anatomy* **51:** 181–249.

PASTEELS, J. L., GAUSSET, P., DANGUY, A. AND ECTORS, F., 1974. Gonadotropin secretion by human foetal and infant pituitaries. *Institut national de la sonté et de recherche médicale* **32:** 13–36.

RISBRIDGER, G. P. AND WEISS, M., 1985. Gonadotrophin and steroid binding to adrenal cortex tissue of female Brushtail Possum, *Trichosurus vulpecula. Australian Journal of Zoology* **33:** 831–35.

ROAF, R., 1935. A study of the adrenal gland of the rabbit. *Journal of Anatomy* **70:** 126–34.

ROSS, M. H., 1960. The adrenal cortex in the human fetus as revealed by electron microscopy. *Anatomical Record* **136:** 268–69.

SHOREY, C. D., 1968. An electron microscope study of pituitary-ovarian changes during oogenesis, and embryonic development in *Trichosurus vulpecula.* Unpubl. M.Sc. Thesis, University of New South Wales, Sydney.

SPAGNOLI, D. B., 1979. Model cytological characteristics of the adrenal medulla of the newborn opossum. *Anatomical Record* **193:** 689 (Abstract).

SPAGNOLI, D. B., FREDERICKSON, R. G., ROBINSON, R. L. AND CARMICHAEL, S. W., 1987. Opossum adrenal medulla: II. Differentiation of the chromaffin cell. *American Journal of Anatomy* **179:** 220–31.

STEWART, F. AND TYNDALE-BISCOE, C. H., 1983. Pregnancy and parturition in marsupials. Pp. 1–33 in *Current topics in experimental endocrinology. Vol. 4* ed by L. Martini and V. H. T. James. Academic Press: New York.

SVALANDER, C., 1974. Ultrastructure of the fetal rat adenohypophysis. *Acta Endocrinologica* (Supplement 188) **76:** 1–113.

TYNDALE-BISCOE, C. H. AND EVANS, S. M., 1980. Pituitary-ovarian interactions in marsupials. Pp. 259–68 in *Comparative physiology: primitive mammals* ed by K. Schmidt-Nielsen, L. Bolis and C. R. Taylor. Cambridge University Press: Cambridge.

VINSON, G. P. AND KENYON, C. J., 1978. Pp. 248–52 in *General comparative and clinical endocrinology of the adrenal cortex. Vol. 2* ed by I. Chester-Jones and I. W. Henderson. Academic Press: London.

WALKER, M. T. AND GEMMELL, R. T., 1983. Organogenesis of the pituitary, adrenal and lung at birth in the wallaby, *Macropus rufogriseus*. *American Journal of Anatomy* **168:** 331–44.

WARING, H., 1935. The development of the adrenal gland of the mouse. *Quarterly Journal of Microscopic Science* **78:** 329–66.

WEISS, M., 1980. Adrenocorticosteroids in protherian, metatherian and eutherian mammals. In *Comparative physiology: primitive mammals* ed by K. Schmidt-Nielsen, L. Bolis and C. R. Taylor. Cambridge University Press: London.

WHEELER, R. S., 1943. Normal development of the pituitary in the opossum and its responses to hormonal treatments. *Journal of Morphology* **73:** 43–87.

Olfactory sense organs in *Isoodon macrourus* in the early pouch young and adult

JEAN E. KRATZING[1]

Studies using light and transmission electron microscopy show that the adult *Isoodon macrourus* has a complex nasal cavity with three areas of chemosensory receptors, the vomeronasal organ (VNO) rostrally, and the main olfactory membrane (MOE) and septal olfactory organ (SOO) in the caudal two-thirds of the cavity. Basic structure of the receptors is similar in all areas, a bipolar sensory neuron with a modified dendritic process, ciliated in the MOE and SOO but not in the VNO, and an axonic process which forms a fibre of the olfactory nerve.

The newborn *I. macrourus* has some receptors in the MOE and VNO with well-differentiated dendritic and axonic processes but their perikarya lack many features of adult cells. The SOO is not apparent until day 5. Olfactory, vomeronasal and respiratory glands are absent at birth but differentiate in the first two weeks of pouch life.

Key words: *Isoodon*, olfactory cells, vomeronasal cell, septal organ.

INTRODUCTION

THE part played by olfaction in the life of mammals has been explored by many authors in recent behavioural and functional studies, revealing its importance in social and reproductive behaviour as well as in food gathering (Vandenburgh 1983). Russell (1985) has recorded its importance in the social life of marsupials, and Sharman (1973) and Hughes and Hall (1984) considered it probable that the olfactory system is sufficiently developed at birth to assist the immature newborn in locating the pouch and teat. However, in at least some marsupials the olfactory system, in common with visual and auditory systems, is poorly developed at birth (Nelson and Harman 1983; Nelson, pers. comm. 1985), and further studies are needed to evaluate the developmental aspects of the peripheral and central olfactory systems. While it may not be representative of marsupials generally, *Isoodon macrourus* is a convenient species in which to follow development. The adult is known to have an extensive peripheral olfactory system (Kratzing 1978), and its short gestation period provides access to early stages of known age.

MATERIALS AND METHODS

Specimens were obtained from individuals of the following ages; Pouch young at days 1, 2, 5, 11, 16, 35 and mature adults. Fixation was by immersion in gluteraldehyde in the case of the young specimens, and by perfusion for the adults. For each stage, appropriate areas of one specimen were dissected out under fixative and prepared for transmission electron microscopy.

A second specimen was processed for light microscopy, using the whole head of young specimens and appropriate areas of the nose of an adult. Serial sections were cut at 7μ, and every tenth section stained and examined.

RESULTS

Light microscopy

In the adult, there are three chemosensory neuroepithelia, the main olfactory epithelium (MOE), the septal olfactory organ (SOO) and the vomeronasal epithelium (VNO). The internal anatomy of the nose is complex, with convoluted dorsal and ventral conchae. The nasal septum is incomplete in the caudal two-thirds of the cavity (Fig. 1), and ends ventrally in a laterally expanded ridge resembling an inverted letter "T". The MOE covers most of the dorsal septum and turbinates in the caudal two-thirds of the cavity and the SOO extends along the widened ventral edge of the septum in the same region. The VNO lines the medial aspect of the blind tubes of the organ in the rostral third of the cavity (Fig. 2). Other surfaces within the cavity are lined by two modifications of ciliated respiratory epithelium, one of which contains mucous secretory cells. Glands under the epithelium include the olfactory glands under the MOE and SOO, vomeronasal glands caudal to the sensory epithelium, lateral nasal glands under the ventral and lateral epithelium, and turbinate, septal and maxillary sinus glands.

At birth, simple right and left nasal cavities without conchae are divided by a wide septum. Rostrally the ventral septum is occupied by the blind tubes of the vomeronasal organ. These open by a short duct to the nasal cavity and to the incisive duct, and thus have access to odours in the oral cavity. The VNO lies medially in the tubes. Caudal to the vomeronasal organ the

[1]Department of Veterinary Anatomy, University of Queensland, St Lucia, Queensland, Australia 4067.
Pages 153–58 *in* BANDICOOTS AND BILBIES, ed by J. H. Seebeck, P. R. Brown, R. L. Wallis and C. M. Kemper. Surrey Beatty & Sons, Sydney, 1990.

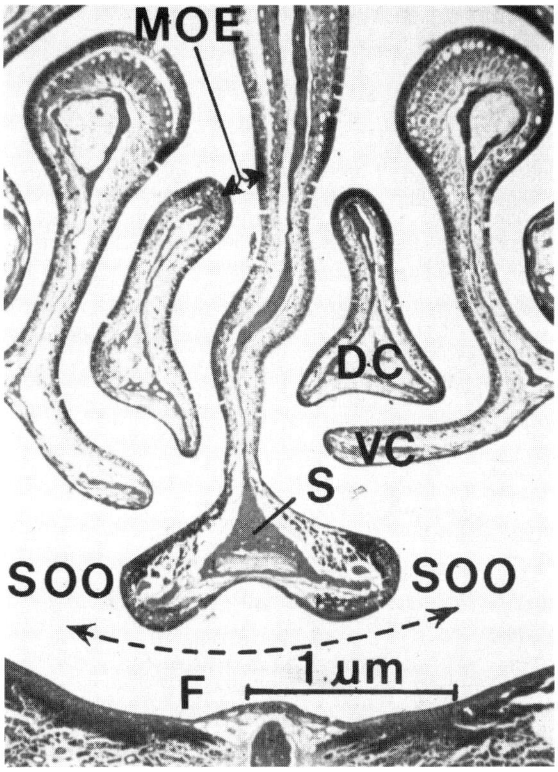

Fig. 1. T.S. Ventral nasal cavity of adult at mid-length. The septal olfactory organ (SOO) lies on the laterally expanded septum (S). Olfactory epithelium (MOE) lines the dorsal septum, dorsal concha (DC) and lateral wall. The ventral septum, ventral concha (VC) and floor (F) are lined by respiratory epithelium. Note the free communication (arrows) between right and left nasal cavities.

septum is incomplete ventrally. MOE lines the dorsal wall of the cavity and the dorsal septum but there is no SOO. There is no glandular tissue under any area of the epithelium though the main duct of the lateral nasal gland (Steno's duct) is present and the nasolacrimal duct is present as a solid cord. At five days the SOO can be recognized on the ventral septum, there are a few secretory cells under the MOE and maxillary sinus epithelium, and in the lateral nasal gland. At 11 days, the ventral and ethmoidal conchae are represented by slight bulges into the cavity. There is little change at 16 days, but by 35 days, conchal development resembles that of the adult.

Fine structure

The chemosensory receptor cells in the three neurosensory areas are peripherally located neurons with modified dendrites reaching to the surface, and proximal processes which leave the base of the epithelium to become axons of the olfactory nerves. In the MOE, at birth, some receptors can be distinguished from the supporting cells by dendritic processes which end in a rounded rod containing a cluster of centrioles (Fig. 3A) and in a few cases a single cilium. Supporting cells carry a few microvilli and form a well-defined terminal web at the junctional complex. By five days after birth, some receptor dendrites have developed one or more cilia, and most have clusters of centrioles and dark granules towards the surface (Fig. 3B). By 11 days, many receptors have dendritic processes which resemble those seen in the adult (Fig. 4), where the olfactory rod carries many cilia projecting amongst the thick fringe of microvilli on supporting cells. However, at 11, 16 and 35 days and in the adult, some dendrites resembling those seen in the first few days after birth can still be seen. In the newborn, MOE receptor cytoplasm is studded with ribosomes and there are a few mitochondria, but even in the perinuclear region there is little differentiation of other organelles (Fig. 5A). At the epithelial base some proximal processes have differentiated neurotubules, but in most cases these are better seen in the small clusters of axons below the basal lamina. Supporting cell cytoplasm differs little from that of receptors, and no basal cells can be distinguished.

Five days after birth there is little change in receptor perinuclear organization (Fig. 5B), but neurotubules are more evident in dendritic and axonic processes. Supporting cells have

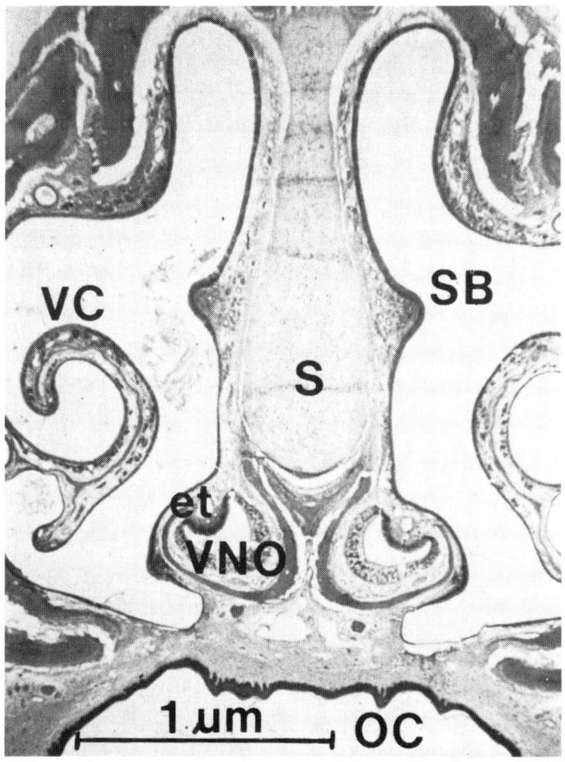

Fig. 2. T.S. Rostral nasal cavity of adult. The vomeronasal organ (VNO) lies ventral to the septal cartilage (S) and is lined by neuroepithelium medially and ciliated epithelium laterally. Erectile vascular tissue (et) under the lateral epithelium acts as a pump to draw in air to the organ. OC — oral cavity; SB — swell body; VC — ventral concha.

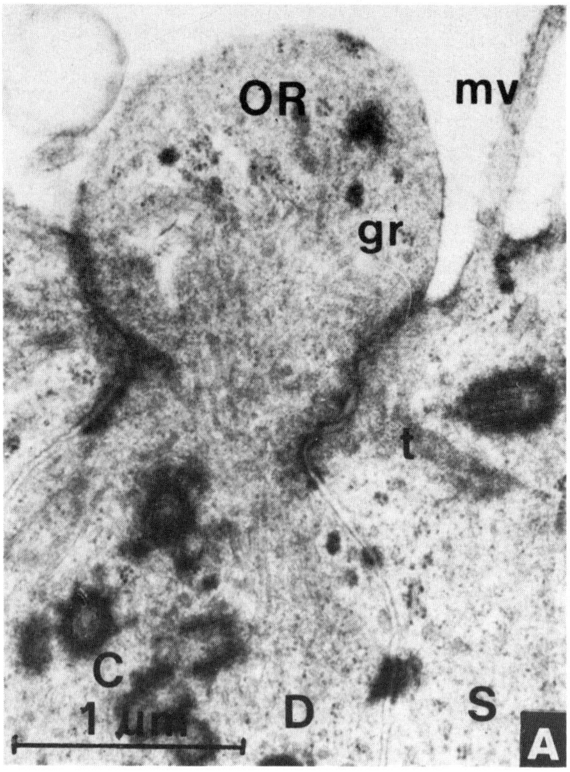

differentiated some rough endoplasmic reticulum, and a few basal cells can be distinguished by the presence of fibrillar bundles. Below the epithelium, some cells contain a few secretion granules. Receptors of the SOO are present on the ventrolateral septum; their fine structure resembles that of the MOE at birth. Secretory cells do not differentiate under the SOO until 16 days after birth. At 16 days many of the receptors in the MOE and SOO have a Golgi apparatus and numerous mitochondria as in the adult perikaryon (Fig. 5C). By 35 days the fine structure of the MOE and SOO closely resemble that of the adult with receptors at different stages of differentiation, supporting cells and two types of basal cells, distinguished by "light" and "dark" cytoplasm. VNO receptor dendrites are less well-differentiated at birth than those in the MOE and are difficult to distinguish from supporting cells except in a few cases where there is a centriole towards the surface (Fig. 6A). Both cell types have sparse microvilli. By five days dendrites have microtubules, more mitochondria and clusters of centrioles towards the surface (Fig. 6B). Receptor dendrites in the adult have more microvilli, clusters of centrioles and dark granules (Fig. 7). Cilia do not usually develop on receptors or supporting cells in the adult

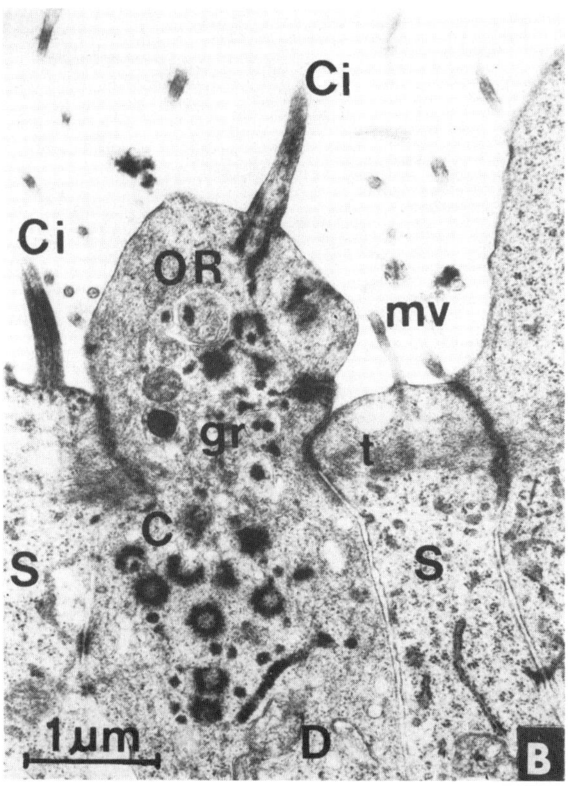

Fig. 3A,B. MOE dendrites (D) at birth (A) and five days (B) end in a rounded olfactory rod (OR). Most rods lack cilia (Ci) at birth but have clusters of centrioles (C) and dark granules (gr). Supporting cells (S) have microvilli (mv) but may have a cilium or centriole. t — terminal web.

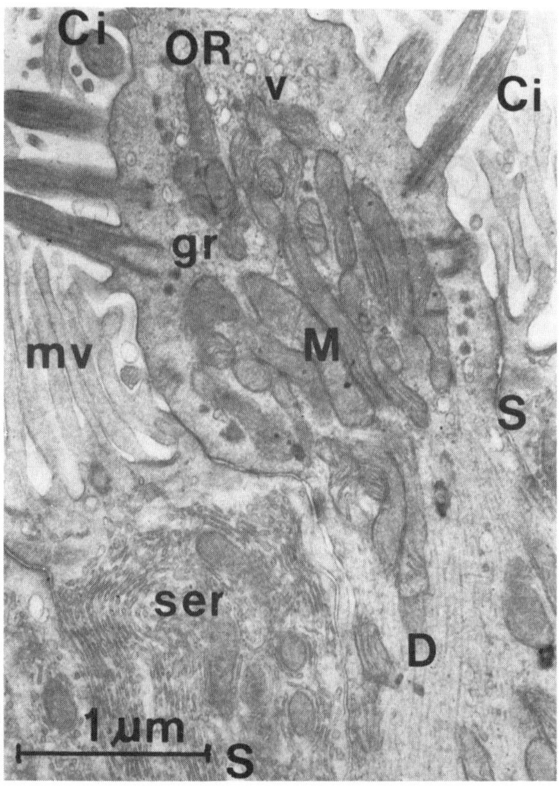

Fig. 4. Receptor surface in The adult MOE. The dendrite (D) ends in a ciliated (Ci) rod (OR) whose cytoplasm contains neurotubules, dark granules (gr), vesicles (v) and mitochondria (M). Supporting cells have extensive smooth reticulum (ser) and numerous microvilli. (M). S — supporting cell.

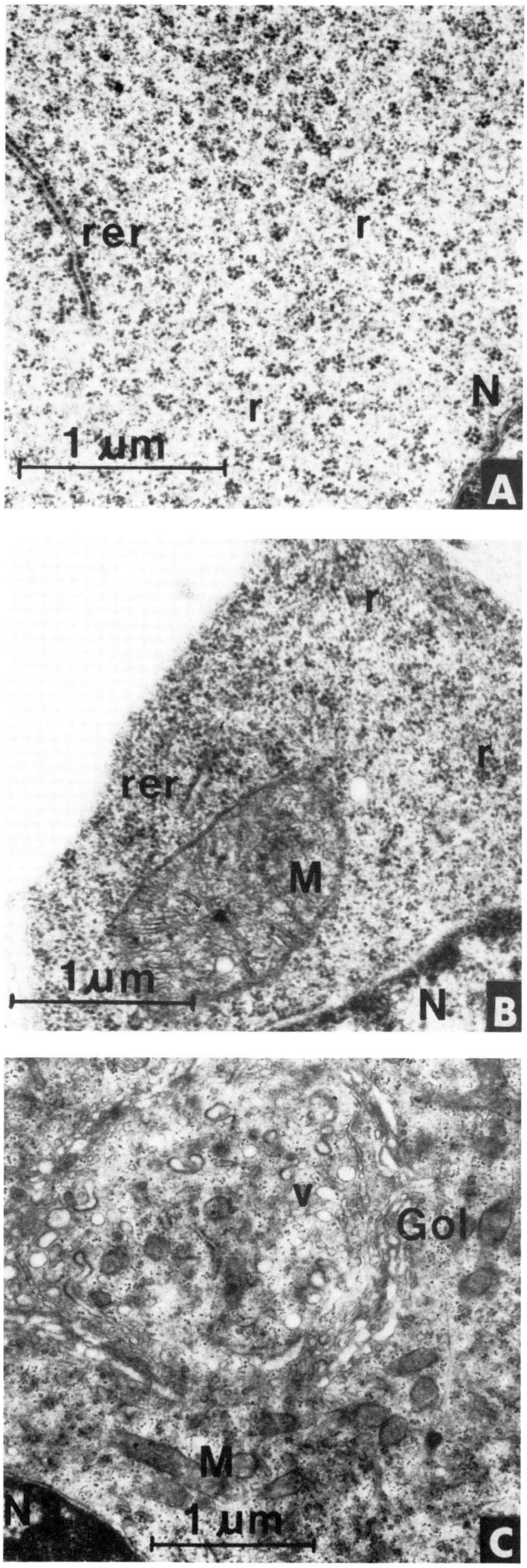

Fig. 5A,B,C. Perikarya of MOE receptors at birth (A) five days (B) and adult (C). Cytoplasm contains ribosome rosettes (r), limited rough reticulum (rer) and mitochondria (M) up to five days after birth. Adult perikarya have Golgi apparatus (Gol), vesicles (v) and numerous mitochondria. N — nucleus.

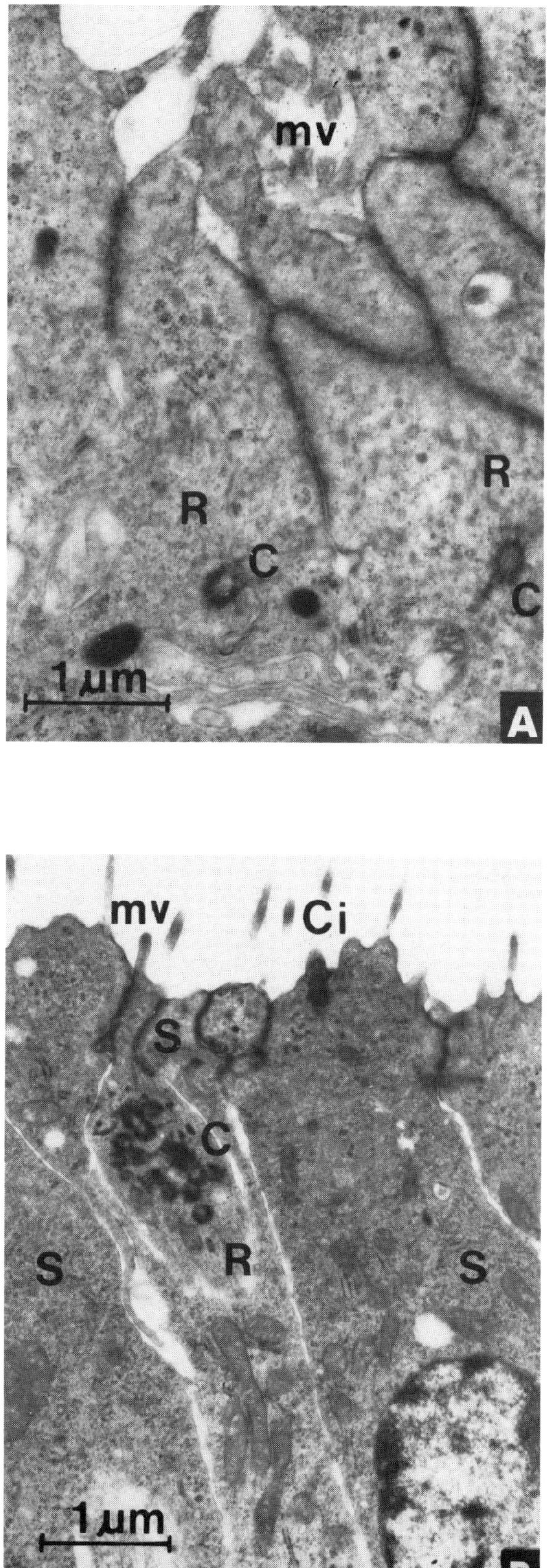

Fig. 6A,B. Surface of VNO at birth (A) and five days (B). Receptor dendrites (R) are difficult to identify at birth but may be indicated by a centriole (C); centriolar clusters are evident at five days. Receptors and supporting cells (S) have microvilli (mv). One supporting cell has a single cilium.

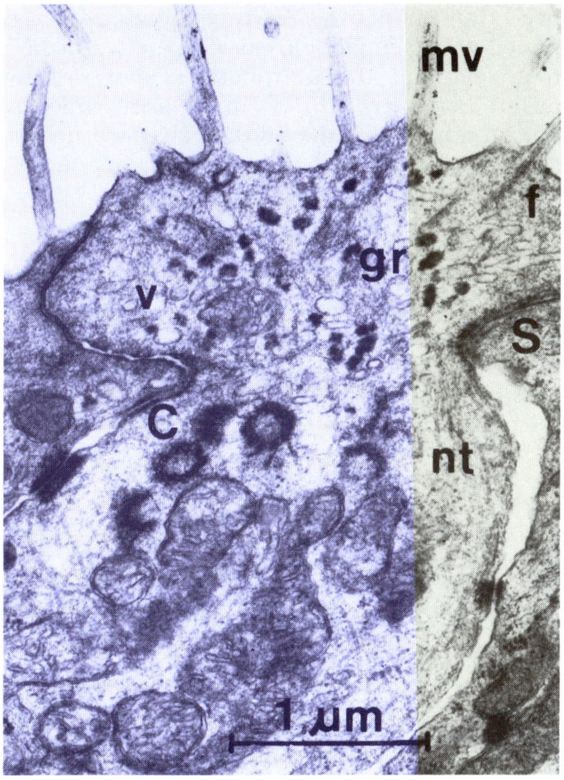

Fig. 7. Adult VNO dendrite. Microvilli (mv) on the dendrite have a fibrillar core (f) extending into surface cytoplasm which contains neurotubules (nt), dark granules (gr) and vesicles (v). Centriolar clusters (C) and mitochondria lie slightly deeper. S — supporting cell.

VNO but may be seen occasionally, particularly in younger specimens up to 16 days. Cytoplasmic differentiation in the VNO perikarya and axonic processes at birth is slightly further advanced than in the MOE but in general the changes at different ages are similar (Figs 8A,B,C). Vomeronasal glands do not show evidence of secretion until later than olfactory glands (between 16 and 35 days). By 35 days after birth, all three neuroepithelia have achieved characteristic adult structure.

DISCUSSION

While it is clear from fine structural studies that the adult *I. macrourus* has an extensive array of chemosensory receptors in its nasal cavity, results are not so conclusive for the newborn. Anatomical studies cannot give an unequivocal guide to functional capacity but the absence of a Golgi apparatus and other organelles in the perikarya of the MOE and VNO receptors suggests that they are not capable of full sensory function in the first few days of pouch life despite the presence of dendritic and axonic processes. Immaturity is also obvious at the central level. Detailed studies of the maturation of the olfactory bulb were not carried out, but the serial sections prepared for light microscopy show that the bulb at birth consists of a dense

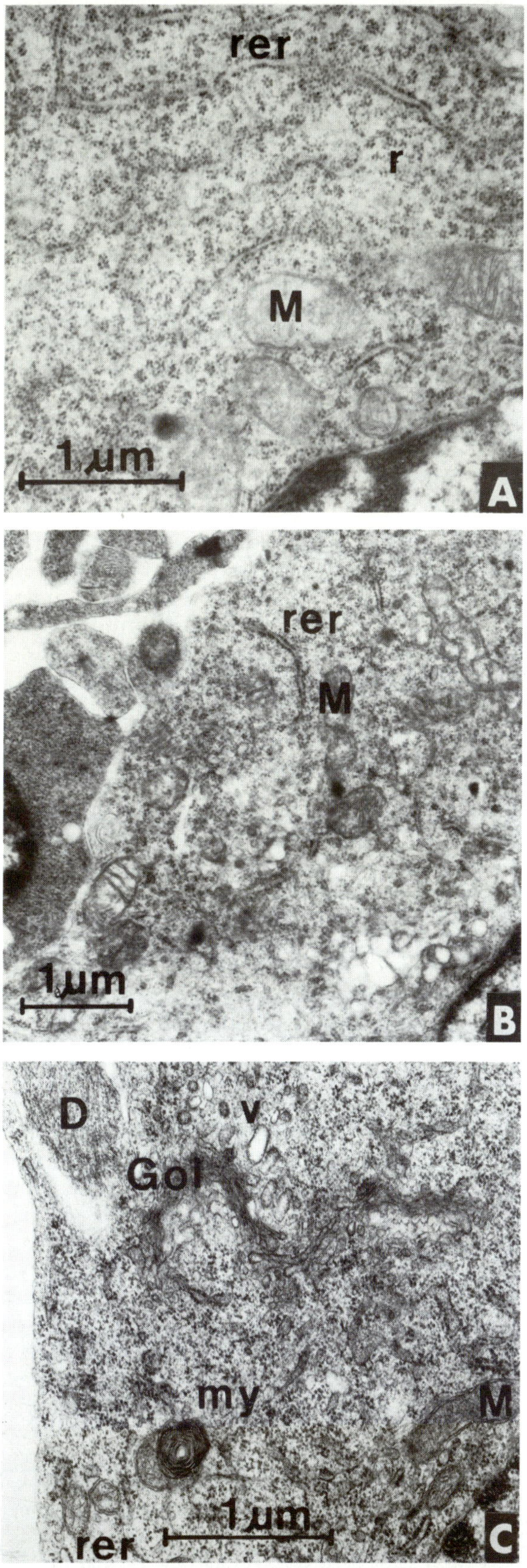

Fig. 8A,B,C. VNO perikarya at birth (A), five days (B) and in the adult (C). Cytoplasm up to five days contains ribosome clusters (r), some rough reticulum (rer) and mitochondria (M). Adult perikarya have a Golgi apparatus (Gol), vesicles (v), rough reticulum (rer) and a myelin body (my). D — receptor dendrite.

layer of neuroblasts around the central canal. Mitral cells and the first indications of a glomerular layer are not distinguishable until five days after birth.

By about 11 days the cytoplasmic organization in many receptors in the MOE and VNO resembles that in the adult. With increasing age the proportion of fully differentiated receptors in the MOE increases, but differentiation of new olfactory cells is a continuing process in most vertebrates (Graziadei and Metcalf 1971) so that immature receptors can still be seen in the adult. It is probable that the "light" basal cells of the adult MOE serve as a stem cell for new receptors (Kratzing 1978). A similar process occurs in the VNO but the stem cells are difficult to identify structurally.

The functional capacity of the peripheral system at birth may also be limited by the absence of glandular secretions. The environment of the pouch probably reduces the need for the respiratory glands which function in humidity and temperature control (Kratzing 1984). However, olfactory (Bowman's) glands occur under the MOE in all terrestrial vertebrates, and under the SOO where this is present. Similarly, the vomeronasal organ receives secretions from specific glands which do not open on to other surfaces. Since the part these secretions play in the olfactory process is not clear, it is uncertain if their absence limits the functional capacity of the receptors (Cuschieri and Bannister 1974).

Uncertainty also remains concerning the differences in the roles of the MOE, SOO and VNO in olfaction. The septal organ or Organ of Rodolfo-Masera (Rodolfo-Masera 1943; Breipohl, Naguro and Miragall 1982) occurs in some rodents and other species. It has not been reported in marsupials other than *I. macrourus*. Though some authors have found minor structural differences, in *I. macrourus* its fine structure is identical with the MOE. From its position in the nasal cavity it is tempting to speculate that it is well-placed to monitor smells in the quiet respiratory air flow in the ventral meatus and perhaps help to detect their direction. Whatever its role in the adult, it is slower to develop than the other two areas and can play no part in location of the pouch and teat for the neonate.

The vomeronasal organ has been shown to play an important role in hormonal and reproductive systems of many species (Wysocki 1982). It is clear that it develops early in pouch life in *Isoodon*, overlapping that of the MOE and has an active part in early sensory development. Superficially, there would appear to be little need for an active olfactory system for the pouch-young individual once securely attached to a teat, but the peripheral system is fully differentiated by 35 days after birth, which is well in advance of the 50 days at which time it should be free to explore away from the pouch.

ACKNOWLEDGEMENTS

I wish to thank Ms Lynn Tolley and Mr Paul Addison for their technical assistance, and my special thanks are due to Mr Leslie Hall, who provided the specimens that made the study possible. The work was supported by an Australian Research Grants Scheme grant.

REFERENCES

Breipohl, W., Naguro, T. and Miragall, F., 1983. Morphology of the Masera Organ in NRMI Mice. *Verhandlungen der Anatomischen Gesellschaft* **77:** 741–43.

Cuschieri, A. and Bannister, L. H., 1974. Some histochemical observations on the mucosubstances of the nasal cavity of the mouse. *Histochemical Journal* **6:** 543–58.

Graziadei, P. P. C. and Metcalf, J. F., 1971. Neuronal dynamics in the olfactory mucosa of the adult vertebrates. *American Journal of Anatomy* **19:** 11.

Hughes, R. L. and Hall, L. S., 1984. Embryonic development in the Common Brushtail Possum *Trichosurus vulpecula*. Pp. 197–212 in *Possums and gliders* ed by A. Smith and I. Hume. Australian Mammal Society: Sydney.

Kratzing, J. E., 1978. The olfactory apparatus of the bandicoot *(Isoodon macrourus)*: fine structure and presence of a septal olfactory organ. *Journal of Anatomy* **125:** 601–13.

Kratzing, J. E., 1984. The structure and distribution of nasal glands in four marsupial species. *Journal of Anatomy* **139:** 553–64.

Nelson, J. S. and Harman, A. E., 1983. Developmental anatomy of *Dasyurus hallucatus*. *The Australian Mammal Society Bulletin* **8**(1): 19 (abstract).

Rodolfo-Masera, T., 1943. Su l'esistanza di un particolare organo olfattivo nel setto nasale della cavia e di altri roditori. *Archivio italiano di anatomia e di embriologia* **48:** 157–213.

Russell, E. M., 1984. Social behaviour and social organization of marsupials. *Mammalogy Review* **14:** 101–54.

Russell, E. M., 1985. The Metatherians: order Marsupialia. Pp. 45–104 in *Social odours in mammals, Vol. 1* ed by D. Macdonald and R. D. Brown. Oxford University Press: Oxford.

Sharman, G. B., 1973. Adaptations of marsupial pouch young for extra-uterine existence. Pp. 67–90 in *The mammalian fetus* in vitro, ed by C. R. Austin. Chapman and Hall: London.

Vandenburgh, J. G. (ed), 1983. *Pheromones and reproduction in mammals.* Associated Press: New York.

Wysocki, C. I., 1982. A review of the neurobehavioural evidence for the involvement of the vomeronasal system in mammalian reproduction. *Neuroscience and Behavioural Reviews* **3:** 301–41.

body stain) or for myelinated axons using Weil-haematoxylin methods and mounted on glass slides for microscopic examination.

The remaining bandicoots were treated for autoradiography as follows. Under ether anaesthesia each bandicoot received an injection of the amino acid tracer, ^{3}H-Proline, into the vitreous chamber of one eye. Following 1–2 days survival, the animals were deeply anaesthetized and perfused through the heart with physiological saline followed by 10% formalin. The region of the brain containing the thalamus and upper midbrain was removed, frozen and sectioned in the frontal plane at 40 μm. The sections were mounted on gelatinized slides and processed as follows. Each slide was dipped in Ilford K2 emulsion, stored in the dark at 4°C for 4–8 weeks, developed in Kodak D-19 developer and lightly counterstained with a Nissl stain (cresyl violet) to enable identification of LGd and other nuclear cytoarchitecture plus the position of label within the LGd. With this procedure the course and terminations of optic nerve fibres are revealed as concentrations of silver grains in the developed autoradiograph.

Counterstained, autoradiographically labelled sections through the LGd were photographed in both dark field and, at higher magnifications, in bright field using a photomicroscope. Magnified outlines of LGd sections, selected at equal intervals through the rostrocaudal extent of the nucleus, were drawn using a microscope drawing tube attachment. The labelled areas, indicating those regions within the LGd which received crossed and uncrossed retinal projections, were also traced on these outlines. The outlines were then traced on a digitising pad which interfaced with a planimetric image analysis programme (*Bioquant,* R & M Biometrics). Through serial reconstruction of these data we were able to calculate the relative proportions of the bandicoot LGd receiving crossed and uncrossed projections. These planimetric procedures were also employed with data obtained previously from other mammalian species.

RESULTS

LGd cytoarchitecture

The LGd in both bandicoots is similar. Cytoarchitecturally, the nucleus is homogenous, having few internal distinguishing features. Nissl stains (Fig. 1A) show that while neuronal packing density is greatest in the lateral part of the nucleus, near the optic tract, and least in the medial portion of the nucleus, there is no separation into distinct lateral or α and medial β segments as is seen in most other Australian marsupials (Sanderson 1985). Nor are cells in any portion of the nucleus segregated into laminae based upon differential cell size, cell shape or packing density differences as seen in many other marsupials (Johnson and Marsh 1969; Sanderson *et al.* 1984; Sanderson 1985). Fibre stains (Fig. 1B) show fascicles, derived from the optic tract, travelling vertically through the LGd and extending into the lateroposterior thalamic nucleus. The projection from the LGd to visual cortex (geniculocortical projection) consolidates in the ventral part of the LGd and courses medially before losing its identity among the fibres of the internal capsule.

Retinogeniculate projections, macro-organization

Injections of ^{3}H-proline into one eye levelled both LGds plus a small part of the contralateral lateroposterior thalamic nucleus immediately dorsal to the LGd and adjacent to the optic tract. In both bandicoots the LGd receives most of its retinal input from the opposite, e.g., crossed, eye (Fig. 2). In *P. gunnii* uncrossed terminals are found in small patches confined to the dorsal half of the LGd which together comprise about 7% of the LGd volume (Table 2). One patch is adjacent to the optic tract; the others, separated by intervening strips of contralateral terminals, are scattered irregularly throughout the dorsal half of the nucleus. The pattern in *I. obesulus* is similar except that, at caudal levels within the LGd, two well-defined uncrossed laminae extend into the ventral half of the nucleus (Fig. 3). Nearly 10% of the nuclear volume is occupied by uncrossed terminals in this species (Table 2). In dark field photographs the labelled regions within the LGd and the optic tract show up as dusty white clouds, while the parts of the LGd and other central structures which do not receive retinal inputs remain dark (Fig. 4C). The drawings in Figures 2 and 3 have been made from a series of these autoradiographs and show the distribution of retinal projections within the LGd for both species.

Retinogeniculate projections, micro-organization

Higher magnification Nissl counterstained autoradiographs are shown for six mammalian species in Figures 6 and 7, including *P. gunnii* (Fig. 7A). In these illustrations the plane of focus is within the photographic emulsion layer and hence is *above* the tissue section. Consequently, the counterstained cells in the background appear, to varying degrees, to be out of focus. The photographs were taken under bright field conditions; the small black spots are developed silver grains located within the emulsion. These spots correspond to the white "dust" seen in the lower magnification views of Figures 4 and 5. With careful technique, nearly every spot corresponds with the position of a radioactive amino acid residue, either within the axon terminal region or within

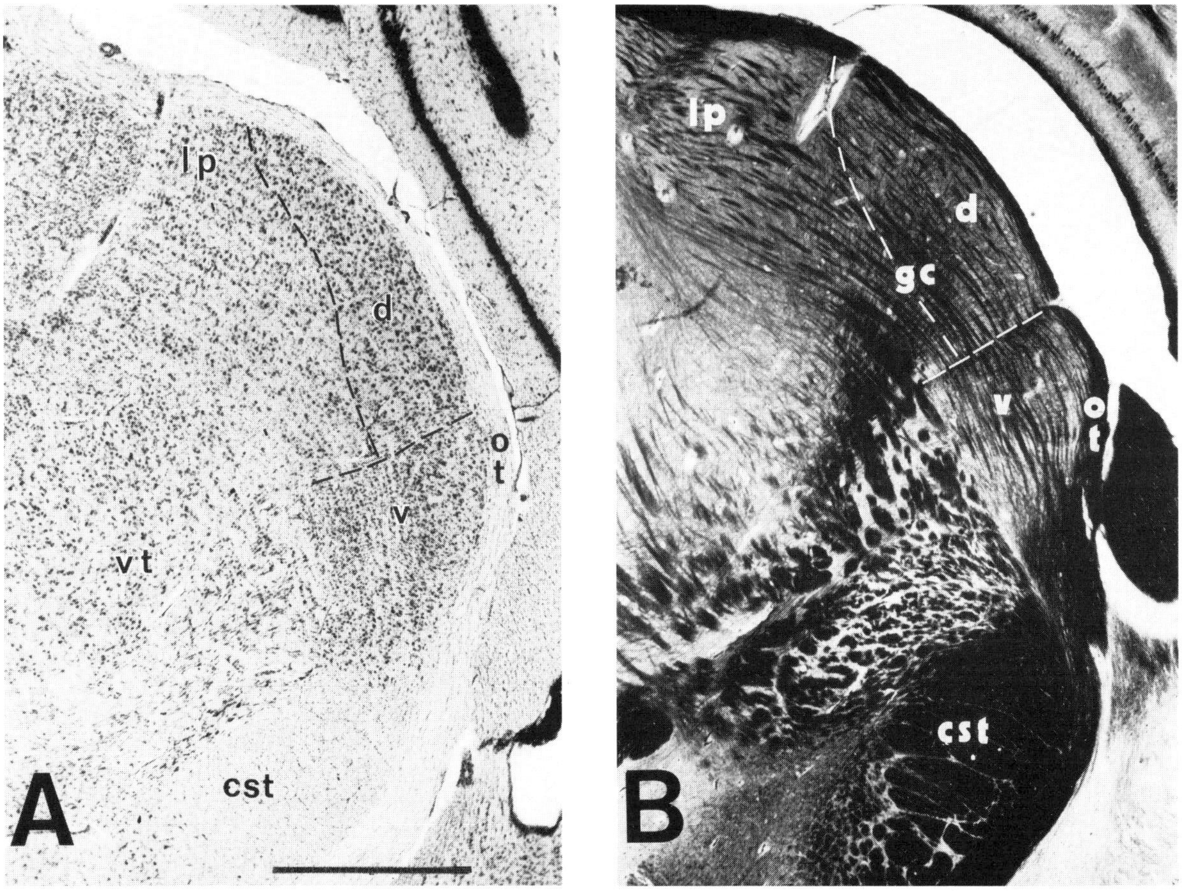

Fig. 1A,B. Frontal sections through mid levels of the right hand LGd in *Isoodon obesulus*. A. Nissl (thionin) stained section. The cytoarchitectural homogeneity of the LGd is apparent. The medial LGd border is poorly defined here though at higher magnifications the lateroposterior nucleus and the LGd are easily separated. B. Myelin (Weil-Heidenhain) stained section, approximately the same level as A. Fascicles of myelinated fibres, derived from the optic tract can be seen coursing vertically through both the ventral and dorsal geniculates into the lateroposterior thalamic nucleus. The geniculocortical radiation is seen consolidating within the ventral part of the LGd and exiting from its ventromedial corner. Abbreviations: *cst*, corticospinal tract; *d*, dorsal lateral geniculate nucleus; *gc*, geniculocortical and corticogeniculate fibres; *lp*, lateroposterior thalamic nucleus; *ot*, optic tract; *v*, ventral lateral geniculate nucleus; *vt*, ventral thalamic tier. Marker bar=1.0 mm.

Table 2. Bilateral, crossed and uncrossed retinal representation in selected LGds.

Class, Order, Species	% LGd volume occupied by: Bilateral input	Crossed input	Uncrossed input	Source: derived from data originally presented by
Marsupialia, Polyprotodonta				
Didelphis virginiana	39.8	60.2 (80.1)[1]	19.9[2]	Royce *et al.* 1976
Dasyuroides byrnei	38.4	61.6 (80.9)[1]	19.3[2]	Haight and Sanderson 1986
Sarcophilus harrisii	43.9	56.1 (78.1)[1]	22.0[2]	Sanderson *et al.* 1979
Perameles gunnii	0	93.1	6.9	Present paper
Isoodon obesulus	0	90.2	9.8	Present paper
Marsupialia, Diprotodonta				
Petaurus breviceps	0	74.0	26.0	Sanderson *et al.* 1987
Trichosurus vulpecula	0	68.4	31.6	Sanderson *et al.* 1978
Lasiorhinus latifrons	0	89.1	10.9	Sanderson and Pearson 1981
Potorous tridactylus	0	75.9	24.1	Sanderson *et al.* 1984
Macropus eugenii	0	80.5	19.5	Sanderson *et al.* 1984
Placentalia, Rodentia				
Rattus villosissimus	0	89.0	11.0	Mayner *et al.* 1980

[1]First figure represents measured volume of LGd occupied solely by crossed terminals. Figure in brackets represents estimate of total crossed input. This was obtained by dividing the "bilateral input" volume in half and adding this to the first figure.
[2]Estimate of uncrossed input obtained by dividing the "bilateral input" volume in half as explained above.

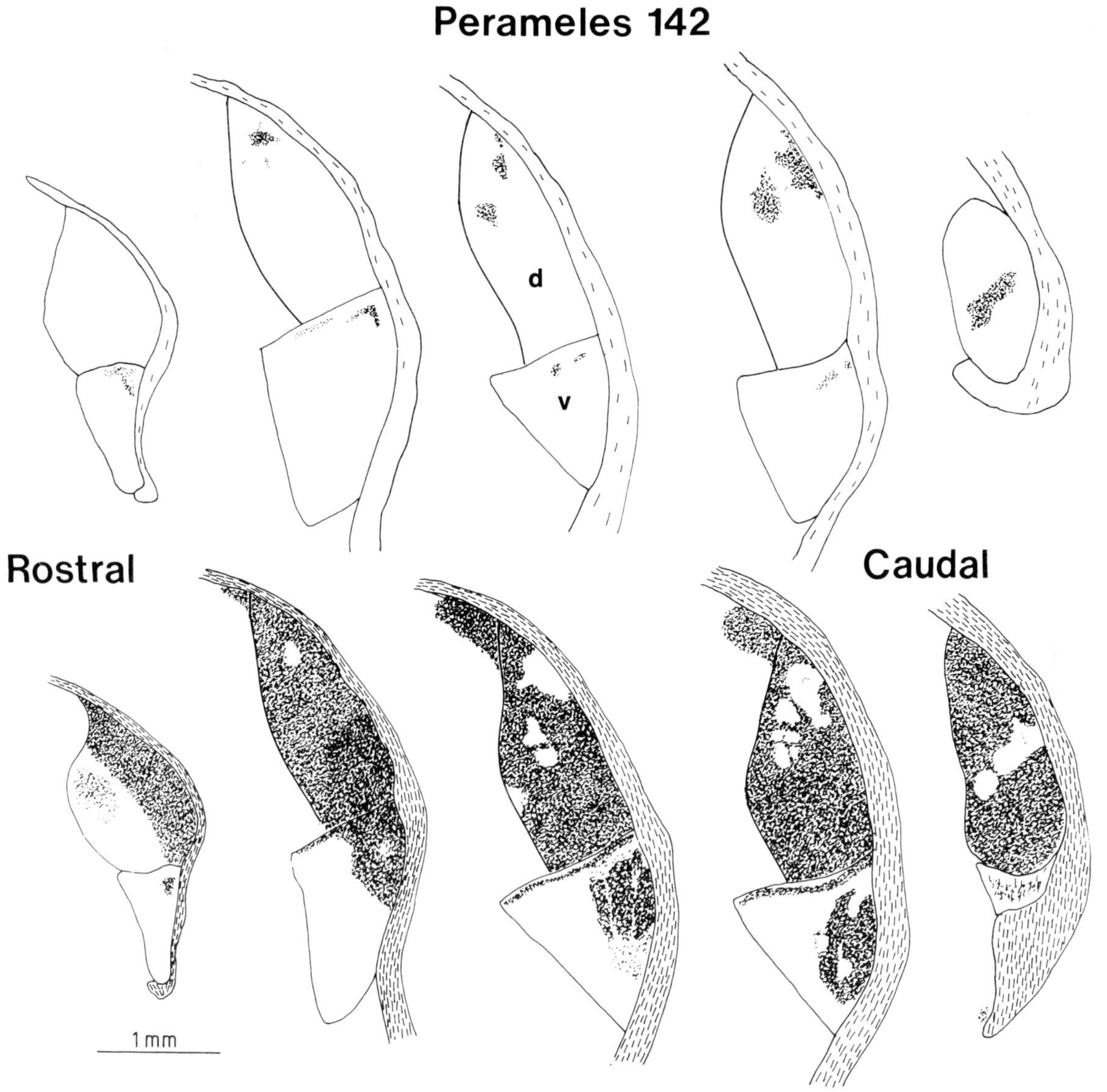

Fig. 2. Drawings of the distribution of uncrossed (top row) and crossed (bottom row) retinal terminals in the *Perameles gunnii* LGd following ^{3}H-proline injection into one eye. Drawings were made from frontal sections, taken at five equally-spaced intervals through the rostrocaudal extent of the LGd. Each vertical pair corresponds to matching levels in each LGd. In order to facilitate comparison of ipsilateral and contralateral terminal patterns, all LGds have been referred to the right side of the brain. Abbreviations: *d*, dorsal lateral geniculate nucleus; *v*, ventral lateral geniculate nucleus.

either within the axon terminal region or within the axon cylinder. As transport of amino acids is fairly rapid, being approximately 250–300 mm/day, within 24 hours most of the label will have accumulated in or near the axon terminals with the remainder being incorporated into the axon cylinder itself. It should also be noted that the weak α radiation emitted from the tritiated tracer can penetrate only a few μm of tissue before being absorbed. Consequently, though the sections shown are 50 μm thick, the data are derived from only the top 10% or so of the tissue thickness (Cowan and Cuénod 1975). Consequently, the neurons which appear least in focus in Figures 4 and 5 are most distant from the layer of exposed silver grains. These somata (though not necessarily their processes) are less likely to be associated with the silver grains overlying them, than are those of cells more clearly in focus.

Examination of the distribution of retinal terminals at higher magnification discloses that the retinal projections to the LGd are asymmetric in both bandicoots, with each eye producing different synaptric distribution patterns within the LGd. In the LGd of *P. gunnii* crossed retinal terminals are both more numerous and more densely packed than uncrossed terminals (Fig. 7A). They tend to be found scattered throughout the intercellular region or neuropil, indicating that most retinal afferents synapse with cell processes rather than with cell bodies. However, some

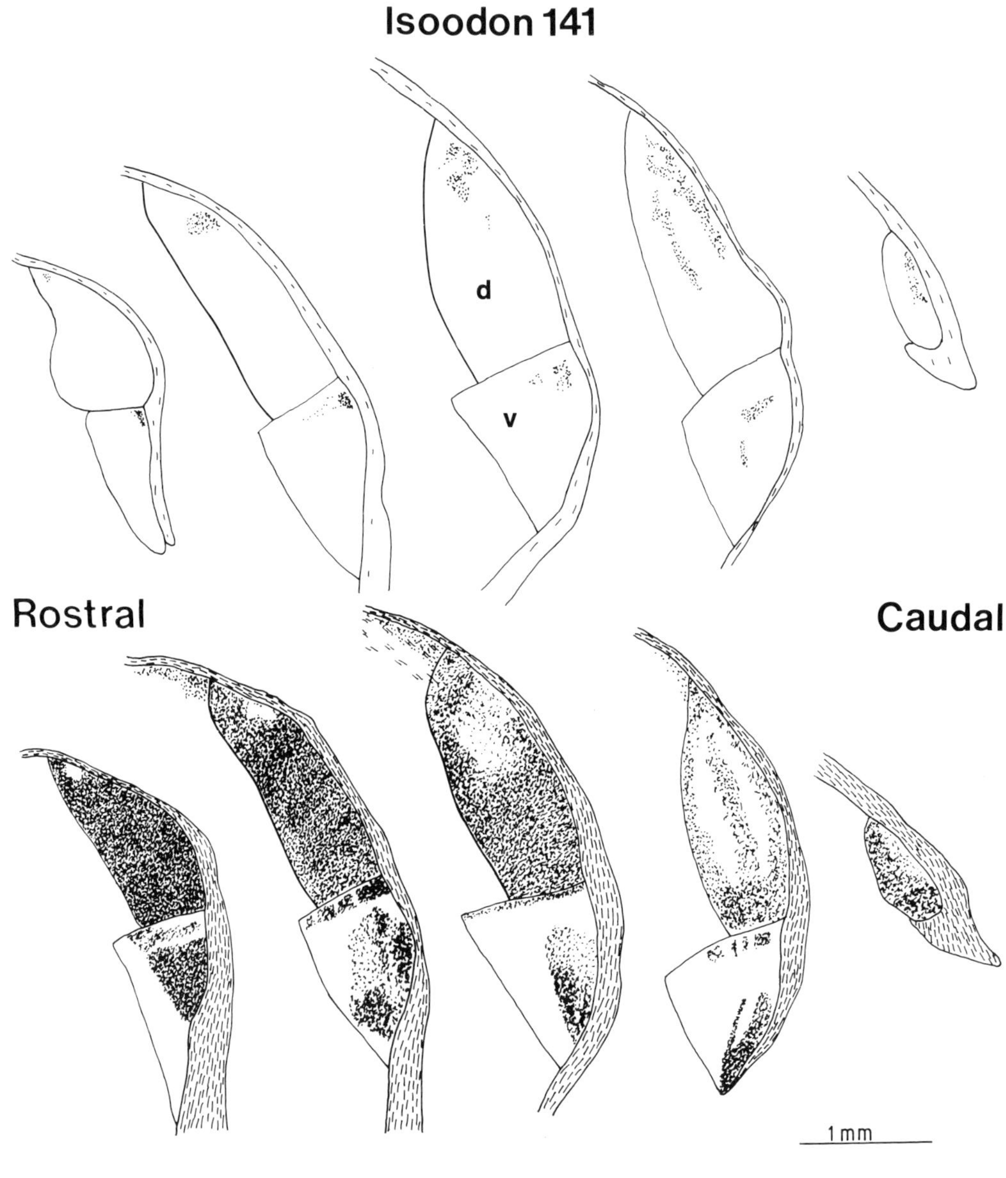

Fig. 3. Drawings of the distribution of uncrossed (top row) and crossed (bottom row) retinal terminals in the *Isoodon obesulus* LGd following ^{3}H-proline injection into one eye. Drawings were made from frontal sections, taken at five equally spaced intervals through the rostrocaudal extent of the LGd. Each vertical pair corresponds to matching levels in each LGd. In order to facilitate comparison of uncrossed and crossed terminal patterns, all LGds have been referred to the right side of the brain. Abbreviations: *d*, dorsal lateral geniculate nuclues; *v*, ventral lateral geniculate nucleus.

Fig. 4A,B,C, (opposite). Darkfield autoradiographs showing uncrossed (left) and crossed (right) retinal terminal distributions within the LGd following intraocular ^{3}H-proline injections in three representative polyprotodont marsupial species. Row A, the Virginia Opossum, *Didelphis virginiana*. Row B, the Kowari, *Dasyuroides byrnei*. Row C, the Eastern Barred Bandicoot, *Perameles gunnii*. Note substantial amount of binocular overlap of retinal terminals in A and B and none in C. All illustrations are taken from frontal sections located within the central one-third of each LGd. The right and left photograph of each pair derived from approximately the same level within the LGd, and all illustrations have been arranged to correspond to a left eye injection. The *P. gunnii* autoradiographs correspond to a level between the third and fourth vertical pair of drawings in Figure 2. Abbreviations: *b*, region of LGd receiving bilateral retinal input; *i*, regions of ipsilateral retinal input, label free areas within the contralateral LGd corresponds to labelled areas in the ipsilateral LGd; *lp*, lateroposterior thalamic nucleus, *m*, portion of LGd which receives input from the contralateral eye only, the corresponding region in the ipsilateral LGd is label free; *ot*, optic tract; *v*, ventral lateral geniculate nucleus. Marker bars=1.0 mm.

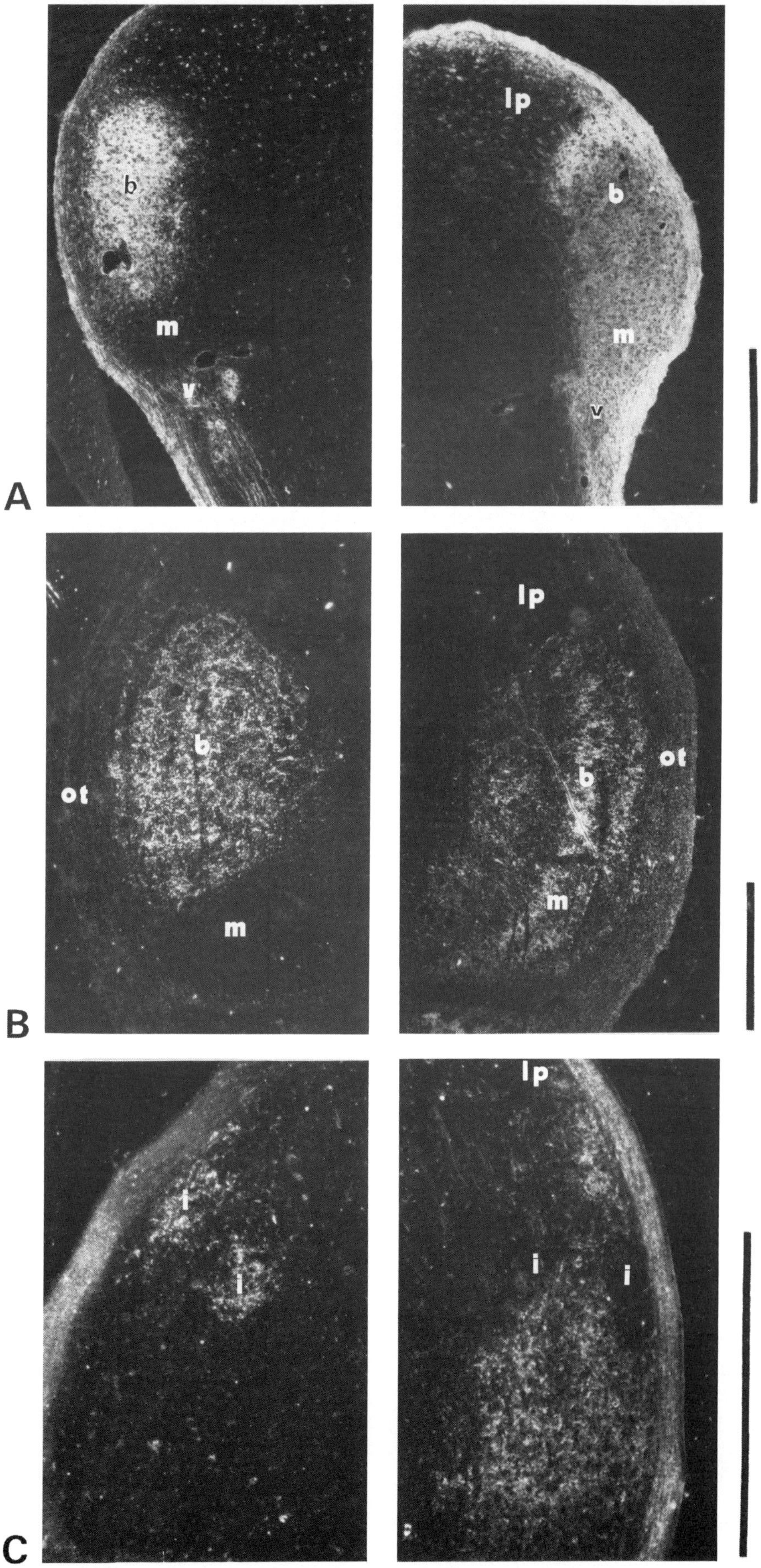
lp
b
b
m
m
v
v
A
lp
ot
b
b
ot
m
m
B
lp
i
i
i
i
C

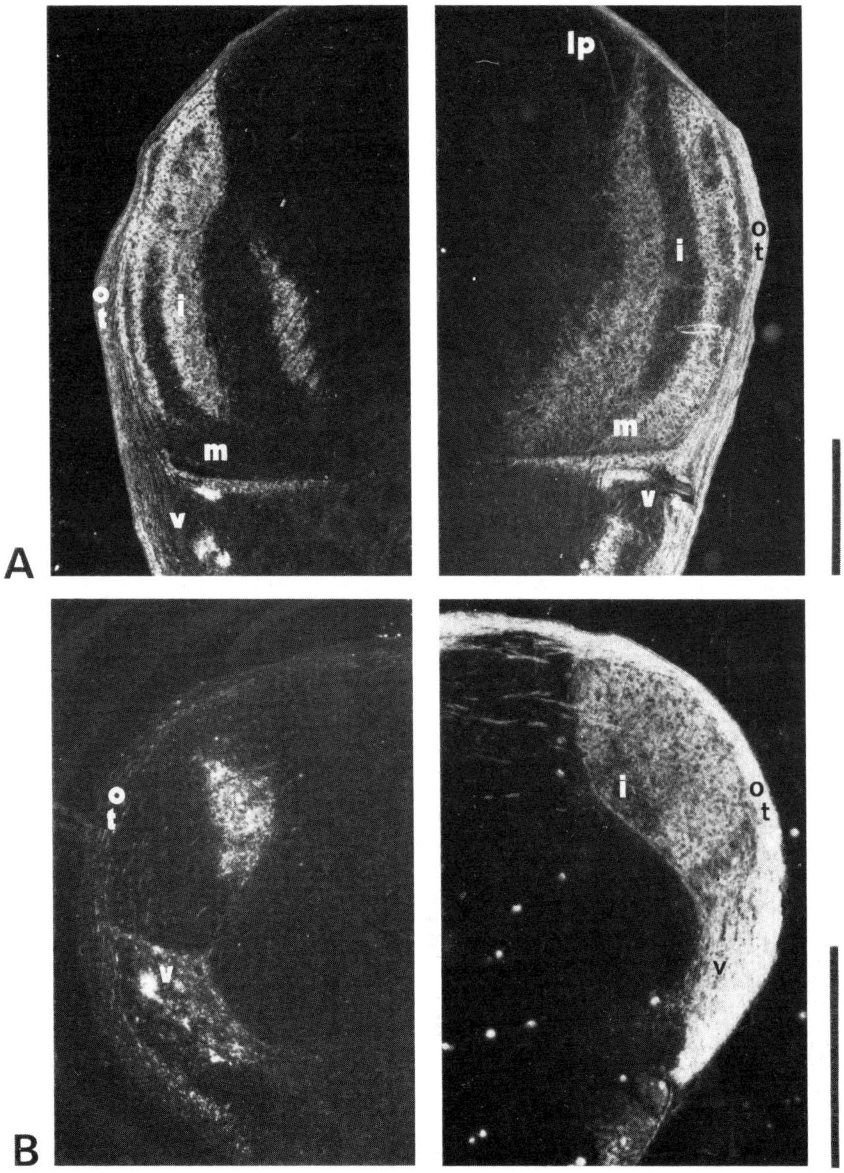

Fig. 5A,B. Darkfield autoradiographs showing uncrossed (left) and crossed (right) retinal terminal distributions within the LGd following intraocular ^{3}H-proline injections in a diprotodont marsupial and in a placental mammal. Row A, the Brushtailed Possum, *Trichosurus vulpecula*; Row B, the Plains Mouse, *Pseudomys australis.* Note absence of binocular overlap in these LGds. All illustrations are taken from frontal sections located within the central one-third of each LGd. Each right and left pair derives from approximately the same level within the LGd, and all illustrations have been arranged to correspond to a left eye injection. For list of abbreviations see caption to Figure 4. Marker bars=1.0 mm.

terminals are seen which appear to be in contact with well-focused cell bodies, suggesting that some axosomatic synapses may be present, an observation needing ultrastructural verification. Within the neuropil the crossed terminals are fairly evenly scattered (Fig. 7A, right), though there is some gathering into loose terminal aggregations.

Uncrossed terminals, as seen in *P. gunnii* (Fig. 7A, left), generally occur deep within the interstitial region. In the uncrossed terminal areas the infocus neurons are surrounded by terminal free "halos", indicating that uncrossed synapses tend to be located at some distance from neuronal cell bodies where they are gathered into dense clusters against a thinly scattered background of individual label found elsewhere in the interstitium. That this scattered label does not represent background artefact is demonstrated by the abrupt cessation of label along the lower left margin of the illustration, this being of portion of the LGd which receives only contralateral projections.

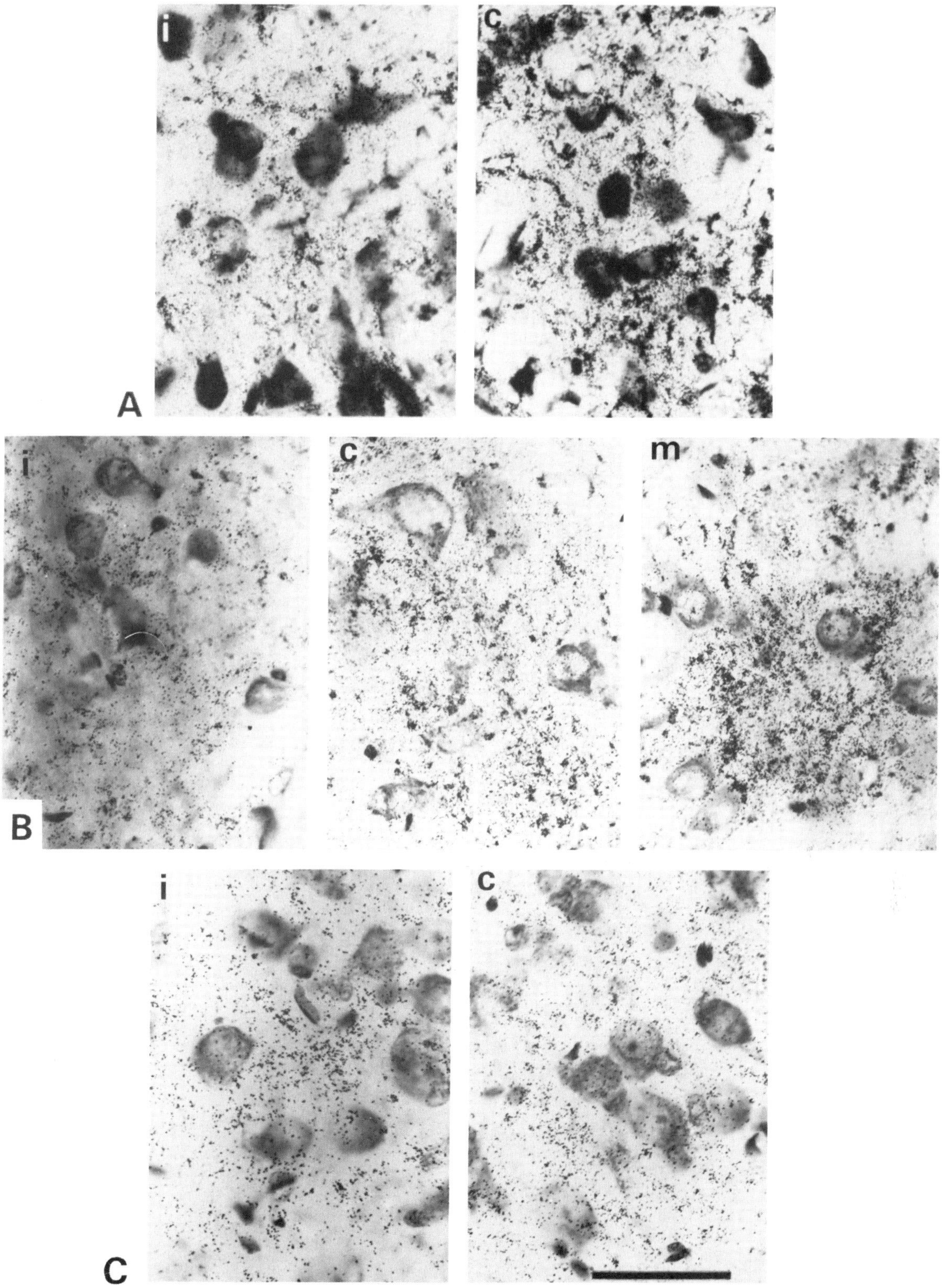

Fig. 6A,B,C. Higher magnification, Nissl counterstained autoradiographs of retinal terminal distribution patterns in didelphid and dasyurid marsupials. Row A, Virginia Opossum, *Didelphis virginiana;* Row B, Tasmanian Devil, *Sarcophilus harrisii;* Row C, Kowari, *Dasyuroides byrnei.* In all of these marsupials large portions of the LGd receive overlapping inputs from both eyes. All photographs have been arranged to correspond to a left eye injection, and each uncrossed or ipsilateral (labelled *i*) and crossed or contralateral (labelled *c*) pair was taken from homologous positions within each LGd. The photograph in Row B, labelled *m*, was taken from the monocular portion of the uncrossed LGd. Note that terminals in A and B display unequal crossed-uncrossed densities but are clustered. In C crossed and uncrossed terminals have equal densities but are not clustered. See text for further details. Marker bar in Row C refers to all illustrations and equals 50 μm.

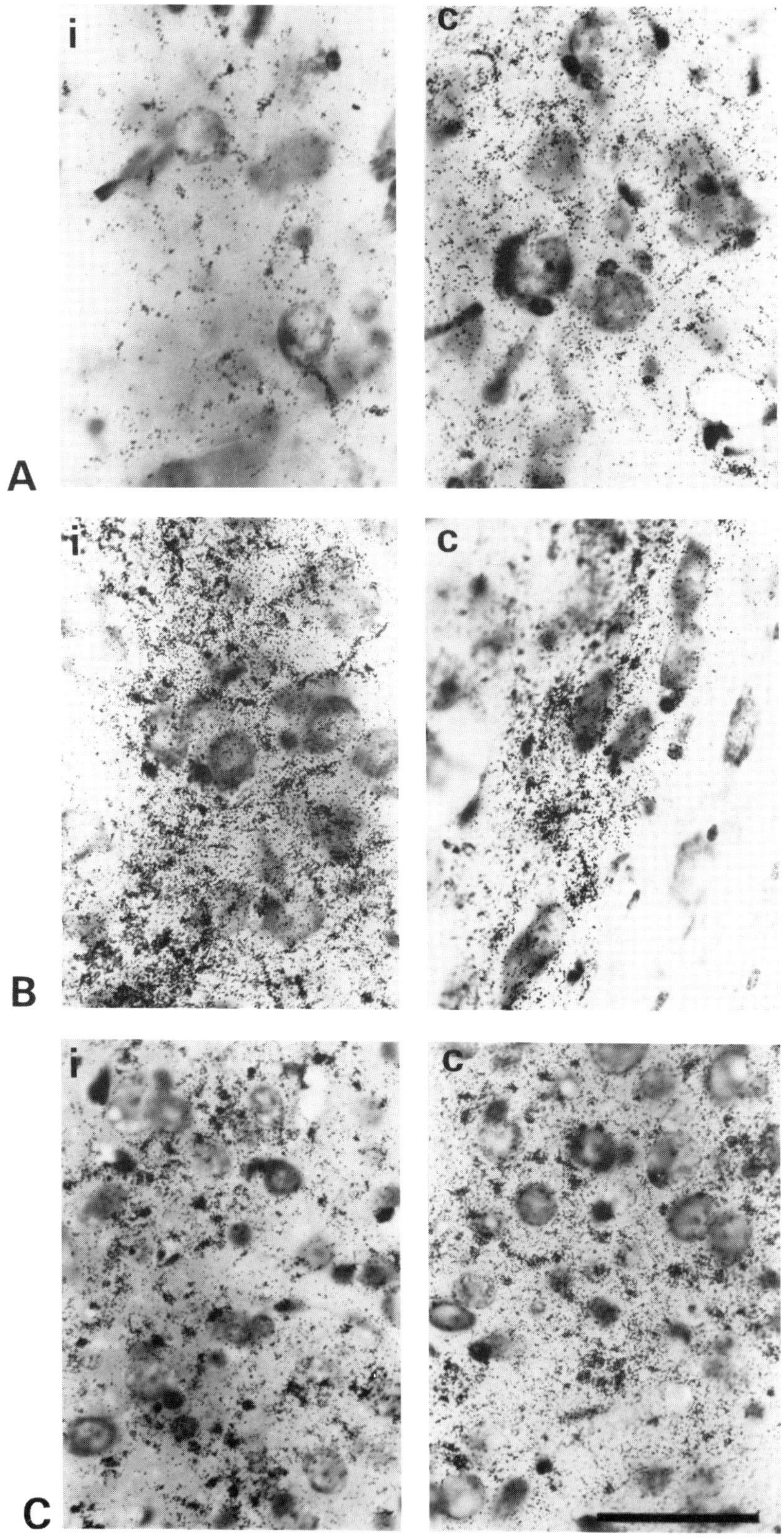

Fig. 7A,B,C. Higher magnification, Nissl counterstained autoradiographs of retinal terminal distribution patterns in a peramelid polyprotodont, a diprotodont marsupial and a placental. Row A, Eastern Barred Bandicoot, *Perameles gunnii;* Row B, Brushtailed Possum, *Trichosurus vulpecula;* Row C, Plains Mouse, *Pseudomys australis.* In these LGds crossed (contralateral) and uncrossed (ipsilateral) retinal terminals are effectively segregated within the nucleus. In A crossed and uncrossed terminal distributions are grossly asymmetric and only the uncrossed terminals are stongly clustered. In B and C the terminals are symmetric and clustered. See text for further details. The photographs have been arranged to correspond to a left eye injection. See text for further details. Marker bar in Row C refers to all illustrations and equals 50 μm.

In bandicoot LGds the distribution of the crossed and uncrossed retinal terminals is highly asymmetric: per unit area, there are many more crossed terminals than uncrossed. The former tend to be distributed evenly throughout the neuropil while the latter are clustered deep within the interstitium.

DISCUSSION

In this discussion we will compare the following anatomical features of the bandicoot LGd with those of other mammals:

a. cytoarchitectural complexity
b. degree of binocular overlap of retinal terminals
c. symmetry of crossed and uncrossed retinal terminal distributions
d. clustering versus non-clustering of retinal terminals.

We will try to assess the functional, behavioural and phylogenetic significance of these features, though it should be realized that these comments will be poorly supported by hard data, especially in regard to bandicoots. As much as we would like to relate visual anatomy and function in marsupials, such correlations remain nebulous because we know so little about marsupial visual physiology or behaviour. However, extensive anatomical and physiological studies of visual organization in placental mammals have led to certain generalizations about visual ecology in mammals; generalizations which, as will be seen, may not be entirely applicable to marsupials. Recent morphological studies of the marsupial visual system have shown that many of these mammals differ considerably from the placental "norm".

Anatomical complexity in the LGd: does it relate to lifestyle?

Placentals

An architecturally simple LGd, as seen in bandicoots, is presumed to be a primitive mammalian character. Complex LGds, i.e., those displaying distinct, multiple cell laminae receiving segregated projections from each eye, are the exception rather than the rule among placental mammals (Sanderson *et al.* 1984; Sanderson 1985 for discussion and review). Complex LGd organization has been associated with active predation, as in the Carnivora (Kaas, Guillery and Allman 1972, 1973; Sanderson 1974), or with colour vision and an arboreal or aerial habitus, as in many primates, tree squirrels and the non-echolocating bats (Weber, Casagrande and Harting 1977; Kaas, Huerta, Weber and Harting 1978; Suthers and Bradford 1980). Placentals which do not occupy these niches, such as terrestial rodents, insectivores, echolocating bats, ruminants, edentates, and, in fact, most placental mammals, display a simple LGd macro-architecture which is combined with a simple pattern of retinal projections to the LGd (Guillery 1979).

Marsupials

In two polyprotodont families, the Didelphidae and the Dasyuridae, active predation may be correlated with an anatomically complex LGd. Among the Didelphidae the arboreal and predatory Mouse Opossum, *Marmosa robinsoni (= mitis),* has a laminated LGd (Royce *et al.* 1976), while the omnivorous and sluggish Virginia Opossum, *Didelphis virginiana,* does not (Royce *et al.* 1976). In Australia most dasyurids examined possess multi-laminated LGds (Sanderson and Pearson 1977; Haight and Sanderson 1986 and authors' observations) whereas the Tasmanian Devil *Sarcophillus harrisii,* a cumbersome scavenger, does not (Sanderson *et al.* 1979). The bandicoots, members of another polyprotodont family, are land-bound, nocturnal, fossicking omnivores and with their anatomically simple LGds, would seem to fit comfortably into the non-complex category.

These behavioural-ecological correlations collapse when the diprotodont marsupials are examined (Table 1). Most of these marsupials have complex LGds, some of which are far more elaborately laminated than those of any other mammal (Sanderson *et al.* 1984). Often, however, there is no obvious relationship between lifestyle and LGd complexity. Consider the gliding possums, animals which would be expected to have excellent visual co-ordination. In the Feathertail Glider, *Acrobates pygmaeus,* the LGd has two cell laminae (Sanderson *et al.* 1987), while that of the Sugar Glider, *Petaurus breviceps,* has five (Johnson and Marsh 1969; Sanderson *et al.* 1987). The arboreal, but non-gliding, Common Ringtail Possum *Pseudocheirus peregrinus* has a two layered LGd (Pearson, Sanderson and Wells 1976) while the less arboreal Brush-tail Possum, *Trichosurus vulpecula,* has six (Hayhow 1967). Members of the kangaroo family, most of which are nocturnal grazers and browsers, display up to eight distinctive laminae within their LGds (Sanderson *et al.* 1984; Sanderson *et al.* 1987). Their placental counterparts, the antelope, deer, sheep and cattle, are restricted to no more than three (see Sanderson *et al.* 1984 for review). The only placentals which approach the diprotodont level of LGd lamination are the primates. These LGds usually contain six cell laminae, three associated with each eye, although eight or more have been reported in some primate species (Kaas *et al.* 1978).

Thus, among the marsupials a complex LGd is the rule rather than the exception; the reverse of the placental situation. In this respect the anatomically simple bandicoot LGd resembles that of a large number of placentals and comparatively few marsupials.

LGd cytoarchitecture and hidden laminae

Retinal terminal patterns within the LGd are often more complex than geniculate cytoarchitecture would suggest. For example, the cytoarchitecturally homogenous LGd of *I. obesulus* contains three distinct zones of retinal input; a large zone of crossed terminals in which two strips of uncrossed terminals are embedded. Such zones form *hidden laminae,* first described in the rat LGd by Hayhow, Sefton and Webb (1962), and found in all marsupial groups and in many placentals (see Kaas *et al.* 1978). In marsupials a particular cytoarchitectural lamina may contain two or three well-defined hidden laminae, as has been demonstrated in the didelphid *M. robinsoni* (Royce *et al.* 1976), the dasyurid *D. viverrinus* (Sanderson and Pearson 1977) and in several members of the kangaroo family (Sanderson *et al.* 1984). In an earlier paper we named the regions within the LGd which received discrete retinal projections *terminal bands,* while the term *lamina* was retained for cytoarchitectural features (Sanderson *et al.* 1984). One result of the lack of correspondence between cytoarchitecture and retinal terminals is that there are often more terminal bands than there are cytoarchitectural laminae (see comparison of terminal bands and laminae Table 1). As will be seen below, the organization of retinal terminals within the bandicoot LGd is more complex than its simple cytoarchitecture would suggest.

Ocular segregation within LGd: where do bandicoots fit?

The fundamental patterns of retinogeniculate macro-organization found in mammals are illustrated in the examples presented in Figures 4 and 5. In didelphid and dasyurid marsupials there is always a substantial area of the LGd which receives a binocular input (Fig. 4A,B) though there may be additional regions in which the ocular inputs are segregated (see Lent *et al.* 1976; Royce *et al.* 1976; Sanderson and Pearson 1977; Haight and Sanderson 1986). To date, binocular overlap of retinal terminals within the LGd has been observed in eight didelphid and dasyurid species and in only one diprotodont, *A. pygmaeus* (Sanderson *et al.* 1987). In the bandicoots (Fig. 4C), as in virtually all diprotodonts (Fig. 5A), and in all placentals studied so far (Fig. 5B), each eye tends to project to separate regions within the LGd (Table 1), displaying no binocular overlap of retinal terminals within the LGd. If binocular overlap, as seen in other polyprotodonts, is a primitive marsupial character, then the segregation of retinal projections in bandicoots and diprotodonts may be convergent upon the placental condition.

Crossed, uncrossed and bilateral areas within the LGd: how do they compare?

Opossums and dasyures

As previously mentioned, these marsupials exhibit a large area of binocular overlap within the LGd, making determinations of the LGd volume attributable to a single eye difficult (Fig. 4). In the Kowari, *Dasyuroides byrnei,* a small dasyurid, approximately 38% of the LGd volume receives overlapping projections from both eyes. Similar figures are obtained for the large dasyurid, *S. harrisii* (44%) and for the didelphid *D. virginiana* (40%). Qualitative examination of crossed and uncrossed terminal densities in higher magnification photomicrographs indicates that these probably equal in *D. byrnei* (Fig. 6C), but that in *D. virginiana* (Fig. 6A) and *S. harrisii* (Fig. 6B) the uncrossed terminal densities are less than those derived from the opposite eye. Nonetheless, we have simply divided the volume of the binocular input region in these LGds by two, assuming thereby that half the bilateral zone of each LGd is devoted to processing information from one eye or the other. This yields estimates of from 19% to 22% of the LGd volume occupied by uncrossed retinal inputs, in these three representative polyprotodonts (Table 2).

Possums, wombats and kangaroos

Apart from that of *A. pygmaeus* these LGds display very little binocular overlap, the volume of the LGd occupied by uncrossed terminals ranges between 11% and 32% (Table 2). In *T. vulpecula,* an animal with demonstrable manipulative skills and eye-hand co-ordination (Rees and Hore 1970; Megirian, Weller, Martin and Watson 1977), the uncrossed terminals occupy 32% of the LGd volume, the highest proportion yet observed in a marsupial. In most other possums and in the kangaroos the figure is 20–26%. An exception is found in the Hairy-nosed Wombat, *Lasiorhinus latifrons,* which has a broad skull and laterally placed eyes and in which only 11% of the LGd receives uncrossed projections.

Bandicoots

LGd macro-organization in bandicoots differs from that of most other marsupials in that the proportion of the LGd receiving uncrossed inputs is only 7% in *P. gunnii* and 10% in *I. obesulus*. With the exception of the wombat, most other marsupials appear to devote 20% or more of their LGd to processing information from the ipsilateral eye. The bandicoot and wombat figures are not unusual among mammals. For example, in the Australian murid, *Rattus villosissimus,* the volume of the LGd occupied by uncrossed projections is 11%, a figure comparable to that observed in wombats and bandicoots

(Table 2). The small input from the uncrossed eye in *P. gunnii* may be related to the longer snout characteristic of this species. The shorter nosed *I. obesulus* presumably has less nose to get in the way of binocular vision and thus has a larger ipsilateral retinal representation within its LGd. Useful binocular vision in bandicoots is likely to be limited to a small solid angle centred on the midline and located just dorsal to the snout.

LGd micro-organization: terminal clusters and crossed-uncrossed terminal asymmetries

Polyprotodonts

In micro-organization didelphid and dasyurid LGds differ markedly from the mammalian norm (Haight, Astheimer and Tennent 1987; Haight and Sanderson 1988). As seen at light microscopic levels these differences take two forms: a. the densities and distribution of crossed and uncrossed terminals can be unequal or asymmetric (Figs 6A,B), whereas in other mammals the distribution is symmetric (Fig. 7C), and b. the terminals can be clustered or non-clustered (cf. Figs 6A,B,C and 7C), while in other mammals they are always clustered. In respect of both these conditions bandicoots display their own unique variations (Fig. 7A).

Bandicoots

Asymmetry of crossed and uncrossed retinal inputs is extreme. The density (=number of terminals per unit of LGd volume) of uncrossed retinal terminals is very much less than that of the crossed terminals (Fig. 7A). Further, the uncrossed terminals are arranged in clusters which tend to be located at a maximum distance from the neuronal somata. Crossed terminals are more evenly distributed throughout the neuropil, and clustering is not marked.

Dasyures

In most Australian dasyurids, as exemplified by *D. byrnei* (Fig. 6C), the density of crossed and uncrossed terminals is approximately the same, and terminal clusters are absent. A similar organization is seen in the Fat-tailed Dunnart, *Sminthopsis crassicaudata* (Haight and Sanderson 1988) and in *D. viverrinus* (authors' observations). A different pattern is seen in *S. harrisii* (Fig. 6B) and *D. virginiana* (Fig. 6A) where both crossed and uncrossed terminals are clustered. Uncrossed terminals tend to be located at greater distances from the somata and their density is lower than that of the crossed terminals, though this difference is not as extreme as in bandicoots (Fig. 7A).

Diprotodont marsupial

In the well laminated LGd of *T. vulpecula* retinal terminals are organized in alternating bands, derived in turn from each eye. In both crossed and uncrossed terminal bands the degree of clustering and overall terminal density are equal (Fig. 7B) though in the example shown the crossed band is narrower than the uncrossed band. The abrupt cessation of label on either side of the crossed band is obvious, these unlabelled areas corresponding to regions which receive uncrossed retinal projections.

Other mammals

The LGd organization in dasyurids and didelphids differs from that seen in all other therian mammals. In the latter crossed and uncrossed retinal terminals are of equal or near equal density and tend to form myriad dense, patchy clusters within the LGd. Further, terminal patterns do not differ qualitatively whether they are derived from the crossed or the uncrossed eye, observations which are borne out by electron microscope studies of geniculate organization in placentals (Guillery and Colonnier 1970) and which are supported by electrophysiological studies (see Rodieck 1979 for review). Hence, in most mammals, inspection of retinal termination patterns within the LGd will not yield information as to whether they are derived from the ipsilateral or contralateral retina. This is clearly seen in the following two examples.

Rodent

Within the LGd of the Plains Mouse, *Pseudomys australis* the similarity in pattern of the crossed and uncrossed terminals is obvious (Fig. 7C). Most of the terminals are organized into clusters which tend to be located deep within the interstitum, a feature also reported in primates (Guillery and Colonnier 1970). Additionally, the LGd neurons in this rodent are considerably smaller than most of those seen in marsupials.

Finally, we draw the reader's attention to the ventral part of the LGd in *S. harrisii* and *D. virginiana*. This region (labelled "m" in Figs 4A and 4B) receives only crossed projections and is probably homologous to the monocular segment (i.e., far lateral visual field) in other mammals (Sanderson and Pearson 1977). Interestingly, in didelphids and dasyurids the density of terminals in this region is greater than elsewhere in the LGd as is seen in the right hand photograph in Figure 6B which was taken from this region. Whether the terminal density in this monocular region equals or exceeds the summed densities of the overlapping crossed and uncrossed terminals in more dorsal parts of the nucleus awaits quantitative measurements. However, in other mammals, including bandicoots, this region of the LGd does not display an increased density of retinal terminals.

These results are of considerable functional and taxonomic interest. Differing LGd anatomies could suggest differing visual physiologies (see below); they may have taxonomic significance as well: *a, S. harrisii* and *D. virginiana* LGds display binocular overlap, crossed-uncrossed terminal asymmetry and terminal clustering; *b,* other dasyurid LGds display binocular overlap, crossed-uncrossed terminal symmetry and no clustering; *c,* bandicoot LGds display no binocular overlap, gross cross-uncrossed terminal asymmetry and their crossed terminals are not clustered while their uncrossed terminals are. On the other hand, *d, T. vulpecula* LGds show no binocular overlap and their terminal distributions are both symmetric and clustered. The evolutionary implications are strong: similarities between diprotodont and placental LGds arose independently and are convergent; the didelphids and *S. harrisii* have probably retained the primitive marsupial configuration; while other marsupials, namely the bandicoots and the remaining dasyurids, have developed their own patterns. Visual system morphology among Australasian marsupials has undergone a number of radiations.

Binocular overlap and asymmetries; are there functional implications?

Morphometric analyses of the pattern and densities of retinal synapses within representative marsupial LGds are needed. What is not known at this stage is whether, at the micro-organizational level, the retinal terminals in the binocular overlap region of the dasyurid and didelphid LGd are discrete or mixed. If discrete, i.e., if there are micro-islands of distinctly separated ipsilateral and contralateral inputs within the geniculate which simply are not segregated into larger macro-zones or laminae, this would represent an anatomical pattern peculiar to certain marsupials, but one which physiologically need not be radically different from that observed in other mammals where crossed and uncrossed terminals become more widely separated. If mixed, i.e., if individual geniculate neurons or functional neuronal clusters within the LGd do receive inputs from both eyes, this would represent a most unusual organization and could suggest that visual physiology in polyprotodonts differs considerably from that of the better known placental mammals and very probably from the bandicoots and diprotodont marsupials as well. Apart from the fact that comparative and developmental studies now show very clearly that the tendency has been toward ocular separation and functional specialization within the LGd, we are little closer to understanding exactly why this should be (Guillery 1979; Kaas *et al.* 1978).

In terms of their LGd anatomy bandicoots are "fence straddlers", poised between the polyprotodonts and the diprotodonts. The complete separation of the crossed and uncrossed retinal terminals within the LGd is clearly an advanced character; the extreme asymmetry of the crossed and uncrossed terminal densities is a more radical manifestation of the presumably less developed condition seen in *S. harrisii* and *D. virginiana*. Perhaps the strangest feature is the clustering of the uncrossed projection and the non-clustering of the crossed projection. In this regard bandicoots seem to waver between the non-clustered condition seen in most dasyurids and the presumably more primitive clustered condition retained in other mammals — but with a difference. Has the uncrossed visual projection remained primitive, while parts of the crossed projection advanced? Why should there be asymmetries in the organization of crossed and uncrossed terminals at all? Are both eyes bringing the same type of information to the brain? Or, is that part of the retina which provides the crossed projections to one side of the brain doing something quite different to that part of the opposite retina which is providing the uncrossed projections? The authors of this paper are firmly in the camp which would maintain that anatomical differences usually have functional significance. When finally understood, the visual ecology of bandicoots and other polyprotodont marsupials will prove very interesting indeed.

ACKNOWLEDGEMENTS

We are indebted to Dr J. K. Harting of the University of Wisconsin who kindly loaned us the *Didelphis* spp. material used in this study. Dr John Nelson of Monash University gave us access to his data on retinogeniculate projections in a number of Australian marsupial species. For this we are very grateful. Thanks also go to the National Parks and Wildlife Service of Tasmania for permission to take bandicoots, *S. harrisii* and possums and to the Waite Institute in Adelaide, South Australia which kindly provided *D. byrnei* specimens. We also wish to thank Dr Lee Astheimer and an unknown reviewer for many helpful comments on the manuscript. This work was supported by grants from the Australian Research Grants Scheme (to K. J. Sanderson), the Australian National Health and Medical Research Council (to J. R. Haight) as well as by research funds from our respective universities. Finally, we wish to acknowledge the photographic assistance of Michael Stranger.

REFERENCES

Brauer, K., Schober, W. and Winkelmann, E., 1978. Phylogenetical changes and functional specializations in the dorsal lateral geniculate nucleus (dLGN) of mammals. *Journal für Hirnforschung* **19:** 177–87.

Cowan, W. M. and Cuenod, M. (eds), 1985. *The use of axonal transport for studies of neuronal connectivity.* Elsevier: Amsterdam.

GUILLERY, R. W., 1979. A speculative essay on geniculate lamination and its development. *Progress in Brain Research* **51:** 403–18.

GUILLERY, R. W. AND COLONNIER, M., 1970. Synaptic patterns in the dorsal lateral geniculate nucleus of the monkey. *Zeitschrift für Zellforschung und Mikroscopische Anatomie* **103:** 90–108.

HAIGHT, J. R. AND NELSON, J. E., 1987. A brain that doesn't fit its skull: a comparative study of the brain and endocranium of the koala, *Phascolarctos cinereus* (Phascolarctidae). Pp. 331–52 in *Possums and opossums, studies in evolution* ed by M. Archer. Zoological Society of New South Wales: Sydney.

HAIGHT, J. R. AND SANDERSON, K. J., 1986. Visual organization in dasyurids and peramelids. *Bulletin of the Australian Mammal Society* **9:** 31.

HAIGHT, J. R. AND SANDERSON, K. J., 1988. Retinal projections in two Australian polyprotodont marsupials: kowari, *Dasyuroides byrnei*, and fat-tailed dunnant, *Sminthopsis crassicaudata* (Dasyuridae). *Brain, Behaviour and Evolution* **31:** 96–110.

HAIGHT, J. R., ASTHEIMER, L. AND TENNENT, R. J., 1987. A cylindrical "glomerulus" in the dorsal lateral geniculate of the quoll, *Dasyurus viverrinus*. *Neuroscience Letters*, Supplement 27: **S29.**

HAYHOW, W. R., 1967. The lateral geniculate nucleus of the marsupial phalanger, *Trichosurus vulpecula*. An experimental study of cytoarchitecture in relation to the intranuclear optic nerve projection fields. *Journal of Comparative Neurology* **131:** 571–604.

HAYHOW, W. R., SEFTON, A. AND WEBB, C., 1962. Primary optic centres of the rat in relation to the terminal distribution of the crossed and uncrossed optic nerve fibres. *Journal of Comparative Neurology* **118:** 295–322.

JOHNSON, J. I., 1977. Central nervous system of marsupials. Pp. 157–278 in *The biology of marsupials* ed by D. Hunsaker. Academic Press: New York.

JOHNSON, J. I. AND MARSH, M. P., 1969. Laminated lateral geniculate in the nocturnal marsupial *Petaurus breviceps* (Sugar Glider). *Brain Research* **15:** 250–54.

KAAS, J. H., GUILLERY, R. W. AND ALLMAN, J. M., 1971. Some principles of organization in the lateral geniculate nucleus. *Brain Behaviour and Evolution* **6:** 253–99.

KAAS, J. H., GUILLERY, R. W. AND ALLMAN, J. M., 1973. Discontinuities in the lateral geniculate nucleus corresponding to the optic disc: A comparative study. *Journal of Comparative Neurology* **147:** 163–80.

KAAS, J. H., HUERTA, M. F., WEBER, J. T. AND HARTING, J. K., 1978. Patterns of retinal terminations and laminar organization of the lateral geniculate nucleus of primates. *Journal of Comparative Neurology* **182:** 517–54.

LENT, R., CAVALCANTE, L. A. AND ROCHA-MIRANDA, C. E., 1976. Retinofugal projections in the opossum. An anterograde degeneration and radioautographic study. *Brain Research* **107:** 9–26.

LEVENTHAL, A. G., 1979. Evidence that the different classes of relay cells of the cat's lateral geniculate nucleus terminate in different layers of the striate cortex. *Experimental Brain Research* **37:** 349–72.

MAYNER, L., PEARSON, L. J. AND SANDERSON, K. J., 1980. An autoradiographic study of visual pathways of the Australian rodents *Notomys alexis, Pseudomys australis* and *Rattus villosissimus*. *Australian Journal of Zoology* **28:** 381–93.

MEGIRIAN, D., WELLER, W. L., MARTIN, G. F. AND WATSON, C. R. R., 1977. Aspects of laterality in the marsupial *Trichosurus vulpecula* (Brushtailed Possum). *Annals of the New York Academy of Sciences* **299:** 197–212.

PEARSON, L. J., SANDERSON, K. J. AND WELLS, R. T., 1976. Retinal projections in the Ringtailed Possum, *Pseudocheirus peregrinus*. *Journal of Comparative Neurology* **170:** 227–40.

REES, S. AND HORE, J., 1970. The motor cortex of the Brushtailled Possum *(Trichosurus vulpecula)*: motor representation, motor function and the pyramidal tract. *Brain Research* **20:** 439–51.

RODIECK, R. W., 1979. Visual Pathways. *Annual Review of Neuroscience* **2:**193–225.

ROYCE, G. J., WARD, J. P. AND HARTING, J. K., 1976. Retinofugal pathways in two marsupials. *Journal of Comparative Neurology* **170:** 391–414.

SANDERSON, K. J., 1974. Lamination of the dorsal lateral geniculate nucleus in carnivores of the weasel (Mustelidae) raccoon (Procyonidae) and fox (Canidae) families. *Journal of Comparative Neurology* **153:** 239–66.

SANDERSON, K. J., 1986. Evolution of the lateral geniculate nucleus. Pp. 183–95 in *Visual neuroscience, festschrift for P. O. Bishop* ed by J. Pettigrew, W. Levick and K. J. Sanderson. Cambridge University Press: London.

SANDERSON, K. J. AND PEARSON, L. J., 1977. Retinal projections in the Native Cat, *Dasyurus viverrinus*. *Journal of Comparative Nuerology* **174:** 347–58.

SANDERSON, K. J. AND PEARSON, L. J., 1981. Retinal projections in the Hairy-nosed Wombat, *Lasiorhinus latifrons*. *Australian Journal of Zoology* **29:** 473–81.

SANDERSON, K. J., PEARSON, L. J. AND DIXON, P. G., 1978. Altered retinal projections in Brushtailed Possum, *Trichosurus vulpecula,* following removal of one eye. *Journal of Comparative Neurology* **180:** 841–68.

SANDERSON, K. J., PEARSON, L. J. AND HAIGHT, J. R., 1979. Retinal projections in the Tasmanian Devil, *Sarcophilus harrisii*. *Journal of Comparative Neurology* **188:** 335–46.

SANDERSON, K. J., HAIGHT, J. R. AND PETTIGREW, J. D., 1984. The dorsal lateral geniculate nucleus of macropodid marsupials: cytoarchitecture and retinal projections. *Journal of Comparative Neurology* **224:** 85–106.

SANDERSON, K. J., NELSON, J. E., CREWTHER, D. P., CREWTHER, S. G. AND HAMMOND, V. E., 1987. Retinogeniculate patterns in diprotodont marsupials. *Brain, Behaviour and Evolution* **30:** 22–42.

SHATZ, C. J. AND SRETAVAN, D. W., 1986. Interactions between retinal ganglion cells during the development of the mammalian visual system. *Annual Review of Neuroscience* **9:** 171–207.

SUTHERS, R. A. AND BRADFORD, M. R., Jr, 1980. Visual systems and the evolutionary relationships of the Chiroptera. Pp. 331–46 in *Proceedings of the 5th International Bat Conference, Lubbock Texas* ed by D. E. Wilson and A. L. Gardner. Texas Technical University Press: Lubbock, Texas.

WEBER, J. T., CASAGRANDE, V. A. AND HARTING, J. K., 1977. Transneuronal transport of (^{3}H)proline within the visual system of the grey squirrel. *Brain Research* **129:** 346–52.

Physiological effects of glucocorticoid hormones and ACTH in the Southern Brown Bandicoot *Isoodon obesulus* (Shaw and Nodder)

LESLIE N. THOMAS[1] and ADRIAN J. BRADLEY[2]

Some marsupials such as the Red Kangaroo and the Quokka are unusual in that they show few metabolic responses following the injection of adrenocortical hormones such as cortisol. The lack of either a diabetogenic or nitrogen mobilizing response has been attributed to an adaptation to life in a stressful environment.

The effects of adrenocorticotrophic hormone ACTH and cortisol on carbohydrate and nitrogen metabolism and haematology of *Isoodon obesulus* were determined during short and long-term administration of these hormones.

The results of these experiments show that *I. obesulus* is similar to the Red Kangaroo and the Quokka in resisting the diabetogenic and nitrogen mobilizing action of glucocorticoids. *I. obesulus*, however, does show the classic haematological disturbances following long-term glucocorticoid elevation.

Key words: Glucocorticoids, ACTH, *Isoodon obesulus,* short-term response, long-term response, glucose, urea nitrogen, free fatty acids, haematology.

INTRODUCTION

ADRENAL corticosteroid hormones such as cortisol and corticosterone were originally termed "glucocorticoids" because of their effect on carbohydrate metabolism in eutherians (Ashmore and Morgan 1967). Basically, plasma glucose concentration is elevated by high concentrations of glucocorticoids in the blood. This effect is biphasic, and is initially due to reduced uptake of glucose by the skin, thymus, lymphatic tissue, adipose tissue and the heart, but is then maintained by decreased protein anabolism and increased protein catabolism (Ashmore and Morgan 1967; Munck 1971). This results in an elevation of plasma free fatty acids concentration and increased nitrogen excretion as a by-product of the protein breakdown. If nitrogen intake is not elevated to balance nitrogen output, a negative nitrogen balance ensues, which leads to loss of weight and, if prolonged, leads to death.

Chronic elevations of glucocorticoid concentration have been shown to have an immunosuppressive effect (Claman 1972). In some species this is due to the increased rate of lymphocytopenia (Cassileth 1983), and in other species it is due to a decrease in the mass of lymphoid tissue, thus reducing the production of lymphocytes (Barker, Beveridge, Bradley and Lee 1978). It should be possible then to use changes in the proportion of lymphocytes to other white cell types as an indicator of corticosteroid activity (Cheal, Lee and Barnett 1976).

Total plasma corticosteroids are composed of three fractions: free steroid, steroid bound to corticosteroid binding globulin (CBG) and albumin-bound steroid (Tait and Burstein 1964). It is generally agreed by most workers that the CBG-bound corticosteroid is biologically inactive whereas the free or unbound corticosteroid is biologically available (Tait 1982). Free corticosteroid has been used in place of TPCS in some studies in natural populations of marsupials (Bradley, McDonald and Lee 1976, 1980; McDonald, Lee, Bradley and Than 1981).

An immediate metabolic response to high plasma glucocorticoid levels, characterized by an elevation of plasma glucose concentration, has been shown to occur in some species of marsupials after infusion of ACTH for just 1 h: *Trichosurus vulpecula* (Than and McDonald 1974) and *Macropus eugenii* (Cooley and Janssens 1977). Short-term infusion of cortisol produced increases in plasma glucose levels but not in other metabolites in *T. vulpecula* and *M. eugenii.*

The response of marsupials to long-term elevation of plasma glucocorticoid levels appear to be more varied than the short-term response. This was first noted by Griffiths, McIntosh and Leckie (1969) who found that daily cortisone injections over eight days had no effect on blood glucose levels or nitrogen balance of *M. rufus.* The classic diabetogenic, nitrogen-mobilizing effect is shown by *T. vulpecula* during eight days of cortisol infusion (Than and McDonald 1974). During six days of cortison injections, *M. eugenii* exhibited a similar plasma glucose and metabolite profile to its short-term response (Cooley and Janssens 1977; Janssens and Hinds 1981). In the long-term experiment, however, both nitrogen intake and output rose so that a

[1]Zoology Department, University of Western Australia, Perth, Western Australia, Australia 6000.
[2]Zoology Department, University of Tasmania, Hobart, Tasmania, Australia 7001.
Pages 175–83 *in* BANDICOOTS AND BILBIES, ed by J. H. Seebeck, P. R. Brown, R. L. Wallis and C. M. Kemper. Surrey Beatty & Sons, Sydney, 1990.

positive nitrogen balance was maintained despite the catabolic effects of cortisol. McDonald and Bradshaw (1981) found that injection of cortisol acetate over a period of five days had no effect on plasma glucose concentrations in *Setonix brachyurus*, nor did this treatment have any effect on nitrogen intake or output of female *S. brachyurus* although it produced a small, but significant, increase in the nitrogen output of males.

Both *S. brachyurus* and *M. rufus* inhabit fairly arid environments where the nitrogen content of the vegetation is seasonally low. Their lack of sensitivity to elevated plasma glucocorticoid levels is an adaptation which allows the conservation of nitrogen (McDonald 1977).

Cortisol is the main glucocorticoid secreted by most of the marsupials studied so far (Oddie, Blaine, Bradshaw, Coghlan, Denton, Nelson and Scoggins 1976). *Vombatus ursinus* and *Phascolarctos cinereus* were the only two species found to produce corticosterone as their main glucocorticoid but after treatment with ACTH, cortisol was found in higher concentration than corticosterone in *P. cinereus* (Oddie *et al.* 1976). Cortisol is the most potent of the naturally occurring glucocorticoids as corticosterone has only 20% of the potency of cortisol for glucocorticoid activity (Liddle 1981). Any change in the ratio of cortisol to corticosterone is important because it determines the effectiveness of the total circulating glucocorticoid concentration.

The variety of responses from the few species of marsupial so far studied has prompted this investigation. The Southern Brown Bandicoot, *Isoodon obesulus,* is a relatively common marsupial found in southeastern and southwestern Australia (Heinsohn 1966; Jones 1924). It is insectivorous and has a high protein diet as opposed to the other species mentioned above, which are mainly herbivorous.

The response of *I. obesulus* to both short-term and long-term administration of ACTH was examined. The short-term experiment was concerned with changes in plasma corticosteroid, glucose, urea nitrogen and free fatty acid concentrations. The long-term administration of ACTH made possible a broader study. As well as the factors measured during the short-term infusion, changes in the nitrogen balance were monitored as well as changes in some haematological parameters. The effects of prolonged stress on steroid-partitioning and the cortisol:corticosterone ratio were also monitored.

METHODS

Blood sampling procedure

Blood samples of 200 to 500 μl were routinely taken via catheters inserted into either the caudal or jugular vein, however, cardiac puncture was sometimes used.

Individuals were anaesthetized during jugular vein catheter insertion by an initial intramuscular injection of 0.45 IU Ketalar kg^{-1} (Parke-Davis, Australia), reinforced by 2% Halothane (CSL, Melbourne, Australia) in a nitrous oxide/oxygen gas mixture. A minimum of two days was allowed for recovery from the operation.

Bard-i-cath inside-needle catheter intravenous placement units (Luer 1619, 203.2 mm long, 19 g needle; Bard, Sunderland, England) were used for caudal vein catheters. They were inserted without anaesthesia in the caudal vein at the base of the tail and secured with a strip of Elastoplast. The catheter and tail were then covered with a 75–125 mm length of hollow electrical conduit (20 mm diameter) with the flanged end directed anteriorly and this was then secured to the base of the tail with Elastoplast.

Jugular catheters were either of medical grade, single lumen, clear vinyl tubing (0.98 mm OD, 0.58 mm ID) or double lumen, clear vinyl tubing (1.5 mm OD, 2×0.71 mm ID; Dural Plastics, NSW). Jugular catheters were passed subcutaneously, using a stainless steel tube as a guide, to emerge dorsally between the shoulder blades. Cannulae tubing was plugged with a steel pin and secured with Elastoplast.

All blood samples taken during the short-term experiment were taken via venous catheters. At the conclusion of all short-term experiments, saline-washed red cells were reinjected into the animals.

The pool of 10 experimental animals used in the long-term experiment was divided into two groups of five: four males and one female per group. Both groups were subjected to the same injection regime, but were sampled on different days. Most samples were obtained within 1.5 min and all samples were obtained within 3 min of disturbing the animal.

Housing

All bandicoots used in these experiments were housed in metabolism cages 450 mm × 700 mm × 600 mm with 13 mm^2 mesh situated above a mesh collecting tray (for faeces) and an aluminium collecting funnel (for urine). Individuals were housed in this manner for three months prior to the start of the experiment. During the acclimatization period they were fed on dry dog food pellets (Luv Tender Chunx) supplemented by apples, hard-boiled eggs, snails and mealworms. Water was supplied *ad libitum.* The cages were in a room kept at 25±1.5°C at 85% humidity.

Experimental procedure

Short-term infusion of ACTH

Three adult male and three adult female *I. obesulus* were used. Four animals had jugular catheters and two had tail vein catheters.

No food was given for 24 h before each experimental infusion. Individuals were restrained in calico bags during the infusion procedure but further disturbance was kept to a minimum by passing the catheter through the mouth of the bag so that it was not necessary to move the animal during either infusion of ACTH or the withdrawal of blood samples. A Braun Infusion Pump (Braun, Melsunger) was used for all infusions.

Six animals (experimental) were infused with ACTH (Synacthen; Ciba-Geigy, Aust.) at a dose rate of 0.45 IU kg^{-1} h^{-1} in 2.6 ml saline for 1 h. Three animals (control) were infused for the same period with saline alone. Blood samples were withdrawn into heparinized syringes every 15 min. This interfered with the infusion for 30 to 50 seconds each time when single lumen tubing was used, but infusion was uninterrupted when double lumen jugular cannulae tubing was used. Care was taken to refill the predetermined dead space in the catheter when infusions were recommenced after blood sampling. Sampling was continued for a further 150 min. after ACTH infusion ceased.

Long-term administration of ACTH

The injection regime began with daily intramuscular injections of 0.25 ml of 0.154 M NaCl solution for four days. It was followed by daily intramuscular injections of Acthar gel (Amour Pharmaceutical Co., England) at a dose rate of 3 IU kg^{-1} day^{-1} for six days. All injections were then terminated though sampling continued until day 14. Blood samples were taken 23 hr after injection.

The effects of ACTH on nitrogen balance was examined during the long-term experiment by comparing nitrogen intake and output. Each animal was given 75 g of dog food pellets and any uneaten food was removed and weighed at the end of 24 h. The pellets contained 2.63% Nitrogen (wet weight). The weight of uneaten food was corrected for weight loss by dehydration before nitrogen intake was calculated.

Urine was collected in glass jars containing 1.00 ml 10% H_2SO_4, as a preservative, and a layer of Ondina oil to prevent evaporation. A small amount of distilled water was used to wash down the walls of the collecting funnel. The washings were collected with the urine and stored frozen until assayed.

Faecal matter was collected from mesh screens positioned beneath each cage and stored until assayed. Faeces and urine were collected every 24 h.

The urine and faeces of each sample were made up to 300 ml with distilled water and blended in a food processor until a uniform texture was obtained. This was immediately subsampled and assayed for total nitrogen by Kjehdahl distillation (Kaplan 1969).

Blood sample analysis

The following measurements were made on blood samples taken during the short-term ACTH infusion experiment: blood glucose, plasma urea nitrogen, plasma free fatty acids (FFAs), total plasma corticosteroid concentration and corticosteroid partitioning into CBG bound, albumin bound and free steroid.

A wider range of parameters were measured during the long-term experiment. As well as those measured during the short-term experiment, various haematological parameters and the cortisol, corticosterone ratio were also measured.

Blood glucose was measured by the glucose oxidase method, using a portable Ames Dextrostix reflectance meter (Miles, USA). McDonald and Bradshaw (1981) found that this gave good agreement with a colorimetric glucose oxidase method (Worthington, USA).

The method outlined by Tietz (1976) was used to determine plasma urea nitrogen concentration. The assay was scaled down to use 5 μl plasma samples.

The method of Laurell and Tibbling (1967) was modified for the determination of plasma-free fatty acids. Polypropylene test-tubes were used instead of glass-stoppered test-tubes. The assay was scaled down to use plasma samples of 20 μl.

Packed Cell Volume (PCV): a 75 μl heparinized haematocrit tube was filled with fresh whole blood and centrifuged at 12 000 rpm for 2 min. The proportion of packed red cells to total volume was expressed as a percentage.

Red Blood Cell Count (RBC): 10 μl of whole mixed unclotted blood was diluted with 2.0 ml Hayem's solution and thoroughly mixed. The count was made using a haemocytometer with a standard Neubauer ruling.

A portable haemoglobinometer (American Optical Co., USA) was used to determine plasma Hb concentration to the nearest 0.1 g dl^{-1}.

Mean Corpuscular Haemoglobin Concentration (MCHC), Mean Corpuscular Haemoglobin (MCH) and Mean Corpuscular Volume (MCV) were all calculated using the method of Cheal *et al.* (1976).

White Cell Count (WCC): 20 μl of whole unclotted blood was diluted with 100 μl Turk's Fluid and mixed well. Leucocytes were counted using a haemocytometer with standard Neubauer ruling.

Differential Cell Count (DCC): fresh, unclotted, whole blood was used to make blood smears on clean glass slides. These were rapidly air-dried by vigorously waving, then, fixed in methanol and stained with May-Grunwald-Giemsa's stain (Disbrey and Rack 1970). Five lots of 100 cells were counted in longitudinal strips running from one end of the smear to the other. The differential totals were expressed as mean percentages (Dacie 1975).

The method of Bassett and Hicks (1969) was used to determine total plasma corticosteroid concentration (TPCS) with the modifications described by McDonald *et al.* (1981).

The ratio of free:protein-bound:albumin-bound steroid was calculated by the method of Tait and Burstein (1964), using the high affinity binding constant (K_T) of 5.44×10^7 l mol^{-1} at 36°C for *I. obesulus* (Bradley, unpublished).

The method described by McDonald *et al.* (1981) was used to determine the MCBC, but with the following modifications: the assay was scaled down to accommodate plasma samples of 30 μl and before being refrigerated, the samples were incubated at 37°C for 5 min.

Plasma albumin was assayed using the method of Rodkey (1965). The assay was scaled down for plasma samples of 10 μl. Bovine serum albumin (Sigma, USA) was used as a standard, but all values so determined were corrected for *I. obesulus* albumin binding characteristics. To compare BSA and *I. obesulus* albumin, the absorbance of known concentrations of each type of albumin was determined when mixed with bromocresol green reagent. In this way a correction factor was generated for each *I. obesulus* sample when BSA was used as a standard. *I. obesulus* albumin was isolated by column chromatography on a 20 cm × 1 cm Sephadex G-200 column using a tris buffer (0.1 M, pH 7.5). The albumin fraction from this column was concentrated in dialysis tubing (20/32 Union Carbide, USA) using polythylene glycol 6000 (BDH, England) and the concentration was determined by the method of Bradford (1976).

Plasma cortisol, corticosterone and 11-deoxycortisol were resolved on 0.8 cm (I.D.) × 16 cm columns of Sephadex LH20 (Pharmacia, Sweden) using dichloromethane/ethanol (99:1) as the column solvent (Segre, Friedrich, Dodek, Lloyd, Lobotsky, Levin and Klaiber 1966). In this system, corticosterone (B) was excluded in the 2 to 4.5 ml fraction, 11-deoxycortisol (S) in the 4.5 to 7 ml fraction and cortisol (F) in the 12 to 18 ml fraction.

The exclusion volumes were reproducible when room temperature was maintained at 20 ± 2°C. This method eliminated high blank values (Murphy 1971). Routine collections were made of corticosterone and cortisol only. The 40 μl plasma samples were extracted with 3 × 5 volumes of dichloromethane, extracts were dried with air and then applied to the columns in 2 × 100 μl column solvent. Recoveries were estimated by the addition of 4 000 dpm of the appropriate tritiated steroid: they were 85 ± 5% SD (n=6) for B and F respectively.

RESULTS

Long-term administration of ACTH

Total plasma corticosteroid (TPCS) levels were maintained at approximately 80 nM during saline treatment and a significant rise in these levels was recorded 24 h after the first ACTH injection (Table 1). Peak TPCS values were recorded on days 8 and 9 (fifth and sixth days of ACTH treatment) and the TPCS levels began to decrease as soon as ACTH treatment ceased. Values recorded on day 13 were not significantly different from the saline treatment values. At no stage did TPCS values exceed the corresponding MCBC values.

Table 1. Total plasma corticosteroid concentration (TPCS) and Maximum Corticosteroid Binding Capacity (MCBC) during long-term administration of ACTH.

Treatment	Day	N	TPCS x̄±SE	MCBC x̄±SE
Daily injections of isotonic saline	0	5	80.0±35.6	290.8±38.07
	2	5	86.6±25.1	309.0±35.9
	3	5	77.8±10.6	310.4±16.8
Daily injections of ACTH (3 IU ACTH kg^{-1} day^{-1})	5	5	132.7±20.0	326.2±16.2
	6	5	181.0±26.2	377.2±33.8
	8	2	226.7±16.2	424.4±16.4
	9	4	222.4±17.5	426.5±27.7
No injections	11	4	186.9±33.8	381.9±36.41
	13	7	115.6±19.5	327.8±15.8

Changes in the maximum corticosteroid binding capacity (MCBC) followed the same pattern as the TPCS values but there was a notable time lag between the initial response of the TPCS level to ACTH and the response of the MCBC. No significant elevation of the MCBC was recorded until day 7 (third day of ACTH treatment) and though MCBC decreased after ACTH treatment ceased, the final values were still slightly, but not significantly, above the saline treatment values.

These changes in MCBC suggest that either the rate of synthesis of binding globulin was increased during the ACTH treatment or release of binding globulin from the liver was increased or that both these responses occurred concurrently. The response of the

Table 2. Steroid partitioning during long-term administration of ACTH (x̄SD).

Treatment	Day	N	Concentration of corticosteroid (nM) Free	CBG-bound	Albumin bound	Percentage of TPCS in each category Free	CBG-bound	Albumin bound
Daily injections of isotonic saline	0	5	4.83±1.64	64.3 ±23.5	10.9±3.40	6.3±0.39	78.5±1.65	15.2±1.47
	2	5	5.42±1.45	70.39±20.1	10.8±2.45	6.2±0.24	81.3±1.42	12.4±2.11
	3	5	4.76±1.08	63.70±14.6	9.3±1.90	6.1±0.18	81.9±1.35	11.9±1.63
Daily injections of ACTH (3 IU ACTH kg^{-1} day^{-1})	5	5	8.82±1.41	105.70±19.0	18.2±2.05	6.6±0.41	79.6±1.62	13.7±1.21
	6	5	11.69±1.30	147.2 ±20.6	25.5±2.82	6.4±0.29	79.2±1.70	14.4±1.41
	8	2	13.55±1.11	180.1 ±11.6	33.1±3.01	6.0±0.21	79.4±1.01	14.0±1.52
	9	4	13.10±1.02	177.4 ±14.1	31.4±2.49	5.9±0.12	79.9±0.35	14.2±0.28
No injections	11	4	11.98±1.31	150.7 ±26.4	24.2±2.22	6.4±0.36	80.6±1.21	13.0±0.99
	13	7	7.22±1.24	98.1 ±15.7	17.3±2.60	6.4±0.10	79.7±1.04	13.9±0.54
One-way analysis of variance df = 8, 33			F=4.51 P<0.05	F=8.39 P<0.05	F=3.81 P<0.05			

MCBC effectively reduced potential increases in plasma free corticosteroid, though not sufficiently to maintain them at pre-ACTH treatment levels (Table 2). The ratio of free:CBG-bound:albumin-bound steroid remained constant throughout the experiment (Table 2). No significant difference between the sexes were found for either TPCS or MCBC.

During the saline treatment cortisol and corticosterone were found in roughly equal proportions (53.3:46.7) but by day 3 of ACTH treatment, cortisol was 71% of the TPCS and by day 6, 77.8%. Four days after ACTH treatment was terminated, the proportion of cortisol had fallen to 56.2% of the TPCS (Table 3).

Throughout the experiment the concentration of corticosterone remained constant and it was the changes in the concentration of cortisol that resulted in the proportional changes.

Mean plasma glucose concentration showed some daily fluctuation (Fig. 3) though there was no significant difference between means either within or between treatments ($F_{[2,42]}$ = 2.61, $P > 0.05$).

Plasma FFAs, on the other hand, showed a significant elevation initially during ACTH treatment ($F_{[3,56]}$ = 17.98, $P < 0.05$). The elevation was not prolonged and FFA concentration had fallen to control level by the end of the experiment (Fig. 1).

Plasma urea nitrogen concentration rose steadily between day 0 and day 13 (Fig. 1). Day to day changes were not significant though the plasma urea nitrogen for day 0 differed significantly from that for day 13 ($F_{[1,10]}$ = 17.93, $P < 0.01$). The rise in plasma urea nitrogen began at the start of the control period and continued after all injections had ceased. It is not possible to determine the precise effect of exogenous ACTH in this case because plasma urea concentration did not return to pre-experimental levels before the experiment was concluded.

Mean white cell count (WCC) rose slightly during ACTH administration and a mild state of lymphocytopenia was noted, however, neither of these proved to be significant (Table 4).

Mean daily nitrogen intake did not vary significantly either between or within treatments ($F_{[2,60]}$ = 0.086, $P > 0.05$). Nitrogen output rose slightly during the experiment (Table 5) but a positive nitrogen balance was maintained throughout the experiment, apart from two instances. On day 10, the nitrogen input and output of animal No. 4 were equal. On day 6, animal No. 9 entered negative nitrogen balance. Neither of these conditions was maintained for more than 24 h.

Table 3. The concentration and proportions of cortisol and corticosterone during the long-term administration of ACTH.

Treatment	N	Cortisol + Corticosterone (Total) (nM) x̄±SD	Cortisol (nM) x̄±SD	% of Total x̄	Corticosterone (nM) x̄±SD	% of Total x̄
Control (daily isotonic saline injection)	5	80.0±35.6	44.2 ±17.08	53.3	35.8±11.43	46.7
Day 3 of ACTH treatment (3 IU ACTH kg^{-1} day^{-1})	5	182.0±22.4	127.8 ±15.37	71.0	54.6±11.43	29.0
Day 6 of ACTH treatment	4	222.0±17.5	169.8 ± 5.25	77.8	52.2±13.48	22.2
Day 4 after cessation of ACTH treatment	6	115.6±18.3	67.15±13.20	56.2	48.5± 7.28	43.7
One-way analysis of variance			$F_{[3,16]}$=13.24, P<0.05		$F_{[3,16]}$=0.51, P<0.05	

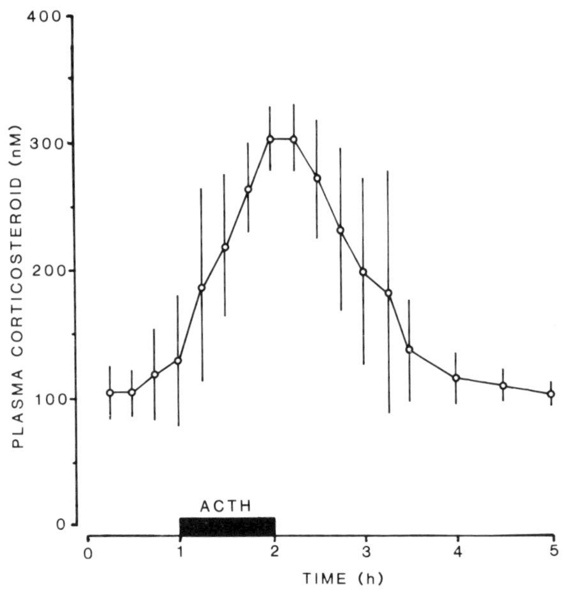

Fig. 1. Total plasma corticosteroid (TPCS) concentration observed in *Isoodon obesulus* during short-term infusion of ACTH. x̄±SE, n=6.

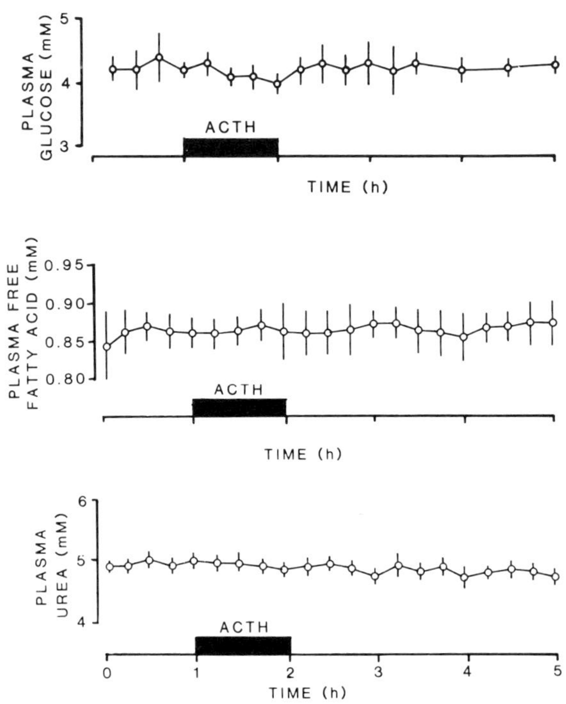

Fig. 2. Plasma glucose (n=6), plasma FFA (n=2) and plasma urea nitrogen (n=3) concentrations observed in *Isoodon obesulus* during short-term infusion of ACTH. x̄±SE.

Short-term effects of ACTH

Infusion of ACTH intravascular at a rate of 0.45 IU kg^{-1} h^{-1} caused a significant rise in TPCS concentration (Fig. 1) from a concentration of 104.8 ± 21.8 nM prior to the commencement of infusion to a maximum of 303.0 ± 26.4 nM of the 1-h infusion period ($F_{[2,140]}$ = 4.50, $P < 0.05$). Within approximately 2 h of the termination of the ACTH infusion TPCS concentration had returned to the pre-infusion level. The large SEs during the post infusion recovery period may be attributed to one animal in which a peak TPCS concentration of 354 nM was achieved 30 min. after the termination of the ACTH infusion and in this same animal TPCS returned to pre-infusion levels more slowly than in the other five animals. In contrast to the significant changes seen in the TPCS concentrations, no significant changes were observed in either plasma glucose, plasma-free fatty acids or plasma urea nitrogen concentrations (Fig. 2) either before, during or after the infusion of ACTH.

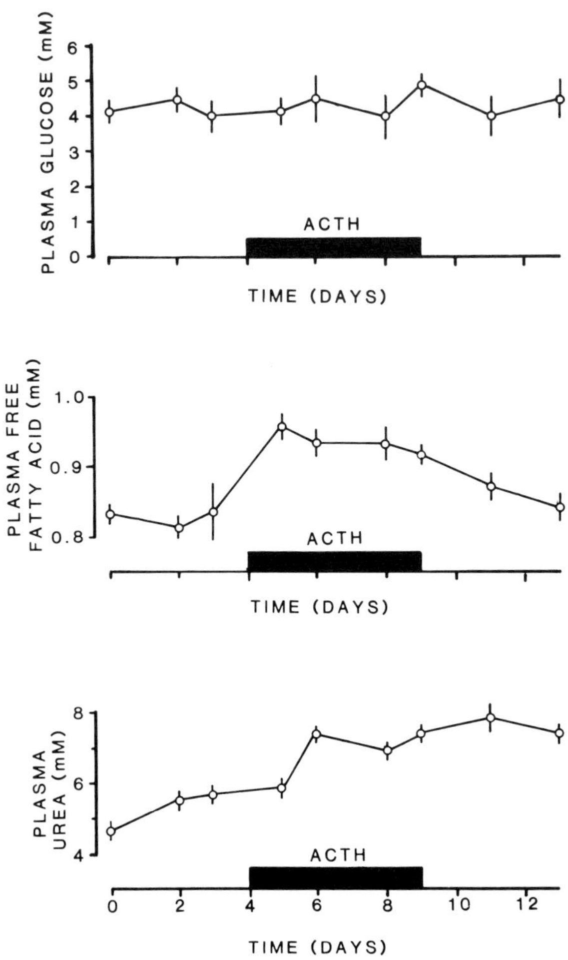

Fig. 3. Plasma glucose, plasma FFA and plasma urea nitrogen concentrations observed in *Isoodon obesulus* during the long-term administration of ACTH. x̄±SE, n=5. Injections of 0.25 ml isotonic saline were given daily from day 0 to day 3.

Table 4. Mean white cell count and lymphocyte/neutrophil ratio and haemoglobin concentration observed during four days of saline treatment followed by six days of ACTH treatment and in samples obtained during the four-day period immediately after the cessation of ACTH treatment.

Treatmernt	N	White cell count (WBC/mm^3) $\bar{x}\pm SE$	Neutrophil/ Lymphocyte ratio $\bar{x}\pm SE$	Haemoglobin (%) $\bar{x}\pm SE$
Daily isotonic saline injections	12	6 125±505	0.56±0.076	16.3±1.24
Daily ACTH injections (3 IU ACTH kg^{-1} day^{-1})	12	7 150±373	0.92±0.081	14.6±1.07
No injections	11	6 290±768	0.69±0.10	12.0±0.87
One way analysis of variance df=2,32		$F=1.00, P>0.05$		

Table 5. Nitrogen intake and output (g N kg^{-1} day^{-1}) during long-term ACTH administration.

Treatment	N	Nitrogen intake $\bar{x}\pm SE$	N	Nitrogen output $\bar{x}\pm SE$
Control (daily isotonic saline injections)	20	0.93±0.054	20	0.35±0.032
Daily ACTH injections (3 IU ACTH kg^{-1} day^{-1})	29	0.91±0.040	28	0.39±0.029
Post-ACTH treatment sampling	14	0.93±0.061	12	0.42±0.055
One way analysis of variance		$F_{[2,60]}=0.086$ $P>0.05$		$F_{[2,57]}=4.47$ $P>0.05$

DISCUSSION

The effect of ACTH on the indices of carbohydrate, fat and protein metabolism in *I. obesulus* resemble those described for *M. rufus* (Griffiths *et al.* 1969) and *S. brachyurus* (McDonald and Bradshaw 1977). These species may be described as being resistant to the diabetogenic and nitrogen mobilizing effects of cortisol. In marked contrast to the metabolic resistance exhibited in these species is the effect of cortisol in *T. vulpecula* which has been shown to be extremely sensitive to this hormone (Than and McDonald 1974), the most notable effects being an elevation in plasma glucose, glycosuria and protein catabolism which is reflected in increased urinary nitrogen excretion and the induction of a negative nitrogen balance.

The response of *M. eugenii* to ACTH and cortisol during short-term administration (Cooley and Janssens 1977) and the administration of cortisol during long-term treatment (Janssens and Hinds 1981) lies somewhere between the categories of sensitive and resistant. While ACTH and cortisol may be shown to mobilize glucose during both short and long-term treatment, the long-term treatment with cortisol did not result in a net negative nitrogen balance since, while nitrogen output increased, reflecting increased protein catabolism, the nitrogen intake of the animals increased to more than compensate for the increased loss (Janssens and Hinds 1981).

Comparison between corticosteroid values observed in the field (Thomas, this volume) and in these experiments show marked similarities. The same range of TPCS values was observed in both studies and this suggests that the effects recorded during both short-term and long-term administration of ACTH are physiological, rather than pharmacological effects. The dosage of ACTH used in each of the experiments was sufficient to cause significant elevations in TPCS and in free steroid concentration.

Short-term infusion of ACTH produced a three-fold increase in TPCS but had no effect on plasma glucose, plasma FFA or plasma urea nitrogen. These results contrast with those of Than and McDonald (1974) for *T. vulpecula* and Cooley and Janssens (1977), both of whom found that short-term infusion of ACTH at the same dose rate as in this experiment produced significant increases in both TPCS and plasma glucose.

Long-term administration of ACTH induced an increase in TPCS of the same magnitude as that of the short-term experiment. There was no significant change in plasma glucose concentration but on the first day of ACTH administration, a significant increase in plasma FFA concentration was observed. This increase was transient, but suggests that ACTH may have a lipolytic effect on *I. obesulus*. If so, then in this respect *I. obesulus* shows similarity to *T. vulpecula* (Than and McDonald 1974). In rabbits, the lipolytic response to ACTH is due to the mobilization of depot fat (Hollenberg, Raben and Astwood 1961). The increase in plasma urea nitrogen during long-term administration of ACTH was not reflected in the nitrogen intake or output, or body weight, all of which remained constant.

The slight lymphocytopenia observed in the long-term experiment is evidence that *I. obesulus* is not resistant to this aspect of the immunosuppressive effect of glucocorticoid hormones. Cheal *et al.* (1976) and Barker *et al.* (1978) found that marked lymphocytopenia was related to high plasma corticosteroid concentrations in *Antechinus stuartii* and Dickens (1978) found that when subjected to stress *P. cinereus* also exhibited lymphocytopenia.

Cortisol was found to be the main glucocorticoid in *I. obesulus,* though when TPCS concentrations were low, cortisol and corticosterone were present in almost equal concentrations. These results are similar to those reported by McDonald and Bradshaw (1977) for *S. brachyurus*. Oddie *et al.* (1976) found that of the nine species of marsupial that they examined, only *V. ursinus* and *P. cinereus* exhibited corticosterone as their main glucocorticoid when in an unstressed condition.

The two species of marsupial previously found to be resistant to the effects of glucocorticoid hormones (Griffiths *et al.* 1969; McDonald and Bradshaw 1981) are both herbivores which commonly inhabit arid environments (Storr 1964; McDonald 1977). Their lack of response to glucocorticoids may be considered as an adaptation to conserve nitrogen and carbohydrates (McDonald 1977). *Isoodon obesulus,* on the other hand, is an insectivore and has a nitrogen-rich diet (Heinsohn 1966). It may be suggested that resistance in *I. obesulus* to the metabolic effects of glucocorticoid hormones arose in response to a different stimulus than in *S. brachyurus* and *M. rufus*. It may be an adaptation to seasonally high levels of social and reproductive stress. Thomas (this volume) found that high concentrations of plasma free corticosteroid hormones were associated with both high population estimates and the recruitment of transient sub-adults into the adult population.

ACKNOWLEDGEMENTS

We would like to thank Iain McCleod for his assistance with the nitrogen analysis, Stan Hopwood for his help with the graphs, Greg Thomas for his assistance with trapping and Rosilyne Briggs for her assistance with typing. This study was assisted by grant number D18015574 to A. J. Bradley from the Australian Research Grants Scheme.

REFERENCES

Ashmore, J. and Morgan, D., 1967. Metabolic effects of adrenal glucocorticoid hormones: carbohydrate, protein, lipid and nucleic acid metabolism. Pp. 249–68 in *The adrenal cortex* ed by A. B. Eisenstein. Little, Brown & Co: Boston.

Barker, I. K., Beveridge, I., Bradley, A. J. and Lee, A. K., 1978. Observations on spontaneous stress-related mortality among males of the dasyurid marsupial *Antechinus stuartii* Macleay. *Australian Journal of Zoology* **26:** 435–47.

Bassett, J. M. and Hicks, N. T., 1969. Micro-determination of corticosteroids in ovine peripheral plasma: effects of venipuncture, corticotrophin, insulin and glucose. *Journal of Endocrinology* **44:** 387–403.

Bradford, M. M., 1976. A rapid and sensitive method for the quantitation of microgram quantities of protein utilizing the principle of protein-dye binding. *Analytical Biochemistry* **72:** 248–54.

Bradley, A. J., McDonald, I. R. and Lee, A. K., 1976. Corticosteroid binding globulin and mortality in a dasyurid marsupial. *Journal of Endocrinology* **70:** 323–24.

Bradley, A. J., McDonald, I. R. and Lee, A. K., 1980. Stress and mortality in a small marsupial (*Antechinus stuartii* Macleay). *General and Comparative Endrocrinology* **40:** 188–200.

Cassileth, P. A., 1983. Lymphocytopenia. Pp. 86–94 in *Haematology* ed by W. J. Williams, E. Beutler, A. J. Erslev and M. A. Lichtman. McGraw-Hill: New York.

Cheal, P. D., Lee, A. K. and Barnett, J. L., 1976. Changes in the haematology of *Antechinus stuartii* (Marsupialia) and their association with male mortality. *Australian Journal of Zoology* **24:** 299–311.

Claman, H. M., 1972. Corticosteroids and lymphoid cells. *New England Journal of Medicine* **287:** 388–97.

Cooley, H. and Janssens, P., 1977. Metabolic effects of infusion of cortisol and adrenocorticotropin in the tammar wallaby (*Macropus eugenii* Desmarest). *General and Comparative Endocrinology* **33:** 352–58.

Dacie, J. V., 1975. *Practical haematology*, 5th ed. Longman: New York.

Dickens, R. K., 1978. A comparison of haematological values from Koala in different locations and the effect of stress on that value. Pp. 183–97 in *The Koala. Proceedings of the Taronga Symposium* ed by T. J. Bergin. Zoological Parks Board of New South Wales: Sydney.

Disbrey, B. D. and Rack, J. H., 1970. *Histological laboratory methods*. Livingstone: London.

Griffiths, M., McIntosh, D. L. and Leckie, R. M. C., 1969. The effect of cortisone on nitrogen balance and glucose metabolism in diabetic and normal kangaroos, sheep and rabbits. *Journal of Endocrinology* **44:** 1–12.

Heinsohn, G. E., 1966. Ecology and Reproduction of the Tasmanian Bandicoots (*Perameles gunni* and *Isoodon obesulus*). *University of California Publications in Zoology* **80:** 1–107.

Hollenberg, C. H., Raben, M. S. and Astwood, E. B., 1961. The lipolytic response to corticotrophin. *Endocrinology* **58:** 97–109.

JANSSENS, P. A. AND HINDS, L. A., 1981. Long-term effects of corticosteroid administration in the tammar wallaby, *Macropus eugenii*. *General and Comparative Endocrinology* **45:** 56–60.

KAPLAN, A., 1969. The determination of urea, ammonia and urease. Pp. 206–59 in *Methods of biochemical analysis*. Vol. 17 ed by D. Glick. Wiley: New York.

LAURELL, S. AND TIBBLING, G., 1967. Colormetric microdetermination of Free Fatty Acids in plasma. *Clinica Chimica Acta* **16:** 57–62.

LIDDLE, G. W., 1981. The adrenals. Pp. 805–63 in *Textbook of endocrinology*. 6th ed. ed by R. H. Williams. Saunders: Philadelphia.

MCDONALD, I. R., 1977. Adrenocortical functions in marsupials. Pp. 345–77 in *The biology of marsupials* ed by B. Stonehouse and D. Gilmore. University Park Press: London.

MCDONALD, I. R. AND BRADSHAW, S. D., 1977. Plasma corticosteroids and the effect of adrenocorticotrophin in a macropodid marsupial (*Setonix brachyurus*, Quoy and Gaimard). *Journal of Endocrinology* **75:** 409–18.

MCDONALD, I. R. AND BRADSHAW, S. D., 1981. Resistance to the nitrogen-mobilizing, diabetogenic action of cortisol in a small wallaby *(Setonix brachyurus)*. *General and Comparative Endocrinology* **43:** 39–41.

MCDONALD, I. R., LEE, A. K., BRADLEY, A. J. AND THAN, K. A., 1981. Endocrine changes in dasyurid marsupials with differing mortality patterns. *General and Comparative Endocrinology* **44:** 292–301.

MUNCK, A., 1971. Glucocorticoid inhibition of glucose uptake by peripheral tissues: Old and new evidence, molecular mechanisms and physiological significance. *Perspectives in Biology and Medicine* **14:** 265–89.

MURPHY, B. E. P., 1971. "Sephadex" column chromotography as an adjunct to competitive protein binding assays of steroids. *Nature* **232:** 21–24.

ODDIE, C. J., BLAINE, E. H., BRADSHAW, S. D., COGHLAN, J. P., DENTON, D. A., NELSON, J. F. AND SCOGGINS, B. A., 1976. Blood corticosteroids in Australian marsupial and placental mammals and one monotreme. *Journal of Endocrinology* **69:** 341–48.

RODKEY, F. L., 1965. Direct spectrophotometric determination of albumin in human serum. *Clinical Chemist* **11**(4): 478–87.

SEGRE, E. J., FRIEDRICH, E. H., DODEK, O. I., LLOYD, C. W., LOBOTSKY, J., LEVIN, J. AND KLAIBER, E. L., 1966. Effects of epinephrine on the production and metabolic clearance of cortisol and in normal men and women with idiopathic hirsutism. *Acta Endocrinologica* **53:** 561–70.

STORR, G. M., 1964. Studies on marsupial nutrition iv. Diet of the Quokka, *Setonix brachyurus* (Quoy and Gaimard), on Rottnest Island, Western Australia. *Australian Journal of Biological Science* **17:** 469–81.

TAIT, J. F., 1982. The biological availability of free and protein-bound steroid. Pp. 65–70 in *Free hormones in blood* ed by A. Albertini and R. P. Ekins. Elsevier: Amsterdam.

TAIT, J. F. AND BURSTEIN, S., 1964. *In vivo* studies of steroid dynamics in man. Pp. 441–557 in *The hormones*. Vol. 5 ed by G. Pincus. K. V. Thimann and E. G. Astwood. Academic Press: New York.

THAN, K. A. AND MCDONALD, I. R., 1974. Effect of cortisol on insulin sensitivity of the marsupial brush-tailed opossum, *Trichosurus vulpecula* (Kerr). *Journal of Endocrinology* **63:** 473–81.

TIETZ, N., 1976. *Clinical chemistry*. Saunders: London.

Tolerance to fluoroacetate of populations of *Isoodon* spp. and *Macrotis lagotis* and its implications for fauna management

LAURIE E. TWIGG[1,3], DENNIS R. KING[2] and ROBERT J. MEAD[1]

Toxic fluoroacetate occurs naturally in southwestern Australian species of the legume genera *Gastrolobium* and *Oxylobium* and in *G. grandiflorum* and *Acacia georginae* in northern Australia. The tolerance to sodium fluoroacetate (1080) of two geographically separate populations of the Southern Brown Bandicoot *(Isoodon obesulus)* and populations of the Northern Brown Bandicoot *(I. macrourus)*, the Golden Bandicoot *(I. auratus barrowensis)* and the Greater Bilby *(Macrotis lagotis)* each with different degrees of exposure to fluoroacetate-bearing vegetation is presented. *Isoodon obesulus* from southwestern Australia and *M. lagotis* which co-exist with the toxic plants tolerated approximately 20 and 15 mg of 1080 kg^{-1} respectively. However, *I. obesulus* from the Franklin Islands, South Australia and *I. macrourus* from Dorrigo, New South Wales, both of which are outside the range of the poisonous plants, were more sensitive to 1080 with approximate LD_{50} values of 8 and 3.5 mg of 1080 kg^{-1} respectively. *Isoodon auratus barrowensis* from Barrow Island, Western Australia, separated from contact with fluoroacetate-bearing vegetation for 6–8 000 years, had an LD_{50} of 8.94 mg kg^{-1}. These tolerances are discussed in relation to diet, past and present distribution of the bandicoots and the hazard posed to bandicoot populations during 1080 baiting programmes.

Key words: 1080, fluoroacetate-tolerance, *Isoodon obesulus I. macrourus, I. auratus, Macrotis lagotis,* non-target species.

INTRODUCTION

In addition to habitat alteration by man, competition and predation by introduced species has been strongly implicated in the decline of many indigenous small to medium-sized mammals in Australia (Jones 1925; Perry 1973; Christensen 1980; King, Oliver and Mead 1981; McKenzie 1981; King and Smith 1985). Predation by the Fox, *Vulpes vulpes* (Watts 1969; Christensen 1980) and Cat, *Felis catus* (Burbidge 1971) and competition from the Rabbit, *Oryctolagus cuniculus* (Watts 1969) have been suggested as major contributing factors in the decline of some species of bandicoots on the Australian mainland. Control of Rabbit, Cat and Fox populations and other mammalian pests, such as feral Pigs, *Sus scrofa* and Dingoes *Canis familiaris dingo* using sodium fluoroacetate (Compound 1080) may therefore be necessary in some management programmes aimed at the conservation of bandicoots. However, until recently, little was known about the effect of such poisoning campaigns on non-target species.

Fluoroacetate is highly toxic to most species of mammals (Atzert 1971) with mammalian carnivores being particularly sensitive (McIlroy 1981). Fluoroacetate is converted within the animal to fluorocitrate which competitively inhibits the tricarboxylic acid cycle enzyme aconitate hydratase (E. C. 4. 2. 1. 3) with resultant citrate accumulation (Morrison and Peters 1954). Elevation of plasma citrate levels in response to dosing has previously been shown to reflect the sensitivity of animals to fluoroacetate poisoning (King, Oliver and Mead 1978, 1981; Oliver, King and Mead 1979).

Fluoroacetate also occurs as a natural plant product and is found in the genera *Oxylobium, Gastrolobium* and *Acacia* in Australia. The distribution of these toxic plants is shown in Figure 1. Thirty-three of the 35 species in Australia known to produce fluoroacetate are confined to the lower south-west of Western Australia and these species have been shown to contain between 100 and 2 650 mg fluoroacetate per kg air-dried sample (Aplin 1971). Air-dried samples of *A. georginae* (Oelrichs and McEwan 1961, 1962) and *G. grandiflorum* (McEwan 1964) contain up to 25 and 185 mg fluoroacetate per kg, respectively.

Animal populations which are currently exposed to this vegetation or have been exposed to it in the past are much less susceptible to fluoroacetate intoxication than are geographically separate, unchallenged populations in southeastern Australia (Oliver, King and Mead 1977, 1979; King *et al.* 1978, 1981). However, populations of *Isoodon obesulus, Perameles nasuta* and *P. gunnii* from southeastern Australia, although not exposed to fluoroacetate-bearing vegetation, all have higher tolerances to fluoroacetate than do most other eastern Australian mammals (McIlroy 1983).

[1]School of Biological and Environmental Sciences, Murdoch University, Murdoch, Western Australia, Australia 6150.
[2]Agriculture Protection Board of Western Australia, Bougainvillea Avenue, Forrestfield, Western Australia, Australia 6058.
[3]Current address: New South Wales Agriculture and Fisheries, Agricultural Research and Veterinary Centre, Forest Road, Orange, New South Wales, Australia 2800.
Pages 185–92 *in* BANDICOOTS AND BILBIES, ed by J. H. Seebeck, P. R. Brown, R. L. Wallis and C. M. Kemper. Surrey Beatty & Sons, Sydney, 1990.

Fig. 1. Distribution of fluoroacetate-bearing vegetation in Australia; vertical lines: 33 toxic species of *Gastrolobium* and *Oxylobium;* horizontal lines: *Gastrolobium grandiflorum;* stippled area: *Acacia georginae.* Distribution after: Aplin (1971, and pers. comm.), Bell *et al.* (1955), Everist (1947), Jessop (1981), R. W. Johnson (pers. comm.), B. Maslin (pers. comm.).

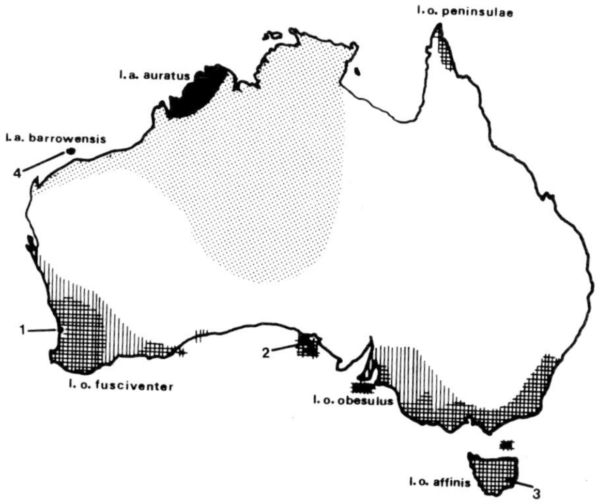

Fig. 2. The range of the Southern Brown Bandicoot *Isoodon obesulus* and the Golden Bandicoot *I. auratus;* vertical hatching: Former range of *I. obesulus;* cross hatching: Current range of *I. obesulus;* stippled area: Former range of *I. auratus*; solid area: Current range of *I. auratus*. Distribution modified from: Braithwaite (1983), McKenzie (1983), A. Baynes (pers. comm.).

Population 1: Perth, WA, exposed to fluoroacetate-bearing vegetation; Population 2: Franklin Islands, SA, not exposed to fluoroacetate-bearing vegetation; Population 3: Tasmanian population studied by McIlroy (1983); not exposed to fluoroacetate-bearing vegetation; Population 4: Barrow Island, WA, not currently exposed to fluoroacetate-bearing vegetation.

Dietary information on bandicoots is limited, but they are generally considered to be omnivorous (Watts 1969; Stoddart and Braithwaite 1979; Hume 1982; Johnson 1983; McKenzie 1983) and may feed directly on the toxic plants or on animals which have fed on them. As the distribution of some populations of *I. obesulus, I. macrourus, I. auratus* and *Macrotis lagotis* coincides with that of the fluoroacetate-bearing vegetation (Figs 1, 2, 3 and 4), it is possible that the tolerance to fluoroacetate of these populations may be even greater than that reported for eastern Australian bandicoots.

This paper presents the results of a comparison of the sensitivity to fluoroacetate of populations of *I. obesulus, I. macrourus, I. auratus barrowensis* and *M. lagotis,* each with differing degrees of past or current exposure to naturally-occurring fluoroacetate. The results obtained are discussed in relation to diet, former distribution of the bandicoots and the hazard posed to these non-target species during 1080 baiting programmes.

MATERIALS AND METHODS

Animals

Isoodon obesulus: Male and female animals (n = 6), with exposure to fluoroacetate-bearing vegetation, were collected around Perth, Western Australia. Adult male and female animals (n = 6) not exposed to the toxic plants were collected from the Franklin Islands, South Australia by A. Robinson and P. Copley, National Parks and Wildlife Service, Adelaide, South Australia.

Isoodon macrourus: Male animals (n = 6), not exposed to the toxic plants, were bred in captivity but had originated from Dorrigo, New South Wales. They were supplied by B. Green, CSIRO Division of Wildlife and Ecology. *Isoodon auratus barrowensis:* Male and female animals, not currently exposed to fluoroacetate-bearing vegetation, were collected from Barrow Island, Western Australia with the permission of the Western Australian Wildlife Authority and with the assistance of W. H. Butler.

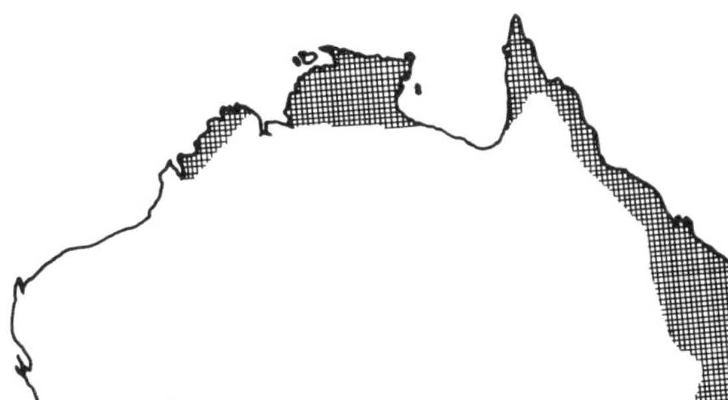

Fig. 3. The current distribution of the Northern Brown Bandicoot *Isoodon macrourus.* (●) Dorrigo, New South Wales; not exposed to fluoroacetate-bearing vegetation. Distribution after Gordon (1983).

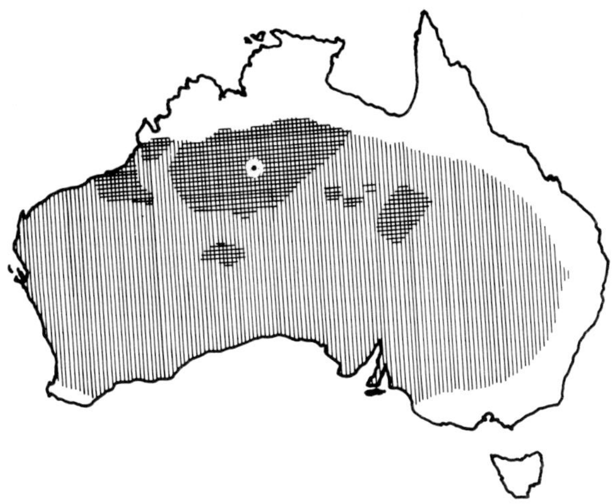

Fig. 4. The past (vertical hatching) and present (cross hatching) range of *Macrotis lagotis*. (●) Rabbit Flat (NT) Distribution modified from: Burbidge and Fuller (1979), Johnson (1983), Watts (1969), R. Southgate (pers. comm.).

Macrotis lagotis: Two male animals from the captive colony at Rabbit Flat, Northern Territory, were provided by K. Johnson, Northern Territory Conservation Commission. *G. grandiflorum* occurs within the range of this population.

Housing

With the exception of those *I. a. barrowensis* from which mortality data were obtained, all other experimental animals were housed indoors in Perth. Bandicoots were kept in individual metabolism cages provided with nest boxes. Ambient temperature was 23 ± 1°C, with 70% relative humidity and a 12:12 h photoperiod. Mortality data on *I. a. barrowensis* were obtained from experiments conducted in a field laboratory on Barrow Island during March 1983. Animals were held in individual cages and were subjected to a natural photoperiod with a temperature range of 21–26°C. All animals were fed on tinned dog food, lamb heart, bananas and mixed seed. Water was provided *ad libitum*.

Dosing and bleeding

Sodium monofluoroacetate (1080) was administered intraperitoneally in aqueous solution. All animals were dosed between 0815 and 0930 h. Those animals administered progressively higher amounts of 1080 were given a minimum of 2–4 weeks between doses to allow recovery.

Blood samples were collected in heparinized syringes, by cardiac puncture from *I. a. barrowensis* and from the lateral caudal vein of *I. obesulus* and *I. macrourus*. The topical anaesthetic Lignocaine was applied prior to venipuncture. Occasionally, when difficulty was experienced in obtaining samples from the caudal vein, blood was collected from the orbital sinus.

Plasma citrate determination

Plasma citrate concentrations were determined using 20–50 μl plasma samples by the enzymatic method of Mollering and Gruber (1966) as modified by Dagley (1974) with the exception that 0.1 M tris-HCl buffer pH 7.6 was used in place of triethanolamine.

Mortality data

To obtain the mortality data on *I. a. barrowensis* and *M. lagotis*, all animals were inspected at hourly intervals for the first 12 h after dosing. *Macrotis lagotis* was then observed daily for 14 days, while daily inspections of *I. a. barrowensis* were conducted for only 3–4 days because of the short duration of the field trip to Barrow Island. Because of this limitation, tolerance to fluoroacetate was further assessed by following the increase in plasma citrate levels of *I. a. barrowensis* (n = 3) after dosing with 1080 under controlled conditions in Perth.

The LD_{50} for *I. a. barrowensis* was calculated according to the moving average method of Thompson (1947).

Statistical analysis

The elevation in plasma citrate concentration of the bandicoots from the two *I. obesulus* populations and of *I. a. barrowensis* administered 5 mg 1080 kg^{-1} (Fig. 8) were compared using a single factor analysis of variance. The harmonic mean was used to compensate for uneven group size and the Levene test employed to test homogeneity of variance (Keppel 1973). Significant differences between means were determined by the Scheffe test (Keppel 1973). However, the data for *I. macrourus* were not included in this analysis because of the extreme response of these bandicoots at this dose level, and because the variance of the response was found to be unequal (Levene test, Keppel 1973).

Chemicals

Commercial grade 1080 (94% sodium monofluoroacetate by HPLC analysis) was obtained from Rentokil Laboratories Ltd. Citrate lyase, malate dehydrogenase and reduced nicotinamide adenine dinucleotide (NADH) were purchased from Sigma Chemical Co., St Louis, Missouri. All other chemicals were of analytical reagent grade.

RESULTS

Elevations in plasma citrate concentration in response to administration of various amounts of 1080 to *I. obesulus* from Perth, Western Australia and from the Franklin Islands South Australia are shown in Figures 5 and 6. The tolerance to fluoroacetate of *I. obesulus* from Perth was approximately two-fold greater than that of the conspecifics from the Franklin Islands. Two of the three Franklin Island bandicoots dosed with 10 mg 1080 kg^{-1} displayed large increases in plasma citrate concentration and subsequently died within 24 h of dosing. Bandicoots from the Perth population, however, survived this dose level and displayed increases in plasma citrate concentration less than those of the Franklin Island animals (Figs 5 and 6). No deaths occurred in Perth bandicoots at any dose level within 48 h of dosing but one animal died nine days after dosing with 15 mg 1080 kg^{-1} and two animals died four to five days after administration of 20 mg 1080 kg^{-1}.

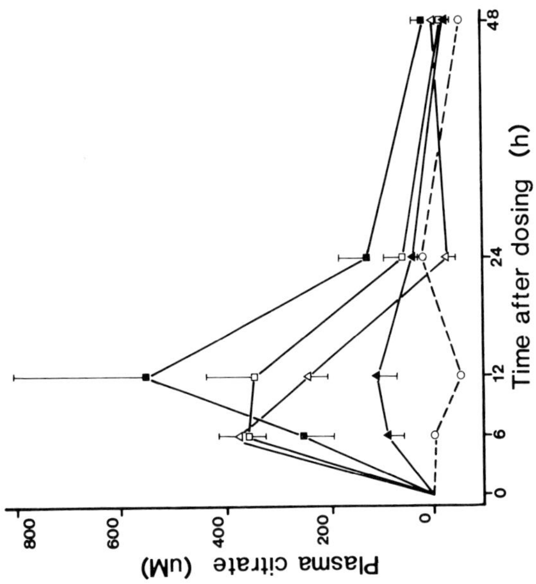

Fig. 5. The mean and standard error of the increase above base-level (time zero) values of plasma citrate concentrations of *Isoodon obesulus* from Perth, WA following administration of various doses of sodium monofluoroacetate (1080). ○ Undosed (n = 1); ▲ 5 mg 1080 kg^{-1} (n = 4); □ 10 mg 1080 kg^{-1} (n = 4); △ 15 mg 1080 kg^{-1} (n = 4); ■ 20 mg 1080 kg^{-1} (n = 4). The four animals in each dose group were randomly selected from the six animals available. Ambient temperature 23 ± 1°C.

Other populations of *I. obesulus* in Western Australia also displayed an enhanced tolerance to fluoroacetate. During preliminary dosing, mortality data were obtained for three *I. obesulus* collected from the lower south-west of Western Australia. When administered 10 mg 1080 kg^{-1} all three bandicoots survived. Two individuals dosed with 15 mg 1080 kg^{-1} also survived but developed symptoms (see McIlroy 1983) of fluoroacetate intoxication. A single bandicoot given 22.5 mg 1080 kg^{-1} died approximately 8 h after dosing. This suggests that the approximate LD_{50} values for *I. obesulus* from the south-west of Western Australia and from the

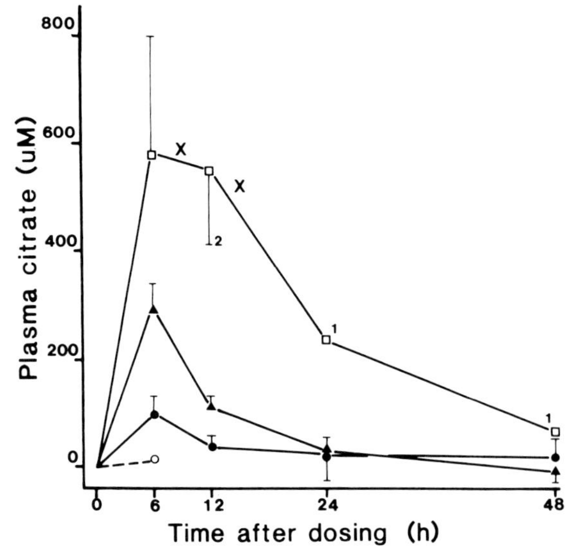

Fig. 6. The mean and standard error of the increase above base-level (time zero) values of plasma citrate concentrations of *Isoodon obesulus* from the Franklin Islands, SA following intraperitoneal administration of sodium fluoroacetate (1080). ○ Undosed (n = 1); ● 2 mg 1080 kg^{-1} (n = 3); ▲ 5 mg 1080 kg^{-1} (n = 5); □ 10 mg 1080 kg^{-1} (n = 4). The animals in each dose group were randomly selected from the six animals available. X indicates death of animal. Ambient temperature 23 ± 1°C.

Franklin Islands, South Australia, as indicated by both mortality and citrate elevation data (Figs 5 and 6) are 20 and 8.0 mg 1080 kg^{-1} respectively.

Of the bandicoots investigated, *I. macrourus* from Dorrigo, New South Wales were found to be the most sensitive to 1080 poisoning (Fig. 7). All four *I. macrourus* dosed with 5 mg 1080 kg^{-1} displayed substantial increases in plasma citrate concentration and subsequently died within 24 h of dosing. The concentration of citrate in the plasma of *I. macrourus* administered 2 mg 1080 kg^{-1} also increased considerably but no deaths occurred at this dose level (Fig. 7). It appears therefore that *I. macrourus* from Dorrigo have an approximate LD_{50} of 3.5 mg 1080 kg^{-1}.

The LD_{50} value calculated from the mortality data obtained for *I. a. barrowensis* from Barrow Island, Western Australia was 8.94 mg 1080 kg^{-1} (Table 1). Increases in plasma citrate concentration of *I. a. barrowensis* dosed with 5 mg 1080 kg^{-1} are shown in Figure 8. At the 5 mg 1080 kg^{-1} dose level, the plasma citrate elevations of *I. a. barrowensis* 6 h after dosing were significantly greater (ANOVA; $P < 0.05$) than those of the Perth population of *I. obesulus* but were not significantly different (ANOVA; $P > 0.05$) from the 6 h levels of the Franklin Island *I. obesulus* (Fig. 8). At this dose level, the 6 h plasma citrate elevations of the Franklin Island *I. obesulus* were also significantly greater

(*ANOVA*; $P < 0.05$) than those from the Perth population. *Isoodon macrourus* accumulated much greater amounts of plasma citrate at both the 2 and 5 mg 1080 kg^{-1} dose levels than did the bandicoots from the other three populations at any dose level (Figs 7 and 8).

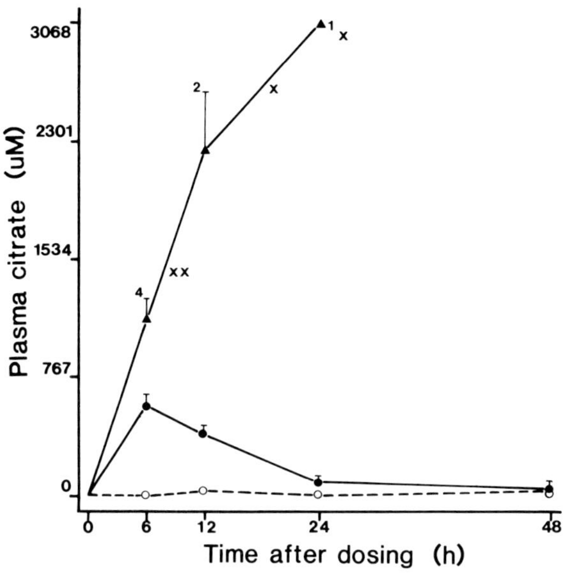

Fig. 7. The mean and standard error of the increase above base-level (time zero) values of plasma citrate concentrations of *Isoodon macrourus* from Dorrigo, NSW, following administration of sodium fluoroacetate (1080). ○ Undosed (n = 1); ● 2 mg 1080 kg^{-1} (n = 4); ▲ 5 mg 1080 kg^{-1} (n = 4). The four animals in each dose group were randomly selected from the six animals available. X indicates death of animal. Ambient temperature 23 ± 1°C.

An estimate of the sensitivity of *M. lagotis* to sodium fluoroacetate was made using two bandicoots and a "leap-frog" dosing technique. Animals were dosed with increasing levels of 1080, one at 0.1, 0.5, and 2.0 mg 1080 kg^{-1} body weight and the other with 0.2, 1.0, and 5.0 mg kg^{-1}. No obvious symptoms of fluoroacetate intoxication were induced at any of these doses. Both bandicoots also survived administration of 10 mg 1080 kg^{-1}, but each developed symptoms of poisoning. When dosed with 15 mg 1080 kg^{-1}, the animals developed symptoms typical of fluoroacetate intoxication but appeared to recover after 48 h. One animal, however, subsequently died 75 h after dosing. These data suggest an estimated LD_{50} for this species of approximately 15 mg 1080 kg^{-1}.

Table 1. Mortality data and LD_{50} with 95% confidence limits for *Isoodon auratus barrowensis* from Barrow Island, WA. (Ambient temperature 21–26°C).

	Dose (mg 1080 kg^{-1})						
	0	5.1	6.4	8.0	10.0	12.5	15.6
Deaths	0/2	0/4	2/4	1/4	2/4	3/4	4/4
		95% LCL 7.26		LD_{50} 8.94		95 % UCL 11.02	

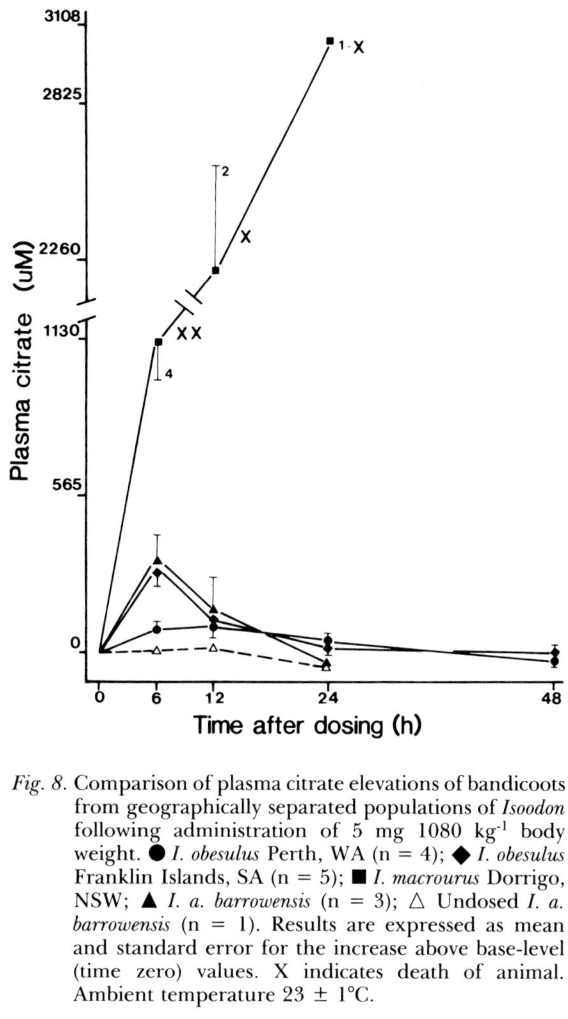

Fig. 8. Comparison of plasma citrate elevations of bandicoots from geographically separated populations of *Isoodon* following administration of 5 mg 1080 kg^{-1} body weight. ● *I. obesulus* Perth, WA (n = 4); ◆ *I. obesulus* Franklin Islands, SA (n = 5); ■ *I. macrourus* Dorrigo, NSW; ▲ *I. a. barrowensis* (n = 3); △ Undosed *I. a. barrowensis* (n = 1). Results are expressed as mean and standard error for the increase above base-level (time zero) values. X indicates death of animal. Ambient temperature 23 ± 1°C.

DISCUSSION

The greater tolerance to fluoroacetate of bandicoots exposed to this toxin in nature is consistent with data obtained for other mammals with past or current exposure to fluoroacetate-bearing vegetation (Oliver *et al.* 1977, 1979; King *et al.* 1978, 1981). However, a relatively high innate tolerance to fluoroacetate is apparent in bandicoots not exposed to fluoroacetate in their diets. For example, the LD_{50} for *Perameles nasuta* from New South Wales is 7.70 mg 1080 kg^{-1} (range 5.28–11.23) and *P. gunnii* from Tasmania, whose range is outside that of the toxic plants, are reported to survive 5 mg 1080 kg^{-1} (McIlroy 1983). The approximate LD_{50} for *I. obesulus* from Tasmania is 7.0 mg 1080 kg^{-1} (n = 4; McIlroy 1983) and is similar to that obtained for *I. obesulus* from the Franklin Islands which have an approximate LD_{50} of 8.0 mg 1080 kg^{-1}. The ranges of all these bandicoots are outside those of the toxic plants. Although the approximate LD_{50} value for *I. obesulus* from eastern Australia is higher than that of most unadapted mammals, it is much less than the approximate LD_{50} of 20 mg 1080 kg^{-1} obtained for *I. obesulus* from Perth, Western Australia.

Isoodon macrourus from Dorrigo, New South Wales, is considerably less tolerant to fluoroacetate than most other species of bandicoots thus far examined (McIlroy 1983). The approximate LD_{50} for this species is 3.5 mg 1080 kg^{-1} (Fig. 7), which is only marginally greater than that of most other species of unadapted mammals. The LD_{50} values for the latter are generally in the region of 0.1–2 mg 1080 kg^{-1} (Atzert 1971; McIlroy 1981, 1982). More northerly populations of *I. macrourus* co-exist with fluoroacetate-bearing vegetation (Figs 1 and 3). These populations were not investigated in this study but may be more tolerant to fluoroacetate than the Dorrigo population which is not exposed to the toxic plants. If this is the case it appears that there has been little gene flow between the northern and more southerly populations of this species.

The LD_{50} for *I. a. barrowensis* was calculated to be 8.94 mg 1080 kg^{-1} (Table 1). Whether this tolerance reflects adaptation to fluoroacetate, or an innate tolerance to the toxin is unclear. *Isoodon a. barrowensis* on Barrow Island is not exposed to fluoroacetate-bearing vegetation but the former, more widespread distribution of *I. auratus* on the adjacent mainland did include areas where fluoroacetate-bearing plants occur (Figs 1 and 2). As populations of *Isoodon* (Figs 6 and 7) and *Perameles* from areas where exposure to fluoroacetate does not occur are less tolerant to fluoroacetate (McIlroy 1983), it is possible that the tolerance to fluoroacetate of *I. a. barrowensis* reflects past adaptation by ancestral mainland bandicoots before Barrow Island and nearby Middle Island were isolated approximately 6 000 to 8 000 years BP (Sharred and King 1981; Buckley 1983).

Retention of tolerance to fluoroacetate in animals from island populations which are no longer exposed to fluoroacetate-bearing vegetation has previously been demonstrated for the Western Grey Kangaroo, *Macropus fuliginosus*, the Tammar Wallaby, *M. eugenii* (Oliver *et al.* 1979), the Burrowing Bettong, *Bettongia lesueur* (King *et al.* 1981) and the Quokka, *Setonix brachyurus* (Mead, Twigg, King and Oliver 1985). In addition, populations of *M. fuliginosus* from eastern Australia which are not exposed to fluoroacetate-bearing plants have tolerances to fluoroacetate similar to that found in Western Australian populations which co-exist with the toxic vegetation. On this basis, it was suggested that *M. fuliginosus* radiated from western to eastern Australia (Oliver *et al.* 1979).

Although the data on *M. lagotis* are based upon only two individuals, it appears that this species is moderately tolerant to fluoroacetate (approximate LD_{50} 15 mg 1080 kg^{-1}). Fluoroacetate-producing plants occur in areas currently inhabited by *M. lagotis* and the former distribution of this species includes areas in the south-west of Western Australia where the more toxic species of *Gastrolobium* and *Oxylobium* occur (Figs 1 and 4). The diet of *M. lagotis* includes insects and their larvae, seeds, bulbs, fruit and fungi (Watts 1969; Johnson 1983). The burrows of *M. lagotis* tend to be located in soft open ground or along watercourses (Watts 1969), and *G. grandiflorum* is most often found on rocky ranges and along dry sandy riverbeds (Jessop 1981). Consequently, it is possible that *M. lagotis* may have developed a primary tolerance to fluoroacetate by feeding on the seeds of the toxic plants and/or may have acquired a secondary tolerance to the toxin by ingesting invertebrates which feed on fluoroacetate-bearing vegetation. Development of tolerance to fluoroacetate by populations of *Isoodon* spp. exposed to this toxin in nature is also likely to have resulted from the incorporation of toxic vegetation in the diet and/or the ingestion of invertebrates which feed on fluoroacetate-bearing plants.

It is apparent that the tolerance to fluoroacetate of these omnivorous marsupials is generally lower than that of most adapted marsupial herbivores. With the exception of *M. eugenii* and *M. irma*, which are predominantly grazing species with LD_{50} values between 5 and 10 mg 1080 kg^{-1}, the approximate LD_{50} values for browsing herbivore populations currently exposed to fluoroacetate-bearing vegetation range from 40–60 mg 1080 kg^{-1} for *S. brachyurus* (Mead *et al.* 1985) and *M. fuliginosus* (King *et al.* 1978) to greater than 100 mg 1080 kg^{-1} for *B. penicillata* (King *et al.* 1981) and *Trichosurus vulpecula* (Oliver *et al.* 1977).

The risk posed to non-target species during pest control programmes is determined by the size of an individual as well as its sensitivity to 1080. To be at risk, a larger, more sensitive animal may have to ingest more 1080 than does a smaller animal with a moderate fluoroacetate tolerance. While bandicoots from eastern Australia appear to have an innate tolerance to fluoroacetate, the size of the animals is such that they are placed at risk during 1080 poisoning programmes directed at feral Pigs and herbivorous mammals (McIlroy 1983). In particular, our study suggests that *I. obesulus* from the eastern Australian mainland and *I. macrourus* could be seriously threatened during such campaigns. However, because of their greater tolerance to fluoroacetate, *I. obesulus* from Western Australia and *M. lagotis* are unlikely to be placed at risk during 1080 poisoning programmes (Table 2).

While caution is required when extending laboratory data to a field situation (McIlroy 1983; Oliver and King 1983), the use of 1080 for

Table 2. Relative susceptibility of three species of bandicoots and four target species to 1080 poison. *Each meat bait and "one-shot" oat bait used in Western Australia contains c. 5–6 mg 1080 (Robinson and Wheeler 1983).

Species	LD_{50} (mg kg^{-1})	Reference	Weight range of adult (kg)	Reference	Amount of* 1080 equivalent to LD_{50} (mg)
Isoodon auratus barrowensis	8.94	1	0.26– 0.46	6, 3	2.3– 4.1
Isoodon obesulus (WA)	c. 20	1	0.40– 1.60	7	8.0–32.0
Isoodon obesulus (SA)	c. 8.00	1	0.40– 0.80	14	3.2– 6.4
Isoodon macrourus	c. 3.5	1	0.50– 3.10	13	1.8–10.9
Macrotis lagotis	c. 15	1	0.80– 2.50	8	12.0–37.5
Felis catus	0.4	2	2.50– 6.40	9	1.0– 2.6
Canis familiaris	0.11	2	9.60–16.00	10	1.1– 1.8
Vulpes vulpes	c. 0.13	3	4.70– 8.30	11	0.6– 1.1
Oryctolagus cuniculus	0.37	4	0.96– 2.42	12	0.4– 0.9
	0.46	5			0.4– 1.1

1. This study; 2. McIlroy (1981); 3. King (unpubl. data); 4. McIlroy (1983); 5. Wheeler and Hart (1979); 6. McKenzie (1983); 7. Braithwaite (1983); 8. Johnson (1983); 9. Jones (1983); 10. Newsome (1983); 11. Coman (1983); 12. Myers (1983); 13. Gordon (1983); 14. P. Copley and A. Robinson (pers. comm.).

control of exotic predators and other mammalian pests may be beneficial in areas where bandicoot populations have adapted to the presence of naturally-occurring fluoroacetate. It may also be possible to re-introduce *M. lagotis* into areas of its former range which contain fluoroacetate-bearing plants or where 1080-baiting is carried out. However, before either practice could be used with confidence, the long-term effects of sodium fluoroacetate on animal populations require further study. Relatively low levels of sodium fluoroacetate have been reported to cause temporary infertility in male *Rattus rattus* (Mazzanti, Lopez and Berti 1965) and may also chronically affect liver function (Twigg, Mead and King 1986) and fertility (Twigg, King and Bradley 1988) in reptiles.

ACKNOWLEDGEMENTS

We are grateful for the assistance of W. H. Butler, B. Green, P. Copley, A. Robinson, K. Johnson, J. Merchant and N. Thomas who supplied some of the animals used in this study. Dosing of *I. a. barrowensis* on Barrow Island was made possible through the provision of transport to, and accommodation on, Barrow Island as part of a WAPET Research Grant to D. R. King to carry out an unrelated study on the island. A. Eastman and B. Green assisted in the handling of experimental animals. R. W. Johnson from the Queensland Herbarium supplied information on the distribution of *G. grandiflorum* and Jean Hitchcock and Carol Hooper typed the manuscript.

REFERENCES

APLIN, T. E. H., 1971. Poison plants of Western Australia: The toxic species of the genera *Gastrolobium* and *Oxylobium*. *Department of Agriculture, Western Australia Bulletin* **3772:** 1–66.

ATZERT, S. P., 1971. A review of sodium monofluoroacetate (compound 1080), its properties, toxicology and use in predator and rodent control. Special Scientific Report. *Wildlife* **146:** 1–34.

BELL, A. T., NEWTON, L. G., EVERIST, S. L. AND LEGG, J., 1955. *Acacia georginae* poisoning of cattle and sheep. *Australian Veterinary Journal* **31:** 249–57.

BRAITHWAITE, R. W., 1983. Southern Brown Bandicoot. Pp. 94–95 in *Complete book of Australian mammals* ed by R. Strahan. Angus & Robertson: Sydney.

BUCKLEY, R. C., 1983. The fauna and flora of Barrow Island, Western Australia. *Journal of the Royal Society of Western Australia* **66:** 91–105.

BURBIDGE, A. A., 1971. The fauna and flora of the Monte Bello Islands. *Department of Fisheries and Fauna, Western Australia, Report* No. **9:** 1–19.

BURBIDGE, A. A. AND FULLER, P. J., 1979. Mammals of the Warburton Region, Western Australia. *Records of the Western Australian Museum* **8:** 57–73.

CHRISTENSEN, P. E. S., 1980. A sad day for native fauna. *Forest Focus* **23:** 1–12.

COMAN, B. J., 1983. Fox. Pp. 486–87 in *Complete book of Australian mammals* ed by R. Strahan. Angus & Robertson: Sydney.

DAGLEY, S., 1974. Citrate UV spectrophotometric determination. Pp. 1562–69 in *Methods of enzymatic analysis*, Vol. 3 ed by M. U. Bergmeyer, Academic Press: New York.

EVERIST, S. L., 1947. Plants poisonous to sheep. *Queensland Agricultural Journal* **64:** 13–25.

GORDON, G., 1983. Northern Brown Bandicoot. Pp. 96–97 in *Complete book of Australian mammals* ed by R. Strahan. Angus & Robertson: Sydney.

HUME, I. D., 1982. *Digestive physiology and nutrition of marsupials (Monographs on Marsupial Biology).* Cambridge University Press: Cambridge.

JESSOP, J., 1981. *Flora of central Australia.* Reed Publishing Co.: Sydney.

JOHNSON, K. A., 1983. Bilby. Pp. 107–08 in *Complete book of Australian mammals* ed by R. Strahan. Angus & Robertson: Sydney.

JONES, E., 1983. Feral Cat. P. 489 in *Complete book of Australian mammals* ed by R. Strahan. Angus & Robertson: Sydney.

JONES, F., 1925. *The mammals of South Australia* (Reprinted 1969). Government Printer: Adelaide.

KEPPEL, G., 1973. *Design and analysis: a researcher's handbook.* Prentice-Hall: Englewood Cliffs, New Jersey.

KING, D. R., OLIVER, A. J. AND MEAD, R. J., 1978. The adaption of some Western Australian mammals to food plants containing fluoroacetate. *Australian Journal of Zoology* **26:** 699–712.

KING, D. R., OLIVER, A. J. AND MEAD, R. J., 1981. *Bettongia* and fluoroacetate: a role for 1080 in fauna management. *Australian Wildlife Research* **8:** 529–36.

KING, D. R. AND SMITH, L. A., 1985. The distribution of the European Red Fox (*Vulpes vulpes*) in Western Australia. **12:** 197–205.

MAZZANTI, L., LOPEZ, M. AND BERTI, M. G., 1965. Atrofia del testicolo prodotta dal monofluoroaceto sodico nel ratto albino. *Experientia* **21:** 446–47.

McEWAN, T., 1964. Isolation and identification of the toxic principle of *Gastrolobium grandiflorum. Queensland Journal of Agricultural Science* **21:** 1–14.

McILROY, J. C., 1981. The sensitivity of Australian animals to 1080 poison. II. Marsupial and eutherian carnivores. *Australian Wildlife Research* **8:** 385–99.

McILROY, J. C., 1982. The sensitivity of Australian animals to 1080 poison. IV. Native and introduced rodents. *Australian Wildlife Research* **9:** 505–17.

McILROY, J. C., 1983. The sensitivity of Australian animals to 1080 poison: VI. Bandicoots. *Australian Wildlife Research* **10:** 507–12.

McKENZIE, N. L., 1981. Mammals of the phanerozoic West Kimberley, Western Australia: biogeography and recent changes. *Journal of Biogeography* **8:** 263–80.

McKENZIE, N. L., 1983. Golden Bandicoot. P. 98 in *Complete book of Australian mammals* ed by R. Strahan. Angus & Robertson: Sydney.

MEAD, R. J., TWIGG, L. E., KING, D. R. AND OLIVER, A. J., 1985. The tolerance to fluoroacetate of geographically separated populations of the quokka (*Setonix brachyurus*). *Australian Zoologist* **21:** 503–11.

MOLLERING, H. AND GRUBER, W., 1966. Determination of citrate with citrate lyase. *Analytical Biochemistry* **17:** 369–76.

MORRISON, J. F. AND PETERS, R. A., 1954. Biochemistry of fluoroacetate poisoning. The effect of fluorocitrate on purified aconitase. *Biochemical Journal* **58:** 473–79.

MYERS, K., 1983. Rabbit. P. 481 in *Complete book of Australian mammals* ed by R. Strahan. Angus & Robertson: Sydney.

NEWSOME, A. E., 1983. Dingo. Pp. 483–85 in *Complete book of Australian mammals* ed by R. Strahan. Angus & Robertson: Sydney.

OELRICHS, P. B. AND McEWAN, T., 1961. Isolation of the toxic principle in *Acacia georginae. Nature (Lond.)* **190:** 808–09.

OELRICHS, P. B. AND McEWAN, T., 1962. The toxic principle of *Acacia georginae. Queensland Journal of Agricultural Science* **19:** 1–16.

OLIVER, A. J. AND KING D. R., 1983. The influence of ambient temperatures on the susceptibility of mice, guinea pigs and possums to Compound 1080. *Australian Wildlife Research* **10:** 297–301.

OLIVER, A. J., KING, D. R. AND MEAD, R. J., 1977. The evolution of resistance to fluoroacetate intoxication in mammals. *Search* **8:** 130–32.

OLIVER, A. J., KING, D. R. AND MEAD, R. J., 1979. Fluoroacetate tolerance, a genetic marker in some Australian mammals. *Australian Journal of Zoology* **27:** 363–72.

PERRY, D. H., 1973. Some notes on the decline and subsequent recovery of mammal populations in the southwest. *West Australian Naturalist* **12:** 128–30.

ROBINSON, M. H. AND WHEELER, S. H., 1983. A radiotracking study of four poisoning techniques for control of the European rabbit, *Oryctolagus cuniculus* (L.). *Australian Wildlife Research* **10:** 513–20.

SHARRAD, R. D. AND KING, D. R., 1981. The geographical distribution of reptile ticks in Western Australia. *Australian Journal of Zoology* **29:** 861–73.

STODDART, D. M. AND BRAITHWAITE, R. W., 1979. A strategy for utilization of regenerating heathland habitat by the Brown Bandicoot (*Isoodon obesulus*; Marsupialia, Peramelidae). *Journal of Animal Ecology* **48:** 165–79.

THOMPSON, W. R., 1947. Use of moving averages and interpolation to estimate median-effective dose. *Bacteriological Reviews* **11:** 115–45.

TWIGG, L. E., MEAD, R. J. AND KING, D. R., 1986. Metabolism of fluoroacetate in the skink (*Tiliqua rugosa*) and the rat (*Rattus norvegicus*). *Australian Journal of Biological Sciences* **39:** 1–15.

TWIGG, L. E., KING, D. R. AND BRADLEY, A. J., 1988. The effect of sodium monofluoroacetate on plasma testosterone concentration in *Tiliqua rugosa* (Gray). *Comparative Biochemistry and Physiology* **91c:** 343–47.

WATTS, C. H. S., 1969. Distribution and habits of the Rabbit Bandicoot. *Transactions of the Royal Society of South Australia* **93:** 135–41.

WHEELER, S. H. AND HART, D. S., 1979. The toxicity of sodium monofluoroacetate to wild rabbits, *Oryctolagus cuniculus* (L.) from three sites in Western Australia. *Australian Wildlife Research* **6:** 57–62.

Toxoplasmosis in wild Eastern Barred Bandicoots, *Perameles gunnii*

DAVID L. OBENDORF[1] and BARRY L. MUNDAY[2]

Infection with *Toxoplasma gondii* is reported in free-ranging Eastern Barred Bandicoots, *Perameles gunnii* from Tasmania. Bandicoots from two localities showed central nervous system abnormalities (incoordination, blindness and staggering). Some animals were also found dead. It is proposed that animals acquired infection through the ingestion of earthworms and insects. These findings are discussed in the light of conservation efforts for the endangered populations of Eastern Barred Bandicoots in Victoria.

Key words: Toxoplasmosis, *Toxoplasma gondii*, Eastern Barred Bandicoot.

INTRODUCTION

THE protozoan parasite, *Toxoplasma gondii* is a well recognized cause of disease in Australian marsupials (Barker, Calaby and Sharman 1963; Attwood, Woolley and Rickard 1975; Arundel, Barker and Beveridge 1977). The organisms normally undergo a two-host life cycle. In Australia, the introduced cat, *Felis catus* acts as the only definitive or final host. A wide range of warm-blooded vertebrates (both mammals and birds) are capable of acting as intermediate hosts. These intermediate hosts may acquire infection by (i) the ingestion of the infective oocysts of *T. gondii* which are shed in cats' faeces, (ii) the ingestion of arthropods or annelids which contain oocysts of *T. gondii* within their digestive tracts, (iii) the ingestion of infective *T. gondii* cysts present in the tissues of mammals or birds or (iv) transplacental infection when a pregnant female becomes infected during gestation.

For herbivorous marsupials (macropods and wombats) ingestion of herbage contaminated with oocysts is the most likely source of infection. In captive populations of these marsupials, enclosures where animals graze or stored feedstuffs are likely to become contaminated (Munday 1978). For carnivorous marsupials (quolls, Kowaris, phascogales and possibly antechinus) infection may be acquired through ingesting cysts of *Toxoplasma* present in the tissues of predated mammals or birds. In captivity, dasyurids fed on diets containing fresh meat are prone to infection (Attwood *et al.* 1975).

For omnivorous and insectivorous marsupials (bandicoots, antechinus and some possums and gliders) a probable source of infection is from eating arthropods and earthworms which contain *Toxoplasma* oocysts passing through their alimentary tracts. Such transport hosts acquire oocysts through coprophagia or by ingesting soil or plant material contaminated with cat faeces (Markus 1973; Ruiz and Frenkel 1980).

The disease, toxoplasmosis, commonly develops in man, domestic or free-ranging animals as a result of lowered immunity (immunosuppressive factors), stress-related conditions environmental, physiological or population-based stressors), debilitation, senility or intercurrent disease states. *Toxoplasma* infection may remain clinically inapparent however, under the abovementioned circumstances, such latent infections may become clinically overt as the parasite undergoes rapid tissue multiplication, dissemination throughout the body and subsequent reinfection of other host cells. The inflammatory response of the host to the parasite coupled with the tissue damage caused by the parasite are responsible for the clinical disease, and the pathology seen at necropsy.

In this brief case study we report the occurrence of toxoplasmosis in free-ranging Eastern Barred Bandicoots *Perameles gunnii*. In the course of our necropsy examinations, numerous parasites were recovered and identified. We therefore present a checklist of the known protozoan, helminth and arthropod parasites reported from *P. gunnii* across their whole range (see Appendices 1 and 2).

DISEASE OCCURRENCE

During the winter of 1984 reports of a central nervous system disease in *P. gunnii* were received from two localities in southern Tasmania. There were several bandicoots observed to be affected at each locality. These animals showed signs of incoordination, apparent blindness, erratic staggery movements and unnatural daytime activity. Despite attempts by members of the public to nurse these affected

[1]Tasmanian Department of Primary Industry, Mt Pleasant Laboratories, P.O. Box 46, Kings Meadows, Tasmania, Australia 7249.
[2]Tasmanian State Institute of Technology, P.O. Box 1214, Launceston, Tasmania, Australia 7250.

Pages 193–97 *in* BANDICOOTS AND BILBIES, ed by J. H. Seebeck, P. R. Brown, R. L. Wallis and C. M. Kemper. Surrey Beatty & Sons, Sydney, 1990.

bandicoots, they eventually died. Some animals were also found dead in farm paddocks adjacent to preferred bandicoot habitat. Three carcasses were submited to Mt Pleasant Laboratories for necropsy examination. Selected tissues were fixed in neutral buffered 10% formol saline for histological examination, paraffin embedded and stained with haematoxylin and eosin. The only gross findings of note were (1) generally poor body condition, (2) absence of ingesta in the gastrointestinal tract and (3) generalized pulmonary consolidation.

Histological examination of the brain revealed numerous focal areas of non-suppurative meningo-encephalitis characterized by tissue necrosis infiltration of lymphocytes and the presence of protozoan cysts typical of *T. gondii* (Fig. 1). The lungs showed evidence of congestion and oedema with infiltrates of lymphocytes into the alveolar wall, again *Toxoplasma* cysts were detected. Severe generalized non-suppurative myocarditis was seen in all animals (Fig. 2). The extent and

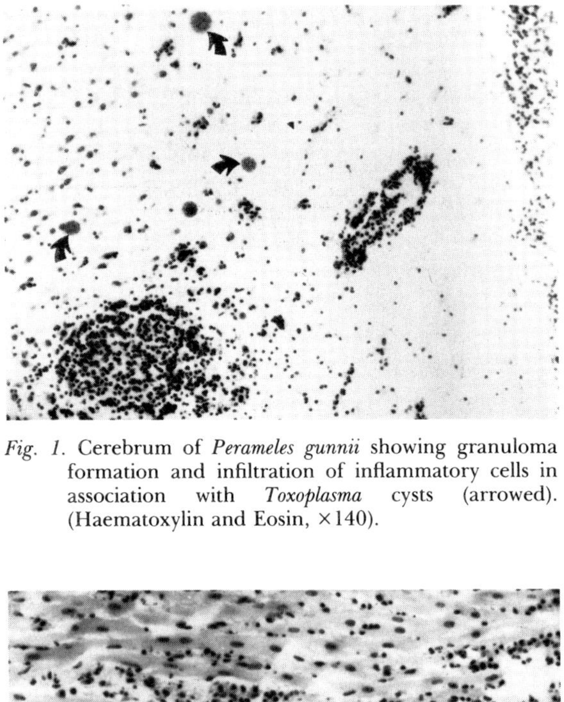

Fig. 1. Cerebrum of *Perameles gunnii* showing granuloma formation and infiltration of inflammatory cells in association with *Toxoplasma* cysts (arrowed). (Haematoxylin and Eosin, ×140).

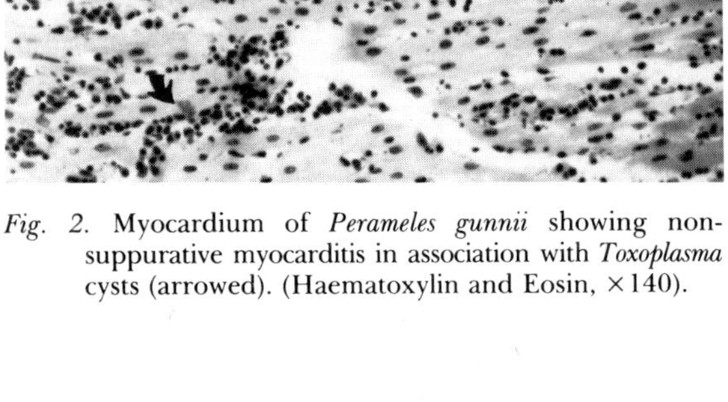

Fig. 2. Myocardium of *Perameles gunnii* showing non-suppurative myocarditis in association with *Toxoplasma* cysts (arrowed). (Haematoxylin and Eosin, ×140).

severity of the pathological findings left no doubt that *T. gondii* had caused the clinical signs and the subsequent deaths of these bandicoots.

The affected bandicoots were collected from agricultural land adjacent to native bushland. One locality was an outer suburban fringe area, the other a farming valley containing several small farms and orchards. *P. gunnii* are commonly found foraging on improved and semi-improved agricultural land as well as in native grasslands (Heinsohn 1966). Although the source of *T. gondii* infection for these bandicoots was not determined, it is highly likely that mechanical transport hosts such as earthworms, arthropods and other soil and pasture invertebrates were involved (see Table 1).

Table 1. Foods taken by *Perameles gunnii.* Source: Heinsohn (1966). +occasional, ++common, +++main food item.

Food	
Earthworms	+++
Slugs (Pulmonata)	+
Adult beetles (Coleoptera) (ground beetles, click beetles, weevils and leaf-eating beetles)	++
Larval insects (wire worms, stag beetles, curl grubs and weevil larvae)	++
Lepidoptera larvae (corbie, cockchafer, armyworms)	+++

Invertebrates which have close association with faeces, soil and herbage are known to act as transport hosts of *T. gondii* oocysts (Smith and Frenkel 1978; Ruiz and Frenkel 1980; Markus 1980). Inapparent, non-clinical *Toxoplasma* infection is commonly encountered in a wide range of Australian native rodents and marsupials (Cook and Pope 1959; Munday 1970; Attwood *et al.* 1975; Glazebrook, Campbell, Hutchinson and Stallman 1978; Obendorf and Smales 1985). The factors which led to the expression of clinical disease in these bandicoots is not known. At the farm valley, wild *Macropus rufogriseus rufogriseus* and *Thylogale billardierii* have also died of toxoplasmosis (Obendorf and Munday 1983). Nutritional and weather stresses were considered possible factors causing latent toxoplasma infection to become clinically obvious and subsequently fatal.

IMPLICATIONS OF TOXOPLASMOSIS TO BANDICOOT CONSERVATION

The disease occurrence may be of interest in relation to the conservation of endangered small populations of Australian bandicoots. The only population of *P. gunnii* known to be extant on mainland Australia exists in and around the city of Hamilton. Predation by foxes, urban dogs and cats (both domestic and feral) is a recognized cause of declining bandicoot numbers at this locality (Seebeck 1979). Habitat contraction and alteration into suburban, commercial or agricultural environments have also taken place.

Under these circumstances, the available foraging habitats for bandicoots are likely to be contaminated with *Toxoplasma* oocysts. Strees factors imposed on the bandicoot population due to a reduction in both preferred habitat and food availability may cause toxoplasmosis to occur. Additionally, bandicoots caught in order to establish captive breeding colonies may have inapparent *T. gondii* infections at the time of capture. Acclimatization to captivity can be sufficiently stressful to allow recrudescence of the latent infection. Dietary rations based on raw sheep meats have caused clinical toxoplasmosis in captive *P. gunnii* (Obendorf, unpublished findings) and in *Antechinus apicalis, Dasyuroides byrnei* and *Dasycercus cristicauda* (Attwood *et al.* 1975). Raw mutton is a common source of *Toxoplasma* cysts (Munday 1970). When necessary, meat products used in diets for captive animals should be cooked (>60°C) for at least 20 min. Freezing at −18°C in a domestic freezer for nine days followed by thawing at 4°C is reported to render *Toxoplasma*-infected goat meat safe (Dubey 1980).

REFERENCES

ARUNDEL, J. H., BARKER, I. AND BEVERIDGE, I., 1977. Diseases of marsupials. Pp. 141–54 in *The biology of marsupials* ed by B. Stonehouse and D. Gilmore. University Park Press: Baltimore.

ATTWOOD, H. D., WOOLLEY, P. A. AND RICKARD, M. D., 1975. Toxoplasmosis in dasyurid marsupials. *Journal of Wildlife Diseases* **11:** 543–51.

BARKER, S., CALABY, J. M. AND SHARMAN, G. B., 1963. Diseases in Australian laboratory marsupials. *Veterinary Bulletin* **33:** 539–44.

BARKER, I. K., MUNDAY, B. L. AND HARRIGAN, K. E., 1975. *Klossiella* spp. in the kidneys of peramelid, petaurid and macropodid marsupials. *Zeitschrift für Parasitenkunde* **46:** 35–41.

COOK, I. AND POPE, J. M., 1959. *Toxoplasma* in Queensland. III. A preliminary survey of animal hosts. *Australian Journal of Experimental Biology and Medical Science* **37:** 253–62.

DOMROW, R., 1961. New and little known Laelaptidae, Trombiculidae and Listrophoridae (Acarina) from Australasian mammals. *Proceedings of the Linnean Society of New South Wales* **86:** 60–95.

DOMROW, R., 1963. New records and species of Austromalayian laelapid mites. *Proceedings of the Linnean Society of New South Wales* **88:** 199–220.

DUBEY, J. P., 1980. Presistence of encysted *Toxoplasma gondii* in caprine livers and public health significance of Toxoplasmosis in goats. *Journal of the American Veterinary Medical Association* **177:** 1203–07.

DUNNET, G. M. AND MARDEN, D. K., 1974. A monograph of Australian fleas (Siphonaptera). *Australian Journal of Zoology.* Supplementary Series No. **30**.

GLAZEBROOK, J. S., CAMPBELL, R. S. F., HUTCHINSON, G. W. AND STALLMAN, N. D., 1978. Rodent zoonoses in north Queensland. The occurrence and distribution of zoonotic infections in north Queensland rodents. *Australian Journal of Experimental Biology and Medical Science* **56:** 147–56.

HEINSOHN, G. E., 1966. Ecology and reproduction of the Tasmanian bandicoot (*Perameles gunnii* and *Isoodon obesulus*). *University of California Publications in Zoology* **80:** 1–96.

MARKUS, M. B., 1973. Earthworms and coccidian oocysts. *Annals of Tropical Medicine and Parasitology* **68:** 247–48.

MARKUS, M. B., 1980. Flies as natural transport hosts of *Sarcocystis* and other coccidia. *Journal of Parasitology* **66:** 361–62.

MAWSON, P. H., 1960. Nematodes belonging to the Trichostrongylidae, Subuluridae, Rhabdiasidae and Trichuridae from bandicoots. *Australian Journal of Zoology* **8:** 261–84.

MUNDAY, B. L., 1970. *The epidemiology of Toxoplasmosis with particular reference to the Tasmanian environment.* M.V.Sc. Thesis, University of Melbourne.

MUNDAY, B. L., 1978. Marsupial disease. Pp. 335–86 in *Fauna — Part B. Proceedings No. 36. Postgraduate Committee in Veterinary Science.* University of Sydney.

MUNDAY, B. L., MASON, R. W., HARTLEY, W. J., PRESIDENTE, P. J. A. AND OBENDORF, D. L., 1978. *Sarcocystis* and related organisms in Australian Wildlife I. Survey findings in mammals. *Journal of Wildlife Diseases* **14:** 417–33.

OBENDORF, D. L. AND MUNDAY, B. L., 1983. Toxoplasmosis in wild Tasmanian Macropods. *Australian Veterinary Journal* **60:** 62.

OBENDORF, D. L. AND SMALES, L. R., 1985. The internal parasites and pathological findings in *Hydromys chrysogaster* (Muridae: Hydromyinae) from Tasmania. *Australian Journal of Zoology* **33:** 33–38.

ROBERTS, F. H. S., 1964. The tick fauna of Tasmania. *Records of the Queen Victoria Museum* No. **17**.

RUIZ, A. AND FRENKEL, J. K., 1980. Intermediate and transport hosts of *Toxoplasma gondii* in Costa Rica. *American Journal of Tropical Medicine and Hygiene* **29:** 1161–66.

SCHMIDT, G. D. AND EDMONDS, S. J., 1989. *Australiformis semoni* (Linstow, 1898) n. gen., n. comb. (Acanthocephala:Moniliformidae) from marsupials of Australia and New Guinea. *Journal of Parasitology* **75:** 215–17.

SEEBECK, J. H., 1979. Status of the barred bandicoot, *Perameles gunnii* in Victoria: with a note on husbandry of a captive colony. *Australian Wildlife Research* **6:** 255–64.

SMITH, D. D. AND FRENKEL, J. K., 1978. Cockroaches as vectors of *Sarcocystis muris* and of other coccidia in the laboratory. *Journal of Parasitology* **64:** 315–19.

VAUCHER, C., BEVERIDGE, I. AND SPRATT, D. M., 1984. Cestodes du genre *Hymenolepis* Weinland, 1858 *(sensu lato)* parasites de Marsupiaux australiens et description de cinq éspèces nouvelles. *Annales de la Société suisse de zoologie et du Muséum d'histoire naturelle de Genève.*

APPENDIX 1

Protozoan and helminth endoparasites reported from *Perameles gunnii.*

Endoparasite	Site	Location and record references
PROTOZOA		
APICOMPLEXA		
Eimeria sp.	small intestine	Vic[1]
Toxoplasma gondii Nicolle and Marceaux, 1909	various tissues	Vic[1],Tas[2]
Sarcocystis sp. Munday *et al.*, 1978	skeletal muscles	Tas[3]
Klossiella sp. Barker *et al.*, 1975	kidney	Vic[1,4],Tas[4]
CESTODA		
CYCLOPHYLLIDEA		
Hymenolepidae		
Hymenolepis perimelidarum Nybelin, 1917	small intestine	Tas[5]
NEMATODA		
TRICHOSTRONGYLOIDEA		
Dromaeostrongylidae		
Peramelistrongylus skedastos Mawson, 1960	stomach	Vic[1],Tas[5]
OXYUROIDEA		
Labiobulura inglisi Mawson, 1960	intestine	Tas[5]
PHYSALOPTEROIDEA		
Physaloptera sp.	stomach	Vic[7]
SEURATOIDEA		
Seuratidae		
Echinonema cinctum (Linstow, 1898); Yorke and Maplestone, 1926	small intestine	Vic[7]
Echinonema sp. n.	small intestine	Vic[7]
RHABDITOIDEA		
Strongyloididae		
Parastrongyloides australis Mawson, 1960	small intestine	Vic[1],Tas[6]
TRICHUROIDEA		
Capillariidae		
Capillaria sp. (undescribed)	tongue	Tas[1]
Capillaria sp. (undescribed)	oesophagus	Tas[1]
Capillaria sp. (undescribed)	small intestine	Vic[1],Tas[7]
METASTRONGYLOIDEA		
Angiostrongylidae		
Marsupostrongylus bronchialis Mackerras and Saunders, 1953	lung	Tas[7]
FILAROIDEA		
Onchocercidae		
Cercopithifilaria johnstoni Bain, Barker and Chabaud, 1982	subcutaneous tissues	Tas[7]
ACANTHOCEPHALA		
Polymorphidae		
Australiformis semoni (Linstow, 1898); Schmidt and Edmonds, 1989	small intestine	Tas[7,8]
TREMATODA		
DIGENEA		
Cathaemasiidae		
Mehlisia sp.	small intestine	Tas[7]

[1]Obendorf unpublished data; [2]Obendorf and Munday, this study; [3]Munday *et al.*, 1978; [4]Barker, Munday and Harrigan, 1975; [5]Vaucher *et al.*, 1984; [6]Mawson, 1960; [7]Spratt, unpublished data; [8]Schmidt and Edmonds, 1989.

APPENDIX 2

Arthropod ectoparasites (mites, ticks and fleas) reported from *Perameles gunnii.*

Ectoparasite	Location and record references
ARACHNIDA	
ACARI: MESOSTIGMATA	
Dermanyssidae	
Haemolaelaps domrowi Wormersley, 1958	Tas[1]
Haemolaelaps marsupialis (Berlese, 1910)	Vic[2],Tas[1]
Mesolaelaps antipodianus (Hirst, 1926)	Tas[3]
Mesolaelaps australiensis (Hirst, 1926)	Vic[2],Tas[1]
ACARI: IXODOIDEA	
Ixodidae	
Ixodes tasmani Neumann, 1899	Tas[4]
Ixodes cornuatus Roberts, 1960	Tas[4]
SIPHONAPTERA	
CERATOPHYLLOIDEA	
Pygiopsyllidae	
Pygiopsylla hoplia Jordan and Rothschild, 1922	Tas[5]
Pygiopsylea zethi Rothschild, 1904	Tas[5]
Stephanocircidae	
Stephanocircus dasyuri Skuse, 1893	Tas[1]

[1]Domrow, 1961; [2]Domrow, pers. comm.; [3]Domrow, 1963; [4]Roberts, 1964; [5]Dunnet and Mardon, 1974.

Reproduction in the Greater Bilby, *Macrotis lagotis* (Reid) — a comparison with other perameloids

H. E. McCRACKEN[1]

The current knowledge of the reproductive biology of the Greater Bilby, *Macrotis lagotis*, is reviewed and compared with that of other bandicoots. *Macrotis lagotis* has an XX/XY_1Y_2 chromosomal sex determination mechanism. The urogenital anatomy of female *M. lagotis* includes bilateral vaginal complexes each composed of a short median vaginal cul-de-sac, a vaginal caecum and a long lateral vagina. The breeding season of *M. lagotis* appears to vary according to the environment, but the species is capable of breeding in all months of the year in arid regions and in captivity. Females are polyoestrous with an oestrous cycle length of 20.6 ± 7.3 days (n = 14). There is a lactational anoestrus terminated close to the time of pouch exit. The gestation period of *M. lagotis* is estimated as 14 ± 1.4 days (n = 4). Litter size is usually one or two with triplets occurring rarely. Pouch young develop rapidly and leave the pouch at 80 ± 2 days (n = 6). Weaning occurs approximately two weeks later.

Despite some minor differences, the reproductive strategy of *M. lagotis* closely resembles that of other perameloids and includes some of the features which are considered unique to bandicoots amongst the Marsupialia.

Key words: *Macrotis*, reproduction, Peramelidae.

INTRODUCTION

BANDICOOTS are unique amongst marsupials in their mode of reproduction. The distinctive anatomical features of the female urogenital tract (Hill 1899), the brief gestation period (Hughes 1962a; Stodart 1966; Lyne 1974) and rapid rate of development of young relative to adult body size (Lyne 1952, 1964; Mackerras and Smith 1960; Heinsohn 1966), the chorio-allantoic placenta (Hill 1897; Padykula and Taylor 1976), and the persistence of the corpora lutea of pregnancy throughout most of lactation (Hughes 1962a; Gemmell 1979; Lyne and Hollis 1979; Hollis and Lyne 1980), all set the bandicoots apart from other marsupials.

The majority of studies revealing these unique reproductive strategies have been of the species *Isoodon obesulus, I. macrourus, Perameles nasuta* and *P. gunnii*. Very little has been published on reproduction in other bandicoot species.

This paper reviews current knowledge of reproduction in one of these lesser known species, the Greater Bilby, *Macrotis lagotis*, and compares this reproductive pattern with that of *I. obesulus, I. macrourus, P. nasuta* and *P. gunnii*. This comparison is interesting in the light of recent controversy concerning the familial classification of *Macrotis* spp. Strong similarities with the typical peramelid mode of reproduction would provide further evidence to support the inclusion of *Macrotis* spp. in the family Peramelidae (Groves and Flannery, this volume) rather than in a distinct taxon, the Thylacomyidae (Archer and Kirsch 1977).

SEX CHROMOSOMES

The karyotype of *M. lagotis* consists of 16 autosomes and an XX/XY_1Y_2 chromosome system (Martin and Hayman 1967). This is very different from the 12 autosomes and XX/XY sex determination mechanism of *Isoodon* spp. and *Perameles* spp. (Sharman 1961).

REPRODUCTIVE ANATOMY

The female reproductive tract of *M. lagotis* (see Figs 1 and 2) consists of the following parts (Hill 1899; McCracken 1983):

(a) two ovaries;

(b) two oviducts;

(c) two distinct uterine horns;

(d) two vaginal complexes each composed of three distinct elements: a short median vaginal cul-de-sac, a large thin-walled vaginal caecum and a long, narrow lateral vagina. The lateral vaginae are closely applied to the medially situated urethra throughout their length and are embedded in a mass of connective tissue which cranially also encloses the median vaginal culs-de-sac.

(e) a short urogenital sinus.

This morphology is very similar to that described for *P. nasuta, I. obesulus* (cited as *P. obesulus*) (Hill 1899) and *I. macrourus* (Tedman, this volume). *Macrotis lagotis* shares the other species' unusual combination of expansive vaginal caeca, very long lateral vaginae and urethra embedded in connective tissue, and short median vaginal cul-de-sac terminating a considerable

[1]Veterinary Department, The Royal Melbourne Zoological Gardens, P.O. Box 74, Parkville, Victoria, Australia 3052.
Pages 199–204 *in* BANDICOOTS AND BILBIES, ed by J. H. Seebeck, P. R. Brown, R. L. Wallis and C. M. Kemper. Surrey Beatty & Sons, Sydney, 1990.

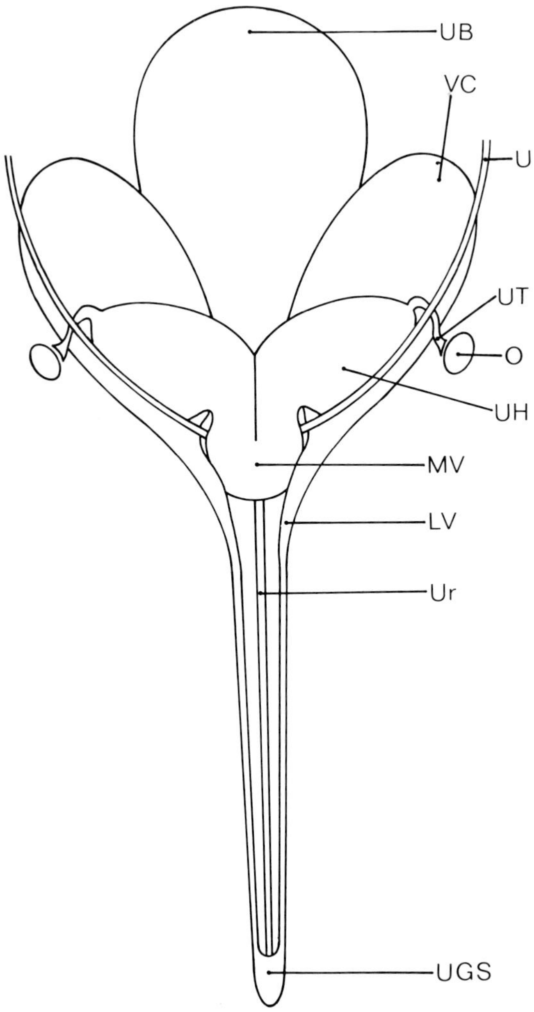

Fig. 1. The urogenital tract of female *Macrotis lagotis* — dorsal aspect. UB, urinary bladder; VC, vaginal caecum; U, ureter; UT, oviduct; O, ovary; UH, uterine horn; MV, median vaginal cul-de-sac; LV, lateral vagina; Ur, urethra; UGS, urogenital sinus.

distance from the cranial end of the short urogenital sinus. This anatomical similarity suggests that *M. lagotis* also shares with these other bandicoot species the unusual mode of parturition in which the young pass to the exterior via a pseudovaginal passage created in the connective tissue between the lateral vaginae (Hill 1899).

The tract of *M. lagotis* differs from that of *P. nasuta, I. obesulus* and *I. macrourus* in only three minor respects. The vaginal caeca of *M. lagotis* are approximately half the length of those of *P. nasuta* and *I. obesulus* (22 mm vs 45 mm) (McCracken 1983; Hill 1899). In the former species, the caeca are shorter than the directly ventrally situated urinary bladder (length=30 mm) (McCracken 1983), whereas in the latter two, the caeca greatly exceed the bladder in length (dimensions not given) (Hill 1899). The morphology of the vaginal caeca also differs in that those of *M. lagotis* are completely separate from each other (McCracken 1983), unlike those of *P. nasuta, I. obesulus* and *I. macrourus* which share a common median partition wall and externally have an undivided bi-lobed appearance (Hill 1899; Tedman, this volume). The third difference is that the median vaginal culs-de-sac of both nulliparous and parous *M. lagotis* are separated by a median septum (McCracken 1983), whereas in the other three species, the septum is lacking in parous animals (Hill 1899; Tedman, this volume).

The pouch of *M. lagotis,* like that of *P. nasuta, P. gunnii, I. obesulus* and *I. macrourus,* opens ventrocaudally and houses eight teats arranged in two crescentic rows (Jones 1924; Stodart 1983; Seebeck 1983; Braithwaite 1983; Gordon 1983).

The male reproductive tract of *M. lagotis* has not been described.

BREEDING SEASONS

Macrotis lagotis is physiologically capable of breeding at all times of the year. In the field, the breeding season appears to vary according to environment. In the temperate zones of South Australia early this century, *M. lagotis* was observed to have a distinct breeding season from about March to May (Jones 1924). In arid regions however, breeding may occur at any

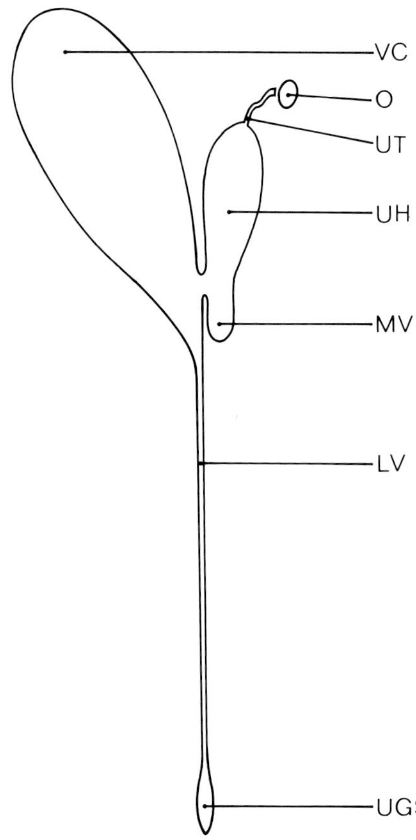

Fig. 2. The urogenital tract of female *Macrotis lagotis* — lateral aspect, showing the relationships of the components of the vaginal complex. (Symbols as in Fig. 1.)

time of the year (Jones 1924; McCracken 1983), perhaps, as suggested by Jones (1924), dependent on rainfall and food availability.

In captivity and fed *ad libitum, M. lagotis* in a colony at the Arid Zone Research Institute, Alice Springs, NT, bred throughout the year (70 litters over 3.5 years) with no significant peaks in reproductive activity (McCracken 1983). These observations differ markedly from those of Hulbert (1972, 1982) who reported the birth of only four litters in a captive colony at the University of New South Wales, all born in the months March to May. The reason for this difference is unknown.

The breeding seasons of other bandicoot species are highly variable. In *P. nasuta* and *I. macrourus,* breeding may occur throughout the year but there is decreased breeding activity from late autumn to mid-winter (Lyne 1979; Mackerras and Smith 1960; Gordon 1983; Gemmell 1982). *Perameles gunnii* and *I. obesulus* are more seasonal, with a non-breeding period of variable length in autumn and early winter (Heinsohn 1966; Seebeck 1983; P. R. Brown, pers. comm. 1987; Braithwaite 1983). As with *M. lagotis* there are environmental variations in breeding seasons within species but it appears that the time of breeding in *Perameles* and *Isoodon* spp. is more strongly correlated with the rate of change of minimum temperature than with rainfall (Barnes and Gemmell 1984).

OESTROUS CYCLE

Macrotis lagotis is polyoestrous with an oestrous cycle length of 12 to 37 days ($\bar{x} \pm$ s.e.: 20.6 ± 7.3 days) (n=14). Stages of the cycle may be detected by the technique of smearing the urogenital sinus. Oestrus, the time of the female's peak sexual attraction, is characterized by a distinctive urogenital smear containing greater than 80% cornified epithelial cells and very few polymorphonuclear leucocytes. The duration of oestrus is 2–11 days ($\bar{x} \pm$ s.e.: 4.3 ± 2.1 days) (n=30). The end of oestrus is detected by a markedly increased number of leucocytes and a decreased number of cornified cells in the urogenitial smear. Lactating *M. lagotis* are acyclic for the major part of lactation. Cycling recommences towards the end of lactation, immediately before, or soon after, the pouch exit of the suckling young (McCracken 1986).

This cyclic pattern closely resembles that described for *I. macrourus* and *P. nasuta* (Hughes 1962a; Lyne 1976). The cycle length and variability both within and between individuals is very similar for all three species (*I. macrourus*: 9–34 days (Lyne 1976), *P. nasuta*: 10–34 days (Hughes 1962a; Lyne 1976)). With all three species, difficulty has been experienced in obtaining normal oestrous cycle lengths by vaginal smearing, with periods of extended anoestrus or ultrashort cycles having been detected (Hughes 1962b; Lyne 1976; McCracken 1986). It may be that intensive handling is highly stressful for bandicoots, causing disturbances to ovarian function.

Macrotis lagotis further resembles the other bandicoots in its display of a lactational anoestrus terminated close to the time of the young's departure from the pouch. It is not yet known for *M. lagotis* whether, as in *P. nasuta* and *I. macrourus,* this phenomenon is due to the persistence of a functional corpus luteum of pregnancy into lactation. In *P. nasuta* and *I. macrourus,* the corpus luteum regresses at a stage approximately two-thirds through lactation, and the animal subsequently recommences cycling (Hughes 1962a; Lyne and Hollis 1979; Gemmell 1979). This is an unusual feature in marsupials, as in most species the corpus luteum begins to regress before parturition (Lyne and Hollis 1979). The phenomenon results in ovulation towards the end of lactation and, therefore, bandicoots can produce a new litter of young as soon as the previous litter is weaned, thus accelerating the potential rate of reproduction (Hughes 1962a; Stodart 1966; Lyne and Hollis 1979; McCracken 1983).

The major difference between the cycles of *M. lagotis* and those of *P. nasuta* and *I. macrourus* is a shorter period of oestrus, lasting only one day, in the latter two species (Lyne 1976).

GESTATION

The gestation period of *M. lagotis* has been estimated as 14 days ($\bar{x} \pm$ s.e.: 14.0 ± 1.4 days) (n=4). This value was determined by measurement of the interval between parturition and the first day of the preceeding oestrus (McCracken 1986). As a successful mating has not been observed in *M. lagotis,* gestation length has not yet been accurately determined. This estimated value is not very different from that of 12.5 days in *P. nasuta* and *I. macrourus* (Hughes 1962b; Stodart 1966; Lyne 1974).

LITTER SIZE AND SEX RATIO

The litter size of *M. lagotis* ranges from one to three (Ride 1970) but triplets occur rarely. All litters reported to be born in captivity (n=77) have had a size of one or two (Watts 1970; Hulbert 1982; McCracken 1983). In the captive colony at the Arid Zone Research Institute, Alice Springs, singletons (n=38) and twins (n=34) occurred at a ratio of 1.1:1 (McCracken 1983).

The sex ratio of pouch young was found to be 1:1 in a survey of the animals born in the Alice Springs captive colony. Of 119 animals born, 48 were male, 49 were female, and 22 were not sexed. This ratio was not found to change significantly with development (McCracken, unpubl.). These results differ from observations of Hulbert (1972, 1982) who reported the captive birth of only seven animals, all male. The bias in the latter result may have been due to the small sample size.

The other bandicoot species are more prolific breeders. *Perameles nasuta, P. gunnii* and *I. obesulus* all bear litters of similar size, ranging from one to five with a mean of 2.2–2.4 (Lyne 1964; Heinsohn 1966; Stodart 1966) The litters of *I. macrourus* are larger, ranging from one to seven, with a mean of four, but in this species the litters generally decrease in size during lactation, to a mean of 2.3 at weaning (Mackerras and Smith 1960; Gemmell, Walker, Johnson and Cepon 1984).

The sex ratios of pouch young of *P. nasuta, P. gunnii, I. obesulus* and *I. macrourus* do not differ from that of *M. lagotis*, all being approximately 1:1 (Lyne 1964; Heinsohn 1966; Gemmell *et al.* 1984).

GROWTH AND DEVELOPMENT

Growth of *M. lagotis* young has been monitored from birth to independence by frequent measurements and observations (Watts 1970; Hulbert 1972; McCracken 1983). Of all body measurements, head length was found to be the most reliable index of pouch young age, being easily taken and showing least variation. Growth of the head was best described by the power curve:

$$X = \left(\frac{Y}{0.18}\right)^{0.78} - 14,$$

where X=age (days) and Y=head length (mm) (see Fig. 3) (McCracken 1983). The rate of development of *M. lagotis* pouch young has been studied by both Hulbert (1972) and McCracken (1983). The findings of only the latter study are presented in Table 1. There are some discrepancies between the results of the two studies with Hulbert (1972) describing a more rapid rate of development with eyes fully open at days 43–64 and initial pouch exit at days 67–68. These discrepancies are likely to have occurred due to differences in experimental method. In the 1983 study, birth dates and developmental stages were determined by pouch checks made daily rather than every six to eight days as in the 1972 study. The results of the later study, therefore, are presumed to be of greater accuracy.

Initial pouch exit occurs at 80 ± 2 days ($\bar{x}$ ± s.e.) (n = 6), followed by a one to two-day period of occasional returns to the pouch (McCracken 1983). The young remain down the burrow for approximately 14 days, dependent on suckling from their mother who makes frequent returns to them during her nocturnal foraging activities

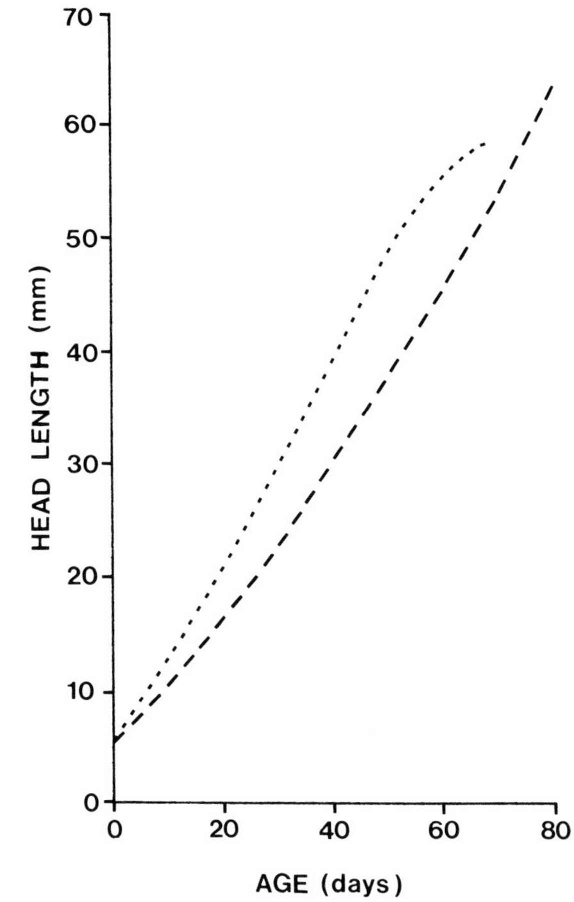

Fig. 3. Growth curves of *Macrotis lagotis* [– – –, from McCracken (1983)] and *Perameles nasuta* [- - -, from Lyne (1964)] pouch young.

Table 1. Comparison of development rates of *Macrotis lagotis* and *Perameles nasuta.*

Developmental characteristic	*M. lagotis*[1] (days)	*P. nasuta*[2] (days)
EARS — unfused, turned backwards	17–21	12
EMERGENCE OF VIBRISSAE	31–41	20–30
PIGMENTATION — initial	37–45	—
— complete	50–58	—
MOUTH — fully open	59–65	45
EYES — fully open	63–66	40–53
PELAGE — body completely covered with short fur	66–69	44–48
POUCH EXIT — initial	78–82	50–54[3]
WEANED	94–160	62–63[3]
SEXUAL MATURITY (female)	150	120

[1]McCracken (1983); [2] Lyne (1964); [3]Stodart (1966).

(Johnson and Johnson 1983). After emergence from the burrow, the young commence eating solids but may continue to share their mother's burrow for a variable period, possibly suckling during the day (McCracken 1983; Johnson and Johnson 1983). If however, the female gives birth to a new litter, which may occur within two weeks of pouch exit, suckling of the previous litter appears to cease immediately (McCracken 1983).

The young female *M. lagotis* reaches sexual maturity at approximately five months of age (McCracken 1983). The age of onset of maturity in the male is unknown.

Studies on growth and development of *P. nasuta, P. gunnii, I. obesulus* and *I. macrourus* have revealed that these animals develop more rapidly and become independent sooner than any other marsupial studied (Lyne 1964; Stodart 1966; Heinsohn 1966; Mackerras and Smith 1960; Gordon 1974; Russell 1982). The rate of development of *P. nasuta,* which is very similar to that of other *Perameles* and *Isoodon* species studied, is compared with that of *M. lagotis* in Table 1 and Figure 3. Although growth in *M. lagotis* is significantly slower than that of the other bandicoots, it is more rapid than that of a number of other marsupials of similar adult size (*Bettongia lesueur, Dasyurus viverrinus, Didelphis virginiana, Petauroides volans, Trichosurus vulpecula* (Russell 1982)). *Macrotis lagotis*, therefore, shares with other bandicoots their characteristic accelerated life history pattern.

CONCLUSION

There is still much to learn about reproduction in *M. lagotis* but the information presented in this review provides an interesting comparison with our more extensive knowledge of reproduction in other bandicoots. Despite several differences, the reproductive biology of *M. lagotis* shows strong similarities to the pattern observed in other bandicoots. These similarities add to those observed by Groves and Flannery (this volume) who proposed that, despite karyotypic, serological and morphological differences (Martin and Hayman 1967; Archer and Kirsch 1977), the genus *Macrotis* should be classified with the other Australian bandicoots in the Peramelidae, rather than in a distinct family, the Thylacomyidae.

REFERENCES

ARCHER, M. AND KIRSCH, J. A. W., 1977. The case for the Thylacomyidae and Myrmecobiidae, Gill, 1872, or why are marsupial families so extended? *Proceedings of the Linnean Society of New South Wales* **102:** 18–25.

BARNES, A. AND GEMMELL, R. T., 1984. Correlations between breeding activity in the marsupial bandicoots and some environmental variables. *Australian Journal of Zoology* **32:** 219–26.

BRAITHWAITE, R. W., 1983. Southern Brown Bandicoot, *Isoodon obesulus.* P. 94 in *The Australian Museum complete book of Australian mammals* ed by R. Strahan. Angus & Robertson: Sydney.

GEMMELL, R. T., 1979. The fine structure of the luteal cells in relation to the concentration of progesterone in the plasma of the lactating bandicoot, *Isoodon macrourus* (Marsupialia: Peramelidae). *Australian Journal of Zoology* **27:** 501–10.

GEMMELL, R. T., 1982. Breeding bandicoots in Brisbane (*Isoodon macrourus;* Marsupialia, Peramelidae). *Australian Mammalogy* **5:** 187–93.

GEMMELL, R. T., WALKER, M. T., JOHNSON, G. AND CEPON, G., 1984. The number of young present in sequential litters of the marsupial bandicoot, *Isoodon macrourus* in captivity. *Australian Journal of Zoology* **32:** 623–29.

GORDON, G., 1974. Movements and activity of the short-nosed bandicoot, *Isoodon macrourus. Mammalia* **38:** 405–32.

GORDON, G., 1983. Northern Brown Bandicoot, *Isoodon macrourus.* Pp. 96–97 in *The Australian Museum complete book of Australian mammals* ed by R. Strahan. Angus & Robertson: Sydney.

HEINSOHN, G. E., 1966. Ecology and reproduction of the Tasmanian bandicoots (*Perameles gunnii* and *Isoodon obesulus*). *University of California Publications in Zoology* **80:** 1–107.

HILL, J. P., 1897. The placentation of *Perameles. Queensland Journal of Microscopic Science* **40:** 385–442.

HILL, J. P., 1899. Contributions to the morphology and development of the female urogenital organs in the Marsupialia. I. On the female urogenital organs of *Perameles,* with an account of the phenomenon at parturition. *Proceedings of the Linnean Society of New South Wales* **24:** 42–82.

HOLLIS, D. E. AND LYNE, A. G., 1980. Ultrastructure of luteal cells in fully formed and regressing corpora lutea during pregnancy and lactation in the marsupials *Isoodon macrourus* and *Perameles nasuta. Australian Journal of Zoology* **28:** 195–211.

HUGHES, R. L., 1962a. Role of the corpus luteum in marsupial reproduction. *Nature* **194:** 890–91.

HUGHES, R. L., 1962b. Reproduction in the macropod marsupial *Potorous tridactylus* (Kerr). *Australian Journal of Zoology* **10:** 193–224.

HULBERT, A. J., 1972. Growth and development of pouch young in the Rabbit-eared Bandicoot, *Macrotis lagotis,* Peramelidae. *Australian Mammalogy* **1:** 38–39.

HULBERT, A. J., 1982. Notes on the management of a captive breeding colony of the Greater Bilby *Macrotis lagotis.* P. 53 in *The management of Australian mammals in captivity* ed by D. Evans. Zoological Board of Victoria: Melbourne.

JOHNSON, C. AND JOHNSON, K., 1983. Behaviour of the Bilby, *Macrotis lagotis* (Reid) in captivity. *Australian Wildlife Research* **10:** 77–87.

JONES, F. W., 1924. *The mammals of South Australia.* Part III. *The Bandicoots and Herbivorous Marsupials.* Government Printer: Adelaide.

LYNE, A. G., 1952. Notes on the external characteristics of the pouch young of four species of bandicoot. *Proceedings of the Zoological Society of London* **122:** 625–49.

LYNE, A. G., 1964. Observations on the breeding and growth of the marsupial *Perameles nasuta* Geoffroy with notes on other bandicoots. *Australian Journal of Zoology* **12:** 322–39.

LYNE, A. G., 1974. Gestation period and birth in the marsupial *Isoodon macrourus. Australian Journal of Zoology* **22:** 303-09.

LYNE, A. G., 1976. Observations on oestrus and the oestrous cycle in the marsupials *Isoodon macrourus* and *Perameles nasuta. Australian Journal of Zoology* **24:** 513–21.

LYNE, A. G., 1979. The bandicoots *Isoodon macrourus* and *Perameles nasuta:* their maintenance and breeding in captivity. Pp. 47–52 in *The management of Australian mammals in captivity* ed by D. D. Evans. Zoological Board of Victoria: Melbourne.

LYNE, A. G. AND HOLLIS, D. E., 1979. Observations on the corpus luteum during pregnancy and lactation in the marsupials *Isoodon macrourus* and *Perameles nasuta*. *Australian Journal of Zoology* **27:** 881–99.

McCRACKEN, H. E., 1983. *Reproduction in the female Bilby,* Macrotis lagotis. B.Sc.(Vet) (Honours) Thesis: University of Sydney, New South Wales.

McCRACKEN, H. E., 1986. Observations on the oestrous cycle and gestation period of the Greater Bilby, *Macrotis lagotis* (Reid) (Marsupialia: Thylacomyidae). *Australian Mammalogy* **9:** 5–16.

MACKERRAS, M. J. AND SMITH, R. H., 1960. Breeding the short-nosed marsupial bandicoot, *Isoodon macrourus* (Gould) in captivity. *Australian Journal of Zoology* **8:** 371–82.

MARTIN, P. G. AND HAYMAN, D. L., 1967. Quantitative comparisons between the karyotypes of Australian marsupials from three different superfamilies. *Chromosoma Berl.* **20:** 290–310.

PADYKULA, H. A. AND TAYLOR, M. J., 1977. Uniqueness of the bandicoot chorioallantoic placenta (Marsupialia, Peramelidae). Cytological and evolutionary interpretations. P. 52 in *Reproduction and evolution, Proceedings of the 4th Symposium on Comparative Biology of Reproduction* ed by J. H. Calaby and C. H. Tyndale-Biscoe. Australian Academy of Science: Canberra.

RIDE, W. D. L., 1970. *A guide to the native mammals of Australia.* Oxford University Press: London.

RUSSELL, E. M., 1982. Patterns of parental care and parental investment in marsupials. *Biological Review* **57:** 423–86.

SEEBECK, J. H., 1983. Eastern Barred Bandicoot, *Perameles gunnii.* P. 100 in *The Australian Museum complete book of Australian mammals* ed by R. Strahan. Angus & Robertson: Sydney.

SHARMAN, G. B., 1961. The mitotic chromosomes of marsupials and their bearing on taxonomy and phylogeny. *Australian Journal of Zoology* **9:** 38–60.

STODART, E., 1966. Management and behaviour of breeding groups of the marsupial *Perameles nasuta* Geoffroy in captivity. *Australian Journal of Zoology* **14:** 611–23.

STODART, E., 1983. Long-nosed Bandicoot, *Perameles nasuta.* P. 99 in *The Australian Museum complete book of Australian mammals* ed by R. Strahan. Angus & Robertson: Sydney.

WATTS, C. H. S., 1970. The Bilby or "Rabbit Bandicoot" *(Macrotis lagotis). Royal Zoological Society of South Australia* 92nd Annual Report 25–26.

The initiation of the breeding season of the Northern Brown Bandicoot *Isoodon macrourus* in captivity

R. T. GEMMELL[1]

Although *Isoodon macrourus* show evidence of annual cycles in their reproductive activity, the factor(s) which initiate breeding activity are still not known. Three aspects of the start of the breeding season were examined: the annual variation in the commencement of breeding, the hormonal events preceding the first births of the season and the role of the male in stimulating ovarian activity.

The results indicate that the commencement of the breeding season is not triggered by any single environmental factor such as daylength, rainfall or minimum temperature. In four of 11 females, rises in plasma progesterone presumably associated with an oestrous cycle preceded the first pregnancy while pregnancy occurred with the first ovulation of the breeding season in the remaining seven females. Oestrous cycles were observed in females housed in enclosures without males, therefore it is unlikely that stimuli from males are involved in the initiation of breeding activity.

Photoperiod is suggested as the basic cue for the initiation of breeding activity in *I. macrourus*. However, closer to the equator other environmental factors such as temperature and rainfall may play an important role in altering the start of the breeding season.

Key words: Breeding season, *Isoodon macrourus,* photoperiod.

INTRODUCTION

In contrast to the growing number of publications concerned with the factors governing seasonal breeding in eutherian mammals (see review Clarke 1981) there is little information about the cues utilized by the marsupials. Apart from the data for *Macropus rufus* (Newsome 1965) and *M. eugenii* (Sadlier and Tyndale-Biscoe 1977; Tyndale-Biscoe 1980) there is scant information concerning factors governing seasonal breeding in marsupials in the wild. Newsome (1965) observed that reproduction in *M. rufus* was opportunistic and that all females bred when green herbage was abundant after good rains. In *M. eugenii,* a wallaby which exhibits embryonic diapause, reactivation of development is related to the time of full moon relative to the summer solstice (Sadlier and Tyndale-Biscoe 1977). Recent studies would suggest that melatonin secretion controlled by photoperiod influences the reproductive activity of *M. eugenii* (McConnell and Tyndale-Biscoe 1985). In other macropodids reproduction occurs continually throughout the year, for example *M. giganteus* (Poole 1975), *M. agilis* (Merchant 1976), *M. rufogriseus* (Merchant and Calaby 1981), *Bettongia lesueur* (Tyndale-Biscoe 1968), *Setonix brachyurus* (Shield 1964) and *Potorous tridactylus* (Hughes 1962). Although all these marsupials are reported to be continuous breeders, in some there is a decrease in the breeding rate during the year. In both *P. tridactylus* and *B. lesueur* there is a decrease in the number of births during November and December (Hughes 1962; Tyndale-Biscoe 1968). Rose and McCartney (1982) reported that *Thylogale billardierii* was a seasonal breeder in Tasmania with most births in the months April, May and June. These authors noticed that in captivity, females came into oestrus throughout the year and they suggested that nutrition may be involved with fertility in the wild. They also noted that seasonal breeding was characteristic of all the more specialized feeding Tasmanian kangaroos, *M. giganteus, M. rufogriseus* and *T. billardierii* whereas the less specialized feeders *P. tridactylus* and *Bettongia gaimardi* bred throughout the year in the wild. With *S. brachyurus,* wallabies on the mainland bred continuously throughout the year, whereas those on Rottnest Island had a restricted breeding period of six months. Shield (1964) surmised that since both populations were in the same latitude and less than 80 km apart, it was unlikely that climatic conditions were responsible for the different breeding patterns. Seasonal breeding on Rottnest Island was thought to be due to nutrition.

Whereas nutrition has been suggested to be important in controlling breeding in the large marsupials, photoperiod is thought to be more important in the small species, for example

[1]Department of Anatomy, University of Queensland, St Lucia, Queensland, Australia 4067.
Pages 205–12 *in* BANDICOOTS AND BILBIES, ed by J. H. Seebeck, P. R. Brown, R. L. Wallis and C. M. Kemper. Surrey Beatty & Sons, Sydney, 1990.

Sminthopsis larapinta (= *macroura*) (Godfrey 1969), *S. crassicaudata* (Smith, Bennett and Chesson 1978) and the genus *Antechinus* (Calaby and Taylor 1981; Dickman 1985; Selwood 1985). However, in other marsupials such as the ubiquitous possum, *Trichosurus vulpecula*, breeding is seasonal but no suggestions of cues or mechanisms which may control breeding have been advanced (Gemmell, Hughes and Jenkin 1987).

Reproductive strategies in the marsupials, like all mammals, maximize the probability that young gain independence in the most favourable environmental conditions. The period from conception to weaning is short, approximately two months in the bandicoot *Isoodon macrourus* (Lyne 1964; Gemmell 1982; Gemmell *et al.* 1984). This species reproduces throughout the year with a substantial decrease in breeding in the months March to June. Bandicoots born in these months leave the pouch during the months May to August, the coldest period of the year and the least conducive for survival. Therefore the bandicoot weans her young mainly in the summer, however if environmental conditions are not harsh some litters are also produced in the winter.

Although reproductive activity is seasonal in bandicoots, the initiation and termination of the breeding season appears to vary annually amongst the various species (Heinsohn 1966; Gordon 1971; Stoddart and Braithwaite 1979; Gemmell 1982; Barnes and Gemmell 1984). Heinsohn (1966) reported that the build-up of breeding activity in *Perameles gunnii* was gradual with about 43% of females bearing pouch young by the end of June, about 80% by the end of July and 100% by the end of August. A similar pattern of gradual build-up in the number of *I. macrourus* breeding was observed by Gordon (1971). Gordon also noted that the start of the breeding season varied from year to year, from mid July (1968) to early September (1970), a period of 55 days, and a similar variation occurred in the time of termination of lactation (late February to late April). Similar variations in the commencement of the breeding season of *I. obesulus* from 1974 to 1977 were reported by Stoddart and Braithwaite (1979). Females were first observed to carry pouch young during August, except for 1977 when the first pouch young were seen in June.

The percentage of lactating females in wild populations, observed during each month of the year (Heinsohn 1966; Gordon 1971; Stoddart and Braithwaite 1979; Gemmell 1982) has been correlated with the environmental variables of daylength, rainfall and temperature, and derivatives of these variables. When all four studies were considered (Barnes and Gemmell 1984), the strongest correlation was between the proportion of lactating females in the populations and the rate of change of minimum temperature. An analysis of a captive population of *I. macrourus* in Queensland showed a similar correlation between births and rate of change of minimum temperature (Barnes and Gemmell 1984).

Although the number of births of *I. macrourus* in captivity correlates well with the rate of change of minimum temperature, it is still not known whether temperature or other environmental cues are used to initiate and terminate breeding in the colony. The aim of this study was to examine the events which may initiate breeding in the bandicoot *I. macrourus.*

MATERIALS AND METHODS

The occurrence of births was monitored in a captive breeding colony of *I. macrourus* from 1980 to 1984. The housing, feeding and methods of recapture have been described previously (Gemmell 1982; Gemmell *et al.* 1984). The number of births and the number of adult females capable of birth (at risk), were noted for each month.

To examine the hormonal events preceding the first births of the breeding season, blood samples were obtained from five parous and six nulliparous females housed with male bandicoots and from two parous and one nulliparous female in an enclosure with no male animals. This female-only enclosure was located 1 km away from the enclosures with males. Blood samples were obtained weekly from a total of 14 bandicoots. Five animals were sampled from November 1983, one animal from February 1984, four animals from April 1984 and four from July 1984 through to the start of the next breeding season. Sampling was terminated during August to October, 1984 as soon as a birth was observed. Samples were collected by cardiac puncture from bandicoots lightly anaesthetized with a halothane (Fluothane — I.C.I.) and oxygen mixture. The blood was placed into

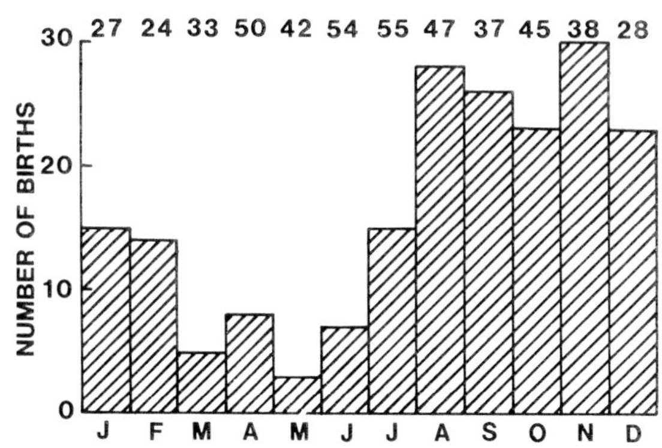

Fig. 1. Number of *Isoodon macrourus* which gave birth within the enclosures from January 1980 to December 1984. Number of animals at risk or capable of giving birth is indicated for each month.

heparinized tubes, centrifuged within 15 min of collection and the plasma frozen until assayed.

Plasma progesterone concentrations were determined by radioimmunoassay using the method described previously for the bandicoot (Gemmell 1979), using sheep anti-progesterone-11-hemisuccinate-bovine serum albumin (anti-serum No. 334) kindly donated by Dr R. I. Cox, CSIRO, Prospect, New South Wales, Australia. The anti-serum is highly specific for progesterone, the only significant cross-reaction being that with 11β-hydroxy-progesterone (11.9%). The limit of the sensitivity of the assay was 25 pg/tube and the intra- and inter-assay coefficients of variation were 13.4% (n = 5) and 14% (n = 10) respectively. The assay buffer blank was less than 0.1 ng/ml and the efficiency of extraction was 87.0% (n = 10).

The contents and the condition of the pouch were also examined weekly. Rainfall, maximum and minimum temperature were monitored daily from January 1982 to December 1984.

RESULTS

Births monitored from January 1980 to December 1984 totalled 197 (Figs 1 and 2). Although the composite histogram of the monthly frequency of births indicated that the breeding rate was depressed from March to June, with gradual increase in births during July and August, examination of the annual record revealed a variation in the start of the breeding season (Fig. 3). The first births, after at least a one-month period of no births, occurred on 1 August 1980, 7 July 1981, 27 May 1983 and 28 July 1984. Births occurred throughout 1982. To further examine the effect of minimum temperature and rainfall on the frequency of births, weekly means of these parameters were examined during 1982, 1983 and 1984 (Figs 4 and 5).

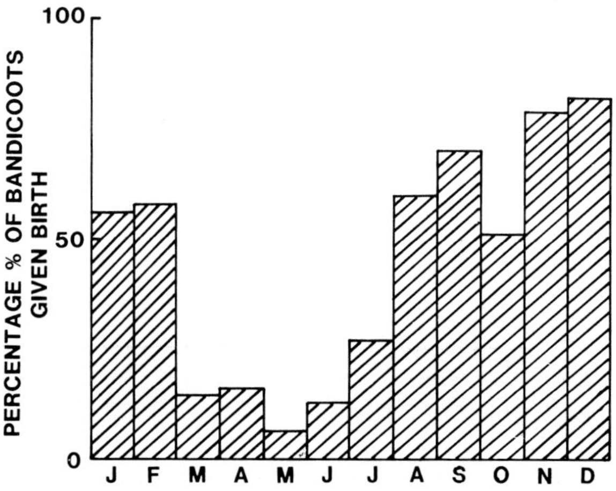

Fig. 2. Percentage of *Isoodon macrourus* giving birth from January 1980 to December 1984.

During 1984, dates of first litters ranged from 28 July to 20 October, a span of three months. There was no significant difference between birth dates of formerly parous and formerly

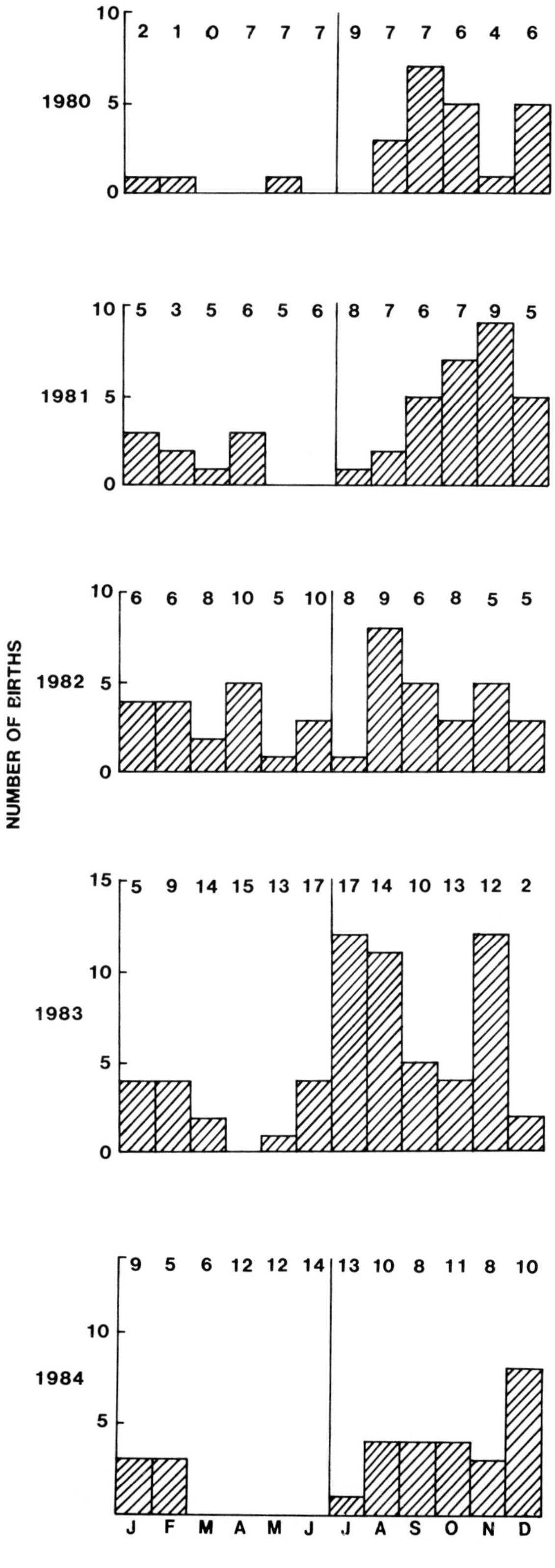

Fig. 3. Number of *Isoodon macrourus* which gave birth within enclosures each year from 1980 to 1984. Number of animals at risk is indicated for each month.

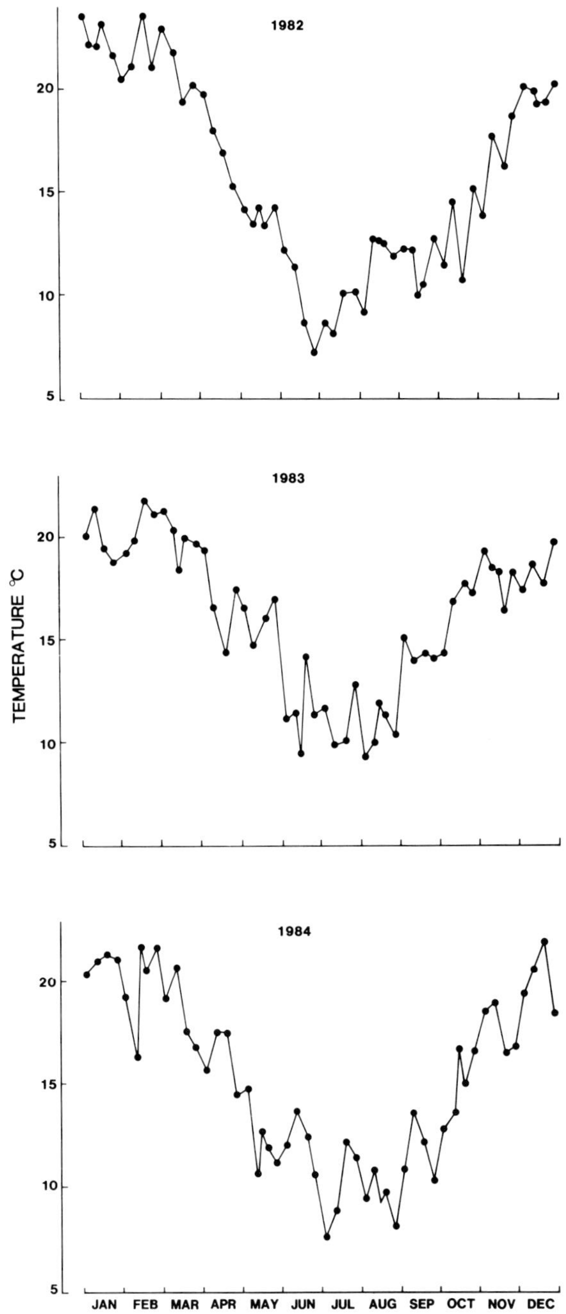

Fig. 4. Mean weekly minimum temperatures for the years 1982 to 1984.

nulliparous females. A similar range of dates for the first rise in plasma progesterone was obtained for parous females (17 July to 30 August, n=5) and for nulliparous females (31 July to 25 September, n=6) (Table 1). Peaks of plasma progesterone, not linked to pregnancy were observed in bandicoot 2 during July and August with birth occurring on 13 September; however, bandicoot 1 gave birth to a litter on 28 July without the occurrence of a preceding oestrous cycle (Fig. 6). In bandicoot 3 plasma progesterone was elevated in July and August and birth was observed on 30 September. Bandicoot 4 gave birth to a litter on 8 September without any evidence of a preceding oestrous cycle and the fifth parous bandicoot gave birth after a presumed oestrous cycle. Most nulliparous bandicoots gave birth to their first litter without showing evidence of a previous oestrous cycle (Fig. 7) the exception being bandicoot 8. In three of the parous bandicoots (2, 3 and 5) and one of the nulliparous bandicoots (8), the plasma progesterone profile indicated that the animals probably entered at least one oestrous

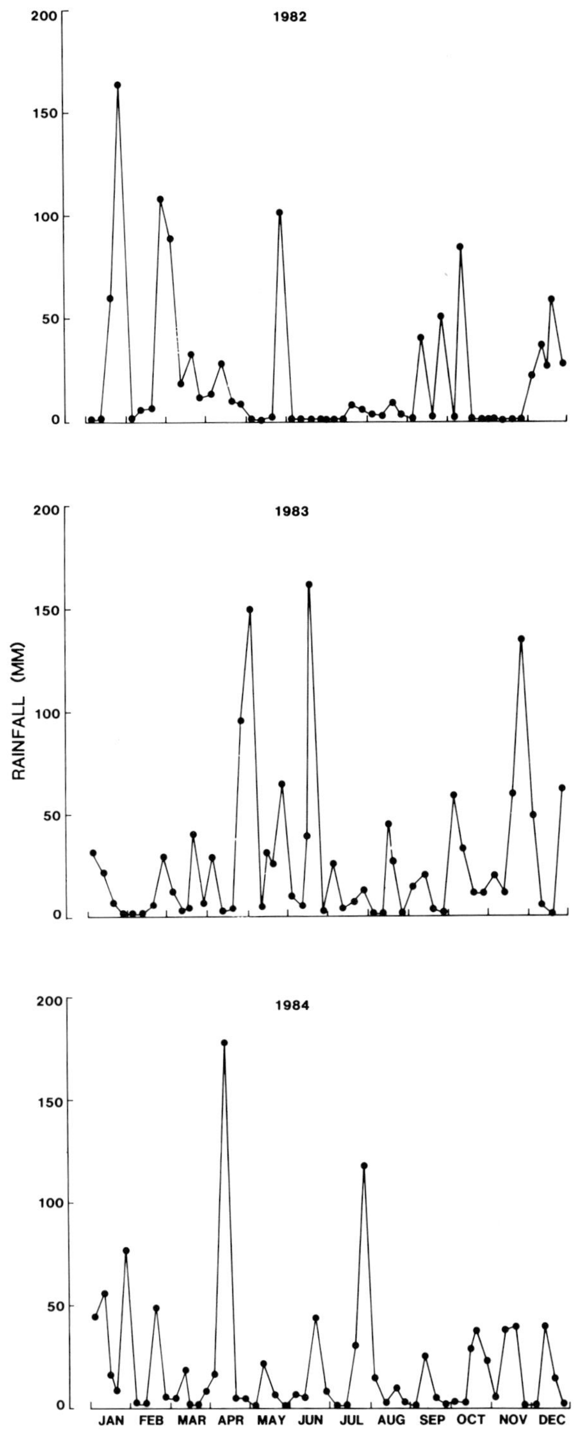

Fig. 5. Total weekly rainfalls for the years 1982 to 1984.

cycle prior to becoming pregnant. However, in two of the parous animals (1 and 4) and five of the nulliparous (6, 7, 9, 10 and 11), pregnancy followed first ovulation.

Evidence of oestrous cycles was observed in two of the three female bandicoots housed in an enclosure without males (Fig. 8). These females were housed in a yard isolated from the presence of males.

Table 1. The dates in 1984 of the first observed rise in progesterone and birth of the first litter in parous and nulliparous female *Isoodon macrourus* with and without males.

Experiment	First rise in progesterone	Day of birth of first litter
With males		
Parous females		
1	17 Jul.	28 Jul.
2	17 Jul.	13 Sep.
3	24 Jul.	30 Sep.
4	30 Aug.	8 Sep.
5	30 Aug.	20 Oct.
Nulliparous females		
6	31 Jul.	9 Aug.
7	7 Aug.	13 Aug.
8	before 14 Aug.	23 Sep.
9	21 Aug.	23 Aug.
10	11 Sep.	26 Sep.
11	25 Sep.	26 Sep.
Without males		
Parous females		
12	21 Aug.	—
13	—	—
Nulliparous female		
14	7 Aug.	—

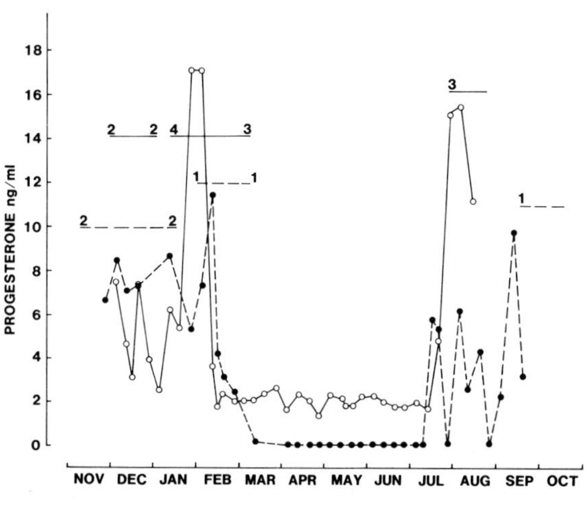

Fig. 6. Plasma progesterone profiles of parous *Isoodon macrourus* 1 (○—○) and 2 (●--●) prior to the birth of the first litter in 1985. Bars designate a period of lactation and the numbers above the bar the number of young present in the pouch.

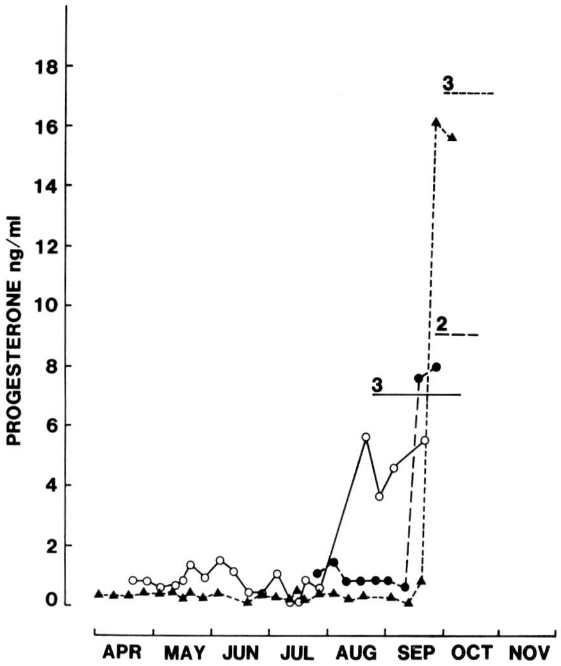

Fig. 7. Plasma progesterone profiles of nulliparous *Isoodon macrourus* 9 (○—○), 10 (●--●) and 11 (▲--▲).

DISCUSSION

Breeding activity at the start of the 1984/85 breeding season gradually increased from no births from March to June, to 7.7% in July, and 40%, 50%, 36%, 38% and 80% giving birth in August, September, October, November and December, respectively. The number of bandicoots giving birth gradually increased in a similar manner to that seen in the composite histogram for the years 1980 to 1984 (Fig. 2). Just as the first litters of the breeding season for individual females were observed from July to October, the first rise in progesterone occurred from July to September. This rise in progesterone would presumably indicate an ovulation and the subsequent formation of corpora lutea, however, the ovulation and rise in progesterone could be associated with either a pregnancy or an oestrous cycle. In the parous animals the first signs of ovarian activity were observed from 17 July to 30 August, and with the nulliparous animals from 31 July to 25 September. Therefore, this gradual build-up in breeding activity in the bandicoot colony could be due to the variability in the occurrence of the first ovulation in individual animals. Ovulation is either inhibited during the anoestrous period and this factor is removed during the breeding season, or ovulation is stimulated by a factor which is released at the start of the breeding season. From the results obtained in these studies it is apparent that the effect of these factors on individual animals in variable with some adult females becoming sexually active before others. A similar variation in the initiation of breeding activity was observed with nulliparous females. Thus, the

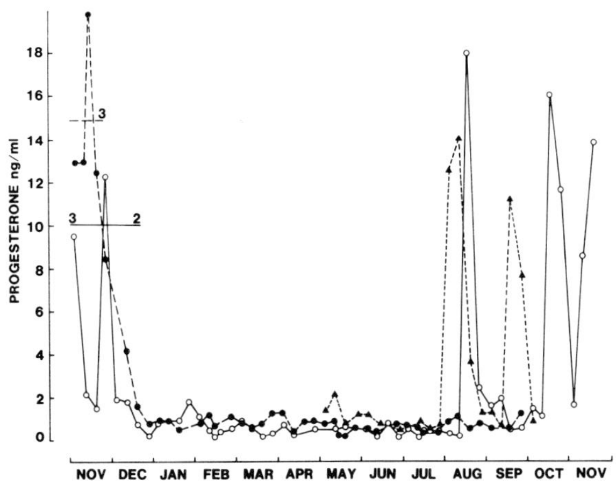

Fig. 8. Plasma progesterone profiles of parous *Isoodon macrourus* 12 (○--○) and 13 (●--●) and nulliparous animals 14 (▲--▲) held in an enclosure devoid of males.

gradual build-up of breeding activity in *I. macrourus* at the start of the breeding season in probably due to several factors. Some females may be more susceptible to the factor(s) which allow ovulation in bandicoots, with some females an oestrous cycle or cycles may precede the first pregnancy and some of the scatter of births may be due to the differing individual onset of sexual maturity in developing females.

When considering the various cues which would be used by *I. macrourus* to initiate breeding, the effect of the sexually active male on the anoestrous female was considered, since the concentration of testosterone in adult male bandicoots inceased in April several months prior to the increase in birth rate (Gemmell, Johnston and Barnes 1985). The presence of elevated progesterone levels in two of the three females in the enclosures without males would indicate that the males do not play a role in activating ovarian activity in the female.

As with a large number of mammals, bandicoots in the southern regions of Australia appear to use the rate of change of day length as the primary cue, and rate of change of minimum temperature and rainfall as additional cues (Barnes and Gemmell 1984). In New South Wales and Queensland, the rate of change of temperature was the primary cue, with the rate of change of day length and rainfall being of secondary importance. When the rate of change of day length is the primary cue, the start of the breeding season should be similar every year. From the data of Heinsohn (1966), the breeding season for *P. gunnii* began in June in the years 1960, 1961 and 1962. Similarly with the Victorian study of Stoddart and Braithwaite (1979) breeding in *I. obesulus* began in July in 1974, 1975 and 1976 and in May in 1977. However, in New South Wales, Gordon (1971) reported that the start of the breeding season varied annually, births of *I. macrourus* beginning in August 1968, July 1969 and September 1970. Gordon did not observe clear peaks in the proportion of females breeding: the pattern was one of gradual build-up in numbers of breeding females. Gordon (1971) noted some correlation between rainfall and breeding, but rainfall was not an absolute determinant of breeding and he suggested that rainfall may modify the length of the season. In our own captive population, breeding occurs throughout the year with a decrease in breeding from March to June. The first births in the breeding season occurred on 1 August, 7 July, 27 May and 28 July during 1980, 1981, 1983, and 1984 respectively and births occurred throughout 1982. Due to this annual variation in the start of the breeding season it is unlikely that day length is the only cue in initiating breeding. An analysis of the numbers of births from January 1977 to February 1983 showed a stronger association between the numbers of births and the rate of change of minimum temperature than to any other environmental variable examined (Barnes and Gemmell 1984). Nevertheless it is unlikely that the rate of change of minimum temperature is the only cue which initiates breeding in the captive colony after an examination of the data presented in this study. Although there is a large variation in the occurrence of births between the years 1982 to 1984, there is no obvious change in minimum temperature which would account for the variation in the start of the breeding season. Similarly, rainfall does not appear to be the cue.

This irregular start to the breeding season, observed in the various species of bandicoot, has been reported for other mammals. Asdell (1926) observed in goats that a cool summer provided an early oestrus while a hot summer had the opposite effect. Similarly, in sheep, although photoperiod is the major cue for the start of the breeding season, other factors such as temperature can modify its effect (Dutt and Bush 1955; Wodzicka-Tomaszewska, Hutchinson and Bennett 1967). Bronson (1985) stated that "when seasonal breeding is required, a mammal may use a predictor such as photoperiod to prepare metabolically for reproduction. However, opportunistic breeding, unenforced by a predictor, may be the most prevalent strategy extant among today's mammals". The decrease in the degree of seasonality in the bandicoot as the latitude decreases would suggest that this marsupial is exhibiting the "high degree of flexibility and opportunism associated with the breeding of most small mammals" (Bronson 1985). If environmental cues are the trigger for initiating breeding in the female bandicoot, a possible mechanism using daylength, temperature and rainfall can be suggested. The testosterone level in male bandicoots, a measure of sexual activity in the male, correlates well with the rate of change of daylength (Gemmell,

Johnston and Barnes 1985). Therefore, the rate of change of daylength may be the primary cue in initiating breeding activity both in the male and female. The rate of change of temperature and rainfall may exert an increasing influence in regions where the rate of change of daylength is not acute. Thus in Tasmania and Victoria, the rate of change of daylength may be the primary cue in controlling the start of the breeding season. In New South Wales and Queensland, although the rate of change of daylength may play some part in initiating the breeding season, the more variable factors such as temperature and rainfall may play a more important role.

The occurrence of ovulation, with or without oestrus, preceding pregnancy at the start of the breeding season, observed in the bandicoot, has been reported for other animal species. Although the ewe has a distinct breeding season with oestrous cycles generally starting as day length is decreasing, in some animals, ovulation without oestrus occurs during the non-breeding season (Thimonier 1981). Walton *et al.* (1977) monitored the changes in the plasma concentration of progesterone from anoestrus through to breeding activity and observed that 13 of 15 ewes had their first ovulation of the season, as judged by a luteal phase pattern of progesterone secretion, independently of oestrus. In some sheep, ovulation without oestrus occurred several times. With the bandicoot ovulation without oestrus probably occurred in four of 11 females.

In the sheep, a suggested hypothesis for the control of seasonal breeding is via "a photoperiodically controlled modulation of the brain and/or pituitary to gonadal steroid feedback" (Hoffman 1973; Legan and Karsch 1980). The occurrence of a breeding season or anoestrus is thought to be governed by the capacity of oestradiol to inhibit the release of the pituitary hormones luteinizing hormone (LH) and follicle stimulating hormone (FSH). During the breeding season, oestrogen released during the cycle cannot inhibit LH secretion and progesterone, which is the major inhibitor of LH, is the regulator of cyclic activity (Hauger, Karsch and Foster 1977). During the non-breeding period, the hypothalamus-hypophyseal axis becomes more sensitive to oestrogen, the estradiol from enlarging follicles inhibiting the release of LH and preventing ovulation (Legan, Karsch and Foster 1977; Legan and Karsch 1980). Presumably, with the transition from anoestrus to the breeding season, in some animals the hormonal events which allow ovulation to occur are active prior to those controlling oestrus and this is the reason for the non-pregnant cycles preceding the first pregnancy. Further study of other animal species is required to ascertain if these events control seasonal breeding in all animals.

ACKNOWLEDGEMENT

This study was supported by a grant from the Australian Research Grants Scheme.

REFERENCES

Asdell, S. A., 1926. Variation in the onset of the breeding year in the goat. *Journal of Agricultural Science* **16:** 632–69.

Barnes, A. and Gemmell, R. T., 1984. Correlations between breeding activity in the marsupial bandicoots and some environmental variables. *Australian Journal of Zoology* **32:** 219–26.

Bronson, F. H., 1985. Mammalian reproduction: an ecological perspective. *Biology of Reproduction* **32:** 1–26.

Calaby, J. H. and Taylor, J. M., 1981. Reproduction in two marsupial-mice *Antechinus bellus* and *A. bilarni* (Dasyuridae) of tropical Australia. *Journal of Mammalogy* **62:** 329–41.

Clarke, J. R., 1981. Physiological problems of seasonal breeding in eutherian mammals. Pp. 244–312 in *Oxford reviews of reproductive biology* Vol. 3 ed by G. A. Finn. Clarendon Press: Oxford.

Dickman, C. R., 1985. Effects of photoperiod and endogenous control on timing of reproduction in the marsupial genus *Antechinus. Journal of Zoology* (London) (A) **206:** 509–24.

Dutt, R. H. and Bush, L. F., 1955. The effect of low environmental temperature on initiation of the breeding season and fertility in sheep. *Journal of Animal Science* **14:** 885–96.

Gemmell, R. T., 1979. The fine structure of the luteal cells in relation to the concentration of progesterone in the plasma of the lactating bandicoot *Isoodon macrourus* (Marsupialia: Peramelidae). *Australian Journal of Zoology* **27:** 501–10.

Gemmell, R. T., 1982. Breeding bandicoots in Brisbane (*Isoodon macrourus:* Marsupialia, Peramelidae). *Australian Mammalogy* **5:** 187–93.

Gemmell, R. T., Hughes, R. L. and Jenkin, G., 1987. Comparative studies on the hormonal profiles of progesterone and prostaglandin F metabolite in the possum, *Trichosurus vulpecula*. In *Possums and opossums* ed by M. Archer. Royal Zoological Society of New South Wales: Sydney.

Gemmell, R. T. Johnston, G. and Barnes, A., 1985. Seasonal variations in plasma testosterone concentrations in the male marsupial bandicoot *Isoodon macrourus* in captivity. *General and Comparative Endocrinology* **59:** 184–91.

Gemmell, R. T., Walker, M. T., Johnston, G. and Cepon, G., 1984. The number of young present in sequential litters of the marsupial bandicoot *Isoodon macrourus*, in captivity. *Australian Journal of Zoology* **32:** 623–29.

Godfrey, G. K., 1969. Reproduction in a laboratory colony of the marsupial mouse *Sminthopsis larapinta* (Marsupialia: Dasyuridae). *Australian Journal of Zoology* **17:** 637–54.

Gordon, G., 1971. *A study of island populations of the Short-Nosed Bandicoot,* Isoodon macrourus *Gould.* Ph.D. Thesis, University of New South Wales: Kensington, New South Wales.

Hauger, R. L., Karsch, F. J. and Foster, D. L., 1977. A new concept for control of the estrous cycle of the ewe based on the temporal relationship between luteinizing hormone, estradiol and progesterone in peripheral serum and evidence that progesterone inhibits tonic LH secretion. *Endocrinology* **101:** 807–17.

HEINSOHN, G. E., 1966. Ecology and reproduction of the Tasmanian bandicoots (*Perameles gunni* and *Isoodon obesulus*). *University of California Publications in Zoology* **80:** 1–107.

HOFFMANN, J. C., 1973. Light and feedback control of gonadotrophin secretion. Pp. 886–90 in *Proceedings of the IVth International Congress of Endocrinology, Washington, D.C. June 18–24, 1972* ed by R. O. Scow. Elsevier: New York.

HUGHES, R. L., 1962. Reproduction in the macropod marsupial *Potorous tridactylus* (Kerr). *Australian Journal of Zoology* **10:** 193–224.

LEGAN, S. J. AND KARSCH, F. J., 1980. Photoperiodic control of seasonal breeding in ewes: modulation of the negative feedback action of estradiol. *Biology of Reproduction* **23:** 1061–68.

LEGAN, S. J., KARSCH, F. J. AND FOSTER, D. L., 1977. The endocrine control of season reproductive function in the ewe: a marked change in response to the negative feedback action of estradiol on luteinizing hormone secretion. *Endocrinology* **101:** 818–24.

LYNE, A. G., 1964. Observations on the breeding and growth of the marsupial *Perameles nasuta* Geoffroy, with notes on other bandicoots. *Australian Journal of Zoology* **12:** 322–39.

MCCONNELL, S. J. AND TYNDALE-BISCOE, C. H., 1985. Response in peripheral plasma melatonin to photoperiod change and the effects of exogenous melatonin on seasonal quiescence in the Tammar, *Macropus eugenii*. *Journal of Reproduction and Fertility* **73:** 529–38.

MERCHANT, J. C., 1976. Breeding biology of the Agile Wallaby, *Macropus agilis* (Gould) (Marsupialia: Macropodidae), in captivity. *Australian Wildlife Research* **3:** 93–103.

MERCHANT, J. C. AND CALABY, J. H., 1981. Reproductive biology of the Red-necked Wallaby *(Macropus rufogriseus banksianus)* and Bennett's wallaby *(M. r. rufogriseus)* in captivity. *Journal of Zoology* **194:** 203–17.

NEWSOME, A. E., 1965. Reproduction in natural populations of the Red Kangaroo, *Megaleia rufa* (Desmarest), in central Australia. *Australian Journal of Zoology* **13:** 734–59.

POOLE, W. E., 1975. Reproduction in the two species of grey kangaroos, *Macropus giganteus* Shaw and *M. fuliginosus* (Desmarest). II. Gestation, parturition and pouch life. *Australian Journal of Zoology* **23:** 333–53.

ROSE, R. W. AND MCCARTNEY, D. J., 1982. Reproduction of the Red-bellied Pademelon, *Thylogale billardierii* (Marsupialia). *Australian Wildlife Research* **9:** 27–32.

SADLIER, R. M. F. S. AND TYNDALE-BISCOE, C. H., 1977. Photoperiod and the termination of embryonic diapause in the marsupial *Macropus eugenii*. *Biology of Reproduction* **16:** 605–08.

SELWOOD, L., 1985. Synchronization of oestrus, ovulation and birth in female *Antechinus stuartii* (Marsupialia: Dasyuridae). *Australian Mammalogy* **8:** 91–96.

SHIELD, J., 1964. A breeding season difference in two populations of the Australian macropod marsupial *(Setonix brachyurus)*. *Journal of Mammalogy* **45:** 616–25.

SMITH, M. J., BENNETT, J. H. AND CHESSON, C. M., 1978. Photoperiod and some other factors affecting reproduction in female *Sminthopsis crassicaudata* (Gould) (Marsupialia: Dasyuridae) in captivity. *Australian Journal of Zoology* **26:** 449–63.

STODDART, D. M. AND BRAITHWAITE, R. W., 1979. A strategy for utilization of regenerating heathland habitat by the brown bandicoot (*Isoodon obesulus;* Marsupialia, Peramelidae). *Journal of Animal Ecology* **48:** 165–79.

THIMONIER, J., 1981. Control of seasonal reproduction in sheep and goats by light and hormones. *Journal of Reproduction and Fertility* Supplement **30:** 33–45.

TYNDALE-BISCOE, C. H. S., 1968. Reproduction and postnatal development in marsupial *Bettongia lesueur* (Quoy and Gaimard). *Australian Journal of Zoology* **16:** 577–602.

TYNDALE-BISCOE, C. H. S., 1980. Photoperiod and control of seasonal reproduction in marsupials. Pp. 277–82 in *Endocrinology 1980* ed by I. A. Cumming, J. W. Funder and F. A. O. Mendelsohn. Australian Academy of Science: Canberra.

WALTON, J. S., MCNEILLY, J. R., MCNEILLY, L. S. AND CUNNINGHAM, G. F., 1977. Changes in concentrations of follicle-stimulating hormone, luteinizing hormone, prolactin and progesterone in the plasma of ewes during the transition from anoestrus to breeding activity. *Journal of Endocrinology* **75:** 127–36.

WODIZICKA-TOMASZEWSKA, M., HUTCHINSON, J. C. D. AND BENNETT, J. W., 1967. Control of the annual rhythm of breeding in ewes: effect of an equatorial day length with reversed thermal season. *Journal of Agricultural Science* **68:** 61–67.

Longevity and reproductive capability of *Isoodon macrourus* in captivity

R. T. GEMMELL[1]

Information on mortality of juvenile and longevity of adult bandicoots was obtained in this study and the results assessed in relation to the survival of *Isoodon macrourus* in captivity and in the wild. To aid in the survival of juvenile bandicoots, young were housed in cages from 60 days post partum until they weighed approximately 600 g. When the young reached this weight they were released into the outside enclosures. Of the 76 juvenile females released 32 survived to give birth to a litter. The life span of the bandicoot was 23.3±1.27 months (mean±SEM, n=7) and 18.9 ± 1.1 months (n=19) for males and females respectively. Analysis of these data suggests that the population of *I. macrourus* can treble each two years in captivity. However, in the wild, the populations are static or declining.

Key words: Longevity, juvenile mortality, *Isoodon macrourus,* life span.

INTRODUCTION

ALTHOUGH 11 extant species of bandicoot are found in Australia (Ride 1970; Kirsch and Calaby 1977; Aitken 1979), data concerning captive breeding colonies are mainly restricted to two species, *Isoodon macrourus* and *Perameles nasuta.* The bandicoot *I. macrourus* was first bred in captivity by MacKerras and Smith (1960) and subsequently by Lyne (1971, 1982) and Gemmell (1982). Information concerning the breeding of *P. nasuta* in captivity is available from the publications of Lyne (1964, 1982), Stodart (1966) and Close (1977). Notes on the maintenance of captive colonies of *Perameles gunnii* (Seebeck 1979) and *Macrotis lagotis* (Johnson and Johnson 1983) have also been presented. In all the above studies, the procedures utilized in maintaining breeding colonies of bandicoots have been outlined, however, details concerning the long-term reproductive capability of the members of the colony have not been presented.

A self-sustaining colony of bandicoots is dependent on numerous factors. Some of these parameters have been examined previously in our colony, for example: seasonality of breeding (Gemmell 1982; Barnes and Gemmell 1984); age at sexual maturity (Gemmell 1984); number of litters per year; number of young per litter; and survival of young to weaning (Gemmell 1982; Gemmell, Walker, Johnson and Cepon 1984). This paper presents information on mortality of juveniles and longevity of adult bandicoots in captivity and assesses the results in relation to the survival of *I. macrourus* in captivity and in the wild.

MATERIALS AND METHODS

All bandicoots used in this study were bred in captivity (Gemmell 1982). Three male bandicoots were housed with approximately seven female bandicoots in each of two large outdoor enclosures measuring 30×30 metres, fed on dog food (Drimeat dog ration, Provincial Traders Pty Ltd, Queensland) and supplied with water *ad libitum.* The animals were examined at least once a week, following recapture, using rectangular wire cages approximately 23×23×52 cm in size, which were baited with apple or carrot smeared with peanut paste. Each bandicoot was lightly anaesthetized with a halogen:oxygen mixture to facilitate examination and weighing. When recovered, the bandicoots were released back into the enclosure. Bandicoots not trapped for several weeks were presumed dead. In the majority of cases the decomposing body was located within the enclosure, however, in some cases the body was not found. These deaths may have been due to predation by either snakes or large goannas.

To aid in the survival of juvenile bandicoots, females with pouch young approximately 45 days post partum were transferred from the outside enclosures to smaller individual indoor wire mesh cages 30×40×50 cm with masonite flooring and nesting materials. The mothers were separated from their young at day 60 post partum, and were returned to the outside enclosures. The young were housed in the individual cages until approximately 600 g in weight, when they were released into the outside enclosures.

[1]Department of Anatomy, University of Queensland, St Lucia, Queensland, Australia 4067.

Pages 213–17 *in* BANDICOOTS AND BILBIES, ed by J. H. Seebeck, P. R. Brown, R. L. Wallis and C. M. Kemper. Surrey Beatty & Sons, Sydney, 1990.

Table 1. Longevity of male *Isoodon macrourus.*

Bandicoot	Date of birth	Date of death	Life span (months)	Maximum weight (g)	Weight at death (g)
1	3 Nov. 1979	7 Jul. 1981	19	2 400	2 080
2	22 Dec. 1980	21 Sep. 1982	21	2 200	2 130
3	22 Dec. 1980	30 Nov. 1983	23	2 400	2 250
4	22 Dec. 1980	21 Sep. 1982	21	2 100	2 000
5	23 Jan. 1981	9 Feb. 1983	25	2 080	1 480
6	27 Jan. 1981	8 Feb. 1983	25	1 510	1 500
7	1 Apr. 1981	24 Aug. 1983	29	2 780	2 780

The date of birth and death for both male and female bandicoots and the age at which sexual maturity was reached in the female bandicoots was noted.

The term life span has been used to describe the days from birth to death for females which had produced young and for males which had fathered at least one litter.

RESULTS

Survival of juveniles

From June 1982 to May 1983, 44 female juveniles weighing approximately 600 g were released into the outside enclosure. Twenty-two of these females gave birth to at least one litter, the remaining 22 died without giving birth. From August 1983 to July 1984, 32 juvenile females were released, 10 of which gave birth to a litter, the remaining 22 died. The proportion of released juvenile females surviving to produce a litter was 50% and 31.2% respectively.

Longevity of male bandicoots

The life span of the male bandicoot in captivity was 23.3±1.27 months (mean±SEM, n = 7), range 19 to 20 months. With three animals, there was a significant reduction in weight before death, whereas in the other three animals, a slight decrease in weight was observed in two animals and no reduction in another animal (Table 1).

Longevity of female bandicoots

The life span of female bandicoots ranges from nine to 28 months, with a mean of 18.9±1.1 months (mean±SEM, n=19). In 13 animals, a significant reduction in weight was observed before death, and the mean life span of these animals was 21.3±1.0 months (mean±SEM, n=13), range 17 to 28 months (Table 2).

DISCUSSION

The male bandicoot has a mean life span of 23.3 months whereas that of the female is 18.9 months in conditions of captivity described in this paper. It is possible that female longevity may be slightly longer than reported here. With 13 females (68%), a significant reduction in weight was observed prior to death and in these animals the life span was 21.3 months. If decrease in weight is indicative of old age, approximately 30% of females did not reach old age, and 21.3 months may be a truer estimation of the female life span.

There is very little accurate information concerning longevity in the bandicoot, as available data are mainly from estimates. *Perameles gunnii*

Table 2. Longevity of female *Isoodon macrourus.*

Bandicoot	Date of birth	Date of death	Life span (months)	Maximum weight (g)	Weight at death (g)
1	22 Dec. 1980	24 Aug. 1982	20	980	690
2	29 Sep. 1981	13 May 1984	19	1 450	1 250
3	17 Feb. 1982	23 Nov. 1982	9	880	880
4	24 Mar. 1982	29 Mar. 1983	12	1 100	1 100
5	10 Apr. 1982	17 Jul. 1984	27	1 300	910
6	1 Aug. 1982	28 Feb. 1984	18	1 080	1 080
7	1 Aug. 1982	17 Jan. 1984	17	1 200	880
8	5 Aug. 1982	3 Jul. 1984	23	1 150	920
9	17 Aug. 1982	25 Sep. 1984	25	1 500	1 122
10	25 Aug. 1982	10 Jan. 1984	16	1 220	1 220
11	22 Oct. 1982	2 Jan. 1984	15	980	980
12	29 Oct. 1982	28 Feb. 1985	28	908	692
13	12 Dec. 1982	26 Jun. 1984	18	1 300	1 100
14	26 Dec. 1982	7 Aug. 1984	18	1 400	1 300
15	13 Feb. 1983	7 Nov. 1984	21	1 100	1 004
16	24 Jul. 1983	12 Mar. 1985	20	1 100	790
17	19 Aug. 1983	10 May 1985	21	1 086	823
18	19 Aug. 1983	2 Apr. 1985	20	1 100	975
19	11 Nov. 1983	11 Dec. 1984	13	1 066	1 066

are thought to live more than two years in the wild and up to three years in captivity (Heinsohn 1966). *Isoodon obesulus* are thought to survive to an age of about three years (Braithwaite 1983). Gordon (1971) observed five *I. macrourus* at 770 days of age and all were lost by 920 days, and in the following year two animals alive at 555 days were not trapped at 750 days. Lyne and Mort (1981) described skulls from wild and captive *I. macrourus* ranging from 39 to 1 249 days. Information concerning the methods of dating the 1 249-day skull or whether the bandicoot was captive or wild was not presented. The majority of studies would suggest that *I. macrourus* live for approximately two to three years in the wild and for a similar period in the semi-captivity conditions of outside enclosures. The present study shows that the animals held in enclosures, the figure is no more than two years. Bandicoots may survive longer in the more stable environment of a laboratory animal house. It is of interest to note that the Virginia Opossum *Didelphis virginiana,* another litter-bearing marsupial, has a life expectancy which seldom exceeds two years (Gottschang 1984).

Gordon (1971) reported that there was a large loss of *I. macrourus* immediately after weaning. Survival of juveniles from weaning into the next breeding season favoured young born early in the breeding season. When studying *I. macrourus* in south-east Queensland, Gordon (1974) observed that only three juveniles out of 26 young reached maturity, a survival rate of 11.5%. Similarly, Stoddart and Braithwaite (1979) reported a loss of 87.5% of young *I. obesulus* produced, on 12 to 18% of the young being trapped presumably as juveniles or young adults within the study area. Heinsohn (1966) also reported a high mortality of newly weaned *P. gunnii.* Only one of 44 large size pouch young and juveniles accompanying their mother were recaptured. Indirect evidence of juvenile mortality is available from the trapping information with very few *I. macrourus* (100–350 g) being trapped in the wild (Hall 1983). Hall suggested that the lack of young bandicoots in traps may indicate a high mortality of juveniles. In our breeding colonies, a similar loss of juvenile bandicoots was observed when the young were allowed to be weaned in the enclosure. Approximately 90% of all pouch young did not reach adulthood. This loss of young may be due to snakes or goannas although it is unlikely that all the loss is due to predation. In order to improve the survival rate of the newly weaned bandicoots, they were kept in cages and "fattened" up to 600 g before release. Although a three- to four-fold increase in survival was obtained, in the present study when juveniles were held in cages until they reached 600 g body weight there was still a 58% loss (44 of a total of 76) of female bandicoots between weaning and sexual maturity, in captivity.

The reproductive potential of a species is dependant on such parameters as length of the breeding season, number of litters per season, survival of young, sexual maturity and longevity. Information concerning these factors is now available for our captive breeding colony and the viability of our population of bandicoots can be assessed.

Although births occur throughout the year, there is a period of reduced breeding activity during the months April, May and June (Gemmell 1982; Barnes and Gemmell 1984). Since the bandicoot is capable of producing a litter every two months, a maximum of four to five litters per year can be produced. An average of 4.33 litters was observed for six bandicoots during 1982 (Gemmell *et al.* 1984). This figure may be slightly greater than normal since there was no decline in breeding activity throughout 1982, an occurrence which happened only once in the seven years from 1976 to 1984. The female bandicoot reaches sexual maturity at 230 days post partum or approximately seven months (Gemmell 1984) and has an average life span of 19 months. Therefore it is capable of producing litters for approximately 12 months. The average female which survives to sexual maturity will produce approximately four to five litters during her life span. Some bandicoots which live to 28 months will give birth to more than five litters and similarly others which live only nine months will give birth to only one litter.

The average litter size observed at birth is four (n=14) in *I. macrourus* (Gemmell 1982). This observation is in accord with the results of Lyne and Hollis (1979) who reported 4.05±0.29 embryos (mean±SEM, n=20) in pregnant *I. macrourus.* Thus, one female bandicoot, in her life span, will give birth to 18 young (four young times 4.5 litters). The survival rate of young from birth to weaning is 72.8% in our colony (Gemmell *et al.* 1984), so only 13.1 young will leave the pouch and 42% or 5.9 young will reach sexual maturity. Since the sex ratio is 1 to 1 (Gemmell *et al.* 1984) each female will produce another three mature females in our semi-captive environment.

In the wild, this production rate of offspring would be lower. Stoddart and Braithwaite (1979) reported that each female *I. obesulus* produced 8.27 young per year, but, recruitment into the trappable population varies between 12 and 18%. Therefore if female *I. obesulus* have a life span of approximately 19 months they will produce 8.27 young of which 1.3 young will reach maturity; presumably 0.65 of these young being female. Gordon (1974) observed an 11.5% survival rate from pouch to young adult; thus, if one *I. macrourus* female gave birth to 18 young in her life time and 13.1 of the young were weaned, only 1.5 of these young would reach maturity, presumably 0.75 of these young are female.

Table 3. Reproductive information concerning bandicoots.

Species	Reproductive information
Isoodon auratus	Breeds continuously throughout the year. Litter size 2–3 (McKenzie 1983)
Isoodon macrourus	Seasonality in breeding, litter size, survival of young in captivity and in the wild (Mackerras and Smith 1960; Gordon 1974; Lyne 1982; Gemmell 1982; Hall 1983; Gemmell *et al.* 1984)
Isoodon obesulus	Seasonality in breeding, litter size, survival of young in the wild (Heinsohn 1966; Stoddart and Braithwaite 1979)
Perameles bougainville	Breeding season May and June, litter size of two (Jones 1924; Ride and Tyndale-Biscoe 1962; Burbidge 1983)
Perameles eremiana	"Apparently two young produced at a time" stated by Baldwin Spencer in Gordon (1983)
Perameles gunnii	Seasonality in breeding, litter size, longevity, sexual maturity in the wild (Heinsohn 1966)
Perameles nasuta	Continuous breeding, litter size, sexual maturity, in captivity and in the wild (Lyne 1964; Stodart 1966; Close 1977; Lyne 1982)
Chaeropus ecaudatus	Two young in litter, breeding season in winter (Ride 1970)
Echymipera rufescens	Seasonal breeder (Hulbert, Gordon and Dawson 1971; Gordon and Laurie 1977; Gordon 1983)
Macrotis lagotis	Breeding throughout the year in captivity, two young per litter (Hulbert 1971; Johnson 1983)
Macrotis leucura	Two young per litter (Johnson 1983)

In conclusion, the population of *I. macrourus* can treble each two years, in captivity. In the wild, the populations are static or on the decline. This observation is supported by the results of several ecological studies. Heinsohn (1966) observed a reduction of 59 *P. gunnii* per 100 acres in March 1961 down to 18 bandicoots per 100 acres in April 1962. In Victoria, the number of *I. obesulus* within a grid area remained constant from 1974 to 1977. The rate of input did not cause an increase in the population because of a high rate of loss of previously known animals (Stoddart and Braithwaite 1979). This steady state of population may be due to excess animals leaving the trapping site or a balance between the young animals maturing and the older dying.

The numbers and species of bandicoots are declining and although various suggestions based on environmental change have been put forward for this, little information is available concerning the biological reasons for this phenomenon. There is very little known about the reproductive biology of the various species of bandicoots (Table 3). The description of the reproductive biology of *Chaeropus ecaudatus* is typical: "reliable accounts state that not more than two young are carried. As far as can be construed, the breeding season seems to be in winter" (Ride 1970). No descriptions of reproductive ability are available for the species *I. auratus, P. bougainville, C. ecaudatus, Echymipera rufescens* and *M. leucura.* Although reproductive information is available for the other species, data concerning longevity, age at sexual maturity and survival of juveniles are still required for a fuller understanding of the reproductive capability of each species.

ACKNOWLEDGEMENT

This study was supported by a grant from the Australian Research Grants Scheme.

REFERENCES

AITKEN, P., 1979. The status of endangered Australian wombats, bandicoots and the marsupial mole. Pp. 61–65 in *The status of endangered Australasian wildlife* ed by M. J. Tyler. Royal Zoological Society of South Australia: Adelaide.

BARNES, A. AND GEMMELL, R. T., 1984. Correlations between breeding activity in the marsupial bandicoots and some environmental variables. *Australian Journal of Zoology* **32:** 219–26.

BRAITHWAITE, R. W., 1983. Southern Brown Bandicoot *Isoodon obesulus.* Pp. 94–95 in *The Australian Museum complete book of Australian mammals* ed by R. Strahan. Angus & Robertson: Sydney.

BURBIDGE, A. A., 1983. Western Barred Bandicoot *Perameles bougainville.* P. 101 in *The Australian Museum complete book of Australian mammals* ed by R. Strahan. Angus & Robertson: Sydney.

CLOSE, R. L., 1977. Recurrence of breeding after cessation of suckling in the marsupial *Perameles nasuta. Australian Journal of Zoology* **25:** 641–45.

GEMMELL, R. T., 1982. Breeding bandicoots in Brisbane (*Isoodon macrourus:* Marsupialia, Peramelidae). *Australian Mammalogy* **5:** 187–93.

GEMMELL, R. T., 1984. The age of sexual maturity in the female bandicoot *Isoodon macrourus:* the effect of season. *Proceedings of the Australian Reproductive Biology Society:* Melbourne University p. 24.

GEMMELL, R. T., WALKER, M. T., JOHNSTON, G. AND CEPON, G., 1984. The number of young present in sequential litters of the marsupial bandicoot, *Isoodon macrourus* in captivity. *Australian Journal of Zoology* **32:** 623–29.

GORDON, G., 1971. *A study of island populations of the Short-nosed Bandicoot,* Isoodon macrourus *Gould.* PhD Thesis, University of New South Wales.

GORDON, G., 1974. Movements and activity of the short-nosed bandicoot *Isoodon macrourus* Gould (Marsupialia). *Mammalia* **38:** 405–31.

GORDON, G., 1983. Northern Brown Bandicoot *Isoodon macrourus* Pp. 96–97, Desert Bandicoot *Perameles eremiana,* P. 102, Rufous Spiny Bandicoot *Echymipera rufescens,* P. 103 in *The complete book of Australian mammals* ed by R. Strahan. Angus & Robertson: Sydney.

GORDON, G. AND LAURIE, B. V., 1977. The rufescent bandicoot, *Echymipera rufescens* (Peters & Doria) on Cape York Peninsula. *Australian Wildlife Research* **5:** 41–45.

GOTTSCHANG, J. L., 1984. The Virginia Opossum, *Didelphis virginiana* in an urban environment. *Abstract of the joint meeting, Australian Mammal Society and American Society of Mammalogists, Sydney, May 1984* (not paginated).

HALL, L. S., 1983. Observations on body weights and breeding of the Northern Brown Bandicoot, *Isoodon macrourus* trapped in South-east Queensland. *Australian Wildlife Research* **10:** 467–76.

HEINSOHN, G. E., 1966. Ecology and reproduction of the Tasmanian bandicoots (*Perameles gunni* and *Isoodon obesulus*). *University of California Publication in Zoology* **80:** 1–96.

HULBERT, A. J., 1971. Growth and development of pouch young in the Rabbit-eared Bandicoot, *Macrotis lagotis* (Peramelidae). *Australian Mammalogy* **1:** 38–39.

HULBERT, A. J., GORDON, G. AND DAWSON, T. J., 1971. Rediscovery of the marsupial *Echymipera rufescens* in Australia. *Nature (Lond.)* **231:** 330–31.

JOHNSON, K. A., 1983. Bilby *Macrotis lagotis,* Pp. 107–08; Lesser Bilby *Macrotis leucura.* P. 109 in *The complete book of Australian mammals* ed by R. Strahan. Angus & Robertson: Sydney.

JOHNSON, C. N. AND JOHNSON, K. A., 1983. Behaviour of the Bilby, *Macrotis lagotis* (Reid), (Marsupialia: Thylacomyidae) in captivity. *Australian Wildlife Research* **10:** 77–87.

JONES, F. W., 1924. *The mammals of South Australia. II. The bandicoots and herbivorous mammals.* Government Printer: Adelaide.

KIRSCH, J. A. W. AND CALABY, J. H., 1977. The species of living marsupials: an annotated list. Pp. 9–25 in *The biology of marsupials* ed by B. Stonehouse and D. Gilmore. Macmillan: London.

LYNE, A. G., 1964. Observations on the breeding and growth of the marsupial *Perameles nasuta* Geoffroy, with some notes on other bandicoots. *Australian Journal of Zoology* **12:** 322–29.

LYNE, A. G., 1971. Bandicoots in captivity. *International Zoology Yearbook* **11:** 41–43.

LYNE, A. G., 1982. The bandicoots *Isoodon macrourus* and *Perameles nasuta:* their maintenance and breeding in captivity. Pp. 47–52 in *Management of Australian mammals in captivity* ed by D. D. Evans. Zoological Board of Victoria: Melbourne.

LYNE, A. G. AND HOLLIS, D. E., 1979. Observations on the corpus luteum during pregnancy and lactation in the marsupials *Isoodon macrourus* and *Perameles nasuta. Australian Journal of Zoology* **27:** 881–89.

LYNE, A. G. AND MORT, P. A., 1981. A comparison of skull morphology in the marsupial bandicoot genus *Isoodon:* its taxonomic implication and notes on a new species, *Isoodon arnhemensis. Australian Mammalogy* **4:** 107–33.

MACKERRAS, M. J. AND SMITH, R. H., 1960. Breeding the short-nosed marsupial bandicoot, *Isoodon macrourus* (Gould), in captivity. *Australian Journal of Zoology* **8:** 371–82.

MCKENZIE, N. L., 1983. Golden Bandicoot *Isoodon auratus.* P. 98 in *The complete book of Australian mammals* ed by R. Strahan. Angus & Robertson: Sydney.

RIDE, W. D. L., 1970. *A guide to the native mammals of Australia.* Oxford University Press: Melbourne.

RIDE, W. D. L. AND TYNDALE-BISCOE, C. H., 1962. Mammals. Pp. 54–97 in *The results of an expedition to Bernier and Dorre Islands, Shark Bay, Western Australia, in July 1959* ed by A. J. Fraser. Fauna Bulletin No. 2. Western Australian Fisheries Department: Perth.

SEEBECK, J. H., 1979. Status of the barred bandicoot, *Perameles gunnii,* in Victoria: with a note on husbandry of a captive colony. *Australian Wildlife Research* **6:** 255–64.

STODART, E., 1966. Management and behaviour of breeding groups of the marsupial *Perameles nasuta* Geoffroy in captivity. *Australian Journal of Zoology* **14:** 611–23.

STODDART, D. M. AND BRAITHWAITE, R. W., 1979. A strategy for utilization of regenerating heathland habitats by the brown bandicoot (*Isoodon obesulus;* Marsupialia, Peramelidae). *Journal of Animal Ecology* **48:** 165–79.

Aspects of lactation in the Northern Brown Bandicoot *Isoodon macrourus*

J. C. MERCHANT[1]

Milk samples from Northern Brown Bandicoots *Isoodon macrourus* were collected at regular intervals throughout the 55–60 days of lactation. Milk composition changed as lactation proceeded; total milk solids at birth were 8–10% (w/w) and increased to more than 40% by 55 days; carbohydrate concentration declined from about 45% of the solids fraction initially to about 7% at the end of lactation. Lipid was the major component in late lactation and reached concentrations as high as 65% by 55 days. Milk quality changed rapidly at the end of lactation so that within one to two days of weaning of a litter newborn young of the next litter could attach to previously sucked teats and survive. Milk intake by the pouch young increased throughout lactation from about 2 ml day^{-1} at 20 days of age to about 18 ml day^{-1} by 55 days. Maternal body weight increased markedly during the first half of lactation to reach a maximum at 34 ± 5 days (mean ± SD). Thereafter body weight declined to its original level by the end of lactation. The magnitude of weight gain during lactation was correlated with season, the lowest increases being in the cooler months.

Key words: Lactation, milk composition, milk intake, maternal weight change, teat selection.

INTRODUCTION

MARSUPIALS stimulated the interest of biologists because of their mode of reproduction and the remarkably small size of their newborn young. The subsequent growth and development of the young in the pouch was likened to the intrauterine phase of pregnancy in the eutherian mammal (Tyndale-Biscoe, Stewart and Hinds 1984). The nourishment of young during lactation has been studied intensively in eutherians but not in marsupials. This is probably due, in part at least, to the small size of both marsupial neonate and mammary gland and the consequent difficulties of obtaining milk samples and measuring milk consumption (Green 1984). In recent years, however, various marsupial groups have been studied to determine their lactational strategies. In terms of its composition the milk of the Tammar Wallaby *Macropus eugenii* is the best known of any marsupial species (Green, Newgrain and Merchant 1980; Findlay and Renfree 1984; Green 1984). Milk of the small (1 kg) dasyurid *Dasyurus viverrinus* has now been analysed (Green, Merchant and Newgrain 1987) and, in recent years, various other marsupial groups have been studied to determine their lactational strategies (Green and Merchant 1988; Merchant 1989).

The bandicoots were considered to be a suitable species for study because of certain unique features. Apart from an exceptionally short gestation of 12.5 days, the shortest of any mammal (Lyne 1974), bandicoot young grow and develop very rapidly (Mackerras and Smith 1960; Lyne 1964). Lactation lasts only 60 days, which is shorter than any other marsupial of comparable size. Oestrus may occur during lactation so that a new litter of young may be born as soon as the previous young vacate the pouch (Gemmell 1982). The Northern Brown Bandicoot *Isoodon macrourus* may breed continuously in the northern parts of its range (Gordon 1974) so that lactation may also be continuous.

The present study was undertaken to elucidate the lactational strategy employed by the bandicoot, which not only allowed the mother to cope with the energy requirements of raising a litter to independence every two months but also provided sufficient nutrients for rapid growth in the young.

METHODS

Isoodon macrourus were trapped near Dorrigo, New South Wales, and transferred to outdoor enclosures (8 m × 10 m) at the Division of Wildlife and Ecology, CSIRO, Canberra, ACT. Although Canberra is about 200 km south of the extreme southern limit of their present range and about 750 km south-west of Dorrigo, they thrived and bred well. Animals were fed *ad libitum* a diet of dog food (Harper's Puppy Chow, Arnott-Harper Pty Ltd) and water was freely available. The animals also had access to insects and earthworms which appeared to be in reasonable numbers in the pens.

Females were either trapped in the wild or born in captivity. The former were assumed to be about one year old on the basis of their weights at capture and two years old in July 1984 when the captive-bred animals were seven to nine months old.

[1]CSIRO Division of Wildlife and Ecology, Canberra, ACT, Australia 2600.

Pages 219-28 *in* BANDICOOTS AND BILBIES, ed by J. H. Seebeck, P. R. Brown, R. L. Wallis and C. M. Kemper. Surrey Beatty & Sons, Sydney, 1990.

At weekly intervals females with young of known age were caught by removing them from the nests they had built and were placed in calico bags for weighing. Mothers' weights were corrected for the weight of pouch young up to 40 days by subtracting the weights of young estimated from a graph of body weight for known-age young (unpublished data). Thereafter young were removed from the pouch prior to weighing. When mothers were to be milked, the young were removed from the pouch and, if naked, were kept in labelled containers in a humid incubator at 35°C. When furred they were simply kept warm in calico bags. Milk was allowed to accumulate in the mammary glands for four to five hours and then collected after an intramuscular injection of oxytocin (Syntocinon; Sandoz, Australia) at a dosage of 0.8 i.u./kg body weight. Anaesthetic was not required as females, gently handled, lay quietly on their sides during milking. Data were derived from 10 females over 29 lactations.

Most females started breeding in July 1984 and gave birth for the first time in August. The breeding season was then divided into two-month periods. These were August–September, October–November, December–January and February–March which corresponded with the 60-day periods of lactation. Lactations which did not fall precisely into these periods were assigned on the basis of when the greater part of the first 30 days fell. Two lactations starting in July were included in the August–September period.

Milk solids were determined gravimetrically to the nearest 0.1 mg by freeze-drying approximately 50 μl or 100 μl of milk weighed in pre-weighed plastic vials. Very small samples, about 10 μl, were weighed in pre-weighed aluminium containers and dried in a desiccator. All samples were subsequently reweighed to determine total solids. All weighings were made to the nearest 0.1 mg on a Sartorius balance (Model 1207 MP2).

Carbohydrate concentrations were determined using the methods of Messer and Green (1979) and crude lipid determinations by the methods of Green, Griffiths and Leckie (1983).

Total nitrogen was estimated using the dried residues of the total solids determinations which were digested using the Kjeldahl procedure. Total nitrogen was estimated by microdiffusion technique (Conway 1962) and converted to protein using a factor of 6.38.

Milk intake was estimated by measuring the dilution over time of the radioisotope ^{22}Na in the blood of the pouch young using the techniques of Green and Newgrain (1979).

The selection of teats by newborn young was recorded for each litter and their subsequent survival noted. The number of lactating teats was also recorded at each observation in case this was greater than the number of young surviving later in lactation as was observed in the Eastern Quoll, *D. viverrinus* (Merchant, Newgrain and Green 1984).

The ages of pouch young of five females trapped in the field to obtain milk samples were determined by measurement and comparison with growth curves of captive-bred young (Merchant unpublished).

RESULTS

Milk solids

Total solids increased throughout lactation from about 8% (w/w) in the first few days of lactation to 40–45% (w/w) in the last week (Fig. 1). From day 50 of lactation total solids in some females began to decline. This was quite rapid so that in one case, for example, solids fell from 42.6% at day 56 to 14.9% at day 59 while in another they fell from 41.2% at day 51 to 13.6% at day 59.

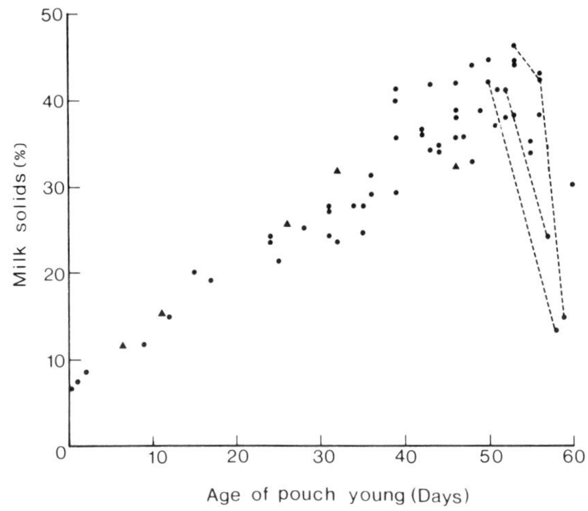

Fig. 1. Changes in total milk solids during lactation in *Isoodon macrourus*. ▲ field samples. ---- joins values for individual females.

Milk solids determined from five milk samples obtained from females trapped and released in the field are included (Fig. 1) and were similar to those of captive bandicoots.

Carbohydrates, lipid and protein

Analyses for carbohydrate, lipid and protein are at a preliminary stage with lipid and protein concentrations having been determined for the latter half of lactation only. When examined as a percentage of the milk solids fraction carbohydrate declined from an initial level around 45% to about 30% at 30 days and 7% at the end of lactation (Fig. 2). Lipid, however, was high in the latter half of lactation with values around 35% at 30 days rising to about 65% by

56 days. Protein declined from around 35% at 30 days to about 20% at 55 days.

As a percentage of whole milk, carbohydrate concentrations ranged from about 4 g 100 ml^{-1} initially to about 8 g 100 ml^{-1} at 25–30 days and then declined to 5 g 100 ml^{-1} at 56 days. Protein was relatively uniform at about 10 g 100 ml^{-1} from 30 to 56 days. Lipid, however, increased from about 10 g 100 ml^{-1} at 30 days to as high as 28 g 100 ml^{-1} at 56 days. The increase in milk solids throughout lactation is, therefore, due mainly to increasing lipid concentration.

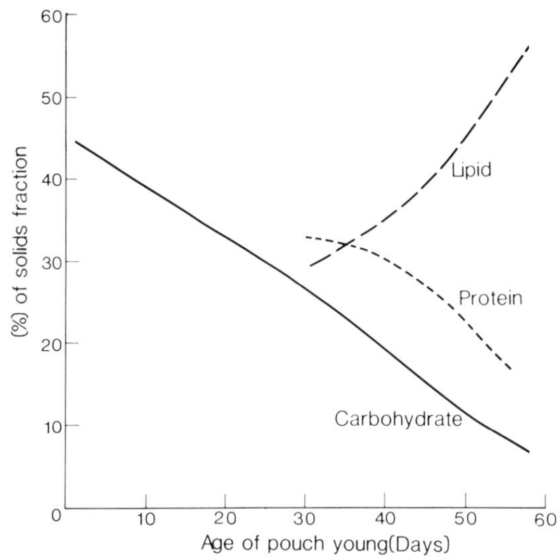

Fig. 2. Changes in the relative concentrations of carbohydrates, proteins and lipids in the solids fraction of the milk during lactation in *Isoodon macrourus*.

Milk intake by pouch young

The values presented (Fig. 3) are based on corrected estimates of sodium flux calculated from the regression equation:

Y (sodium intake) = 0.203 + 0.8264X (sodium influx)
(r = 0.9061)

which was derived from the results of two validation experiments in which young were fed known amounts of sodium and their sodium fluxes measured using the techniques of Green and Newgrain (1979). Bandicoot pouch young drank about 2 ml^{-1} day^{-1} at 20 days of age and a body weight of around 6 g. Milk intake reached about 10 ml and 18 ml^{-1} day^{-1} by 45 and 55 days respectively.

Weights of females during lactation

The pattern of weight change in nine females from the start of breeding in August 1984 to May 1985, by which time breeding had ceased, is presented (Fig. 4). Females were synchronized on their first lactation, about August. Each lactation was then assigned a 60-day period which

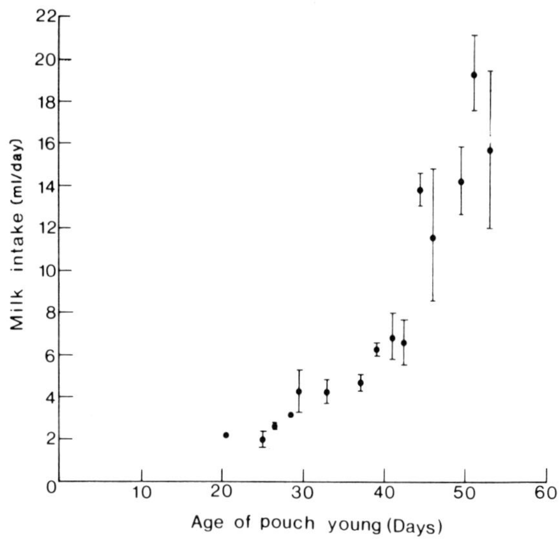

Fig. 3. Milk consumption by young *Isoodon macrourus* during lactation (mean ± 1 SD).

represents both the duration of lactation and the age of the pouch young. Data for old and young animals are presented separately.

Young and old females showed a consistent pattern of weight increase during the first 30–40 days of lactation which was followed by a decrease over the remaining period (Fig. 4). The mean age of pouch young at maximum maternal weight was 34 ± 5 days (SD) in 29 litters. In the younger group, however, superimposed on this pattern, there was an upward trend in both minimum and maximum weights.

Exceptions to this pattern occurred when females lost entire litters, either deliberately removed from the pouch (RPY) or found to have disappeared from the pouch (DPY) between one examination and the next. Most females lost one or two young during lactation so that commonly only two or three young were weaned (mean 2.44 ± SD 1.19, n = 25). The observed pattern of weight change, therefore, occurred under partial but not complete loss of litters.

The major changes seen during lactation were not shown by females when not lactating (Fig. 4).

Factors affecting body weight increase

Using the data for all females there were no significant differences between the mean body weights of females at the time of parturition in each period ($F_{3.25}$ = 0.737) (Table 1). The maximum weight increase for each female was calculated by subtracting her weight at parturition from her maximum weight during lactation. Maximum weight increase varied significantly among periods ($F_{3.25} = 6.58$; $P < 0.01$) with that for August–September being lowest (Table 1). When maximum weight increase was

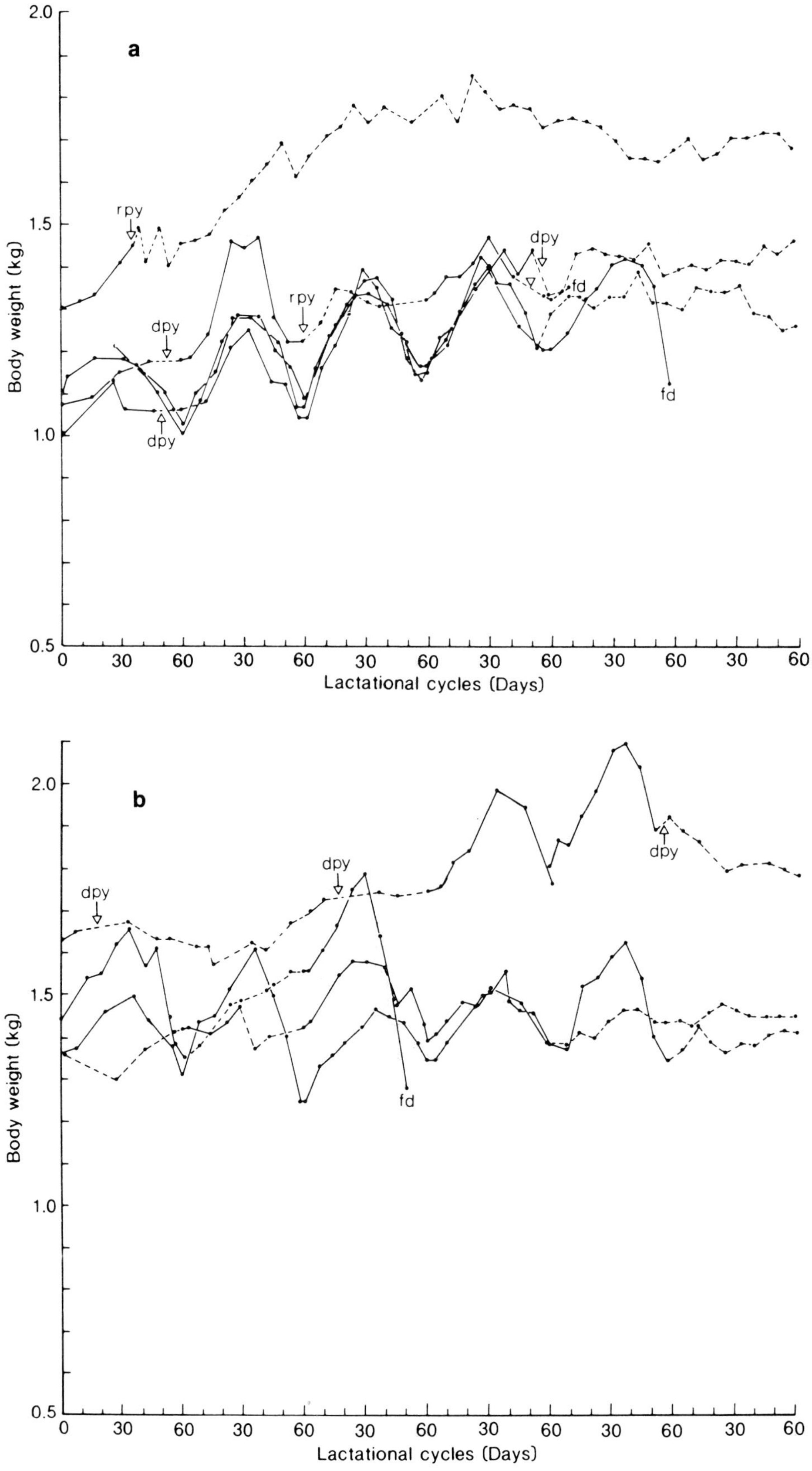

Fig. 4. Weights of female *Isoodon macrourus* in the younger age group (a) and in the older age group (b) over the period August 1984 to May 1985; first lactational cycles synchronized (see text). ——— lactating; ---- non-lactating; rpy=removal of pouch young; dpy=disappearance of entire litter; fd=female died.

Table 1. Mean maternal weight (g) of *Isoodon macrourus* at the time of birth, mean maximum weight increase (g) and maximum weight increase (%).

		August/ September	October/ November	December/ January	February/ March	Significance
Maternal wt. at birth	Mean	1 207.1	1 204.6	1 289.8	1 347.0	
	SD	164.7	195.2	236.8	247.5	n.s.
	N	7	9	7	6	
Maximum wt. increase (g)	Mean	129.86	241.1	218.9	251.5	
	SD	44.3	73.3	60.7	31.3	$P<0.01, F_{3.25}=6.58$
	N	7	9	7	6	
Maximum wt. increase (%)	Mean	10.6	20.8	17.7	19.1	
	SD	2.6	7.9	6.8	3.5	$P<0.05, F_{3.25}=4.16$
	N	7	9	7	6	

expressed as a percentage of maternal weight at parturition the differences between periods were again significant ($F_{3,25} = 4.16$; $P<0.05$) with August–September lowest (Table 1). Percentage increases ranged from 7.8% in the August–September period to 33.6% in October–November.

There was no significant correlation between body weights at parturition and the subsequent maximum weight increase ($r = 0.038$; n.s.).

The mean weight increase within 10-day intervals during lactation was calculated for both age groups for each seasonal period and are presented (Fig. 5). Figure 5 shows the overall tendency for August–September weight increases to be smaller than in other periods throughout lactation. The differences in mean weight increases within 10-day intervals between two-month periods for the older group were not significant. In the younger group, however, significant differences occurred as early as 10–19 days into lactation (Table 2).

The possibility that the magnitude of weight increase during lactation might be correlated with litter size at birth was also considered. Using all data, irrespective of age or season, a positive correlation was obtained ($r = 0.523$; $P<0.01$). Mean litter sizes were not significantly different between two-month periods either at birth ($F_{3.25} = 1.80$) or at weaning ($F_{3.23} = 1.58$) but were significantly different at the time of maximum maternal body weight in mid-lactation ($F_{3.25} = 4.16$; $P<0.05$). The October–November mean of 4.11 young per litter was significantly higher (SNK*-test) than the 2.43 of August–September ($P<0.02$) and 2.67 of February–March ($P<0.05$) but not the 3.14 recorded for December–January (Table 3).

Teat selection by pouch young

Because the bandicoot may mate late in lactation (Lyne 1976) new young are born and enter the pouch within a few days of the exit of the previous litter. They may thus have a choice of either small, previously unsucked teats (PU) or large, previously sucked teats (PS). I found that some newborn young attached to the previously sucked teats (Fig. 6) and that their survival was not significantly different to those young attached to previously unsucked teats ($\chi^2 = 0.62$; n.s.) (Table 4A). When the number of previously unsucked teats was either greater than or equal to the number of new young entering the pouch the young clearly preferred the smaller previously unsucked teats ($P<0.01$) (Table 4B).

Although there was little difference in the survival of young on the two types of teat, I observed retarded growth in three out of 18 young of different mothers when attached to previously sucked teats. One, which was recorded as "half the size" of its littermates at seven days of age, had disappeared from the pouch by 14 days. The second was recorded as "half the size" of its littermates at 15 days, "about three-quarter size" by 21 days, "indistinguishable" by 30 days and subsequently survived. The third young is shown in Figure 7 at the age of eight days; at 15 days of age it weighed 1.2 g while its apparently normal sibling weighed 3.5 g. Occasionally, therefore, young on previously sucked teats may not be initially supplied with milk of the appropriate quality but, as noted above, may subsequently survive.

Table 2. Significance of the differences between mean weight increases within 10-day intervals during lactation for August–September against the three other time periods for young female *Isoodon macrourus.*

	Days from birth					
	0–9	10–19	20–29	30–39	40–49	50–59
October/November	n.s.	n.s.	$P<0.01$	$P<0.001$	n.s.	n.s.
December/January	n.s.	$P<0.01$	$P<0.05$	$P<0.01$	n.s.	n.s.
February/March	n.s.	$P<0.05$	$P<0.02$	$P<0.01$	$P<0.01$	n.s.

*SNK = Student-Newman-Keuls.

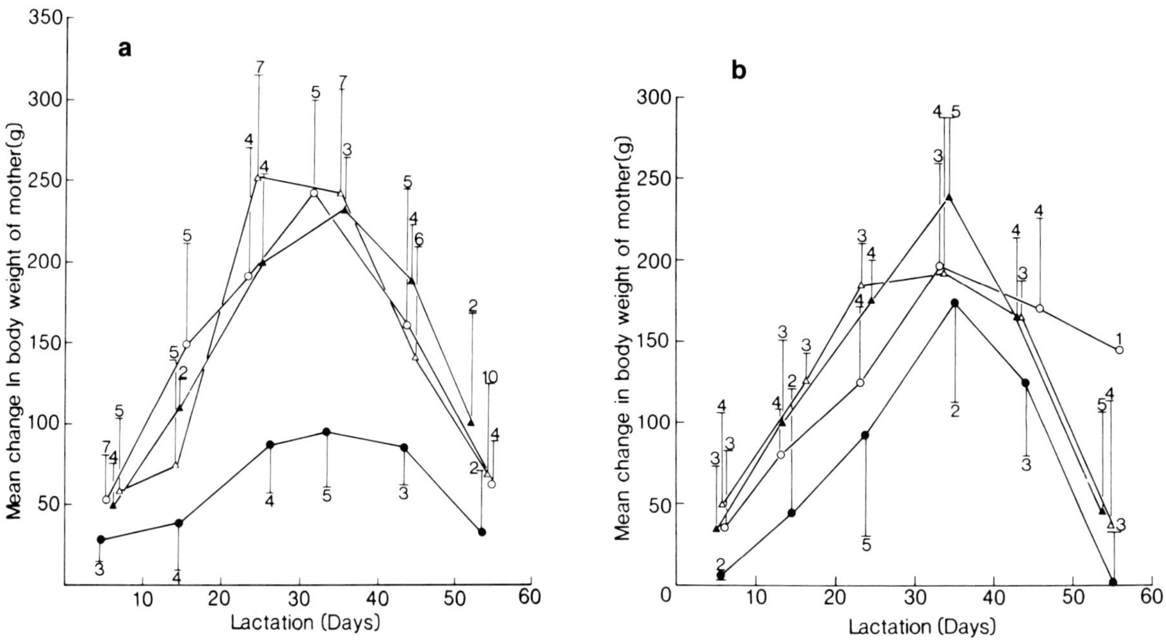

Fig. 5. Mean change in weight of female *Isoodon macrourus* in the younger age group (a) and in the older age group (b) within 10-day periods throughout lactation in August-September (●); October-November (△); December-January (○) and February-March (▲). S.D. and N are given for each mean.

Table 3. Mean litter sizes of *Isoodon macrourus* at birth, at the time of maximum maternal weight and at weaning.

		August/ September	October/ November	December/ January	February/ March	Significance
Litter size at birth	Mean	3.00	4.33	4.28	4.33	
	SD	1.29	1.22	1.11	1.63	n.s.
	N	7	9	7	6	
Litter size max. weight	Mean	2.43	4.11	3.14	2.67	
	SD	1.13	1.27	0.89	0.52	$P<0.05, F_{3,25}=4.16$
	N	7	9	7	6	
Litter size at weaning	Mean	2.00	3.00	2.00	1.83	
	SD	1.67	0.76	1.00	1.17	n.s.
	N	6	8	7	6	

Table 4. A. Attachment of young to previously sucked (PS) and unsucked *Isoodon macrourus* teats (PU); B. Choice of teat when the number of unsucked teats available is equal to or greater than the number of new young.

A.	Number of litters	Number of females	Number of young	Attachment PS	Attachment PU	Survival PS	Survival PU	Significance
All data	17	8	74	18	56	—	—	—
Survival known	11	6	37	10	27	8	18	n.s.

B.	Number of young	Teats available PS	Teats available PU	Choice PS	Choice PU	
	42	32	56	7	35	$P<0.01$

Pouch young mortality

In 25 litters from eight females, 61 of 98 young (62.2%) present in the pouch at birth survived to independence, an average of 2.44 young per litter. The 37 which disappeared from the pouch comprised four age groups — 0–14, 15–29, 30–44 and 45–59 days. Since mothers were examined only weekly the exact day of the death of the pouch young was not known. Deaths were not evenly distributed over age groups ($\chi^2 = 8.94$; $P<0.005$) with 17 of the 37 (45.9%) occurring between 30–44 days. In this age group young detach from the teats and may move around freely and their eyes open. These changes may make them more vulnerable to loss from the pouch.

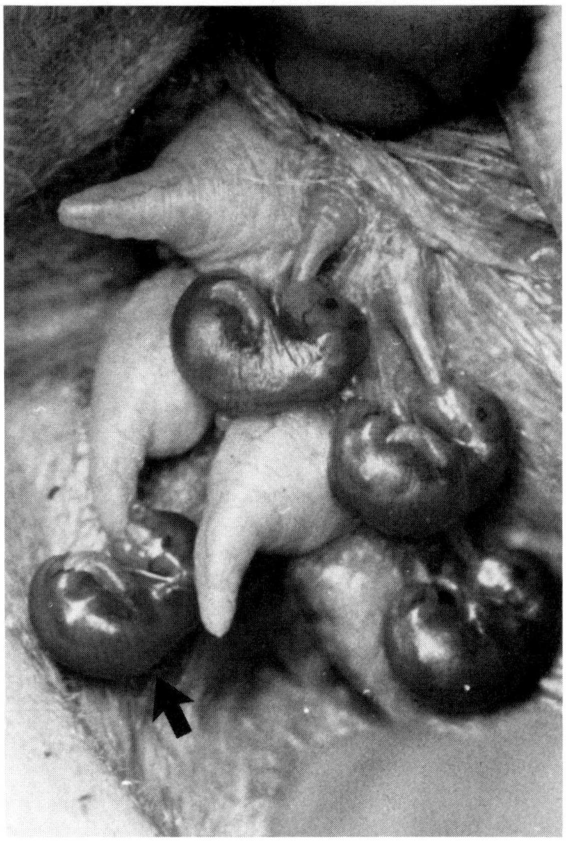

Fig. 6. Newborn young *Isoodon macrourus* attached to three previously unsucked teats (right of photo) and a previously sucked teat (arrowed) ×2. (Photo G. Chapman).

DISCUSSION

Duration of lactation in the bandicoots is short when compared to other marsupials of similar size but is still longer than eutherian species of similar size. Blaxter (1971) related the duration of lactation to body mass (W, grams) in eutherians by the expression

$$\text{Lactational Duration (days)} = 12W^{0.15}.$$

Using this formula the duration of lactation in a comparable-sized eutherian should be about 35 days, much less than the observed 55 to 60 days of the bandicoot. Lactation in *I. macrourus* is effectively over by about 55 days and by 60 days the mammary glands may be again producing milk suitable for newborn young. This dramatic change in milk quality, from late to early milk, is in contrast to most macropodids which, in late lactation, may produce milk of two types from separate mammary glands: one for the older young and one for the newborn (Sharman 1970).

During this short lactation bandicoot milk quickly reaches total solids levels as high as 45% (w/w) by 55 days. *Macropus eugenii* and *D. viverrinus*, however, may take 200 and 100 days respectively to reach 35% total milk solids (Green *et al.* 1980; Green *et al.* 1987).

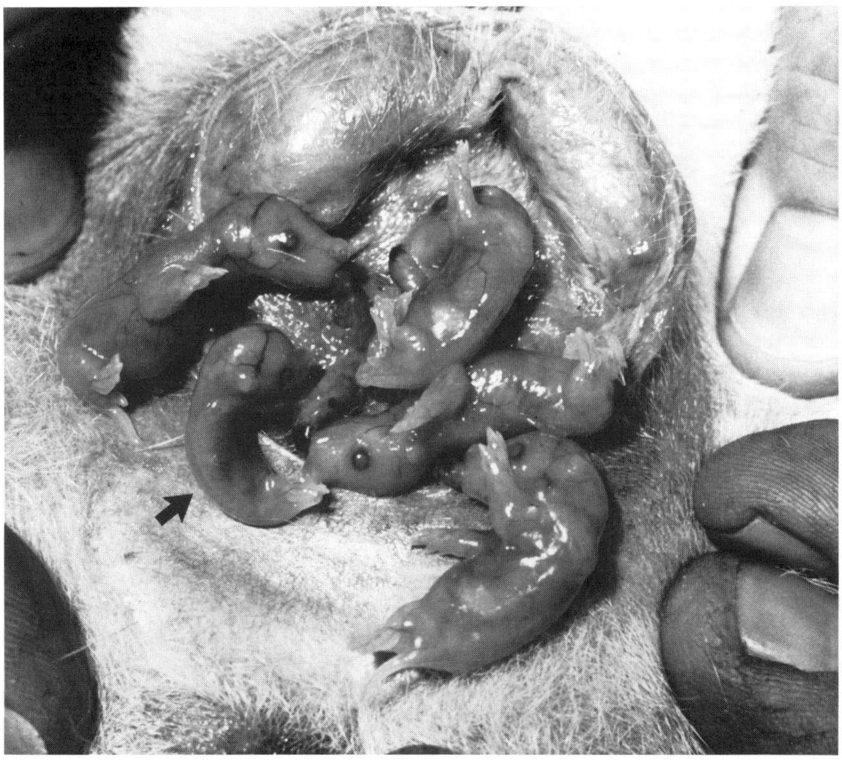

Fig. 7. Five young *Isoodon macrourus*, aged eight days, showing the small size of one young (arrowed) which attached to a previously sucked teat × 1.46. (Photo G. Chapman).

Although analyses of early milk are as yet incomplete, the probable pattern of change in carbohydrate and lipid concentrations is evident. Since the concentrations of these two components are similar at about day 30 of lactation and widely divergent in late lactation, the pattern of change is apparently similar to that seen in *M. eugenii* and *D. viverrinus*, and therefore may prove to be common among the marsupials.

As lactation progresses milk intake increases so that by 50 days of age estimated milk intake may be as high as 18 ml per young per day. Whether litter size affects individual milk intake or overall milk production is not known. Estimations for *D. viverrinus* (Green *et al.*, unpublished), a species of similar adult size suggest that at 50 days of age the young quoll is consuming about 2 ml milk per day, much less than the bandicoot. Quoll milk is also less concentrated at 20% of total solids compared to the 40 to 45% of the bandicoot. An adult *M. eugenii*, of about 4 kg, supplies 2.86 ml milk per day to young aged 50 days (Green 1984). These comparisons do not take into account the quite different life histories and breeding strategies and that the young are very different in size at this age. On the other hand, details of milk composition and its rate of ingestion help to explain why the young bandicoot has so often been described as exhibiting very rapid growth (Mackerras and Smith 1960; Lyne 1964). The bandicoot, therefore, benefits from both a more concentrated and greater supply of milk. Green (1984) argued that there was considerable evidence from studies of lactation in *M. eugenii* and transfer experiments involving the same and different species (Merchant and Sharman 1966) to suggest that the growth rates of young marsupials are limited by the milk supply. The bandicoot data appear to support this hypothesis.

Lactation in eutherians is associated with changes in the mother such as increases in organ weights, food consumption and blood flow as well as a breakdown of body stores and a great inequity in the proportion of some metabolic substrates removed by the mammary glands (Hanwell and Peaker 1977; Linzell 1974; CIBA 1976). The body weights of female bandicoots change quite markedly during lactation. Some preliminary results from unpublished estimations of total body water suggest that fat storage and its mobilization may represent a significant proportion of the observed weight changes. Weight loss in the latter part of lactation is considered to be due to the increased demands of milk production which rises quickly after 30 days. Lipid is the major component of milk at this stage so that fat mobilization may represent the simplest pathway to its production. Some females in the younger age group also showed an overall upward trend in body weight which is considered to be due to a growth component.

Body weights of mothers at the end of lactation are generally similar to or even slightly higher than at parturition (Fig. 4). The initial phase of weight increase following parturition, therefore, cannot be simply a period of recovery. Comparison of patterns of body weight change in lactating and non-lactating females suggests that weight increase is a response stimulated by some aspect of pregnancy, parturition or suckling. For example, the heaviest female (Fig. 4B), which was subjectively assessed as fat, weighed 1 750 g at a parturition which followed a period of about 180 days of almost continuous non-lactation. Her body weight then increased by 242 g in the next 33 days. At her next lactation, which commenced a few days after weaning her last litter, her weight increased by 291 g from 1 812 g in 36 days. The significant correlation between litter size at birth and the magnitude of weight increase during lactation further supports this idea. Weight increase, therefore, might well be under hormonal control and mediated via the suckling stimulus. *Isoodon obesulus* (Stoddart and Braithwaite 1979) and *I. macrourus* (Hall 1983) also show a positive correlation between litter size and the body weight of the mother, further support for the above suggestion. Stoddart and Braithwaite (1979), however, also found that the oldest females produced the largest litters. In the present study there was no significance in the difference between litter size at birth in the older (mean 4.27 ± SD 1.35, n = 11) and younger females (mean 3.83 ± SD 1.38, n = 18).

The magnitude of weight increase was correlated with season, with the August–September increase being smallest. Since this study was made in captivity food was always available in excess so that food shortage may be discarded as a possible cause of lower weight increase at the start of the breeding season. A reasonable explanation is that during August–September in Canberra, when overnight temperatures may be below zero, the demands of thermoregulation may reduce the energy available for storage.

The present study did not show any significant differences in the numbers of young at birth or at weaning between August and March. Gordon (1971), however, observed that smaller litters of 2.75 young (n = 4) were born from April to June than in the period October to December when 4.10 young were born (n = 7), a value very similar to that in this study. Gemmell, Walker, Johnston and Cepon (1984) also found smaller litters in the colder months due to a greater loss of young then. They proposed that in the colder months females ceased to breed and those carrying young stopped lactating and entered anoestrus at this time. Barnes and Gemmell (1984) found a high correlation between both the number of births and the percentage of females lactating, and the rate of change of minimum temperature. My results also suggest that females are adversely affected in the colder months and show this in a reduced ability to increase weight during lactation compared to the warmer months, a possible reason for failure to breed or lactate successfully in August–September.

The magnitudes of weight increments in the periods from October through to March were not significantly different. Therefore smaller weight increases cannot be implicated in the failure of females to become pregnant in the final lactations of February–March nor the loss of three out of an overall total of seven complete litters lost in the later stages of lactation. Therefore a reduced ability to increase body weight to a sufficient level may not be the only mechanism involved in successful breeding.

Heinsohn (1966), Stoddart and Braithwaite (1979), Gemmell (1982) and Barnes and Gemmell (1984) have all implicated photoperiod as a controlling factor in breeding. The latter authors also considered additional factors and found daylength and rainfall was of lesser importance than minimum temperature. It may well be that a delicate balance of several factors, changing in importance with time, control the initiation and cessation of breeding. Hall (1983) and Gemmell *et al.* (1984) found no evidence to suggest preferential selection of any one of the eight teats in the pouch or group of teats by newborn young, and no selective mortality arising from choice of attachment, but they did not remark on the use of previously sucked teats. Hall (1983) suggested that the average litter size of less than four seemed to be advantageous since newborn young had available to them previously unsucked teats, especially since previously sucked teats were too large to allow the attachment of the newborn for about a month after the cessation of lactation. I found, however, that newborn young could successfully attach to these large teats within a few days of the previous lactation and that this was not an uncommon occurrence. Although newborn young showed a distinct preference for the smaller unsucked teats there was no significant difference in survival on the two types of teat. Whether this preference is due to the ability of the new young to detect a difference in the size of the teats or in the smell or taste of the milk is not known. The olfactory sense organs are apparently non-functional at birth (Kratzing, this volume). It is quite clear that the recently active mammary gland of late lactation is capable of very quickly undergoing the changes necessary for the renewed production of milk appropriate to the needs of the newborn young.

In general, I concluded that the continuous breeding strategy of the bandicoot does not limit litter size by rendering some teats unavailable. Instead, the bandicoot appears to have adapted well to coping with successive litters through its ability to change its milk quality very rapidly. As well as this, the body weight increases of females during the first half of lactation, when milk production is low, would appear to be an excellent way of preparing for the forthcoming drain of producing large quantities of highly concentrated milk. Whether wild bandicoots exhibit weight changes such as those observed here is not known but their milk solids are similar to those of captive females at the same stage of lactation. The study of milk composition is continuing as well as more detailed work on the energetic costs of lactation and the control and mechanism of weight change.

ACKNOWLEDGEMENTS

I would like to thank John Libke for specialized technical assistance and stimulating discussion. I would also like to thank Brian Green and Keith Newgrain for their assistance and criticism. Rod Dorey and Megan Stewart also gave able assistance with handling and feeding animals. The New South Wales National Parks and Wildlife Service gave permission to collect animals to establish the captive colony.

REFERENCES

BARNES, A. AND GEMMELL, R. T., 1984. Correlations between breeding activity in the marsupial bandicoots and some environmental variables. *Australian Journal of Zoology* **32:** 219–26.

BLAXTER, K. K., 1971. The comparative biology of lactation. Pp. 51–69 in *Lactation* ed by I. R. Falconer. Butterworths: London.

CIBA, 1976. *Breast-feeding and the mother.* Associated Scientific Publishers: Amsterdam.

CONWAY, E. J., 1962. Microdiffusion analysis and volumetric error. Crosby, Lockwood and Son: London.

FINDLAY, L. AND RENFREE, M., 1984. Growth, development and secretion of the mammary gland of macropodid marsupials. *Symposium of the Zoological Society of London.* **51:** 403–32.

GEMMELL, R. T., 1982. Breeding bandicoots in Brisbane. *Australian Mammalogy* **5:** 185–93.

GEMMELL, R. T., WALKER, M. T., JOHNSTON, G. AND CEPON, G., 1984. The number of young present in sequential litters of the marsupial bandicoot, *Isoodon macrourus,* in captivity. *Australian Journal of Zoology* **32:** 623–29.

GORDON, G., 1971. A study of island populations of the short-nosed bandicoot, *Isoodon macrourus* Gould. Ph.D. Thesis: University of New South Wales: Sydney.

GORDON, G., 1974. Movements and activity of the short-nosed bandicoot *Isoodon macrourus* Gould (Marsupialia). *Mammalia* **38:** 405–31.

GREEN, B., 1984. Composition of milk and energetics of growth in marsupials. *Symposium of the Zoological Society, London* **51:** 369–87.

GREEN, B. AND MERCHANT, J. C., 1988. The composition of marsupial milk. Pp. 41–54 in *The developing marsupial. Models for biomedical research* ed by C. H. Tyndale-Biscoe and P. A. Janssens. Springer-Verlag: Berlin.

GREEN, B. AND NEWGRAIN, K., 1979. Estimation of the milk intake of sucklings by means of ^{22}Na. *Journal of Mammalogy* **60:** 556–59.

GREEN, B., NEWGRAIN, K. AND MERCHANT, J., 1980. Changes in milk composition during lactation in the tammar wallaby *(Macropus eugenii). Australian Journal of Biological Sciences* **33:** 35–42.

GREEN, B., GRIFFITHS, M. AND LECKIE, R. M. C., 1983. Qualitative and quantitative changes in milk fat during lactation in the tammar wallaby *(Macropus eugenii). Australian Journal of Biological Sciences* **36:** 455–61.

HALL, L. S., 1983. Observations on body weights and breeding of the northern brown bandicoot, *Isoodon macrourus,* trapped in south-east Queensland. *Australian Wildlife Research* **10:** 467–76.

HANWELL, A. AND PEAKER, M., 1977. Physiological effects of lactation on the mother. *Symposium of the Zoological Society, London* **41:** 297–312.

HEINSOHN, G. E., 1966. Ecology and reproduction of the Tasmanian bandicoots (*Perameles gunni* and *Isoodon obesulus*). *University of California Publications in Zoology* **80:** 1–96.

LINZELL, J. L., 1974. Mammary blood flow and methods of identifying and measuring precursors of milk. Pp. 143–225 in *Lactation* ed by B. L. Larson and V. R. Smith. Academic Press: New York and London.

LYNE, A. G., 1964. Observations on the breeding and growth of the marsupial *Perameles nasuta* Geoffroy, with notes on other bandicoots. *Australian Journal of Zoology* **12:** 322–39.

LYNE, A. G., 1974. Gestation period and birth in the marsupial *Isoodon macrourus. Australian Journal of Zoology* **22:** 303–09.

LYNE, A. G., 1976. Observations on oestrus and the oestrous cycle in the marsupials *Isoodon macrourus* and *Perameles nasuta. Australian Journal of Zoology* **24:** 513–21.

MACKERRAS, M. J. AND SMITH, R., 1960. Breeding the short-nosed bandicoot *Isoodon macrourus* (Gould) in captivity. *Australian Journal of Zoology* **8:** 371–82.

MERCHANT, J. C., 1989. Lactation in macropodoid marsupials. Pp. 355–64 in *Kangaroos, wallabies and rat-kangaroos* ed by G. Grigg, P. Jarman and I. Hume. Surrey Beatty & Sons: Sydney.

MERCHANT, J. C., NEWGRAIN, K. AND GREEN, B., 1984. Growth of the eastern quoll, *Dasyurus viverrinus* (Shaw), (Marsupialia) in captivity. *Australian Wildlife Research* **11:** 21–29.

MERCHANT, J. C. AND SHARMAN, G. B., 1966. Observations on the attachment of marsupial young to the teats and on the rearing of pouch young by foster-mothers of the same or different species. *Australian Journal of Zoology* **14:** 593–609.

MESSER, M. AND GREEN, B., 1979. Milk carbohydrates of marsupials. II. Quantitative and qualitative changes in milk carbohydrates during lactation in the tammar wallaby *(Macropus eugenii). Australian Journal of Biological Sciences* **32:** 519–31.

SHARMAN, G. B., 1970. Reproductive physiology of marsupials. *Science, N.Y.* **167:** 1221–28.

STODDART, D. M. AND BRAITHWAITE, R. W., 1979. A strategy for utilization of regenerating heathland by the brown bandicoot (*Isoodon obesulus*; Marsupialia, Peramelidae). *Journal of Animal Ecology* **48:** 165–79.

TYNDALE-BISCOE, C. H., STEWART, E. AND HINDS, L. A., 1984. Some factors in the initiation and control of lactation in the tammar wallaby. *Symposium of the Zoological Society of London* **51:** 389–401.

The seminiferous tubules, rete testis and efferent ducts in peramelid marsupials

P. A. WOOLLEY[1]

The anatomy of the seminiferous tubules, rete testis and excurrent duct system of the testis of bandicoots of the family Peramelidae has been investigated and comparisons made with these structures in members of other families of marsupials, both Australian and American. In peramelids the seminiferous tubules are of large diameter, there are relatively low numbers of tubuli recti and the rete is a simple duct. The urinogenital connection appears to occur outside the body of the epidiymis, in the single long excurrent duct.

Key words: Anatomy, seminiferous tubules, tubuli recti, rete testis, efferent ducts, peramelid marsupials.

INTRODUCTION

STUDIES on the anatomy of the rete and the excurrent duct system of the testis of bandicoots of the family Peramelidae are few. They include those of Fraser (1919) on the development of the rete in pouch young of *Perameles* sp., de Burlet (1921) on the form of the seminiferous tubules and their connections with the duct draining the testis in a pouch young of *Perameles obesula* (= *Isoodon obesulus*) and Hughes (1965), who noted the presence of a single relatively long efferent duct connecting the testis and epididymis in *P. nasuta* and *I. macrourus*. The observations made on these species suggest that the rete in peramelids consists of a simple duct system and that a single excurrent duct connects the testis and epididymis. In these respects peramelids are similar to Australian marsupials of the families Dasyuridae, Thylacinidae, Myrmecobiidae and Tarsipedidae, but different from the Phalangeridae and Macropodidae. They also share similarities in rete testis structure with the American families Caenolestidae and Didelphidae, but not the Microbiotheriidae (Woolley 1987).

The present study was undertaken to provide information on a wider range of species in five of the eight currently recognized genera of the Peramelidae. Observations have been made on the diameter of the seminiferous tubules, the number of tubuli recti and the form of the rete and the duct connecting the rete with the epididymis.

MATERIALS AND METHODS

Seven species of peramelids were examined. The collection numbers and source of the 10 individuals studied follow: two *Peroryctes papuensis* (M 21337, M 21338—Western Australian Museum) collected by the author in Central Province, Papua New Guinea; one *Echymipera kalubu* (M 25417 — Western Australian Museum) collected by G. L. Grossek in Morobe Province, Papua New Guinea; one *Perameles gunnii* (FWD 13101), a road kill, collected at Hamilton, Victoria and donated by P. R. Brown, Arthur Rylah Institute for Environmental Research; one *P. bougainville* (M 4516 — Western Australian Museum) collected on Bernier Island, Western Australia; one *P. nasuta* (M 11830 — Australian Museum) collected 60 km north-west of Grafton, New South Wales; two *Macrotis lagotis* (M 18067 — Western Australian Museum), a road kill from the Kimberley region of Western Australia and (NT 17), a captive-bred specimen donated by K. A. Johnson, Conservation Commission of the Northern Territory; two *I. macrourus*, (LTU 5) donated by S. Armstrong, La Trobe University, and (S 63) donated by J. C. Merchant, CSIRO Division of Wildlife and Ecology, both captive-bred specimens derived from stock obtained from the vicinity of Dorrigo, New South Wales.

Body weight, testis length and width, and length of the excurrent duct, from point of emergence from the testis to point of entry into the epididymis, were obtained for most specimens. Two testes, one from *P. gunnii* (FWD 13101), and one from *I. macrourus* (LTU 5), were partially dissected while fresh to expose the rete. Testis and epididymal material from *P. papuensis*, *E. kalubu* and *I. macrourus* was fixed while fresh in aqueous Bouin's solution. The post-mortem history and method of fixation of the material from *P. bougainville*, *P. nasuta* and *M. lagotis* was not known. The *P. gunnii* had been frozen. The anterior half of one testis and a portion of the excurrent duct from nine individuals (representing all species except

[1]Department of Zoology, La Trobe University, Bundoora, Victoria, Australia 3083.

Pages 229–34 *in* BANDICOOTS AND BILBIES, ed by J. H. Seebeck, P. R. Brown, R. L. Wallis and C. M. Kemper. Surrey Beatty & Sons, Sydney, 1990.

P. gunnii), and the head of one epididymis and remainder of the excurrent duct from three individuals (*P. papuensis, P. bougainville* and *I. macrourus*) were serially sectioned at 8 or 10 μm and stained with haematoxylin and eosin.

A drawing mirror attached to a microscope was used to project the sections of each testis to make outline drawings of the rete and the connections of the seminiferous tubules to it. From these drawings a count of the number of tubuli recti was obtained.

The diameter of the seminiferous tubules in sections was measured using a microscope with a measuring eyepiece calibrated with a stage micrometer. Only tubules containing spermatozoa as well as other stages of the spermatogenic cycle were measured. The form of the excurrent duct between the testis and epididymis was analysed visually from the serial sections.

RESULTS

In all species of peramelids examined, a single non-coiled excurrent duct connecting the testis with the epididymis could be seen macroscopically. The duct emerged from the tunica albuginea of the testis on the dorsal surface close to the anterior pole, in close association with the testicular blood vessels. It continued, suspended in the mesorchium, to enter the caput epididymis. The length of the duct was approximately 4 mm in *P. papuensis*, 5 mm in *E. kalubu*, 8 mm in *I. macrourus* and *P. bougainville*, and 10 mm in *P. gunnii*.

When the tunica albuginea was removed from the unfixed testes of *P. gunnii* and *I. macrourus* the duct could be seen to lie free for a short distance before entering the tubular tissue alongside the testicular artery. Partial separation of the seminiferous tubules showed that they

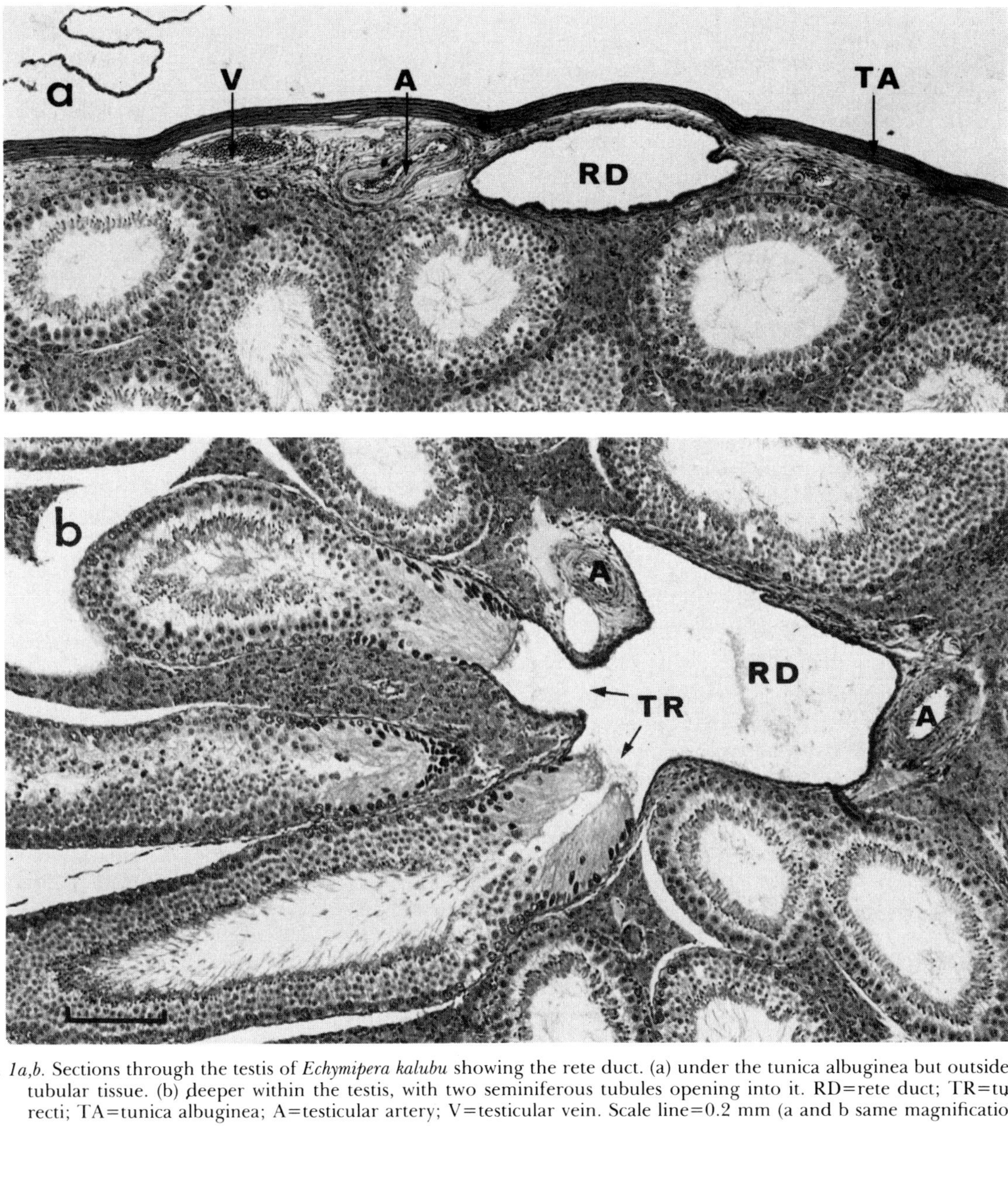

Fig. 1a,b. Sections through the testis of *Echymipera kalubu* showing the rete duct. (a) under the tunica albuginea but outside the tubular tissue. (b) deeper within the testis, with two seminiferous tubules opening into it. RD=rete duct; TR=tubuli recti; TA=tunica albuginea; A=testicular artery; V=testicular vein. Scale line=0.2 mm (a and b same magnification).

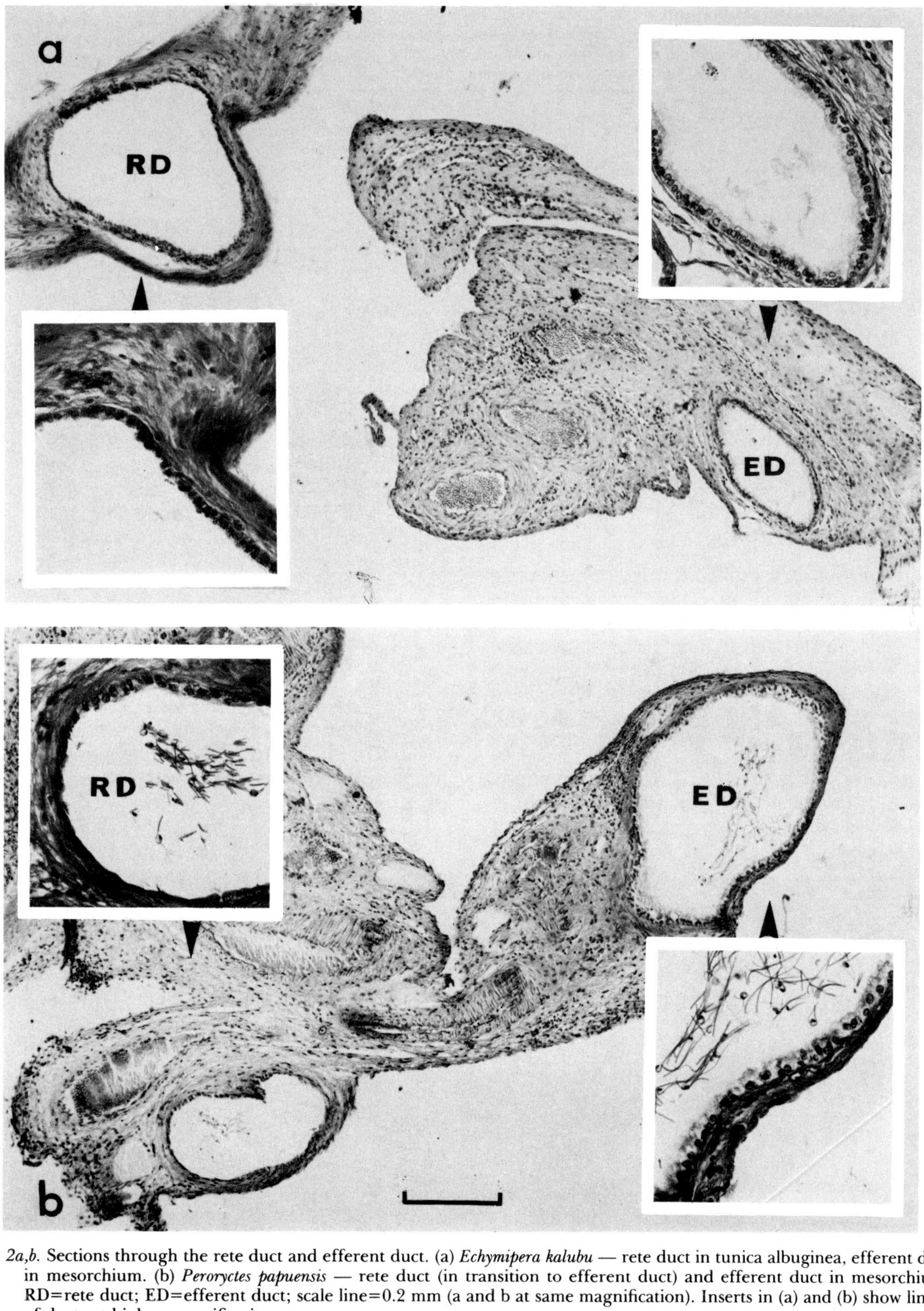

Fig. 2a,b. Sections through the rete duct and efferent duct. (a) *Echymipera kalubu* — rete duct in tunica albuginea, efferent duct in mesorchium. (b) *Peroryctes papuensis* — rete duct (in transition to efferent duct) and efferent duct in mesorchium. RD=rete duct; ED=efferent duct; scale line=0.2 mm (a and b at same magnification). Inserts in (a) and (b) show lining of ducts at higher magnification.

made connection with the duct in the anterior pole of the testis, as in dasyurid marsupials (Woolley 1975). The ends of the tubules tapered before joining the short terminal branches of the rete duct.

In the testes of the six species examined histologically the rete duct was seen to be lined with a single layer of squamous to very low columnar non-ciliated cells with dark-staining nuclei. (Figs 1a,b). Slight folding and a few villous projections of the epithelium of the rete duct were seen in some specimens. Spermatogenic cells were absent from the tapered ends of the seminiferous tubules, which were lined with tall, pale-staining cells (Fig. 1b). No distinct

Table 1. Data on the seminiferous tubules and rete testis in peramelid marsupials. Tissues of some specimens (asterisked) were poorly fixed and shrinkage of the tissues, evidenced by folding of the basement membrane and collapse of the seminiferous tubules, had occurred. Body weight obtained from preserved specimens shown in parentheses.

Species	No.	Body weight (g)	Testis (mm)	Diameter of tubules (mm)	No. of tubuli recti
Peroryctes papuensis	M 21337	150	13.2×9.6	0.45	12
Peroryctes papuensis	M 21338	145	13.0×9.2	0.46	12
Echymipera kalubu	M 25417	380	15.6×10.3	0.38	20
Perameles bougainville	M 4516*	(200)	14.5×10.5	—	24
Perameles nasuta	M 11830*	(680)	—	—	22
Macrotis lagotis	M 18067*	—	—	—	16
Macrotis lagotis	NT 17*	(1 850)	19.0×11.5	—	18
Isoodon macrourus	LTU 5	—	18.0×14.5	—	40
Isoodon macrourus	S 63	2 884	22.5×17.5	0.37	44

mediastinum was found in any species. Before entering the tubular tissue the rete duct was surrounded by a layer of loose connective tissue (Fig. 1a) but deeper within the testis there was no distinct layer (Fig. 1b).

The number of tubuli recti ranged from 12 in *P. papuensis* to 44 in one *I. macrourus*, and the diameter of the seminiferous tubules, in well-fixed material, from 0.37 to 0.46 mm (Table 1). Although a few spermatozoa were seen in the epididymis of *I. macrourus* LTU 5, spermatogenesis was no longer occurring and the epithelium lining the tubules appeared abnormal.

The cells lining the rete duct were generally flattened, non-ciliated and dark-staining where the duct passed through the tunica albuginea, but a short distance from the testis the lining cells were taller, ciliated and pale-staining (Figs 2a,b). Between the testis and epididymis the duct followed a straight course and no further change in the epithelium was seen as the duct approached (Fig. 3a) and entered (Fig. 3b) the caput epididymis. Within the caput epididymis it followed a convoluted course. There was a reduction in diameter but whether the duct remained single or divided into two or more branches before it connected with the epididymal duct(s) was not determined. No marked change in the lining epithelium was associated with the reduction in diameter.

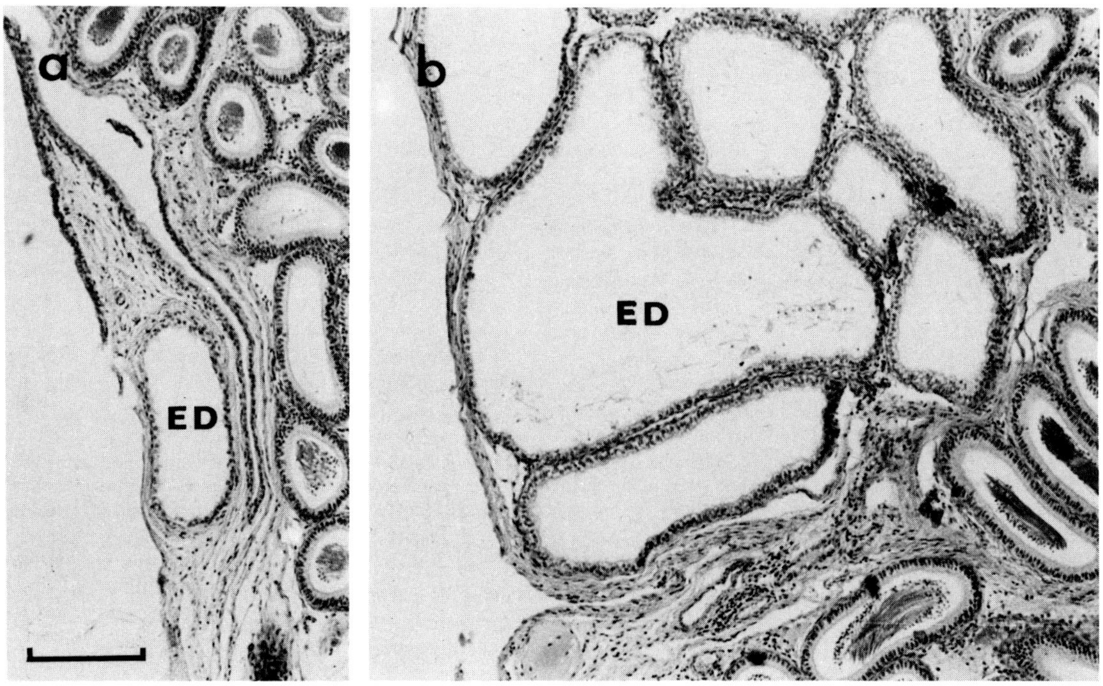

Fig. 3a,b. Sections through the anterior portion of the epididymis of *Peroryctes papuensis.* (a) the single efferent duct external to the epididymis. (b) efferent duct within the caput epididymis. ED=efferent duct; scale line=0.2 mm (a and b at same magnification).

DISCUSSION

The gross morphology of the seminiferous tubules was not directly investigated because in both species (*P. gunnii* and *I. macrourus*) from which fresh testes were available for dissection the testes were large and the seminiferous tubules delicate. De Burlet (1921), by modelling serial sections, found that the testis of a pouch young of *P. obesula* (=*I. obesulus*) contained six single loops and one three-limbed loop, with 15 connections between the tubules and the duct draining the testis. With the exception of *I. macrourus* the species examined in the present study had from 12 to 24 tubule-duct connections, suggesting that the form of the seminiferous tubules may be relatively simple as in *I. obesulus*, at least in the species with the low numbers of tubuli recti.

In peramelids the number of tubuli recti ranged from 12 in the smallest of the species examined (*P. papuensis*) to 44 in the largest (*I. macrourus*), providing further support for the general relationship between body size and number of tubuli recti seen in dasyurid and didelphid marsupials (Woolley 1975, 1987).

The diameter of active seminiferous tubules in well-preserved material from the peramelids examined ranged from 0.37 to 0.46 mm. The diameter has also been recorded for *P. nasuta* (0.355 mm, Setchell 1977) and *Perameles* sp. (0.39 mm, Benda 1906). The seminiferous tubules of peramelids thus are among the widest known in mammals (see Woolley 1975, 1987) and are comparable in width to those of some other marsupials e.g., dasyurids (0.36 to 0.52 mm, Woolley 1975); the thylacinid, *Thylacinus cynocephalus* (0.43 mm, Woolley unpublished) and the tarsipedid, *Tarsipes rostratus* (0.35 mm, Woolley and Scarlett 1984).

In all species of peramelids examined the rete was found to be a simple duct, as in dasyurids (Woolley 1975), *Tarsipes rostratus* (Woolley and Scarlett 1984) and didelphids and caenolestids (Woolley 1987). A single duct emerges from the testis and follows a straight course to enter the head of the epididymis. The nature of the excurrent duct in peramelids has not previously been studied. If, as stated by Gier and Marion (1970), rete ducts are lined with non-ciliated cells and efferent ducts (of mesonephric origin) with ciliated cells, the change in the lining epithelium of the duct of peramelids which occurs after its emergence from the testis suggests that the rete duct connects with a single efferent duct. If this is so the urinogenital connection occurs outside the epididymis in peramelids, as it does in American marsupials (Woolley 1987) and not within it, as in Australian dasyurid (Woolley 1975) and tarsipedid (Woolley and Scarlett 1984) marsupials. Fraser (1919) concluded that the urogenital connection in *Perameles* sp. involved a single rete canal and a small number of mesonephric tubules from which the epididymis arose but she did not establish where the connection occurred. The recent study by Ullman (1981) on the sexual differentiation of the gonads in bandicoots does not extend to observations on the rete duct and the urinogenital connection, and further studies are required to elucidate events during development of the system in bandicoots and other marsupials.

Among the medium sized marsupials, peramelids, with a long (up to 10 mm) single excurrent duct, would seem to offer advantages over species with multiple ducts for the collection of rete fluid from the testis by cannulation of the duct.

ACKNOWLEDGEMENTS

I wish to thank Dr D. J. Kitchener, Miss L. Gibson, Mr K. A. Johnson, Mr P. R. Brown, Mr J. C. Merchant, Dr S. Armstrong and Mr G. L. Grossek for providing most of the material upon which this paper is based. Assistance from Jennifer Cheney, especially in the careful preparation of serial sections, and from David Walsh for photomicrography is gratefully acknowledged.

NOTE ADDED IN PROOF

An eighth species, *Microperoryctes* sp. (M 25375 — Papua New Guinea Museum), collected by I. Craven in the Arfak Mountains in Irian Jaya, which may be distinct from *M. murina* (J. Menzies, pers. comm.), has been examined. One testis and epididymis were available for examination and the anterior half of both were serially sectioned. Body weight of the individual was not obtained. The intact testis measured 17 × 10 mm and there were 18 tubule-duct connections. Spermatogenesis was occurring but because the tissues were poorly fixed and shrunken the diameter of the seminiferous tubules was not measured. Features of the rete duct were similar to those in the testes of the other species examined, and the urinogenital connection occurred outside the body of the epididymis.

REFERENCES

BENDA, C., 1906. Die spermiogenese der marsupialier. *Denkschriften der Medizinisch — Naturwissenschaftlichen Gellschaft (Jena)* **6:** 441–58.

DE BURLET, H. M., 1921. Zur Entwicklung und Morphologie des Saugerhodens. II. Marsupialer. *Zeitschrift für Anatomie und Entwicklungsgeschichte* **61:** 19–31.

FRASER, E. A., 1919. The development of the urogenital system in the Marsupialia, with special reference to *Trichosurus vulpecula*. Part II. *Journal of Anatomy* **53:** 97–129.

GIER, H. T. AND MARION, G. B., 1970. Development of the mammalian testis. Pp. 1–45 in *The testis* ed by A. D. Johnson, W. R. Gomes and N. L. VanDemark. Academic Press: New York.

HUGHES, R. L., 1965. Comparative morphology of spermatozoa from five marsupial families. *Australian Journal of Zoology* **13:** 533–43.

SETCHELL, B. P., 1977. Reproduction in male marsupials. Pp. 411–57 in *The biology of marsupials* ed by B. Stonehouse and D. Gilmore. Macmillan: London.

ULLMAN, S. L., 1981. Sexual differentiation of the gonads in bandicoots (Peramelidae, Marsupialia). Pp. 41–51 in *Development and function of reproductive organs* ed by A. G. Byskov and H. Peters. Elsevier North-Holland: Amsterdam.

WOOLLEY, P. A., 1975. The seminiferous tubules in dasyurid marsupials. *Journal of Reproduction and Fertility* **45:** 255–61.

WOOLLEY, P. A., 1987. The seminiferous tubules, rete testis and efferent ducts in didelphid, caenolestid and microbiotheriid marsupials. Pp. 217–27 in *Possums and opossums: studies in evolution* ed by M. Archer. Royal Zoological Society of New South Wales: Sydney.

WOOLLEY, P. A. AND SCARLETT, G., 1984. Observations on the reproductive anatomy of male *Tarsipes rostratus* (Marsupialia: Tarsipedidae). Pp. 445–50 in *Possums and gliders* ed by A. Smith and I. Hume. Australian Mammal Society: Sydney.

△ PLATE 1: *Perameles bougainville*, Bernier Island, Shark Bay, Western Australia, 1975.
Photo: A. C. Robinson

▽ PLATE 2: *Perameles gunnii*, Hamilton, Victoria, 1984.
Photo: P. R. Brown

△ PLATE 3: *Isoodon auratus*, Barrow Island, Western Australia.

Photo: K. A. Johnson

▽ PLATE 4: *Isoodon macrourus*, Townsville, Queensland, 1975.

Photo: A. C. Robinson

△ PLATE 5: *Isoodon macrourus,* Kimberleys, Western Australia, June 1989.

Photo: D. F. Gibson

▽ PLATE 6: *Isoodon obesulus,* Cranbourne, Victoria, May 1982.

Photo: B. Lobert

△ PLATE 7: *Isoodon obesulus*, Franklin Islands, South Australia.

Photo: H. J. Aslin

▽ PLATE 8: *Macrotis lagotis*, Northern Territory, 1984.

Photo: R. I. Southgate

△ PLATE 9: *Macrotis lagotis,* Northern Territory, 1983.

Photo: H. McCracken

▽ PLATE 10: *Echymipera clara,* West Irian, June 1988.

Photo: T. F. Flannery

△ PLATE 11: *Echymipera kalubu*, Wau, Morobe Province, Papua New Guinea.
Photo: P. A. Woolley and D. Walsh

▽ PLATE 12: *Peroryctes broadbenti*, Port Moresby, Papua New Guinea, July 1969.
Photo: Roy D. Mackay

△ PLATE 13: *Peroryctes raffrayanus*, Baiyer River Sanctuary, Papua New Guinea, December 1985. Photo: Roy D. Mackay

▽ PLATE 14: *Microperorctes longicauda*, Papua New Guinea, April 1987. Photo: H. Cogger

△ PLATE 15: *Microperoryctes longicauda,* Neon Basin, Papua New Guinea, December 1981.
Photo: T. F. Flannery

▽ PLATE 16: *Chaeropus ecaudatus,* possibly from Ryan Wells, Northern Territory, 1891.
Photo: South Australian Museum

Ultrastructure of spermatozoa and epididymal sperm maturation in some perameloids

H. R. HARDING[1], C. D. SHOREY[2] and K. W. CLELAND[2]

Spermatozoa of *Perameles nasuta* and *Isoodon macrourus* are very similar to one another and show a number of unique features compared with those of non-perameloid marsupials. These include: a distinctive and complex sperm head shape; rostral location of the acrosome; the very extensive ventral nuclear groove; location of the sperm neck close to the anterior of the head, in a hollowed-out portion of the ventral nucleus which forms an extension of the ventral groove; marked heterogeneity of the nuclear contents; a distinctive diamond-shaped anterior principal piece as seen in transverse section; lack of the well-developed mid and principal piece fibre networks typical of marsupials. In common with the dasyurids and *Tarsipes rostratus* the dense outer fibres are widely separated from the axoneme doublets, but in contrast to these groups the midpiece is not flattened.

The ultrastructural changes apparent during epididymal sperm maturation are relatively slight in these peramelids compared with most other marsupials. Most unusual is the very marked relocation of the neck from its position in the implantation fossa in immature spermatozoa to one close to the anterior of the nucleus within the extension of the ventral groove in mature spermatozoa.

Preliminary observations of spermatozoa from *Perorcytes longicauda* and *Echymipera kalubu* suggest that there is consistency in sperm form within the Perameloidea. However, sperm head structure in *Macrotis lagotis* has yet to be examined.

The distinctive features of perameloid spermatozoa coupled with their large size (intermediate between those of the diprotodonts and the giant spermatozoa of *Tarsipes* and the dasyurids) raises interesting structure-function and phylogenetic questions. Little can be concluded regarding the phylogenetic significance of perameloid sperm structure at present, but it is suggested that the combination of distinctive features and intermediate sperm size, relative to other Australian marsupials, make spermatozoa of this group potentially useful for structure-function studies.

Key words: Perameloidea (Marsupialia), bandicoots, *Perameles nasuta, Isoodon macrourus,* spermatozoa, ultrastructure, epididymal sperm maturation.

INTRODUCTION

ALTHOUGH the information available on spermiogenesis in *Perameles nasuta* almost certainly exceeds that for any other mammal, a comprehensive description of the mature spermatozoon and epididymal sperm maturation has not yet been published for any perameloid species.

Since the papers of Sapsford, Rae and Cleland (1967, 1969a,b, 1970) provide a very detailed and comprehensive description of all stages of spermiogenesis in *P. nasuta* through to the movement of the spermatids into the lumen of the seminiferous tubules (Sapsford and Rae 1969), these phases of sperm development will not be discussed in this paper.

In contrast, the information available on mature sperm ultrastructure and epididymal sperm maturation is very incomplete. Using light microscopy Cleland (1955) described the sperm tail of *P. nasuta,* and Hughes (1965) included *P. nasuta* and *Isoodon macrourus* in a comparative study of marsupial spermatozoa. At the ultrastructural level Cleland and Rothschild (1959) have provided a detailed description of the principal piece of the sperm tail in *P. nasuta*, Harding (1977) and Harding, Carrick and Shorey (1977, 1979, 1983) have briefly described sperm structure and maturation in *P. nasuta* and *I. macrourus* as part of comparative studies, and Cummins (1980) included *I. macrourus* in a comparative study of nuclear stability and has briefly described epididymal sperm maturation in this species (Cummins 1979). Spermatozoa from peramelids other than *P. nasuta* and *I. macrourus* have not been described in any detail. Hughes (1965) made preliminary light microscope observations on spermatozoa from *P. gunnii, I. obesulus* and *Echymipera rufescens* and found them to be comparable in morphology to those of *P. nasuta* and *I. macrourus.*

These studies have shown *P. nasuta* and *I. macrourus* spermatozoa to be very similar to one another and to have unique features of the sperm head and the mid- and principal pieces which set them apart from other marsupials. Certain similarities in flagellar structure with the

[1]Centre for Liberal and General Studies, University of New South Wales, P.O. Box 1, Kensington, New South Wales, Australia 2033.
[2]Department of History and Embryology, University of Sydney, New South Wales, Australia 2006.
Pages 235-50 *in* BANDICOOTS AND BILBIES, ed by J. H. Seebeck, P. R. Brown, R. L. Wallis and C. M. Kemper. Surrey Beatty & Sons, Sydney, 1990.

dasyurids and *Tarsipes rostratus* are also evident, and the possible phylogenetic implications of these have been discussed previously (Harding 1987; Harding, Carrick and Shorey 1981; Harding, Woolley, Shorey and Carrick 1982; Temple-Smith 1987).

In this paper we will examine, in greater detail than previously, mature sperm ultrastructure and epididymal sperm maturation in *P. nasuta* and *I. macrourus,* and provide preliminary comparative comments for *Macrotis lagotis, E. kalubu* and *Perorcytes longicauda.*

MATERIALS AND METHODS

Epididymal samples have been examined from seven *P. nasuta* and two *I. macrourus*. For one animal of each species, sperm samples were obtained from the caput, corpus and cauda epididymis and processed using the methods described in Harding, Carrick and Shorey (1975). For the remaining *I. macrourus,* samples were obtained from two additional intermediate regions, the caput-corpus and corpus-cauda epididymis. These samples were removed and processed using methods described previously (Harding *et al.* 1982). Samples from this animal were also examined by light microscopy. The specimens for light microscopy were prepared by placing small pieces of fixed tissue which had been stored in cacodylate/sucrose buffer on a slide and teasing the tissue apart to release the sperm. Remaining pieces of tissue were removed and the wet slide covered with a cover slip. The slides were examined using a Leitz Dialux 20 light microscope employing positive phase contrast. The spermatozoa were photographed and a calibrated grid slide (10 μm intervals) was photographed at the same magnifications. Measurements were made following printing of the calibrated grid and spermatozoa negatives at the same enlargement.

The six remaining *P. nasuta* were animals sampled by Cleland in 1964. Cauda epididymidal samples were taken from five of these animals and caput samples from the sixth. This material was fixed in Palade's (1% osmium tetroxide in veronal-acetate buffer at pH 7.4) for periods ranging from one to two hours at 0°C. It was embedded in araldite.

Two *M. lagotis* have been sampled, but in both cases the spermatozoa were very degenerate. For one animal the samples were removed whilst the animal was still alive under anaesthetic using the methods described by Harding *et al.* (1982). The animal was old and in poor condition and presumably this accounted for the degenerate condition of the spermatozoa. The other animal had been preserved whole in alcohol for an unknown period of time (some years) before sampling. Sampling was carried out as for the other *M. lagotis*. In this case inadequate, and for electron microscopy purposes, inappropriate fixation was probably responsible for the poor sperm condition.

Preliminary observations have been made on epididymal samples from two *E. kalubu* and one *P. longicauda* from New Guinea. The epididymis of *P. longicauda* was frozen in liquid nitrogen before thawing and processing using the fixation methods described in Harding *et al.* (1982). The samples from *E. kalubu* were fixed in a 3%/2% mixture of glutaraldehyde/formaldehyde. Post-fixation techniques were as for the *Perorcytes* specimens. Sample from *E. kalubu* were also examined by light microscopy using the methods described above.

For transmission electron microscopy, specimens were examined using a Philips EM 200, EM 201, EM 300 or JEOL 100 CX.

OBSERVATIONS AND DISCUSSION

The information below is based on *P. nasuta* and *I. macrourus*. Only very preliminary observations have been made on spermatozoa from *Perorcytes, Echymipera* and *Macrotis* and these will be summarized at the end of this section.

Mature epididymal spermatozoa

Perameloid spermatozoa are large by mammalian standards. Among the eutherian mammals only some members of the families Cricetidae (hamsters) and Muridae (rats) have larger spermatozoa (Cummins and Woodall 1985). Table 1 shows that, compared to other Australian marsupials, they are exceeded in size only by spermatozoa of the dasyurids and *Tarsipes rostratus; Tarsipes* spermatozoa are by far the largest reported to date for mammals (Cummins and Woodall 1985).

Spermatozoa of *P. nasuta* and *I. macrourus* are very similar to one another and display a number of unique features in both head and flagellar structure compared with non-peramelid marsupials. Despite their close similarity, spermatozoa of these species can be distinguished at least by size differences (Table 1) and probably also by minor differences in head form.

Viewed dorsally or ventrally or in longitudinal or transverse sections, the sperm heads of *P. nasuta* and *I. macrourus* have a unique and complex form compared to other marsupials. As is typical of marsupials, they are dorsoventrally flattened with the neck inserted into what is conventionally described as the ventral surface of the nucleus rather than into its caudal pole as in eutherians (Harding *et al.* 1979).

Except for the hooked heads of the vombatiforms (Koala and wombats) the sperm heads of other Australian marsupials all display

Table 1. Dimensions of perameloid spermatozoa compared with those of other Australian marsupials.

	Head length μm	Midpiece length μm	Total sperm length μm	Reference
Perameles nasuta	5.7	14	199.8	Hughes 1965
Isoodon macrourus	6.0	10.7	171.1	Hughes 1965
Echymipera kalubu	5.8	19.3	—	1
Dasyurids	Range 9.5–12.7	Range 34–40	Range 218–255	Hughes 1965 Harding *et al.* 1982
Tarsipes rostratus	10.6	91	342.6	Woolley and Scarlett 1984
	12.4			Harding *et al.* 1984
	13.97	86	356.3	Cummins and Woodall 1985
Macropodoids–phalangeroids[2]	Range 4–9	Range 8–16	Range 106–124	Hughes 1965 Cummins and Woodall 1985

[1]Harding, Archer, Aplin *et al.* unpublished, very approximate preliminary measurements.
[2]Excluding *Distoechurus pennatus* which has a total length of approximately 150 μm (Harding, Archer, Aplin *et al.* unpublished observations).

proportional varations of a similar theme. In most cases they have a fairly simple profile and viewed dorsally or ventrally they are pointed or rounded anteriorly and rounded or blunt posteriorly and the lateral margins if showing curvature at all, are convex. By contrast, the profile of perameloid spermatozoa is far from simple and viewed dorsally or ventrally the head is rounded anteriorly but widens fairly abruptly (Figs 1, 7). The lateral margins are concave, although this is more pronounced in *I. macrourus* than in *P. nasuta,* and posteriorly the head flanges out such that the lateral width close to its anterior and posterior margins is equivalent (Figs 1, 7).

Sequences of longitudinal (in both the dorso-ventral (Figs 2, 9, 10) and lateral (Fig. 7) planes) and transverse sections (Figs 3–6, 8) show that the profile of the sperm head is a complex series of depressions and protuberances rather than the generally simple contours seen in other marsupial spermatozoa (Harding 1977, 1987).

To some degree the complex profile is illusory. Close examination of Figures 2–10 shows that the nucleus has two components, the electron-dense chromatin typical of sperm nuclei, and a granular relatively electron-lucent component which surrounds the dense chromatin, particularly on the dorsal surface posteriorly from the region of the implantation fossa, in the posterior ventral region and within the anterior ventral groove. The uneven profile displayed by the nucleus is to some extent due to the uneven profile of its very visible electron dense component while the electron-lucent component fills in many of the gaps giving a more even profile to the sperm head than is readily obvious. Additionally, the electron-dense component is pitted or vesiculated, particularly in its posterior and ventral extremities and anteriorly at the point where the sperm head widens laterally (Figs 2–8).

The only other Australian marsupial group in which the nuclear contents consistently appear unevenly condensed in mature spermatozoa is the Dasyuridae. However, the relatively electron-lucent granular component seen in the perameloids is not found. Instead the surface of the nucleus, and particularly the ventral surface, is frequently more electron-dense than its interior. The demarcation between the two zones is distinct and the more electron dense zone is divided by apparently empty, electron-lucent channels (Harding *et al.* 1982). It is not clear whether these unusual nuclear modifications in the perameloids and dasyurids are similar in nature or function.

Sapsford *et al.* (1969a) observed zones of differing electron density in the nucleus of *P. nasuta* during spermiogenesis and considered possible functions of the parachromatin. They found that pores in the nuclear membrane only occurred in the regions underlain by parachromatin and suggested that the parachromatin may represent "dispensable" nuclear material whose disposal is facilitated by its proximity to nuclear pores. Relatively electron lucent regions associated with nuclear pores in the ventral or caudal regions of the nucleus are common in spermatids of other marsupials (Sapsford *et al.* 1969a; Rattner 1972; Harding 1977) and eutherians (Franklin 1968). Although positive proof is lacking, it seems likely that these electron-lucent zones adjacent to pore-containing regions of the nuclear envelope are functionally similar in eutherian and marsupial spermatids, serving to facilitate dispersal of redundant nuclear material. However, this gives no insight into possible functions of the parachromatin in mature spermatozoa, since presumably reduction in nuclear volume is complete by this stage.

In all Australian marsupial species except the vombatiforms, the ventral nuclear surface is grooved over at least its posterior half and the neck is inserted near the anterior extremity of the groove (Hughes 1965; Harding 1977). This groove makes possible the streamlined form of the mature spermatozoon since it accommodates the flagellar structures in the region in which the nucleus overlies the neck and midpiece (Figs 2, 10). As previously reported by Hughes (1965) using light microscopy the ventral groove, which is broad posteriorly and narrow anteriorly, is very extensive in the perameloids reaching almost to the anterior tip of the nucleus (Figs 2–7).

Among mammals the depression in the nucleus for articulation of the neck with the sperm head is called the implantation fossa. This is characteristically deep in marsupials (Figs 8, 9) compared with eutherians (Harding 1977). In eutherians the implantation fossa is lined with a thick layer of electron dense material called the basal plate, and a large number of fine filaments connect the basal plate to the articular surface of the neck. It is believed that these structures effect head-tail attachment (Fawcett 1975). The implantation fossa in Australian marsupials is likewise lined with structures (Figs 8, 9) which appear to perform a similar attachment function in spermatozoa from the seminiferous tubules of the testis and from the proximal regions of the epididymis (Harding 1977). As discussed below, the neck becomes dislocated from the implantation fossa as the head takes on a parallel alignment with the flagellum during maturation (Figs 2, 10). Whilst dislocation occurs to some extent in all Australian marsupials studied to date except *Phascolarctos cinereus* in which realignment of the head relative to the flagellum does not occur (Harding 1977), it is most extreme in the perameloids. In mature spermatozoa of *P. nasuta* and *I. macrourus* the neck is located almost at the anterior extremity of the extensive ventral nuclear groove (Figs 2, 7, 10). In this region the groove is both ventrally and laterally bounded by the nucleus and hence more precisely forms a hollowing-out of the nucleus (Figs 2, 4).

This dislocation of the neck from the implantation fossa raises the question of how head-flagellar union is maintained in mature marsupial spermatozoa, other than by containment of both structures within the sperm plasma membrane. Figure 2 shows that in the peramelids the neck is very deeply inserted into the hollow section of the anterior nucleus. Even though specific connecting structures are not apparent, it seems likely that over most directions of stress this deep insertion would maintain head-tail attachment. The situation in the other Australian marsupials which lack this pronounced anterior development of the ventral groove has not been examined in detail, but at least in the dasyurids a secondary depression in the nucleus is evident, and this accommodates the neck in its mature position (Harding 1977). However, as in the perameloids, connecting structures are not evident in association with the neck in the secondary fossa. In terms of specific connecting structures, the ultrastructural evidence suggests that the head-flagellar connection is not as secure in mature marsupial spermatozoa as it is in immature spermatozoa. Perhaps this makes good functional sense. It is likely that the head and flagellum are not as subject to separation in their mature streamlined alignment as they are in immature spermatozoa in which the long axis of the sperm head is angled at varying degrees to the long axis of the flagellum (Fig. 9). Sapsford *et al.* (1969a) previously reported the changing neck insertion in *P. nasuta* and suggested that a secure head-neck joint may be necessary during certain stages of spermiogenesis, whilst the dislocation is required to facilitate streamlining.

Description of the complex array of structures which comprise the sperm neck is beyond the scope of this paper. Brief comparative observations and discussion have previously suggested that the neck in many Australian marsupial species is more complex than in most eutherians and that among Australian marsupials, the neck structures are particularly well developed in *P. nasuta* and *I. macrourus* (Harding 1977).

The acrosome in mature spermatozoa of *P. nasuta* and *I. macrourus* covers the rostrum of the nucleus which is very visible at the light microscope level as a rounded projection in this region (Fig. 1). This location is unique among Australian marsupials in which the acrosome is characteristically restricted to the dorsal nuclear surface except in certain pseudocheirid species in which it may also extend down the lateral surfaces of the nucleus (Harding 1987). In longitudinal sections cut in a dorsal-ventral plane the acrosome is seen to have (at least over its anterior two-thirds) a fairly smooth convex profile (Figs 2, 9, 10) as is common among other Australian marsupials, but laterally cut longitudinal sections (Fig. 7) and transverse sections (Fig. 3) show that in these dimensions the acrosome conforms to the underlying nuclear profile and has a more complex saddle shape.

Like the sperm head, the flagellum in *P. nasuta* and *I. macrourus* spermatozoa also has a number of unique features. As reported previously (Harding 1977, 1987; Harding *et al.* 1977, 1979, 1981, 1982), perhaps most interesting from both phylogenetic and functional perspectives is the wide separation of the dense outer fibres from their underlying axoneme doublets over much of the length of the flagellum (Figs 11, 13). At the anterior extremity of the midpiece where the nucleus overlies the mitochondrial sheath, the dense outer fibres lie fairly close the the underlying axoneme (Fig. 6), but in all sections of the midpiece posterior to the head, and over most of the principal piece, the dense outer fibres are fairly widely separated from the axoneme and each is connected to its corresponding axoneme doublet by paired fibrils (the connecting laminae). The separation is greatest for dense outer fibres 1, 5 and 6 and least for 3 and 8, and for the latter two the connecting laminae appear single or absent rather than double[1] (Figs 11, 13).

Cleland and Rothschild (1959) have provided a detailed description of the arrangement of these connecting laminae in *P. nasuta.* The paired components of the connecting laminae are each fibrils of between 20–25 nm in diameter running unbroken from the dense outer fibre to the axoneme. Longitudinal sections (Fig. 15) show that the fibrils are discrete structures and not continuous in the longitudinal dimension. As Hughes (1982 and pers. comm.) has pointed out, the term "connecting lamina" is therefore not altogether appropriate since the structures are fibrils rather than plates or sheets. However, in order to maintain consistency with Cleland and Rothschild's original usage the term connecting lamina will be used in this paper.

A wide separation of the dense outer fibres and axoneme is also found in the dasyurids and *T. rostratus* (Harding *et al.* 1982; Harding, Carrick and Shorey 1984), but in most Australian marsupials the dense outer fibres fairly closely overlie their corresponding axoneme doublets (Harding 1977, 1987). Among American marsupial species a slight radial displacement is evident in *Didelphis virginiana* (Fawcett 1970; Temple-Smith and Bedford 1980; Temple-Smith 1987). We have discussed in detail elsewhere the likely significance of this unusual feature shared by the perameloids, dasyurids and *Tarsipes,* and whilst our initial conclusion was that it indicated phylogenetic links between these groups (Harding *et al.* 1981, 1982, 1984), on the basis of more recent evidence, we and others have concluded that at least for *Tarsipes* independent derivation of radially displaced dense outer fibres provides the most economic explanation for this shared feature at present (Harding 1987; Temple-Smith 1987). As discussed above, spermatozoa of these three groups are large, and we have suggested that the radial displacement of the dense outer fibres may represent a structural solution to facilitate motility in these large spermatozoa (Harding *et al.* 1981; Harding 1987). In this regard it is perhaps significant that the radial displacement is not as great in the perameloids as in the larger spermatozoa of the dasyurids and *Tarsipes.*

The connecting laminae in *Tarsipes* like those in the perameloids are paired structures, but they are single in the dasyurids (Harding *et al.* 1982).

In terms of midpiece shape, the perameloids conform to the usual mammalian circular profile seen in transverse sections, in contrast to the markedly flattened midpieces found in the large spermatozoa of *Tarsipes* and the dasyurids (Harding *et al.* 1982). However, as in the latter groups the mitochondria in the perameloids are sculptured on their inner surfaces to accommodate the displaced dense outer fibres (Fig. 11). Surrounding the mitochondrial sheath is a layer of granular material which in some superficial longitudinal sections is seen to have a faint fibrous substructure. In some such sections this substructure appears to resolve as incomplete longitudinal sections of microtubules (Fig. 12). The perameloid midpiece is unusual in lacking a well-defined fibre network underlying the plasma membrane. Varying forms of a helically wound network of fibrous bands have been reported in all other Australian (Olson 1975; Harding *et al.* 1975, 1979, 1982, 1984; Harding 1977, 1987; Temple-Smith and Bedford 1976) and American (Olson and Hamilton 1976; Olson *et al.* 1977; Temple-Smith and Bedford 1980; Temple-Smith 1987) marsupials examined so far, except for *Phascolarctos cinereus* (Harding *et al.* 1987). The function of the fibre network is not clear but it seems likely that it may have some passive stiffening role in motility. Given the ubiquity of a midpiece fibre network among marsupials it also seems likely that the faint fibrous substructure in the midpiece granular layer of perameloids may represent either a poorly developed or a degenerate fibre network. In the Common Brushtail Possum, *Trichosurus vulpecula* (Temple-Smith 1981), and the Virginia Opossum, *Didelphis virginiana* (Olson *et al.* 1977) freeze fracture studies have revealed highly ordered intramembranous particle rows which at least in *T. vulpecula* are associated with the underlying midpiece fibre network. Similar studies in the perameloids would seem worthwhile.

At its anterior extremity the granular layer surrounding the mitochondrial sheath is seen in longitudinal sections to be bounded by a wedge-shaped mass of granular material of slightly greater electron density than the midpiece granular layer. This material forms an annular structure surrounding the mitochondrial sheath at the posterior margin of the nucleus (Figs 7, 10). Apparently associated with this granular wedge and lying exterior to it and just underneath the plasma membrane, it a thin plate of material of similar electron density to the nucleus (Figs 2, 7, 10). Longitudinal sections cut in the lateral dimension show these structures to surround the midpiece at the posterior margin of the nucleus (Fig. 7) but longitudinal sections

[1]The numbering system for the dense outer fibres used in this paper follows that in Fawcett (1962), since this is the conventional system in use. It should be noted that Cleland and Rothschild (1959) have used a different method. The conversion of one to the other is as follows. Cleland and Rothschild's number is given first and the appropriate number in Fawcett's system is given second: 1=3, 2=4, 3=5, 4=6, 5=7, 6=8, 7=9, 8=1, 9=2.

cut in a dorso-ventral plane do not show the very electron dense plate on the ventral side (Fig. 10). Hence this electron dense plate appears to surround the posterior margin of the nucleus on all sides but the ventral where the ventral groove is present. Cummins (1979) previously reported the existence of an electron dense sheath in this position in *I. macrourus* spermatozoa, and the present observations show that these structures occur in both *P. nasuta* and *I. macrourus*, but are more prominent in the latter species (compare Figs 2 and 10). There have been no reports of similar structures in other marsupials.

The midpiece is separated from the principal piece by the annulus. In *P. nasuta* and *I. macrourus* this forms a fibrous ring caudal to the most posterior mitochondria and joined to the fibrous sheath of the principal piece by an array of fine fibres (Fig. 15). The latter appear to be an uniquely perameloid feature. In most Australian (Harding 1977, 1987) and in American marsupials (Fawcett 1970; Phillips 1970; Olson and Hamilton 1976; Olson *et al.* 1977; Temple-Smith and Bedford 1980; Temple-Smith 1987) the plasma membrane at the annulus is indented, and, as well as forming a layer of fibrous material separating the mitochondria and fibrous sheath the annulus lines the indented portion of the midpiece plasma membrane which forms a small posteriorly directed fold over the principal piece. In some cases it may also line the indentation on the principal piece side. The perameloids resemble the dasyurids and *Tarsipes* in that the plasma membrane is not indented to the extent that it forms a fold, and in longitudinal sections the annulus appears as a fairly unobtrusive wedge of fibrous material separating the mitochondria and fibrous sheath.

The principal piece in *P. nasuta* and *I. macrourus* also displays unique features compared with other marsupials. Since our observations basically agree with the detailed description and discussion of this region of the spermatozoon given by Cleland and Rothschild (1959) and Sapsford *et al.* (1970), a brief summary will suffice here. In transverse sections (Fig. 13) the principal piece has a diamond shape, but with the points of the diamond rounded and the sides concave in profile. The "diamond" is not totally symmetrical since a major axis in the plane at right angles to a line joining the central axoneme pair, and a minor axis through the plane joining this axoneme pair, can be recognized. In addition the "diamond" is mostly slightly skewed. At the points on the minor axis are the longitudinal columns which are shown by longitudinal sections (Fig. 16) to be continuous down the length of the flagellum. Connecting the two columns are circumferentially arranged fibrous bands (the ribs) which are shown by longitudinal sections not to be continuous longitudinally (Fig. 16). These bands are bifurcated where they join the longitudinal columns such that in transverse sections double fenestrations are evident on either side of each column (Fig. 13). As described by Cleland and Rothschild (1959), surrounding the fibrous sheath is a continuous membrane which has a fuzzy electron dense appearance on its outer surface (described by Sapsford *et al.* (1970) as the accessory sheath). The plasma membrane at least in anterior regions of the principal piece fits very loosely over the fibrous sheath (Fig. 13).

As is typical in other mammals, the principal piece tapers posteriorly, and in the perameloids takes on an oval and finally a circular shape in transverse section (Fig. 14). The double fenestrations adjacent to the longitudinal columns are replaced by single gaps on either side of each column. Accompanying these changes the dense outer fibres decrease in size, move close to the axoneme and differentially terminate (Fig. 14). This combination of changes is typical of other mammals with large and/or displaced dense outer fibres. As reported by Cleland and Rothschild (1959) no end piece is present in the perameloid flagellum since the termination of the axoneme occurs before that of the fibrous sheath (Fig. 14).

The chief distinguishing feature of the principal piece in *P. nasuta* and *I. macrourus* is its shape. Although a generalized diamond shape for the principal piece in transverse section is not uncommon among marsupials, that of the perameloids is distinctive due to the concave rather than convex profile of the sides joining the points of the diamond and in that the minor axis is the one on the plane joining the central axoneme pair, whereas in other marsupials with similarly-shaped principal pieces, this forms the major axis. The other groups, with distinctively large spermatozoa — the dasyurids and *Tarsipes* — have a similar arrangement to the perameloids of major and minor axes relative to the central axoneme pair, but their principal pieces are oval shaped in transverse section. The perameloid fibrous sheath is further distinguished by the presence of double fenestrations in each rib on either side of the longitudinal columns, rather than the single cavities typical of other Australian marsupials (Harding 1977). Finally, perameloids lack a principal piece fibre network, an irregular array of fibrous bands between the fibrous sheath and plasma membrane typical of all other marsupials examined so far except for the dasyurids and possibly *Phascolarctos* (Harding 1977).

Epididymal sperm maturation

Epididymal sperm maturation has been described for a number of Australian and American marsupials (Hughes 1965; Harding *et al.* 1975, 1976, 1977, 1979, 1982, 1983, 1984;

Cummins 1976, 1979, 1980; Temple-Smith and Bedford 1976, 1980; Olson and Hamilton 1976; Harding 1977; Olson *et al.* 1977; Temple-Smith 1984; Temple-Smith 1987). By comparison with this information, the maturation process in perameloid spermatozoa is seen to be not as morphologically obvious as it is in most other marsupials. In common with other marsupials the prominent cytoplasmic droplet contracts anteriorly, becomes eccentrically placed to one side of the anterior midpiece and sperm head, and is shed as the spermatozoa pass through the corpus epididymis (Fig. 1). In contrast the cytoplasmic droplet in eutherians migrates posteriorly before being shed (Bedford 1975).

In most marsupials the nucleus lies with its long axis perpendicular to the long axis of the flagellum at spermiation and a prominent feature of epididymal maturation is a rotation of the nucleus such that its long axis comes to lie close to a plane parallel with the long axis of the flagellum. Such a rotation was first described by Fürst (1887) for *Philander opossum* and *Sminthopsis murina* and has since been described for many other marsupials by several researchers including Hughes (1965) for a range of Australian marsupials. In *P. nasuta* (Sapsford *et al.* 1969a) and *I. macrourus* the angle between the long axes of the head and flagellum is already much reduced from the perpendicular at spermiation and hence nuclear rotation is not as obvious a part of epididymal maturation as in many marsupials (Harding *et al.* 1979). A notable feature in the perameloids however, is the marked change in the position of neck insertion (as described above) which accompanies the final reduction in the angle between the head and flagellar long axes (Figs 9, 10). In caput epididymidal spermatozoa of *P. nasuta* and *I. macrourus* the neck is inserted into the deep implantation fossa and an apparently secure ball and socket type articulation joins the head and neck (Fig. 8). In spermatozoa from the cauda epididymis in which the head very tightly overlies the flagellum, the neck structures have lost this ball and socket articulation with the nucleus and moved anteriorly into the hollowed-out portion of the anterior nucleus which forms an extension of the ventral groove. The implantation fossa remains to indicate the former position of the neck (Figs 2, 10). Although a similar movement of the neck structures occurs in most other marsupials (Harding 1977) it is most extreme in the perameloids.

Pronounced modification of the acrosome is a prominent feature in most Australian marsupials during epididymal transit (Harding *et al.* 1976, 1979, 1983, 1984; Cummins 1976; Temple-Smith and Bedford 1976) but this process is absent in the perameloids. Possible reasons for the apparent absence of acrosomal maturation in the perameloids have been discussed elsewhere (Harding *et al.* 1983).

The most notable change in the perameloid sperm head during maturation is a progressive tightening of the fit of the nuclear and plasma membranes. In spermatozoa from the anterior half of the epididymis in both *P. nasuta* and *I. macrourus* the nuclear and plasma membranes are typically ballooned out from the electron dense nuclear material over parts of the sperm head (Fig. 8). This expanded region approximately matches the regions in which parachromatin is found in the mature sperm head and patches of parachromatin-like granular material are evident in the "ballooned" regions at this stage (Fig. 8). In spermatozoa from the posterior half of the epididymis, the nuclear and plasma membranes fit fairly closely over the sperm head and the parachromatin evenly fills these areas (Figs 2, 5, 7).

Changes in the plasma membrane and closely underlying regions of the midpiece constitute a prominent part of epididymal maturation in most American and Australian marsupials (Harding *et al.* 1975, 1977, 1979, 1982, 1984; Cummins 1976, 1980; Temple-Smith and Bedford 1976, 1980; Olson and Hamilton 1976; Harding 1977; Olson *et al.* 1977; Temple-Smith 1981, 1984). The development of a helically-wound midpiece fibre network is widespread and in some species flask-like invaginations of the midpiece plasma membrane also develop. Neither of these structures develop in the perameloids and instead the mitochondrial sheath is surrounded by a layer of granular material. Although this layer is present in caput epididymidal spermatozoa it is possible that it increases slightly in thickness during epididymal transit. However, further analysis is required to determine this with certainty. An unique feature of epididymal maturation in *P. nasuta* and *I. macrourus* is the development of the electron dense layer and underlying "wedge" of granular material, at the posterior margin of the nucleus. This is first seen in corpus-cauda epididymidal spermatozoa in *I. macrourus* and although present in cauda epididymidal spermatozoa of *P. nasuta,* its site of first appearance has not been determined for this species. Cummins (1980) previously reported this feature of maturation in *I. macrourus.*

Other perameloid species

Preliminary observations on admittedly poorly preserved material from the New Guinean species *P. longicauda* and *E. kalubu* indicate that spermatozoa from these species conform to the distinctive perameloid pattern described from *P. nasuta* and *I. macrourus.* There are probably minor size and shape differences which would enable separation of the species, but detailed examination of additional material will be required to define these. It is clear from measurements made using light microscopy, that spermatozoa of *E. kalubu* have considerably longer midpieces than those of *P. nasuta* or *I. macrourus* (Table 1) and the lateral margins of the head are not as distinctively concave as in *I. macrourus* (Fig. 1).

The material available from *M. lagotis* is so poor that only sections of the principal and, less commonly, the midpiece are present. No sperm heads have been observed. *Macrotis lagotis* sperm flagella clearly conform to the peramelid pattern in terms of wide separation of the dense outer fibres from the axoneme, presence of paired connecting laminae joining these structures, and in the form of the annulus. However, assuming that sections from the full length of the principal piece are represented in the material examined, marked differences from other perameloids are the presence of single rather than double fenestrations on either side of the longitudinal columns of the fibrous sheath, and a notable lack of the concave profile of the sides of the diamond shaped fibrous sheath in transverse sections which typifies other perameloid sperm (Fig. 13).

CONCLUDING REMARKS

Mature epididymal spermatozoa of *P. nasuta* and *I. macrourus* are very similar to one another and display a number of distinctive features compared with those of other marsupials. That these unusual features are characteristic of the perameloids as a group is indicated by the preliminary observations of *P. longicauda* and *E. kalubu* spermatozoa. In this regard the perameloids resemble the dasyurids in which there is a high degree of intra-family consistency in sperm form rather than the diprotodonts which generally show much greater intra-family diversity in this regard (Harding 1987).

These remarks are based on examination of spermatozoa from four of the seven recognized extant perameloid genera and brief observations of flagellar structure in a fifth *(Macrotis)*. *Macrotis* is the genus which has typically been separated by systematists from the other perameloids at family level. Since this separation has recently been challenged (Groves and Flannery, this volume) and since our very preliminary observations suggest a possible difference between *M. lagotis* and the other perameloids in principal piece structure, ultrastructural examination of well-preserved sperm material from this species is an important priority.

The perameloid spermatozoon is undeniably marsupial (see Harding *et al.* 1979 for features of marsupial compared with eutherian spermatozoa) but there are features of the head, midpiece and principal piece which are unique compared with other marsupials examined to date. These have been described above and it remains to briefly consider their phylogenetic and/or functional significance.

Sperm head shape is of phylogenetic interest because it is generally a useful distinguishing feature between species or higher taxonomic ranks and because it seems likely that the variability between species may be less functionally correlated than other aspects of sperm structure. Sperm head form in the perameloids is both complex and distinctive compared with other Australian marsupials. Unfortunately, phylogenetic analysis is hampered by the lack of information on *Macrotis* spp. sperm head form, but if this is shown to resemble that of the other perameloids then sperm head form is best regarded as synapomorphic within the Perameloidea and as an autapomorphy for the group. As such this feature is of little use in elucidating extra-perameloid phylogenetic relationships.

The other unusual sperm head features include the rostral position of the acrosome and the heterogeneous appearance of the nuclear contents. Those pose interesting structure-function questions for future research.

The flagellum of perameloid spermatozoa also possesses unusual features. Most interesting is the wide radial displacement of the dense outer fibres from the axoneme and the sharing of this feature with the dasyurids and *Tarsipes*. As mentioned above, we have previously discussed the possible phylogenetic and functional significance of this feature in some detail and on the basis of recent evidence concluded that at least for *Tarsipes* convergence provides the most economic explanation at present (Harding 1987). Whether the same conclusion should be reached regarding the sharing of this feature in the dasyurids and perameloids is unresolved. Certainly non-convergence would fit with some previous suggestions regarding phylogenetic links between the perameloids and dasyurids (Kirsch 1977). On the other hand there are distinct differences in flagellar structure between these groups. Seen in transverse sections the midpiece is rounded in the perameloids, but flattened into an oval shape in the dasyurids (as in *Tarsipes*) and the connecting laminae are paired in the perameloids (as in *Tarsipes*), but single in the dasyurids. In accepting the possibility of convergence between the dasyurids/perameloids and *Tarsipes* in this feature, we stressed the necessity for an underlying structure-function requirement concerning motility in these large spermatozoa (Harding 1987). If indeed this is the case then presumably a three-way convergence is equally acceptable. Hence the phylogenetic significance of this shared feature between the dasyurids and perameloids is not clear at present. However, it is of interest from a structure-function perspective that the three groups with the largest spermatozoa among marsupials (dasyurids, perameloids, *Tarsipes*) share not only radial displacement of the dense outer fibres, but also a unique form for the annulus compared with other marsupials. Comparative motility studies of these large spermatozoa with those of other marsupials possessing smaller spermatozoa may be rewarding.

Observations in this study suggest that perameloid spermatozoa may be very useful for further research. From a phylogenetic perspective, examination of *Macrotis* spermatozoa is essential for understanding intra-perameloid relationships. However, assessment of the potential utility of sperm form for interpreting the broader relationships of the perameloids must await further analysis of the plesiomorphic or apomorphic status of various sperm characters.

Perameloid spermatozoa may offer a great potential for comparative structure-function studies of mammalian spermatozoa. An obvious example concerns sperm motility and the possible adaptive structural features related to sperm size. Perameloid spermatozoa being intermediate in size between the giant spermatozoa of *Tarsipes* and the dasyurids and the smaller diprotodontan spermatozoa may provide key information. It is interesting to note in this regard that the radial displacement of the dense outer fibres in peramelids is greater than in any diprotodontan spermatozoa but less than that seen in the dasyurids and *Tarsipes*. Likewise it is interesting that peramelid spermatozoa conform to the diprotodontan pattern of a circular midpiece as seen in transverse sections rather than the flattened midpiece found in *Tarsipes* and the dasyurids.

We suggest that by virtue of their many unique features and also their unique combination of the features seen in other marsupial groups, the perameloids may be a key group for comparative structure-function spermatozoan studies.

ACKNOWLEDGEMENTS

We are grateful to the following people for providing animals and/or collecting samples from them: Ken Aplin, Frank Carrick, Les Hall, Tony Hulbert and Gordon Lyne. The *Echymipera* and *Peroryctes* material was collected by Tim Flannery on a trip funded by a National Geographic Society Grant No. 2699-83 to Michael Archer. We also thank Leisa Manning for her invaluable technical assistance and Anne Woods for efficiently typing the manuscript.

REFERENCES

BEDFORD, J. M., 1975. Maturation, transport and fate of spermatozoa in the epididymis. Pp. 303–17 in *Handbook of physiology, male reproductive system* Vol. 5, Sect. 7 ed by R. O. Greep and E. B. Astwood. American Physiological Society: Washington.

CLELAND, K. W., 1955. Structure of the bandicoot sperm-tail. *Australian Journal of Science* **18:** 96–97.

CLELAND, K. W. AND ROTHSCHILD, LORD, 1959. The bandicoot spermatozoon: an electron microscope study of the tail. *Proceedings of the Royal Society* **150:** 24–42.

CUMMINS, J. M., 1976. Epididymal maturation of spermatozoa in the marsupial *Trichosurus vulpecula:* changes in motility and gross morphology. *Australian Journal of Zoology* **24:** 499–511.

CUMMINS, J. M., 1979. Sperm maturation in the epididymis of the Brindled Short-nosed Bandicoot, *Isoodon macrourus*. *Proceedings of the Australian Society of Reproductive Biology* (Perth): 84.

CUMMINS, J. M., 1980. Decondensation of sperm nuclei of Australian marsupials: effects of air drying and of calcium and magnesium. *Gamete Research* **3:** 351–67.

CUMMINS, J. M. AND WOODALL, P. F., 1985. On mammalian sperm dimensions. *Journal of Reproduction and Fertility* **75:** 153–75.

FAWCETT, D. W., 1962. Sperm tail structure in relation to the mechanism of movement. Pp. 147–69 in *Spermatozoon motility* ed by D. W. Bishop. American Association for the Advancement of Science. Publication No. 72.

FAWCETT, D. W., 1970. A comparative view of sperm ultrastructure. *Biology of Reproduction* 2, suppl. **2:** 90-127.

FAWCETT, D. W., 1975. The mammalian spermatozoon. *Developmental Biology* **44:** 394–436.

FRANKLIN, L. E., 1968. Formation of the redundant nuclear envelope in monkey spermatids. *Anatomical Record* **161:** 149–62.

FÜRST, C. M., 1887. Uber die Entwicklung der Samenkörperchen bei den Beutelthieren. *Arkiv für Mikroscopische Anatomie und Entwicklungsmechanik* **30:** 336–65.

HARDING, H. R., 1977. *Reproduction in male marsupials: a critique, with additional observations on sperm development and structure.* Ph.D. Thesis. University of New South Wales: Sydney.

HARDING, H. R., 1987. Interrelationships of the families of the Diprotodonta — a view based on spermatozoon ultrastructure. Pp. 195–216 in *Possums and opossums: studies in evolution* ed by M. Archer. Surrey Beatty & Sons and the Royal Zoological Society of New South Wales: Sydney.

HARDING, H. R., CARRICK, F. N. AND SHOREY, C. D., 1975. Ultrastructural changes in spermatozoa of the Brush-tailed Possum, *Trichorurus vulpecula* (Marsupialia), during epididymal transit. Part I: the flagellum. *Cell and Tissue Research* **164:** 11–32.

HARDING, H. R. CARRICK, F. N. AND SHOREY, C. D., 1976. Ultrastructural changes in spermatozoa of the Brush-tailed Possum, *Trichosurus vulpecula* (Marsupialia) during epididymal transit. Part III: the acrosome. *Cell and Tissue Research* **171:** 61–73.

HARDING, H. R., CARRICK, F. N. AND SHOREY, C. D., 1977. Spermatozoa of Australian marsupials: ultrastructure and epididymal development. Pp. 151–52 in *Reproduction and evolution: fourth international symposium on comparative biology of reproduction* ed by J. H. Calaby and C. H. Tyndale-Biscoe. Australian Academy of Science: Canberra.

HARDING, H. R., CARRICK, F. N. AND SHOREY, C. D., 1979. Special features of sperm structure and function in marsupials. Pp. 289–303 in *The spermatozoon* ed by D. W. Fawcett and J. M. Bedford. Urban and Schwarzenberg Inc.: Baltimore–Munich.

HARDING, H. R., CARRICK, F. N. AND SHOREY, C. D., 1981. Marsupial phylogeny: new indications from sperm ultrastructure and development in *Tarsipes spenserae? Search* **12:** 45–47.

HARDING, H. R., CARRICK, F. N. AND SHOREY, C. D., 1983. Acrosome development during spermiogenesis and epididymal sperm maturation in Australian marsupials. Pp. 411–14 in *The sperm cell* ed by J. André. Proceedings of the Fourth International Symposium on Spermatology, Seillac, France, 1982. Martinus Nijhoff Publishers: The Hague.

HARDING, H. R., CARRICK, F. N. AND SHOREY, C. D., 1984. Sperm ultrastructure and development in the Honey Possum, *Tarsipes rostratus*. Pp. 451–61 in *Possums and gliders* ed by A. P. Smith and I. D. Hume. Australian Mammal Society: Sydney.

HARDING, H. R., CARRICK, F. N. AND SHOREY, C. D., 1986. The affinities of the Koala, *Phascolarctos cinereus* (Marsupialia), on the basis of sperm ultrastructure and development. Pp. 353–64 in *Possums and opossums: studies in evolution* ed by M. Archer. Surrey Beatty and Sons and the Royal Zoological Society of New South Wales: Sydney.

HARDING, H. R., WOOLLEY, P. A., SHOREY, C. D. AND CARRICK, F. N., 1982. Sperm ultrastructure, spermiogenesis and epididymal sperm maturation in dasyurid marsupials: phylogenetic implications. Pp. 659–73 in *Carnivorous marsupials* ed by M. Archer. Royal Zoological Society of New South Wales: Sydney.

HUGHES, R. L., 1965. Comparative morphology of spermatozoa from five marsupial families. *Australian Journal of Zoology* **13:** 533–43.

HUGHES, R. L., 1982. Reproduction in the Tasmanian Devil *Sarcophilus harrisii* (Dasyuridae, Marsupialia). Pp. 49–63 in *Carnivorous marsupials* ed by M. Archer. Royal Zoological Society of New South Wales: Sydney.

KIRSCH, J. A. W., 1977. The classification of marsupials. Pp. 1–50 in *The biology of marsupials* ed by D. Hunsacker. Academic Press: New York.

OLSON, G., 1975. Observations on the ultrastructure of a fiber network in the flagellum of sperm of the Brushtailed Phalanger, *Trichosurus vulpecula*. *Journal of Ultrastructure Research* **50:** 193–98.

OLSON, G. E. AND HAMILTON, D. W., 1976. Morphological changes in the midpiece of Wooly opossum spermatozoa during epididymal transit. *Anatomical Record* **186:** 387–404.

OLSON, G. E., LIFSICS, M., FAWCETT, D. W. AND HAMILTON, D. W., 1977. Structural specializations in the flagella plasma membrane of opossum spermatozoa. *Journal of Ultrastructure Research* **59:** 207–21.

PHILLIPS, D. M., 1970. Ultrastructure of spermatozoa of the Wooly Opossum *Caluromys philander*. *Journal of Ultrastructure Research* **33:** 381–97.

RATTNER, L. B., 1972. Nuclear shaping in marsupial spermatids. *Journal of Ultrastructure Research* **40:** 498–512.

SAPSFORD, C. S. AND RAE, C. A., 1969. Ultrastructural studies on Sertoli cells and spermatids in the bandicoot and ram during the movement of mature spermatids into the lumen of the seminiferous tubule. *Australian Journal of Zoology* **17:** 415–45.

SAPSFORD, C. S., RAE, C. A. AND CLELAND, K. W., 1967. Ultrastructural studies on spermatids and Sertoli cells during early spermiogenesis in the bandicoot *Perameles nasuta* Geoffroy (Marsupialia). *Australian Journal of Zoology* **15:** 881–909.

SAPSFORD, C. S., RAE, C. A. AND CLELAND, K. W., 1969a. Ultrastructural studies on maturing spermatids and on Sertoli cells in the bandicoot *Perameles nasuta* Geoffroy (Marsupialia) *Australian Journal of Zoology* **17:** 195–292.

SAPSFORD, C. S., RAE, C. A. AND CLELAND, K. W., 1969b. The fate of residual bodies and degenerating germ cells and the lipid cycle in Sertoli cells in the bandicoot *Perameles nasuta* Geoffroy (Marsupialia). *Australian Journal of Zoology* **17:** 729–53.

SAPSFORD, C. S., RAE, C. A. AND CLELAND, K. W., 1970. Ultrastructural studies on the development and form of the principal piece sheath of the bandicoot spermatozoon. *Australian Journal of Zoology* **18:** 21–48.

TEMPLE-SMITH, P. D., 1981. Redistribution of intramembranous particles in the midpiece plasma membrane of the Brushtail Possum spermatozoon during epididymal transit. *Journal of Anatomy* **132:** 301.

TEMPLE-SMITH, P. D., 1984. Reproductive structures and strategies in male possums and gliders. Pp. 89–106 in *Possums and gliders* ed by A. P. Smith and I. D. Hume. Australian Mammal Society: Sydney.

TEMPLE-SMITH, P. AND BEDFORD, J. M., 1976. The features of sperm maturation in the epididymis of a marsupial, the Brushtailed Possum, *Trichosurus vulpecula*. *American Journal of Anatomy* **147:** 471–500.

TEMPLE-SMITH, P. D. AND BEDFORD, J. M., 1980. Sperm maturation and the formation of sperm pairs in the epididymis of the opossum, *Didelphis virginiana*. *Journal of Experimental Zoology* **214:** 161–71.

TEMPLE-SMITH, P. D., 1987. Sperm structure and marsupial phylogeny. Pp. 171–93 in *Possums and opossums: studies in evolution* ed by M. Archer. Surrey Beatty & Sons and the Royal Zoological Society of New South Wales: Sydney.

WOOLLEY, P. A. AND SCARLETT, G., 1984. Observations on the reproductive anatomy of male *Tarsipes rostratus* (Marsupialia: Tarsipedidae). Pp. 445–50 in *Possums and gliders* ed by A. P. Smith and I. D. Hume. Australian Mammal Society: Sydney.

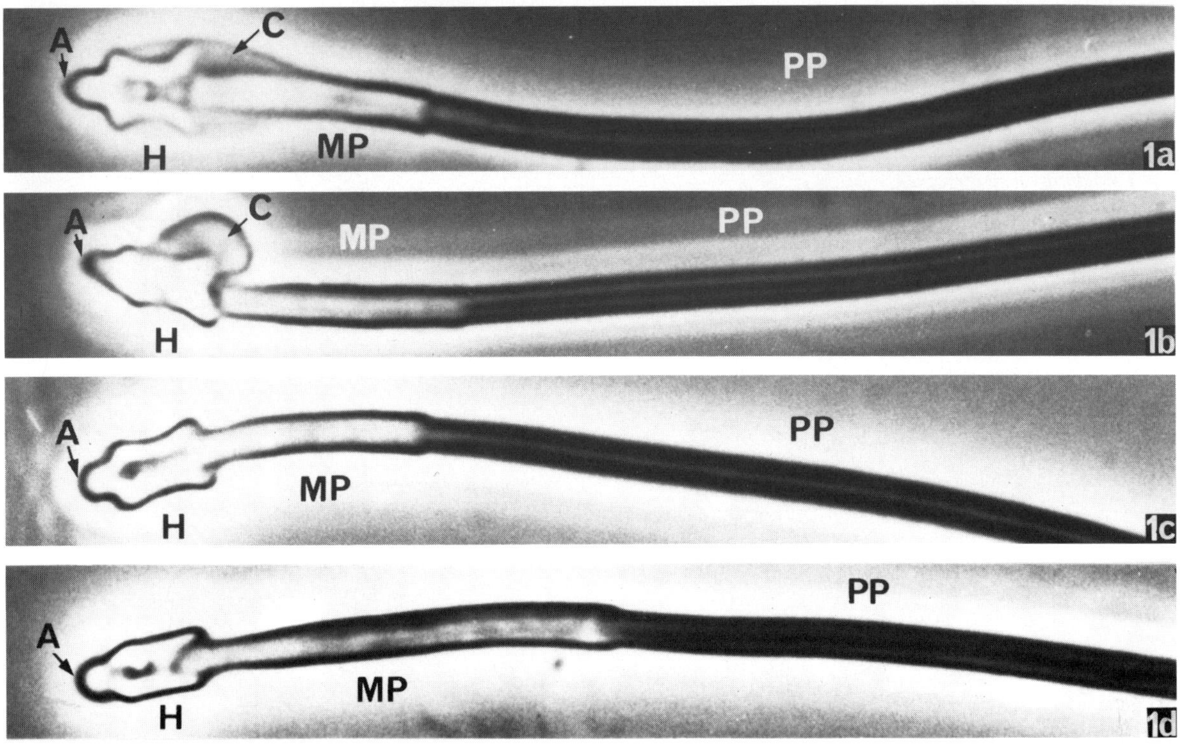

Fig. 1a,b,c,d. Phase contrast light micrographs all to the same magnification X 2 900. (a) Spermatozoon from the caput epididymis of *Isoodon macrourus* viewed ventrally. The cytoplasmic droplet (C) surrounds the posterior part of the sperm head (H) and the anterior midpiece (MP). The ventral groove in the head (H) is obvious and the acrosome (A) forms a rostral protrusion. (b) Spermatozoon from the caput-corpus epididymis of *I. macrourus* viewed dorsally. The cytoplasmic droplet (C) has contracted anteriorly and become eccentrically placed to one side of the head (H) and midpiece (MP). (c) Spermatozoon from the corpus-cauda epididymis of *I. macrourus* viewed ventrally. Note that the cytoplasmic droplet has been shed. (d) Spermatozoon from the cauda epididymis of *Echymipera kalubu*. The head (H) lacks the distinctive concave lateral profile seen in *I. macrourus* and the midpiece (MP) is much longer relative to head length than in this species. A=Acrosome; PP=principal piece.

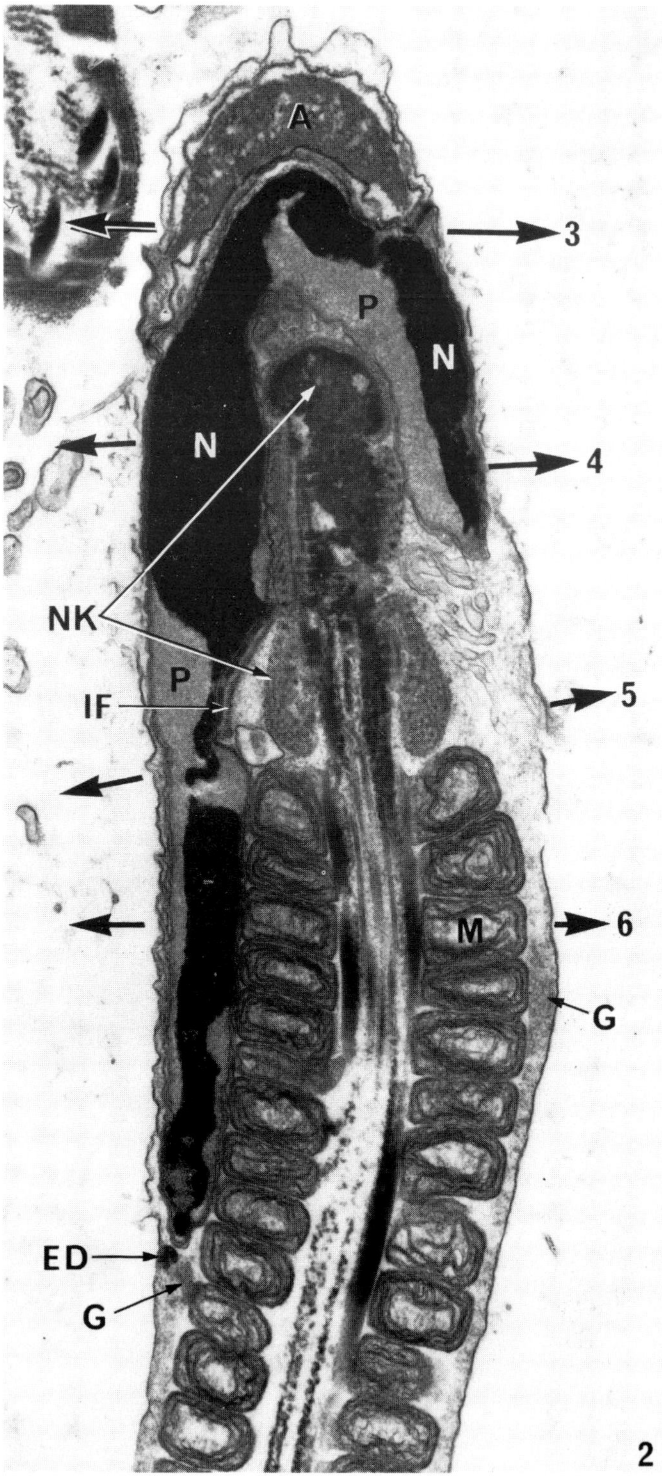

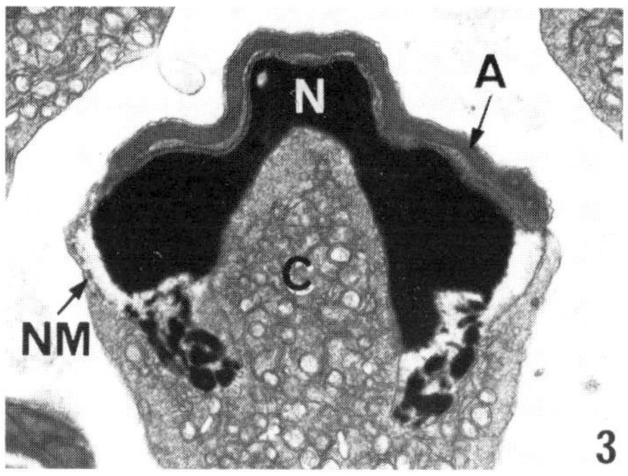

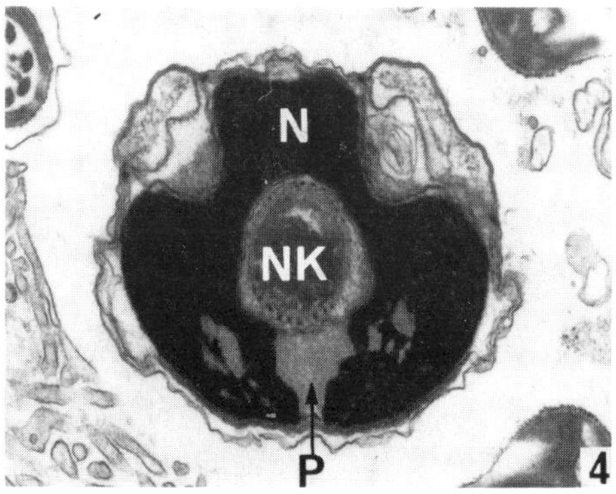

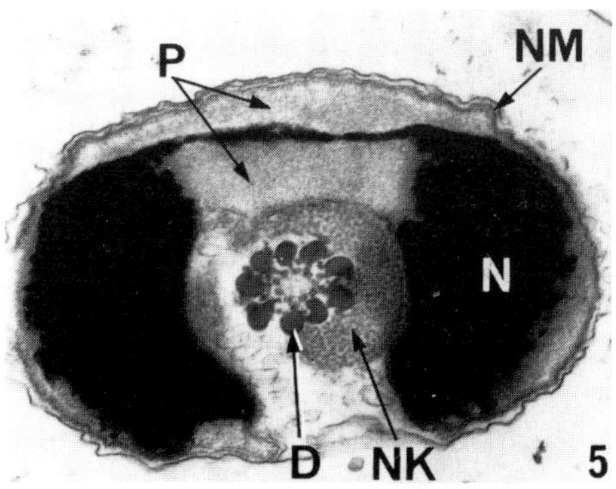

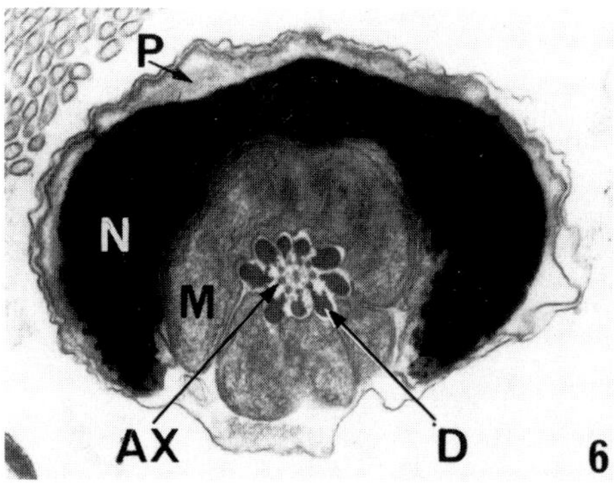

Fig. 2. Longitudingal section in a dorso-ventral plane through the head and anterior midpiece of a cauda epididymidal spermatozoon of *Perameles nasuta.* The neck (NK) has moved anteriorly from the implantation fossa (IF) in which it is located in immature spermatozoa. The arrows labelled 3, 4, 5 and 6 mark the approximate positions of the transverse sections shown in Figs 3–6 respectively. Neither the electron dense structure (ED) partially surrounding the midpiece posterior to the nucleus, nor the annular granular wedge (G) associated with it, are as well developed as in *Isoodon macrourus* (compare with Fig. 10). A=acrosome, M=mitochondria, N=nucleus, P=parachromatin. X 33 500.

Fig. 3. Transverse section through the anterior region of the head in a corpus epididymidal spermatozoon of *Isoodon macrourus.* Note the pitting of the nucleus (N) ventrally. A=acrosome, C=cytoplasmic droplet, NM=nuclear membrane. X 17 000.

Fig. 4. Transverse section through the head of a cauda epididymidal spermatozoon of *Perameles nasuta* in the region of the neck. N=nucleus, NK=neck, P=parachromatin. X 22 000.

Fig. 5. Transverse section through the head of a cauda epididymidal spermatozoon of *Perameles nasuta* in the posterior region of the neck. D=dense outer fibres, N=nucleus, NM=nuclear membrane, P=parachromatin. X 26 500.

Fig. 6. Transverse section through the head of a cauda epididymidal spermatozoon of *Perameles nasuta* in the posterior region of the head which overlies the anterior midpiece. AX=axoneme, D=dense outer fibres, M=mitochondria, N=nucleus, P=parachromatin. X 22 000.

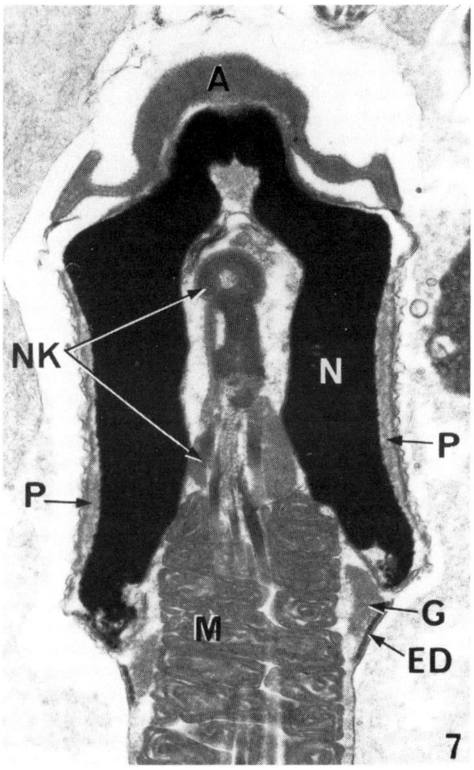

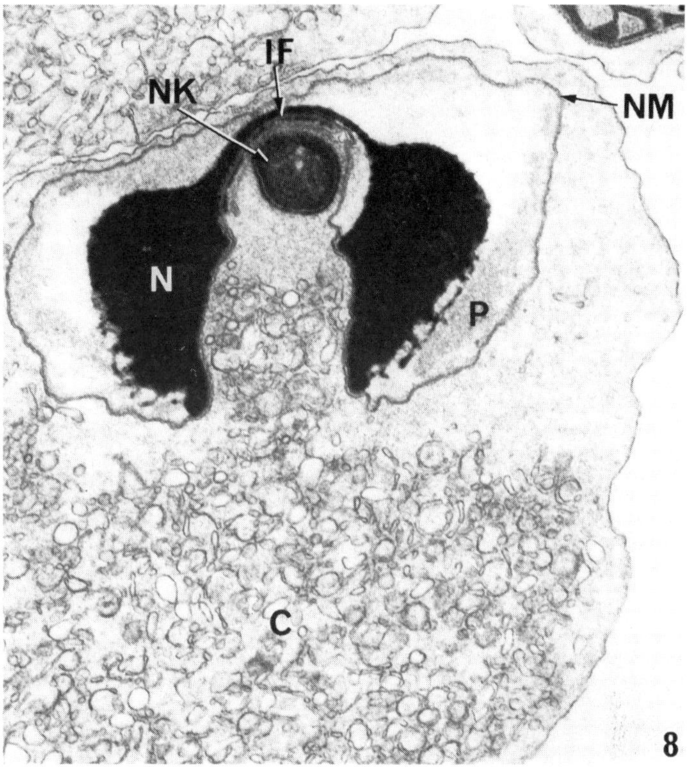

Fig. 7. Longitudinal section in a lateral plane through the head and anterior midpiece of a corpus-cauda epididymidal spermatozoon of *Isoodon macrourus*. Note that the neck (NK) lies close to the anterior of the nucleus (N). A= acrosome, ED=electron dense plate surrounding the midpiece, G= granular wedge underlying the electron dense plate, M=mitochondria, P=parachromatin. X 17 000.

Fig. 8. Transverse section through the head of a caput epididymidal spermatozoon of *Perameles nasuta* in the region of the neck (NK). Note the close connection between the neck and the electron dense layer lining the deep implantation fossa (IF). In contrast to the mature spermatozoon in Fig. 7 the nuclear membrane is ballooned out from the main body of the nucleus (N) and contains patches of parachromatin (P). The nucleus is pitted ventrally. C=cytoplasmic droplet. X 23 500.

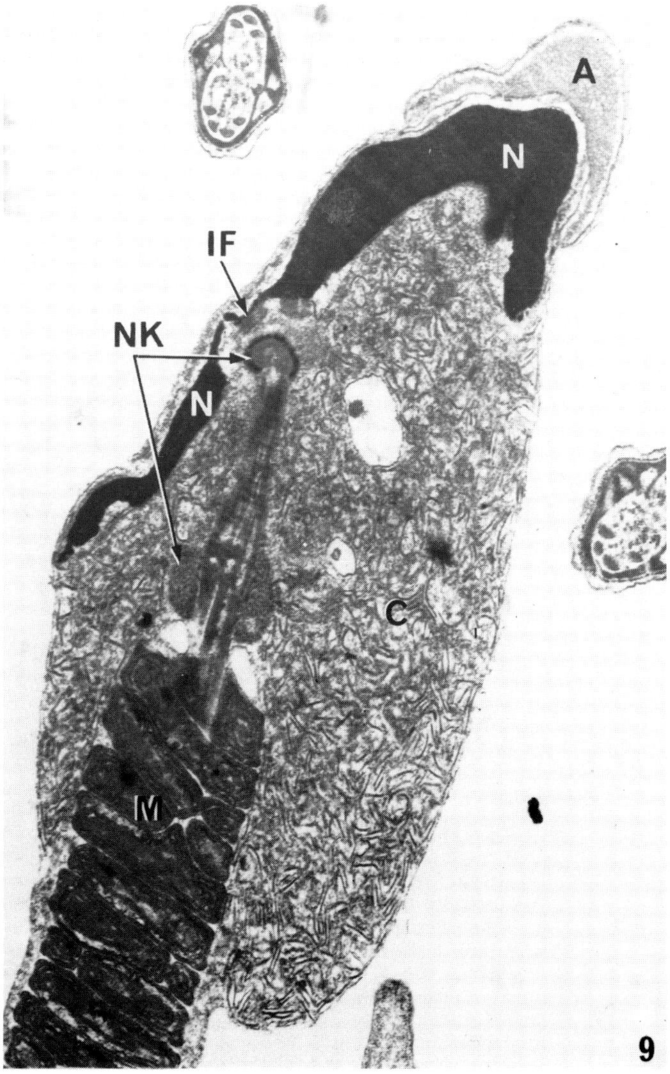

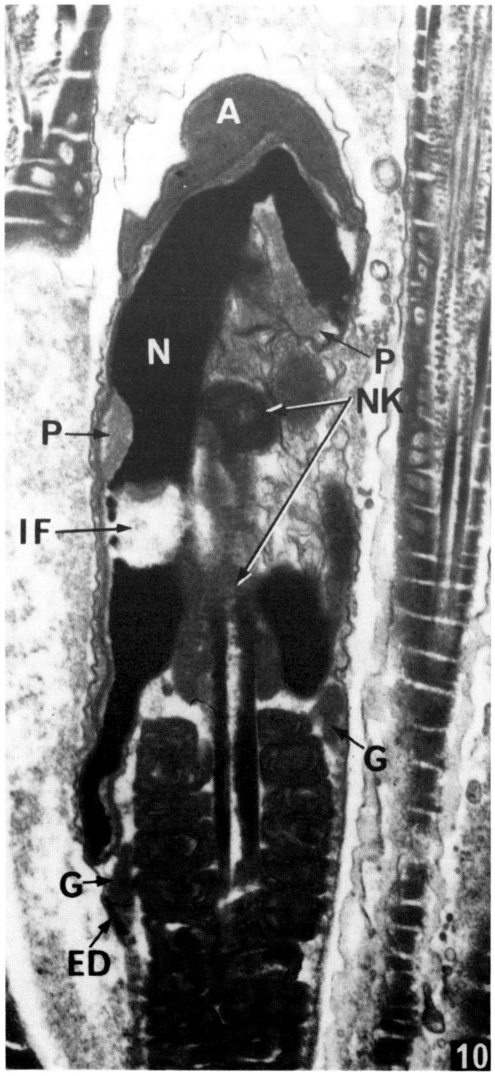

Fig. 9. Longitudinal section in a dorso-ventral plane through the head and anterior midpiece of a corpus epididymidal spermatozoon of *Isoodon macrourus*. Note that the nucleus (N) does not so closely overlie the neck (NK) and mitochondria (M) as in the mature spermatozoon (see Fig. 10). The anterior articular portion of the neck (NK) is housed in the deep implantation fossa (IF). The cytoplasmic droplet (C) is contracted around the head and anterior midpiece. A=acrosome. X 19 500.

Fig. 10. Longitudinal section in a dorso-ventral plane through the head and anterior midpiece of a corpus-cauda epididymidal spermatozoon of *Isoodon macrourus*. The nucleus (N) now tightly overlies the neck (NK) and mitochondria (M) of the anterior midpiece. The neck (NK) has moved anteriorly from the implantation fossa (IF) and the cytoplasmic droplet has been shed. A=acrosome, ED=the electron dense plate is seen only dorsally posterior to the nucleus, G=granular wedge underlying the electron dense plate is seen both dorsally and ventrally, P=parachromatin. X 19 000.

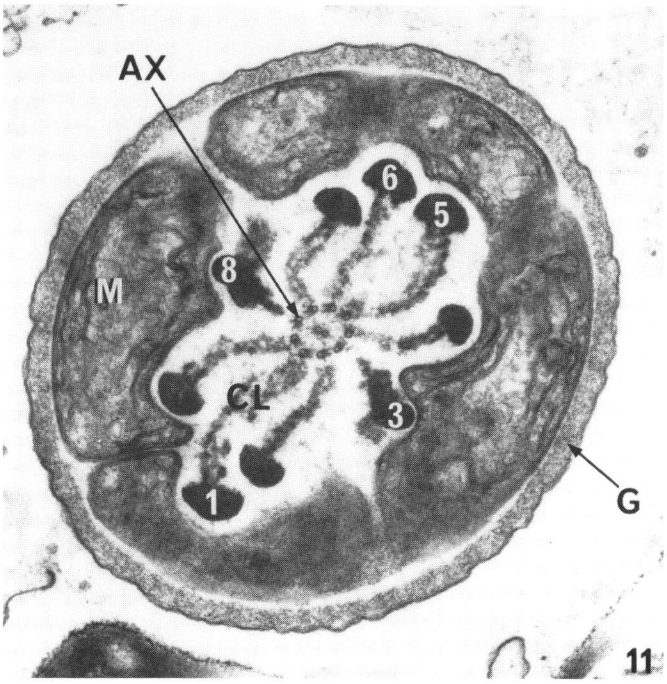

Fig. 11. Transverse section through the midpiece of a cauda epididymidal spermatozoon of *Perameles nasuta.* AX=axoneme, CL=connecting lamina, G=granular layer surrounding mitochondria (M). Dense outer fibres 1, 3, 5, 6 and 8 and numbered accordingly. X 43 000.

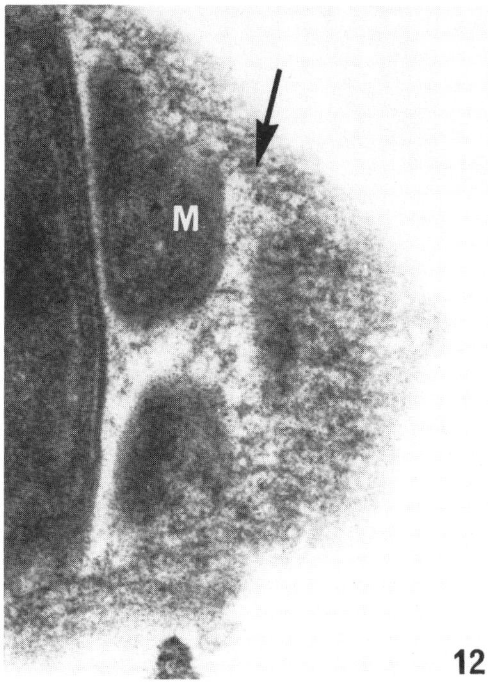

Fig. 12. Superficial longitudinal section through portion of the midpiece of a corpus epididymidal spermatozoon of *Isoodon macrourus.* The section cuts through the granular layer seen surrounding the mitochondria in Fig. 11 and its fibrous or micro-tubular substructure (arrowed) is apparent. Key: M=mitochondria. X 96 000.

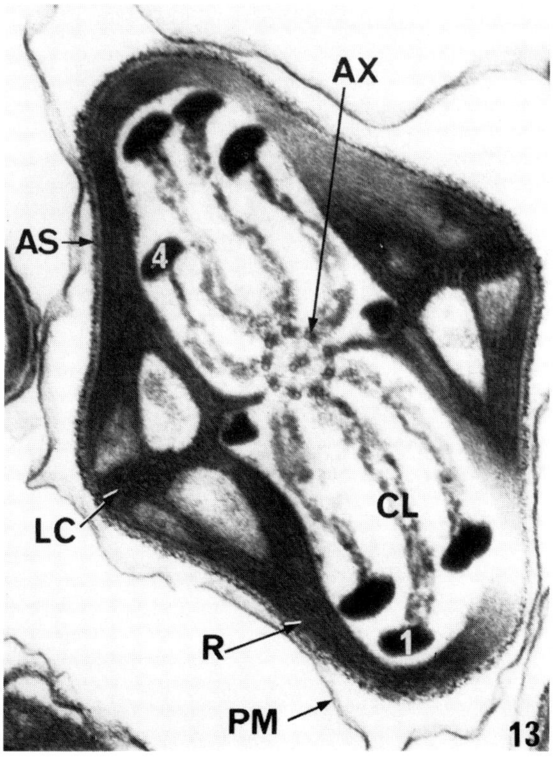

Fig. 13. Transverse section through the anterior principal piece of cauda epididymidal spermatozoon of *Perameles nasuta.* AS=accessory sheath, AX=axoneme, CL=connecting lamina, LC=longitudinal column of the fibrous sheath, PM=plasma membrane, R=rib of the fibrous sheath. Dense outer fibres 1 and 4 are labelled accordingly. X 52 500.

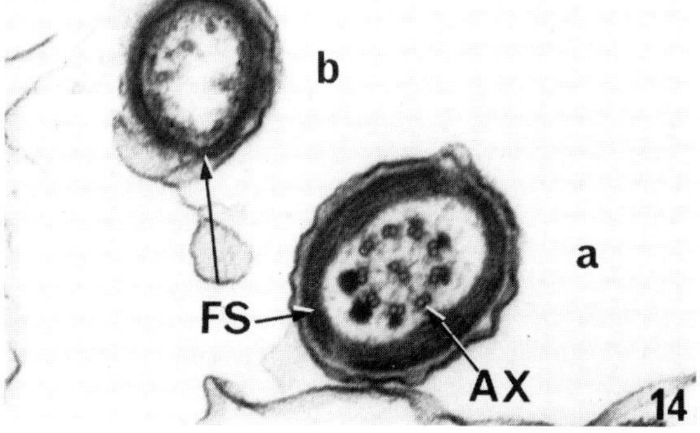

Fig. 14. Transverse sections through the posterior principal piece of a cauda epididymidal spermatozoon of *Perameles nasuta* at the same magnification as Fig. 13. Section a shows that the fibrous sheath (FS) is now oval in transverse section and the ribs are no longer bifurcated. Only dense outer fibres 1, 2, 4, 5, 6 and 9 remain and they are much reduced in size and closely abut the axoneme doublets (AX). Section b is posterior to a and shows that the axoneme and dense outer fibres have terminated before the fibrous sheath (FS). X 52 500.

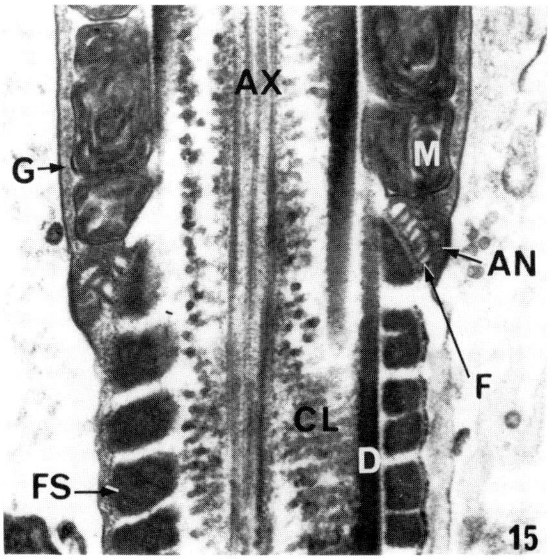

Fig. 15. Longitudinal section through the flagellum of a corpus-cauda epididymidal spermatozoon of *Isoodon macrourus* in the region of the annulus (AN). AX=axoneme, CL=connecting laminae, D=dense outer fibre, F=fibres of annulus attached to fibrous sheath (FS), G=granular layer surrounding mitochondria (M). X 37 000.

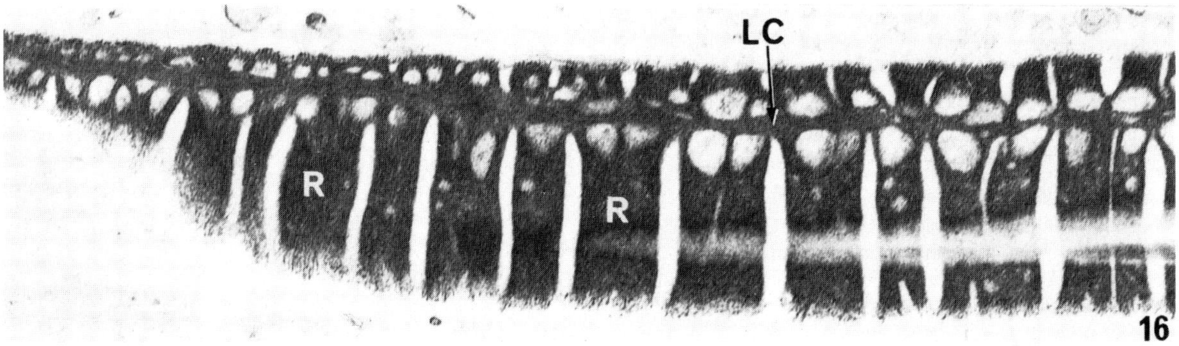

Fig. 16. Longitudinal section through the fibrous sheath of a cauda epididymidal spermatozoon of *Perameles nasuta*. LC=longitudinal column of fibrous sheath, R=ribs of fibrous sheath. X 31 000.

Seasonal effects on androgen secretion in the male Northern Brown Bandicoot, *Isoodon macrourus*

J. R. McFARLANE[1], F. N. CARRICK[1], R. T. GEMMELL[2] and B. MacDONALD[3]

The release of testosterone (T), androstenedione (A), 17β-estradiol (E2), and progesterone (P) in the male bandicoot *(Isoodon macrourus)* was found to be highly episodic. Mean concentrations of 8–12 ngT/ml and 1–2 ngA/ml were measured in peripheral blood samples during the breeding season. In the non-breeding season, concentrations dropped to about 2 ngT/ml and 0.28 ngA/ml. Gonadectomy resulted in the apparent concentration of T and A rapidly decreasing to a few picograms, whereas the peripheral concentrations of P and E2 showed little change following gonadectomy.

Key words: Androgens, bandicoot, episodic release, gonadectomy, *Isoodon macrourus*, seasonal effects, radioimmunoassay.

INTRODUCTION

It is now evident that the release of testosterone from the testis of mammals that have been adequately studied typically occurs in an episodic manner, rather than at a steady rate. However, most of the reports of this phenomenon result from studies on eutherian mammals.

The pattern of episodic release is superimposed on an apparent diurnal cycle in the release of these hormones in some species such as the horse (Kirkpatrick, Vail, Devons, Schwend, Baker and Weisner 1976; Sharma 1976), man (Rowe, Lincoln, Racey, Lehane, Stephenson, Shenton and Glover 1974), miniature pig (Ellendorff, Parvizi, Pomerantz, Hartjen, Konig, Smidt and Elsaesser 1975), Owl Monkey *Aotus trivirgatus* (Dixon and Gardner 1981) and Rhesus Monkey *Macaca mulatta* (Goodman, Hotchkiss, Karsch and Knobil 1974). Other species do not appear to exhibit a diurnal component, as in the bull (Katongole, Naftolin and Short 1971), cat (Johnstone, Bancroft and McFarlane 1984), dog (DePalatis, Moore and Falvo 1978) and ram (Katongole, Naftolin and Short 1974; Purvis, Illius and Haynes 1974).

Many seasonally breeding species have been shown to have a corresponding variation in the concentration of androgens: high in the breeding season and low in the non-breeding season. This has been reported for a number of species such as the hibernating bat, *Myotis lucifugus lucifugus* (Gustafson and Shemesh 1976), Common Brushtail Possum (Carrick and Cox 1977), Koala (Carrick, McFarlane, Bancroft and Brown 1981), Platypus (Carrick and Hughes 1978), ram (Sanford, Winter, Palmer and Howland 1974) and Tammar Wallaby (Catling and Sutherland 1980).

In all of the above reports involving marsupial species, diprotodontid marsupials were studied. To determine whether the findings of these studies were applicable to the Marsupialia in general, these phenomena needed to be studied in a representative species of the Polyprotodonta, thereby extending the work of Gemmell, Johnson and Barnes (1985) on the Northern Brown Bandicoot, *Isoodon macrourus*.

This paper reports an examination of the seasonal changes in 24 h secretion profiles of four steroid hormones — testosterone (T), androstenedione (A), estradiol (E2) and progesterone (P) in *I. macrourus*. It also establishes that these bandicoots have an episodic release pattern for a number of steroid hormones.

MATERIALS AND METHODS

Experimental animals

Adult male *I. macrourus* were captured from bush areas within 20 km of Brisbane and housed under natural daylight conditions in rabbit cages (0.5 × 0.5 × 0.5 m with a solid floor and straw nesting material) for two to three weeks prior to commencement of the experiments. Between 24 and 36 h prior to the initial collection of blood samples, an indwelling catheter (1.0 mm ID, 1.5 mm OD) made from medical grade vinyl catheter tubing (Dural Plastics NSW) was surgically placed in the jugular vein while the bandicoot was under Fluothane (ICI)/oxygen anaesthesia. The animals' necks were then bandaged with a crepe bandage, which secured the end of the catheter behind the animals' heads. The bandicoots tolerated the catheters well and attemps to disoldge the catheters were rare, even after several months.

[1]Reproductive Biology Group, Department of Veterinary Anatomy, University of Queensland, St Lucia, Queensland, Australia 4067.
[2]Department of Anatomy, University of Queensland, St Lucia, Queensland, Australia 4067.
[3]Department of Public Health and Nutrition, Queensland Institute of Technology, North Quay, Queensland, Australia 4000.
Pages 251-57 *in* BANDICOOTS AND BILBIES, ed by J. H. Seebeck, P. R. Brown, R. L. Wallis and C. M. Kemper. Surrey Beatty & Sons, Sydney, 1990.

Animal handling

Blood samples were drawn without removing the animals from their cages. An extra length of catheter tubing was connected to the implanted tubing at each bleed interval, in order not to restrict or interfere with the animals whilst samples were being taken. The animals appeared to be aware of the sampling process but showed little reaction to the procedure, provided care was taken not to frighten them or to provoke aggressive behaviour. There was sufficient low-level light from sources external to the animal house to permit samples to be taken at night without the need for additional illumination.

Blood sampling

Blood samples (1 ml) were taken at hourly or two-hourly (October/November) intervals over a 24 h period, using a tuberculin syringe. Each of these samples was accurately dispensed in duplicate 0.5 ml aliquots into a clean extraction vial (20 ml scintillation vial) containing 5 ml of hexane/toluene (May & Baker) mixture (2:1 v/v), 0.5 ml of distilled water and 1 000 CPM of tritiated T (1,2,6,7,16,17 ^{3}H T Amersham). The vials were shaken by hand for 2 min. and then stored at 4°C pending further processing.

Experimental treatments

Seasonal effects: Adult male bandicoots (11) were bled over a 12-month period on four occasions — March (2), June (2) , July (5) and November (2) — to determine any seasonal variation in the steroid hormone profiles.

Gonadectomy: The opportunity arose to collect blood samples from two bandicoots that were being castrated as part of another study. A further three animals were used as controls for this experiment by sampling during an anaesthetic procedure similar to that employed with the castrated animals. All the bandicoots were bled hourly for 10 h before the onset of anaesthesia and to determine pre-experimental hormone profiles; they were then sampled after castration or sham operation every 15 min. for the first hour and then hourly for the next 3 h. Single samples were taken once daily for the next three days.

Steroid assay

A phosphate buffer (0.05 M at pH 7.4, 0.15 M NaCl) with 1% gelatin (Difco) was used for preparing dilutions of antisera, tracer (1,2,6,7,16,17 ^{3}H-testosterone, 1,2,6,7 ^{3}H-androstenedione, 2,4,6,7,16,17 ^{3}H-estradiol and 1,2,6,7 ^{3}H-progesterone, Radiochemical Centre, Amersham), and 1% gamma globulin (bovine, Calbiochem). A 22% solution of PEG (polyethylene glycol 6000 Merck) was prepared using the phosphate buffer without gelatin. Non-radioactive steroids were obtained from Sigma Chemical Co., USA, for use as authentic standards. The standard curves for T, A and P covered the range of 1–1 000 pg; and for E2, the range 0.8–100 pg. Antisera were prepared using testosterone-3-CMO-BSA, androstenedione-3-CMO-BSA, 17β-estradiol-6-CMO-BSA, and progesterone-11-HS-BSA, and were supplied through the generosity of Dr Ron Cox, CSIRO, Division of Animal Production, Prospect, New South Wales. The antisera were tested for cross reactivity with a wide range of steroidal compounds, especially those containing 17-hydroxy groups (T antiserum) or 17-keto groups (A antiserum). The T antiserum had a significant cross reaction only with 5α-dihydrotestosterone (31%), and 4-androsten-3β, 17β-diol (30%). The only significant cross reaction with the A antiserum was 5-androstan-3-ol-17-one (28%). There was no significant cross reactivity with the P or E2 antisera.

Within 24 h of the conclusion of blood sample collection, the samples were further processed as follows. The extraction vials were shaken for 15 min., the aqueous phase frozen on dry ice, and after the solvent was decanted into a clean scintillation vial, the remaining aqueous phase was discarded. The solvent phase was then evaporated under dry nitrogen or vacuum. The extract was redissolved in 1 ml of ethanol (Merck AR), and aliquots were then used to determine the concentration of the four steroid hormones using radioimmunoassay (RIA) procedures.

For each blood sample, nine aliquots of the extract were pipetted into polythene tubes (10 × 50 mm Kayline Plastics) and evaporated under vacuum or dry nitrogen. Scintillant (0.3% PPO, 0.03% POPOP, 1.5% methanol in toluene) was dispensed into one tube and counted in a Packard 300C beta counter, to estimate recovery efficiency (recoveries averaged 85%) for every sample (the exact extraction efficiency for the other three steroids will differ from that calculated using T: however, the proportional differences will remain constant). The other eight tubes were assayed for the four steriods in duplicate. The antisera (300 μl) were used at final dilutions of 1:25 000 T, 1:30 000 A, 1:150 000 E2 and 1:2 500 P, and together with the appropriate radioactive tracer (100 μl containing approximately 10 000 cpm) was added to each sample and standard curve tube. The tubes were then vortexed and incubated overnight at 4°C. Separation of bound from free steroid was achieved by adding 100 μl of gamma globulin and 1 ml PEG to all tubes, which were then vortexed and given a further incubation at 4°C for 30 min prior to centrifugation for

30 min at 2 500 rpm. The supernatant was discarded by wedging the assay tubes in custom made plastic racks which allowed the assay tubes to be inverted and drained for 10 min. Scintillant (1.6 ml) was added to each tube and the pellet resuspended by vortexing. The bound radioactive ligand was then counted after 4–6 h equilibration.

The sensitivity of the assay was 10 pg for both T and A; 0.8 pg for E2; and 2.5 pg for P. The nonspecific binding for all assays was 2.5%. The inter-assay variation for T and A was 11%, 7% for E2 and was 15% for P. The intra-assay variation for all assays was 4–6%.

A log/logit transformation (Rodbard, Bridson and Rayford 1969) of the RIA standard curves was used to calculate a linear equation, from which the steroid concentrations of the unknowns were calculated.

The mean hormone concentrations for the breeding and non-breeding seasons were derived from the hormone concentrations measured in the individual hourly blood samples, and not from pooling the hourly samples and then assaying a single composite sample. These means were tested for significant differences using Student's *t* tests.

RESULTS

A release episode of a particular steroid has been defined as an increase in steroid concentration over a local basal, level which is greater than the intra-assay variation followed some time later by a decrease, which is greater than the intra-assay variation (Sodersten, Eneroth and Pettersson 1983). Thus in this study if a rise in steroid concentration occurred which was greater than 4–6% of the basal steroid concentration, then the event was a real variation in the peripheral plasma steroid concentration and not a fluctuation due to experimental errors. It is beyond the scope of this paper to analyse episode frequency and amplitude in detail, but it is apparent from a visual appraisal of the data shown in Figures 1, 2 and 4 that the release of these steroids is episodic rather than a constant steady rate.

Non-breeding season

Typical 24 h secretion profiles for a male *I. macrourus* in the non-breeding season (March) are shown in Figure 1. The mean concentration of T was 2.14 ng/ml, with a less variable release of T relative to the breeding season; variability in peripheral blood concentrations of A (mean 0.28 ng/ml) was also secreted at a relative low. However, the production of E2 (mean 1.80 ng/ml) and P (mean 1.90 ng/ml) exhibited a much more episodic pattern.

Breeding season

The secretion profiles typical of a male *I. macrourus* during the breeding season (July) are indicated in Figure 2. These samples were

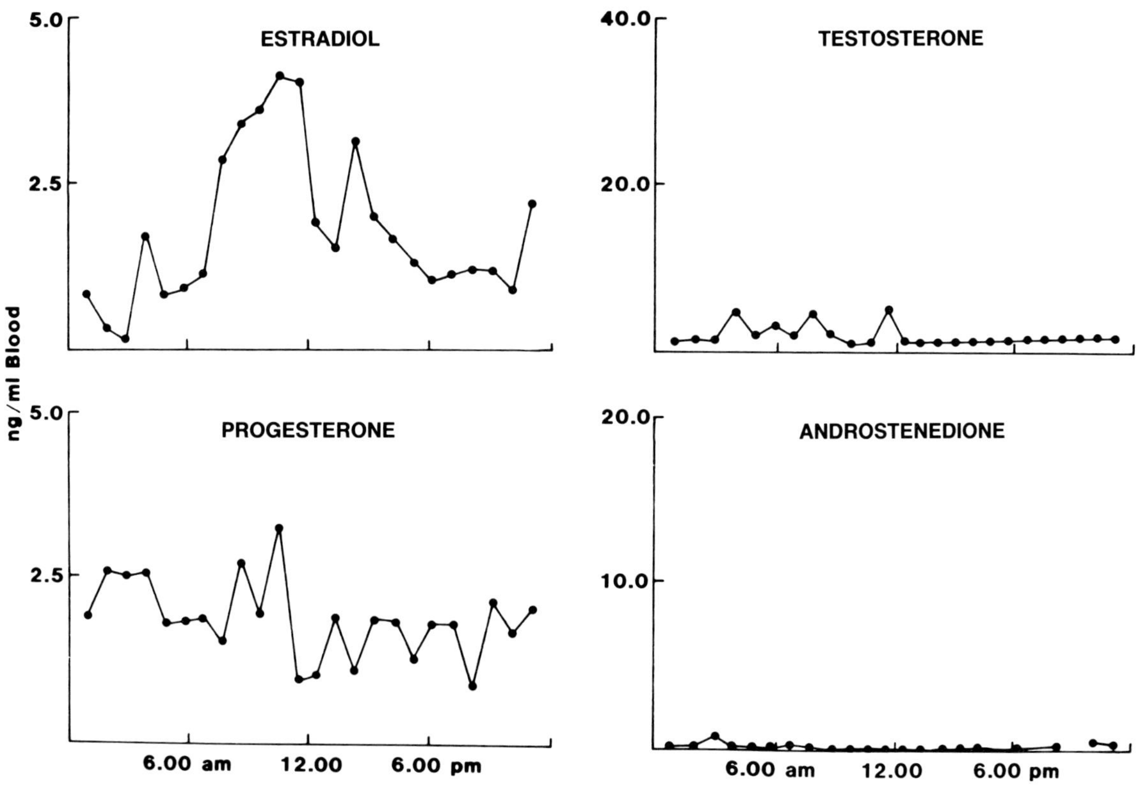

Fig. 1. Non-breeding season (March) 24 h secretion profiles of T, A, E2 and P for a typical male *Isoodon macrourus*.

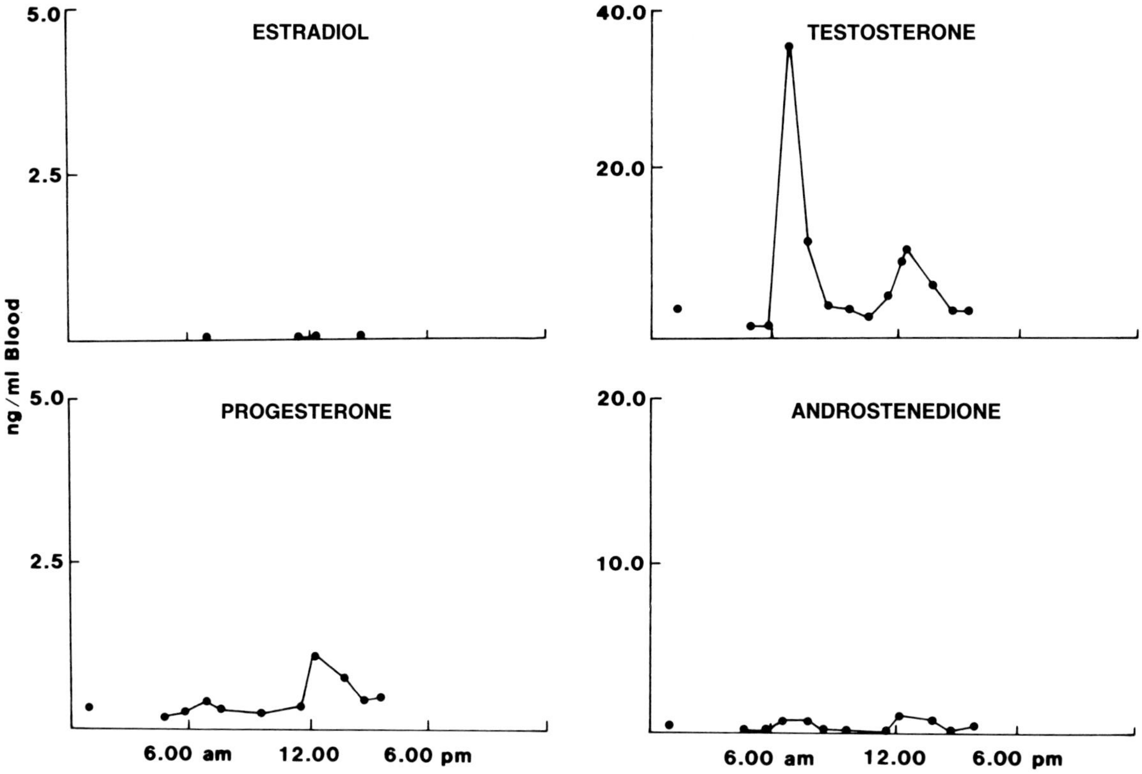

Fig. 2. Breeding season (July) 24 h secretion profiles of T, A, E2 and P for a typical male *Isoodon macrourus*.

collected prior to a castration experiment and thus a full 24 h series of blood samples was not collected. Although a complete 24 h profile was unavailable, the means indicated for July in Figure 3 were derived from five animals and thus should provide a conservative estimation of the true mean typical of animals at this time of year. Even with only a partial profile it is clear that the secretion of T was more episodic than in the non-breeding season. The mean concentration of T in this animal was 5.50 ng/ml and showed an episodic release pattern, however, A was secreted at a relatively constant rate with a mean of 0.43 ng/ml. The mean concentrations of E and P were 0.01 ng/ml and 0.36 ng/ml respectively.

Seasonal changes

A comparison of mean concentration of the four steroids over the four periods of the year during which samples were taken is shown in Figure 3. Several features are evident in these

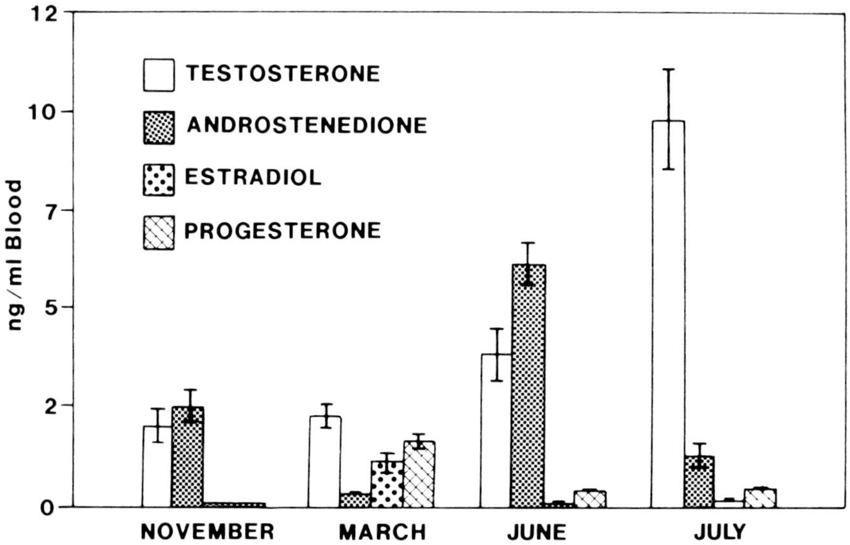

Fig. 3. Mean concentrations and standard errors of steroids in the peripheral plasma of male *Isoodon macrourus* at different times of the year.

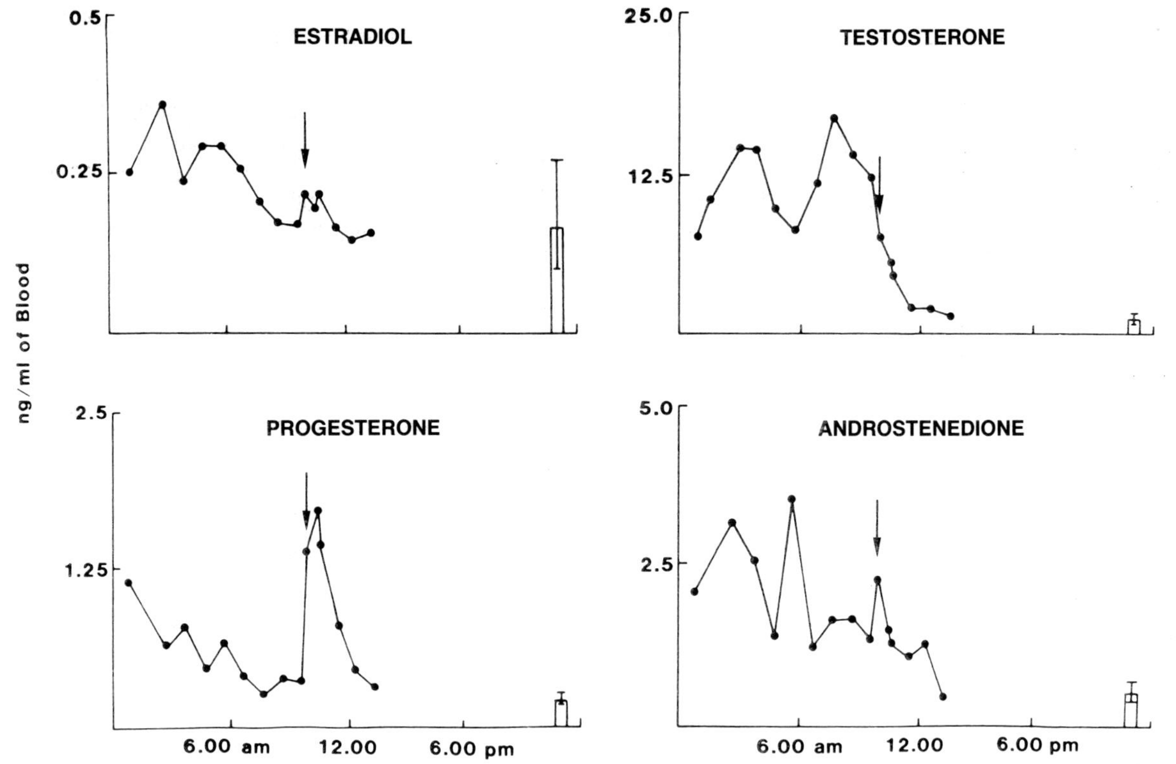

Fig. 4. The effect of gonadectomy on the peripheral concentration of four steroid hormones assessed by hourly bleeds and the mean and range calculated on three single daily samples after castration. The time of castration is indicated by the arrow.

data. The overall concentration of T rose from a basal level in the non-breeding season (March mean 2.18 ng/ml, 48 hourly samples from two animals) to a peak in the breeding season (July mean 9.47 ng/ml, 47 hourly samples from five animals, $P<0.001$). At the start of the breeding season A was at maximum concentration (June mean 5.93 ng/ml, 39 hourly samples from two animals) and then declined during the breeding (July mean 1.84 ng/ml, 47 hourly samples from five animals, $P<0.001$) and non-breeding season (March mean 0.35 ng/ml, 48 hourly samples from two animals, $P<0.001$). On the other hand E2 and P were at maximal concentrations during the non-breeding season (March mean 1.10 ng E2/ml, $P<0.001$, and 1.59 ng P/ml, $P<0.001$, 48 hourly samples from two animals) and were almost non-detectable by the end of the breeding season (November).

Gonadectomy

The effect of castration on the four steroid hormones in an adult male *I. macrourus* during the breeding season is shown in Figure 4. Although the concentration of T was declining at the time of castration, the overall effect can be seen by comparing the mean concentration (12.14 ng/ml) before castration with the mean concentration (1.13 ng/ml) of the three samples taken over the next three days.

These results indicate that most of the circulating T and A are released by the testis, and that a larger proportion of A than T is derived from sources other than the testis. However, in the case of E2 and P, a far larger proportion is from another source, possibly the adrenal glands. With respect to E2, the large range in concentration after castration, although based on only three single samples, indicates that an episodic release pattern for this hormone may still have persisted, which contrasts with the other three hormones which show much less variable peripheral concentrations following castration.

DISCUSSION

Although it is clear that release of testicular steroids in eutherian mammals is typically of an episodic nature, recent studies on the secretion of androgens in male non-eutherian mammals show some inconsistencies. Studies in this laboratory have shown that both the Koala (Carrick *et al.* 1981) and the Platypus (unpublished data) exhibit episodic release of T, A and E2, which indicate that these species, together with *I. macrourus,* follow the eutherian pattern of episodic release. Other studies on several species of macropods (Lincoln 1978) and the Common Brushtail Possum (Curlewis and Stone 1985) are not so conclusive, since it is likely that the animals involved were stressed due to animal handling procedures. This factor was acknowledged by the authors, and is likely to have resulted in suppression of the response of hypothalamo/hypophysial/testicular axis. The data of both Allen and Bradshaw (1980) and Curlewis and Stone (1985) on the Common Brushtail Possum do provide indications of episodic release, although in both cases these effects are masked by insufficiently frequent blood sampling. Additionally, more frequent blood samples would have been required to confirm the pattern of hormone release in these studies.

The consensus that episodic release of steroids is a generalized mammalian trait lends support to our contention that the use of single blood samples taken from a few animals cannot be used to gain a true picture of the endocrine status of the animal. However, if there are large differences in concentrations of hormones between breeding and non-breeding seasons, seasonal differences can be identified using single blood sample, but this often requires that large numbers of animals be studied.

Free ranging *I. macrourus*, as well as those in captivity, have been shown to have a seasonal breeding cycle (Gordon 1971; Gemmell 1982). Births occur throughout the year (Gemmel *et al.* 1985), however, there is a distinct reduction in birthrate during late autumn to mid-winter. The direct causal agent for this seasonal cycle has not yet been determined; however, the bandicoot probably responds to a physical environmental cue.

Male bandicoots are reported to produce sperm throughout the year (*I. macrourus*, Gordon 1971; *Isoodon obesulus* and *Perameles gunnii*, Heinsohn 1966) and this can lead to the conclusion that males are always capable of successful reproductive activity. However, there is some evidence from a study using single blood samples taken weekly throughout the year (Gemmell *et al.* 1985), that male *I. macrourus* exhibit a seasonal change in the concentration of circulating plasma testosterone.

The results from the present study also support this finding, showing that in the non-breeding months the testosterone concentration is low, with apparently reduced episodic activity: during the breeding season the mean concentration of T rises and the episodic release pattern becomes more pronounced. While it is said that sperm are present all year in male bandicoots, no publication provides quantitative data on sperm production rates and possible seasonal variation. The large changes of androgen concentration from breeding season to non-breeding season may also effect other variables influencing reproductive success such as behaviour or the secretions of the accessory sex glands. It has been shown that the prostate in the male Common Brushtail Possum varies markedly with season (Gilmore 1969; McFarlane, Carrick and Brown 1985) and these changes are likely to be under androgen control.

The effects of gonadectomy on T and A concentrations are similar to the responses of these hormones to gonadectomy in other eutherian (Hodges 1978; Sodersten *et al.* 1983; Johnstone *et al.* 1984) and marsupial species (Catling and Sutherland 1980). There was a rapid drop to basal level immediately after gonadectomy, and the means of the single sample bleeds taken over the next few days showed that the hormone concentration remained depressed, although somewhat higher residual concentrations were detected than those reported for the Tammar Wallaby (Catling and Sutherland 1980).

The high concentration of E2 and P found in the non-breeding season can not be explained without further study, but it is possible that these hormones may play a role in the onset of seasonal breeding activity.

REFERENCES

ALLEN, N. T. AND BRADSHAW, S. D., 1980. Diurnal variation in plasma concentrations of testosterone, 5α-dihydrotestosterone, and corticosteroids in the Australian Brush-Tailed Possum, *Trichosurus vulpecula* (Kerr). *General Comparative Endocrinology* **40:** 455–58.

CARRICK, F. N. AND COX, R. I., 1977. Testicular endocrinology of male marsupials and monotremes. Pp. 137–41 in *Reproduction and evolution* ed by J. H. Calaby and C. H. Tyndale-Biscoe. Australian Academy of Science: Canberra.

CARRICK, F. N. AND HUGHES, R. L., 1978. Reproduction in male monotremes. *Australian Zoology* **20:** 211–31.

CARRICK, F. N., MCFARLANE, J. R., BANCROFT, B. J. AND BROWN, A. S., 1981. Androgen production in the Koala. Proceedings of the 2nd Koala Symposium p. 5.

CATLING, P. C. AND SUTHERLAND, R. L., 1980. Effect of gonadectomy, season and the presence of female Tammar Wallabies *(Macropus eugenii)* on concentrations of testosterone, luteinizing hormone and follicle-stimulating hormone in the plasma of male Tammar Wallabies. *Journal of Endrocrinology* **86:** 25–33.

CURLEWIS, J. D. AND STONE, G. M., 1985. Peripheral androgen levels in the male Brush-Tail Possum *(Trichosurus vulpecula). Journal of Endocrinology* **105:** 63–70.

DEPALATIS, L., MOORE, J. AND FALVO, R. E., 1978. Plasma concentrations of testosterone and LH in the male dog. *Journal of Reproduction and Fertility* **52:** 201–07.

DIXON, A. F. AND GARDNER, J. S., 1981. Diurnal variations in plasma testosterone in a male nocturnal primate, the Owl Monkey *(Aotus trivirgatus). Journal of Reproduction and Fertility* **62:** 83–86.

ELLENDORFF, F., PARVIZI, N., POMERANTZ, D. K., HARTJEN, A., KONIG, A., SMIDT, D. AND ELSAESSER, F., 1975. Plasma luteinizing hormone and testosterone in the adult male pig: 24 hour fluctuations and the effect of copulation. *Journal of Endocrinology* **67:** 403–10.

GEMMELL, R. T., 1982. Breeding bandicoots in Brisbane *(Isoodon macrourus*: Marsupialia, Peramelidae). *Australian Mammalogy* **5:** 187–93.

GEMMELL, R. T., JOHNSTON, G. AND BARNES, A., 1985. Seasonal variations in plasma testosterone concentrations in the male marsupial bandicoot *Isoodon macrourus* in captivity. *General and Comparative Endocrinology* **59:** 184–91.

GILMORE, D. P., 1969. Seasonal reproductive periodicity in the male Australian Brush-tailed Possum *(Trichosurus vulpecula). Journal of Zoology, London* **157:** 75–98.

GOODMAN, R. L., HOTCHKISS, J., KARSCH, F. J. AND KNOBIL, E., 1974. Diurnal variation in serum testosterone concentration in the adult male rhesus monkey. *Biology of Reproduction* **11:** 624-30.

GORDON, G., 1971. A study of island populations of the short-nosed bandicoot, *Isoodon macrourus* (Gould). Ph.D. Thesis. University of New South Wales: Sydney.

GUSTAFSON, A. W. AND SHEMESH, M., 1976. Changes in plasma testosterone levels during the annual reproductive cycle of the hibernating bat, *Myotis lucifugus lucifugus* with a survey of plasma testosterone levels in adult male vertebrates. *Biology of Reproduction* **15:** 9–24.

HEINSOHN, G. E., 1966. Ecology and reproduction of the Tasmanian bandicoots (*Perameles gunni* and *Isoodon obesulus*). *University of California Publications in Zoology* **84:** 1–107.

HODGES, J. K., 1978. Effect of gonadectomy and oestradiol treatment on plasma luteinizing hormone concentration in the marmoset monkey, *Callithrix jacchus. Journal of Endocrinology* **76:** 271–81.

JOHNSTONE, I. P., BANCROFT, B. J. AND McFARLANE, J. R., 1984. Testosterone and androstenedione profiles in the blood of domestic tom-cats. *Animal Reproduction Science* **7:** 363–75.

KATONGOLE, C. B., NAFTOLIN, F. AND SHORT, R. V., 1971. Relationship between blood levels of luteinizing hormone and testosterone in bulls and the effects of sexual stimulation. *Journal of Endocrinology* **50:** 457–66.

KATONGOLE, C. B., NAFTOLIN, F. AND SHORT, R. V., 1974. Seasonal variations in blood luteinizing hormone and testosterone levels in rams. *Journal of Endocrinology* **60:** 101–06.

KIRKPATRICK, J. F., VAIL, R., DEVONS, S., SCHWEND, S., BAKER, C. B. AND WEISNER, L., 1976. Diurnal variation of plasma testosterone in wild stallions. *Biology of Reproduction* **15:** 98–101.

LINCOLN, G. A., 1978. Plasma testosterone profiles in male macropodid marsupials. *Journal of Endocrinology* **77:** 347–51.

PURVIS, K., ILLIUS, A. W. AND HAYNES, N. B., 1974. Plasma testosterone concentrations in the ram. *Journal of Endocrinology* **61:** 241–53.

McFARLANE, J. R., CARRICK, F. N. AND BROWN, A. S., 1985. Seasonal changes in the size of the prostate in the Brushtail Possum *Trichosurus vulpecula. Australian Mammal Society Bulletin* **9:** 56.

RODBARD, D., BRIDSON, W. AND RAYFORD, P. L., 1969. Rapid calculation of radioimmunoassay results. *Journal of Laboratory and Clinical Medicine* **74:** 770–81.

ROWE, P. H., LINCOLN, G. A., RACEY, P. A., LEHANE, J., STEPHENSON, M. J., SHENTON, J. C. AND GLOVER, T. D., 1974. Temporal variations of testosterone levels in the peripheral blood plasma of men. *Journal of Endocrinology* **61:** 63–73.

SANFORD, L. M., WINTER, J. S. D., PALMER, W. M. AND HOWLAND, B. E., 1974. The profile of LH and testosterone secretion in the ram. *Endocrinology* **95:** 627–31.

SHARMA, P. O., 1976. Diurnal variations of plasma testosterone in stallions. *Biology of Reproduction* **15:** 158–62.

SODERSTEN, P., ENEROTH, P. AND PETTERSSON, A., 1983. Episodic secretion of luteinizing hormone and androgen in male rats. *Journal of Endocrinology* **97:** 145–53.

Observations on placentation and development in *Echymipera kalubu*

R. L. HUGHES[1], L. S. HALL[1], M. ARCHER[2] and K. APLIN[2]

Observations were made on a collection of 14 female *Echymipera kalubu* and one sexually immature *E. rufescens* from New Guinea. The breeding season occupies an extended period. The corpus luteum of pregnancy continues to function into at least early lactation and females may become reimpregnated before the completion of weaning. The medullary plate stage of embryonic development occurs at an early stage of blastocyst expansion. A discoidal chorio-allantoic placenta is formed during the foetal portion of the gestation period and a yolk-sac placenta is also of significance. Birth is via a marsupial-type pseudovaginal passage and the chorio-allantoic placenta detaches from the uterine endometrium after birth.

Key words: *Echymipera,* placentation, foetal membranes, development, reproduction.

INTRODUCTION

THE first description of the foetal membranes of the Peramelidae was that of Hill (1895) who concluded from the morphological arrangement in dissected specimens that the peramelids were unique amongst the marsupials in that they possessed a true chorio-allantoic placenta which developed in the terminal stages of gestation. Hill (1897, 1899) subsequently gave an account of the histology of the chorio-allantoic placenta and foetal membranes of *Perameles obesula (=Isoodon obesulus)* and concluded that the placental arrangement was unique amongst mammals. Unlike eutherians, the trophoblast in the region of the true chorion completely disappeared during the invasion of maternal tissues. From this he concluded that *Perameles (=Isoodon)* exhibited a primitive mode of placentation not represented in mammals described to that date.

Subsequently, Flynn (1923) made a detailed study of placentation in *P. obesula (=I. obesulus)* and re-examined Hill's material from *Perameles gunnii* and *I. obesulus*. Flynn (1923), completely rejected Hill's (1897) contention of the disappearance of the chorion and described the formation of a mixed foetal/maternal syncytium in which the foetal chorionic cells were incorporated. Flynn (1923) interpreted these events not as a primitive structure, but as a bandicoot specialization, and although the events he described had features which he believed were unique to the Peramelidae, the persistance of the giant foetal chorionic cells was thought to be in keeping with that found in the formation of the chorio-allantoic placenta found in a variety of eutherians. Flynn (1923) also compared his findings with placental structures and foetal membranes in other marsupials. From this he highlighted the uniqueness of the peramelid placenta where the discoidal invasive chorio-allantoic placenta formed the dominant placental structure in the terminal stages of gestation (Flynn 1923). This distanced the Peramelidae from other marsupials where the yolk-sac placenta remains as the dominant respiratory and nutrient organ up to the time of birth.

Subsequent observations on placentation in the Peramelidae have been extended by Hughes (1974) on *Perameles nasuta* and *Echymipera rufescens*, and Padykula and Taylor (1976, 1977, 1982) on *Isoodon macrourus*. These more recent observations, although confirming Flynn's (1923) light microscopy, have added relatively little to the understanding of cellular specializations in the histogenesis of the mixed foeto/maternal chorio-allantoic placenta. However, Padykula and Taylor (1976) recorded the presence of desmosomes between chorionic cells and luminal epithelium and described homo and heterokaryons, in the context of similar structures reported for a diversity of eutherian tissue, as an interesting cellular phenomena in *I. macrourus*.

The present paper describes the cellular details of an earlier and later stage in the formation of the chorio-allantoic placenta than has been previously recorded for the Peramelidae. These observations include the state of organogenesis of a pre-implantation stage and the fully developed chorio-allantoic placenta of an immediately post partum female. Other material studied includes pre-primitive streak stages which are correlated with the condition of the reproductive tract. This material, together with information on the condition of non-pregnant females, is used to give a comparison of the reproductive ecology of *Echymipera*

[1]Department of Anatomy, University of Queensland, St Lucia, Queensland, Australia 4067.
[2]School of Zoology, University of New South Wales, Kensington, New South Wales, Australia 2033.

Pages 259-70 *in* BANDICOOTS AND BILBIES, ed by J. H. Seebeck, P. R. Brown, R. L. Wallis and C. M. Kemper. Surrey Beatty & Sons, Sydney, 1990.

Table 1. Details of specimens used in this study. Specimen NGM695 is *Echymipera rufescens*. All others are *E. kalubu*. In the reproductive condition column, CL = corpus luteum; CA = corpus albicans; followed by diameter in mm. In the details of pouch young column, first number = number of pouch young; HL = head length; HB = head and body length; BW = body weight in g; FL = foot length; followed by measurement in mm. All specimens collected by K. Keith except for 343 which was collected by K. Aplin.

Specimen	Locality/Date	Body weight	Reproductive condition	Details of pouch young
NGM695	Aupik, 12/5/1965	200 g	Immature, no CL	No pouch young
NGM329	Japandai, 13/6/1963	450 g	Lactating; 2CL, 2.2	2; HL7.7; sex indifferent
NGM405	Korogo, 9/3/1964	580 g	Not pregnant; 3CL, 2.3	3; BW 13.1; 2 male, 1 female
NGM406	Korogo, 9/3/1964	510 g	Not pregnant; 2CL, 1.9	2; HL16; HB31.5; 1 male, 1 female
NGM411	Korogo, 10/3/1964	470 g	Not pregnant; 3 lactating teats; 3CL, 2.0	1; no fur, partly eaten; HB45
NGM656	Amanab, 30/4/1965	690 g	Not pregnant, 2 enlarged teats	2; HL25; BW 105; HB74; 1 male, 1 female
NGM399	Korogo, 8/3/1964	500 g	Not pregnant	1; HB37; FL6.7; fur just appearing; eyes closed
NGM352	Maprik, 27/2/1964	780 g	Lactating; no embryos found; 2CL, 2.3	1; HB95; BW18; FL18; furred and off teat in pouch
NGM264	Chambri Lake, 30/5/1963	480 g	Lactating; distended caecum; no embryos found; 3CA, 2.0; 3 enlarged teats 15 mm long	Young out of pouch
NGM407	Korogo, 9/3/1964	430 g	Lactating; 2CA, 2.0; 2 enlarged teats	No pouch young
NGM304	Maprik, 7/6/1963	490 g	Lactating; CL, 2.2; unilaminar blastocyst of several hundred cells 0.5 mm dia.; used teats 15–20 mm long	No pouch young
NGM387	Kundangai, 3/3/1964	470 g	2 uterine blastocysts 2.1 mm dia.; CL, 3.0; medullary plate stage; lactating	1; HB123; BW42
NGM301	Japanaut, 7/3/1963	730 g	2 uterine blastocysts 2.3 mm dia.; 2CL, 3.0; medullary plate; 2 stages; lactating; teats 15–20 mm long	2; HL 42.7; BW56; 2 males both furred
NGM424	Maprik, 20/3/64	410 g	Not lactating; pregnant in both uteri; pre-implantation stage; 2CL, 2.9; greatest embryonic length 5 mm	No pouch young
343	Haia, 16/5/1984		Post partum; undegenerated placenta in each uterus; uterine body 14.5 × 8 mm; 2CL, 2.9	2; Newborn young; HL6.0; BW 0.207

Fig. 1. Edge of medullary plate area from 2.1 mm blastocyst of specimen NGM387. Expanding medullary plate cells exhibiting mitosis (arrows) and large cells of yolk-sac ectoderm are also undergoing mitosis (YS). Magnification ×227.

Fig. 2. True chorion from 5.0 mm prefoetal stage of specimen NGM424. The thickened endometrial epithelial nests (EN) contain maternal blood vessels (BV) and overlie endometrial glands (EG) which are in a secretory phase. No intracellular endometrial leucocytes are present at this stage. Note the thin layer of non-vascularized somatic mesoderm (SM) attached to the chorion ectoderm (CE). Uterine lumen (UL). Magnification ×227.

Fig. 3. Uterine endometrium near trilaminar yolk-sac from specimen NGM424. Thickened luminal nests of epithelial cells (EN) contain blood vessels (BV) which lie close to the luminal surface. Endometrial glands (EG) penetrate to the luminal surface and are in a secretory phase. The foetal trilaminar yolk-sac (TYS) consists of flattened ectodermal and endodermal cells enclosing a rich foetal blood vessel net containing nucleated erythrocytes (FBV). Magnification ×227.

Fig. 4. Uterine endometrium adjacent to bilaminar yolk-sac from specimen NGM424. There is a marked reduction in the thickness of the luminal nests of epithelial cells (EN) which contain and overlie a rich vascular network (BV). Numerous capillary loops reach the luminal surface. The endometrial glands (EG) are in a secretory phase. The foetal ectodermal cells are large and vacuoalated (FE) with an irregular border suggesting the presence of microvilli. The foetal endodermal cells (FED) are cuboidal. Magnification ×227.

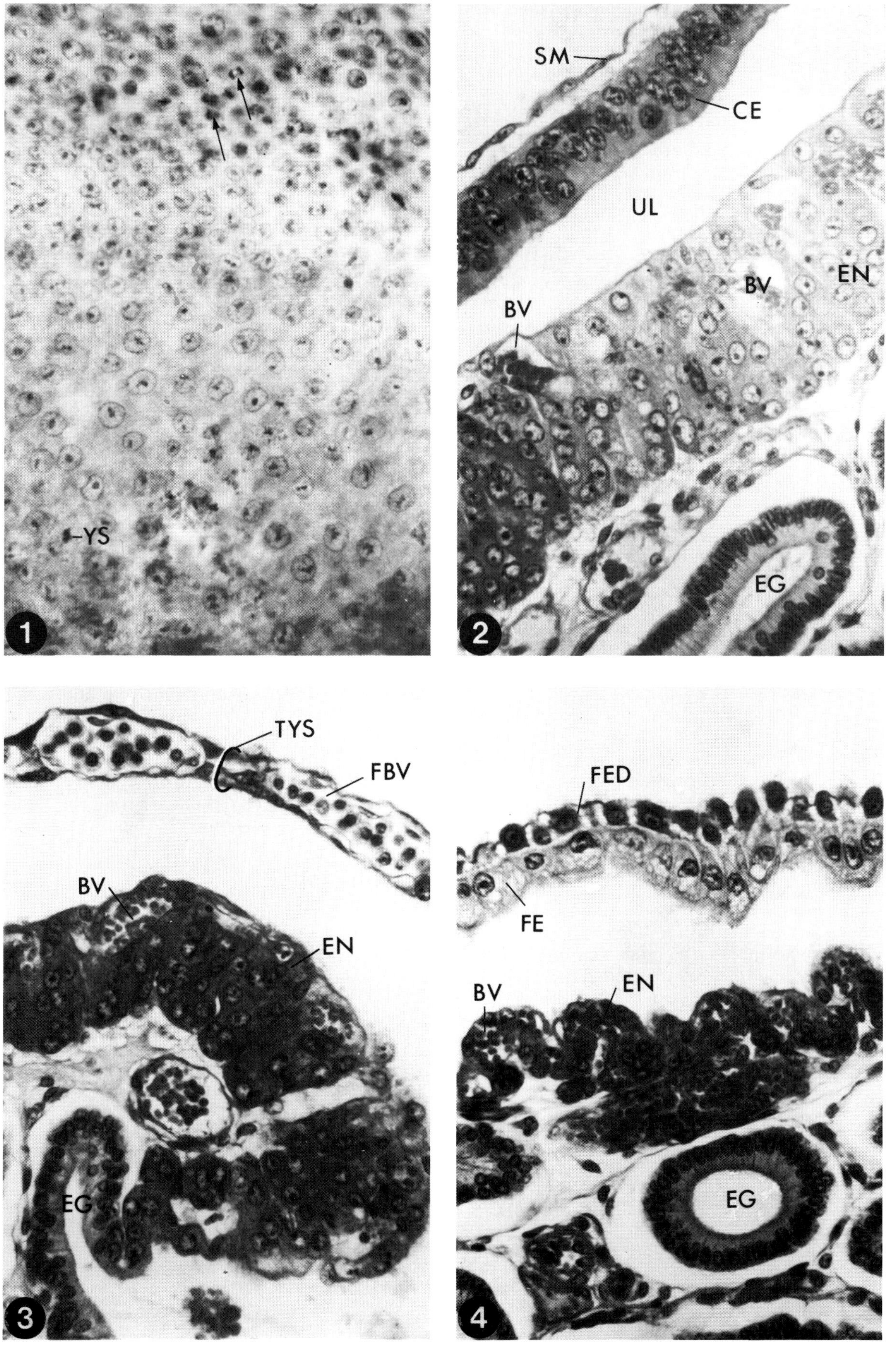
1
YS
2
SM
CE
UL
BV
BV
EN
EG
3
TYS
FBV
BV
EN
EG
4
FED
FE
BV
EN
EG

with other members of the Peramelidae, for which there exists an extensive and growing literature.

MATERIAL AND METHODS

Table 1 shows the details of the 14 *E. kalubu* and one sexually immature *E. rufescens* specimens used in this study. The specimens were preserved in the field in Bouin's or 10% buffered formalin and were returned to the laboratory for detailed dissection. Uteri were removed and examined under a stereo microscope. Blastocycts were recovered from dissected uteri using a dissecting microscope. Measurements of placentae and other structures were taken prior to further processing. The chorio-allantoic placenta with attached umbilical cords was removed from the uterine lumen and processed separately. Part of the ovary bearing the large corpus luteum was dissected and processed separately.

Material for histological examination was dehydrated through graded alcohols, embedded in wax, serially sectioned at 5 μm, and stained with haematoxylin and alcoholic eosin (H and E).

RESULTS

Reproductive ecology

A perspective of the reproductive ecology of *E. kalubu* is outlined in Table 1. The major features to be noted are: (i) the breeding season, as indicated by the presence of pouch young, extends from February to June. It should be noted that this includes both dry and monsoon periods. Although specimens are lacking from July to January, it appears that the breeding season is not sharply defined; (ii) an immature female weighed 200 g and breeding females ranged in weight from 410–536 g; (iii) mean litter size, as indicated by pouch young, is 1.82 (n = 11) and the eggs ovulated per set, as indicated by corpora lutea was 2.27 (n = 11); (iv) the corpus luteum of pregnancy persists during the early stages of lactation. Although the preservation of material was not ideal, the undegenerated luteal cells of the corpus luteum were well vascularized by small blood vessels which passed through a fine reticular net of connective tissue, and on this basis the corpora lutea were considered to be functional; (v) females may become re-impregnated while still lactating, and this occurred in March and June; (vi) as in other marsupials, teats enlarge with suckling young and can be used as an index of reproductive activity as described for *I. macrourus* by Hall (1983).

Foetal membranes

Table 1 shows there were three early pregnancy stages. In specimen NGM304, a unilaminar blastocyst of several hundred cells and with an overall diameter (including shell membrane) of 0.5 mm was recovered from the uterus. In NGM387 there were two medullary plate stages with an overall diameter, that included the shell membrane, of 2.1 mm. Figure 1 shows a portion of the blastocyst at the junctional region between the expanding medullary plate and ectodermal wall of the primitive yolk-sac. It is important to note that the medullary plate stage is present in blastocysts of small diameter. In NGM301 there were also two medullary plate stages with an overall diameter of 2.3 mm. It is also of interest that at the medullary plate stage the simple columnar luminal epithelium of the uterine endometrium has not yet commenced to form the luminal endometrial nests characteristic of the immediate pre-attachment stage. A pre-implantation stage was exhibited by NGM424 which was pregnant in both uteri. The greatest length of the definitive embryos was 5 mm, with the head facing the cervix in one uterus and the utero-tubal junction in the other. A description of the arrangement of the extra-embryonic membranes and uterine endometrium will be followed by a brief account of the level of organogenesis in the embryo.

Fig. 5. Embryonic pole of uterus from specimen NGM424. The myometrium (MYM), glandular endometrium (GUE) and luminal epithelium (LUE) are markedly thickened. The true chorion (TC) is a thickened multicellular layer which follows the contours of the endometrium. Vascularized trilaminar yolk-sac (TYS), two layered amnion (AM), extra embryonic coelom (EEC), yolk-sac splanchnopleure (YSP), limb bud (LB), closed neural tube (NT), dorsal aorta (DA) and somite (S). Magnification ×40.

Fig. 6. Abembryonic pole of uterus from specimen NGM424. The myometrium (MYM), glandular endometrium (GUE), and luminal epithelium (LUE) are noticeably thinner than in Figure 5. Bilaminar yolk-sac (BYS) is unattached to endometrium. Magnification ×40.

Fig. 7. Pre-implantation stage embryo from specimen NGM424 closed neural tube (NT), differentiating neuroblasts (NB), spinal ganglia (SG), notochord (NO), precartilaginous sclerotome (PS), dorsal aorta (DA), posterior cardinal vein (PC), urogenital ridge with mesonephric tubule (UR) and myotome (MY). Magnification ×103.

Fig. 8. Pre-implantation stage embryo from specimen NGM424. Open neural tube (NT) covered by amnion (AM), neural crest cells (NC) and foetal erythrocyte (NE). Magnification ×103.

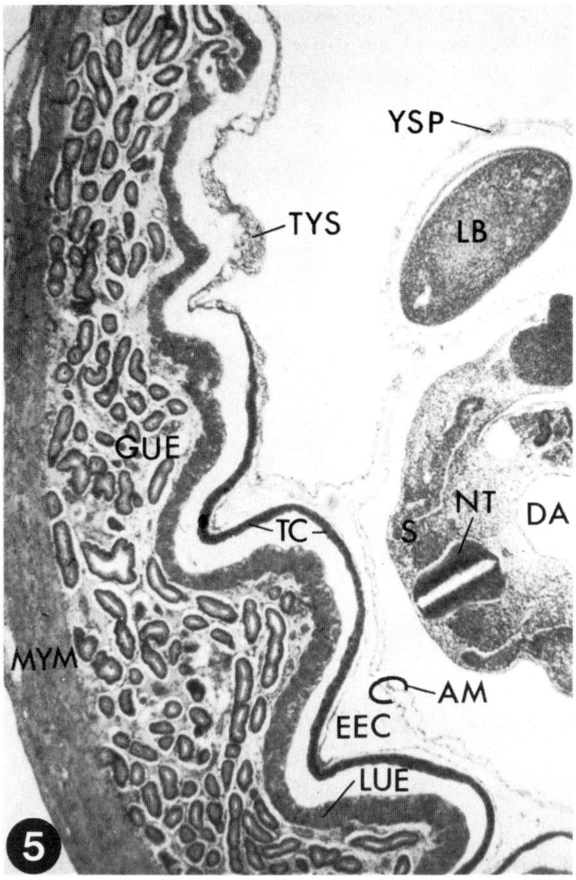
YSP
TYS
LB
GUE
NT
DA
S
TC
MYM
AM
EEC
LUE
5

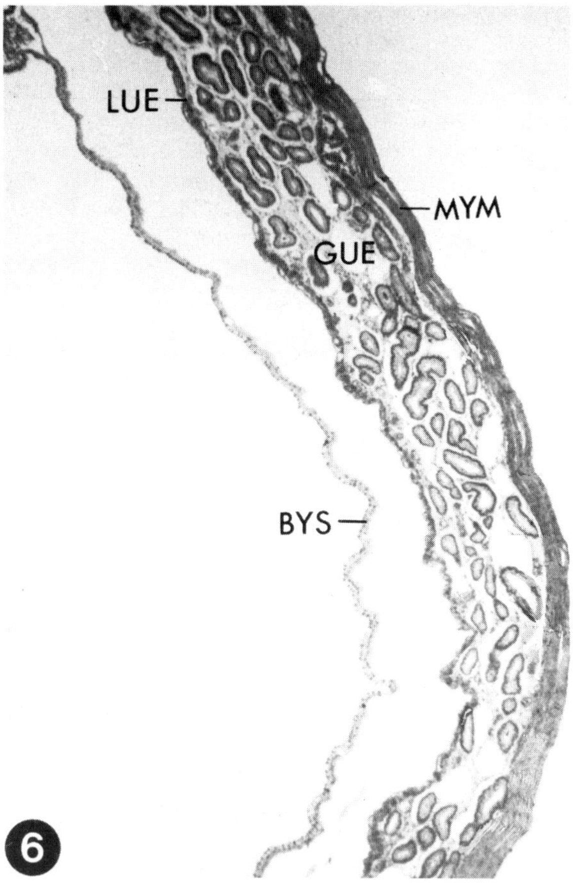
LUE
MYM
GUE
BYS
6

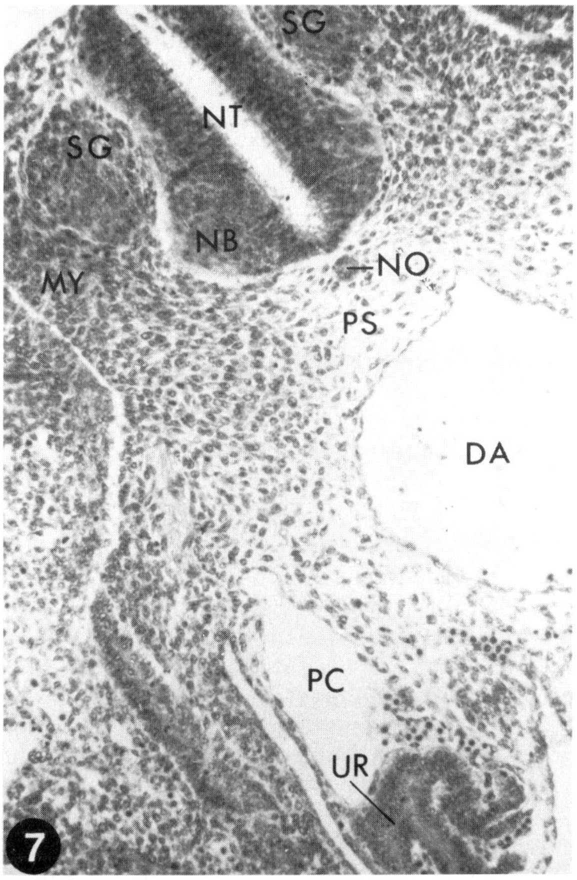
SG
NT
SG
NB
MY
NO
PS
DA
PC
UR
7

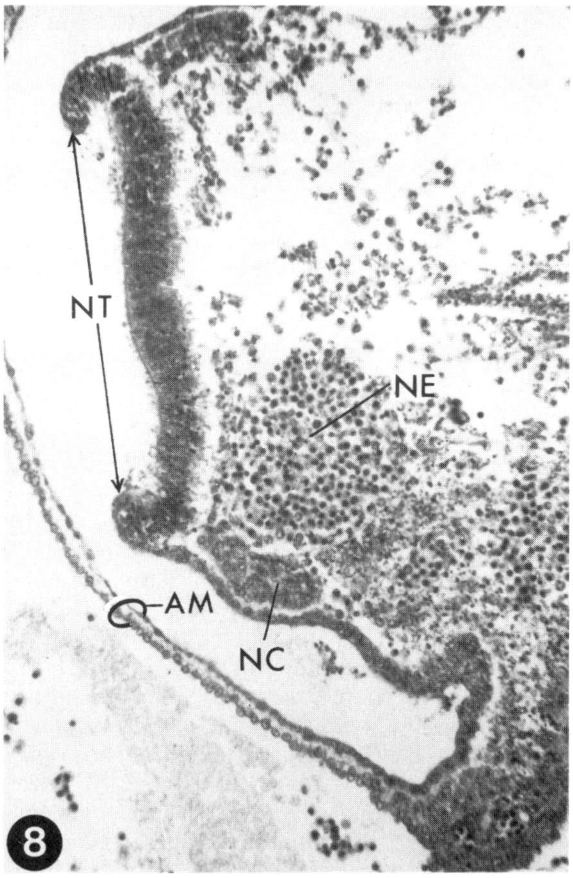
NT
NE
AM
NC
8

The chorion

The arrangement of the chorion and the immediately adjacent associated uterine endometrium is shown in Figure 2. The chorionic ectoderm is the thickest ectodermal component of all the extra-embryonic membranes abutting the uterine lumen. The nuclei are enlarged and arranged in a thickened layer of three to four cells. This area was immediately underlain by a thin non-vascular layer of somatic mesoderm. Under oil immersion the cells of the thickened chorionic ectoderm appeared to have frequent cellular boundaries. The immediately adjacent luminal epithelium of the uterine endometrium was arranged as thickened groups of cell nests (Fig. 2). Each section of a luminal nest contained the nuclei of up to 24 cells. These nuclei are vesicular with a prominent nucleolus and nest exhibit a scalloped appearance at the interface with the deeper endometrial tissue. The luminal surface was relatively smooth and at all points was widely separated from the foetal chorionic ectoderm. A rich sub-epithelial vascular net of blood vessels was found to penetrate the epithelial nests and at several points was separated from the surface by only a thin layer of cytoplasm.

The endometrium exhibited a marked polarization of thickening with the maximum thickness of myometrium, luminal epithelium and glandular endometrium occurring in the area immediately adjacent to the chorion (Fig. 5). The minimum dimensions of this tissue were located at the opposite poles in the region of the bilaminar yolk-sac (Fig. 6). This polarization of maternal tissue is shown diagrammatically in Figure 9.

Trilaminar yolk-sac

In the region of the trilaminar yolk-sac of the tissue from NGM424 the surface ectoderm of the foetal extra-embryonic membrane consisted of a single layer of flattened cells (Fig. 3). Immediately below this layer the rich network of foetal capillaries constituted an almost unbroken layer of vessels distended with nucleated erythrocytes. The endodermal layer also consisted of a single layer of flattened cells (Fig. 3). The trilaminar yolk-sac is widely separated from the luminal epithelium which consisted of luminal nests of approximately 12 nuclei per section. Its total thickness was approximately two-thirds of the surface epithelium of the endometrium in the chorionic area of the embryo. A rich and slightly more developed sub-epithelial vascular network lies between and among luminal endometrial nests close to the luminal surface. In some places the uterine glands penetrated to the surface (Fig. 3).

The foetal ectodermal cells of the bilaminar yolk-sac were arranged as enlarged vacuolated columnar cells with vesicular nuclei immediately overlying slightly enlarged cuboidal cells of the endodermal layer (Fig. 4). The endometrium adjacent to the distal extremity of the bilaminar yolk-sac was markedly thinned (Fig. 6) in comparison to other areas (Fig. 5). In this area the bilaminar yolk-sac was widely separated from the luminal epithelium which was the most highly vascularized part of the luminal epithelium and exhibited the minimal development of the luminal nests. The luminal endometrial nests consisted of groups of up to six nuclei per section and were markedly reduced in diameter.

The above observations indicated a level of polarity manifested in a gradation of thickness in maternal components with a maximum degree of development adjacent to the chorionic area. There was a reciprocal level of maternal uterine luminal vascularization with a maximum occurring at the most distal region of the bilaminar yolk-sac. The vacuolation in the immediately adjacent foetal ectodermal cells suggests that the level of physical interchange between maternal and foetal tissues is greatest at this region.

The double-layered amnion closely invests the embryo and follows the major surface contours of its body. There is a separate pocket which encloses the large forelimb (Figs 5 and 8).

Pre-implantation embryo

The torted embryo obtained from specimen NGM424 exhibited forelimb buds and was at the branchial cleft stage with unfused neural folds in the caudal region. Neural crest cells were seen (Fig. 8). Organ systems present included the primordia of liver, early mesonephric tubules without mesonephric duct (Fig. 7), heart, fore, mid and hind gut. The partly vascular yolk-sac was serviced by an early but functional cardiovascular system. Somites (Fig. 5), spinal ganglia (Fig. 7) and primary brain vesicles were present. The orientation of the embryo in relation to its extra-embryonic membranes is shown diagrammatically in Figure 9.

The head region of the embryo was deeply buried within the yolk-sac splanchnopleure. The close apposition of the amnion and the yolk-sac splanchnopleure over the cephalic and dorsal regions of the embryo eliminated the cavity of the extraembryonic coelom. The caudal region of the embryo occupied a relatively superficial position immediately adjacent to a conspicuously wide component of the extra-embryonic coelom. The umbilical region with its vascularized yolk-sac stalk consequently occupied a fairly superficial position and was

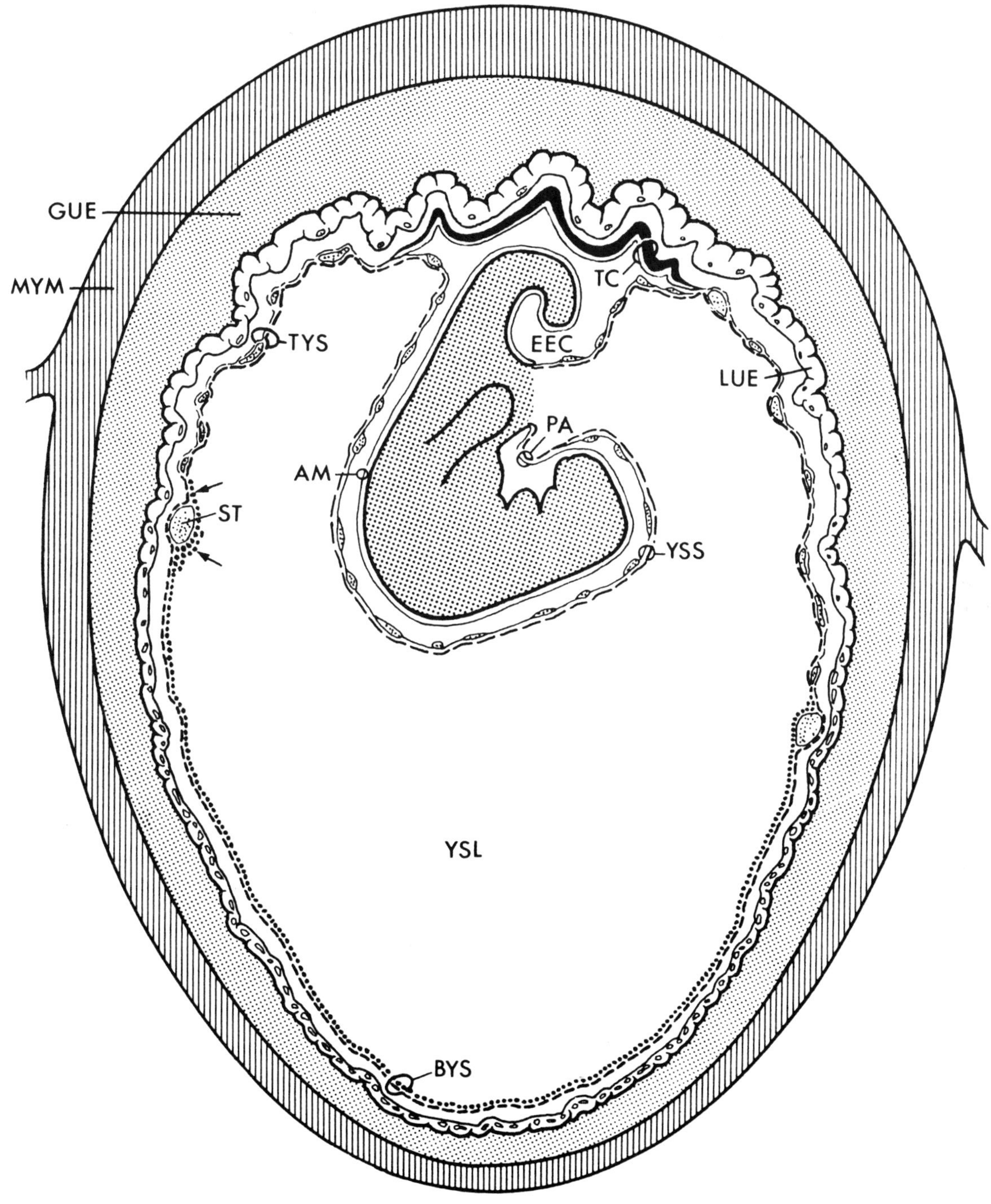

Fig. 9. Schematic diagram of foetus and foetal membranes in specimen NGM424. This is a pre-implantation specimen and the myometrium (MYM), glandular endometrium (GUE) and luminal epithelium (LUE) of the uterus show a pronounced thickening adjacent to the embryonic pole of the foetal membranes. These three tissues grade to a minimum thickness at the abembryonic pole of the foetal membranes in the most distal region of the bilaminar yolk-sac (BYS). The membrane closely investing the embryo is the amnion (AM) which is composed of an inner layer of somatic mesoderm and an outer layer of ectoderm (a single line represents both these layers). A small component pro-amnion, consisting of ectoderm and endoderm is situated in a region between the head fold and forelimb (PA). The cephalic region of the embryo is buried in a deep fold of the yolk-sac splanchnopleure (YSS) which consists of a vascularized splanchnic mesoderm and yolk-sac endoderm. At this stage the allantois is absent, however the region of the future allantois is situated in close proximity to a markedly thickened area of true chorion (TC) and a spacious area of extra embryonic coelom (EEC). The extra embryonic foetal membranes immediately adjacent to the luminal uterine epithelium are composed of true chorion (TC), which contains a thickened ectoderm underlaid by a non-vascular somatic mesoderm, and the vascularized trilaminar yolk-sac (TYS), which consists of ectoderm, vascularized splanchnic mesoderm and endoderm. In a region of the trilaminar yolk-sac, immediately adjacent to the sinus terminalis (ST) is an area lined by cuboidal endodermal cells (arrows). In other areas these cells are more squamous. Yolk-sac lumen (YSL).

relatively close to a conspicuous area of true chorion. An allantoic vesicle was absent and its intra-embryonic primordium was as yet undifferentiated. However, the close proximity of the umbilical region to the true chorion would obviously facilitate the future formation of the chorion-allantois.

Fully developed chorio-allantoic placenta

The immediately post-partum specimen NGM343 possessed two new-born young in her pouch. Subsequent histological examination of uterine tissue revealed a detached discoidal chorio-allantoic placenta free in each of the uterine lumina. Attached to these were umbilical cords which extended through the cervix and into the vaginal complex. The dimensions of these structures were ovoid placental discs of 4 mm × 6 mm and umbilical cords of 8 mm in length.

The preservation and preparation of this tissue revealed that the endometrium and chorio-allantoic placental tissues were in a relatively undegenerated state (Figs 10, 11, 12 and 13). The endometrium in the placental area exhibited remnants of giant placental cells (Fig. 10). The tissue was relatively loose and was infiltrated with numerous polymorpholeucocytes. The rich maternal vessels contained erythrocytes (Fig. 10).

In the peri-placental endometrial area there was a rich vascular maternal blood supply which extended beneath the luminal epithelium. The luminal epithelial nests consisted of groups of relatively intact cells but there were signs of slight degenerative changes. These signs included a disorganization of tissue and a reduction in the height of luminal nests of cells as well as a reduction in the diameter of their vesicular nuclei (compare Fig. 3 with Fig. 11).

An examination of the discoidal chorio-allantoic placenta showed a rich interwoven foetal and maternal network of tissue with numerous polymorpho-leucocytes invading the surrounding tissue (Figs 12 and 13). In some sections foetal and maternal capillaries had an apparent immediately adjacent location which indicated little or no separation between their endothelial layers as was previously reported for bandicoots by Flynn (1923).

DISCUSSION

Details of the reproductive ecology of the Peramelidae were unable to be constructed from the small amount of material reported in Hill (1895, 1897) and Flynn (1923) and waited, more or less, until modern times. The first observations on the breeding of the Peramelidae were by Mackerras and Smith (1960) who captured and bred *I. macrourus* in captivity. This was followed by the monograph of Heinsohn (1966) who recorded the reproductive ecology of wild-caught *P. gunnii* and *I. obesulus* in Tasmania. Lyne (1964, 1974, 1976), Lyne and Hollis (1977, 1979) and Hollis and Lyne (1980) have made considerable contributions from observations on captive *P. nasuta* and *I. macrourus*. Further publications by Stodart (1966), Gordon (1971, 1974a and b), Close (1977), Stoddart and Braithwaite (1979), Gemmell (1981) and Hall (1983) have contributed to the understanding of reproduction in the female peramelids. From the previous studies a general pattern can be established. Female bandicoots have a short gestation of about 12.5 days, being considerably less than the oestrous cycle of about 26 days; the corpus luteum persists as a functional organ into lactation; there is a chorio-allantoic placenta; growth of pouch young is rapid and lasts about two months; and females become re-impregnated approximately one gestation period before they wean their pouch young. The present study confirms this general pattern in the genus *Echymipera.*

Despite the extensive studies on the reproductive ecology of the Peramelidae, few studies have added to the classical observations on the chorio-allantoic placenta of the Peramelidae made by Flynn (1923), which is among the most distinctive of the reproductive characters of the Peramelidae. There is only one previous observation on placentation in a species of *Echymipera.* Hughes (1974) reported that in an early foetal stage of *E. rufescens* (= *E. kalubu* reidentified GL = 5.0 mm) the chorion and

Fig. 10. Uterine endometrium in placental area from recent post partum specimen NGM343. Note the darkly stained giant chorio allantoic placental cells (GS). The tissue is highly invaded by polymorphs (PM) and the uterine glands which are still intact are highly vascularized (GUE). The endometrial tissue is very disorganized. Magnification ×227.

Fig. 11. Peri-placental area of post partum female NGM343. Note highly folded luminal endometrium (LUE), with a rich sub-epithelial vascular network (MBV). The luminal epithelium is multicellular and has been invaded by polymorphs (PM). Magnification ×227.

Fig. 12. Detached luminal chorio-allantoic placenta from specimen NGM343. Note the close apposition of the endothelia of foetal (FBV) and maternal (MBV) capillary networks. Numerous polymorphs (PM). Rich foetal network of blood vessels contain nucleated erythrocytes. Magnification ×545.

Fig. 13. Detached luminal chorio-allantoic placenta from specimen NGM343. Numerous polymorphs (PM) have infiltrated the tissue which features a very well developed capillary network of foetal blood vessels (FBV). The enlarged pale cells (PC) are foeto-maternal syncytical cells. Magnification ×545.

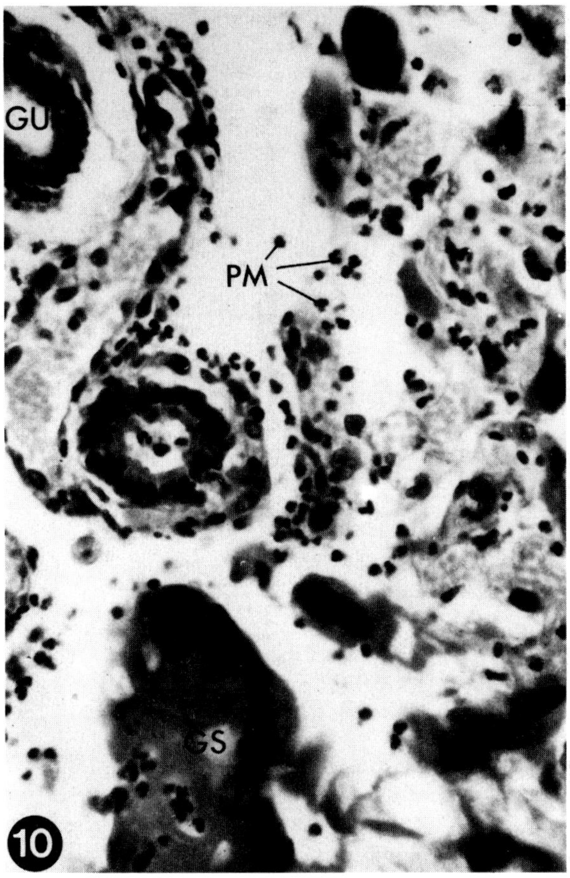
GU
PM
GS
10

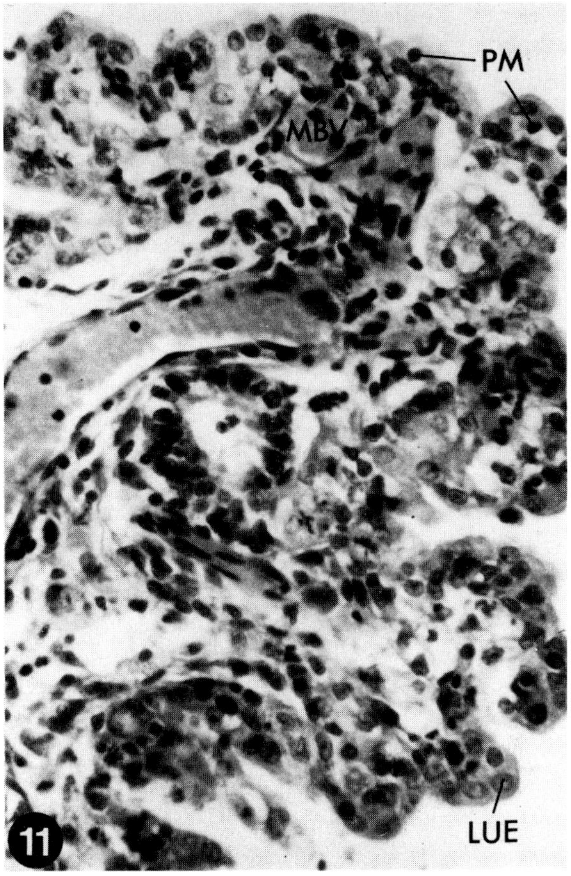
PM
MBV
LUE
11

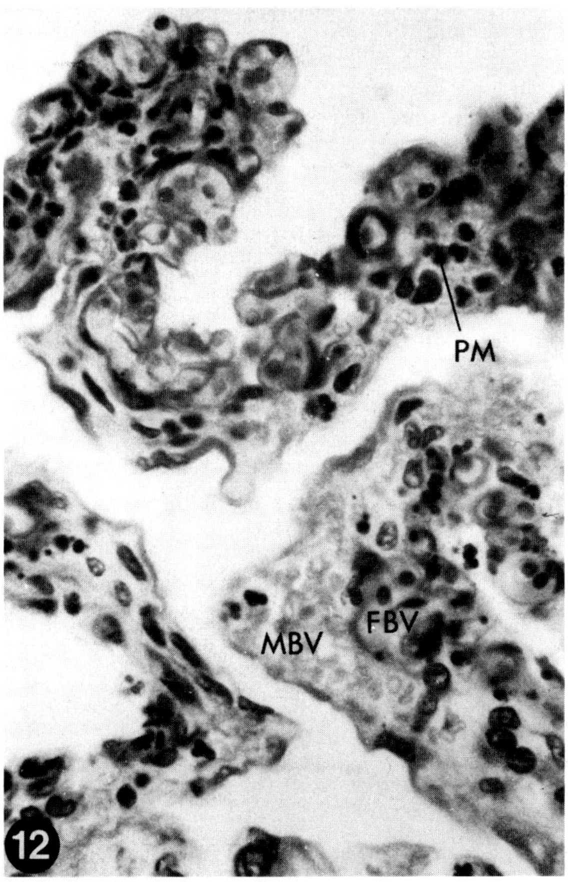
PM
MBV
FBV
12

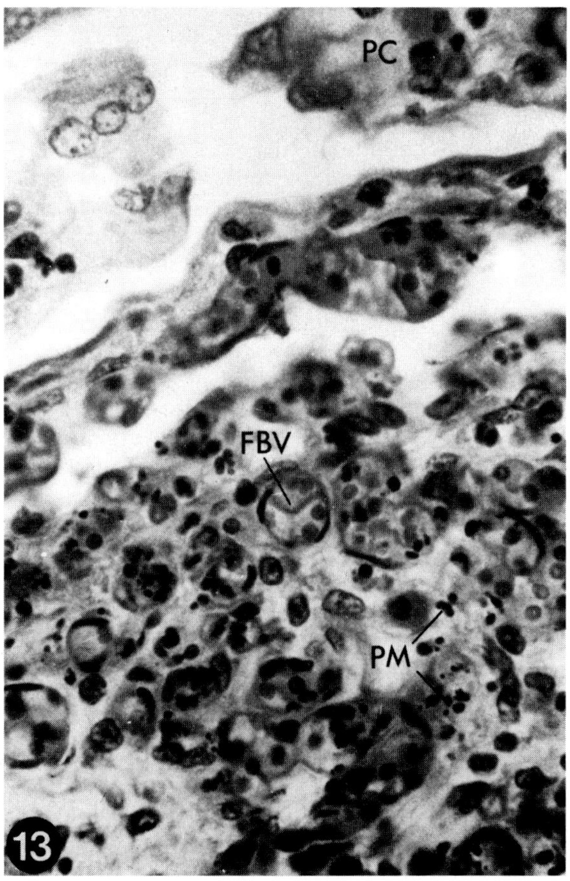
PC
FBV
PM
13

adjacent luminal epithelium of the uterus exhibited localized thickening and only remnants of the shell membrane remained. The present paper includes a more detailed examination of the specimen previously reported in Hughes (1974).

Lyne and Hollis (1977) reported that the medullary plate stage, in the process of endoderm formation, was attained in *P. nasuta* and *I. macrourus* in small blastocysts of 1.00 mm to 1.5 mm in diameter. In comparison to other Australian marsupials Lyne and Hollis (1977) note that an equivalent stage in embryonic development was attained in larger blastocysts of over 4.5 mm in diameter but that in the American marsupial *Didelphis virginiana* endoderm formation occurred at an even earlier blastocyst stage than in *P. nasuta* and *I. macrourus*. Our observations on the two medullary plate stages of *E. kalubu* show that this bandicoot, like *P. nasuta* and *I. macrourus*, also exhibits a medullary plate stage in small blastocysts of slightly larger diameter (2.1 mm and 2.3 mm respectively). It will consequently be of interest to determine for other peramelids whether the medullary plate stage is attained at an early stage of blastocyst expansion and whether this represents yet another distinctive peramelid reproductive feature. The early formation of the medullary plate reflects rapid embryonic development of peramelids and is a factor that contributes to the short gestation.

In our immediate pre-implantation stage of *E. kalubu* the development of the maternal subluminal endometrial capillary net was slightly more prominent in regions immediately adjacent to the richly vascularized, but non-invasive portion of the yolk-sac. In the absence of an allantoic rudiment the vascularized yolk-sac at this stage constitutes the sole structure for foetal-maternal vascular exchange. In this 5 mm embryo of *E. kalubu* the enlarged vacuolated foetal ectodermal cells of the non-vascularized bilaminar yolk-sac and perhaps also the slightly thickened endodermal cells of this bilaminar region are morphologically consistent with significant placental function. On similar grounds Flynn (1923) considered that in bandicoots the bilaminar yolk-sac was a significant placental structure. This is consistent with a finding of maximum luminal endometrial vascularization of our 5 mm pre-attachment embryo in the region of the bilaminar yolk-sac.

In this pre-implantation stage we described the uterine wall as consisting of three components: superficial vascularized luminal epithelial nests; a central region of secretory endometrial glands; and a peripheral layer of smooth muscle. A morphological polarization in thickness was exhibited by each of these three uterine wall components. The maximum thickness was at the embryonal pole of the uterus in the region near the true chorion and this progressively diminished towards the region immediately adjacent to the embryonic vascularized yolk-sac. These tissue components were thinnest at the abembryonal pole at the distal extremity immediately adjacent to the non-vascularized bilaminar yolk-sac. The functional significance of this polarity of uterine tissues obviously deserves further investigation. In this context a gradation in the dimensions of the luminal endometrial syncytial nests consistent with our findings was reported by Flynn (1923) to be evident at the early attachment phase in the formation of the bandicoot chorio-allantoic placenta. It is of interest to note that at the immediately pre-implantation stage of development of *E. kalubu* the prominence of the endometrial subepithelial capillary net exhibits an inverse polarity to other endometrial tissue components with descending orders of prominence adjacent to the bilaminar yolk-sac: vascularized; and true chorion respectively.

Our account of the chorio-allantoic placenta in *E. kalubu* is based on material from only two pregnant females. These specimens constitute an earlier and a later stage than those previously described for other bandicoots. Despite the small amount of material at our disposal it is absolutely certain that the distinctive peramelid-type of chorio-allantoic placenta as originally described by Hill (1895, 1897, 1899) and later in more detail by Flynn (1923) has now been shown by us to occur in the rather specialized bandicoot *E. kalubu*. We consequently believe that it might reasonably be presumed that this type of placental arrangement will be found to constitute a distinctive feature of the Peramelidae.

The major features in the formation of the chorio-allantoic placenta of *E. kalubu* will now be summarized.

Our earliest specimen consisted of an immediate pre-attachment stage. Syncytialization was not as yet evident in either the thickened true chorion or in the multicellular maternal luminal endometrial nests. However, in other respects the cellular detail of both the foetal and the maternal tissues closely paralleled that reported by Flynn (1923) for the chorio-allantois of the early attachment phase of *Perameles*.

In regard to the cellular detail of our terminal post-parturient stage in the formation of the chorio-allantoic placenta it is of interest that a detached discoidal chorio-allantoic placenta was found free in the lumen of each uterus. This in conjunction with the relatively undegenerated cellular structure is not consistent with Hill's

(1897, 1899) belief that the chorio-allantoic placenta of the bandicoots he studied was non-deciduate (contra-deciduate) and consequently degenerated while still attached to the uterine endometrium.

The numerous polymorphonuclear leucocytes we have found to be associated with both the detached chorio-allantoic placentae and also within the uterine endometrium in our immediately post-partum specimen of *E. kalubu* should not necessarily be interpreted as evidence for birth associated with immunological rejection of the placenta as suggested by Lillegraven (1975) and Tyndale-Biscoe (1973). In this context the classical studies of the bandicoot placenta by Flynn (1923) indicated that abundant leucocytes appear at the early embryonic attachment stage within the maternal endometrial tissues that are being subjected to destructive invasion by the syncytiochorion. That this earliest appearance of leucocytes is not at a terminal stage of gestation is illustrated by the figures of Hughes (1974, 1984) where a *P. nasuta* embryo at this time showed a pre-foetal embryo at the early branchial cleft stage of development. Flynn's (1923) studies also indicated that the high endometrial concentrations of leucocytes were not a pre-parturient phenomenon but leucocytic invasion initiated at the earliest stages of embryonic attachment persisted throughout the entire histogenesis of the chorio-allantois. His data suggest that the abundance of leucocytes was continuously fuelled by the progressive tissue destruction culminating at the terminal stages of pregnancy in a relatively simple chorio-allantoic placental arrangement where the separate foetal and maternal capillary beds were closely intertwined. According to Flynn (1923) the demise of the luminal endometrial syncytial nests occurred at their interface with the syncytiochorionic expansion. Involution of the majority of the giant invasive syncytiochorial cells occurred at a later stage behind the invasive progress zone and is well advanced by the terminal stages of gestation so that in the fully formed chorio-allantoic placenta little tissue remained to prevent the extremely close apposition of the endothelia of the foetal and the maternal capillary nets. Consequently our findings of massive leucocytic invasion in combination with the few remnants of presumptive giant foetal cells in both the vacated endometrial placental area and also within the detached discoidal chorio-allantoic placentae, as well as the rich, closely intertwined foetal and maternal capillaries, can be interpreted in a parallel histological manner that is consistent with the description of the terminal placental stages of *Perameles* as outlined by Flynn (1923). It is also of interest that in our material the umbilical cords could be traced in the histological sections of the vaginal complex to confirm the presence of a pseudovaginal birth canal similar to that originally described for *Perameles* by Hill (1897) and recognized by him as universal in the mode of marsupial parturition.

The relatively recent ultrastructural studies of the chorio-allantoic placenta of *P. nasuta* and *I. macrourus* by Padykula and Taylor (1976, 1977, 1982) showed that during the attachment phase of the foetal chorionic cells, these formed desmosomes with the cells of the maternal luminal endometrial syncytial nests and were a prerequisite to cell fusion and the consequent formation of mixed nests of foetal-maternal syncytium they designated as heterokaryons. The pure uninvaded luminal endometrial syncytial nests they designated homokaryons. Padykula and Taylor (1976) also showed that the foetal and maternal capillaries of the fully formed placenta although extremely closely opposed were separated by extensions of the syncytial masses and were not immediately opposed as suggested by Flynn (1923).

The important but limited ultrastructural contributions of Padykula and Taylor (1976, 1977, 1982) in the understanding of the cellular dynamics of the peramelid chorio-allantoic placenta indicates the need for ultrastructural investigations to parallel and extend the classical investigations of Flynn (1923). In this regard aspects of particular interest would be further investigations on the ultrastructure of syncytialization; cell fusion; cell destruction; and studies of aspects of exchange between the foetal and the maternal capillaries in the developing and fully functional chorio-allantois and yolk-sac. Likewise the brevity of pregnancy in bandicoots and the speed of development in pouch life suggest that mitotic rate as well as other matters of the cell cycle deserve further attention.

REFERENCES

CLOSE, R. L., 1977. Recurrence of breeding after cessation of suckling in the marsupial *Perameles nasuta*. *Australian Journal of Zoology* **25:** 641–45.

FLYNN, T. T., 1923. The yolk-sac and allantoic placenta in *Perameles*. *Quarterly Journal of Microscopic Science* **67:** 123–82.

GEMMELL, R. T., 1981. The role of the corpus luteum of lactation in the bandicoot, *Isoodon macrourus* (Marsupialia: Peramelidae). *General and Comparative Endocrinology* **44:** 13–19.

GORDON, G., 1971. *A study of island populations of the short-nosed bandicoot,* Isoodon macrourus *Gould.* Ph.D. Thesis University of New South Wales, Sydney.

GORDON, G., 1974a. Movements and activity of the short-nosed bandicoot *Isoodon macrourus* Gould (Marsupialia). *Mammalia* **38:** 405–31.

GORDON, G., 1974b. Queensland bandicoots. *Queensland Agricultural Journal* **100:** 16–20.

HALL, L. S., 1983. Observations on body weights and breeding of the Northern Brown Bandicoot, *Isoodon macrourus*, trapped in south east Queensland. *Australian Wildlife Research* **10:** 467–76.

HEINSOHN, G. E., 1966. Ecology and reproduction of the Tasmanian bandicoots (*Perameles gunni* and *Isoodon obesulus*). *University of California Publications in Zoology* **80:** 1–96.

HILL, J. P., 1895. Preliminary note on the occurrence of a placental connection in *Perameles obesula*, and on the foetal membranes of certain macropods. *Proceedings of the Linnean Society of New South Wales* **10:** 578.

HILL, J. P., 1897. The placentation of *Perameles*. Contributions to the embryology of the marsupials — 1. *Quarterly Journal of Microscopic Science* **40:** 385–446.

HILL, J. P., 1899. Contributions to the morphology and development of the female urogenital organs in the Marsupialia. I. On the female urogenital organs of *Perameles* with an account of the phenomenon of parturition. *Proceedings of the Linnean Society of New South Wales* **14:** 42–82.

HOLLIS, D. E. AND LYNE, A. G., 1980. Ultrastructure of luteal cells in fully formed and regressing corpora lutea during pregnancy and lactation in the marsupials *Isoodon macrourus* and *Perameles nasuta*. *Australian Journal of Zoology* **28:** 195–211.

HUGHES, R. L., 1974. Morphological studies on implantation in marsupials. *Journal of Reproduction and Fertility* **39:** 173–86.

HUGHES, R. L., 1984. Structural adaptations of the eggs and the fetal membranes of monotremes and marsupials for respiration and metabolic exchange. In *Respiration and metabolism of embryonic vertebrates* ed by R. S. Seymour. Junk: Dordrecht.

LILLEGRAVEN, J. A., 1975. Biological considerations of the marsupial-placental dichotomy. *Evolution* **29:** 707–22.

LYNE, A. G., 1964. Observations on the breeding and growth of the marsupial *Perameles nasuta* Geoffroy, with notes on other bandicoots. *Australian Journal of Zoology* **12:** 322–39.

LYNE, A. G., 1974. Gestation period and birth in the marsupial *Isoodon macrourus*. *Australian Journal of Zoology* **22:** 303–09.

LYNE, A. G., 1976. Observations on oestrus and the oestrous cycle in the marsupials *Isoodon macrourus* and *Perameles nasuta*. *Australian Journal of Zoology* **24:** 513–21.

LYNE, A. G. AND HOLLIS, D. E., 1977. An unusual blastocyst stage of the marsupial *Isoodon macrourus*. *Australian Journal of Zoology* **25:** 225–31.

LYNE, A. G. AND HOLLIS, D. E., 1979. Observations on the corpus luteum during pregnancy and lactation in the marsupials *Isoodon macrourus* and *Perameles nasuta*. *Australian Journal of Zoology* **27:** 881–99.

MACKERRAS, J. AND SMITH, R. H., 1960. breeding the short-nosed bandicoot, *Isoodon macrourus* (Gould) in captivity. *Australian Journal of Zoology* **8:** 371–82.

PADYKULA, H. A. AND TAYLOR, J. M., 1976. Ultrastructural evidence for loss of the trophoblastic layer in the chorioallantoic placenta of Australian bandicoots (Marsupialia: Peramelidae). *Anatomical Record* **186:** 357–85.

PADYKULA, H. A. AND TAYLOR, M. J., 1977. Uniqueness of the bandicoot chorioallantoic placenta (Marsupialia: Peramelidae). Cytological and Evolutionary Interpretations. In *Reproduction and evolution* ed by J. H. Calaby and C. H. Tyndale-Biscoe. Australian Academy of Science: Canberra.

PADYKULA, H. A. AND TAYLOR, M. J., 1982. Marsupial placentation and its evolutionary significance. *Journal of Reproduction and Fertility* Suppl. **31:** 95–104.

STODART, E., 1966. Management and behaviour of breeding groups of the marsupial *Perameles nasuta* Geoffroy, in captivity. *Australian Journal of Zoology* **14:** 611–23.

STODDART, D. M. AND BRAITHWAITE, R. W., 1979. A strategy for utilization of regenerating heathland habitat by the brown bandicoot (*Isoodon obesulus*; Marsupialia, Peramelidae). *Journal of Animal Ecology* **48:** 165–79.

TYNDALE-BISCOE, H., 1973. *Life of marsupials*. Arnold: London.

Unusual circadian activity rhythms in captive bandicoots *Isoodon obesulus* and *Isoodon macrourus*

S. M. ARMSTRONG[1], M. J. CHESWORTH[1], R. WILLIG[1], G. J. COLEMAN[1] and V. M. CASSONE[2]

Circadian rhythms in running wheel activity, time spent in a nest box, and activity associated with feeding and drinking were recorded in 16 *Isoodon obesulus* and nine *I. macrourus* under a variety of laboratory environmental lighting manipulations. These included: (a) reversal of the light-dark cycle, (b) increases and (c) decreases in photoperiod. Previous work by Lyne (1981) which reported unusually unstable circadian rhythms in bandicoots was confirmed and extended. Unstable rhythms in bandicoots appear to be due to a combination of three factors: (1) increase in photoperiod and concomitant compression of nocturnal activity, (2) reversal of photoperiod and (3) previous history of lighting manipulations.

Key words: Circadian rhythms, entrainment, photoperiod, light pulses, free-running, pacemaker.

INTRODUCTION

A DAILY rhythm is classified as circadian if it persists under temporal isolation such as in conditions of constant light (LL) or constant darkness (DD). In DD or LL the rhythm manifests its own endogenous periodicity which is usually slightly longer or shorter than 24 h and is described as free-running (Aschoff 1960). Under natural conditions or under artificial light-dark (LD) cycles in the laboratory, circadian rhythms are reset each day so that they are entrained or synchronized to a particular phase of the lighting conditions. The circadian activity pattern of mammals may be diurnal, nocturnal or crepuscular.

With few exceptions, such as the Numbat *Myrmecobius fasciatus)* which is diurnal (Friend 1985), Australian marsupials are nocturnal or crepuscular, irrespective of other aspects of their ecological niche. Therefore, nocturnal activity rhythms represent a fundamental feature of marsupial biology and behaviour, which has received little empirical research under laboratory conditions. Circadian rhythm research is integral to knowledge of photoperiodic mechanisms and seasonal breeding (Zucker, Johnston and Frost 1980) and these may characterize many marsupial species. Furthermore, since most of our knowledge of circadian mechanisms in mammals is based on rodents (Pittendrigh and Daan 1976a), marsupials provide a range of uninvestigated species whose study could generate fresh insight and expand current knowledge of such mechanisms. Circadian activity rhythms of *Isoodon macrourus* and *Perameles nasuta* appear to fall into the latter category (Lyne 1981).

In the laboratory, when the LD cycle is reversed, over half of the *P. nasuta* and 15% of the *I. macrourus* studied either failed to entrain or showed unstable entrainment of the circadian activity rhythm (Lyne 1981). This instability in bandicoots contrasts with rodent species where re-entrainment to the reversed LD cycle is complete within 14 days (Quay 1970). When double plots along a 48 h time scale are produced from previously published bandicoot activity records (Lyne 1981) two phenomena, (i) free-running and (ii) bouncing (a form of unstable entrainment) become apparent (see Pittendrigh and Daan 1976b for description of phenomena). Neither of these phenomena would be expected to occur under conventional LD cycles and so these data are unusual. Therefore, bandicoots may represent a unique experimental mammal for investigating the stability of the circadian pacemakers which are believed to generate rhythmicity.

The aim of the present studies was to confirm the earlier report of unusual circadian wheel-running activity rhythms in bandicoots (Lyne 1981) and then to attempt to isolate the conditions responsible for the unstable entrainment, with the ultimate aim of describing the characteristics of circadian pacemakers in bandicoots.

MATERIALS AND METHODS

Animals

Thirteen adult females and three juvenile male *I. obesulus* were trapped near Seaford, Victoria in June 1981. Traps were opened late in the afternoon and inspected, then closed the following dawn. A mixture of peanut butter,

[1]Department of Psychology and Brain-Behaviour Research Institute, La Trobe University, Bundoora, Victoria, Australia 3083.
[2]Department of Neurology, Health Sciences Center, State University of New York, Stony Brook, New York 11794, USA.
Pages 271-83 *in* BANDICOOTS AND BILBIES, ed by J. H. Seebeck, P. R. Brown, R. L. Wallis and C. M. Kemper. Surrey Beatty & Sons, Sydney, 1990.

rolled oats and honey wrapped in muslin was used as bait. Six adult male *I. macrourus* were obtained from Queensland and a further three from Canberra.

Bandicoots were fed daily at approximately 1600 hrs with 60 g of a mixture of minced meat, rolled oats and a supplement of vitamins and minerals.

Apparatus

A maximum of 12 bandicoots at any one time was housed in a laboratory maintained at 21°± 1°C air temperature and a 12:12 LD cycle with lights on at 0600 hrs. Room lighting, provided by two ceiling fluorescent lights, was 150 lux, while two incandescent red lights provided a dim illumination of 0.4 lux during the dark period. Bandicoots were housed individually in boxes measuring 1 220 mm long, 630 mm wide and 430 mm high, made of grey acrylic sheeting. The floor of each box was covered in dry grass. Each box contained a running-wheel of 570 mm radius, a nest box, a compartment providing food and a compartment providing water. At the opening of the nest box, food compartment and water compartment infra-red cells and monitors were mounted so that entrance and exit of the bandicoot would break the infra-red beam and the position of the animal could be determined via a logic circuit. Running wheel revolutions, time spent in the nest box, and time spent in food and water compartments were monitored and sampled every 15 min by a Computer Automation LSI-2 computer in an adjacent room. Data were plotted to simulate actograms generated by paper event recorders.

Procedure

Responses to several manipulations of photoperiod were examined in both *I. obesulus* and *I. macrourus*. These included: (1) reversal of photoperiod, (2) increase in photoperiod and (3) decrease in photoperiod. These are explained in detail in the figure legends.

In a preliminary attempt to establish the properties of the bandicoot circadian pacemaker, the procedure successfully used by Daan and Pittendrigh (1976a) in rodents to establish a phase response curve was followed. Five *I. macrourus* were allowed to free-run in DD until their period was stable. Once every 14 days a single 30 min light pulse of 150 or 1 300 lux was given at different times in the individual's circadian cycle. The subsequent amount of advance or delay phase-shift in the activity rhythm induced by a particular pulse at a particular phase of the circadian cycle was then calculated.

RESULTS

Under a 12:12 LD cycle all bandicoots except one (Im 2, Fig. 6) showed stable entrainment. In all cases running wheel activity provided the best record of circadian rhythms. Running in the wheel (Fig. 1), drinking and feeding behaviour, occurred predominantly during the dark period. All three behaviours commenced immediately upon onset of darkness but the duration of activity was restricted almost entirely to the first 6 to 8 h of darkness. This contrasts with, for example, rodents such as hamsters and mice whose records typically show a second burst of activity before lights-on (Pittendrigh and Daan 1976c). In laboratory rats feeding too may show this bimodal distribution in the dark (Armstrong 1980). In the rodent, these two bands of activity are thought to reflect the output from two oscillators, which are linked to dawn and dusk (Pittendrigh and Daan 1976c).

With the exception of one animal, all bandicoots used their nest boxes but did so intermittently throughout the experiment. When the nest box was being used regularly, time out of the nest box displayed a pattern complementary to the other three behaviours combined. When the nest box was not used, a make-shrift nest was built under the running wheel, but sometimes animals slept in the drinking or feeding compartments. It can be seen (Fig. 1) that much of the time spent outside the nest box was spent in the running wheel and therefore running wheel activity provides a good measure of general activity.

Reversal of photoperiod

Following LD cycle reversals all *I. obesulus* showed delaying transients to re-entrainment which took place within 8 to 15 days (e.g., Io 7 in Fig. 2) with the exception of two animals (Io 1 and Io 5, Fig. 2). From the running wheel records re-entrainment often appeared to be rapid (Figs 1 and 2) but this was due to a "masking effect" of light on running in the wheel. Masking refers to the suppression of overt activity induced in a nocturnal mammal by exposure to light i.e., a photophobic reaction. Nest box (Fig. 1) and drinking and feeding records (not shown) demonstrated that entrainment in fact took place from the onset of darkness prior to photoperiod reversal. The rapid appearance of activity at the beginning of the dark period within a few days of LD reversal corresponds to a band of activity that comes from the end of the previous dark period. This is well illustrated in Fig. 2 where the record of animal Io 5 has been double plotted for this purpose. Thus, although there is little behavioural evidence for a morning oscillator in the activity records of bandicoots during stable entrainment its presence is indicated in the activity rhythm during the days taken to re-entrain.

Two *I. obesulus* showed unstable entrainment patterns after their first photoperiod reversal. One of these, Io 1 (Fig. 2) showed the bouncing phenomenon similar to that found in the running wheel records published by Lyne (1981). Even after 50 days, when the LD cycle was again reversed, bandicoot Io 1 failed to attain a stable phase relationship with the reversed LD cycle. Bandicoot Io 5 showed a much moderated version of this phenomenon (Fig. 2). However, after the second reversal of the LD cycle both these bandicoots showed stable re-entrainment to the LD cycle. Similarly, all other *I. obesulus* re-entrained within 14 days (range 6–14 days) to the second LD cycle reversal by phase delaying transients.

Photoperiod increases and decreases

Behavioural rhythms of all *I. obesulus* except two (Io 2 and Io 5, Fig. 3) maintained a stable phase-relationship with the decreasing scotoperiod. Onset of activity immediately followed lights-off as exemplified by Io 9 (Fig. 3). For bandicoot Io 5 the activity pattern started to break away from the LD cycle just after LD 14.5:9.5 was reached. A similar pattern was displayed by Io 2 during LD 15:9. It is uncertain whether these two bandicoots had begun to free-run because the subsequent reversal of the LD cycle and decrease in photoperiod confound the issue. Nevertheless, it appeared that the increasing photoperiod and decreasing scotoperiod instigated unstable entrainment in two *I. obesulus*.

The behavioural rhythms of Io 5 commenced to bounce under decreasing photoperiod (Fig. 3) in a similar manner to that reported by Lyne (1981).

All *I. macrourus* except one (Im 2) showed stable entrainment which persisted under conditions of increasing photoperiod and decreasing scotoperiod as exemplified by the records of animals Im 1 and Im 9 (Figs 4 and 5). In contrast to the stable pattern of entrainment shown by the majority, animal Im 2 showed virtually no running wheel activity and no rhythm in drinking and feeding under LD 12:12 (Fig. 6). Under increasing photoperiod bursts of running wheel activity at irregular intervals of several days occurred during feeding time and during the dark, indicating arrhythmicity. However, it became apparent later in the experiment (Fig. 8) that this animal was in fact free-running. The amount of activity, particularly during the first half of the light period was reduced and the manifestation of the free-running rhythm therefore hard to discern, probably due to masking by light.

In a situation where the 16:8 LD cycle was reversed and the increase of photoperiod continued, six *I. macrourus* showed stable re-entrainment by phase-delaying transients, again typified by Im 1 and Im 9 (Figs 4 and 5). Activity in the food and water compartments, associated with food replenishment, now occurred at subjective dawn instead of subjective dusk (not shown). Im 6 free-ran for one complete cycle before re-entraining (Fig. 7) while animal Im 3 free-ran with a period greater than 24 h after reversal of the LD cycle (Fig. 6). However, with animal Im 2, the running wheel activity record appeared arrhythmic at this stage, and the free-running rhythm was not readily discernable until later (Fig. 8).

Following the reversal of a 17:7 LD cycle and the reduction in photoperiod to 14:10 LD, five *I. macrourus* showed stable re-entrainment by delaying transients as depicted by Im 8 (Fig. 8). One bandicoot (Im 9) showed bouncing with re-entrainment eventually stabilizing at the end of the 14 days of the 14:10 LD cycle while Im 10 free-ran with a period greater than 24 h (Fig. 8). As indicated above Im 2 and Im 3 showed free-running rhythms with periods longer than 24 h but the free-running had already commenced before the LD reversal and decrease in photoperiod (Fig. 8).

Light pulses

All of the five *I. macrourus*, none of which had shown unusual rhythms under LD cycle and photoperiod manipulation, free-ran with periods less than 24 h. Thirty minute duration light pulses of 150 lux produced phase delays which were small and no discernable phase advances. Light pulses of the same duration but of 1 300 lux produced large enough phase shifts with which to begin to calculate a phase-response curve. Figure 9 shows a large phase delay and change in free-running period followed two weeks later by a small phase advance when light pulses were applied at circadian time 14 and 20 respectively to Im 12.

DISCUSSION

Lyne (1981) described peculiar circadian activity rhythms in the bandicoots *I. macrourus* and *P. nasuta*. The present work confirms the findings in *I. macrourus*. Therefore, it is reasonable to expect a similar finding for *P. nasuta* but with a greater incidence of individuals with unstable circadian pacemakers. It would appear that the incidence of instability is less likely in *I. obesulus* than *I. macrourus* although in the present investigation the different manipulations of photoperiod and the LD cycle between experiments makes a species comparison tenuous. Furthermore, the majority of *I. obesulus* tested were female whereas all the *I. macrourus* were male, although there is no reason to expect a sex difference.

Our study indicates that the unstable circadian activity rhythms in some species of bandicoots are due to the combination of three factors: (1) reversal of the LD cycle, (2) reduction of scotoperiod, increase in photoperiod and resulting compression of α (the active portion in the circadian cycle) and (3) history dependence of the circadian pacemaker (see Pittendrigh and Daan 1976a for description of terms and phenomena). The latter factor is simply an acknowledgement that reversal of the LD cycle and photoperiod change alone are insufficient to produce instability under the present lighting regimen. Under Lyne's (1981) experimental conditions, particularly for *P. nasuta,* photoperiod increase alone appears to have been sufficient to cause instability in some individuals.

The data of Im 10 shown in Figure 8 are of immediate importance since they illustrate and help elucidate the phemonenon first discovered by Lyne (1981). Clearly, the period of the free-running rhythm of Im 10 lengthens as it crosses the light period and becomes closer to 24 h as it crosses the dark period. This change in free-running period is consistent with Aschoff's Rule (Pittendrigh 1960) which states that the free-running period of a nocturnal animal lengthens with increasing light intensity. In the case of Lyne's animals the free-running rhythm is not manifest during the light period although the change in period of the circadian pacemaker can be inferred by extrapolating the onset of activity seen in the dark into the light period (see Fig. 13b in Lyne 1981). Since the lighting levels in Lyne's experiments were generally higher (e.g., 280–350 lux with fluorescent lamps) than those employed in the present experiments this would account for his observations of (1) the absence of activity due to masking in the light period and (2) the even greater increase in period length than noted here.

All *I. macrourus* with unstable ryhthms died before they could be tested in DD in the present experiment. Therefore, whether bandicoots which show unstable activity rhythms under LD cycles free-run with periods greater than 24 h in DD or free-run with periods less than 24 h in DD but have a larger delay portion to their phase response curve than stable bandicoots remains to be determined.

ACKNOWLEDGEMENTS

We are grateful for the help of Hans Brunner in trapping *I. obesulus* and to Bob Gemmell and Jim Merchant for supplying us with *I. macrourus.* This work was supported by an ARGS grant to S. Armstrong.

REFERENCES

ARMSTRONG, S., 1980. A chronometric approach to the study of feeding behaviour. *Neuroscience and Behavioral Reviews* **4:** 27–53.

ASCHOFF, J., 1960. Exogenous and endogenous components in circadian rhythms. *Cold Spring Harbor Symposia on Quantitative Biology* **25:** 11–26.

DAAN, S. AND PITTENDRIGH, C. S., 1976. A functional analysis of circadian pacemakers in nocturnal rodents. II. The variability of phase response curves. *Journal of Comparative Physiology* **106:** 253–66.

FRIEND, J. A., 1985. Diel and seasonal patterns of activity in the Numbat (*Myrmecobius fasciatus*). *Australian Mammal Society Bulletin* **9(1):** 47 (abstract).

LYNE, A. G., 1981. Activity rhythms in the marsupials *Isoodon macrourus* and *Perameles nasuta* in captivity. *Australian Journal of Zoology* **29:** 821–38.

PITTENDRIGH, C. S., 1960. Circadian rhythms and the circadian organization of living systems. *Cold Spring Harbor Symposia on Quantitative Biology* **25:** 159–84.

PITTENDRIGH, C. S. AND DAAN, S., 1976a. A functional analysis of circadian pacemakers in nocturnal rodents. I. The stability and lability of spontaneous frequency. *Journal of Comparative Physiology* **106:** 223–52.

PITTENDRIGH, C. S. AND DAAN, S., 1976b. A functional analysis of circadian pacemakers in nocturnal rodents. IV. Entrainment: pacemaker as clock. *Journal of Comparative Physiology* **106:** 291–331.

PITTENDRIGH, C. S. AND DAAN, S., 1976c. A functional analysis of circadian pacemakers in nocturnal rodents. V. Pacemaker structure: a clock for all seasons. *Journal of Comparative Physiology* **106:** 333–55.

QUAY, W. B., 1970. Precocious entrainment and associated characteristics of activity patterns following pinealectomy and reversal of photoperiod. *Physiology and Behavior* **5:** 1281–90.

ZUCKER, I., JOHNSTON, P. G. AND FROST, D., 1980. Comparative physiology and biochronometric analyses of rodent seasonal reproductive cycles. *Progress in Reproductive Biology* **5:** 102–33.

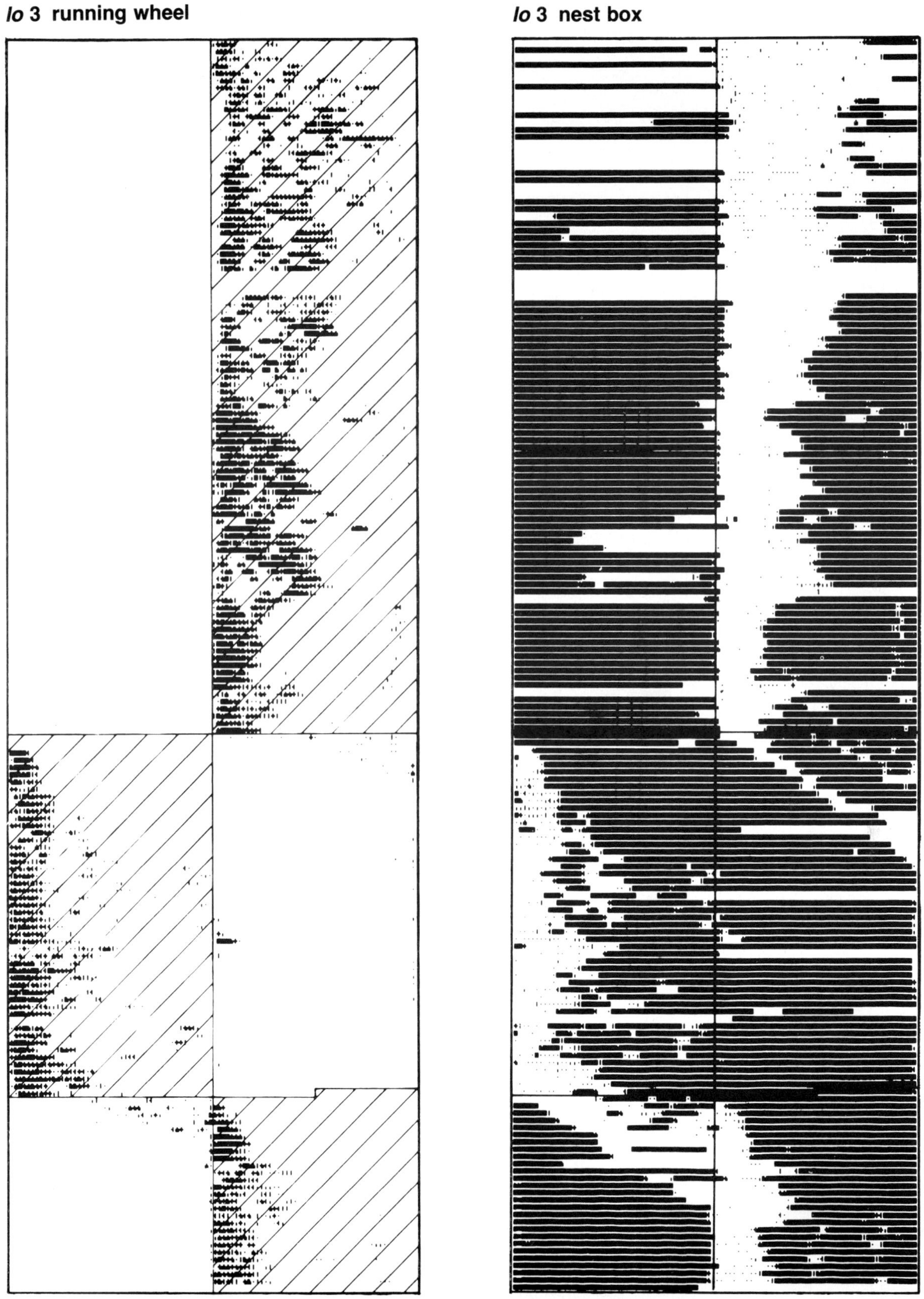

Fig. 1. Running wheel activity and time spent in nest box for an *Isoodon obesulus* under a 12:12 LD cycle for 96 days, reversal of the LD cycle for 50 days and for 23 days after a second reversal of the LD cycle. This animal showed stable entrainment to the LD cycle. Hatching indicates darkness.

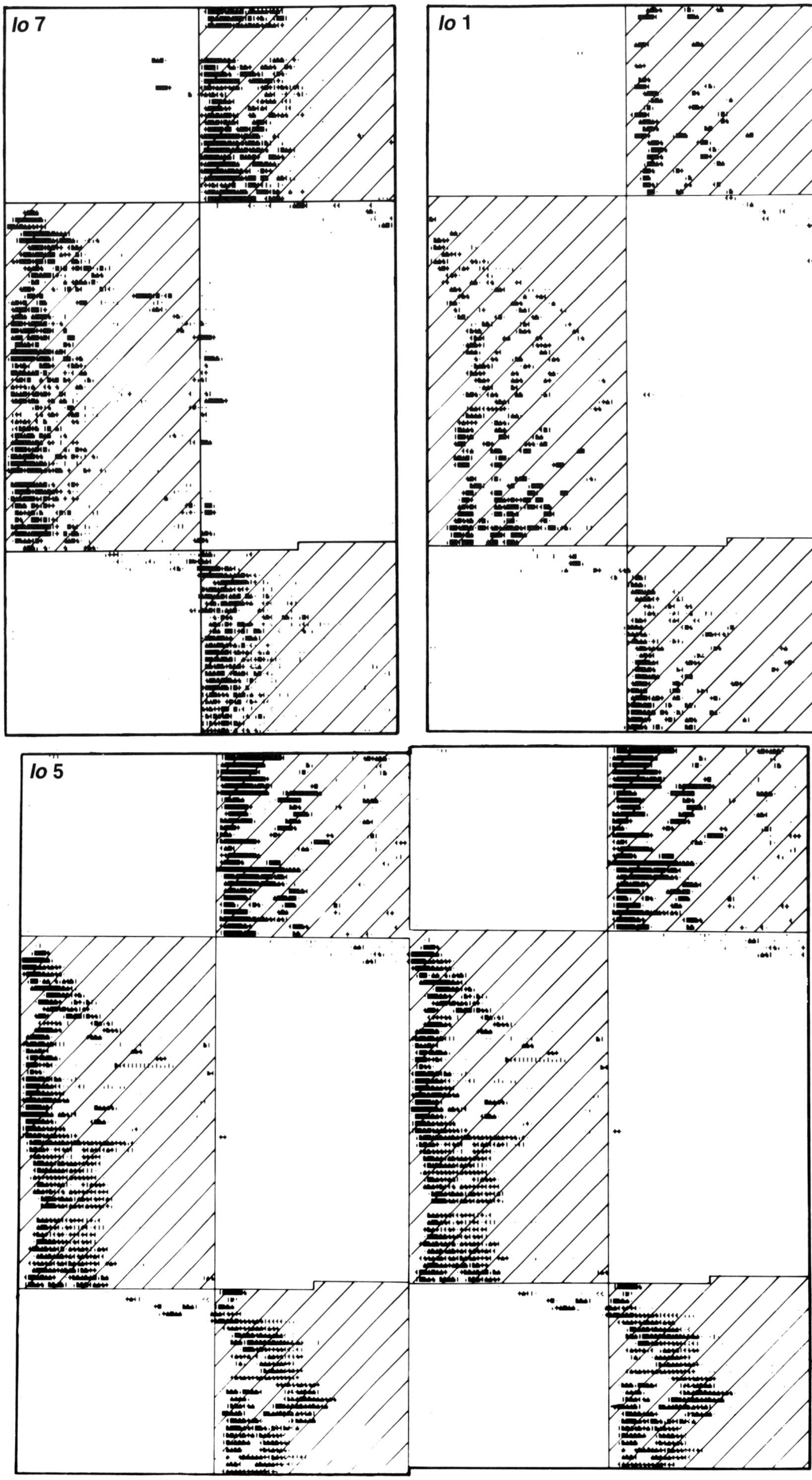

Fig. 2. Running wheel activity patterns for three *Isoodon obesulus* under a 12:12 LD cycle for 26 days, reversed LD cycle for 50 days and reversal of LD cycle back to normal for 23 days. Hatching indicates darkness. Note the unstable entrainment and bouncing of the activity pattern in Io 1. Even after 50 days of the reversed LD cycle stable entrainment has not been achieved but re-entrainment is normal after the second reversal of the LD cycle. The data for Io 5 has been double plotted along a 48 h time scale so that the data from day 1 runs into day 2 and so forth. This method aids visualization of the re-entrainment pattern.

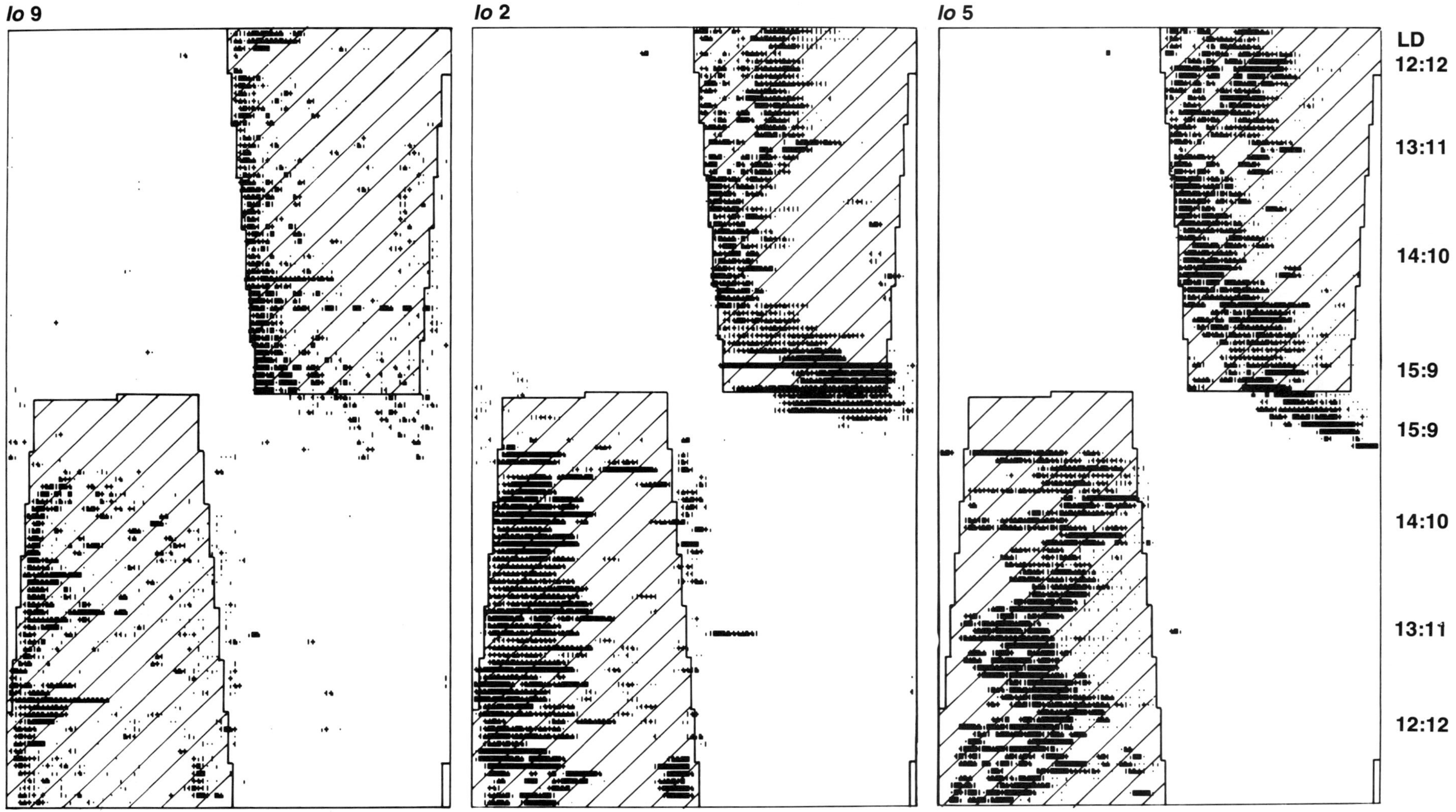

Fig. 3. Running wheel activity patterns for three *Isoodon obesulus* under increasing photoperiod (15 min extra light at lights-on and 15 min extra light at lights-off per week), reversal of LD cycle, and decreasing photoperiod. Hatching indicates darkness. Onset of actvity for Io 9 immediately followed the lights-off signal down to an LD 15:9 photoperiod whereas Io 2 and Io 5 broke away from lights-off at LD 15:9 and LD 14.5:9.5 respectively. While Io 2 subsequently re-entrained to the reversed LD cycle with its decreasing photoperiod Io 5 did not, but showed the unstable entrainment known as bouncing. Instead of the onset of activity becoming aligned with the onset of darkness, the rhythm overshoots by continuing to phase delay until the end of the active period enters the beginning of the light period (LD 14.5:9.5). The rhythm then changes direction and phase advances back across the dark until onset of activity nears the end of the light period (13:11 LD), at which point it again starts to phase delay across the dark until LD 12:12 is reached. The rhythm then phase advances again to reach a steady state with alignment to the LD cycle at 11:13 (not shown).

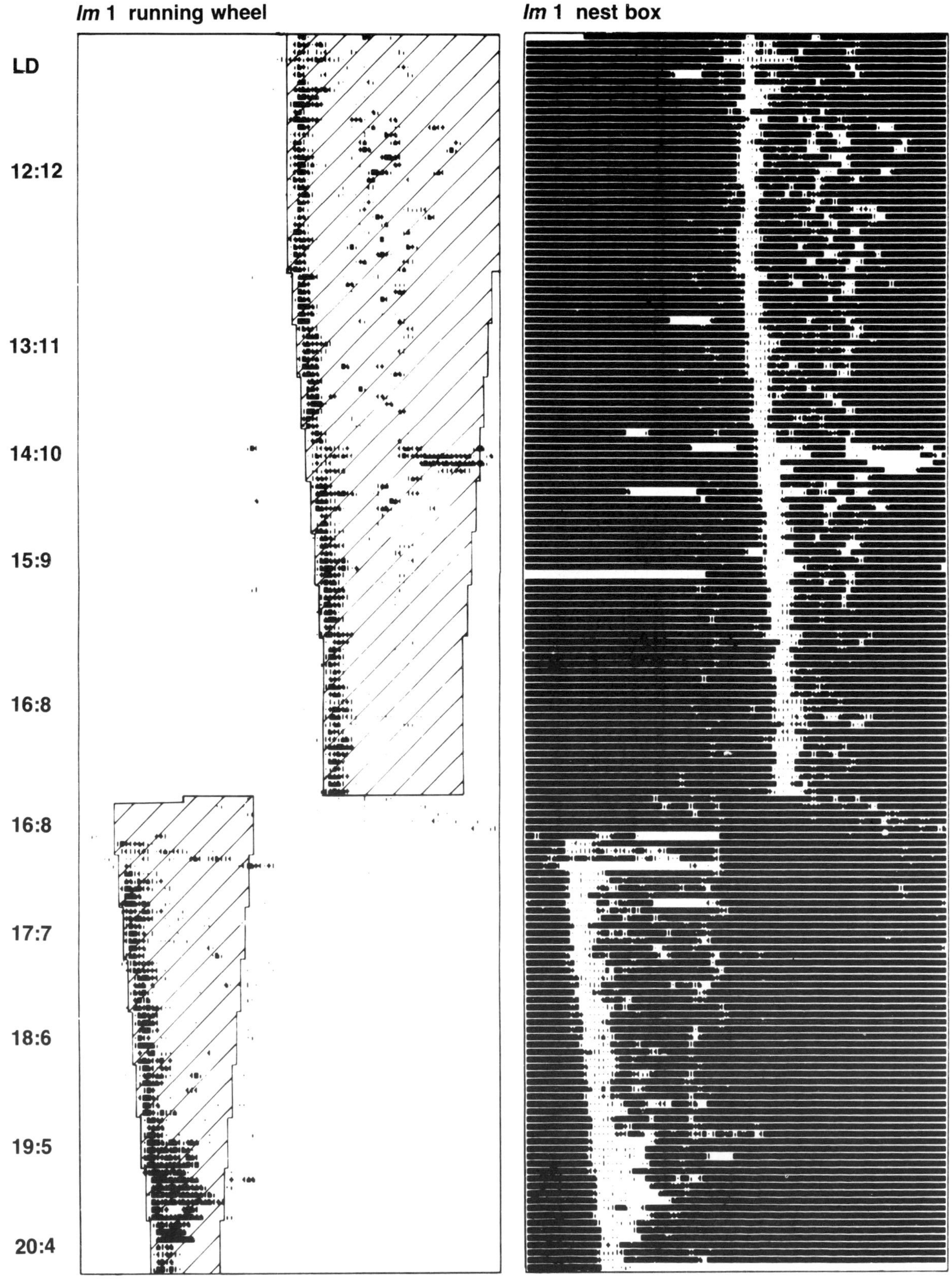

Fig. 4. Running wheel activity and time spent in nest box for *Isoodon macrourus* Im 1. Hatching indicates darkness. After 32 days in a 12:12 LD cycle, darkness was decreased by 30 min per week (15 min at lights-off and 15 min at lights-on) until a 16:8 LD photoperiod was reached. This was maintained for 14 days and then the LD cycle was reversed. The increase in photoperiod was continued until LD 20:4 was reached. This animal showed a very stable entrainment to the reduction and reversal of photoperiod.

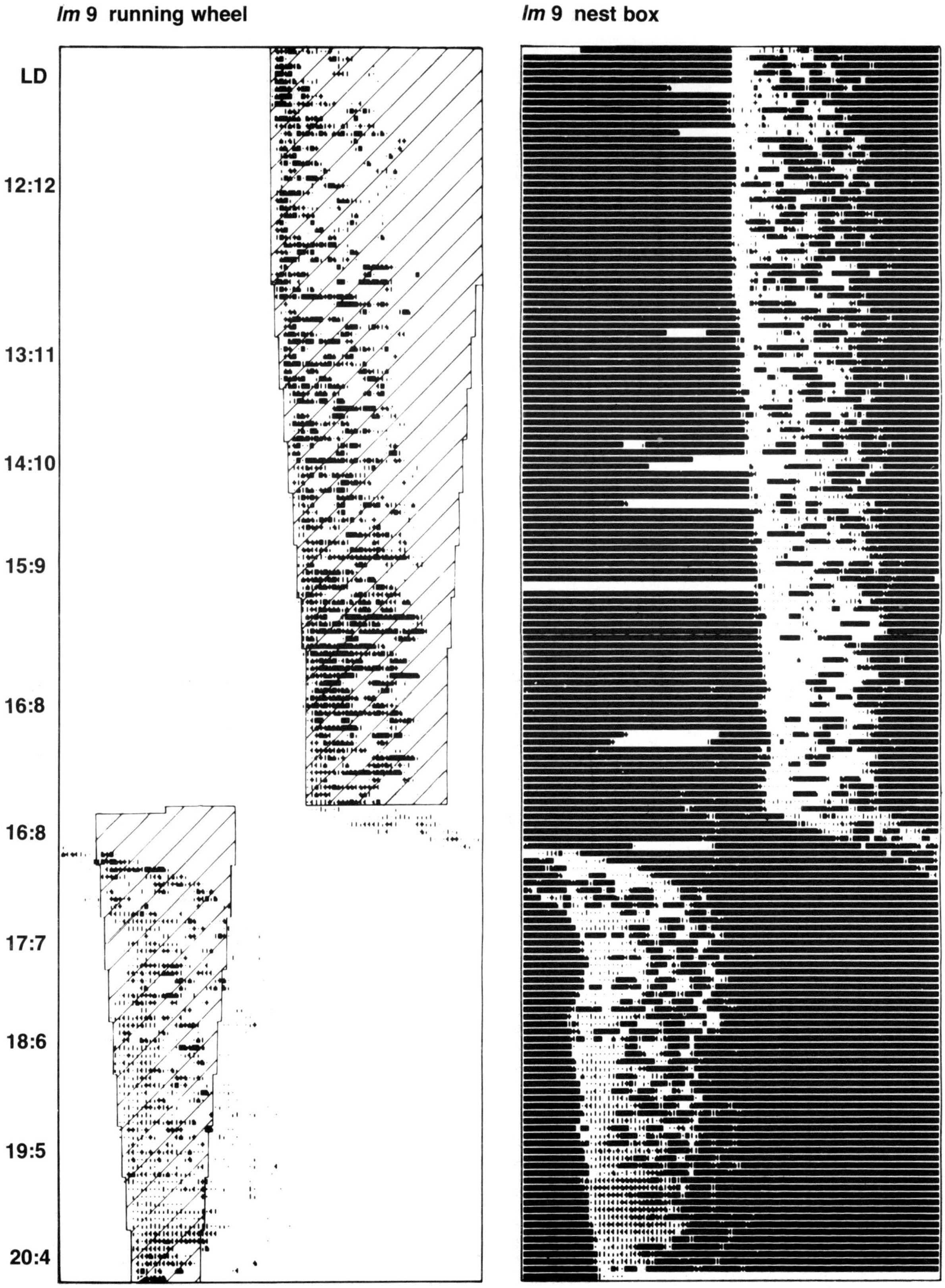

Fig. 5. Data for Im 9 under the same conditions as for *Isoodon macrourus* Im 1 in Fig. 4.

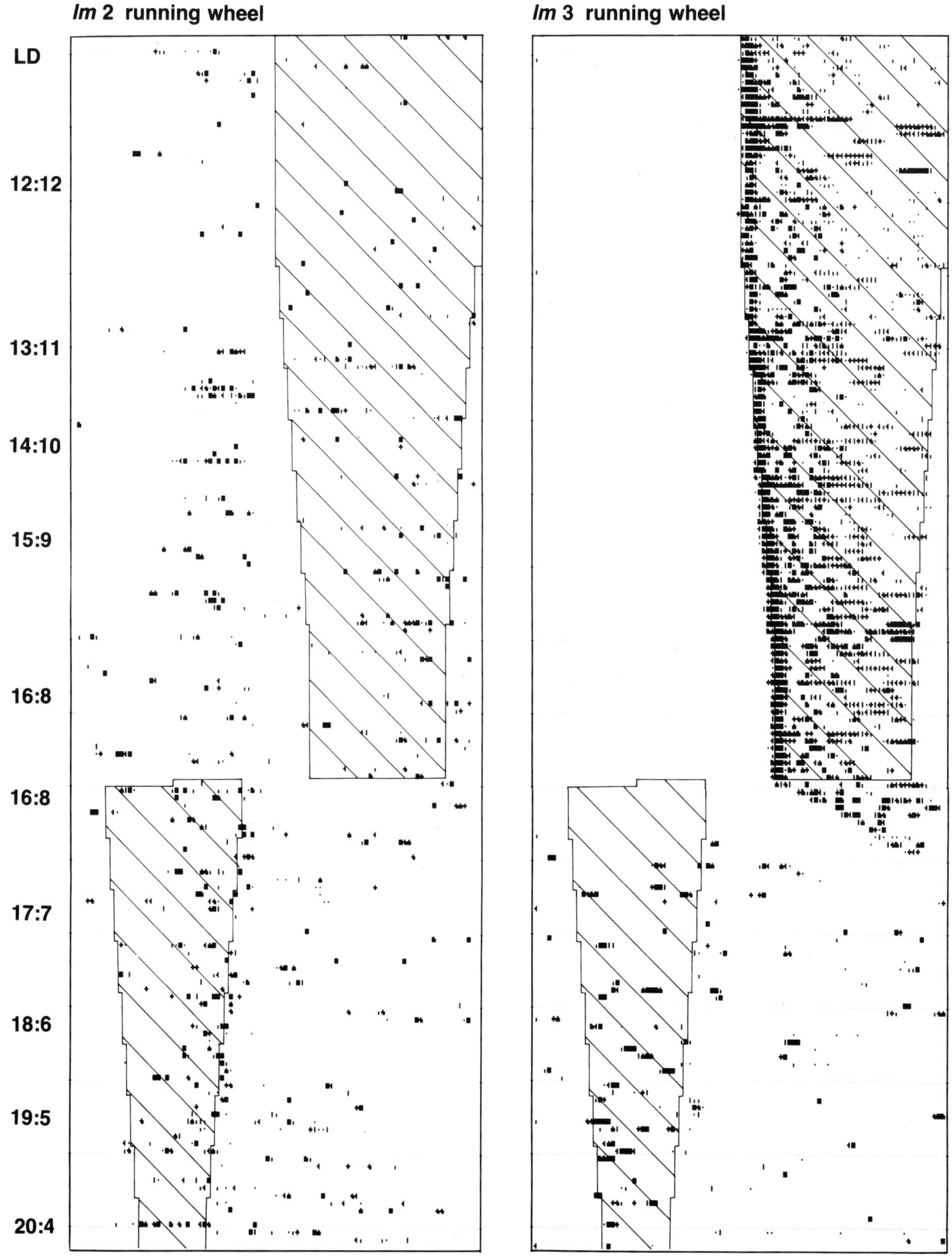

Fig. 6. Running wheel data for *Isoodon macrourus* Im 2 and Im 3. Experimental procedure is described in Fig. 4, i.e., increasing photoperiod, reversal of 16:8 LD cycle and further increases in photoperiod. Activity data of Im 2 indicates it was free-running with a period greater than 24 h throughout the experiment. Im 3 failed to re-entrain after the LD reversal and free-ran with a period of greater than 24 h at a reduced activity level (see also Fig. 8).

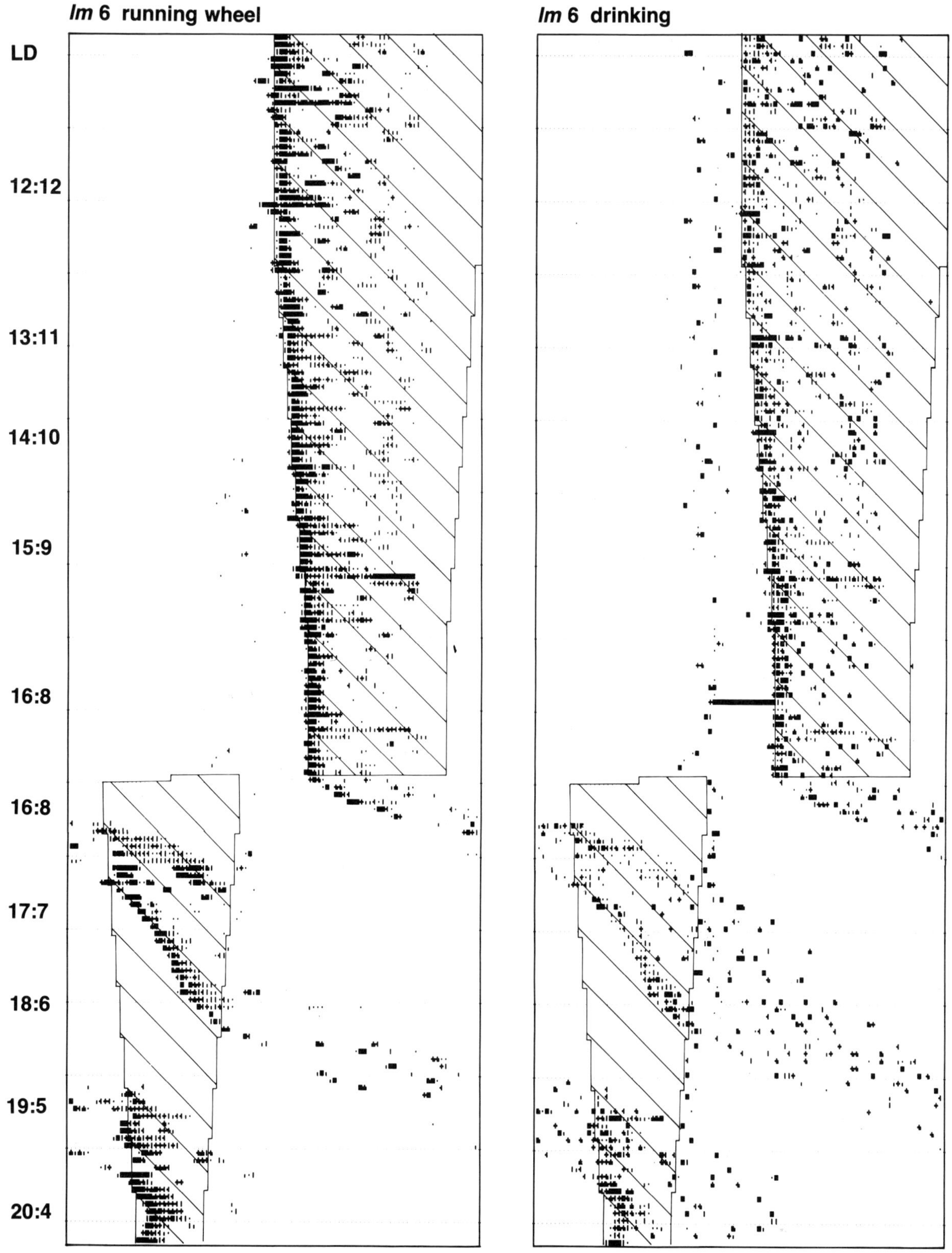

Fig. 7. Running wheel and drinking data for *Isoodon macrourus* Im 6 under the same conditions as described in Fig. 6. Im 6 failed to re-entrain after the reversal of the LD cycle but re-entrained successfully the second time.

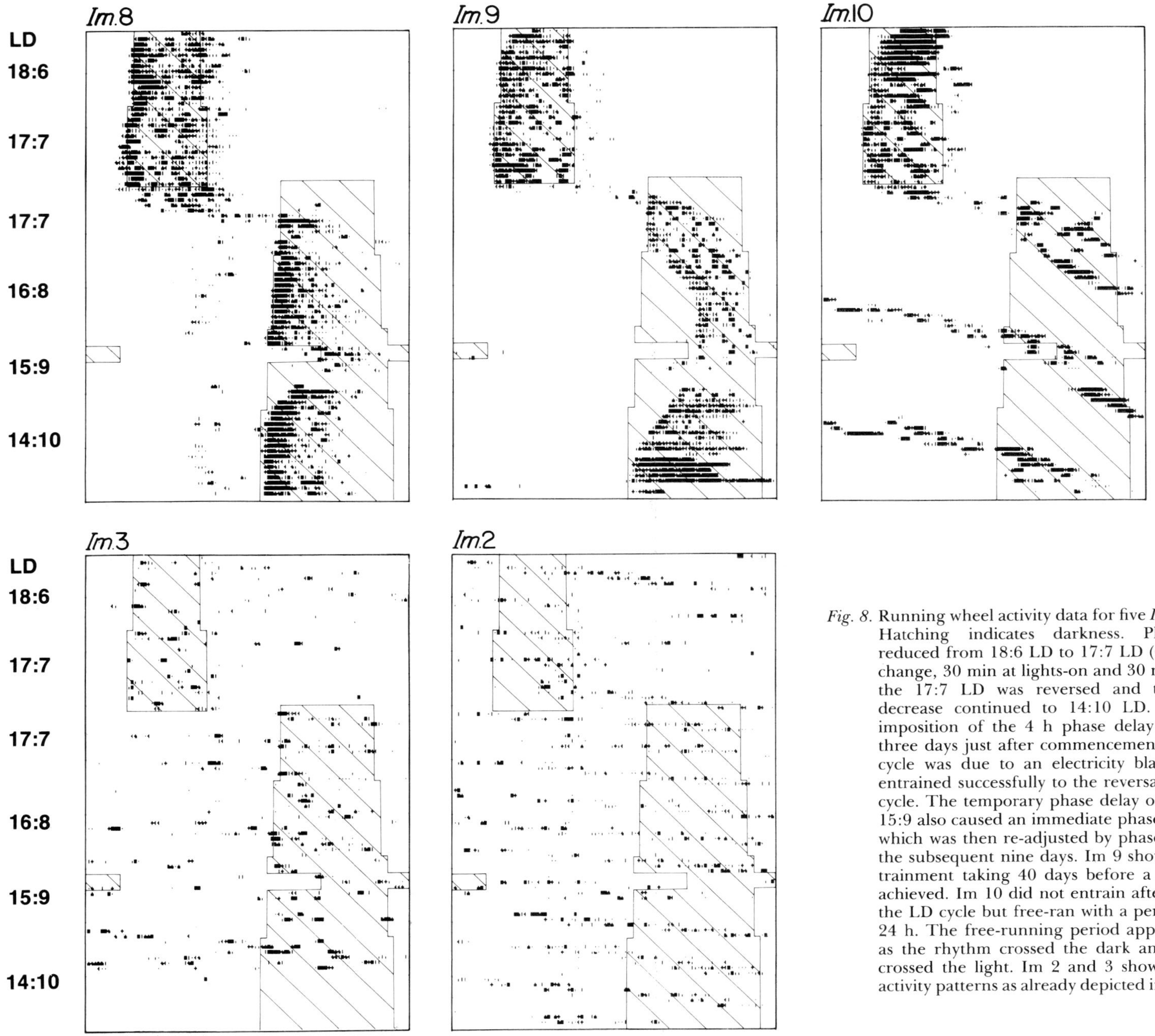

Fig. 8. Running wheel activity data for five *Isoodon macrourus*. Hatching indicates darkness. Photoperiod was reduced from 18:6 LD to 17:7 LD (60 min per week change, 30 min at lights-on and 30 min at lights-off). the 17:7 LD was reversed and the photoperiod decrease continued to 14:10 LD. The temporary imposition of the 4 h phase delay in darkness for three days just after commencement of the 15:9 LD cycle was due to an electricity blackout. Im 8 re-entrained successfully to the reversal of the 17:7 LD cycle. The temporary phase delay of darkness at LD 15:9 also caused an immediate phase delay in activity which was then re-adjusted by phase advancing over the subsequent nine days. Im 9 showed unstable entrainment taking 40 days before a steady state was achieved. Im 10 did not entrain after the reversal of the LD cycle but free-ran with a period greater than 24 h. The free-running period appeared to shorten as the rhythm crossed the dark and lengthen as it crossed the light. Im 2 and 3 showed free-running activity patterns as already depicted in Fig. 6 for Im 2.

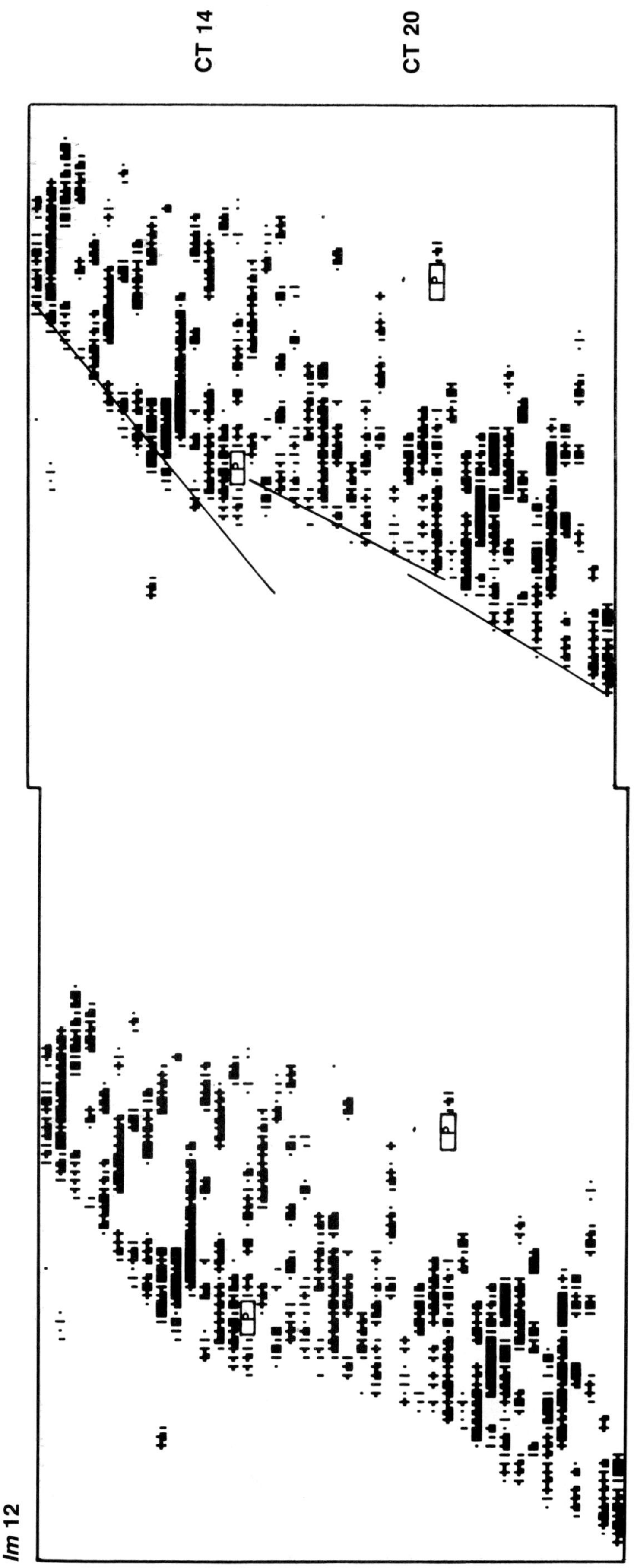

Fig. 9. A 48-hour, double plot of *Isoodon macrourus* Im 12 in DD free-running with a period less than 24 h. At the two times indicated by the 'p' in rectangles, an hour light pulse of 1 300 lx was given and the amount of phase shift calculated. CT 14 and CT 20 refer to circadian times when light pulses were given. Right-hand side of plot has eye-fitted lines drawn to indicate phase and period changes.

Life history of the bandicoots: developmental rigidity and phenotypic plasticity

ANDREW COCKBURN[1]

The Peramelidae are an ancient radiation of marsupials which show little variation in morphology and "programmed" life history traits. They differ from all other marsupials for which data are available by extreme acceleration of ontogeny, and associated reproductive adaptations. This acceleration is manifested in the shortest gestation period of any mammal, a very brief period of lactation for a marsupial of their size, and rapid growth and puberty. Rapid development and reproduction are most strongly favoured in expanding populations, and facilitate exploitation of habitats inaccessible to other marsupials, such as temporally patchy environments. Here bandicoots respond to the rapidly changing life history optima with behavioural plasticity manifested by enforced dispersal of juveniles and brood reduction.

There is a correlation between morphological conservatism, behavioural plasticity, and accelerated ontogeny in several taxa. It is proposed that selection for high fecundity in early bandicoots led to the crossing of an important adaptive threshold, but precluded subsequent morphological diversification and radiation.

Key words: Bet-hedging, development, heterochrony, life history theory, plasticity, radiation, succession.

INTRODUCTION

ALTHOUGH controversy continues to surround the phylogenetic position of the Peramelidae within the marsupials, evidence presented in this volume suggests that bandicoots can be best regarded as a sister taxon to the Dasyuridae (Baverstock, Aplin, Flannery, Birrell and Krieg, this volume), and are thus suitable for use in a comparison of the evolution of life history variability within these higher taxa. Lee, Woolley and Braithwaite (1982) and Lee and Cockburn (1985) have demonstrated that within the Dasyuridae it is possible to recognize at least six distinct strategies of reproduction, characterized by variation in the oestrous pattern, number of periods of rut exhibited by males, the duration of the mating period, the seasonality of breeding, and the age at sexual maturity. The occurrence of these patterns is well correlated with habitat changes (Lee and Cockburn 1985). In addition, there is substantial intrafamilial variation in litter size, and the duration of both gestation and lactation. Much, though not all of this variability, can be explained in terms of allometric variation in body size (Lee and Cockburn 1985), as is true for patterns of variation in many mammalian life history traits (Eisenberg 1981; Russell 1982; Peters 1983; Calder 1985). In this review, I show that by contrast, the extant bandicoots are characterized by negligible variation in body size, and consequently show little allometric variation in life histories. Further, the programmed aspects of life history also appear to show little intrataxonomic variation. However, the bandicoots do differ significantly from the allometric reference curves for the marsupials provided by Russell (1982) and Lee and Cockburn (1985), invariably in the direction of accelerated ontogeny and reproduction. Here I argue that accelerated ontogeny and limited variability are closely related, and that selection for the former had led to the latter. In order to establish this correlation, I initially review the trophic position, habitat utilization, and reproductive variation in the extant bandicoots. Because data on bandicoot life histories are still scanty, particularly for the tropical species, and as I believe that bandicoots afford a number of unusual opportunites to examine questions of general biological importance, I seek to identify profitable lines of research which require our attention.

DIET

Good data on diet are available for very few bandicoot species. The species studied thus far appear to be omnivores. A variety of surface and soil dwelling invertebrate species are consumed, including earthworms, dipterans, earwigs, spiders, adult beetles, and larvae and pupae from beetles, moths, ants and termites (Heinsohn 1966; Johnson 1980; Lobert and Opie 1988). There are no data which suggest that bandicoots exhibit preferences for particular types of invertebrates. Plant tissue comprises a variable proportion of the diet, and does appear to be consumed quite selectively, reflecting the lack of dental and alimentary adaptation for digestion of fibrous foodstuffs (Hume 1982). Use of plants appears to be greatest in *Macrotis lagotis*, where seeds of *Panicum australiense* (Gramineae) and corms of *Cyperus bulbosus* (Cyperaceae) often comprising up to two-thirds

[1]Department of Zoology, Australian National University, GPO Box 4, Canberra City, Australian Capital Territory, Australia 2601.
Pages 285–92 *in* BANDICOOTS AND BILBIES, ed by J. H. Seebeck, P. R. Brown, R. L. Wallis and C. M. Kemper. Surrey Beatty & Sons, Sydney, 1990.

of faecal material (Johnson 1980). In *Perameles* and *Isoodon* hypogeous sporocarpic fungi (mainly *Endogone* and *Glomus*), and fruits from both native and introduced species are prominent in the diet during seasons when they are available (Heinsohn 1966; Lobert and Opie 1988). Among the marsupials, the groups most similar to the bandicoots in diet are some of the potoroines, and the terrestrial didelphids (Lee and Cockburn 1985). McNab (1980a) has suggested that bandicoots closely resemble the tenrecs and armadillos ecologically, and that the low basal metabolism of these taxa is a reflection of the combination of burrowing, and the nature of the prey they consume (see also McNab 1980b). However, there is no evidence that bandicoots have unusually low metabolic rates for marsupials (Hulbert and Dawson 1974; Hinds and MacMillen 1984).

HABITAT PREFERENCES

Bandicoots are extremely widely distributed throughout almost all habitats in the Australo-Papuan region, although many of the Australian species appear to have undergone massive range reductions since European settlement, and at least two species have become extinct (see section on distribution and status, this volume).

On a microgeographic scale, several studies suggest that many bandicoot species exhibit preferences for habitat which is only likely to be available for short periods. Bandicoots are capable of rapid exploitation of alienated areas such as farmland and rubbish tips (Heinsohn 1966; Gordon 1974; Seebeck 1979; Brown 1988). Some of the New Guinean bandicoots are sufficently abundant near human settlements to be a valuable source of food (George and Maynes, this volume). Even in areas of native vegetation, several species appear to prefer ecotonal habitat, with its inherent patchiness (e.g., *Echymipera rufescens,* Gordon and Lawrie 1977), and early seral habitat (e.g., *Isoodon obesulus* Stoddart and Braithwaite 1979), which has been either recently burnt (Heinsohn 1966) or cleared (Stoddart and Braithwaite 1979; Lobert, this volume). In one case this preference has been shown to be correlated with the availability of food (Lee and Cockburn 1985; Lobert and Lee, this volume). Changes in the availability of early seral habitat since European settlement have been implicated in the decline of several species which are now either extinct, or have undergone massive range reductions (Johnson and Southgate, this volume; Kemper, this volume), particularly where patches of habitat are isolated and dispersal between patches of different successional status is frustrated (Kemper, this volume; Opie, Gullan and Mansergh, this volume). Although there are a number of exceptions to this pattern, for example *Echymipera clara,* which is confined to primary rainforests (George and Maynes, this volume), we have no information on the ecology of such species. Research into the population biology of these species would be invaluable.

HOME RANGE AND SOCIAL ORGANIZATION

Bandicoots appear to be solitary, and local aggregations of individuals appear to reflect the dispersion of food resources rather than social bonding (Heinsohn 1966; Gordon 1974). Even juveniles do not associate with their mothers for long periods (Gordon 1974), and local recruitment is very low (Gordon 1974; Stoddart and Braithwaite 1979; Gemmell 1982), suggesting that very high levels of juvenile dispersal prevail (Lee and Cockburn 1985).

Males range over much larger areas than females. In an extreme instance, Heinsohn (1966) found that male ranges were ten times greater than female ranges in Tasmanian *Perameles gunnii.* These data suggest that the mating system is polygynous or promiscuous, but there are no studies of social organization. *Macrotis lagotis* uses scent-marking around burrows (Johnson and Johnson 1983), but the establishment of true territories is unknown. Home ranges in dense island populations may be small in comparison to mainland populations (Copley, Read, Robinson and Watts, this volume), repeating a pattern well known in other mammals (Gliwicz 1980).

BANDICOOT REPRODUCTION IS DIFFERENT

Gestation

It has been known for some time that bandicoot reproduction differs from that of all other marsupials (or mammals generally). For example, the gestation period of *Perameles* and *Isoodon* is the shortest known for any mammal (11.5 to 12.5 days) (Lyne 1974). Short gestation appears to be characteristic of the Peramelidae. McCracken (1986) has recently reported a gestation period of 14 days in *M. lagotis.* Reduction in the duration of gestation in contrast to other mammals appears to involve restriction of both the pre-attachment and organogenesis phases of gestation (Renfree 1980), and the embryonic stasis characteristic of some other marsupials (for an extreme example, see Selwood 1980), has not been documented.

Despite the brevity of gestation, there is no evidence of reduction in the size of neonates, which in fact are substantially larger than would be expected for a dasyurid or didelphid of similar size (Russell 1982).

Lactation

Russell (1982) provided excellent evidence that lactation is extremely brief in bandicoots in contrast to other marsupials, particularly when allometric factors are taken into account (Fig.1). Bandicoot mothers produce milk which is much richer than in other marsupials, and exhibit a particularly sharp rise in lipid concentration as lactation progresses (Merchant, this volume). It is uncertain whether production of concentrated milk facilitates rapid growth and a short period of maternal dependence, or whether the rapid development permits bandicoot pouch young to accept a richer diet.

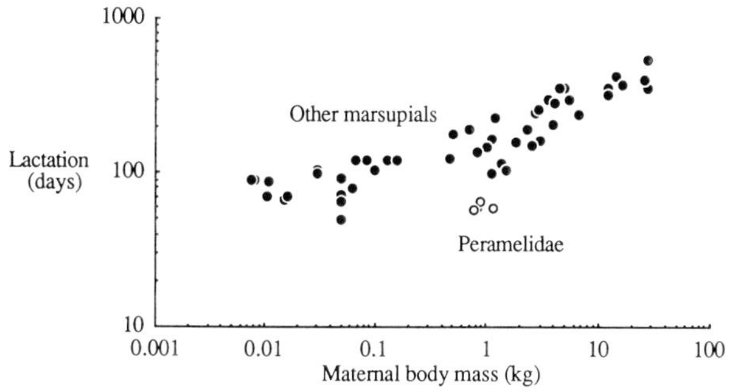

Fig. 1. Relation between duration of lactation in marsupials and maternal body mass. Data are from Russell (1982). Bandicoots have the lowest relative length of lactation, and close to the shortest lactation of any marsupial, regardless of body size effects.

Growth

The most comprehensive analysis of allometric variation in mammalian growth is provided by Zullinger, Ricklefs, Redford and Mace (1984) who concluded that the Gompertz growth constant, K, could be shown to vary with body mass according to the formula,

$$K = 0.045\ M^{-0.30} (R^2=0.72)$$

where M is asymptotic mass in kilograms. Although growth data for four bandicoot species *(P. gunnii, P. nasuta, I. obesulus, I. macrourus)* were considered in this analysis, the values used by Zullinger *et al.* (1984) are suspiciously low, and may have been influenced by the lack of tendency of bandicoot growth trajectories towards an asymptote. Lee and Cockburn (1985) used the algorithms proposed by Kaufmann (1981) to calculate K for all the complete growth curves available for marsupials by October 1983, and concluded that bandicoots had a faster relative growth rate than any other marsupials (Fig. 2). Like Zullinger *et al.* (1984) they argued that bandicoot growth is considerably slower than that predicted for same-sized eutherians. It is worth noting that the results of Zullinger *et al.* (1984) suggest that the differences between marsupials and eutherians are greater than those implied by Lee and Cockburn (1985).

Litter size

Most bandicoots have eight teats, setting an upper limit to the number of young they can rear. However, unlike many dasyurids, the teats are rarely saturated, and litter sizes less than five are most common (Table 1). The modal litter size is three for the species in southeastern Australia, while the remaining species for which data are available generally have two young. There are no good data for tropical species, but it will not be surprising if these species also have litters of two, as similar latitudinal and climatic trends have been noted in many mammalian amd avian taxa (Cockburn, Lee and Martin 1983).

The reduction in litter size below the teat number in bandicoots has led to speculation that the rapid succession of litters means that juveniles in second and subequent litters within a season would be incapable of attaching to the swollen teats from which large young of earlier teats had suckled (Heinsohn 1966, and see photographic evidence presented by Lyne 1974). Gemmell, Walker, Johnston and Cepon (1984) showed that although several litters containing five young were produced in a captive colony of *I. macrourus* the sum of young in consecutive litters never exceeded the number of teats (eight). However, Merchant (this volume) has demonstrated a number of instances of successful attachment by neonatal bandicoots to large teats, casting doubt on the validity of the consecutive litter hypothesis.

Gemmell (1982) and Gemmell *et al.* (1984) also presented evidence that the frequency of brood reduction in bandicoots is extremely high, although the effects may be exaggerated in this and other studies by the trapping and handling techniques employed. Loss of young during the

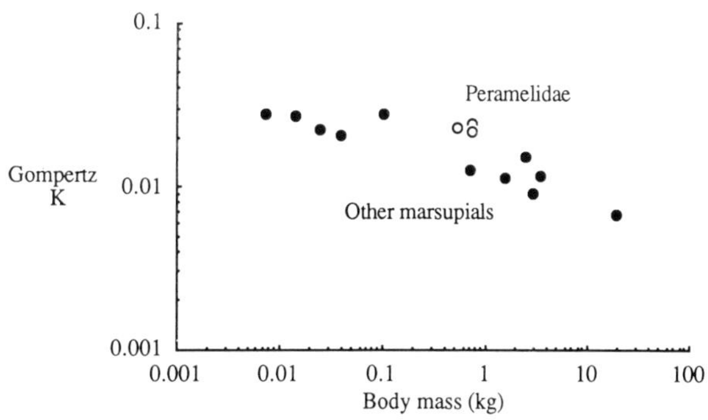

Fig. 2. Relation between the Gompertz growth constant (K) and asymptotic body mass in female marsupial species for which data are available. Data are from Lee and Cockburn (1985), and were calculated using the algorithm provided by Kaufmann (1981). The allometric relation for all mammals derived by Zullinger *et al.* (1984) is provided for comparison. Marsupials grow slowly, but bandicoots have the fastest relative growth rate of any marsupial.

Table 1. Litter size in the bandicoots. Means are provided for studies with large sample sizes.

Species	Mean	Mode	Source
Chaeropus ecaudatus — historical records	—	1or2	Aitken (1983)
Macrotis lagotis — captive colony	1.5	2	Johnson (1983a), McCracken (1986)
Macrotis leucura — historical records	—	2	Johnson (1983b)
Isoodon obesulus nauticus — Franklin I., S.A.	2.1	2	Copley *et al.* (this volume)
Isoodon obesulus obesulus — Tasmania	2.8	3	Heinsohn (1966)
Isoodon obesulus obesulus — Cranbourne, Victoria	3.0	3	Stoddart and Braithwaite (1979)
Isoodon macrourus — south-east Queensland	3.4	3	Gordon (1974)
Isoodon macrourus — Brisbane, Queensland	3.6	4	Gemmell (1982)
Isoodon macrourus — captive colony	2.8	3	Gemmell *et al.* (1984)
Perameles nasuta — captive colony	2.6	3	Stodart (1977)
Perameles gunnii — Tasmania	2.3	2	Heinsohn (1966)
Perameles gunnii — Victoria	2.3	2	Seebeck (1979)

course of lactation raises a number of intriguing questions concerning the allocation of parental investment in this group of animals. Young are not lost only from large litters. Indeed, the data of Gemmell *et al.* (1984) suggest the opposite may be true, as 80% of the large litters (five young) were suckled to weaning. Brood reduction has recently attracted considerable theoretical interest, particularly because mortality in early development often appears to be sex biased (Widdowson 1976; McLure 1981; Clutton-Brock, Albon and Guinness 1985). Re-examination of existing data sets, and future studies which investigate this perspective are likely to be illuminating.

Most authors agree that there are seasonal trends in litter size, with peaks often being recorded in late spring (Heinsohn 1966; Gordon 1974; Stoddart and Braithwaite 1979; Gemmell *et al.* 1984; Copley *et al.*, this volume). Interpretation of these results is confounded by brood reduction, which cannot be assessed in many field studies. Field data suggest that primiparous females have smaller litters than previparous ones (Copley *et al.*, this volume), and that large females have larger litters (Stoddart and Braithwaite 1979), but data from captive colonies do not always support these observations (Gemmell *et al.* 1984).

Duration of the breeding season

All bandicoots are polyoetrous and are capable of producing several litters each year. Bandicoots develop rapidly and may breed in the season of their birth, with females achieving maturity at three months, and males at 4.5 months (Heinsohn 1966). Indeed, it appears that bandicoots may conceive at first oestrus (Gemmell, this volume). The extent of the breeding season determines the maximum possible annual fecundity. While some populations have individuals which give birth in each month of the year (Gordon 1974; P. R. Brown, pers. comm.), there is generally a decline in the frequency of pregnant and lactating females in late summer and autumn. The onset of breeding after this seasonal anoestrus may be very abrupt (Stoddart and Braithwaite 1979), or much more gradual (Gemmell 1982). Although the causes of onset of breeding are poorly understood, there are convincing data linking the cessation of breeding to declining food and water availability, particularly in response to drought. Heinsohn (1966) showed that births in a Tasmanian population of *P. gunnii* stopped as early as January in a drought year, but persisted until May when conditions were more benign. P. R. Brown (pers. comm.) noted complete cessation of breeding in a mainland population of *P. gunnii* during a severe drought, with births occurring within 17 days after widespread heavy rain terminated the drought. Lee and Cockburn (1985) suggested that declines in breeding in *P. nasuta* and *I. obesulus* populations noted by Close (1977) and Stoddart and Braithwaite (1979) may have also been associated with the effect of the drying of soil on prey numbers or accessibility.

Morphological and endocrinological correlates of rapid reproduction

All these data suggest that reproductive rate and ontogeny in the Peramelidae are enhanced relative to the other marsupials.

The means by which bandicoots achieve this acceleration remain uncertain. They differ from other marsupials through the possession of a chorioallantoic placenta in which the chorionic villi are highly invasive (Padykula and Taylor 1976; Hughes, Alpin, Hall and Archer 1986), and the persistence of the corpora lutea of pregnancy through approximately 45 days into lactation (Hughes 1962; Gemmell 1979; Lyne and Hollis 1979; Hollis and Lyne 1980). The corpora lutea continue to secrete progesterone into lactation (Gemmell, Jenkin and Thorburn 1980; Gemmell 1981), though there is no evidence of uterine receptivity to progesterone in bandicoots analogous to that observed in eutherians (Gemmell 1981; Rothchild 1981). As several authors have pointed out, bandicoot reproduction appears to possess some characteristics which are intermediate between that of eutherians and other marsupials (Taylor and Padykula 1978; Rothchild 1981; Gemmell *et al.* 1984). By contrast, the truncation of gestation makes the bandicoots the most "marsupial" of all marsupials, with a heavy emphasis on lactation as the principal source of parental investment. Although reconciliation of these patterns must remain the province of developmental biologists and endocrinologists, in the following sections I explore the evolutionary and ecological consequences of the bandicoot life history. I suggest that the reproduction and life history of this group may be representative of a pattern of extreme interest in understanding the process of morphological diversification.

DISCUSSION

The bandicoot life history

Rapid and widely spread reproduction and extremely low recruitment of bandicoots are consistent with a life history strategy which involves the dispersal of large numbers of young in the hope of colonization of habitat some distance from the mothers, home range (Giesel 1976; Gill 1978; Cockburn, Braithwaite and Lee 1981; Cockburn 1988). This capacity is likely to favour the bandicoots in Australian environments which have proved resistant to colonization by other marsupials, such as heathlands which are regularly perturbed by burning (e.g., Braithwaite 1979). It is certainly true that most of the species exclusively associated with Australian heathlands are rodents (Braithwaite, Cockburn and Lee 1978; Kikkawa, Ingram and Dwyer 1979; Newsome and Catling 1979; Fox 1983), and that the life history of the heathland bandicoot is more similar to that of fire-adapted rodent species than that of many other marsupials (Cockburn *et al.* 1981; Lee and Cockburn 1985).

In summary, it is possible to conclude that the range of reproductive traits exhibited by bandicoots is rather conservative in contrast to the dasyurids, with plasticity in behavioural responses such as brood reduction and natal dispersal likely to be of great importance in facilitating exploitation of the enormous range of habitats bandicoots now occupy. The bandicoots show reproductive patterns which suggest that they have crossed a barrier in marsupial evolution which has led to much higher fecundity than would be anticipated for other species of similar size (e.g., small potoroines).

Why are there no marsupial bears?

Given the impressive reproductive adaptations of the bandicoots, it seems profitable to speculate on the causes of the restricted morphological and reproductive diversity in this taxon. Large marsupial omnivores are absent from Australia, and current phylogenetic interpretation does not suggest that bandicoots have given rise to any daughter taxa, except perhaps the dasyurids. Their morphological conservatism stands in contrast to Cope's Law suggesting that morphological diversification leading to phyletic size increase is a near universal characteristic of the evolutionary history of higer taxa (for an amusing account see Gould (1984)). It should be noted that fossil faunas contain tiny plesiomorphic perameloids which may represent the ancestral stock of the larger forms, suggesting that Cope's Law is indeed applicable. However, given the reinterpretation of marsupial phylogeny provided by Baverstock (pers. comm.) and Baverstock *et al.* (this volume), it seems conceivable that both perameloids and dasyuroids are descendants of a small perameloid ancestor. In this case the lack of diversity in omnivorous bandicoots remains puzzling.

The apparent resistance of many taxa to morphological change across massive periods of time is termed *morphological stasis,* and interpretation of the significance and causes of stasis lies at the heart of some of the most controversial discussion in modern evolutionary biology (for reviews and alternative opinions see Gould 1980, 1983; Charlesworth, Lande and Slatkin 1982; Wake, Roth and Wake 1983; Barnard 1984; Cheverud 1984; Stenseth 1985). Some biologists have argued that morphological change (and to some extent morphological diversification) is a comparatively rare feature in the fossil record, and that stasis is the usual condition of most species. Morphological change is thought to be associated with the rapid genetic restructuring that takes place at the time of speciation, and which is possible because speciation normally takes place in small isolated populations (see however Barton and Charlesworth 1984). This view has been challenged by Wake *et al.* (1983)

who reviewed evidence on the morphological and genetic diversification in the plethodontid salamanders, and showed that, despite massive genetic differentiation (morphologically similar species in eastern and western North America diverged about 42 million years ago), and frequent speciation, all species of the genus *Plethodon* show a conservative and simple morphology. Wake *et al.* argue that this persistence is indicative of fitness and adaptation in a group of organisms which are buffered against the extreme environmental changes which must have occurred during the existence of the genus, through behavioural plasticity. In order to emphasize the magnitude of these changes, Wake *et al.* point out that all terrestrial vertebrate genera which co-occurred with *Plethodon* 60 million years ago are now extinct. The evolutionary history of *Plethodon* is in stark contrast to the diversity exhibited by the genus *Bolitoglossa*, which has undergone prolific speciation and comparative morphological diversification (Alberch 1980; Alberch and Alberch 1981). There is substantial evidence that the morphological diversification in *Bolitoglossa* is achieved through heterochrony, or changes in the timing of development. Indeed, there have been a number of claims that the majority of major morphological shifts are consequence of developmental changes (Gould 1977, 1982). It is therefore worth considering the opportunity for development changes in bandicoots.

Selection for increased fecundity and developmental rate, like other paths of evolution, is likely to follow the line of least resistance (Stebbins 1970; Harper 1982). In bandicoots, this appears to have involved the compression of all phases of maternal investment relative to that observed in other marsupials, rather than any fundamental alteration to the marsupial developmental programme. Even substantial reproductive changes such as persistence of the corpus luteum into lactation are unaccompanied by uterine receptivity to secretions from the corpus luteum (Rothchild 1981).

In discussing the consequences of short gestation, Lee and Cockburn (1985) have proposed that, relative to eutherians, the extremely short period of uterine attachment exhibited by marsupials has caused a high level of synchrony in marsupial ontogeny which has restricted morphological diversification. They argued that this is best interpreted as specialization, rather than primitiveness, or lack of adaptation. It does, however, impose limits on the scope for morphological diversification through a paedomorphic ancestor, which Gould (1977) argued is a common feature underlying the origin and diversification of new radiations. Here I propose that the exaggeration of the marsupial truncation of gestation in the bandicoots means that they are also specialized and that this specialization may impose even stronger developmental constraints on diversification than occurs elsewhere in the marsupials. Thus while the ancestral bandicoots have crossed an important ecological threshold facilitating exploitation of seral and ephemeral habitats, the pathway to accelerated ontogeny has precluded subsequent morphological diversification. Bandicoots are to marsupials, what marsupials are to the eutherian radiation.

These fascinating animals afford an excellent opportunity for future research at the interface of ecology and developmental biology, an area which offers to provide new insights into the evolutionary process (Stearns 1982). The rapid development and reproduction of the bandicoots has been exploited by Gordon Lyne and his co-workers to produce a better picture of the reproduction of these species than is available for any marsupial but the *Macropus eugenii* and *Trichosurus vulpecula* (Lyne 1974, 1976; Lyne and Hollis 1979; Hollis and Lyne 1980). This provides a clear challenge to ecologists to do at least as well.

ACKNOWLEDGEMENTS

This manuscript was prepared while my research was supported by the Australian Research Grants Scheme.

REFERENCES

AITKEN, P. F., 1983. Pig-footed Bandicoot *Chaeropus ecaudatus*. Pp. 104–05 in *The Australian Museum complete book of Australian mammals* ed by R. Strahan. Angus & Robertson: Sydney.

ALBERCH, P., 1980. Ontogenesis and morphological diversification. *American Zoologist* **20:** 653–67.

ALBERCH, P. AND ALBERCH, J., 1981. Heterochronic mechanisms of morphological diversification and evolutionary change in the neotropical salamander, *Bolitoglossa occidentalis* (Amphibia: Plethodontidae). *Journal of Morphology* **167:** 249–64.

BARNARD, C. J., 1984. Stasis: a coevolutionary model. *Journal of Theoretical Biology* **110:** 27–34.

BARTON, N. H. AND CHARLESWORTH, B., 1984. Genetic revolutions, founder effects, and speciation. *Annual Review of Ecological Systems* **15:** 133–64.

BRAITHWAITE, R. W., 1979. Social dominance and habitat utilization in *Antechinus stuartii* (Marsupialia). *Australian Journal of Zoology* **27:** 517–28.

BRAITHWAITE, R. W., COCKBURN, A. AND LEE, A. K., 1978. Resource partitioning by small mammals in lowland heath communities of south-eastern Australia. *Australian Journal of Ecology* **3:** 423–45.

CALDER, W. A., 1985. *Size, function, and life history*. Harvard University Press: Cambridge, Massachusetts.

CHARLESWORTH, B., LANDE, R. AND SLATKIN, M., 1982. A Neo-Darwinian commentary on macroevolution. *Evolution* **36:** 474–98.

CHEVERUD, J. M., 1984. Quantitative genetics and developmental constraints on evolution by selection. *Journal of Theoretical Biology* **110:** 155–71.

CLOSE, R. L., 1977. Recurrence of breeding after cessation of suckling in the marsupial *Perameles nasuta. Australian Journal of Zoology* **25:** 641–45.

CLUTTON-BROCK, T. H., ALBON, S. D. AND GUINNESS, F. E., 1985. Parental investment and sex differences in juvenile mortality in birds and mammals. *Nature (Lond.).* **313:** 131–33.

COCKBURN, A., 1988. *Social behaviour in fluctuating populations.* Croom Helm: London.

COCKBURN, A., BRAITHWAITE, R. W. AND LEE, A. K., 1981. The response of the heath rat, *Pseudomys shortridgei,* to pyric succession: a temporally dynamic life-history strategy. *Journal of Animal Ecology* **50:** 649–66.

COCKBURN, A., LEE, A. K. AND MARTIN, R. W., 1983. Macrogeographic variation in litter size in *Antechinus* (Marsupialia: Dasyuridae). *Evolution* **37:** 86–95.

EISENBERG, J. F., 1981. *The mammalian radiations: an analysis of trends in evolution, adaptation and behavior.* Athlone Press: London.

FOX, B. J., 1983. Mammal species diversity in Australian heathlands: the importance of pyric succession and habitat diversity. Pp. 473–89 in *Mediterranean-type ecosystems. Ecological studies.* Vol. 43 ed by F. J. Kruger, D. T. Mitchell and J. U. M. Jarvis. Springer-Verlag: Berlin.

GEMMELL, R. T., 1979. The fine structure of the luteal cells in relation to the concentration of progesterone in the plasma of the lactating bandicoot, *Isoodon macrourus* (Marsupialia: Peramelidae). *Australian Journal of Zoology* **27:** 501–10.

GEMMELL, R. T., 1981. The role of the corpus luteum of lactation in the bandicoot, *Isoodon macrourus* (Marsupialia: Peramelidae). *General and Comparative Endocrinology* **44:** 13–19.

GEMMELL, R. T., 1982. Breeding bandicoots in Brisbane *(Isoodon macrourus)* Marsupialia, Peramelidae. *Australian Mammalogy* **5:** 187–93.

GEMMELL, R. T., JENKIN, G. AND THORNBURN, G. D., 1980. Plasma concentrations of progesterone and 13,14-dihydro-15-keto-prostoglandin F2 at parturition in the bandicoot, *Isoodon macrourus. Journal of Reproduction and Fertility* **60:** 253–56.

GEMMELL, R. T., WALKER, M. T., JOHNSTON, G. AND CEPON, G., 1984. The number of young present in sequential litters of the marsupial bandicoot, *Isoodon macrourus* in captivity. *Australian Journal of Zoology* **32:** 632–39.

GIESEL, J. T., 1976. Reproductive strategies as adaptations to life in temporally heterogeneous environments. *Annual Review of Ecology and Systematics* **7:** 57–79.

GILL, D. E., 1978. On selection at high population density. *Ecology* **59:** 1289–91.

GLIWICZ, J., 1980. Island populations of rodents: their organization and functioning. *Biological Review* **55:** 109–38.

GORDON, G., 1974. Movements and activity of the short-nosed bandicoot, *Isoodon macrourus* (Gould) (Marsupialia). *Mammalia* **38:** 405–31.

GORDON, G. AND LAWRIE, B. C., 1977. The rufescent bandicoot, *Echymipera rufescens* (Peters and Doria) on Cape York Peninsula. *Australian Wildlife Research* **5:** 41–45.

GOULD, S. J., 1977. *Ontogeny and phylogeny.* Belknap: Cambridge, Massachusetts.

GOULD, S. J., 1980. Is a new and general theory of evolution emerging? *Paleobiology* **6:** 119–130.

GOULD, S. J., 1982. Change in the timing of development as a mechanism of macroevolution. Pp. 333–46 in *Evolution and development* ed by J. T. Bonner. Springer-Verlag: Berlin.

GOULD, S. J., 1983. The meaning of punctuated equilibrium and its role in validating a hierarchical approach to macroevolution. *Scientia* **118:** 135–57.

GOULD, S. J., 1984. *Hen's teeth and horse's toes.* Penguin: Middlesex.

HARPER, J. L., 1982. After description. Pp. 11–25 in *The plant community as a working mechanism* ed by E. I. Newman. Blackwell Scientific Publications: Oxford.

HEINSOHN, G. E., 1966. Ecology and reproduction of the Tasmanian bandicoots *(Perameles gunni* and *Isoodon obesulus). University of California Publications in Zoology* **80:** 1–107.

HINDS, D. S. AND MACMILLEN, R. E., 1984. Energy scaling in marsupials and eutherians. *Science* **225:** 335–37.

HOLLIS, D. E. AND LYNE, A. G., 1980. Ultrastructure of luteal cells in fully formed and regressing corpora lutea during pregnancy and lactation in the marsupials *Isoodon macrourus* and *Perameles nasuta. Australian Journal of Zoology* **28:** 195–211.

HUGHES, R. L., 1962. Role of the corpus luteum in marsupial reproduction. *Nature (Lond.)* **194:** 890–91.

HULBERT, A. J. AND DAWSON, T. J., 1974. Standard metabolism and body temperature of perameloid marsupials from different environments. *Comparative Biochemistry and Physiology* **47A:** 583–90.

HUME, I. D., 1982. *Digestive physiology and nutrition of marsupials.* Cambridge University Press: Cambridge.

JOHNSON, C. N. AND JOHNSON, K. A., 1983. Behaviour of the bilby, *Macrotis lagotis* (Reid), (Marsupialia: Thylacomyidae) in captivity. *Australian Wildlife Research* **10:** 77–87.

JOHNSON, K. A., 1980. Diet of the Bilby, *Macrotis lagotis,* in the western desert region of central Australia. *Bulletin of the Australian Mammal Society* **6:** 46–47.

JOHNSON, K. A., 1983a. Bilby *Macrotis lagotis.* Pp. 107–08 in *The Australian Museum complete book of Australian mammals* ed by R. Strahan. Angus & Robertson: Sydney.

JOHNSON, K. A., 1983b. Lesser Bilby *Macrotis leucura.* Pp. 109 in *The Australian Museum complete book of Australian mammals* ed by R. Strahan. Angus & Robertson: Sydney.

KAUFMANN, K. W., 1981. Fitting and using growth curves. *Oecologia (Berl.)* **49:** 293–99.

KIKKAWA, J., INGRAM, G. J. AND DWYER, P. D., 1979. The vertebrate fauna of Australian heathlands — an evolutionary perspective. Pp. 231–79 in *Heathlands and related shrublands. Descriptive studies. Ecosystems of the world Vol. 9A* ed by R. L. Specht. Elsevier: Amsterdam.

LEE, A. K. AND COCKBURN, A., 1985. *Evolutionary ecology of marsupials.* Cambridge University Press: Cambridge.

LEE, A. K., WOOLLEY, P. AND BRAITHWAITE, R. W., 1982. Life history strategies of dasyurid marsupials. Pp. 1–11 in *Carnivorous marsupials* ed by M. Archer. Royal Zoological Society of New South Wales: Sydney.

LYNE, A. G., 1974. Gestation period and birth in the marsupial *Isoodon macrourus. Australian Journal of Zoology* **22:** 303–09.

LYNE, A. G., 1976. Observations on oestrus and the oestrous cycle in the marsupials *Isoodon macrourus* and *Perameles nasuta*. *Australian Journal of Zoology* **24:** 513–21.

LYNE, A. G. AND HOLLIS, D. E., 1979. Observations on the corpus luteum during pregnancy and lactation in the marsupials *Isoodon macrourus* and *Perameles nasuta*. *Australian Journal of Zoology* **27:** 881–99.

LYNE, A. G. AND VERHAGEN, A. M. W., 1957. Growth of the marsupials *Trichosurus vulpecula* Kerr and *Perameles nasuta* Geoffroy. *Growth* **21:** 167–95.

MCCRACKEN, H. E., 1986. The oestrous cycle and gestation period of the Greater Bilby, *Macrotis lagotis* (Reid) (Marsupialia: Thylacomyidae). *Australian Mammalogy* **9:** 5–16.

MCLURE, P. A., 1981. Sex-biased litter reduction in food-restricted wood rats (*Neotoma floridana*). *Science* **211:** 1058–60.

MCNAB, B. K., 1980a. Energetics and the limits to a temperate distribution in armadillos. *Journal of Mammalogy* **61:** 606–27.

MCNAB, B. K., 1980b. Food habits, energetics and the population biology of mammals. *American Naturalist* **116:** 106–24.

MARLOW, B. J., 1961. Reproductive behaviour of *Antechinus flavipes* (Waterhouse) (Marsupialia) and the development of the pouch young. *Australian Journal of Zoology* **9:** 203–18.

NEWSOME, A. E. AND CATLING, P. C., 1979. Habitat preferences of mammals inhabiting heathlands of warm temperate coastal, montane and alpine regions of south-eastern Australia. Pp. 301–19 in *Heathlands and related shrublands. Descriptive studies. Ecosystems of the world Vol. 9A* ed by R. L. Specht. Elsevier: Amsterdam.

PADYKULA, H. A. AND TAYLOR, J. M., 1976. Ultra structural evidence for loss of the trophoblastic layer in the chorioallantoic placenta of Australian bandicoots (Marsupialia: Peramelidae). *Anatomical Record* **186:** 357–86.

PETERS, R. H., 1983. *The ecological implications of body size.* Cambridge University Press: Cambridge.

RENFREE, M. B., 1980. Placental function and development in marsupials. Pp. 269–84 in *Comparative physiology: primitive mammals* ed by K. Schmidt-Nielsen, L. Bolis and C. R. Taylor. Cambridge University Press: Cambridge.

ROTHCHILD, I., 1981. The regulation of the mammalian corpus luteum. *Recent Progress in Hormone Research.* **37:** 183–298.

RUSSELL, E. M., 1982. Patterns of parental care and parental investment in marsupials. *Biological Review* **57:** 423–86.

SEEBECK, J. H., 1979. Status of the barred bandicoot, *Perameles gunnii* in Victoria: with a note on husbandry of a captive colony. *Australian Wildlife Research* **6:** 255–64.

SELWOOD, L., 1980. A timetable of embryonic development of the dasyurid marsupial, *Antechinus stuartii* (Macleay). *Australian Journal of Zoology* **28:** 649–68.

STEARNS, S. C., 1982. The role of development in the evolution of life histories. Pp. 237–58 in *Evolution and development* ed by J. T. Bonner. Springer-Verlag: Berlin.

STEBBINS, G. L., 1970. Adaptive radiation of reproductive characteristics in angiosperms. I. Pollination mechanisms. *Annual Review of Ecology and Systematics* **1:** 307–26.

STENSETH, N. C., 1985. Darwinian evolution in ecosystems: The Red Queen view. Pp. 55-72 in *Evolution: essays in honour of John Maynard Smith* ed by P. H. Greenwood, P. H. Harvey and M. Slatkin. Cambridge University Press: Cambridge.

STODART, E., 1977. Breeding and behaviour of Australian bandicoots. Pp. 179–91 in *The biology of marsupials* ed by B. Stonehouse and D. Gilmore. Macmillan: London.

STODDART, D. M. AND BRAITHWAITE, R. W., 1979. A strategy for utilization of regenerating heathland habitat by the brown bandicoot (*Isoodon obesulus:* Marsupialia, Peramelidae). *Journal of Animal Ecology* **48:** 165–79.

TAYLOR, J. M. AND PADYKULA, H. A., 1978. Marsupial trophoblast and mammalian evolution. *Nature (Lond.)* **271:** 588.

WAKE, D. B., ROTH, G. AND WAKE, M. H., 1983. On the problem of stasis in organismal evolution. *Journal of Theoretical Biology* **101:** 211–24.

WIDDOWSON, E. M., 1976. The responses of the sexes to nutritional stress. *Proceedings of the Nutrition Society* **35:** 175–80.

ZULLINGER, E. M., RICKLEFS, R. E., REDFORD, K. H. AND MACE, G. M., 1986. Fitting sigmoidal equations to mammalian growth curves. *Journal of Mammalogy* **65:** 607–36.

Distribution and abundance of the Greater Bilby *Macrotis lagotis* Reid (Marsupialia: Peramelidae)

R. I. SOUTHGATE[1]

The Greater Bilby *Macrotis lagotis* was once widely distributed throughout arid and semiarid Australia. It has suffered a rapid decline since European settlement and is now restricted to parts of the Tanami, Great Sandy and Gibson Deserts with outlying populations in south-west Queensland and north of Alice Springs, Northern Territory. Its distribution is examined in relation to historic and scientific records, climatic and soil features, land use and the distribution of rabbits and foxes. The initial contraction of the range of *M. lagotis* occurred a decade or so behind the arrival of the Fox and European Rabbit. The data indicate that *M. lagotis* distribution is still declining and no single factor can be held responsible for its disappearance.

Key words: *Macrotis lagotis*, distribution, abundance, decline.

INTRODUCTION

THE Greater Bilby *(Macrotis lagotis)* was once distributed throughout arid and semiarid Australia. Since European settlement an alarming decline in its range and the abundance has occurred. Concern for its demise was expressed as early as the late 1850s (Kreft 1866) and restated by Jones (1924), Troughton (1932) and Finlayson (1961). Watts (1969) compiled data on the past and present distribution and discussed factors associated with its decline. These factors included predation by foxes, habitat alteration by rabbits and pastoral activities. Changed burning regimes have also been advanced as a possible cause.

This paper updates and expands upon Watt's assessment of changes in the status of *M. lagotis.* To help interpret its past distribution and abundance, the rate of discovery or recognition of the species is first reviewed in relation to exploration, settlement and scientific records. Secondly, its distribution at the time of European settlement is outlined and its decline is followed to the present range. Finally, this paper investigates factors which may be limiting the overall distribution of *M. lagotis* and causing populations to become isolated. Its present distribution is considered in relation to climate and soil factors, land use and the distribution of foxes and rabbits.

METHODS

Past distribution of *M. lagotis* was determined from museum data, published literature and communication with long-standing residents of various areas. As its present distribution was largely unknown, extensive survey work was undertaken. The general procedure used during field work was to interview residents of pastoral stations, mining camps or Aboriginal communities. The area visited is shown in Figure 1. Some 60 000 km were travelled and 250 days spent in the field.

A puppet skin specimen of *M. lagotis* was shown to the people interviewed and a poster and information leaflet were issued. Discussions with Aboriginal people proved to be very fruitful as *M. lagotis* was an important totemic and food item throughout its range. Any recent sightings were noted and these areas were investigated in greater detail. *Macrotis lagotis* activity could be confidently identified from careful examination of feeding activity, tracks, burrows and scats. Frequent night spot-lighting resulted in only two sightings.

A representative set of accurately located and dated records was extracted from all known records, and examined in relation to other data sets for factors which were thought may explain the distribution pattern of *M. lagotis.* Records were used in BIOCLIM, a model for predicting animal or plant distribution in relation to climate (Busby 1985). Soil data extracted from the *Atlas of Australian soils* (Northcote 1960–68) at a scale of 1:2 000 000 were attached to the reliable set of distribution records. To guard against mapping inaccuracy, the soil types 6 km to the north, south, east and west were also recorded. If no difference was found it was assumed with confidence that *M. lagotis* occupied that particular soil type. Where a discrepancy was found between the soil types, the soil type was not included in the analyses. Mapping of land-use, rabbit and fox distribution

[1]Conservation Commission of the Northern Territory, P.O. Box 1046, Alice Springs, Northern Territory, Australia 5750.

Pages 293–302 *in* BANDICOOTS AND BILBIES, ed by J. H. Seebeck, P. R. Brown, R. L. Wallis and C. M. Kemper. Surrey Beatty & Sons, Sydney, 1990.

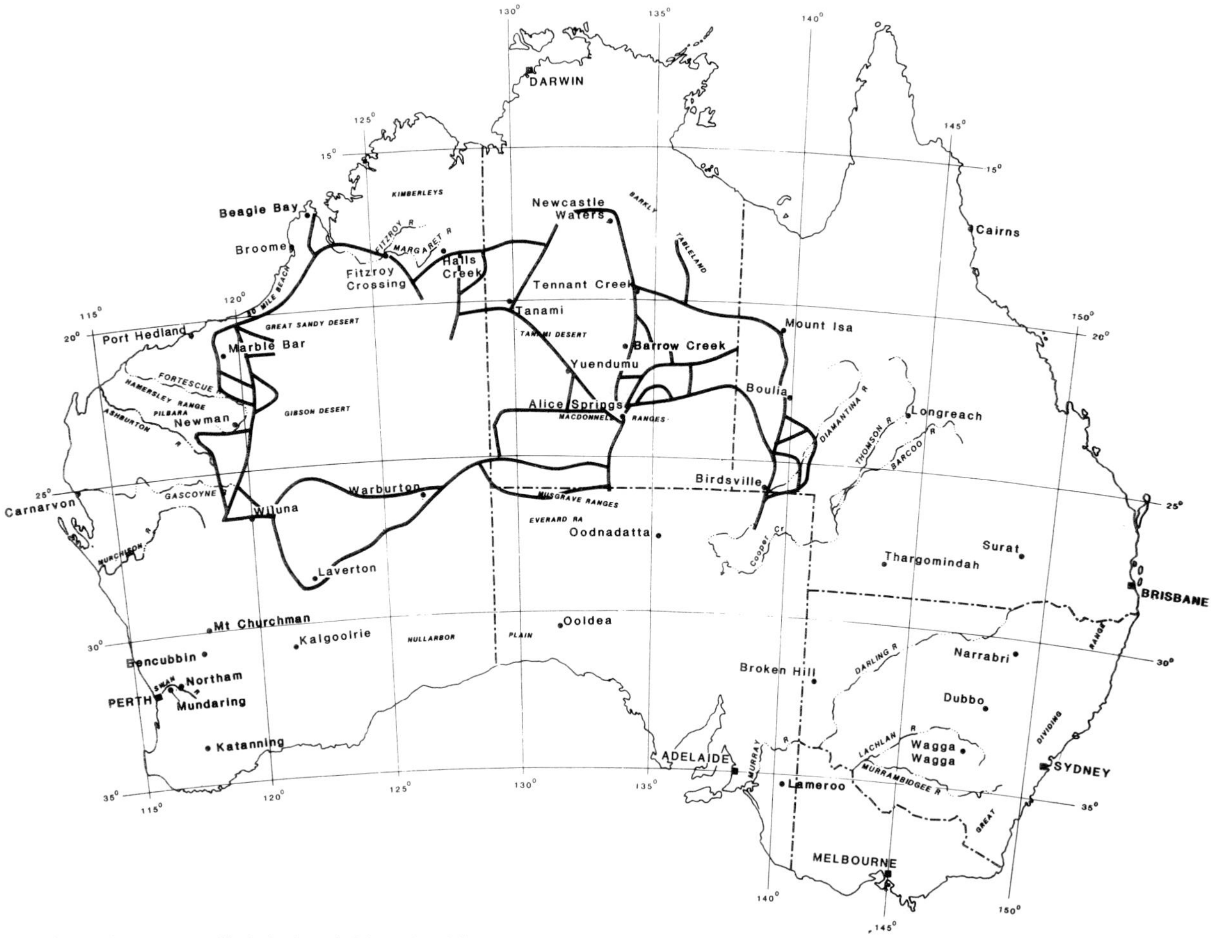

Fig. 1. Routes travelled during field work while studying *Macrotis lagotis*, 1983–85.

was adapted from a range of reports (Mitchell 1982; Low 1983; Strong and Low 1983; King and Smith 1985; D. R. King, pers. comm.) and the *Atlas of Australian resources* (1975 series).

RESULTS

Past distribution and abundance of M. lagotis

The former distribution of *M. lagotis,* both as based on accurate records, and as predicted by BIOCLIM (using 415 localities), is shown in Figure 2. When the historic record of *M. lagotis* was examined, distinct anomalies became apparent. Rarely did observation of the animal correspond with the extent of exploration or settlement. Initially, *M. lagotis* was even wrongly described by Reid (1837) as an inhabitant of Van Diemen's Land (Tasmania) where the species has not existed in historic times.

Reid (1837) also mentioned that an animal of this description had been found beyond the mountains of the Swan River (Western Australia). The Swan River Colony was established in 1829; as the surrounding country was explored, it was reported several times (Glauert 1950; Dixon 1974) that *M. lagotis,* called the "Dol-goitch" or "Dal-gyte" by Aborigines, was widespread and abundant through parts of the interior.

The occurrence of *M. lagotis* outside Western Australia was not confirmed until some 20 years later when specimens were collected by members of the Blandowski expedition near the junction of the Darling and Murray Rivers in 1857 (Wakefield 1966). By this stage at least two-thirds of what is now New South Wales had been explored and overlanders, driving sheep and cattle, had linked the colonies of Adelaide and Melbourne.

The reason for the apparent absence of early *M. lagotis* records from the eastern states is unclear. Krefft (1866), a member of the Blandowski expedition, noted, however, that "... the nocturnal habits of many of the local fauna accounted for apparent scarcity of the animal life seen, except for the occasional kangaroo. Two-thirds of the smaller Mammalia collected were new to the old residents and even the natives, who in many parts had acquired habits different to their former mode of life, had almost forgotten the existence of some species." Krefft, in a specific reference to *M. lagotis*, added "... this beautiful animal like many other

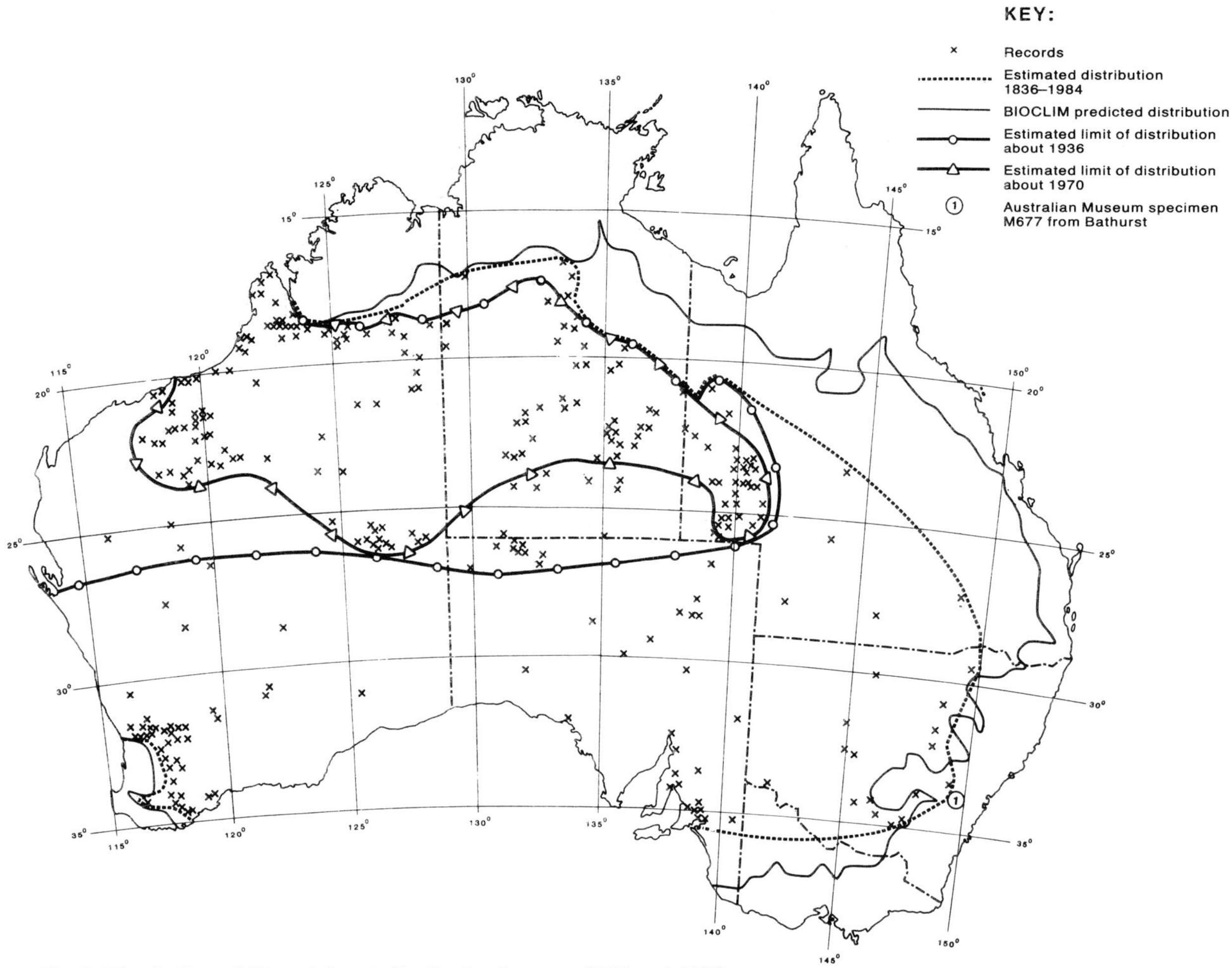

Fig. 2. The decline of *Macrotis lagotis* distribution between 1836 and 1970.

species has long ago retreated north of the Murray." This indirect reference is the primary record of the species' former presence in Victoria.

By the time *The mammals of Australia* was published (Krefft 1871) the range of *M. lagotis* was described as New South Wales, northern Victoria, South Australia, Queensland and Western Australia. Troughton (1932) pointed out that this accurate representation was ignored by Thomas (1888), who stated that it occurred in only South and Western Australia. Troughton continued, "This omission of the eastern states misled subsequent authors, and so we find New South Wales excluded from the range of the genus by Ogilby in his catalogue of 1892, Baldwin Spencer in the Horn Report of 1896, and all systematic and popular works of the time."

At the time, when confusion over the correct distribution was at its height, a rapid decline of *M. lagotis* had commenced in South Australia. Jones (1924) reported that, by 1900, it had disappeared from the more fertile portions of South Australia. He noted, however, that in the 1890s "... it was not unusual for rabbiters, even in the immediate neighbourhood of Adelaide, to take more Bilbies ... than Rabbits in their traps." Up to 1907, some 36 *M. lagotis* were donated to Adelaide Zoo from localities within a few hundred kilometres radius of Adelaide; however, not one of these animals was registered in the Zoo (Rix 1978) indicating that they were considered of minor importance.

In the inland and drier areas *M. lagotis* were still common. Le Souef and Burrell (1926) estimated there were 21 to the square mile (eight per km^2) by surveying their burrow distribution on the Nullarbor Plain in 1921. "Heavy concentrations" of bilbies were also reported by Finlayson (1961) in the 1930s along the Everard and Musgrave Ranges on the South Australian and Northern Territory border.

After the Blandowski expedition there was a 20 year break before there was another authentic *M. lagotis* record from New South Wales with the collection of a museum specimen. Troughton (1932) surmised that, from the museum specimens collected between 1879 and 1908, *M. lagotis* was fairly evenly distributed west of the Great Dividing Range although it was probably never plentiful. He found indications that it was more abundant in some years. Rolls (1969, 1981) also inferred from his historical research that *M. lagotis* was in great numbers in parts of New South Wales in the late 1870s and late 1880s. The last authentic report of *M. lagotis* in New South Wales was from Wagga Wagga in 1912 (Troughton 1932). It is apparent that the collapse of the population occurred rapidly and that all forms of habitat were affected.

By 1875, at least half of Queensland had been settled, and by 1900 the remaining part was taken up. Despite this, museum specimens have been collected from only five different localities in this state and even fewer published reports of *M. lagotis* exist. According to a reliable informant (Kimber 1981) and other unverified accounts, it was found throughout central and western Queensland but in lower densities than in parts of the Northern Territory. In the decade or so after 1900, populations contracted to the south-west of the state.

M. lagotis was first documented in the Northern Territory by the Horn Scientific Expedition (Spencer 1896) in 1894. Well before this time the overland telegraph line had successfully traversed the Northern Territory, and rabbits had already entered from South Australia (Strong 1983). In some areas *M. lagotis* was apparently numerous. Spencer and Gillen (1912) noted that "judging from the supply of alpita (tail tips) in every camp, this bandicoot . . . must abound . . . in many parts of the centre, the burrow this animal makes are very extensive and the animal lives in colonies." Finlayson (1961) referred to them once being one of the most plentiful and universally distributed of the central Australian mammals. He recognized that they were rapidly being reduced to a rare animal and had already been eliminated from much of the southern part of the Northern Territory.

After the initial impetus of collecting in Western Australia between 1840 and 1860, further specimens of *M. lagotis* were not collected until 1900. Shortridge (1909) stated that the species was widely distributed and most plentiful through the south-west (except near the coast). To the north-west it extended its distribution to the coast but it was rarer. He continued, "within recent years it has become extremely rare in the far interior."

In his account of mammals from the south-west Western Australia wheat belt, Leake (1962)

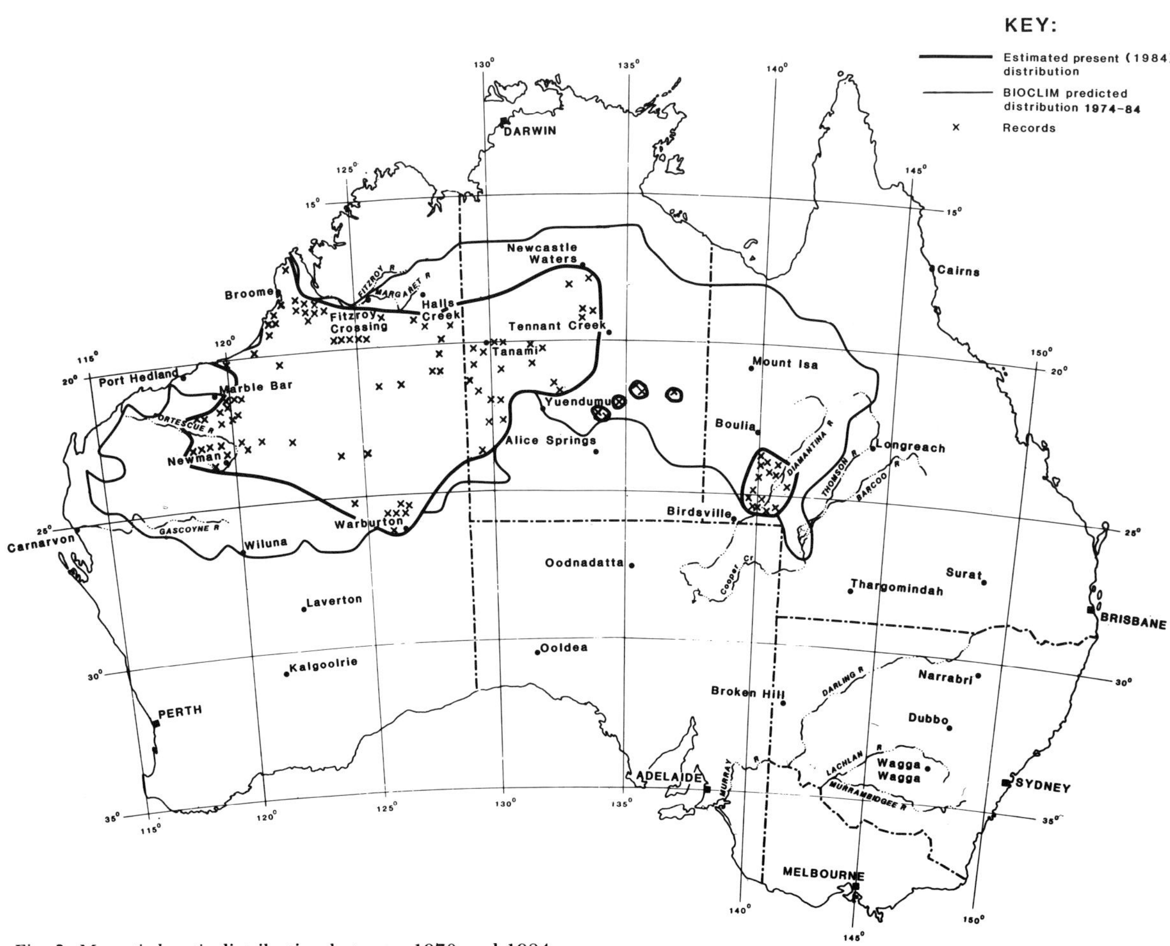

Fig. 3. Macrotis lagotis distribution between 1970 and 1984.

indicated that numbers of *M. lagotis* could vary considerably between years. He stated that "by about 1899 they had probably all gone from the eastern wheat belt, but in 1918 after three very wet years they returned and rapidly increased in numbers . . . at one period their population was dense and shortly afterward their numbers decreased rapidly and by 1929 they had disappeared again and have not returned." Jenkins (1974) similarly discussed the rapid disappearance of *M. lagotis*. He recorded the species as common in Northam in 1926 and plentiful in the Katanning district until 1935.

Since this time authentic reports of *M. lagotis* have only been received from the more northern and drier parts of the state.

Decline and present distribution

It is evident that an extensive and rapid decline of *M. lagotis* occurred between the late 1800s and 1935. In 1935 it was still present in the northern parts of South Australia, north of the Murchison River in Western Australia, in southwestern Queensland and was widespread through parts of the Northern Territory (Fig. 2). Isolated populations probably persisted in some areas to the south of this range, as reasonably reliable reports of *M. lagotis* are on record (West Australian Fisheries and Wildlife files) from areas north-east of Perth until the early 1960s (e.g., Mundaring district, Bencubbin and Mt Churchman). There have been no authenticated reports since. By 1970, the distribution had again altered substantially along its southern boundary with the exception of populations persisting around Warburton (Western Australia) and in south-west Queensland. No longer was it present in the pastoral country south of the Ashburton River in Western Australia, and, in the Northern Territory, *M. lagotis* populations had disappeared from south of the MacDonnell Ranges, and had dwindled in pastoral country to the north. Both Newsome (1962) and Watts (1969) reported extremely low densities in this general area. Watts reported one colony (comprising two adults and offspring) per 78 km^2 near Yuendumu, and Newsome estimated there were four or perhaps six colonies in an area of about 3 885 km^2 of Mulga woodland just north of the MacDonnell Ranges near Alice Springs. This contrasts strongly with the "endless mounds" caused by *M. lagotis* burrowing activity reported by Spencer and Gillen (1912) around Barrow Creek which has a similar mix of land systems (Perry, Mabbut, Lichfield and Quinlan 1963).

The present distribution of *M. lagotis* and that predicted by BIOCLIM (based on 125 records between 1974 and 1984) is shown in Fig. 3.

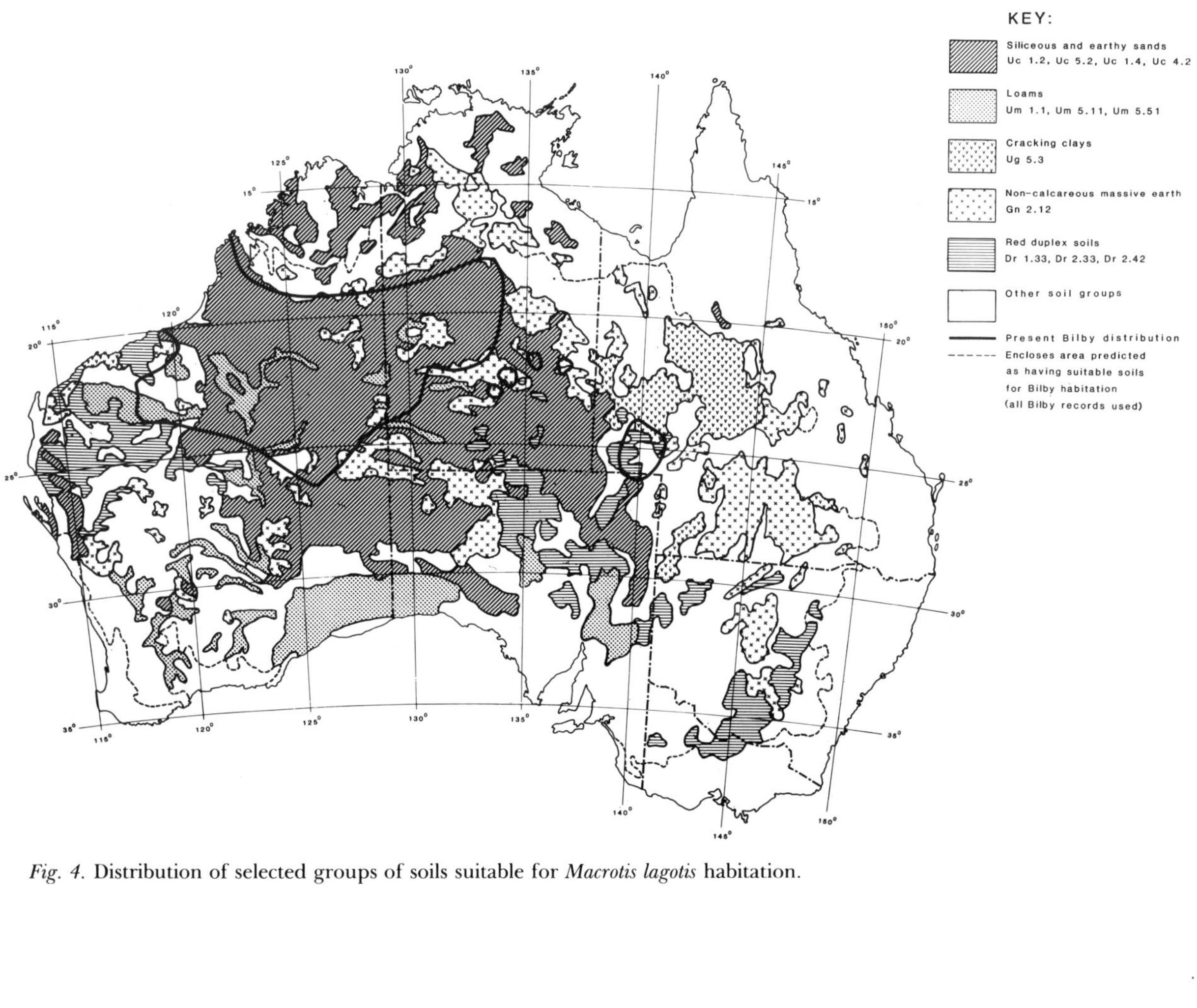

Fig. 4. Distribution of selected groups of soils suitable for *Macrotis lagotis* habitation.

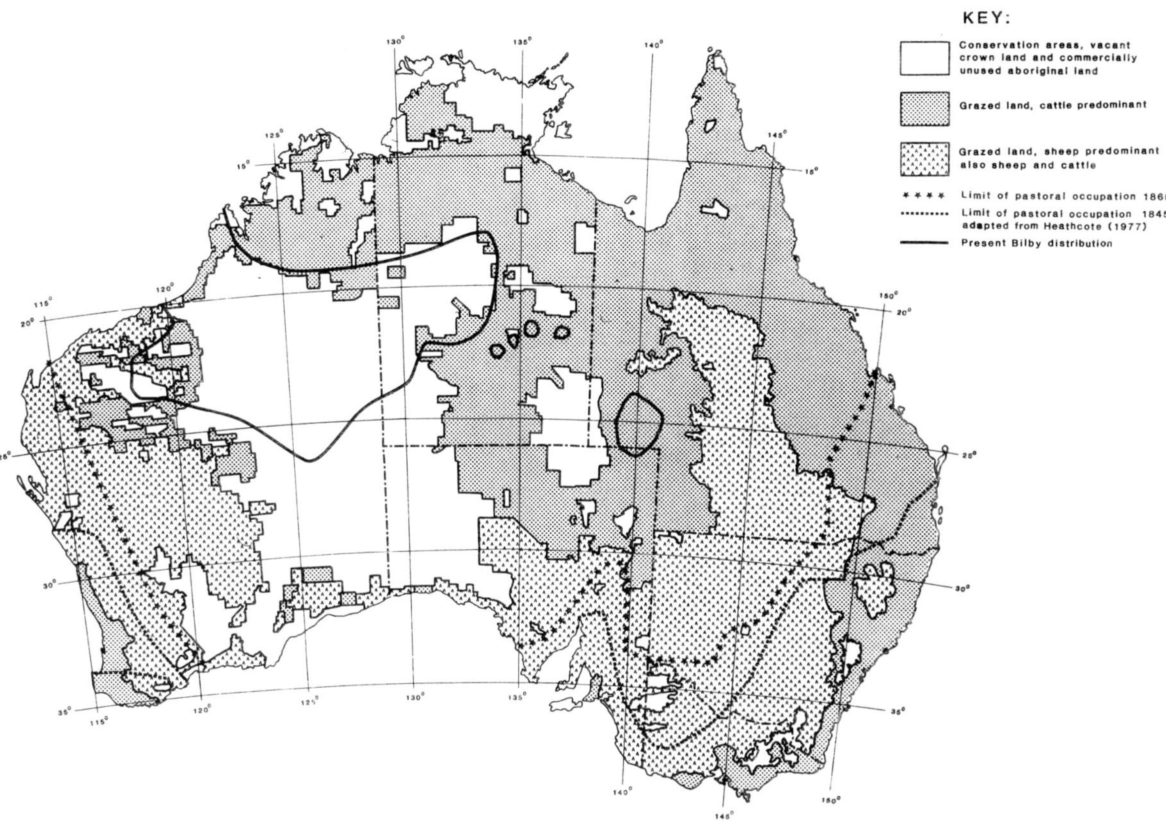

Fig. 5. Landuse in Australia and its relationship with present *Macrotis lagotis* distribution.

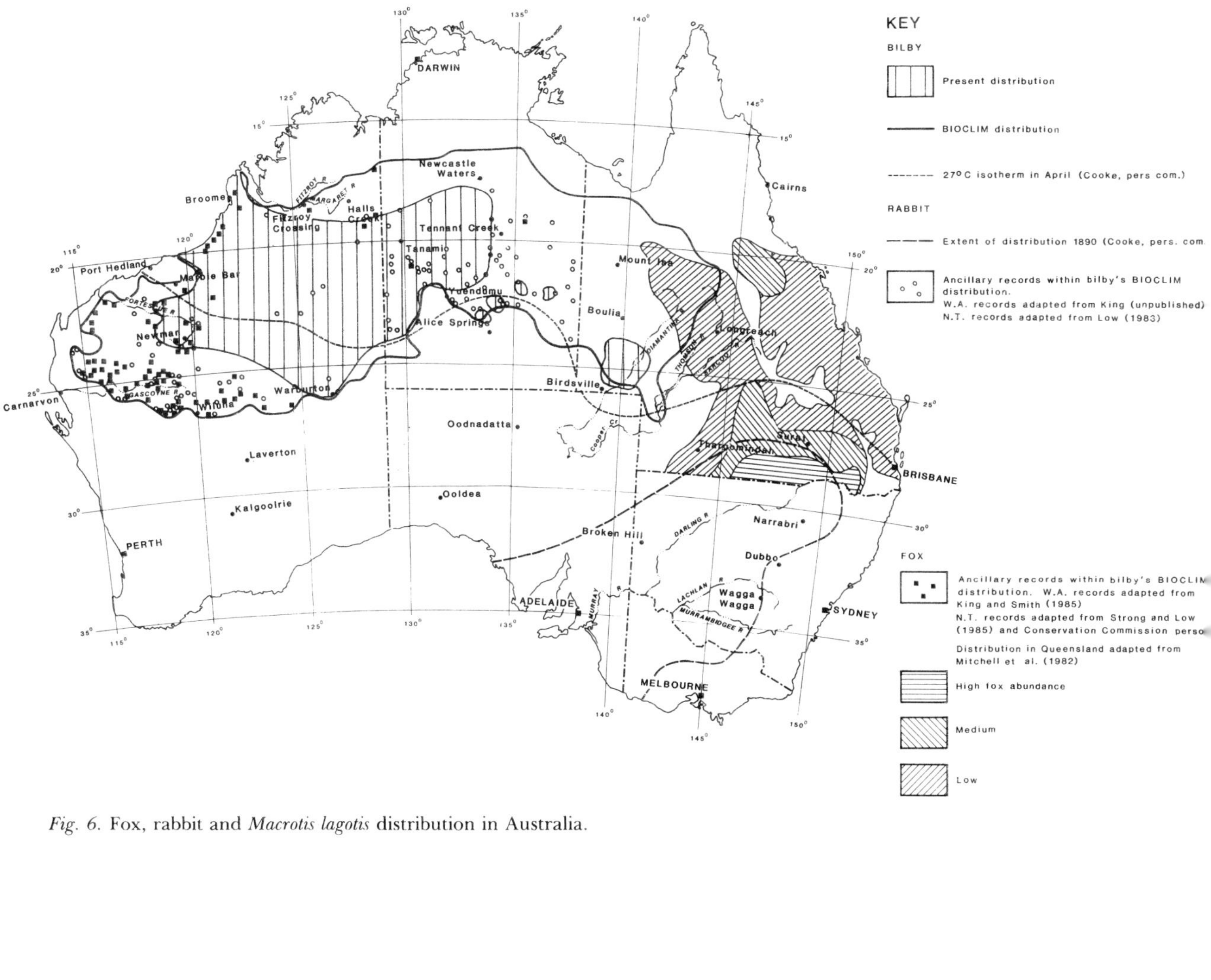

Fig. 6. Fox, rabbit and *Macrotis lagotis* distribution in Australia.

Throughout this distribution a wide range of soil, vegetation and landform types are occupied. Suitable soil types (Fig. 4) are well represented outside this range and this resource cannot be considered limiting to its past or present distribution. Most of the land now occupied by *M. lagotis* is ungrazed and in no areas does it overlap with current sheep grazing activity although this type of pastoralism may have occurred in the past (Fig. 5). Rabbits and foxes overlap with present distribution of *M. lagotis* but are most abundant outside its range (Fig. 6).

The most evident change since 1970 is the continued reduction of *M. lagotis* distribution in the pastoral districts of the Northern Territory. A reduction in distribution has also occurred along the south of their range between Warburton and the Northern Territory border: both Philpott and Smyth (1967) and Burbidge and Fuller (1979) found *M. lagotis* in this area, but Aboriginal people whom I interviewed stated that populations could no longer be found in these parts. In south-west Queensland a recent decline in the distribution of *M. lagotis* has occurred; however, it is unclear whether this change is permanent.

In Western Australia, *M. lagotis* is still present in the Gibson Desert as far south as Warburton, and in desert country to the north-west as far as Newman. Extremely scattered populations exist throughout the Hammersley Range area, along the Fortescue River and the marginal pastoral country in the eastern Pilbara. From the evidence available it is likely that animals are scattered throughout the Great Sandy Desert. The western edge of their range includes some pastoral leases along the 80 Mile Beach and north to Beagle Bay in Dampier Land. In the Kimberley, *M. lagotis* still occurs south of the Fitzroy and Margaret Rivers in the desert areas on the edge of pastoral leases.

In the Northern Territory, scattered populations occur across the northern end of the Tanami Desert and they are widespread and scattered throughout the desert itself (pers. obs., 1980–81; Gibson 1986). There is also evidence of remnant *M. lagotis* populations persisting in pastoral lease country to the north-east and north-west of Alice Springs, and possibly in the non-pastoral areas east of Tennant Creek.

Finally, *M. lagotis* still persists in a block of pastoral country between Birdsville and Boulia in Queensland. Populations are generally sparse but they are locally abundant in some areas.

DISCUSSION

Since Watts' (1969) assessment of *M. lagotis* distribution this study has been able to confirm its presence in parts of the Gibson and Great Sandy Desert. There is evidence that a decline in distribution has occurred in the period since 1969 as it no longer extends to 24°S in the Northern Territory and they have disappeared from a number of other areas.

The use of BIOCLIM enabled the comparison of a known distribution with a hypothetical extreme based on climatic factors. A good match between the historical and predicted distribution was obtained in the south of the range of *M. lagotis* (except for part of Victoria and a possible mislocated record in New South Wales) as shown in Figure 2. An extensive area was also delineated in the northern part of the Northern Territory and in central and eastern Queensland which is outside the known former distribution. *Macrotis lagotis* never occupied the Barkly Tableland because of soil types (Finlayson 1961) which explains the discrepancy in this area. Suitable oils, however, were found in the other mismatched parts (Fig. 4). The possibility that it was present in these parts and disappeared without record cannot be ruled out and requires further historical investigation.

The area predicted by BIOCLIM using recent records was also far more extensive than their present known distribution. Parts of western Queensland, the central Northern Territory and the Pilbara and eastern Gasgoyne appeared to be climatically suitable for occupation, but populations were extremely scarce. Along the southern limits of the assumed distribution there was a close alignment with the BIOCLIM distribution, especially near Warburton and Birdsville. This suggests that resources associated with climate could be important in limiting the distribution of *M. lagotis* in this part of their range.

From historical records it appears that *M. lagotis* was fairly uniformly distributed through its range, although more abundant in certain habitats. It is evident that populations were prone to large fluctuations in size; and that these were a result of seasonal conditions or possibly disease. In the last 10 years fluctuations of populations have been observed in the Tanami Desert (P. K. Latz and K. A. Johnson, pers. comm., 1985) and pockets of high abundance have been noted. This situation has also been observed in *M. lagotis* populations of south-west Queensland (R. Atherton, pers. comm. 1985).

The paucity of early records of *M. lagotis* from parts of Australia is unlikely to have been related to its distribution and abundance. During this survey it was found that its presence, as indicated by burrows and diggings, in certain habitats was easily overlooked by locals or confused with *Rattus* spp. or goanna activity. At the other extreme, when the animal was so common in South Australia, records were probably treated with indifference. Collaboration between early workers and Aboriginal people on the subject appeared to benefit the process of collection and recording of distribution. Aboriginal knowledge does not seem to have been investigated in New South Wales, Victoria and Queensland (except for Krefft's work) and this may explain the lack of records of *M. lagotis* from these states.

The former extent of the distribution of *M. lagotis* in Victoria is uncertain. On one hand we have Krefft's (1866) remark about its early decline. On the other hand, anthropological evidence (Scarlett 1969) is considered inadequate, supportive subfossil material is still lacking and there are no other descriptions. Nevertheless there is a report from Lameroo in South Australia near the Victorian border about 1900 (Rix 1978) and it is difficult to imagine that *M. lagotis* would not have been present in the mallee districts of Victoria at the same time. The north-west of Victoria is clearly delineated by BIOCLIM as part of the climatically suitable range. If Krefft's (1866) remark is correct, only land use, disease or climatic changes can be surmised for the decline of *M. lagotis*, as foxes and rabbits posed no threat at that stage.

In discussing the decline of *M. lagotis*, Watts (1969) dismissed close settlement and the resultant alteration of habitat, and suggested that it was more closely correlated with the arrival of rabbits and foxes. Rabbits were well distributed through Australia by 1935 and they had reached their distribution limits as early as 1910 (Long 1972; Strong 1983). The spread and type of rabbit distribution is shown in Figure 6. In parts of New South Wales, South Australia and Queensland foxes arrived concurrently with rabbits (Frith 1977; Schlager 1981). In other parts they took slightly longer to become established, but by the 1930s they were reported from points along the northern part of the range of *M. lagotis* (King and Smith 1985; Strong and Low 1983). Feral cats preceded both the fox and rabbit in their colonization and were probably widespread throughout the range of *M. lagotis* before 1900. They have been reported as taking *M. lagotis* (Southgate, unpublished data), and may have affected abundance but their long-term and broad scale co-habitation does not correlate well with the decline of *M. lagotis* or its present distribution.

From the records available it is evident that the decline of *M. lagotis* in the south of its range occurred some 10–15 years after the arrival of the rabbit; however, it is difficult to propose a likely mechanism (B. D. Cooke, pers. comm., 1985).

Rabbits and *M. lagotis* do share foods in common, such as the sedge, *Cyperus bulbosus*. Rabbits can also change vegetation composition and structure (Wood 1982); this could have reduced the availability of some food resources such as seeds, witchetty grubs from some *Acacia* species, and *Solanum* spp. fruits. It might have taken 10–15 years for food resources to become disrupted, or this delay may have been because rabbits did not compete directly with adult *M. lagotis* but rather prevented juvenile recruitment into the population so that the population dwindled and died out.

Predation is another suggested mechanism which would act less directly. Rabbit populations undergo periodic booms or busts: predator populations which build up during the plague phase may be forced to switch very heavily to native animals when the rabbit numbers crash. Foxes, in particular, rely largely on rabbits for food and both their initial spread and their present distribution are closely related with those of rabbits. It has been reported (Leake 1962; Jarman and Johnson 1977) that fox populations became abundant when rabbits first assumed a prominent status in an area. This was generally a decade after the rabbits' first arrival; this timing corresponds with the decline of *M. lagotis*.

Thus it is likely that rabbits and foxes are affecting the distribution of *M. lagotis* and abundance in some combination, although it is difficult to determine their relative importance in this effect. Further circumstantial support for this hypothesis can be found by examining the present southern limit of distribution in relation to that predicted by BIOCLIM, and to the northern limit of rabbit establishment. The northern limit to rabbit establishment correlates well with the 27° isotherm in April (B. D. Cooke, pers. comm., 1985) in the Northern Territory, South Australia and Queensland. In Western Australia the northern limit of rabbit distribution lies further south (D. R. King, pers. comm., 1985) as shown in Figure 6. Rabbits and foxes do occur in scattered permanent populations to the north, however rabbit populations can survive only if there is access to free water, succulent vegetation, shaded warrens or warrens in calcareous soils (Parer and Libke 1985). The present southern distribution closely parallels that predicted by BIOCLIM, especially in the Northern Territory, South Australia and Queensland; this implies that a climate-related resource is limiting the distribution of *M. lagotis* in this part of their range. The northern limit of rabbit establishment also coincides neatly with the present southern limit of *M. lagotis* and it is suggested that rabbits (and foxes) are the climate-related resource limiting the southern distribution.

Rabbit and fox distributions cannot explain the decline of *M. lagotis* distribution in the northern part of its range so simply. *Macrotis lagotis* colonies are absent from most pastoral districts and have suffered a gradual decline over a number of decades in these areas. Since the 1940s, when *M. lagotis* was still widespread, there has been a multi-fold increase in stocking rates (Griffin and Friedel 1985) but this has been accompanied by a number of other changes such as a change in the burning regime. Both stocking (Newsome 1971) and fire (K. A. Johnson, pers. comm., 1985) have been suggested to affect the availability of bilby food resources. *Macrotis lagotis*, however, has also disappeared from areas where fire was particularly rare, as in mulga lands, and they still persist in some stocked regions.

Clearly, no single factor has affected the persistence of *M. lagotis*. Work now needs to focus on assessment of the relative impact of disease, foxes and other predators, rabbits, stocking rates and changing fire regimes in causing the decline in different landscapes.

ACKNOWLEDGEMENTS

This survey has been funded by the World Wildlife Fund (Australia) and administered by the Conservation Commission of the Northern Territory.

I would like to thank Drs K. A. Johnson and A. A. Burbidge for their support and guidance throughout the project. Dr B. D. Cooke (South Australian Vertebrate Pests Control Authority) and Dr D. King (Agricultural Protection Board of Western Australia) provided access to unpublished data and invaluable comment. I also extend special thanks to Dr M. R. Fleming, G. Griffin, P. K. Latz and especially Dr M. Stafford Smith who kindly read and criticized the draft of this paper.

REFERENCES

ATLAS OF AUSTRALIAN RESOURCES: maps with accompanying booklet commentaries, 1975. (Scale 1:5 000 000.) AGPS: Canberra.

BUSBY, J. R., 1985. *Bioclimatic prediction system — users manual.* Bureau of Flora and Fauna: Canberra.

BURBIDGE, A. A. AND FULLER, P. J., 1979. Mammals of the Warburton Region, Western Australia. *Records of the Western Australian Museum* **8:** 57–73.

DIXON, J., 1974. *Australian marsupials and monotremes John Gould.* Macmillan: Melbourne.

FINLAYSON, H. H., 1961. On central Australian mammals. Part IV. The distribution and status of Central Australian species. *Records of the South Australian Museum* **14:** 141–91.

FRITH, H. J., 1977. *Wildlife conservation.* Angus & Robertson: Sydney.

GIBSON, D. F., 1986. *A biological survey of the Tanami Desert in Northern Territory.* Government Printer: Darwin.

GLAUERT, L., 1950. The development of our knowledge of the mammals of Western Australia. *Journal of the Royal Society of Western Australia* **34:** 115–34.

GRIFFIN, G. E. AND FRIEDEL, M. H., 1985. Discontinuous change in Central Australia: some implications of major ecological events for land management. *Journal of Arid Environments* **9:** 63–80.

HEATHCOTE, R. L., 1977. Pastoral Australia. Pp. 254–58 in *Australia: a geography* ed by D. N. Jeans. Sydney University Press: Sydney.

JARMAN, P. J. AND JOHNSON, K. A., 1977. Exotic mammals, indigenous mammals, and land use. *Proceedings of the Symposium of the Ecological Society of Australia* **10:** 1–30.

JENKINS, C. F. H., 1974. The decline of the Dalgite *Macrotis lagotis* and other wildlife in the Avon Valley. *West Australian Naturalist* **12:** 169–72.

JONES, F. WOOD, 1924. *The mammals of South Australia.* Government Printer: Adelaide.

KIMBER, R., 1981. Unpublished notes on historical references to rare and feral animals in central Australia, prepared for the Conservation Commission of the Northern Territory.

KING, D. R. AND SMITH, L. A., 1985. The distribution of the European Red Fox *Vulpes vulpes* in Western Australia. *Records of the Western Australian Museum* **12**(2)**:** 197–205.

KREFFT, G., 1866. On the vertebrated animals of the lower Murray and Darling, their habits, economy, and geographical distribution. *Transactions of the Philosophical Society of New South Wales* **(1862–65):** 1–33.

KREFFT, G., 1871. *The mammals of Australia.* Government Printer: Sydney.

LE SOUEF, A. S. AND BURRELL, H., 1926. *The wild animals of Australia.* George Harrap & Co.: Sydney.

LEAKE, B. W., 1962. *Eastern wheatbelt wildlife — experiences of a W.A. naturalist.* Docket Books: Perth.

LONG, J. L., 1972. Introduced birds and mammals in Western Australia. Agricultural Protection Board Technical Series No. **1**.

LOW, W. A., 1983. Distribution of Rabbits in the Northern Territory. Pp. 2–3 in *Rabbit biology and control in arid and semi-arid lands* ed by W. A. Low. Conservation Commission of the Northern Territory: Alice Springs.

MITCHELL, J. L., 1982. *Vertebrate pests of Queensland.* Stock Routes and Rural Lands Protection Board, Department of Lands: Brisbane.

NEWSOME, A. E., 1962. Rabbit-eared Bandicoots or Bilbies. *Australian Natural History* **97:** 98.

NEWSOME, A. E., 1971. Competition between wildlife and domestic stock. *Australian Veterinary Journal* **47:** 577–86.

NORTHCOTE, K. H., 1960–68. *Atlas of Australian soils.* Sheets 1–10. CSIRO: Melbourne.

PARER, I. AND LIBKE, J. A., 1985. Distribution of Rabbit, *Oryctolagus cuniculus,* warrens in relation to soil type. *Australian Wildlife Research* **12:** 387–405.

PERRY, R. A., MABBUT, J. A., LICHFIELD, W. H. AND QUINLAN, T., 1963. General report on the lands of the Alice Springs area, Northern Territory, 1956–57. *CSIRO Land Research Series* No. **6**.

PHILPOTT, C. M. AND SMYTH, D. R., 1967. A contribution to our knowledge of some native mammals from inland Australia. *Transactions of the Royal Society of South Australia* **92:** 115–34.

REID, J., 1837. Description of a new species of the genus *Perameles (P. lagotis). Proceedings of the Zoological Society of London* (**1836**): 129–31.

RIX, C. E., 1978. *Royal Zoological Society of South Australia 1878–1978.* Royal Zoological Society of South Australia: Adelaide.

ROLLS, E. C., 1969. *They all ran wild.* Angus & Robertson: Sydney.

ROLLS, E. C., 1981. *A million wild acres.* Nelson: Melbourne.

SCARLETT, N., 1969. The Bilby, *Thylacomys lagotis,* in Victoria. *Victorian Naturalist* **86:** 292–94.

SCHLAGER, F. E., 1981. The distribution, status and ecology of the rufous rat-kangaroo, *Aepyprymnus rufescens,* in northern New South Wales. M.Sc. Thesis: University of New England: Armidale.

SHORTRIDGE, G. C., 1909. An account of the geographical distribution of the marsupials and monotremes of south-western Australia, having special reference to specimens collected during the Balston Expedition of 1904–1907. *Proceedings of the Zoological Society, London* **(1909):** 803–48.

SPENCER, W. B., 1896. Mammalia Part II Zoology. Pp. 7–17 in *Report on the work of the Horn Scientific Expedition to Central Australia* ed by W. B. Spencer. Dulau: London.

SPENCER, W. B. AND GILLEN, F. J., 1912. *Across Australia.* Macmillan & Co.: London.

STRONG, B. W., 1983. The invasion of the Northern Territory by the European Rabbit *Oryctolagus cuniculus. Conservation Commission of the Northern Territory Technical Report* No. **3**.

STRONG, B. W. AND LOW, W. A., 1983. A short note on the distribution of the Fox *Vulpes vulpes* in the southern Northern Territory. *Northern Territory Naturalist* **6:** 20–23.

THOMAS, O., 1888. *Catalogue of the Marsupialia and Monotremata in the collection of the British Museum (Natural History).* British Museum: London.

TROUGHTON, E. LEG., 1932. Revision of the Rabbit-bandicoots. *Australian Zoologist* **7:** 219–35.

WAKEFIELD, N. A., 1966. Mammals of the Blandowski Expedition to north-western Victoria, 1856–57. *Proceedings of the Royal Society of Victoria* **79:** 379–91.

WATTS, C. H. S., 1969. Distribution and habits of the rabbit bandicoot. *Transactions of the Royal Society of South Australia* **93:** 135–41.

WOOD, D. H., 1984. The Rabbit *(Oryctolagus cuniculus)* as an element in the Arid Biome of Australia. Pp. 273–86 in *Arid Australia* ed by H. G. Cogger and E. E. Cameron. Australian Museum: Sydney.

Habitats and diet of the Greater Bilby *Macrotis lagotis* Reid (Marsupialia: Peramelidae)

R. I. SOUTHGATE[1]

The Greater Bilby *Macrotis lagotis* is still comparatively widespread in Australia although the species has undergone a massive reduction in range since European settlement. Within the present range the distribution of populations is patchy and abundances vary considerably between habitat types. Several broad habitat types were identified with distinct vegetation, fire histories and landform characteristics. Colonies of *M. lagotis* in the south of their range occupied relatively sparse, inhospitable habitats on ridges or rises and fed predominantly on insect material. In the north of their range colonies were found to occupy a broader range of habitat types which included ridges, plains and alluvial areas and fire appeared necessary to reduce the density of ground cover in some of these habitat types. The diet of *M. lagotis* in the north was broader than in the south, and notably included items which were promoted by fire.

Key words: *Macrotis lagotis,* habitats, diet.

INTRODUCTION

THE Greater Bilby *Macrotis lagotis* is an omnivorous, nocturnally active marsupial. Of the five bandicoot species known to have existed in the arid and semiarid parts of Australia, only it has managed to survive. The others have not been positively recorded for at least 30 years and have disappeared without any understanding of their habitat requirements. Similarly, virtually no information exists on the diet and habitats used by *M. lagotis*. This paper presents an assessment of the range of habitat types where *M. lagotis* is presently found and of the important components of their diet.

MATERIALS AND METHODS

Distribution and habitat assessment

Habitat data was collected at occupied and formerly occupied sites from throughout the former and present range of *M. lagotis*. Precise localities for sites were drawn from museum and literature records. In addition, localities were obtained during extensive field surveys in 1983 and 1984 and from discussions with long standing residents of various areas. The distribution of these sites is shown in Figure 1 and the locality information in Table 1.

At each site, vegetation structure, age and floristics were described using a form of Muir's (1977) classification. This classification was broad enough to encompass the range of conditions expected. A similar format was used to map the vegetation of Australia (Beard 1974–79; Carnahan 1976).

Soils were described using Northcote's (1974) notation. Samples were taken to a depth of 1 m and compared with soil types for the area mapped in the *Atlas of Australian soils* (Northcote 1960–68). Morphology and topographical information on the site was described using McDonald, Isbell, Speight, Walker and Hopkins (1984) as a guide. The occurrence of fire and the type of land use and tenure were also noted.

The data collected at each site were used in clustering analysis in the programme TAXON (maintained on CSIRONET). For clustering, binary and numeric attributes were used and an information statistic was the metric. A comparison was made between the sites where recent activity, in the form of faecal material and diggings, was observed and those which were now inactive.

Diet analysis

During field trips dietary information was collected opportunistically. Feeding areas where *M. lagotis* had been digging in the soil and at the base of shrubs were inspected and faecal material found nearby was collected. At least two groups of faecal pellets were collected when an active site was located. The contents of approximately half of each group were reviewed under low powered magnification and the principal food items scored on their abundance. It was relatively easy to identify the remains of termites and ants and larger insects, as well as seeds and bulbs.

[1]Conservation Commission of the Northern Territory, P.O. Box 1046, Alice Springs, Northern Territory, Australia 5750.

Pages 303–09 *in* BANDICOOTS AND BILBIES, ed by J. H. Seebeck, P. R. Brown, R. L. Wallis and C. M. Kemper. Surrey Beatty & Sons, Sydney, 1990.

Table 1a. Location of sites and habitat attributes.

Site	Location	Latitude (S)	Longitude (E)	Activity	Soil texture	Mapped soil	Landform pattern	Landform unit	Unit morphology
LB1	Dampier Downs	18°21′	123°10′	0	LS	Dr2.42	lp	pl	f
LB2	Dampier Downs	18°26′	123°17′	1	S	Uc5.21	lp	pl	f
LG	La Grange	18°42′	121°53′	1	S	Uc5.21	gp	pl	f
RB	Wolf Creek	19°10′	127°40′	1	LS	Uc5.21	up	ri	c
XC	Christmas Creek	19°14′	125°43′	1	S	Uc4.2	up	ri	c
GD	Gordon Downs	18°46′	128°35′	1	CS	Gn2.13	gr	pe	m
BA	Balgo	20°22′	128°06′	1	SC	Uc5.21	gp	pl	f
SB	Salt Beef Lake	20°55′	130°23′	1	S	Um1.1	lp	pl	v
ME	Mount Edgar	21°11′	120°15′	1	S	Um5.51	up	pl	v
CR	Corruna Downs	21°28′	119°50′	1	S	Um5.51	up	pl	f
HO	Hooley Station	22°00′	118°14′	0	SCL	Um6.23	ur	pe	m
AN	Angkula Out-station	22°30′	134°10′	1	SCL	Gn2.12	gr	ri	c
SH	Sand Hill Camp	22°51′	119°41′	0	SCL	Gn2.11	lp	pi	f
TB1	Coorabulka Station	24°02′	140°05′	1	SC	Ug5.3	gr	pe	m
TB2	Coorabulka Station	24°03′	140°09′	0	SC	Ug5.3	gr	pe	m
DD	Davenport Downs	24°10′	140°39′	1	LMC	Ug5.3	up	pl	f
PP	Palparara	24°44′	141°06′	1	LMC	Ug5.3	up	pl	v
CC	Cacoory Ruins	25°11′	139°31′	0	SC	Dr1.33	ur	pe	f
DU	Durrie Downs	25°26′	139°37′	0	SC	Dr1.33	up	pl	f
BD	Mt Beadell	25°31′	125°10′	1	LS	Uc5.21	gr	pe	m
FZ	Warburton	25°56′	126°26′	1	LS	Uc5.21	gr	ri	m
TJ1	Tjirrkali Out-station	25°58′	125°33′	1	LS	Uc5.21	gr	ri	c
TJ2	Tjirrkali Out-station	25°59′	125°33′	1	LS	Uc5.21	gr	ri	c
PU1	Warburton	26°02′	126°30′	1	LS	Gn2.12	gr	ri	c
PU2	Warburton	26°02′	126°30′	1	LS	Gn2.12	gr	ri	c
PU3	Warburton	26°02′	126°30′	1	LS	Gn2.12	gr	rl	c

Table 1b. Location of sites and habitat attributes (cont'd).

Site	Land disturbance	Land tenure	Fire occurrence	Top strata	Bottom strata	Top strata (adapted Carnahan 1976)	Bottom strata
LB1	1	4	2	eLBp.a,ylSCr	gGLc	aSAc	tH
LB2	1	4	2	a,eSBr	gGLi	eLBr	tH
LG	1	2	1	a,lSBi	t,gHi	aSAc	tH
RB	1	4	2	a,lSBr	t,gHi	eLBr	tH
XC	1	4	2	a,ıSBi	t,gHi	eLBr	tH
GD	1	4	3	eLBr.mSBi	t,gGLi	eLBr	tH
BA	0	1	1	a,lSBi	t,gHi	aSAr	tH
SB	0	1	3	mSBi	tHi	aSAi	tH
ME	1	4	1	aSAi.zSCi	tHi	aSAr	tH
CR	1	4	1	aSBb	tHi	aSAr	tH
HO	1	4	4	sSAc	t,gHi	aSAi	gGL
AN	1	3	3	aSAr.a,ySBi	gGLi	aSAi	gGL
SH	1	4	4	aSAi.a,ySBi	t,gGLb	aSAi	gGL
TB1	1	4	5	—	z,gGLi	—	zGLr
TB2	1	4	5	—	z,gGli	—	zGLr
DD	1	4	5	—	gGLi	—	gGLi
PP	1	4	3	—	gGLi	—	gGLi
CC	1	4	4	—	z,gGLi	—	zGLr
DU	1	4	4	—	z,gGLi	—	zGLr
BD	0	1	5	aSBr	t,gHr	aSAi	tH
FZ	0	1	5	aSBi	t,gHi	aSAi	tH
TJ1	0	1	5	aSAp.aSBr	t,gHr	aSAi	tH
TJ2	0	1	5	aSAp.aSBr	t,gHr	aSAi	tH
PU1	0	1	5	aSAp.aSBr	gGLr	aLBi	gGL
PU2	0	1	5	eLBp.aSCi	t,gHi	aLBi	gGL
PU3	0	1	5	aSAi.aSBi	t,gHi	aLBi	gGL

Activity: 1=recent sign, 0=formerly active. Soil texture: S=sand, LS=loamy sand, CS=clayey sand, SCL=sandy clay loam, SC=sandy clay, LMC=light medium clay. Mapped soil from Northcote (1960–68). Landform pattern: lp=level plain, gr=gently undulating rises, gp=gently undulating plain, up=undulating plain, ur=undulating rises. Landform unit: pl=plain, pe=pedimont, ri=ridge. Unit morphology: pl=plain, c=crest, m=mid slope, v=vale. Land disturbance: 0=no effective disturbance, 1=grazed by ungulates. Land tenure: 1=Aboriginal land undisturbed, 2=Aboriginal land disturbed, 3=European lease abandoned, 4=European lease active. Fire occurrence: 1=<1 year, 2=1–2 years, 3=2–4 years, 4=4–10 years, 5=>10 years. Floristic component: a=*Acacia* spp., l=*Grevillea* spp., m=*Melaleuca* spp., e=*Eucalyptus* spp., y=mixed shrub species, z=mixed forb species, g=mixed soft grasses. t=*Triodia* or *Plectrachne* spp. Life form/height: LA=trees 5–15 m, LB=trees <5 m, SA = shrubs >2 m, SB=shrubs 1–2 m, SC=shrubs <1 m, H=hummock grasses, GT=bunch grass >0.5 m, GL=bunch grass <0.5 m, —=no life form, .=combination. Density of canopy cover: d=dense canopy projective foliage cover 70%, c=mid dense canopy p.f.c. 30–70%, i=incomplete canopy p.f.c. 10–30%, r=rare but conspicuous p.f.c. <10%, b=barren vegetation largely absent p.f.c. <<10%, p=scattered groups no definite foliage cover. Formulae examples: tree savannah=eLAr gGLc, shrub steppe=aSBr tHi, open grassland= — gGLi.

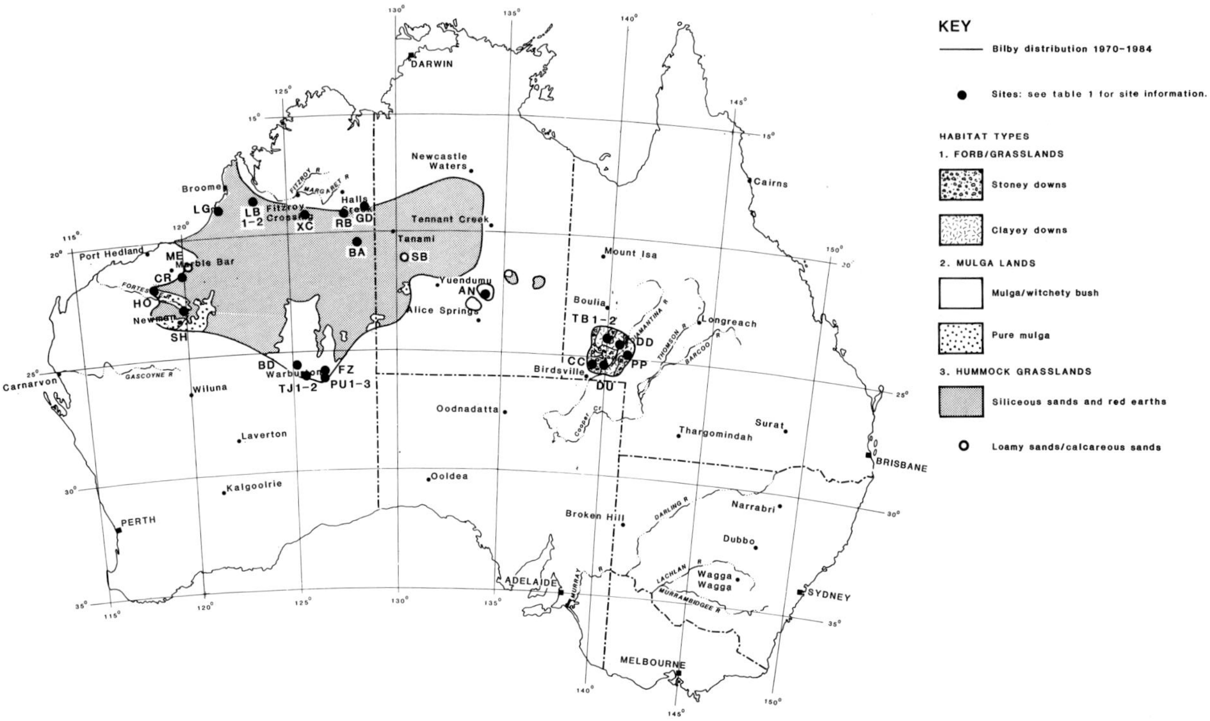

Fig. 1. Present distribution of *Macrotis lagotis*, including habitat types and sites used in analysis.

RESULTS

Habitat description

Macrotis lagotis populations were found in areas with landform patterns ranging from level plains to undulating plains and rises (Table 1). The slope at a colony site never exceeded about 6% (gently inclined). Evidence of the presence of *M. lagotis* was found on crests of ridges, slopes, flats, depressions and watercourses.

Burrows of *M. lagotis* were located in soils ranging in depth from 0.6 m to the deep siliceous sands (>2 m) on some plains. Textures of these soils ranged from coarse sand to light medium clay. Samples were taken from uniform texture profiles, non-calcareous gradational soils and duplex soils. At a number of sites, soils described differed from those mapped for the same area and presented at a scale of 1:2 000 0000 in the *Atlas of Australian soils* (Northcote 1960–68). Soils were generally more uniform in texture where samples were taken.

Vegetation types ranged from a woodland of low trees (<10 m) with an *Acacia* spp. rich understorey in the north of *M. lagotis* distribution, through a variety of shrub steppe communities, to tussock/forb grasslands in

Table 2. Occurrence of food items in *Macrotis lagotis* scats.

Site	n	Insect larvae	Other insect	Termite	Ant	Bulb CYBUL	Seed YAAUS	Other seed	Plant fibre
LB	3	2	4	3			1		
LG	3	1	2					3	
RB	2	4	5	3			1	2	
XC	3		5	4			1	2	3
BA	5	2	4	3				1	5
SB	5		2			1			
ME	4		3			1		2	
CR	2		1	3					4
AN	3	2	3	4				1	
TB	5		2	1				3	4
DD	4	1	2					3	4
PP	5	4	2			1		3	
BD	5	1			3			4	2
FZ	4	1	2	3				4	
TJ	4	1	3	2				5	4
PU	14	1	4	5				2	3

1=most common food item. 5=least common food item. CYBUL=*Cyperus bulbosus*. YAAUS=*Yakirra (Panicum) australiense*.

south-west Queensland. Floristic and structural composition varied considerably throughout the range of vegetation types.

Alteration of habitat type by burning was most evident in the north of the present range of *M. lagotis*. In the south, it occurred in some habitats which were so sparsely vegetated that fires were rare; this included sites near Warburton and some in south-west Queensland. In some areas it overlapped with livestock. These were always cattle and always in low numbers.

Variation between sites

Clustering analysis separated the sites into three main groups. The separation of group 1 from group 2 was in response to topography, soil pH and the presence of *Acacia* spp. in the shrub component. This effectively divided the sites which were rich in *Acacia* spp. on ridges or rises from the other sites.

Separation of group 2 from group 3 was based mainly on soil pedality, ground and shrub cover and soil carbonate reaction. Sites with sandy textured soils and higher shrub and ground cover in parts of the Tanami and Great Sandy Desert were thus distinguished from more clayey soils; the latter had little cover especially in the shrub component. There were no features in the analysis which separated occupied sites from the formerly occupied sites.

Diet

The results of faecal analysis are shown in Table 2. Scats from a particular site usually contained a mixture of food items, but in most cases one food item predominated.

Colonies from Warburton were primarily feeding on Cossidae larvae from the roots of *Acacia kempeana* and *A. rhodophloia*. In south-west Queensland, the main food items varied. *Cyperus bulbosus* bulb remains predominated in scats from a site in a plateau depression area, terrestrial insect remains were most important at another site, and termites at the third. Where *M. lagotis* occurred on the plains in the northern parts of their distribution, seeds, especially from *Yakirra (Panicum) australiense,* appeared to play a more important role in their diet. The remains of *C. bulbosus* were most commonly found in scats from sites in alluvial areas. Large terrestrial insect remains were present at nearly all the sites but generally only in minor proportions.

DISCUSSION

The habitat characteristics and diet described above indicate that *M. lagotis* is a highly flexible animal and many forms of habitat must have been acceptable to it in the past. This is supported by earlier reports of their habitat selection (Spencer and Gillen 1912; Jones 1924; Le Soeuf and Burrell 1926; Troughton 1932; Gould 1845) and is not surprising considering their former broad distribution throughout semiarid and arid Australia. Gould (in Dixon 1974) found they were "tolerably abundant" over the grassy districts of the Swan River (Western Australia) interior where ever the ground was suitable for their burrows. In South Australia, *M. lagotis* was once a familiar species in the more fertile regions (Jones 1924) but was also abundant in habitats in the more arid and less fertile regions of the Nullarbor (Le Soeuf and Burrell 1926) and northern ranges (Finlayson 1961). In the eastern part of Australia they were found throughout the woodlands and plains on the slopes of the Great Dividing Range (Troughton 1932; Marlow 1958).

In more recent work on colonies of *M. lagotis* around Yuendumu, Watts (1969) considered that two-thirds of a 777 km^2 area was suitable for its habitation, the rest being rocky and unsuitable. There was a tendency for burrows to be located in the more open areas and along water courses. He found that their diet consisted of underground tap roots, bulbs (especially of *C. bulbosus*) and fungi *(Endergone)*.

At Warburton, Smyth and Philpott (1968) found that *M. lagotis* showed some preference for certain habitats, but concluded that it only used a small percentage of large homogeneous areas which could support them. Similar conclusions have been reached by broad scale surveys which have come into contact with its populations (Johnson 1979; Youngson, Henry and McKenzie 1981; McKenzie and Youngson 1983; Gibson 1986).

Smyth and Philpott (1967) reported that *M. lagotis* was consuming primarily termites and Lepidoptera larvae. In the Tanami Desert, Johnson (unpublished, 1980) found that *Yakirra australiense* was the principal food source.

A factor common to all the habitats examined during the present survey was the low amount of ground cover, which is thought to be necessary to allow *M. lagotis* to have high mobility for foraging. Sparseness of vegetation appeared to be induced by topography and soils and fire. Bilbies were restricted to fewer habitat types in the southern part of their range compared to the north. In the south, most of the country occupied was elevated and away from major drainage systems.

On the basis of landform, soils, vegetation characteristics, fire frequency and important food resources, the habitat presently occupied by *M. lagotis* was divided into six habitat types in three major associations as presented in Table 3.

Table 3. Separation of habitat types presently used by *Macrotis lagotis.*

Vegetation and soil	Characteristics	Representative area
1. Sparse grass/forbland a. Gravelly desert loams b. Deep brown calcareous clays	Uplands and hills, low fire frequency	South-west Queensland
2. Mulga lands a. Mulga/witchetty bush, lateritic red earths b. Pure mulga stands	Fire sensitive ridges and rises	Gibson Desert, scattered north of Alice Springs; in Pilbara, Great Sandy and Tanami Desert
3. Hummock grasslands a. Siliceous sands and red earths b. Calcareous and siliceous loams	Plains and alluvial areas, high fire frequency	Tanami, Great Sandy Desert, Dampier Land, Kimberley; East Pilbara

Each of the associations is briefly described below and their extent is shown in Figure 1.

1. Sparse grass/forbland

Recent evidence of *M. lagotis* occupation was found in two distinct habitat types in parts of south-west Queensland between Birdsville and Boulia. Both these habitat types were found away from the major drainage systems. Fire was infrequent and unlikely to happen as often as once every 10 years due to the normal sparseness of ground cover or effect of grazing. Foxes *Vulpes vulpes* and rabbits *Oryctolagus cuniculus* were reported to be rare in the vicinity.

a. Low lateric hills: Isolated populations of *M. lagotis* were found in low hills covered with a mixture of ironstone, siliceous and lateritic gravels. Soils are shallow to deep desert loams and support seasonally variable short grasses and forbs with very sparse low shrubs. The areas where *M. lagotis* was found were generally surrounded by gently undulating plains which showed no evidence of occupation. Signs of burrow systems, probably that of *M. lagotis* indicated that occupation of this habitat was once much more extensive than it is now.

b. Raised clayey plains: More continuous populations of *M. lagotis* were found on flat to very gently undulating plains with occasional internal drainage channels. The plains were slightly elevated and composed of deep, brown, self-mulching, calcareous cracking clays with occasional sandstone outcrops. Vegetation was predominantly mitchell grass *Astrebla* sp. with a mix of other grasses and forbs.

2. Mulga scrub/woodlands

This association was characterized by the presence of Mulga *Acacia aneura.* Mulga is very susceptible to fire and relies on seed for regeneration. Wildfire was thus considered an infrequent event (intervals in the order of 20–50 years). Rabbits and foxes were reported in low numbers from the surrounding areas. Two types of habitat have been identified based on the presence of other shrubs and the ground cover.

a. Mulga/witchetty bush: Isolated or continuous but sparse *M. lagotis* populations were found on the rises or ridges of gently undulating plains in areas north of Warburton and north of Alice Springs. Soils are red earths with large quantities of lateritic gravel. Mulga was sparse and stunted, and accompanied by shrubs which provided a source of witchetty grubs, such as *A. kempeana, A. hilliana* or *A. rhodophloia.* Spinifex (*Triodia* spp.) and soft grasses were clumped in distribution, providing little fuel to carry a fire. The unpalatability of the spinifex and the fodder species growing between tussocks has meant that these areas are of little use for grazing stock. This community was distinct from surrounding spinifex dominated communities on the sandier soils and from the denser belts of mulga occurring along the drainage lines.

b. Pure mulga stands: There was evidence that *M. lagotis* occupied almost pure stands of mulga along the Hammersley Range and areas north of Alice Springs. Remains of what could have been burrows of *M. lagotis* were examined in three areas and there were other reliable reports. In the areas examined the mulga was sparse with an understorey of soft grasses and spinifex. Surrounding the belts of mulga in the Pilbara were stony ranges with mallee woodland. These areas offer more potential for stock grazing than the mulga/witchetty association; however, there was little sign of grazing at the locations examined.

3. Hummock grassland mixed shrub or woodland steppe

This association extended over a vast proportion of the Tanami, Great Sandy, southern

Kimberley and Dampier Land. *Macrotis lagotis* populations were generally isolated, but sometimes continuous over an area of several square kilometres.

The effects of fire and rapid biomass accumulation after rainfall are important influences in the creation of suitable habitat. The mean interval between wildfires would be in the order of 4–10 years in this area. Summer rainfall predominates in the parts of the hummock grasslands where *M. lagotis* occurs, and substantial ground coverage can develop with the high plant growth rates. Fire acts to reduce ground cover and create a successional stage where a different range of food resources can be produced. Seed production from plants such as *Yakirra australiense* which grow after fire are probably critical to make habitat suitable for occupation by *M. lagotis* throughout the shrub steppe and woodlands. Two habitat types were delineated primarily on their ability to produce foods important to *M. lagotis*.

a. Siliceous and earthy sands and red earths: This habitat type includes the sandplains and dunefields where soils have high permeability and are of low fertility, and where there is an absence of a defined drainage system. Seeds were of great importance in the diet.

b. Calcareous and siliceous loamy soils: Creek beds and flats, drainage systems, salt lake systems and any alluvial areas which support an abundance of *Cyperus bulbosus* are included in this habitat type. These systems generally retain soil moisture for longer than the previous habitat type and, although the area they cover is negligible compared to that of the hummock grasslands, they appear to be an important refuge for *M. lagotis*.

SUMMARY AND CONCLUSION

The present distribution of *M. lagotis* indicates that it is still comparatively widespread although it has undergone a massive reduction in range since European settlement.

Much speculation has occurred as to why the range has diminished. Predation by foxes and habitat alteration due to rabbits (Jones 1924; Troughton 1932; Watts 1969) have been implicated as have the effects of pastoral activities (see Southgate, this volume). It has become apparent during this study that other factors which dictate the density of ground cover and food resources are also important to consider.

Within its present distribution, *M. lagotis* is patchily distributed and occupies a range of landforms and habitat types. Fire, landform and soil are implicated as agents necessary to create suitable habitat. In the southern part of the range it occupies relatively sparse, inhospitable habitats on ridges and rises, and feeds predominantly on insects. In the north of its range *M. lagotis* can occupy a broader range of habitats which includes ridges, plains and alluvial areas but fire is often necessary to clear the understorey. The diet in the northern part of the range is similarly broad but notably contains items which are promoted by fires.

ACKNOWLEDGEMENTS

I would like to thank the World Wildlife Fund for financial support of the project and Drs K. A. Johnson and A. A. Burbidge for their supervision. I am grateful to P. Fuller who assisted with some of the field work and Dr A. Kerle who commented on the initial draft of this paper and also Dr M. Stafford Smith who provided valuable comments on the final draft.

REFERENCES

BEARD, J. S., 1974–79. *Vegetation survey of Western Australia.* University of Western Australia Press: Nedlands.

CARNAHAN, J. A., 1976. Natural vegetation. In *Atlas of Australian resources* 2nd Series. Division of National Mapping: Canberra.

DIXON, J., 1974. *Australian marsupials and monotremes John Gould.* Macmillan: Melbourne.

FINLAYSON, H. H., 1961. On central Australian mammals. Part IV. The distribution and status of Australian species. *Records of the South Australian Museum* **14:** 141–91.

GIBSON, D. F., 1986. *A biological survey of the Tanami Desert in the Northern Territory.* Government Printer: Darwin.

JOHNSON, K. A., 1979. Report on a survey for Bilby *Macrotis lagotis* in the Tanami Desert 7–13 August, 1979. Conservation Commission of the Northern Territory Internal Report 61/199.

JONES, F. WOOD, 1924. *The mammals of South Australia.* Government Printer: Adelaide.

LE SOEUF, A. S. AND BURRELL, H., 1926. *The wild animals of Australia.* George Harrap & Co: Sydney.

MARLOW, D., 1958. A survey of the marsupials of N.S.W. *CSIRO Wildlife Research* **3:** 71–114.

MCDONALD, R. C., ISBELL, R. F., SPEIGHT, J. G., WALKER, J. AND HOPKINS, M. S., 1984. *Australian Soil and Land Survey.* Inkata Press: Melbourne.

MCKENZIE, N. L. AND YOUNGSON, W. K., 1983. Mammals. Pp. 62–93 in *Wildlife of the Great Sandy Desert, W.A.* ed by A. A. Burbidge and N. L. McKenzie. Research Bulletin of Western Australia **12**.

MUIR, B. G., 1977. Biological survey of the Western Australian wheatbelt. *Records of the Western Australian Museum.* Supplement No. 13.

NORTHCOTE, K. H., 1974. *A factual key for the recognition of Australian Soils.* Rellim Technical Publications: Adelaide.

NORTHCOTE, K. H., 1960–68. *Atlas of Australian soils.* Sheets 1-10. CSIRO: Melbourne.

SMYTH, D. R. AND PHILPOTT, C. M., 1967. A field study of the Rabbit Bandicoot. *Transactions of the Royal Society of South Australia* **93:** 135–41.

SPENCER, W. B. AND GILLEN, F. J., 1912. *Across Australia* Vol. 1 and Vol. 2. Macmillan & Co.: London.

TROUGHTON, E. LEG., 1932. Revision of the rabbit-bandicoots. *Australian Zoology* **7:** 219–35.

WATTS, C. H. S., 1969. Distribution and habits of the rabbit bandicoot. *Transactions of the Royal Society of South Australia* **93:** 135–41.

YOUNGSON, W. K., HENRY, J. AND MCKENZIE, N. L., 1981. Pp. 39–45 in *Wildlife of the Edgar Ranges, S.W. Kimberlies, W.A.* Wildlife Research Bulletin of Western Australia No. **10**.

Reproduction and life history of *Isoodon obesulus* in Victorian heathland

B. LOBERT[1] and A. K. LEE[1]

Most populations of *Isoodon obesulus* breed seasonally, with births occurring between July and December in Victoria, and July to February in Tasmania. Onset and cessation of breeding are highly predictable in southern Victoria, but less so in Tasmania. Litter size varies from one to six. Gestation period is less than 15 days and lactation about 60 days. Females produce up to three litters per breeding season in Victoria, and four litters per breeding season in Tasmania. Litter size and the numbers of litters produced per year appear to be related to food abundance. Dispersal and mortality in Victorian heathlands are highest among juveniles and young adults of both sexes. Maximum longevity of wild animals is about 3.5 to 4 years. In Victoria, sexual dimorphism becomes apparent at about 12 months of age. Females at Cranbourne first breed in the breeding season following birth, whereas in Tasmania females may breed in the season of birth. Tasmanian animals grow faster and weigh more than Victorian animals as adults.

Tactical responses of *I. obesulus* include facultative variations in length of the breeding season, the number of litters produced by females within the breeding season and in litter size.

Key words: Demography, breeding season, reproduction, life history.

INTRODUCTION

THE Southern Brown Bandicoot *(Isoodon obesulus)* occurs in a variety of habitats in coastal southeastern Australia, most of which are prone to frequent disturbance by fire. It is for this reason that the life history strategy of this species has attracted some attention (Stoddart and Braithwaite 1979). Knowledge of this strategy is based upon two substantial studies, Heinsohn (1966) and Stoddart and Braithwaite (1979), and a number of smaller contributions, including McKenzie (1967), Watts (1974), Opie (1980) and Maloney (1982). This paper provides additional information for the Victorian population of *I. obesulus* investigated by Stoddart and Braithwaite (1979) and Opie (1982), and reviews our knowledge of the life history strategy and tactics of this species.

STUDY AREA AND METHODS

This study was conducted at the Royal Botanic Gardens and National Herbarium Annexe (38°07′S, 145°14′E, elevation approximately 85 m), 2 km SE of Cranbourne, 50 km SE of Melbourne.

The area experiences mild winters and warm summers, with monthly minimum and maximum temperatures of 4.6–13.6°C in June and 15.2–26.3°C in February. The mean annual rainfall is 800 mm and rainfall is evenly distributed throughout the year, but varies considerably between years.

The study involved live trapping a roughly rectangular grid (5.1 ha) enclosing 3–4 and 14–18 year-old heath vegetation (Lobert 1985). The grid was trapped for four nights each month. All bandicoots had their weight, left pes length and location on the grid recorded on each capture. In males the size (length × width) of the scrotum, and in females the condition of the pouch, the size of the nipples and size and number of pouch young, were recorded at each capture.

Animals caught prior to the breeding season following their birth were classified as juveniles. Most animals could reliably be assigned to a year cohort using the suite of characters described in Table 1.

Litter size was determined by examining the pouch for young, and by counting the number of enlarged teats present after weaning. J. C. Merchant (pers. comm.) found that, in some instances, newborn young of *Isoodon macrourus* attached to enlarged teats which had been suckled by the previous litter, and also that young can suckle more than one teat. So, counts of enlarged teats are estimates of litter size.

Individual survival was expressed as the percentage of trapped animals of known age surviving at successive breeding and non-breeding seasons following their season of birth. Minimum survival of the 1981 cohort was expressed as the percentage of trapped animals born in 1981 which survived at each subsequent

[1]Department of Zoology, Monash University, Clayton, Victoria, Australia 3168.

Pages 311–18 *in* BANDICOOTS AND BILBIES, ed by J. H. Seebeck, P. R. Brown, R. L. Wallis and C. M. Kemper. Surrey Beatty & Sons, Sydney, 1990.

Table 1. Criteria used to age *Isoodon obesulus* at Cranbourne based on individuals of known age resident on the study grid.

		Weight (g)	Pes (mm)	Age (years)
Males:	juveniles	<500	<53	<1
	adults	450–600	52–55	1–1.5
	”	550–750	54–56	1.5–2
	”	700–950	55–59	2–2.5
	”	900–1 200	58–61	>2.5
Females:	juveniles	<400	<52	<1
	adults	350–500	51–53	1–1.5
	”	450–650	52–55	1.5–2
	”	550–750	54–57	2–2.5
	”	700–850	>56	>2.5

trapping session. A further measure of survival was the period of residency on the grid, which was calculated as the period between the first and last capture. All *t* tests were two-tailed.

RESULTS

Size and growth

Growth rates obtained from a hand-reared litter (Table 2) show that 60-day old animals (age at weaning) weigh between 105 and 140 g. The mean weights of juveniles at first capture were 225.7 g (SE±29.0, n=10) for males, and 229.0 g (±37.0, n=7) for females. Males in the trappable population appeared to grow throughout life whereas growth in females tended to asymptote at about 600 g, when females were 18 to 24 months of age (Figs 1 and 2). Dimorphism between the sexes was apparent in the cohort born in 1981 by April 1982, three months prior to this cohort's first breeding season (Figs 1 and 2). In males there was little or not overlap in weight between successive cohorts (Fig. 1). However, females born in 1981 had similar body weights to females born in 1980 by 12 to 18 months of age (Fig. 2). The heaviest female and male caught weighed 730 g and 1 150 g, respectively. There was no indication of seasonal change in the body weights of males or females.

Table 2. Growth rates of a hand reared litter of *Isoodon obesulus*. Pouch young were about 2 months old at the end of December 1982.

Sex	Weight (g) 14 Dec 1982	20 Dec 1982	4 Jan 1983	Pes (mm) 22 Dec 1982
Male	47	59	140	38
Female	48	52	105	36
Female	43	51	130	38

Minimum survival

Survival of males was higher than females during the first breeding season, giving them an advantage in survival which persisted until the fourth non-breeding season (Fig. 3). Disapppearance of males and females from the population was highest prior to the commencement of the first breeding season. The minimum survival rates of the 1981 cohort, shown in Fig. 4, are similar to survival rates for the population over the entire study period, with a higher male survival rate initially, and survival of both sexes lowest during the non-breeding seasons. No animals were known to survive into their fourth non-breeding season, that is, maximum known longevity was 3.5 years for both males and females.

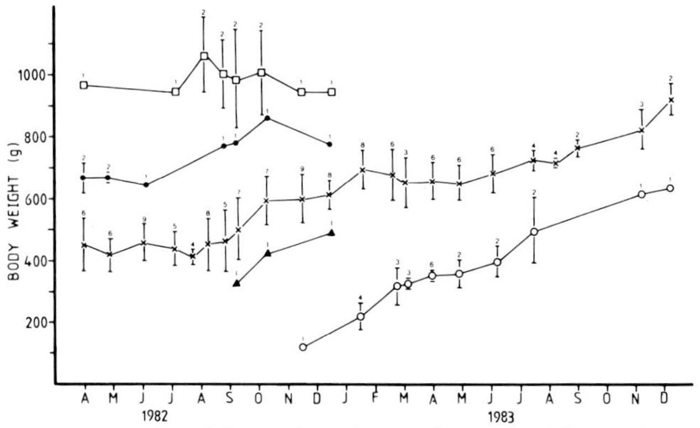

Fig. 1. Mean weight (g±SD) of year cohorts of resident male *I. obesulus* at Cranbourne. Numbers above standard deviation bars represent the number of individuals. Male 21 was part of the 1982 cohort, but was born much earlier then the other members of this cohort. □ 1979 cohort; ● 1980 cohort; × 1981 cohort; ○ 1982 cohort; ▲ male 21 of 1982 cohort.

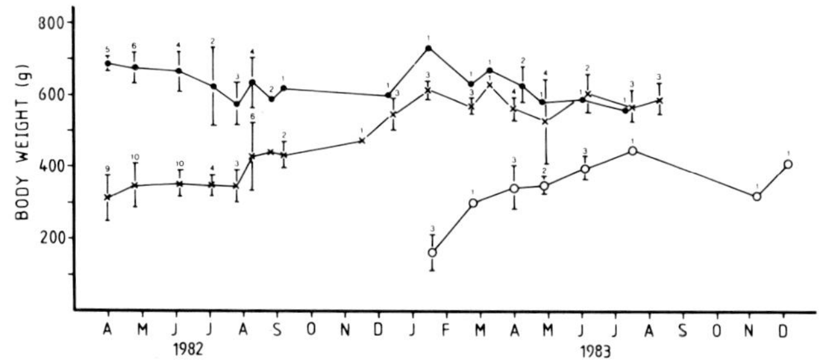

Fig. 2. Mean weight (g±SD) of year cohort of resident female *Isoodon obesulus* without pouch young, at Cranbourne. Numbers above standard deviation bars represent the number of individuals. See Fig. 3 for explanation of symbols.

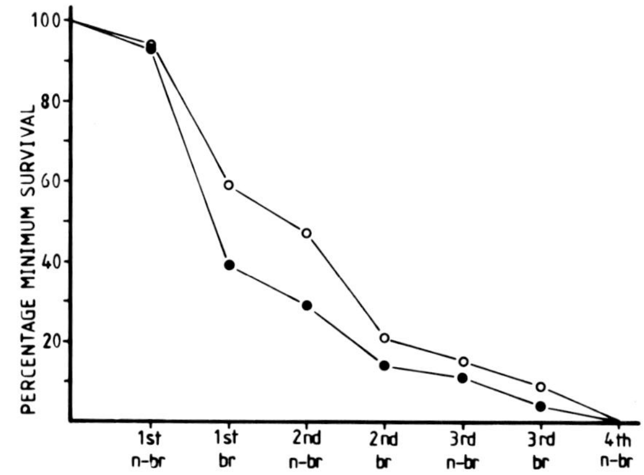

Fig. 3. Percentage minimum survival of *Isoodon obesulus* at Cranbourne. ○ males, ● females, n-br refers to the non-breeding season and br refers to the breeding season. Initial number of individuals was 34 males, 28 females.

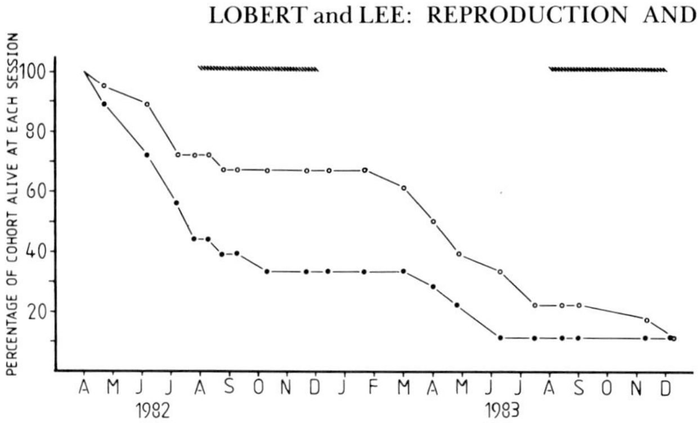

Fig. 4. Percentage survival of 1981 cohort of *Isoodon obesulus* after May 1982. ○ males, ● females. Horizontal hatched bars denote the breeding season. Initial number of individuals was 18 males, 18 females.

Both sexes remained in the trap record for similar periods of time, with mean values of 192.7 days (SE±30.8, n=31) for males, and 171.4 days (±34.7, n=25) for females (t=0.46, NS).

Reproduction

Female *Isoodon obesulus* reached reproductive maturity in the breeding season following their birth and at a minimum of about seven months of age. Reproduction was seasonal with females first observed carrying pouch young during August of each year (Fig. 5). With a gestation period of less than 15 days (B. Lobert, pers obs.), similar to the 12.5 day gestation period of *I. macrourus* (Lyne 1974), and a pouch life lasting two months (Stoddart and Braithwaite 1979; B. Lobert, pers. obs.), mating probably began in late July or early August. However, a juvenile male weighing 324 g was caught on 5 September 1982, indicating that at least one female had produced a litter in late May or early June. The latest that pouch young were observed was in mid December 1982 and early December 1983. No trapping occurred after early December 1983, when only two females carried pouch young (approximately 7–8 mm crown–rump length and 50 mm head–body length, respectively).

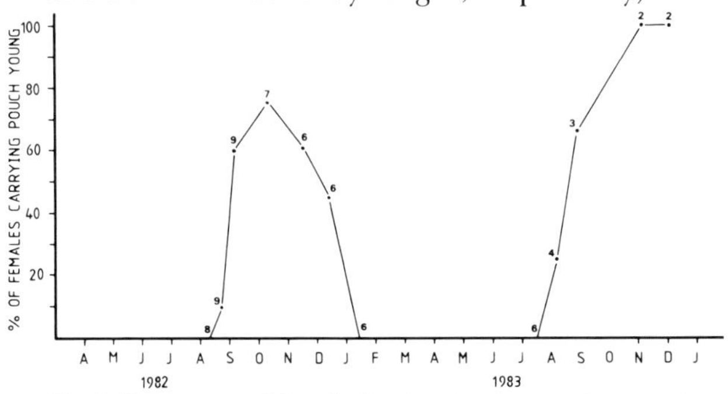

Fig. 5. Percentage of female *Isoodon obesulus* carrying pouch young. Numbers above points denote the number of resident adult females on the grid.

Table 3. Relationship between litter size and time of birth during the breeding season of *Isoodon obesulus*.

Time of	1982							1983			
year	No.2	5	10	13	15	19	20	5	13	25	31
Jul–Aug	3			3		3		1	3	1	
Sept–Oct		2	2	3	3	2	3,3	2			2
Nov–Dec					3			1			

It was estimated that 31 young were weaned in 1982 by eight females from 12 litters, while in 1983, it was estimated that 10 young were weaned by four females from six litters. Hence, the mean litter size of 2.58 (SE±0.19, n=12) neonates in 1982 was significantly larger (t=2.99, df=17, P<0.05) than the mean litter size of 1.57 (±0.32, n=6) neonates in 1983. There was no significant difference in the number of litters produced per female between 1982 (1.50±0.19, n=8) and 1983 (2.00±0.58, n=3). There was no indication that litters produced early in the breeding season were larger than those produced later in the season (Table 3), nor that heavier or older females produced larger or more litters in a breeding season than younger females (Table 4, 5 and 6), although the sample sizes were small.

Table 4. Relationship between female body weight of *Isoodon obesulus* prior to a birth of a litter and litter size [r=0.53 (NS) for 1982, r=0.44 (NS) for 1983].

Year	Litter size	Body weight (g)			
1982	1	450			
	2	450	540	550	
	3	400	550	550	580
		620	690	700	700
1983	1	400	570	620	
	2	550	650		
	3	640			

Table 5. Relationship between female body weight of *Isoodon obesulus* prior to birth of a litter and the number of litters produced in 1982 and 1983 [r=0.51 (NS) for 1982, r=0.34 (NS) for 1983].

Year	Number of litters per female	Body weight (g)			
1982	1	450	540		
	2	475	565	580	700
1983	1	500	640		
	3	620			

Table 6. Relationship between the age of the mother and litter size and the number of litters produced during a breeding season of *Isoodon obesulus*.

Age (years)	Litter size	Number of litters
1	1 1 1 2 2 2 3 3 3	1 1 1 1 1 2 2
2	1 1 2 3 3 3 3 3 3	1 1 2 2 3
	t=1.08 (NS)	t=1.35 (NS)

Scrotum lengths and width were strongly positively correlated (r=0.89). Scrotum length showed marked seasonal fluctuations, reaching maximum size in the middle of the breeding season (Fig. 6). The scrotum of juvenile males attained adult size by May or June in the year after birth, that is, just prior to the breeding

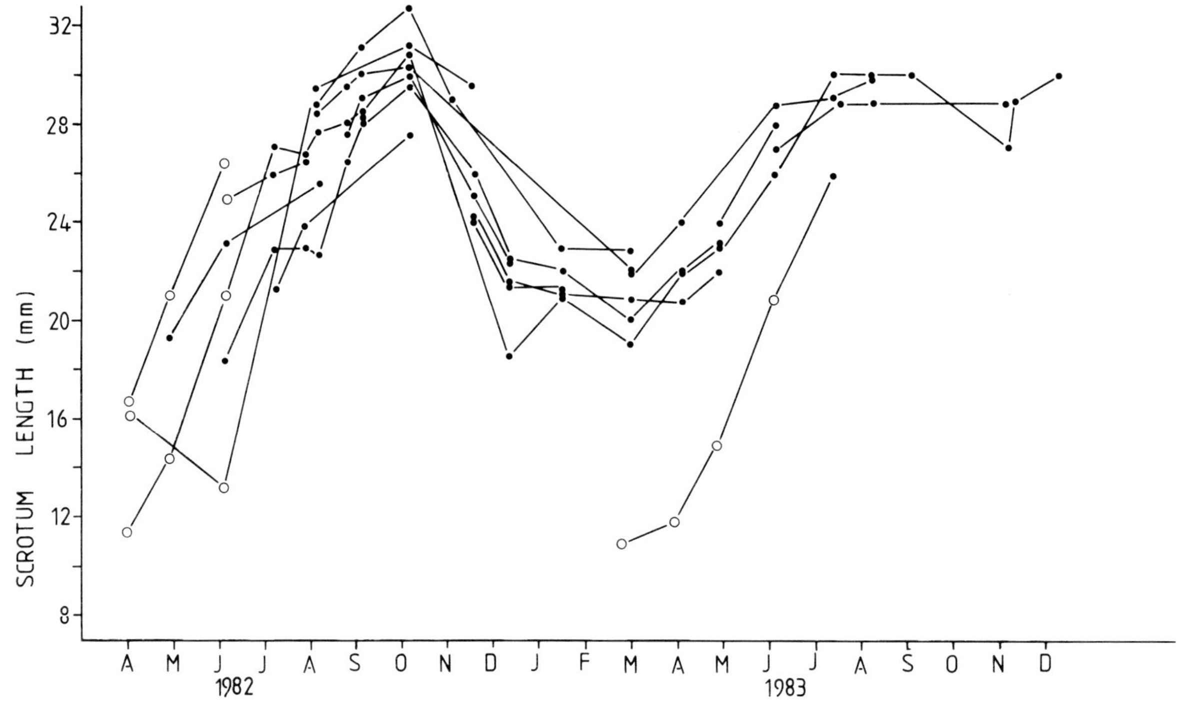

Fig. 6. Seasonal fluctuation in scrotum length of 13 male *Isoodon obesulus.* Asterisks represent juveniles and closed circles represent adults.

season. Scrotum length decreased again after about November and reached a minimum size between January and April. All males except one (male 5) showed this pattern of seasonal fluctuation in scrotum length. This individual had only a single descended testis.

DISCUSSION

Breeding season

Breeding of *I. obesulus* at Cranbourne occurs between July and December, coinciding with the annual peak in abundance of major food items in the soil (Fig. 7). At Healesville, Victoria, Taylor (1965) found breeding in *I. obesulus* restricted to the period August to November (the study ended in November), that is, within the period of breeding at Cranbourne. In Tasmania, the breeding season is 2–3 months

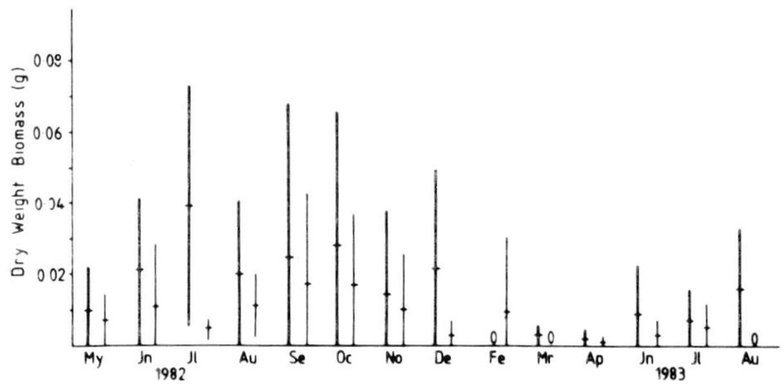

Fig. 7. The level of food abundance per soil core for two structurally distinct vegetation types: low, open 3–4 year-old heath (thin lines) and tall, dense 14–18 year-old heath (thick lines). Horizontal bars represent the mean and vertical bars represent ± one standard deviation (from Lobert 1985).

longer than at Cranbourne, with pouch young present between July and February; breeding again appears to coincide with the time of maximum food abundance (Heinsohn 1966). The pattern of breeding in *I. obesulus* in heathland in southwestern Victoria is not clear; breeding has been reported as year round (R. W. Braithwaite in Stoddart and Braithwaite 1979) and seasonal (A. Cockburn, pers. comm.). An interesting pattern of breeding exists in populations in The Grampians in central western Victoria. *Isoodon obesulus* inhabiting heathland in the Victoria Valley breed all year round, whereas on Mt William (1 167 m), *I. obesulus* breeds seasonally (A. Cockburn, pers. comm.). These observations suggest that some populations in southeastern Australia breed year-round, whereas others breed seasonally, and that seasonal breeding, where it occurs, coincides with peaks in food abundance.

In seven years of study, the breeding season at Cranbourne has always commenced in July/August and finished in December/January (Stoddart and Braithwaite 1979; A. Opie, pers. comm., this study). During that time there have been years of above average rainfall (e.g., 1974) and below average rainfall (e.g., 1982; Fig. 8). Thus, despite marked variation in rainfall, and in the seral stage of the heathland occupied by the bandicoots, the timing of breeding in this population was highly predictable between years.

Seasonality of breeding in bandicoots tends to increse with latitude. Barnes and Gemmell (1984) examined the relationship between

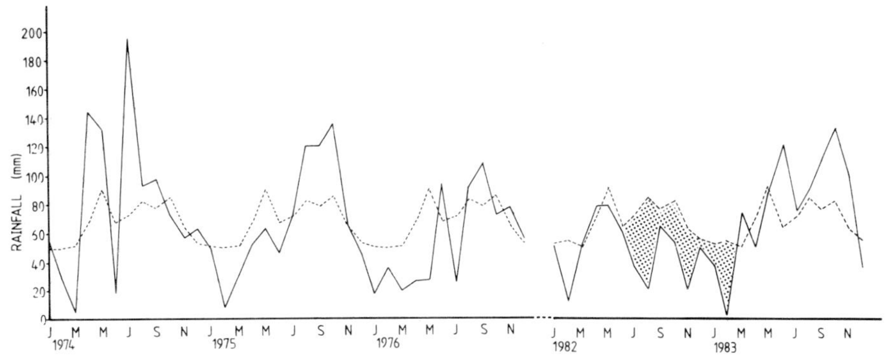

Fig. 8. Rainfall at Cranbourne 1974–1977. The solid line represents monthly totals, the broken line represents the mean monthly totals and the dotted portion of the graph represents the 1982–1983 drought. Source, Dandenong Valley Authority.

different environmental variables and the seasonality of breeding in bandicoot populations from Tasmania *(Perameles gunnii)*, Victoria *(I. obesulus)*, New South Wales *(I. macrourus)* and Queensland *(I. macrourus)*. They found that the proportion of lactating females in a population was strongly correlated with the rate of change of temperature, day length and, to a lesser extent, the pattern of rainfall. These relationships were particularly evident in Tasmania, Victoria and New South Wales, where a period of anoestrus separates periods of breeding.

It is also possible that the timing of births in females may be dependent upon the initiation of breeding activity by males in some populations. There was a marked increase and decrease in scrotum length of males at Cranbourne which coincided with the beginning and end of the breeding season, respectively (Fig. 6). In contrast, female *I. macrourus* in New South Wales (Gordon 1971), and female *P. gunnii* and *I. obesulus* in Tasmania (Heinsohn 1966) breed seasonally; but the testes of males in these populations showed no signs of seasonal fluctuation in amount of sperm production. There is also no seasonal change in testes size in *P. gunnii* in western Victoria (P. R. Brown, pers. comm.), where females breed the year round. This suggests that it is the influence of proximate factors on females, rather than males, which determines the timing of breeding in most populations of these species. The Cranbourne population of *I. obesulus* could be an exception.

Litter size

Investigation of variation in litter size in *I. obesulus* is fraught with difficulty because of the propensity of females to eject pouch young when captured and handled. Litter size in *I. obesulus* at Cranbourne was determined either by counting the number of young in the pouch or by counting the number of enlarged teats in the pouch after weaning. Gemmell (1982) and Gemmell, Walker, Johnston and Cepon (1984) demonstrated that litter size decreased during lactation in *I. macrourus*. A decrease in litter size does not necessarily result in a corresponding decrease in the number of enlarged teats (J. C. Merchant, pers. comm.), so that our estimates of litter size at weaning may overestimate the real value. Because of these sources of error, comparisons of litter size should be approached cautiously.

Although the onset and conclusion of breeding at Cranbourne are highly predictable between years there was variation in the proportion of females carrying pouch young (Fig. 5), mean litter size, number of litters produced per female per season, population density and, in particular, mean, annual reproductive output, between 1974–76 (Stoddart and Braithwaite 1979) and 1982–83 (this study). Onset of breeding likely occurs in response to a highly predictable cue, whereas the between year variation in fertility in individual females suggests that this is influenced by some less predictable factor, such as food abundance.

The abundance of invertebrates in the sandy soil at Cranbourne is strongly influenced by soil-moisture (rainfall and temperature): generally, abundance is highest during late winter to early summer (Fig. 7). The sandy soil dries out quickly during the summer months and most invertebrates either die, emerge or burrow deeper into the soil, thus decreasing the amount of food available to bandicoots (Lobert 1985). The years 1974, 1975 and 1976 (the period of Stoddart and Braithwaite's study) received approximately average monthly and annual rainfall (Fig. 8), and population density of bandicoots varied relatively little between years (Stoddart and Braithwaite 1979). Good spring rains in each of these years suggests that soil moisture and soil invertebrate abundance remained consistent between years. The abundance of soil invertebrates at Cranbourne varied considerably during 1982 and 1983 (Fig. 7). During early to mid 1982, Cranbourne received approximately average rainfall, but between July 1982 and February 1983, Cranbourne received only 52% of the expected rainfall (Fig. 8). This severe drought, and the mass emergence from the soil of what is probably the most important food species *(Diphucephala colaspedoides* larvae) in late November and December 1982 (Lobert 1985), resulted in very low abundance levels of soil invertebrates throughout most of 1983 (Fig. 7).

The marked increase in population density between 1981 and 1982 was associated with animals using for the first time an area of heath that had been slashed in 1979. This population increase may have resulted in increased competition for food between individual bandicoots during 1982, and may have contributed to the small litter sizes during the 1982 breeding season and the population crash of mid to late 1983 (Table 7). The inclusion of certain food items in the diet of *I. obesulus* at Cranbourne (millipedes, scorpions, centipedes, bull-ants), which have been found to be largely ignored by bandicoots (Heinsohn 1966, Maloney 1982; A. Opie, pers. comm.), also suggests that food may have been limited during 1982 and 1983.

Unlike Stoddart and Braithwaite (1979), we found no indication of a relationship between (i) litter size and female body weight prior to the birth of a litter, (ii) litter size and the age of females, or (iii) litter size and the time of birth of the litter during the breeding season. However, the differences in mean litter size with female body weight and timing of birth of the litter during the breeding season, detected by Stoddart and Braithwaite (1979), were often small.

Although Hall (1983) found a positive relationship between litter size, and weight and age (pes length) of wild female *I. macrourus* from Queensland, he did not state at what stage during lactation litter size was estimated. Gemmell *et al.* (1984) found no relationship between litter size and weight of the mother either early or late in lactation in a captive colony of *I. macrourus*, also from Queensland. Gordon (1966, in Gordon 1971) noted that the smallest mean litter size in *I. macrourus* in Queensland coincided with the time of least breeding, suggesting that small litter size was a response to suboptimal breeding conditions. Similarly, Heinsohn (1966) attributed the smaller mean litter size in *P. gunnii* late in the breeding season to diminishing food resources. However, neither Hall (1983), nor Gemmell *et al.* (1984), found strong relationships between litter size and time of litter production. Clearly, the relationships between female condition and age, and litter size require further clarification.

In some species of bandicoots there may be a reduction in litter size during lactation, but for reasons explained above, we have no data on this phenomenon. Both food shortage (Heinsohn 1966; Gemmell *et al.* 1984) and crowding in the pouch (Heinsohn 1966) have been invoked to explain decreases in litter size during lactation in bandicoots. Gemmell (1982) and Gemmell *et al.* (1984) suggest that there is no relationship between the number of young born and the condition of the mother in *I. macrourus*, but that litter size decreases during lactation if the mother is unable to meet the nutritional requirement of the pouch young.

With a mean litter size of between about 2 and 3.9 at weaning and an average of two to three litters per breeding season (Table 7), female *I. obesulus* at Cranbourne which survive into their third breeding season wean between 12 and 35 young during their lifetime. Smith (1980, Tables 9.1 and 9.2) estimated lifetime fecundity of 13 species of arboreal phalangerids (range: 5–32 young) and 10 species of dasyurid marsupials (range: 15–32 young). Estimates of maximumn fecundity of *I. obesulus* at Cranbourne fall in the upper half of the range of these estimates, indicating that *I. obesulus*

Table 7. Mean litter size, mean number of litters produced per female per season, reproductive output and population density of *Isoodon obesulus* in Tasmania (Heinsohn 1966) and Victoria (Stoddart and Braithwaite 1979; Lee and Martin, unpubl. data; Lobert this study). Units: mean reproductive output = number of young produced per female per breeding season, population density = number of animals ha^{-1} on the study grid.

Source	year	Mean litter size	Mean no. of litters per female per season	Mean reproductive output	Population density per ha
Heinsohn (1966)	1960 1961–62	2.8	—	—	0.1
Stoddart and Braithwaite (1979)	1974	3.9		10.2	1.8
	1975	3.5	2.6	9.1	1.8
	1976	3.4		8.9	2.7
Lee and Martin (unpubl. data)	1979–81	3.0	—	—	1.4
Lobert	1982	2.6	1.5	3.9	5.0
	1983	2.0	2.0	3.1	1.0

and other bandicoots are among the most fecund and fertile of all marsupials. Smith's estimates were calculated using maximum captive longevity and assumed maximum rate of litter production, and therefore, probably overestimate fecundity in natural populations.

In summary, *I. obesulus* at Cranbourne, and perhaps in other heathland habitats, has a highly predictable breeding season. During the breeding season, all female bandicoots breed at least once irrespective of the level of food abundance. However, between-year fertility may vary considerably. If food abundance is high, females appear to produce large litters and a number of litters during the breeding season. If food abundance is low, both litter size and the number of litters produced by females appear to decrease.

Growth and longevity

Isoodon obesulus in Tasmania reaches reproductive maturity at a minimum of approximately four and six months of age in females and males, respectively (Heinsohn 1966), compared to a minimum of seven months of age in Victoria. This difference is a consequence of the length of the breeding season. The eight month breeding season in Tasmania allows growth to maturity of juveniles during the season of their birth, whereas the six month breeding season in Victoria is too short to permit this, except perhaps for some animals born very early in the year (May or early June; e.g., male 21, Fig. 1).

Weights of adult male and female *I. obesulus* at Healesville (Taylor 1965; McKenzie 1967) were similar to those at Cranbourne. However, Tasmanian animals grow faster and weigh more than Victorian animals as adults. In Tasmania males weight about 800 g at 12 months of age and about 1 300 g at 18 months of age, while females reach a maximum weight of about 800 g at 12 months of age (Heinsohn 1966; Maloney 1982).

Some animals at Cranbourne lived to be at least 3.5 years old. Thus, maximum longevity of *I. obesulus* at Cranbourne is greater than Heinsohn's (1966) minimum estimate of two years. The weights of animals obtained by Taylor (1965), McKenzie (1967) and Maloney (1982) also suggest that *I. obesulus* live for more than two years in the wild in other populations.

The life history strategy of I. obesulus

The life history strategies displayed by bandicoots have been discussed by Lee and Cockburn (1985) who pointed out that all Australian bandicoot species are broadly similar in their litter size, food habits, body size, rate of growth and duration of parental investment. All are polyoestrous and produce several offspring in a litter and several litters in a year. In short, variation in many of their life history traits appears conservative compared to other marsupial groups (e.g., dasyurids). They summarize by suggesting that bandicoots show mainly facultative differences in their life history strategies, manifested through changes in the length of the breeding season and the production of dispersive offspring.

Most populations of *I. obesulus* breed seasonally, with litters being born from July/August to December/January. Litter size recorded during lactation varies from one to six (Heinsohn 1966; Stoddart and Braithwaite 1979; this study), with an average litter size at weaning ranging from two, in 1983 of this study to 3.9, in 1974 of Stoddart and Braithwaite's (1979) study (Table 7). An oestrous period during late lactation allows pregnancy to progress so that a new litter may be born as soon as the pouch is vacated (Heinsohn 1966; Stoddart and Braithwaite 1979; B. Lobert, pers. obs.). Females produce up to three litters per breeding season in Victoria, and four litters per breeding season in Tasmania (Heinsohn 1966). The gestation period of *I. obesulus* is not precisely known, but is less than 15 days (B. Lobert, pers. obs.) and most likely to be similar to 12.5 days observed in *I. macrourus* (Lyne 1974). Despite the short gestation the young are relatively large at birth compared to other marsupials which produce several young in a litter (Russell 1982). Lactation is also short for a marsupial of this size, lasting approximately two months (Heinsohn 1966; Stoddart and Braithwaite 1979; Russell 1982). Growth is rapid for a marsupial (Lee and Cockburn 1985). Young vacate the pouch when about 47 to 55 days old (Heinsohn 1966). Weaning occurs at approximately 60 days (Heinsohn 1966; Stoddart and Braithwaite 1979; B. Lobert, pers. obs.). Dispersal of young probably occurs as soon as the young are independent of the mother, that is, soon after weaning, or just before the birth of the next litter.

Although the litter size at weaning is not large when compared to dasyurids (Lee and Cockburn 1985), the production of three or four litters per breeding season (in seasonally breeding populations), at short intervals, results in a potentially high reproductive rate. Weaning appears to be followed by a high rate of dispersal (and mortality) of juveniles. This dispersal is seen as being advantageous for species living in unpredictable and patchy habitats such as the heathlands of southeastern Australia (Stoddart and Braithwaite 1979; Cockburn *et al.* 1981).

Stoddart and Braithwaite (1979) suggested that the reproductive strategy of females changes with age (and body weight): young females commenced breeding earlier than old females, litters produced late in the breeding season tended to be larger than litters produced early in the breeding season, and a positive relationship existed between the mean litter size and female body weight. During our study there was no indication of this variation, although the effect of the drought may have masked these trends. The conclusions reached by Stoddart and Braithwaite (1979), although supported by their data, need cautious interpretation, as they are based primarily on small changes in litter size. We suggest that, at this stage, too little is known about the causes of reduction in litter size during lactation in this species to accept that small variations in estimates of litter size are naturally significant. However, this does not detract from the conclusion of Stoddart and Braithwaite (1979) that the strategy of producing small litters of rapidly growing young at frequent intervals and forcing these to disperse, appears suited for habitats in which unoccupied patches are frequently, but unpredictably, available.

The life history tactics of *I. obesulus* at Cranbourne include facultative variation in the annual reproductive output of females through variation in litter size and numbers of litters per female, but the length of the breeding season is highly predictable and is not influenced by the seral stage of the heath occupied by the bandicoots, or by annual variation in food abundance, or rainfall.

ACKNOWLEDGEMENTS

This work was completed to fulfill part of the requirements for the degree of Master of Science (B. L.). We wish to thank Dr D. Churchill for granting permission for the use of the study site in the Royal Botanic Gardens Annexe at Cranbourne. Bruce Fuhrer helped with photography. Animals were trapped and handled under the provisions of a permit issued by the Fisheries and Wildlife Division of Victoria.

REFERENCES

BARNES, A. AND GEMMELL, R. T., 1984. Correlations between breeding activity in the marsupial bandicoots and some environmental variables. *Australian Journal of Zoology* **32:** 219–26.

CLOSE, R. L., 1977. Recurrence of breeding after cessation of suckling in the marsupial *Perameles nasuta. Australian Journal of Zoology* **25:** 641–45.

COCKBURN, A., BRAITHWAITE, R. W. AND LEE, A. K., 1981. The response of the heath rat, *Pseudomys shortridgei* to pyric succession: a temporally dynamic life-history strategy. *Journal of Animal Ecology* **50:** 649–66.

GEMMELL, R. T., 1982. Breeding bandicoots in Brisbane (*Isoodon macrourus* Marsupialia, Peramelidae). *Australian Mammalogy* **5:** 187–93.

GEMMELL, R. T., WALKER, M.T., JOHNSTON, G. AND CEPON, G., 1984. The number of young present in sequential litters of the marsupial bandicoot, *Isoodon macrourus*, in captivity. *Australian Journal of Zoology* **32:** 623–29.

GORDON, G., 1971. A study of island populations of the short-nosed bandicoot *Isoodon macrourus* Gould. Ph.D. Thesis, University of New South Wales: Kensington.

HALL, L. S., 1983. Observations on the body weights and breeding of the northern brown bandicoot, *Isoodon macrourus*, trapped in south-east Queensland. *Australian Wildlife Research* **10:** 467–76.

HEINSOHN, G. E., 1966. Ecology and reproduction of the Tasmanian bandicoots (*Perameles gunni* and *Isoodon obesulus*). *University of California Publications in Zoology* **80:** 1–96.

LEE, A. K. AND COCKBURN, A., 1985. *Evolutionary ecology of marsupials*. Cambridge University Press: Cambridge.

LOBERT, B., 1985. The ecology of the southern brown bandicoot in south-east Australian heathland. M.Sc. Thesis, Monash University, Clayton.

LYNE, A. G., 1974. Gestation period and birth in the marsupial *Isoodon macrourus. Australian Journal of Zoology* **22:** 303–09.

MCKENZIE, N. L., 1967. Some ecological aspects of the behaviour of the short-nosed bandicoot *(Isoodon obesulus)*. Honours Thesis, Monash University: Clayton.

MALONEY, D., 1982. A comparison of the behaviour and ecology of the Tasmanian bandicoots *Perameles gunnii* and *Isoodon obesulus*. Honours Thesis, University of Tasmania: Hobart.

OPIE, A. M., 1980. Habitat selection and the diet of *Isoodon obesulus. Australian Mammal Society Bulletin* **6**(2): 56.

RUSSELL, F. M., 1982. Patterns of parental care and parental investment in marsupials. *Cambridge Philosophical Society Biological Reviews* **57:** 423–86.

SMITH, A., 1980. The diet and ecology of Leadbeaters possum and the sugar glider. Ph.D. Thesis, Monash University: Clayton.

STODDART, D. M. AND BRAITHWAITE, R. W., 1979. A strategy for utilization of regenerating heathland habitat by the brown bandicoot (*Isoodon obesulus;* Marsupialia, Peramelidae). *Journal of Animal Ecology* **48:** 165–79.

TAYLOR, P., 1965. Home range in Australian mammals. Honours Thesis, Monash University: Clayton.

WATTS, C. H. S., 1974. The Nuyts Island Bandicoot. *South Australian Naturalist* **49**(2): 20–24.

Home range and activity period of the Southern Brown Bandicoot (*Isoodon obesulus*) in a Victorian heathland

BERTRAM LOBERT[1]

A comparison of the trap and radio-track determined home ranges of *Isoodon obesulus* suggest that trap-derived data are not suitable for home range studies in heathland and that results of habitat selection studies of this species using trapping data need cautious interpretation. *Isoodon obesulus* inhabiting the heathland were not territorial and some individuals had more than one nest. The location of the nest(s) appeared to influence the level of selection exhibited by an individual for different vegetation types. In this heathland *I. obesulus* was predominantly diurnal.

Key words: Home range, activity period, bandicoot, diurnal.

INTRODUCTION

MANY investigations of home range and habitat selection of Australian small mammals rely on information obtained from mark-recapture studies conducted using a trapping grid or trap lines. All these studies assume that trapping results accurately reflect dispersion of individuals and the population within the study area.

Recent comparisons of home ranges based upon trap-derived data and radio-tracking show that these may differ, and that home ranges based upon radio-tracking data are free of certain biases inherent in trapping studies (Trevor-Deutsch and Hackett 1980). Trap-derived home ranges also depend upon the technique used to calculate the home range (Trevor-Deutsch and Hackett 1980).

This study of the Southern Brown Bandicoot *(Isoodon obesulus)* in Victorian heathland compares home range estimates based on grid trapping with results from radio-tracking. The radio-tracking also allowed examination of the activity period of *I. obesulus* in this heathland.

STUDY AREA

This study was conducted at the Royal Botanic Gardens and National Herbarium Annexe (38° 07′S 145° 14′E; elevation approximately 85 m) near Cranbourne, 50 km south-east of Melbourne.

The area experiences mild winters and warm summers, with a mean monthly minimum temperature of 16°C in July and a mean monthly maximum of 25°C in February. The mean annual rainfall is 800 mm, with autumn, winter and spring typically the wettest months.

The trapping grid was located in Silky Teatree (*Leptospermum myrsinoides*) dominated heathland. The area bounded by the grid included 3–4 year-old heath surrounded by 14 to 18 year-old heath (Fig. 1).

The two main soil types on the grid are sandy podzols and humus podzols, with the latter confined to depressions subject to flooding (Gullan 1978). These soil types support floristically distinct vegetation types: wet heath (humus podzol) and dry heath (sandy podzol) (Table 1; Braithwaite and Gullan 1978; Lobert 1985). However, only five of the 103 trap sites on the grid were situated in wet heath (Fig. 1). The major vegetation discontinuity on the grid was structural. The height of the younger 3–4 year-old low heath varied between 75 and 150 cm, while the height of the older 14–18 year-old high heath varied between 200 and 250 cm. The younger heath was also less dense, with patches of bare ground and herbs between larger clumps of vegetation. The low and high heath were floristically indistinguishable from each other (Lobert 1985).

METHODS

Grid trapping

The presence, distribution and abundance of *I. obesulus* was assessed between March 1982 and August 1983 by live trapping a 5.1 ha grid. Of the 103 trap sites on the grid, 58 were situated in high heath, 43 in low heath and two in isolated patches of high heath (Fig. 1). One Elliott folding trap (32 × 10 × 8 cm) and one wire-mesh cage trap (36 × 13 × 13 cm) were placed at each site and baited with a mixture of rolled oats, peanut butter and golden syrup.

The grid was trapped for four nights each month. In each trapping session traps were

[1]Department of Zoology, Monash University, Clayton, Victoria, Australia 3168.
Pages 319–25 *in* BANDICOOTS AND BILBIES, ed by J. H. Seebeck, P. R. Brown, R. L. Wallis and C. M. Kemper. Surrey Beatty & Sons, Sydney, 1990.

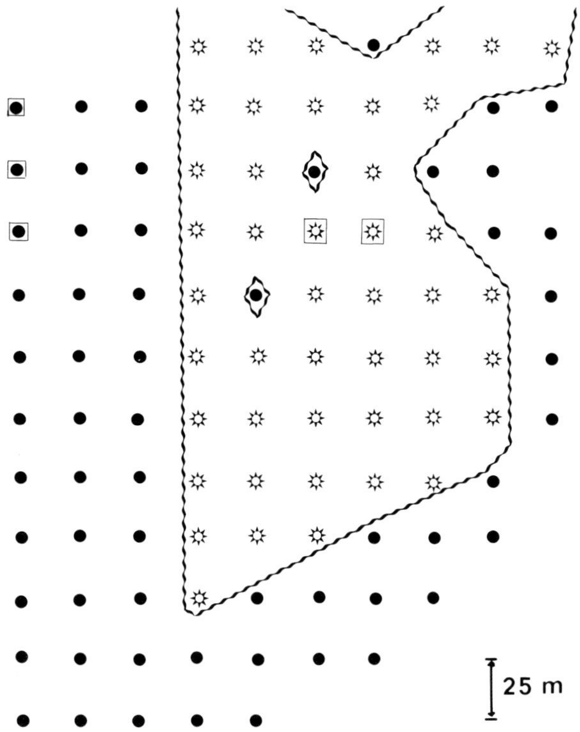

Fig. 1. Cranbourne *Isoodon obesulus* trapping grid. Circles represent sites in the 14–18 year-old heath (high heath), open stars represent sites in the 3–4 year-old heath (low heath), squares represent wet heath sites and the wavy line represents the low/high heath boundary.

open for two consecutive nights, then closed for one night, then open for a further two consecutive nights. Occasionally during the breeding season the two periods of consecutive trapnights were separated by one or two weeks to reduce trap-induced mortality of pouch young. Traps were cleared and reset each morning, except during summer when it was necesary either to clear the traps twice each day, or to close them during the day to avoid mortality.

For all bandicoots the following was recorded: weight, left pes length and location on the grid.

Radio-tracking

Five adult bandicoots (four males and one female) were each radio-tracked for between 21.5 and 67.5 h, from June to September 1983. The duration of radio-tracking depended on how long the transmitters remained on the animal and battery life. A transmitter (approximately 145 MHZ, Austec Inc., Edmonton, Canada) with whip antenna, weighing 7 g, was glued to the rump of the animal in the early morning after its capture. Tracking usually started 2 to 3 h after the transmitter was attached. Animals with transmitters were located by triangulation from two 4 m long directional antennae, mounted on 6 m high masts, using battery operated receivers (A.V.M.). The masts were positioned outside the trap-determined home ranges of the individuals under study. Compass bearings of the signal

Table 1. Vascular plant species characteristic of sandy podzol and humus podzol soils on the *Isoodon obesulus* trapping grid (from Gullan 1978).

Sandy podzol soil (dry heath)
Leptospermum myrsinoides
Epacris impressa
Hypolaena fastigata
Monotoca scoparia
Leucopogon virgata
Humus podzol soil (wet heath)
Melaleuca squarrosa
Gahnia sieberana
Leptospermum juniperinum

direction were taken simultaneously at each station every 5 min, and the duration of radio-tracking sessions was between 1 and 17 h. Signal direction, tested by taking bearings on transmitters with known positions, was found to be consistently accurate to within ± 3°, and so allowed relatively accurate location of animals both on and off the grid (Fig. 2). The dispersion of these animals with regard to the different vegetation types was then examined. The locations of nests were confirmed using a handheld antenna. Radio-tracking and grid trapping were not run concurrently.

Periods of activity (locomotion) were established by monitoring the movements of the animals.

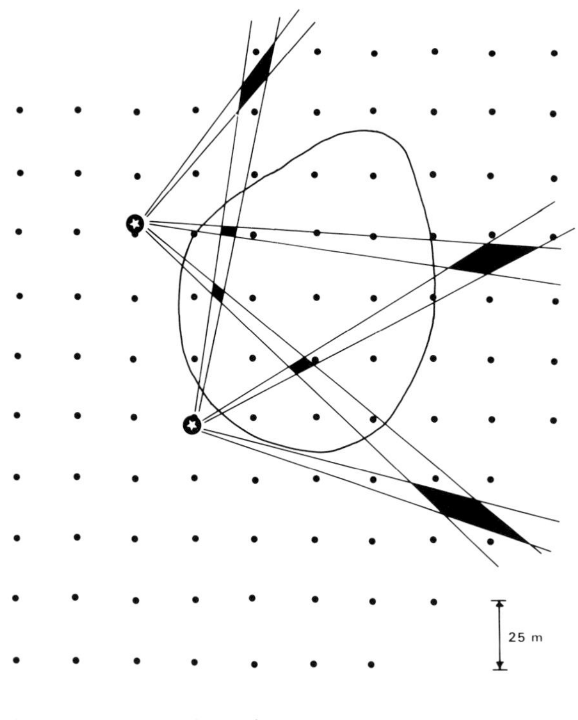

Fig. 2. Trapping grid for *Isoodon obesulus* showing error (shaded areas) involved in locating position of an animal when direction of signal varies by ± 3°. Dots represent trap sites and stars represent position of antennas. Solid line represents a home range boundary.

Home range determination

Radio-tracking home ranges were determined using the method of Dixon and Chapman (1980), incorporating modifications by Spencer and Barrett (in press). This method produces isopleths based on reciprocal mean distance deviations, and provides a close approximation to the true centre of activity, defines home ranges of any shape, relates home range directly to the intensity of activity and can be used to define core areas of activity. Thus, it is of particular value when the habitat is heterogeneous, as is the case in this study. The isopleth enclosing 90% of the loci was chosen as the home range boundary. This procedure excludes single forays away from the principal centres of activity. The method of computation followed Spencer and Barrett (in press).

The home range boundary calculated from radio-tracking data included loci gathered during the day and the night i.e., active and inactive periods. The four animals used in the comparison of the two techniques did not alter their trap-determined home ranges while resident on the grid. Therefore, calculations of home range based on trapping data used data from all seasons for each animal.

Home ranges of grid trapped animals were determined using the method of Dixon and Chapman (1980) when the number of loci exceeded 14, or the minimum area of a polygon method if the number of loci was between five and 14. The minimum area of a polygon method was used to determine home range when loci were few in number because this method was likely to give a more conservative estimate of home range size than the centre of activity method of Dixon and Chapman (1980).

RESULTS

Home range

The home range of all *I. obesulus* resident on the grid are shown in Fig. 3. Although the level of accuracy of some of the trap-derived home range estimates is probably low, it is clear that there was a high degree of home range overlap within and between the sexes.

Comparisons of home ranges based upon trapping and radio-tracking were only possible for the four animals (three males and one female) for which there were sufficient radio-tracking data. Animals with a nest or nests on or close to the grid showed similar trap determined and radio tracked home ranges (Fig. 4). However, when the nest or one of the nests was located some distance from the grid, there was considerable variation in both shape and size between the ranges (Fig. 4, Table 2).

Fig. 3. Home range of all resident *Isoodon obesulus* on grid. a. females resident March–June 1982, b. females resident July–December 1982, c. females resident January–June 1983, d. males resident March–June 1982, e. males resident July–December 1982, f. males resident January–June 1983.

Two of the animals (male 5 and female 5) had two known nests, while the other two (male 18 and 28) each had one known nest. The nests of males 5, 18 and 28 were located in high heath, while the nest of female 5 was located in dense vegetation in a *Gahnia sieberana* dominated swamp in the low heath. Nests were oval shaped mounds of leaf litter and soil, 0.5 to 0.75 m long. An opening at each end of the nest leads to a small, dry, chamber lined with twigs and leaves. The roof and walls of the two openings collapse

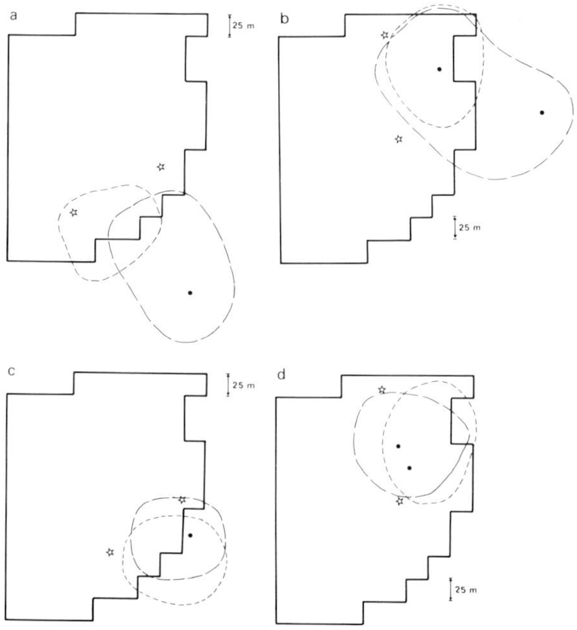

Fig. 4. Trap (short broken line) and radio-tracking (long broken line) determined *Isoodon obesulus* home ranges. a. male 18, b. male 5, c. male 28 and d. female 5. Dots represent nests and stars represent locations of antennas.

after the animal enters or emerges, leaving the opening inconspicuous and sealing the nest.

The proportion of time the four radio-tracked *I. obesulus* spent in the high heath while active ranged from 75–100% (Table 3). These values differ from the proportion of captures of each animal in the two vegetation types (Table 3). The proportion of time spent in the different vegetation types appears to be influenced by the proximity of the nests to the low heath (Table 3, Fig. 4).

Table 2. Home range estimates (m^2) of trapped and radio-tracked *Isoodon obesulus*. Radio-tracked home range was calculated from loci obtained during both active and inactive periods.

Animal No.	Trapped	Radio-tracked
Female 5	10 790	11 016
Male 5	11 003	25 052
Male 18	7 900	13 677
Male 28	8 441	7 875

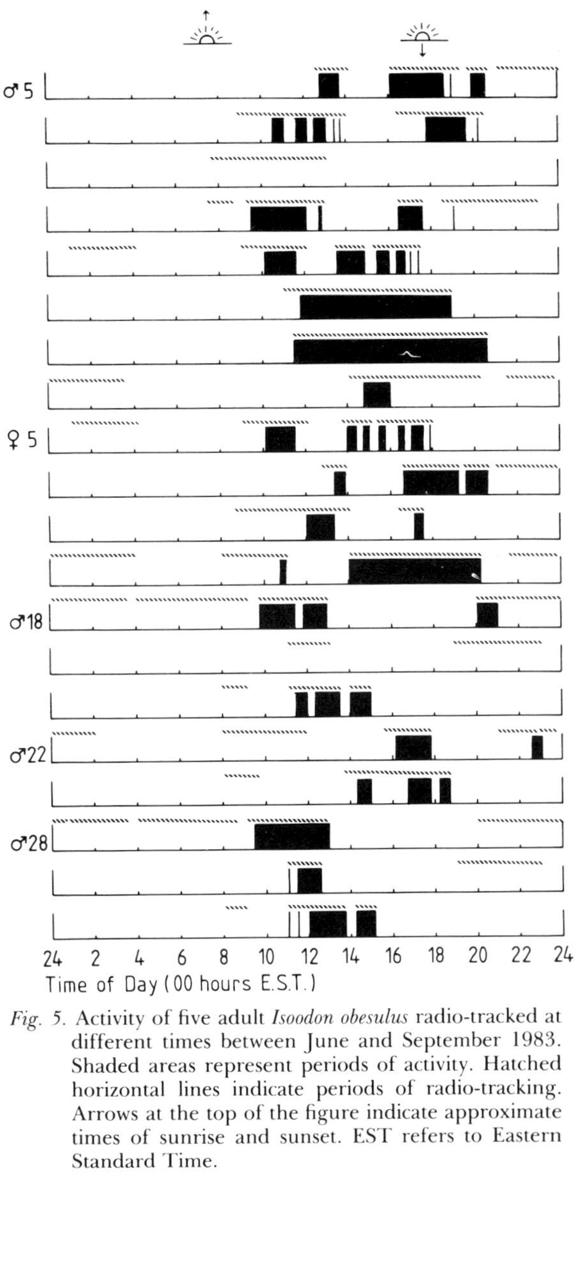

Fig. 5. Activity of five adult *Isoodon obesulus* radio-tracked at different times between June and September 1983. Shaded areas represent periods of activity. Hatched horizontal lines indicate periods of radio-tracking. Arrows at the top of the figure indicate approximate times of sunrise and sunset. EST refers to Eastern Standard Time.

Table 3. Proportion of captures of radio-tracked *Isoodon obesulus* in high heath and the proportion of time the radio-tracked animals spent active in this habitat.

Animal No.	Total No. captures	% captures in high heath	Time spent active (min)	% time in high heath
Female 5	54	50	775	75
Male 5	45	53	1 395	79
Male 18	37	97	420	100
Male 28	33	88	510	86

Activity period

Of the total time spent radio-tracking (139 h 25 min), 65 h 40 min (47.1%) took place between sunset and sunrise. Most of the activity of the five radio-tracked animals occurred between 0900 and 2100 hrs EST (Fig. 5). Only one of the 12 sessions which occurred between 2100 and 0900 hrs included activity. There was no obvious change in activity period with day length, but animals usually returned to, and remained in their nests during heavy rain.

DISCUSSION

Comparison of grid trapping and radio-tracking

Recent investigations comparing home ranges estimated from trapping and radio-tracking suggest that grid trapping reveals trap determined home range as distinct from "actual home range", and that methods of estimating home range from trap data vary in the extent to which they approximate "actual home range" (Trevor-Deutsch and Hackett 1980). The results of the present study, although of limited extent, support the view that home range estimates based on trap data reveal "trap-determined home range". In this study the two techniques produced home ranges of different size and shape. The trap-derived data tended to underestimate the size of *I. obesulus* home range, particularly if one or more of an animal's nests were located near the edge of the grid, or off the grid. The frequency of captures of the four radio-tracked animals in the low heath implies a higher level of selection for the low heath than is suggested by radio-tracking.

There are several potential biases which may account for these findings. *Isoodon obesulus* was readily trapped, and some individuals were repeatedly caught two or three times per 4-night trapping session. High trappability suggests that the animals may be attracted to traps and this may have increased the probability that animals were trapped close to their nest, or near the site where they entered the grid if they nested off the grid. Also, large numbers of Swamp Rats (*Rattus lutreolus*) were caught in the high heath. Their high trappability may have partially excluded *I. obesulus* from traps in the high heath and increased their probability of being trapped in the low heath, in which *R. lutreolus* was virtually absent. Conversely, *I. obesulus* without access to low heath may have been trapped less often than animals which had access to this habitat type. Individual *I. obesulus* may also have preferred directions or paths of movement and this may have influenced which traps were encountered first and most often. Well worn paths used by *I. obesulus* were found near some trap sites, and some of the more frequently trapped animals quickly and repeatedly sought out and followed these paths when released. Most resident animals were trapped at some sites at, or close to the edge of the trapping grid, suggesting that their home ranges included areas off the grid.

Nest location also appears to influence the level of selection of the two vegetation types (Fig. 4, Table 3). Animals with nests in high heath close to the low-high heath interface spent more time in the low heath than animals with nests in high heath removed from the interface. Thus, the distance from the low heath to an animals nest influenced the relative use of both habitat types by that animal.

Although the individual causes of bias could not be examined separately, together the biases appear to have had a marked effect on the trapping results. Consequently, I consider that conclusions reached from trapping on home range and habitat use by *I. obesulus* in this and perhaps other heathlands must be interpreted cautiously.

Home range

The sizes of the radio-tracked home ranges of animals in this study (0.82–3.15 ha) are similar to the home ranges of *I. obesulus* calculated from trapping data by Heinsohn (4.9±1.7 ha, n=5, 1966), McKenzie (2.2±1.5 ha, 1967) and Maloney (1.04±1.1 ha, n=14, 1982), but vary considerably from those reported by Sampson ($\bar{x}$=20.5 and 1.8 ha for males and females, respectively; from Maloney 1982) and Buchmann ($\bar{x}$=0.5 and 0.4 ha for males and females, respectively; from Maloney 1982). These studies used either the minimum area of a polygon (Heinsohn 1966; McKenzie 1967), or the boundary strip method (Maloney 1982), as well as different numbers of loci to determine home range size. The methods used by Sampson and Buchmann (both in Maloney 1982) were not stated.

However, as a result of the different methods employed and the results of the present study, interpretation of the differences in home range sizes of *I. obesulus* in different habitats, although interesting, is presently of dubious value.

This study shows that in heathland there may be considerable individual variability in home range size of *I. obesulus*, and that there was a high degree of home range overlap within and between sexes. The second observation is supported by the trapping results (Fig. 3). Although Heinsohn (1966) and McKenzie (1967) concluded that *I. obesulus* was territorial, the density of animals in the relatively open habitat of their study areas was very low in comparison to the density in heathland at Cranbourne (densities at Cranbourne reached 5 ha^{-1}). Maloney (1982) suggested that territories of *I. obesulus,* if they exist at all, are in the form of spatio-temporal systems based on avoidance. The pugnacious behaviour of individuals of *I. obesulus* shown towards conspecifics (Jones 1924; McKenzie 1976; Maloney 1982), and the overlap of home ranges observed in my study are consistant with this mode of spacing.

Activity period

The radio-tracking data indicate that *I. obesulus* was primarily diurnal during the period they were radio-tracked (May to September). This contrasts with the conclusions reached by Heinsohn (1966), McKenzie (1967), Watts (1974) and L. Thomas (pers. comm.) in Tasmania, Victoria, South Australia and south-western Western Australia, respectively, which suggest that *I. obesulus* is strictly nocturnal. Braithwaite (1977), who also worked in the Cranbourne heathland, found that one-third of *I. obesulus* captures occurred during the day, during September and October 1975. Other bandicoot species, including *Isoodon macrourus* (Mackerass and Smith 1960; Gordon 1974), *Perameles nasuta* (Stodart 1966) and *P. gunnii* (Heinsohn 1966; P. Brown, pers. comm.), are all considered nocturnal, and in fact only one Australian marsupial, the Numbat (*Myrmecobius fasciatus*), is known to be totally diurnal (reviewed by Lee and Cockburn 1985).

Among the factors thought to influence activity patterns are the activity rhythms of predators, competitors for food and other resources, the availability of prey (Park 1940; Cloudsley-Thompson 1961; Ashby 1972), the relationship between energy expenditure and food consumption (Crowcroft 1954), the nature of ground cover (Southern 1964; Crowcroft and Godfrey 1968), rainfall, cloud cover and other weather influences (Doucet and Bider 1974), and high daytime temperatures and low humidity (Cloudsley-Thompson 1961).

Weather in coastal southeastern Australia is rarely extreme, and is similar in Tasmania and southern Victoria, and so is unlikely to account for the observed difference in activity rhythms between sites. Energy expenditure and food consumption are also probably similar in species from different sites as well as between the different species of bandicoot in southeastern Australia, and so is unlikely to account for differences in activity rhythms.

Interspecific competition for resources also does not seem a plausible explanation. Small mammals occurring in the Annexe at Cranbourne are the Swamp Rat (*R. lutreolus*) Black Rat (*Rattus rattus*) and House Mouse (*Mus musculus*), and are specialist herbivore, omnivore and soil fossicking insectivore/small scansorial insectivore, respectively (Braithwaite *et al.* 1978). They are thus unlikely to compete directly with *I. obesulus* which is a soil fossicking generalist omnivore (see Braithwaite *et al.* 1978, Lobert 1985). Where two species of bandicoots are sympatric, as for example *I. obesulus* and *P. gunnii* in Tasmania, both are nocturnal (Maloney 1982), and remain nocturnal where they occur in allopatry in Victoria (McKenzie 1967; P. R. Brown, pers. comm.).

Although comparisons of abundance, availability and dispersion of food resources in different habitats is not possible, it seems unlikely that the activity cycles of the invertebrate prey would vary dramatically between habitats and contribute to different patterns of activity.

Predators of *I. obesulus* at Cranbourne include foxes (*Vulpes vulpes*), diurnal birds of prey (pers. obs.), and perhaps also cats (*Felis cattus*), Tiger Snakes (*Notechis scutatus*) and owls. Tiger Snakes are common in the study area and active during the warmer months, but would probably only pose a threat to recently weaned or juvenile *I. obesulus*. Predation by feral cats and owls may occur, but has not been observed. Although Black-shouldered Kites (*Elanus notatus*) and Australian Kestrels (*Falco centroides*) have been observed hovering over the low heath, and Brown Falcons (*Falco berigora*), Whistling Kites (*Haliastur sphenurus*) and Swamp Harriers (*Circus aeruginosus*) have been seen flying over the heathland, the amount of predation suffered by *I. obesulus* from these potential predators is again unknown. Examination showed that eight fox scats collected on the trapping grid were composed entirely of *I. obesulus* and *R. lutreolus* remains. At Cranbourne, foxes have been seen active during the day and are presumably active at night, although it is not known at which times they hunt. Thus, there is no clear pattern of activity which could account for the shift in activity period of *I. obesulus* at Cranbourne.

One possible explanation for the shift to diurnal activity of *I. obesulus* at Cranbourne is the effect of vegetation type. Much of the heath vegetation occupied by *I. obesulus* at Cranbourne is virtually impenetrable to animals any larger than a bandicoot and conceivably offers greater protection from mammalian and avian predators than the more open vegetation types examined by Heinsohn (1966), McKenzie (1967), Watts (1974), Gordon (1974) and P. Brown (pers. comm., in western Victoria). Reduced predation may in turn allow a shift to diurnality, if such a shift is advantageous to the species. It may also be energetically advantageous to be diurnal in autumn and winter when night time temperatures are low. If this is the case, a shift to nocturnal activity might be expect in summer during periods when diurnal temperatures are high and humidities low. However, at this stage it is not possible to identify the cause of the different activity pattern at Cranbourne.

ACKNOWLEDGEMENTS

Thanks are extended to the many friends who helped with radio-tracking. Tony Lee gave his support throughout the study. Tony, Roger Martin and Gordon Sanson made helpful suggestions on an earlier draft of this paper. Animals were trapped and handled under the provisions of a permit issued by the Fisheries and Wildlife Division of Victoria.

REFERENCES

ASHBY, K. R., 1972. Patterns of daily activity in mammals. *Mammal Review* **1:** 171–85.

BRAITHWAITE, R. W., 1977. Preliminary observations on the activity patterns of *Rattus lutreolus* and other Victorian small mammals. *Victorian Naturalist* **94:** 216–18.

BRAITHWAITE, R. W., COCKBURN, A. AND LEE, A. K., 1978. Resource partitioning by small mammals in lowland heath communities of south-eastern Australia. *Australian Journal of Zoology* **3:** 423–45.

BRAITHWAITE, R. W. AND GULLAN, P. K., 1978. Habitat selection by small mammals in a Victorian heathland. *Australian Journal of Ecology* **3:** 109–27.

CLOUDSLEY-THOMPSON, J. L., 1961. Adaptive functions in circadian rhythms. *Cold Spring Harbor Symposium on Quantitative Biology* **25:** 345–54.

CROWCROFT, P., 1954. The daily cycle of activity in British shrews. *Proceedings of the Zoological Society of London* **123:** 715–29.

CROWCROFT, P. AND GODFREY, G. K., 1968. Daily cycle of activity in two species of *Sminthopsis* (Marsupialia: Dasyuridae). *Journal of Animal Ecology* **37:** 63–73.

DIXON, K. R. AND CHAPMAN, J. A., 1980. Harmonic mean measure of animal activity areas. *Ecology* **61:** 1040–44.

DOUCET, G. J. AND BIDER, J. R., 1974. The effects of weather on the activity of the masked shrew. *Journal of Mammalogy* **55:** 348–63.

GORDON, G., 1974. Movements and activity of the short-nosed bandicoot *Isoodon macrourus* Gould (Marsupialia). *Mammalia* **38:** 405–31.

HEINSOHN, G. E., 1966. Ecology and reproduction of the Tasmanian bandicoots *(Perameles gunnii* and *Isoodon obesulus*). *University of California Publications in Zoology* **80:** 1–96.

JONES, F. WOOD, 1924. *The mammals of South Australia: Part II. The bandicoots and herbivorous marsupials.* Government Printer: Adelaide.

LEE, A. K. AND COCKBURN, A., 1985. *Evolutionary ecology of marsupials.* Cambridge University Press: Cambridge.

LOBERT, B., 1985. *The ecology of the southern brown bandicoot* (Isoodon obesulus) *in south-east Australian heathland.* Masters Thesis, Monash University: Melbourne.

LYNE, A. G., 1974. Gestation period and birth in the marsupial *Isoodon macrourus. Australian Journal of Zoology* **22:** 303–09.

LYNE, A. G., 1981. Activity rhythms in the marsupials *Isoodon macrourus* and *Perameles nasuta* in captivity. *Australian Journal of Zoology* **29:** 821–38.

McKENZIE, N. L., 1967. *Some ecological aspects of the behaviour of the short-nosed bandicoot (*Isoodon obesulus). Honours Thesis, Monash University: Melbourne.

McKENZIE, N. L., 1976. *Behaviour units and functional communication in* Isoodon obesulus *(Marsupialia; Peramelidae).* Masters Thesis, Monash University: Melbourne.

MACKERASS, M. J. AND SMITH, R. H., 1960. Breeding the short-nosed bandicoot, *Isoodon macrourus* (Gould), in captivity. *Australian Journal of Zoology* **8:** 371–82.

MALONEY, D., 1982. *A comparison of the behaviour and ecology of the Tasmanian bandicoots* Perameles gunnii *and* Isoodon obesulus. Honours Thesis, University of Tasmania: Hobart.

PARK, O., 1940. Nocturnalism — the development of a problem. *Ecological Monographs* **10:** 485–536.

SOUTHERN, N. H., 1964. *The handbook of British mammals.* Blackwell: Oxford.

SPENCER, W. D. AND BARRETT, R. H., 1984. An evaluation of the harmonic mean measure for defining carnivore activity areas. *Acta Zoologica Fennica.* **171:** 755–59.

STODART, E., 1966. Management and behaviour of breeding groups of the marsupial *Perameles nasuta* Geoffroy in captivity. *Australian Journal of Zoology* **14:** 611–23.

TREVOR-DEUTSCH, B. AND HACKETT, D. F., 1980. An evaluation of several grid trapping methods by comparison with radio telemetry in a home-range study of the eastern chipmunk (*Tamias striatus* L.). Pp. 375–86 in *A handbook on biotelemetry and radio tracking* ed by C. J. Amlaner and D. W. Macdonald. Pergamon Press: Oxford/New York.

WATTS, C. H. S., 1974. The Nuyts Island Bandicoot. *South Australian Naturalist* **49:** 20–24.

Prediction of the geographic range and habitat preferences of *Isoodon obesulus* and *Perameles nasuta* in Gippsland

ANNE OPIE[1], PAUL GULLAN[1] and IAN MANSERGH[2]

Two procedures were adopted to predict the geographic range and habitat of *Isoodon obesulus* and *Perameles nasuta* from the few known capture sites in Gippsland. One uses floristic vegetation data collected at capture sites compared with that of several thousand other sites distributed throughout Gippsland. The other uses a climatic description of the capture sites compared with predicted climatic data over the entire region. There was considerable overlap in both predicted analyses with the exception of an area of about 2 000 km^2 in south Gippsland. In this area the climatic profiles were considered by be suitable for *P. nasuta* but the vegetation data suggested otherwise. The area has suffered considerable disturbance and clearing for agriculture and other types of development and may represent a local extinction for the species.

The geographic ranges of the two bandicoot species overlap considerably but the habitats differed significantly. These can be summarized as follows: *Isoodon obesulus* favours lowland sites, within 50 km of the coast, supporting sclerophyllous woodlands and forests and heathlands growing on siliceous sandy soils. *Perameles nasuta* has a much broader geographic and habitat range. This species favours a range of forests, woodlands and heathlands on a variety of soils from the sub-alps to the coast. In most cases, however, the known and predicted habitats are associated with watercourses or sites where soils are frequently moist.

Key words: Bandicoots, *Isoodon obesulus, Perameles nasuta,* habitat preference, floristics, distribution, Gippsland, BIOCLIM, vegetation community.

INTRODUCTION

ISOODON OBESULUS (Southern Brown Bandicoot) and *Perameles nasuta* (Long-nosed Bandicoot) are the only species of bandicoot native to Gippsland. Neither is considered rare or endangered in Victoria (Ahern 1982), yet despite extensive surveys throughout Gippsland over an 8-year period (more than 50 000 trap-nights from over 1 000 sites) *I. obesulus* has been captured from only seven localities and *P. nasuta* at 11 (Norris and Mansergh 1981; Suckling 1980). Certainly the animals appear to be more abundant than these results suggest as characteristic diggings and hair samples in predator scats have been observed at many sites at which bandicoots have not been trapped (Carr, Horrocks and Isles 1984; Horrocks, Opie, Carr, Cherry, Brown, Craig and Triggs 1984; Opie, Cherry, Horrocks, Carr, Schulz and Triggs 1984, Chesterfield, Macfarlane, Allen, Hutchinson, Triggs and Barley 1983; Norris, Mansergh, Ahern, Belcher, Temby and Walsh 1983).

This anomaly has created a problem in developing programmes for the conservation and management of these species. The problem centres on the absence of reliable data on their habitat. If the animals are indeed widespread and relatively abundant then it is important to identify their habitats so that monitoring of population densities and habitat change can be planned. If, however, the animals are truly rare (as the trapping data suggest) it is more critical to establish both their actual and potential habitat ranges to ensure their conservation.

This paper deals with this problem by examining two types of habitat data from the sites where *I. obesulus* and *P. nasuta* have been captured in Gippsland. One is floristic information collected during the course of combined flora and fauna surveys conducted in Gippsland between 1977 and 1985. The other is climatic information derived from CSIRO BIOCLIM databases. Both of these data types will be compared with floristic and climatic data throughout Gippsland as a means of predicting habitats for each species at sites where the animals have not been captured.

STUDY AREA

For the purposes of this paper Gippsland is defined as the Gippsland Lakes catchment (i.e., the combined catchment of the Latrobe, Avon, Mitchell, Nicholson and Tambo Rivers) and all the land in Victoria east of this catchment and south of the Great Dividing Range, an area of approximately 35 000 km^2 (Fig. 1). The western region of this area (the Gippsland Lakes

[1]Flora and Fauna Survey and Management Group, Department of Conservation and Environment, 378 Cotham Road, Kew, Victoria, Australia 3101.
[2]Wildlife Management Branch, Department of Conservation and Environment, 123 Brown Street, Heidelberg, Victoria, Australia 3080.
Pages 327–34 *in* BANDICOOTS AND BILBIES, ed by J. H. Seebeck, P. R. Brown, R. L. Wallis and C. M. Kemper. Surrey Beatty & Sons, Sydney, 1990.

catchment) has had a varied history of development including agriculture, open-cut coal mining (in the Latrobe Valley), hardwood and softwood timber production and urban development. Approximately 35% of the 22 000 km^2 catchment is alienated land and the rivers in the area have suffered considerable disturbance both directly (from dams, diversions and snaggings) and indirectly (from runoff and invasion). The eastern region (east of the Tambo River), an area of approximately 13 500 km^2, is predominantly native vegetation with only 10% cleared for agriculture or housing. Eighty-five per cent is Crown land designated as state forest (subject to timber harvesting), parks or other types of reserve.

EASTERN VICTORIA

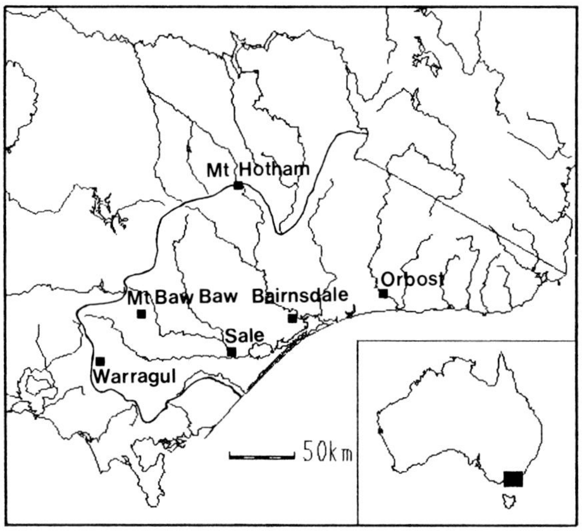

Fig. 1. The study area for predicting the geographic range and habitat preferences of *Isoodon obesulus* and *Perameles nasuta*.

METHODS

Habitat identification

Broad scale flora and fauna surveys were conducted in Gippsland between 1977 and 1985 by the survey teams of the Department of Conservation, Forests and Lands (Fig. 2), (Forbes, Gullan and Walsh 1981; Norris *et al.* 1983).

During these surveys floristic assessments were made at 1 312 sites. At these sites, each 30 m in diameter, all species of vascular plant were recorded and given an estimated cover/abundance value using Braun-Blanquet ratings. The final analysis of the data from these sites and descriptions of the vegetation in Gippsland and contained in Gullan, Walsh and Forbes (1981) and Forbes, Walsh and Gullan (1982). The names used for vegetation communities in this paper are based on those works.

Of these sites 964 were trapped using 20 wire cage traps (36×20×16 cm) or Elliott aluminium

EASTERN VICTORIA

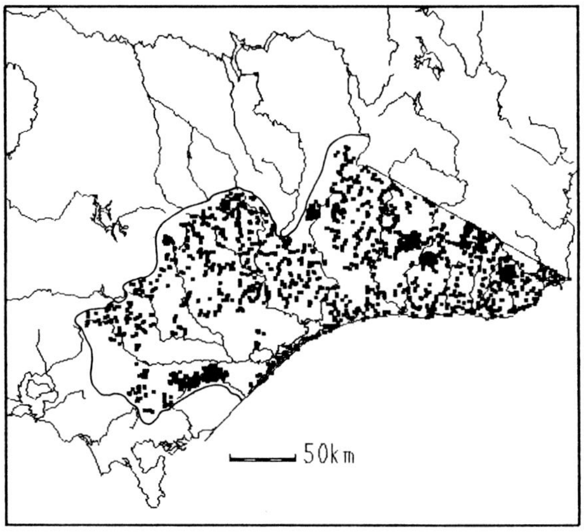

Fig. 2. Distribution of flora and flora/fauna sites in the study area of *Isoodon obesulus* and *Perameles nasuta*.

box traps (30×10×10 cm) and baited with rolled oats, honey and peanut butter. Traps were set at each site for two nights.

More intensive flora and fauna surveys have also been carried out in smaller areas. One was in a 200 km^2 area north of Yarram (Suckling 1980) in which he collected floristic data from 10×10 m quadrats at 275 mammal trapping sites in the Gormandale region. Each site was trapped for three nights using Elliott traps and wire cage traps (35×13×13 cm or 46×20×20 cm) baited with peanut butter, honey and rolled oats.

Others were in 13 forest blocks scattered throughout East Gippsland (Carr, Horrocks, Cherry, Opie, Triggs and Schulz 1984; Chesterfield *et al.* 1984; Horrocks *et al.* 1984; MacFarlane, Loyn, Chesterfield, Traill and Triggs 1984; Opie *et al.* 1984; Brown, Carr, Cherry, Craig, Horrocks, Menkhorst, Opie and Triggs 1986; Carr, Brown, Cherry, Horrocks, Opie and Triggs 1986; Cherry *et al.* 1987; MacFarlane, Parkes, Schulz, Traill and Triggs 1987; Opie, Brown, Carr, Cherry, Horrocks, Menkhorst and Triggs 1987; Parkes, MacFarlane, Schulz and Traill 1987; Schulz, MacFarlane, Parkes and Traill 1987). The forest blocks ranged in size from 100 to 150 km^2 and were part of an area designated for hardwood production.

During these surveys floristic data were collected from 672 sites. Of these sites 163 were trapped using up to 20 traps (including Elliott traps and wire cage traps 36×20×16 cm or 54×36×34 cm) at each site for 1–3 nights.

Carr *et al.* (1981) carried out a combined floristic and fauna survey in the Driffield region. Data were collected from 75 mammal sites and 120 vegetation sites.

Parr-Smith (1978) collected from 84 sites on the Holey Plains. Mammal data were not collected in this survey.

Prediction of habitat using floristics

The technique used successfully by Gullan and Norris (1984) and Mansergh (1983) to predict potential habitat for *Burramys parvus* in the alpine region was applied to the two bandicoot species, *I. obesulus* and *P. nasuta,* in Gippsland.

For the *B. parvus* work the preferred habitat of the animals was determined from an intensive flora and fauna survey of an area where the species was known to occur. Subsequently, a wider area was then surveyed and traps set at sites where similar habitat/vegetation was found. For the present study the wider search was of a large number of vegetation sites stored in a computer data bank rather than a search for sites in the field.

The technique draws on the two main sources of habitat information available. These are:

1. The floristic descriptions for all sites where the bandicoots had been trapped (33 sites). These sites were called the reference sites.
2. The floristic descriptions for the sites at which bandicoots had not been trapped or at which trapping had not been conducted (>2 600 sites).

In order to find potential new habitats the floristic composition of the reference sites was compared with that of all the other sites in Gippsland using a computer programme called GROUPSIM (Gullan, unpublished data) which incorporated the Jaccard Coefficient — a numerical similarity coefficient. A site was considered to be potential new habitat if it exhibited a similarity value greater than 0.5 (on a scale of 0 to 1) with a reference site. This meant that a reference site and potential habitat site usually shared more than two-thirds of their plant species, i.e., they were very similar floristically. In all cases sites with a similarity of 0.5 or greater were classified as part of the same floristic unit in Gullan *et al.* (1981) and Forbes *et al.* (1982).

Prediction of habitat using climate

A bioclimatic prediction system (Busby 1985) was used to determine climate profiles (based on 12 climate parameters) for the reference sites for each bandicoot species and to extrapolate from this information to determine other geographic localities in Gippsland with similar climate profiles. Information provided for each species was the latitude-longitude co-ordinates and altitude of each reference site. This information was then analysed using the BIOCLIM programme and a map of Gippsland was produced with predicted, potential sites marked.

RESULTS

Prediction of habitat using floristics

Isoodon obesulus

This species was trapped at 18 sites from seven localities in Gippsland (Fig. 3). All sites are within 30 km of the coast and at altitudes of less than 100 m ASL. Four vegetation communities are represented by these sites but most captures (14 out of 17) are in lowland sclerophyll forest in the east, or *Leptospermum myrsinoides* heathland, in the west of the study area.

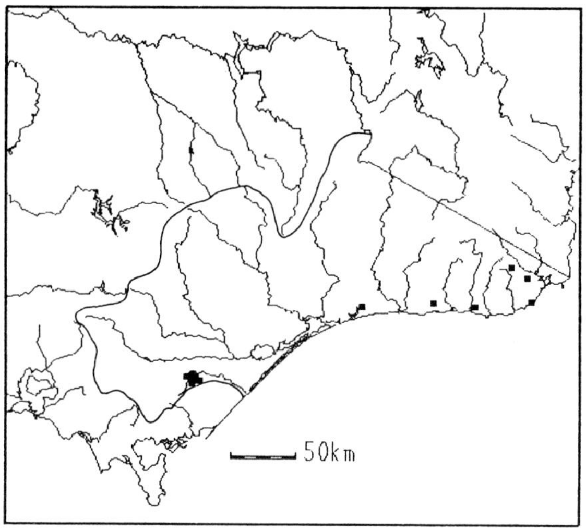

Fig. 3. Location of sites (reference sites) at which *Isoodon obesulus* was rapped in Gippsland. Sites in the south-west of the study area were not used in the bioclimatic analysis.

A further 204 sites from over 27 localities were found to have similar floristics to one or more of the reference sites. These predicted sites are all within 40 km of the coast and occur at altitudes of 300 m ASL or less (mean = 80 m ASL; Fig. 4). Six vegetation communities are represented by these sites although the majority of predictions (80%) are for lowland sclerophyll forest or *Leptospermum myrsinoides* heathland (Table 1).

Two communities in which *I. obesulus* has not been trapped are represented by predicted sites. The first is damp sclerophyll forest of a form which has close floristic affinities with components of *Leptospermum myrsinoides* heathland (Gullan, Forbes, Earl, Barley and Walsh 1985). The second is coastal sclerophyll forest which has close floristic and geographic affinities with lowland sclerophyll forest (Forbes *et al.* 1982). Most of the sites are on relatively well-drained, sandy soils.

Perameles nasuta

This species was trapped at 15 sites from 11 localities in Gippsland (Fig. 5). The sites are geographically more widely spread than those of

Table 1. Vegetation communities (as described by Gullan *et al.* 1985; Forbes *et al.* 1982 and Gullan *et al.* 1985) and the number of sites in each community at which *Isoodon obesulus* and *Perameles nasuta* were trapped (Ref.) or predicted (Pred.).

Vegetation community	*I. obesulus* Ref.	*I. obesulus* Pred.	*P. nasuta* Ref.	*P. nasuta* Pred.
Montane riparian forest	—	—	—	6
Riparian forest	—	—	1	1
Wet sclerophyll forest	—	—	—	1
Damp sclerophyll forest	—	17	1	10
Dry sclerophyll forest	—	—	—	3
Lowland sclerophyll forest	4	46	1	47
Grassy woodland	—	—	2	7
Leptospermum myrsinoides Heathland	10	117	2	60
Warm temperate rainforest	—	—	1	5
Banksia woodland	1	13	—	—
Coastal banksia woodland	—	—	2	—
Coastal heathland	1	1	—	—
Coastal sclerophyll forest	—	2	3	2
Unclassified	2	7	1	6

I. obesulus and occur at altitudes ranging from sea level to about 900 m ASL. Many of the sites are closely associated with watercourses. Nine vegetation communities are represented by these sites with no apparent preference for any one (Table 1).

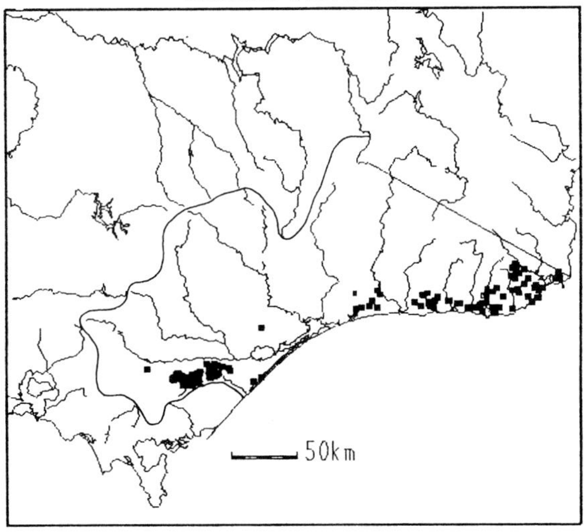

Fig. 4. Location of predicted sites for *Isoodon obesulus* in Gippsland using floristics.

A further 148 sites from over 42 localities were found to have similar floristics to one or more of the reference sites. These predicted sites cover a similar geographic range to the reference sites and occur at altitudes of up to 1 000 m ASL (mean = 200 m ASL). As with the reference sites, the distribution of the predicted sites show a close association with watercourses (Fig. 6). Ten vegetation communities are represented by the predicted sites and 72% of these are in lowland sclerophyll forest of *Leptospermum myrsinoides* heathland. Of the other communities represented (by either the reference or

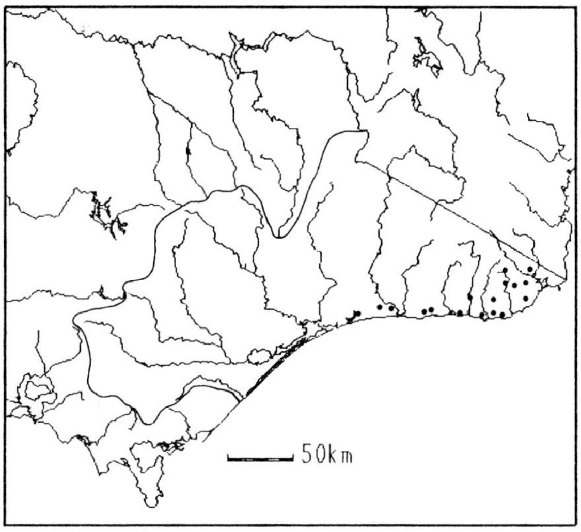

Fig. 5. Location of predicted sites for *Isoodon obesulus* in Gippsland using climate.

predicted site) six are associated with watercourses or moist soils (montane riparian forest, riparian forest, wet sclerophyll forest, damp sclerophyll forest, coastal banksia woodland, warm temperate rainforest). The most unusual of the communities represented is grassy woodland which typically lacks a dense shrub layer and is often heavily disturbed by grazing.

Habitats

Descriptions of the vegetation communities represented by references and predicted sites for *I. obesulus* and *P. nasuta* follow.

Montane riparian forest

A closed scrub to tall open forest of alluvial soils on riparian sites in hilly regions between 500 m and 1 300 m ASL throughout Gippsland.

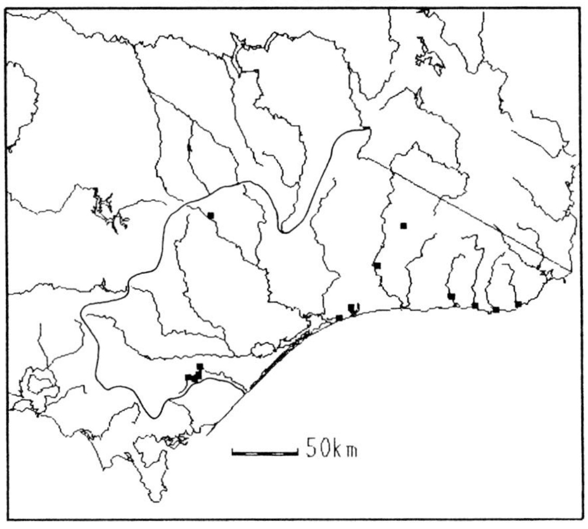

Fig. 6. Location of sites (reference sites) at which *Perameles nasuta* was trapped in Gippsland. The sites in the southwest of the study area were not used in the bioclimatic analysis.

The major tree species is usually *Eucalyptus delegatensis* but *E. pauciflora* and *E. dives* may also be present, often the tree stratum is very sparse. By contrast the shrub layer is usually dense and dominated by *Leptospermum grandifolium* to 10 m. The ground layer is largely made up of grasses, sedges and forbs.

Riparian forest

An open forest or woodland common along watercourses on alluvial soils throughout Gippsland at altitudes ranging from 20 m to 700 m ASL. This is floristically the richest community in Gippsland and often the most disturbed and weedy. The tree layer is usually made up of combinations of *E. bridgesiana, E. viminalis, E. radiata* and *E. cypellocarpa*. The shrub layer is varied and dense. The ground layer is made up of grasses, sedges and forbs.

Wet sclerophyll forest

A tall open forest on deep kraznozem soils in sheltered hillsides and gullies between 500 m and 1 100 m ASL. The principal tree species are tall *E. regnans, E. nitens, E. obliqua, E. cypellocarpa, E. viminalis* and *E. fastigata*. The shrub layer is tall and sparse and consists of broad-leafed species and tree ferns. The ground layer is dominated by rhizomatous and tussock forming ferns.

Damp sclerophyll forest

An open forest of sandy soils in south-west Gippsland at altitudes from 80 m to 300 m ASL. The principal tree species are *E. consideniana* and *E. radiata*. The shrub and ground layers consist of small-leafed sclerophyllous shrubs and herbs.

Dry sclerophyll forest

An open forest of rocky soils between 200 m and 700 m ASL in south-east Gippsland. The principal tree species is *E. globoidea*. The shrub layer is sparse and consists of small-leafed sclerophyllous species. The ground layer is dominated by tussock grasses, sclerophyllous semi-shrubs and forbs.

Lowland sclerophyll forest

An open forest of sandy soils in near coastal regions of East Gippsland at altitudes from sea level to 300 m ASL. The principal tree species are *E. sieberi* and *E. globoidea*. The shrub layer is usually sparse and consists of a range of small and medium-leafed sclerophyllous species. The ground layer is usually dominated by grasses and sclerophyllous semi-shrubs and forbs.

Grassy woodland

An open forest to woodland of clay-loam soils on flat to gently undulating terrain in south-west Gippsland at altitudes of 20 m to 220 m ASL. The principal tree species are *E. radiata* and *E. viminalis*. The shrub layer is sparse. The ground layer is dominated by grasses and other monocotyledons and herbaceous dicotyledons.

Leptospermum mysinoides *heathland*

A woodland with a heathland understorey on sandy soils in south-west Gippsland at altitudes from sea level to 300 m ASL. The understorey is always dominated by the shrub *L. myrsinoides* but the tree layer varies species. In all areas *Banksia serrata* is an abundant species.

Warm temperate rainforest

A closed forest on deep loamy soils in sheltered gullies in East Gippsland at altitudes between 40 m and 500 m ASL. The dominant tree species is the broad-leafed *Acmena smithii* (Lilly Pilly) which is often covered by a range of broad-leafed climbers. The understorey is often sparse and dominated by ferns.

Banksia woodland

A woodland of near coastal, sandy soils throughout East Gippsland from sea level to 200 m ASL. The principal tree species is *B. serrata* although a sparse canopy of *E. botryoides, E. globoidea, E. gummifera, E. consideniana, E. sieberi* or *E. muelleriana* is usually present. The shrub layer may be quite dense and support a range of small-leafed sclerophyllous species. The ground layer is dominated by grasses, sedges and lilies.

Coastal banksia woodland

A woodland of coastal to near coastal sites on sandy soils in East Gippsland at altitudes from sea level to 40 m ASL. The chief tree species is *B. integrifolia* although a few *B. serrata* and *E. botryoides* may be present. The shrub layer may be very dense and consist of thickets of *Melaleuca* spp. and *Ghania clarkei* tussocks.

Coastal heathland

An open forest to closed heath of sandy soils on near coastal sites in East Gippsland at altitudes of 20 m to 150 m ASL. This vegetation is dominated by small-leafed, sclerophyllous shrubs, semi-shrubs and sedges. In the wetter sites dense thickets of *Melaleuca* spp. or plains of *Xanthorrhoea* spp. may develop, while on drier sites grasses and forbs may become abundant.

Coastal sclerophyll forest

An open forest of sandy soils on near coastal, often riparian, sites of East Gippsland at altitudes from sea level to 80 m ASL. The principal tree species is *E. botryoides*. The shrub layer is usually sparse while the ground layer is dominated by grasses and lilies.

Habitat prediction using climate

Nineteen sites in the study area have been predicted by the bioclimatic analysis as climatically suitable habitat for *I. obesulus* (Fig. 7). They are all within 40 km of the coast and contained within the region of potential habitat predicted using floristics.

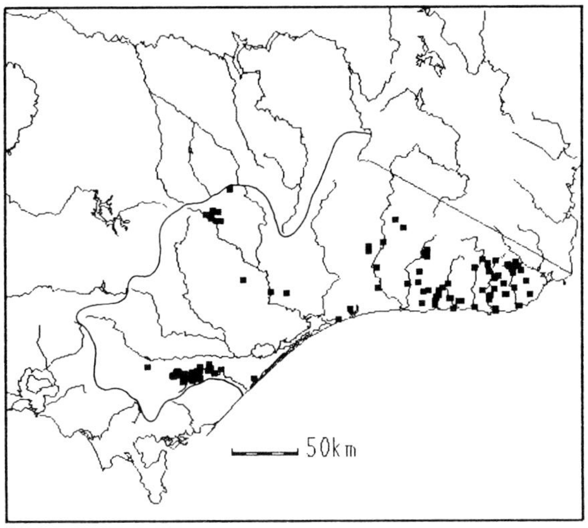

Fig. 7. Location of predicted sites for *Perameles nasuta* in Gippsland using floristics.

The bioclimatic analysis predicts a much broader geographic range for *P. nasuta* than for *I. obesulus*. A total of 149 sites were predicted in the study area and these extend from the coast inland (Fig. 8). The region of potential habitat predicted using florstics is contained within that predicted by the biolcimatic analysis.

The bioclimatic predictions for *P. nasuta* add an extensive range of potential habitat in the south-west of the study area — an area which has suffered considerable disturbance in the past. Although climatic conditions in this area may be suitable for *P. nastua* it appears that the extant vegetation is not.

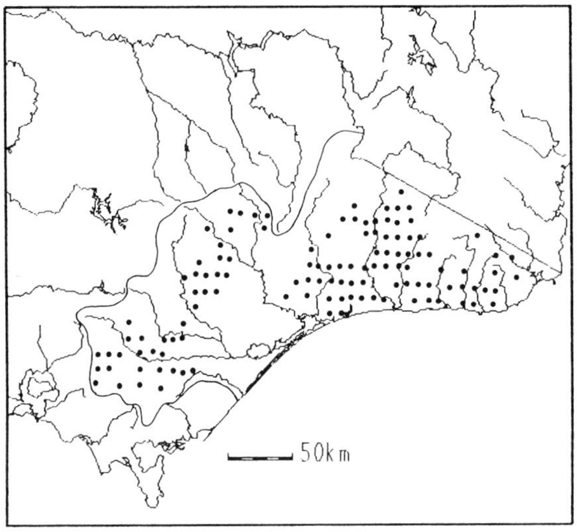

Fig. 8. Location of predicted sites for *Perameles nasuta* in Gippsland using climate.

DISCUSSION

The similarity of the results of the bioclimatic and floristic predictions adds support to the habitat descriptions presented earlier in this paper. It is clear from these descriptions that although the two bandicoot species overlap considerably in their geographic range there are substantial differences in their habitat preferences.

Perameles nasuta occurs throughout a wide geographic range but shows a marked preference for habitats where water is an important feature (e.g., riparian sites or sites with moist soils). *Isoodon obesulus* is more restricted in its geographic, climatic and habitat range and tends to favour sites with well-drained sandy soils. Both species are probably most abundant in *Leptospermum myrsinoides* heathland in the west and lowland sclerophyll forests in the east. However, within these vegetation communities *P. nasuta* was trapped or predicted most often in wetter sites or those nearer watercourses, whereas *I. obesulus* data did not reflect this preference.

An interesting difference between the bioclimatic and floristic analyses is the prediction of *P. nasuta* in parts of western Gippsland by the former but not by the latter. This area, centred south of Sale, consists of large expanses of alienated or at least heavily disturbed land which no longer supports the original native vegetation. This clearly suggests that *P. nasuta* was once more widespread in this area but, although the climate is suitable for the animal, the habitat can no longer support it. The grassy woodland community, in which *P. nasuta* has been trapped is the principal habitat to have suffered in this region and was the first vegetation community to have been utilized by domestic stock.

Both *I. obesulus* and *P. nasuta* are considered by most mammal trappers in Victoria to be trap shy. This hypothesis is supported in two ways by the trapping data collected from the combined flora and fauna surveys described in this paper. First, the trap success for both species is extremely low, even by Australian standards (0.01% for *I. obesulus*, 0.04% for *P. nasuta* — National Parks and Wildlife unpublished data). Second, vegetation similar to that in which bandicoots have been trapped is widespread in Gippsland.

The bioclimatic analysis of trapped sites further supports this hypothesis by predicting a large number of sites (other than the reference sites) where the climate may be suitable for each species.

However, the other unmistakable conclusion that could be drawn from these results is that both species are rare in Victoria. The alarming part of this conclusion is that if they are rare we do not know why. With both the climatic and floristic predictions suggesting an apparent abundance of suitable habitat it could only be that environmental factors more subtle than those we have measured in this paper are the causal factors for either the failure of these species in many sites, or their decline from those sites. If a decline in numbers in many previously favourable sites is a reason for the animals' rarity then we must examine our management programmes for those sites.

Whatever conclusions one draws from the data presented in this paper it is obvious that a programme for the conservation and management of these species is needed. The first step in this programme must be to find out which conclusion is true. If it is the former, then better methods of trapping and assessment must be developed before an understanding of their ecological requirements can be gained. If it is the latter then the problem becomes much more urgent and a closer examination of sites with similar vegetation and climate where the animals do or do not occur will be necessary. Here the emphasis should be on the differences in history and management practices for the sites as well as the assessment of potentially competing or predatory fauna. Either way it is hoped that the predictions from this paper will provide future workers with a habitat framework within which to carry out these investigations.

ACKNOWLEDGEMENTS

The authors gratefully acknowledge John Busby (Bureau of Flora and Fauna, Canberra) for carrying out the BIOCLIM analysis and Geoff Carr, Geoff Parr-Smith and Graeme Suckling for access to their raw data.

REFERENCES

AHERN, L. D., 1982. Threatened wildlife in Victoria and issues related to its conservation. *Fisheries and Wildlife Paper, Victoria No. 27.*

BROWN, G. W., CARR, G. W., CHERRY, K. A., CRAIG, S. A., HORROCKS, G. F. B., MENKHORST, K. A., OPIE, A. M. AND TRIGGS, B. E., 1986. Flora and fauna of the Quadra Forest Block, East Gippsland, Victoria. *Department of Conservation, Forests and Lands, Ecological Survey Report No. 6.*

BUSBY, J. R., 1985. Bioclimatic prediction system — users manual version 1.2. Bureau of Flora and Fauna: Canberra.

CARR, G. W., HORROCKS, G. F. B. AND ISLES, A., 1981. Biological environment and effects of the proposal. Pp. 11.1–11.26 in *Proposed Driffield Project environmental effects statement.* Kinhill Pty Ltd: Melbourne.

CARR, G. W., HORROCKS, G. F. B., CHERRY, K. A., OPIE, A. M., TRIGGS, B. E. AND SCHULZ, M., 1984. Flora and fauna of the Coast Range Forest Block, East Gippsland, Victoria. *Department of Conservation, Forests and Lands, Ecological Survey Report No. 4.*

CARR, G. W., BROWN, G. W., CHERRY, K. A., HORROCKS, G. F. B., OPIE, A. M. AND TRIGGS, B. E., 1986. Flora and fauna of the Buldah Forest Block, East Gippsland, Victoria. *Department of Conservation, Forests and Lands, Ecological Survey Report No. 7.*

CHERRY, K. A., BROWN, G. W., CARR, G. W., HORROCKS, G. F. B., MENKHORST, K. A., OPIE, A. M. AND TRIGGS, B. E., 1987. Flora and fauna of the Freezeout and Pyke Forest Blocks, East Gippsland, Victoria. *Department of Conservation, Forests and Lands, Ecological Survey Report No. 13.*

CHESTERFIELD, E. A., MACFARLANE, M. A., ALLEN, D., HUTCHINSON, M. N., TRIGGS, B. AND BARLEY, R., 1983. Flora and fauna of the Rodger Forest Block East Gippsland, Victoria. *Department of Conservation, Forests and Lands, Ecological Survey Report No. 1.*

FORBES, S. J., GULLAN, P. K. AND WALSH, N. G., 1981. Sites of botanical significance in East Gippsland. *Ministry for Conservation, Victoria, Environmental Studies Series No. 322.*

FORBES, S. J., WALSH, N. G. AND GULLAN, P. K., 1982. Vegetation of East Gippsland. *Muelleria* **5:** 53–113.

GULLAN, P. K., WALSH, N. G. AND FORBES, S. J., 1981. Vegetation of the Gippsland Lakes catchment. *Muelleria* **4:** 333–83.

GULLAN, P. K. AND NORRIS, K. C., 1984. The habitat of *Burramys parvus* (Broom) in Victoria. Pp. 417–21 in *Possums and gliders* ed by A. P. Smith and I. D. Hume. Surrey Beatty & Sons: Sydney.

GULLAN, P. K., FORBES, S. J., EARL, G. E., BARLEY, R. H. AND WALSH, N. G., 1985. Vegetation of South and Central Gippsland. *Muelleria* **6:** 97–145.

HORROCKS, G. F. B., OPIE, A. M., CARR, G. W., CHERRY, K. A., BROWN, G. W., CRAIG, S. A. AND TRIGGS, B. E., 1984. Flora and fauna of the Ellery Forest Block, East Gippsland, Victoria. *Department of Conservation, Forests and Lands, Ecological Survey Report No. 3.*

MACFARLANE, M. A., LYON, R. H., CHESTERFIELD, E. A., TRAILL, B. J. AND TRIGGS, B. E., 1984. Flora and fauna of the Scorpion and Dawson Forest Blocks, East Gippsland, Victoria. *Department of Conservation, Forests and Lands, Ecological Survey Report No. 5.*

MACFARLANE, M. A., PARKES, D. M., SCHULZ, M., TRAILL, B. J. AND TRIGGS, B. E., 1987. Flora and fauna of the Buckland Forest Block, East Gippsland, Victoria. *Department of Conservation, Forests and Lands, Ecological Survey Report No. 8.*

MANSERGH, I. W., 1983. Attempts to locate the Mountain Pygmy-possum *Burramys parvus* in the Mt Hotham–Mt Lock-western fall of the Bogong High Plains areas, November 1982–February 1983. *Arthur Rylah Institute, Environmental Research Technical Report Series 6.*

NORRIS, K. C. AND MANSERGH, I. W., 1981. Sites of zoological significance in East Gippsland. *Ministry for Conservation, Victoria, Environment Studies Series No. 321.*

NORRIS, K. C., MANSERGH, I. M. AHERN, L. D., BELCHER, C. A., TEMBY, I. D. AND WALSH, N. G., 1983. Vertebrate fauna of the Gippsland Lakes catchment, Victoria. *Arthur Rylah Institute, Occasional Paper Series No. 1.*

OPIE, A. M., CHERRY, K. A., HORROCKS, G. F. B., CARR, G. W., SCHULZ, M. AND TRIGGS, B. E., 1984. Flora and fauna of the Yalmy Forest Block, East Gippsland, Victoria. *Department of Conservation, Forests and Lands, Ecological Survey Report No. 2.*

OPIE, A. M., BROWN, G. W., CARR, G. W., CHERRY, K. A., HORROCKS, G. F. B., MENKHORST, K. A. AND TRIGGS, B. E., 1987. Flora and fauna of the Splitters Range Forest Block, East Gippsland, Victoria. *Department of Conservation, Forests and Lands, Ecological Survey Report No. 9.*

PARKES, D. M., MACFARLANE, M. A., SCHULZ, M. AND TRAILL, B. J., 1987. Flora and fauna of the Nunniong North Forest Block, East Gippsland, Victoria. *Department of Conservation, Forests and Lands, Ecological Survey Report No. 11.*

PARR-SMITH, G. A., 1978. The vegetation of the Holey Plains State Park. *Ministry for Conservation, Victoria, Environmental Studies Series No. 208.*

SCHULZ, M., MACFARLANE, M. A., PARKES, D. M. AND TRAILL, B. J., 1987. Flora and fauna of the Mount Murray Forest Block, East Gippsland, Victoria. *Department of Conservation, Forests and Lands, Ecological Survey Report No. 14.*

SUCKLING, G. C., 1980. The effects of fragmentation and disturbance of forest on mammals in a region of Gippsland, Victoria. Ph.D. Thesis: Monash University, Clayton: Victoria.

Stress and population regulation in *Isoodon obesulus* (Shaw and Nodder)

LESLEY N. THOMAS[1]

Isoodon obesulus is a solitary species whose population growth is controlled by territoriality at low population densities. However, when suitable habitat is limited, the ratio of transient to resident individuals increases thus reducing the effectiveness of this mechanism. In this situation, the high reproductive potential of this species suggests that populations of *I. obesulus* could increase significantly in both size and density in a short period of time.

Research on microtine species has shown that the level of stress induced by high population density is sufficient to induce intrinsic population regulation mechanisms.

An enclosed population of bandicoots was studied for 15 months, during which data was collected on population size, breeding, haematology and plasma free corticosteroid levels. The data showed significant seasonal fluctuations.

Key words: Stress, population regulation, *Isoodon obesulus*, population density, ACTH, corticosteroids, haematology.

INTRODUCTION

PREDICTABLE behavioural and demographic phenomena are associated with chronic stress (Novak 1978; Christian 1978). These phenomena form the basis of current theories on small mammal populations which suggest that regulation in many species is derived from density-dependent intrinsic, rather than extrinsic factors (Christian 1950; Krebs and Myers 1974; Christian 1978). The basis of this concept is that as population density increases, social contact between conspecifics increases in frequency and is often associated with a higher frequency of aggressive interactions (Louch 1956; Gains and McClenaghan 1980). Where suitable habitat is limited and population density is high, the individual home range or territory is smaller (Brant 1962). There is greater strain on the natural resources of the population and greater competition between individuals for limited resources (Louch 1956; Windberg and Keith 1976). Under these circumstances the population is said to be experiencing chronic social stress.

During a period of chronic physiological or psychological stress, adrenocorticotrophic hormone (ACTH), secreted by the pituitary, stimulates the release of glucocorticoid hormones from the adrenal cortex. Chronically high concentrations of plasma glucocorticoids may result in loss of body weight and condition due to increased protein catabolism, and supress immune responses by reducing antibody production and inhibiting inflammatory and allergic reactions (Liddle 1981; Cassileth 1983). Erythrocyte number, volume and haemoglobin content are also affected by acute stress and changes in plasma glucocorticoid concentration (Cheal, Lee and Barnett 1976; Brasmer 1984). Dispersion, higher rates of mortality in all age classes but especially juveniles, and a decrease in the number of live births are associated with chronic stress in many species and are important means of population regulation in these species (Fairbairn 1977; Christian 1978; Andrews 1979). Most of this research has centred on microtine rodents in the northern hemisphere.

Using criteria such as changes in the concentration of plasma glucocorticoids, glucose, free fatty acids, plasma urea and amino acids, the sensitivity of some species of Marsupialia to chronic or acute stress has been determined by manipulative experiments (Griffiths *et al.* 1969; Than and McDonald 1974; Cooley and Janssens 1977; McDonald and Bradshaw 1977; Janssens and Hinds 1981) but few species have been studied in the field (Barnett 1979; Bradley, McDonald and Lee 1976, 1980; Miller and Bradshaw 1979).

Isoodon obesulus (Marsupialia: Peramelidae) is a species with low social tolerance (Heinsohn 1966), a high reproductive potential (Stoddart and Braithwaite 1979) and a high frequency of aggressive interactions between conspecifics (Jones 1924; O'Callaghan 1974). This evidence suggests that this species has the potential to increase size rapidly under optimum conditions though its non-gregarious nature suggests intolerance of high population densities.

In order to examine the role of stress in population regulation in *I. obesulus,* an enclosed population, free of mammalian predators, was monitored for 15 months. Demographic indices, such as changes in population size and density,

[1]Zoology Department, University of Western Australia, Western Australia, Australia 6009.
Present address: External Studies Unit, Murdoch University, Murdoch, Western Australia, Australia 6150.
Pages 335–43 in BANDICOOTS AND BILBIES, ed by J. H. Seebeck, P. R. Brown, R. L. Wallis and C. M. Kemper. Surrey Beatty & Sons, Sydney, 1990.

patterns of reproductive activity and intraspecific aggression, were examined. Stress was assessed by examining changes in certain physiological and haematological indices.

MATERIALS AND METHODS

Field procedures

The Harry Waring Marsupial Reserve (32° 1′S, 115° 50′E) is composed of 254 ha of open Jarrah/Banksia woodland and sclerophyllous heathland. It contains an ephemeral "cladium" peat lake, Lake Banganup, and past studies have shown that *I. obesulus* mainly inhabits the dense vegetation around this lake (Monzu 1970; Wright 1975; Craven 1981).

In this study, I intensively trapped a small area of 3 ha at the southern end of Lake Banganup to maximize recaptures. Trapping intensity varied between 70 and 120 trapnights per four day period each month between March, 1982 and June, 1983 inclusive. Treadle operated National traps (National Trap Co., USA), 50 × 20 × 25 cm, were used throughout the study. All population estimates were computed using the Manly-Parr Method (Manly and Parr 1968) which assumes neither a constant survival rate nor that survival is independent of age.

On the morning following capture, a blood sample was obtained by orbital sinus puncture (Halpern and Pachaud 1951). These samples were obtained within 3 min of first disturbing the animals. Blood samples were stored at 4°C for 1–12 min before being centrifuged, then the plasma was frozen and stored. Subsamples of whole blood needed to determine haematological values were taken from mixed blood within 10–15 min of the sample being obtained.

Number 3 fingering ear-tags (Salt Lake Stamp Co., Utah, USA) were used to tag each animal caught. Each bandicoot was weighed to the nearest 10 g and its sex, and the location and extent of any scars, were recorded. Where possible, the number, age and sex of any pouch young were noted using Heinsohn's (1966) data on the development of external features in pouch young of *I. obesulus*.

In order to determine the level of *Salmonella* infection in the population, rectal swabs were taken from each animal caught in 1983. Sterile cotton buds were used, obtaining a faecal sample where possible. The swabs were broken into a vial of 4 ml strontium chloride B broth (Iveson 1971, 1973) and sent to the Department of Public Health for detection and identification of any *Salmonella* infections.

Each animal was briefly searched for external parasites.

Blood sample analysis

Packed cell volume (PCV): a 75 μl heparinized haematocrit tube was filled with fresh whole blood and centrifuged at 12 000 rpm for 2 min. The proportion of packed red cells to total volume was expressed as a percentage.

Red blood cell count (RBC): 10 μl of whole mixed unclotted blood was diluted with 2.0 ml Hayem's solution and thoroughly mixed. The count was made using a haemocytometer with a standard Neubauer ruling.

Plasma haemoglobin concentration was determined by the cyanomethaemoglobin method described in Cheal *et al.* (1976).

Mean corpuscular haemoglobin concentration (MCHC), mean corpuscular haemoglobin (MCH) and mean corpuscular volume (MCV) were all calculated by standard methods.

White cell count (WCC): 20 μl of whole unclotted blood was diluted with 100 l Turk's Fluid and mixed well. Leucocytes were counted using a haemocytometer with standard Neubauer ruling.

Differential cell count (DCC): fresh, unclotted, whole blood was used to make blood smears on clean glass slides. These were rapidly air-dried by vigorous waving, fixed in methanol and stained with May-Grunwald-Giemsa's stain (Disbrey and Rack 1970). Five lots of 100 cells were counted in longitudinal strips running from one end of the smear to the other. The differential totals were expressed as mean percentages (Dacie 1975). Blood smears were also examined for the presence of haemoparasites.

Total plasma protein was determined to the nearest 0.1 g dl^{-1} using a portable refractometer (Atago, Japan).

Analysis of total plasma corticosteroid concentration, maximum corticosteroid binding capacity and plasma albumin concentration are as for Thomas and Bradley (this volume).

RESULTS

Population size and density

Peak population estimates for 1982 occurred between June and September and for 1983 from April until the end of the study (Fig. 1a). Population estimates were lowest during November 1982. Capture rates during November and December were low, only two and three individuals were caught during each month. Changes in population estimates in 1983 followed the pattern recorded in 1982 (Fig. 1a).

Changes in population density mirrored fluctuations in population size since the size of the study area remained constant. Population density estimates were extrapolated from the

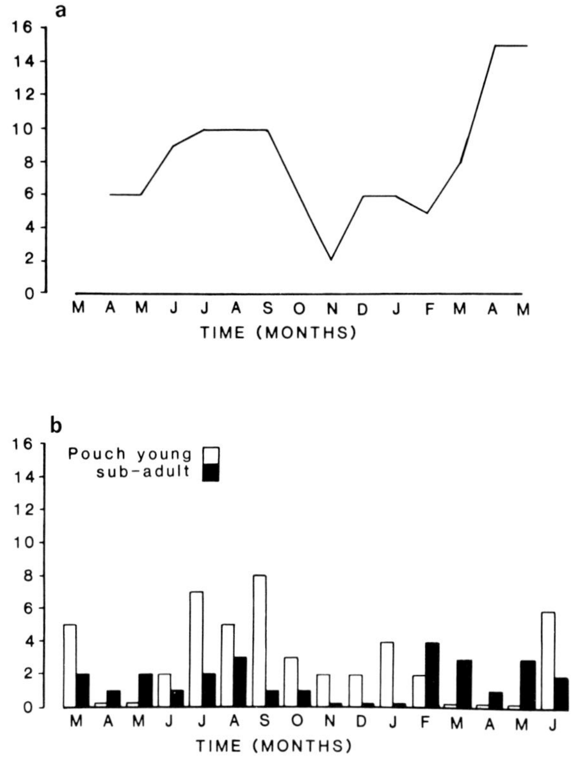

Fig. 1. (a) Population numbers of *Isoodon obesulus* estimated using the Manly-Parr Method (Manly and Parr 1968). (b) Number of subadult *Isoodon obesulus* caught and the number of pouch young observed in the population each month during the study period.

Manly-Parr estimates after including a 50 m wide buffer zone around the study site. The buffer zone was included to allow for movements of individuals within, and adjacent to, the study area. Craven (1981) found that the average width of home ranges in this particular population was 100 metres. Population density ranged from 0.14 to 1.00 animal ha^{-1}.

Isoodon obesulus reach sexual maturity at five to six months of age, at which time they weigh approximately 600 g (Heinsohn 1966). On this basis the trappable population can be divided into adults (>650 g) and subadults (<650 g). Individuals weighing less than 400 g were rarely caught. Subadults were caught only in those months corresponding to high population estimates and density (Fig. 1b).

Breeding data

Litter sizes during 1982 and 1983 were noticeably smaller than in 1981 from the same population (Craven 1981) when a maximum of seven pouch young in a single litter was recorded and litters of three and four young were common. The maximum litter size recorded in this study was three young which was recorded only twice (Table 1). During the study, in-pouch mortality was low, only one litter of two was lost before being weaned.

Between March and September 1981 (inclusive) 67% of all adult females caught were carrying pouch young whereas for the same period in 1982 only 45% of all adult females caught had young. Trapping ceased before the end of the 1983 breeding season.

A maximum of four litters of pouch young a month was noted during July and September 1982 (Table 1). All adult females caught during these months had young present. No pouch young were recorded in April or May 1982 or during March, April or May 1983.

The amalgamation of all *I. obesulus* trapping records from the reserve from 1975 to March 1985, shows a definite July to September peak in the percentage of females bearing young and

Table 1. Litter size and distribution of births of *Isoodon obesulus* during 1982/83 study period. Age of all pouch young was determined using the criteria of Heinsohn (1966). Approximate date of birth of each litter could then be calculated.

Month	Total no. females caught	No. litters "caught"	Individual litter size	No. litters born
March	3	2	2,3	—
April	2	0		—
May	3	0		—
June	4	1	2	2
July	6	4	1,1,1,2	4
August	7	2	1,1,1	1
September	4	4	2,2,2,2	3
October	4	1	3	2
November	2	1	2	1
December	2	1	2	1
January	3	2	2,2	2
February	4	1	2	—
March	6	0		—
April	3	0		—
May	3	0		—
June	5	3	2,2,2	3

Table 2. Amalgamation of all breeding records *Isoodon obesulus* from the Harry Waring Reserve (1975–1984) with additional records courtesy of B. T. Clay.

Month	No. females caught	x=0	No. females with x no. young x=1	x=2	x=3	x=4	x=5	Litter size x̄±SE
January	11	6	1	3	1			2.0±0.35
February	18	10	1	6	1			2.0±0.25
March	38	31	1	3	3			2.28±0.32
April	34	29	1	3		1		2.20±0.44
May	30	23	3	3		1		1.86±0.37
June	22	12	1	7	1	1		2.23±0.27
July	25	10	7	5	2		1	1.86±0.27
August	24	6	4	7	3	2	2	2.5±0.27
September	26	4	2	8	3	8	1	2.95±0.23
October	7	4		1	2			2.67±0.39
November	3	2		1				1
December	14	7	1	5	1			2.0±0.28

in the average litter size (Table 2) $F_{[11,86]}=2.01$, $P<0.05$). This seasonal trend appears to be constant for this reserve and no relationship was found between litter size and the body weight of the mother.

Intraspecific aggression

Fighting in unconfined *I. obesulus* results in loss of fur, here referred to as "scarring". These "scars" are not permanent as the pelage shows partial regrowth after four weeks and is nearly fully regrown at eight weeks. Thus the monthly average area of scarring per individual may be used as an indicator of changing levels of intraspecific agressiveness. A positive correlation ($r=0.73$, $\upsilon=90$, $P<0.05$) was found between population estimates and average scar area (Fig. 2b). During the first five months of the breeding season in 1982, average scar area was larger than for the rest of the year (Fig. 2a). A positive correlation was also found between body weight and scar area ($r=0.401$, $\upsilon=33$, $P<0.05$). The average area of scars on male bandicoots was significantly larger than that of females ($t=3.26$, $\upsilon=82$, $P<0.001$). The averages for each sex ($\bar{x} \pm$ SE) were as follows: males 4.89 ± 0.21 cm^2 (n=32), females 1.63 ± 0.76 (n=52).

Ectoparasites

Two species of ectoparasite were observed on *I. obesulus* during the study. Five immature ticks (Leeuwenhoekiidae) were found on a female bandicoot in January 1983. No other ticks were found.

Seasonal infestations of immature harvest mites, *Odontacarus* sp. (Acari: Trombiculidae) occurred from December to May (Fig. 3). The density of these small orange mites was estimated qualitatively rather than quantitatively since they were densely clustered on sparsely furred areas, particularly around the pouch and on the testicular stalk. On heavily infested animals, mites were also found scattered over the back and the chest.

No difference was found in the degree of infestation between the sexes. Nor was a relationship established between density of mites and either TPCS or plasma-free corticosteroid concentration, though a significant positive correlation was found between WCC and the density of mites on bandicoots caught between December and February ($r=0.683$, $\upsilon=12$, $P<0.05$).

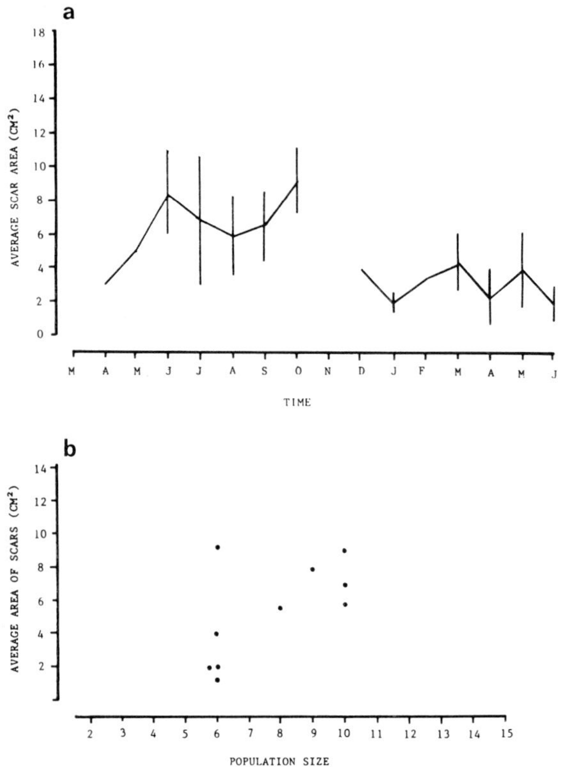

Fig. 2. (a) Average scar area (cm^2) recorded from the population of *Isoodon obesulus* each month. No animals bearing scars were caught during November (x̄±SE). (b) Relationship between average scar area and population density ($r=0.73$, $\upsilon=9$, $P = 0.05$).

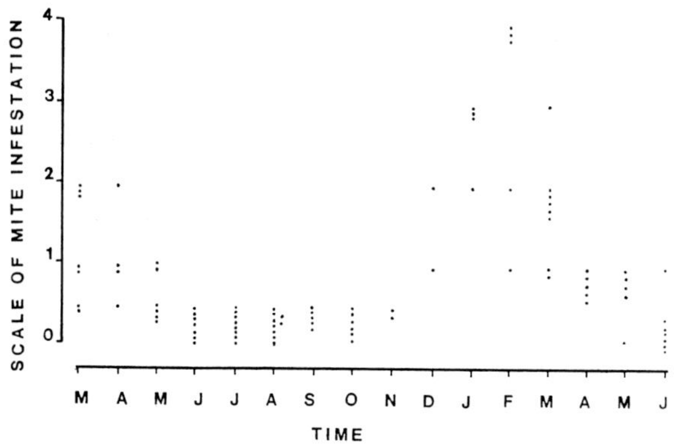

Fig. 3. Seasonal mite (Acari: *Odontocaria*) infestation on *Isoodon obesulus* during the study period. ●=one individual. Scale of infestation 0 =no mites, 1=very few, 2=light, 3=moderate, 4=heavy.

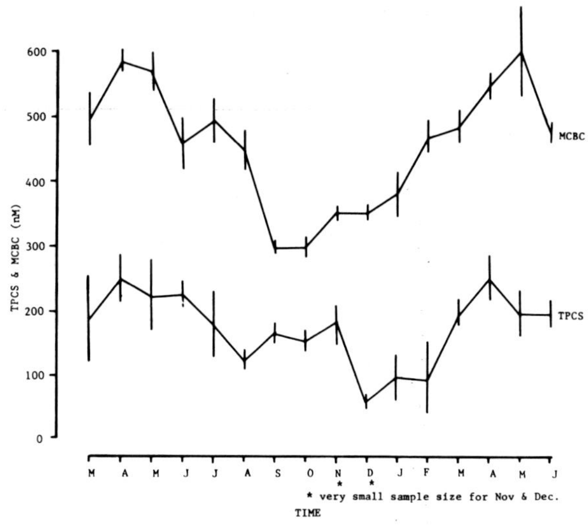

Fig. 4. Total plasma corticosteroid concentration (TPCS) and maximum corticosteroid binding capacity (MCBC) of *Isoodon obesulus*, (x̄±SE) for each month.

Salmonellae and blood parasites

No *Salmonella* infections were detected from any *I. obesulus* captured during this study despite the isolation of 11 species of *Salmonella* on the reserve since 1970 (unpublished annual reports).

Blood parasite concentrations were very low during the study. Microfilaria (species unknown) were found in two blood smears, one from a female *I. obesulus* caught in May 1982 and one from a male caught in August of the same year. Blood smears taken from these animals in September and October 1982 respectively showed no sign of microfilaria. Microfilaria have been recorded from several Macropodidae on this reserve (Ford 1975).

Blood sample analysis

Both the total plasma corticosteroid concentration (TPCS) and the maximum corticosteroid binding capacity (MCBC) showed significant seasonal variation (Fig. 4). Minimum TPCS values (60.5 to 63.0 nM), recorded in December, January and February, were only 25% of the peak values (1982: 249.6 ± 35.2 nM, 1983: 263.1 ± 36.5 nM) which were recorded in April 1982 and May 1983. MCBC values were highest in April and May of both years (1982: April 584 ± 25 nM, May 572 ± 30 nM, 1983: April 557 ± 20 nM, May 613 ± 72 nM) and were lowest during September and October 1982 (303 ± 11 nM, 302 ± 16 nM respectively). TPCS values

Table 3. Steroid partitioning in blood samples taken from unstressed male and female *Isoodon obesulus* caught between March 1982 and June 1983 (x̄ ± SD). Samples were pooled to give adequate sample sizes for statistical analysis.

Months	Sample size ♂	Sample size ♀	Concentration of free steriod (nM) ♂	Concentration of free steriod (nM) ♀	*t*-test	Percentage of free steriod ♂	Percentage of free steriod ♀
March–April	6	9	14.3±2.20	8.2±2.64	16.32 υ=17 *P*<0.001	5.18±0.60	4.18±0.37
May–June	8	8	17.9±1.75	8.9±2.39	9.48 υ=14 *P*<0.001	4.79±0.84	4.20±0.37
July–August	10	9	6.1±1.06	6.7±2.15	1.04 υ=17 *P*>0.05	4.68±0.46	4.17±0.64
Sept–Oct	2	7	9.8±0.24	12.7±1.47	3.02 υ=7 *P*<0.01	5.27±0.47	5.69±0.53
Nov–Dec–Jan	5	4	4.8±2.17	4.6±0.73	0.097 υ=7 *P*>0.05	4.83±0.52	5.36±0.71
Feb–March	5	6	8.6±0.75	7.9±2.06	0.43 υ=9 *P*>0.05	4.18±0.18	4.20±0.32
Apr–May–June	5	11	10.2±0.64	7.4±1.90	4.58 υ=14 *P*<0.001	4.05±0.22	4.32±0.41

Table 4. Red blood cell values recorded from *Isoodon obesulus* from June 1982 to June 1983 (x̄ ± SE). See text for definitions of abbreviations. *Due to the small sample size, results for November and December were combined.

		RBC	HB	PCV	MCV	MCHC	MCH
	N	x 10^{-6}cells/mm^3	%	%	m^3	pg	g 100ml^{-1}RBC
June	5	6.43±0.97	14.51±0.59	44.7±3.9(7)	77.6±18.4	32.6±3.4	25.1±5.3
July	6	6.66±0.60	14.80±0.72(8)	45.9±5.4(8)	71.4±14.3	32.2±3.0	22.1±2.3
August	5	6.73±0.44(9)	14.65±0.67(7)	44.6±4.4	66.5±8.6	32.8±3.4	22.4±2.2
September	5	7.23±0.95	15.15±0.65	45.8±3.0	65.3±9.5	33.2±3.2	20.8±2.9
October	4	6.73±0.67(5)	14.78±0.74	46.4±4.2(5)	61.9±10.0	32.3±4.1	22.4±2.8
November	2	6.92±0.06					
December	2	6.90±0.44	14.96±0.26	46.2±4.8	65.2±7.2	31.8±1.9	19.4±3.4
January	3	7.22±0.72(4)	14.66±0.69(4)	50.0±2.0	66.0±8.9	29.9±2.0	19.6±2.0
February	4	7.08±0.64	15.22±0.80(5)	43.9±6.4(5)	62.2±6.2	33.4±4.1	21.0±2.2
April	4	7.00±0.45(5)	14.62±0.67(5)	46.9±3.6	67.5±4.9	31.3±3.3	20.9±1.6
May	4	6.82±0.67	14.33±0.88	44.1±4.5	69.8±8.1	29.9±2.3	23.5±3.4
June	5	6.78±0.58	14.57±0.71	47.0±4.9	68.5±9.0	31.7±4.3	22.9±2.7

remained well below the corresponding MCBC values in all samples. Significant differences between TPCS concentrations of each sex were recorded only during July and August 1982 when the mean TPCS for females was 202 ± 30 nM and the meant TPCS for males was 81 ± 16 nM (t=8.64, υ=19, P<0.001).

During autumn and early winter of both years, the plasma-free corticosteroid concentration of male *I. obesulus* was found to be significantly higher than the concentrations observed in female *I. obesulus* (Table 3). Females were found to have significantly higher plasma-free corticosteroid concentrations than the males in September and October 1982.

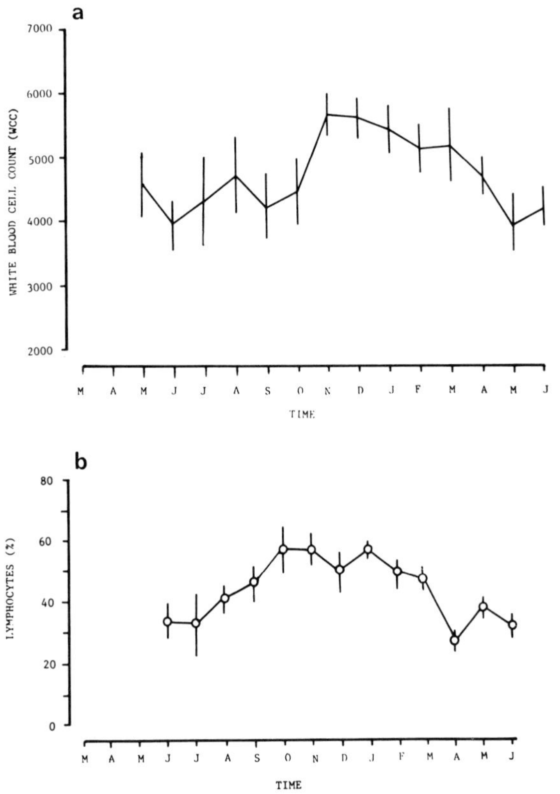

Fig. 5. (a) Mean white cell count (WCC), (x̄±SE). (b) Percentage of lymphocyte in the total WCC (x̄±SE).

The concentration of plasma-free corticosteroid recorded during August, November and December 1982, and January 1983 were lower than values recorded during the rest of the study (Table 3). The proportion of plasma-free to bound corticosteroid remained constant between 4 and 5% throughout the study period with no significant variation.

It may be argued that due to the rapid response time of plasma glucocorticoid concentration to any stressful stimulus, any blood samples taken more than a few minutes after capture would not truly reflect the values to be found in the population at that time, thus invalidating these results. It is not feasible to sample each animal within 2–3 min of capture and the procedure of holding the animals in bags, in a quiet room for 12 h before obtaining a blood sample, has been used successfully with several species (Lee and McDonald 1985).

No significant variation was found in the red blood cell parameters RBC, %Hb, PVC or their derivations MCH, MCHC and MCV (Table 4). No significant correlation was found between these and any other variable examined. It was noted, however, that during the breeding season the %HB of mature female bandicoots (both with and without pouch young) was slightly but not significantly higher than the rest of the population (t=1.43, υ=48, P<0.10).

WCC recorded during the study showed significant variation ($F_{[12,49]}$=3.12, P<0.025), and was markedly higher between November 1982 and March 1983 than during the rest of the study period (Fig. 5a). The elevation in WCC was positively correlated with heavy infestations of trombiculid mites (r=0.644, υ=12, P<0.05). Low neutrophil and high lymphocyte percentages were recorded during this period (Fig. 5b). The percentage of lymphocytes in the total WCC was negatively correlated with TPCS concentration (r=−0.5864, υ=11, P<0.05).

DISCUSSION

The population of *I. obesulus* studied exhibited several fluctuations in population estimates and density over the study period. The seasonal pattern noted was similar to that observed by Copley *et al.* (this volume) in a dense population of *I. o. nauticus* from Nuyts Archipelago. Stoddart and Braithwaite (1979) showed that *I. obesulus* is adept at utilizing newly available habitats, a view which is supported by Lee and Cockburn (1985). The low population size noted in summer was probably due more to emigration than to an increase in the mortality rate. This supposition is supported by the reappearance in the population in January and February of several bandicoots which had been caught repeatedly before November, but not during November or December.

Both TPCS and plasma-free corticoseteroid concentrations were low when the population size was at its estimated minimum. White blood cell parameters recorded during this period also indicate low stress levels since both WCC and the proportion of lymphocytes were high. The changes in white blood cell parameters noted in *I. obesulus* are similar to those associated with changing levels of stress levels in *Antechinus stuartii* (Cheal *et al.*) and *Phascolarctos cinereus* (Dickens 1978).

Despite the homeostatic stability of the free steriod fraction, elevated corticosteroid concentrations were associated with significantly depressed white cell counts and lymphocytopenia in the field population. Even though these immunosuppressive effects were observed, parasite and pathogen loads were very low in both number and diversity, apart from seasonal trombiculid mite infestations. These data suggest that either the immunosuppression noted was not of sufficient magnitude to be deleterious or that the reserve is depauperate in parasite or pathogen fauna.

It is interesting to note that the red blood cell parameters measured during this study showed little seasonal variation but high variability within each monthly sample. This variability appeared to be due to individual differences and was not related to any sub-section of the population, such as male or female, adult or sub-adult, scarred or unscarred individuals, apart from the slightly elevated haemoglobin concentration of all adult females during the breeding season. It should be noted though that the sample sizes of many of the population sub-groups were small. Significant variation in red blood cell parameters has been noted in several species of marsupial such as *P. cinereus* (Dickens 1978), *A. stuartii* (Cheal *et al.* 1976) and *Trichosurus vulpecula* and *T. caninus* (Barnett, How and Humphreys 1979a, b). In these species, variation in red blood cell parameters has been shown to relate to season, age and sex (Cheal *et al.* 1976; Dickens 1978; Barnett *et al.* 1979a, b).

One possible reason for the low-level response to elevated plasma glucocorticoid concentrations is the consistently low proportion of plasma free steroid. Not all of the plasma corticosteroid is active, most of it being bound to a high-affinity corticosteroid-binding globulin (CGB) and is biologically unavailable. Approximately 10–15% is loosely bound to albumin and only a small fraction of free corticosteroid is considered to be biologically available (Tait and Burstein 1964; Tait 1982). Studies of several species show that this fraction of plasma-free corticosteroid is usually around 10% (Sandberg, Rosenthal and Slaunwhite 1967) yet in this population of *I. obesulus* the proportion of free steriod in the TPCS remained constant between 4 and 6% with no significant variation.

Using scarring as a criterion, the effects of intraspecific aggression were greatest on males rather than females. From this it may be suggested that in mature bandicoots, sex is possibly related to an individual's position in the hierarchy and that young females may be more acceptable as recruits than are young males. Because average scar area is larger during the first half of the breeding seasons, it may be that competition between males is a major cause of fighting, but population density was relatively high during this period and this may have also contributed to increased levels of aggression. Concentration of free steroid in males at this time (May–June 1982, April–June 1983) was also significantly higher than that of female bandicoots.

Differences in TPCS and plasma free corticosteroids between the sexes were mainly apparent at the peak of the breeding season, the concentration of both being significantly higher in females than males, though at the beginning of the breeding season the reverse is true. All females caught at this time exhibited this trend, even sexually immature females. It would be interesting to compare these results with data from a peak or pre-peak population of *I. obesulus* to see if this phenomenom was an annual occurrence related to mating, pregnancy and post-natal care of young, or whether it was more related to the combined effects of high population density and the breeding season coinciding.

ACKNOWLEDGEMENTS

I would like to thank Dr Adrian Bradley for all of his invaluable help with this project and for his critical comments on the manuscript. Many people gave me their assistance in the field, these include Brian Clay, Greg Thomas, David Algar and Margaret Heel.

REFERENCES

ANDREWS, R. V., 1979. The physiology of crowding. *Comparative Biochemistry and Physiology* **63A:** 1–6.

BARNETT, J. L., 1979. A stress response in *Antechinus stuartii* Macleay. *Australian Journal of Zoology* **27:** 913–26.

BARNETT, J. L., HOW, R. A. AND HUMPHREYS, W. F., 1979a. Blood parameters in natural populations of *Trichosurus* species (Marsupialia: Phalangeridae) 1. Age, sex and seasonal variation in *T. caninus* and *T. vulpecula. Australian Journal of Zoology* **27:** 913–26.

BARNETT, J. L., HOW, R. A. AND HUMPHREYS, W. F., 1979b. Blood parameters in natural populations of *Trichosurus* species (Marsupialia: Phalangeridae) 2. Influence of habitat and population strategies in *T. caninus* and *T. vulpecula. Australian Journal of Zoology* **27:** 927–38.

BRADLEY, A. J., MCDONALD, I. R. AND LEE, A. K., 1976. Corticosteroid binding globulin and mortality in a dasyurid marsupiial. *Journal of Endocrinology* **70:** 323–24.

BRADLEY, A. J., MCDONALD, I. R. AND LEE, A. K., 1980. Stress and mortality in a small marsupial (*Antechinus stuartii*, Macleay). *General and Comparative Endocrinology* **40:** 188–200.

BRANT, D. H., 1962. Measures of the movements and population densities of small rodents. *University of California Publications in Zoology* **62:** 105–84.

BRASMER, T. H., 1984. The acutely traumatised small animal patient. Saunders: Sydney.

CASSILETH, P. A., 1983. Lymphocytopenia. Pp. 955–57 in *Haematology* ed by W. J. Williams, E. Beutler, A. J. Erslev and M. A. Lichtman. McGraw-Hill: New York.

CHEAL, P. D., LEE, A. K. AND BARNETT, J. L., 1976. Changes in the haematology of *Antechinus stuartii* (Marsupialia) and their association with male mortality. *Australian Journal of Zoology* **24:** 299–311.

CHRISTIAN, J. J., 1950. The adreno-pituitary system and population cycles. *Journal of Mammalogy* **31:** 247–59.

CHRISTIAN, J. J., 1978. Neurobehavioral endocrine regulation of small mammal populations. Pp. 84–91 in *Pymatuning Laboratory of Ecology, Special Publication No. 5 — Small Mammal Populations.*

COOLEY, H. AND JANSSENS, P., 1977. Metabolic effects of infusion of cortisol and adrenocorticotropin in the tammar wallaby (*Macropus eugenii* Desmarest). *General and Comparative Endocrinology* **33:** 352–58.

CRAVEN, L. N., 1981. Ecology of a population of Quendas (*Isoodon obesulus*). M.Sc. prelim. thesis, Department of Zoology, University of Western Australia: Perth.

DACIE, J. V., 1975. *Practical haematology*, 5th ed. Longman: New York.

DICKENS, R. K., 1978. A comparison of hematological values from Koalas in different locations and the effect of stress on those values. Pp. 183–97 in *The koala: Proceedings of the Taronga Symposium* ed by T. J. Bergin. John Sands: Sydney.

DISBREY, B. D. AND RACK, J. H., 1970. Histological laboratory methods. Livingstone: London.

FAIRBAIRN, D. J., 1977. The spring decline in deer mice: death or dispersal? *Canadian Journal of Zoology* **55:** 84–196.

FORD, B. J., 1975. Mammal studies at Jandakot: filariasis and viral infections. Hons. Thesis, Department of Zoology, University of Western Australia: Perth.

GAINES, M. S. AND MCCLENAGHAN, L. R., 1980. Dispersal in small animals. *Review of Ecological Systems* **11:** 163–96.

GRIFFITHS, M., MCINTOSH, D. L. AND LECKIE, R. M. C., 1969. The effect of cortisone on nitrogen balance and glucose metabolism in diabetic and normal kangaroos, sheep and rabbits. *Journal of Endocrinology* **44:** 1–12.

HALPERN, B. H. AND PACHAUD, A. 1951. Technique de prélèvement d'échantillons de sang chez les petits animaux de laboratoire par tonection du plexus opthalmique. *Comptes rendus Seances de la Societé Biologie* **145:** 1465–66.

HEINSOHN, G. E., 1966. Ecology and reproduction of the Tasmanian bandicoots (*Perameles gunni* and *Isoodon obesulus*). *University of California Publications in Zoology* **80:** 1–107.

IVESON, J. B., 1971. Strontium chloride B and E.E. enrichment broth media for the isolation of *Edwardsiella, Salmonella* and *Arizona* species from tiger snakes. *Journal of Hygiene Cambridge* **69:** 323.

IVESON, J. B., 1973. Enrichment procedures for the isolation of *Salmonella, Arizona, Edwardsiella* and *Shigella* from faeces. *Journal of Hygiene Cambridge* **71:** 349–52.

JANSSENS, P. A. AND HINDS, L. A., 1981. Long-term effects of corticosteroid administration in the tammar wallaby, *Macropus eugenii. General and Comparative Endocrinology* **45:** 56–60.

JONES, F. W., 1924. *The mammals of South Australia: Part II The bandicoots and herbivorous marsupials.* Government Printer: Adelaide.

KREBS, C. AND MYERS, J., 1974. Population cycles in small mammals. *Advances in Ecological Research* **8:** 267–399.

LEE, A. K. AND COCKBURN, A., 1985. *Evolutionary ecology of marsupials.* Cambridge University Press: Melbourne.

LEE, A. K. AND MCDONALD, I. R., 1985. Stress and population regulation in small mammals. Pp. 261–304 in *Oxford review of reproductive biology, Vol. 7,* ed by C. A. Finn. Oxford University Press: Oxford.

LIDDLE, G. W., 1981. The adrenals. Pp. 249–92 in *Textbook of endocrinology* ed by R. H. Williams, 6th ed. Saunders: Philadelphia.

LOUCH, C. D., 1956. Adrenocortical activity in relation to the density and dynamics of three confined populations of *Microtus pennsylvannicus. Ecology* **37:** 701–13.

MANLY, B. F. J. AND PARR, M. J., 1968. A new method of estimating population size, survivorship, and birth-rate from capture-recapture data. *Transactions of the Society of British Entomology* **18:** 81–89.

MCDONALD, I. R. AND BRADSHAW, S. D., 1977. Plasma corticosteroids and the effect of adrenocorticotrophin in a macropodid marsupial (*Setonix brachyurus,* Quoy and Gaimard). *Journal of Endocrinology* **75:** 409–18.

MILLER, T. AND BRADSHAW, S. D., 1979. Adrenocortical function in a field population of a macropodid marsupial (*Setonix brachyurus,* Quoy and Gaimard). *Journal of Endocrinology* **82:** 159–70.

MONZU, 1970. A survey of the mammal fauna at Jandakot Reserve. Honours Thesis. Department of Zoology, University of Western Australia, Perth.

NOVAK, M. A., 1978. Social stress and population cycles: an assessment. Pp. 92–96 in *Pymatuning Laboratory of Ecology, Special Publication No.5 — Small Mammal Populations.*

O'CALLAGHAN, W., 1974. Studies on the Behaviour of *Isoodon obesulus.* Unpublished Psychology Hons. Thesis, University of Tasmania.

SANDBERG, A. A., ROSENTHAL, H. AND SLAUNWHITE, W. R., 1967. Transcortin: a corticosteroid-binding protein of plasma. VIII. Parameters affecting cortisol metabolism. Pp. 707–16 in *Proceedings of the second international congress on hormonal steroids* ed by L. Martini, F. Fraschini and M. Motta. Excerpta Medica Foundation: Amsterdam.

STODDART, D. M. AND BRAITHWAITE, R. W., 1979. A strategy for utilization of regenerating heathland habitat by the brown bandicoot (*Isoodon obesulus;* Marsupialia, Peramaelidae). *Journal of Animal Ecology* **48:** 165–79.

TAIT, J. F., 1982. The biological availability of free and protein-bound steroid. Pp. 65–70 in *Free hormones in blood* ed by A. Albertini and R. P. Ekins. Elsevier: Amsterdam.

TAIT, J. F. AND BURSTEIN, S., 1964. In *vivo* studies of steroid dynamics in man. Pp. 441–558 in *The hormones*. Vol. 5 ed by G., Pincus, K. V. Thimann and E. G. Astwood. Academic Press: New York.

THAN, K. A. AND MCDONALD, I. R., 1974. Effect of cortisol on insulin sensitivity of the marsupial brush-tailed opossum, *Trichosurus vulpecula* (Kerr). *Journal of Endocrinology* **63:** 473–81.

WINDBERG, L. A. AND KEITH, L. B., 1976. Snowshoe hare population response to artificial high densities. *Journal of Mammalogy* **57:** 523–53.

Preliminary studies of the Nuyts Archipelago Bandicoot *Isoodon obesulus nauticus* on the Franklin Islands, South Australia

P. B. COPLEY[1], V. T. READ[1, 3], A. C. ROBINSON[1] and C. H. S. WATTS[2]

A mark-release-recapture study of the Nuyts Archipelago Bandicoot, *Isoodon obesulus nauticus* on the Franklin Islands was begun in December 1983. Females were found to have pouch young present throughout the year with a breeding peak in spring. Litters of 1–4 (usually 1–3) young were recorded and at least four litters could be produced annually. Mean litter size ranged from 1.6 in summer to 2.4 in spring. Consistently more female pouch young were recorded (mean females:males; 1.4:1.0). A similar excess of females over males was recorded for independent juveniles captured. Conversely, the sex ratio of adults captured was nearer parity but favoured males slightly in most trapping periods. Mean home range areas of 2.1–2.2 ha and 1.4–1.6 ha were calculated for males and females respectively. Average range lengths changed seasonally with a decline during May and June and a subsequent increase by November which continued through to February. No strong preference for open or dense cover could be demonstrated from the recapture data. The bandicoots occur at mean densities of about 1.3–1.4 animals per hectare on the two main trapping areas and despite a steady rate of recruitment the population size remains essentially static. Extrapolation of mean density estimates suggests that West Franklin Island supports a population of about 550 bandicoots.

Key words: *Isoodon obesulus nauticus,* Franklin Islands, population, reproduction, mortality, home range, recruitment.

INTRODUCTION

In 1922, whilst visiting the Franklin Islands of Nuyts Archipelago, South Australia, Professor F. Wood Jones made the first collections of a relatively small, insular form of the Southern Brown Bandicoot. Later that year Oldfield Thomas (1922) described this form as a distinct species, *Isoodon nauticus.* This he distinguished from the mainland species *I. obesulus* by its markedly smaller size, comparatively paler colour (especially on the ventral surface which is white rather than yellowish grey) and differences in skull morphology including "peculiarly shortened" bullae and shorter narrower nasals. Jones (1924) also pointed out that "*I. nauticus*" has a far wider braincase and less prominent cranial ridges than *I. obesulus.*

Subsequently, the Nuyts Archipelago bandicoot has been little studied and its taxonomic status made uncertain due to lack of nomenclatural conformity. Tate (1948) in his review of the Peramelidae stated, on the basis of an examination of the type specimen, that it is essentially only a small-sized race of *I. obesulus*. Watts (1974) in his brief description of the West Franklin Island population gave it subspecific status as *I. obesulus nauticus.* Robinson and Smyth (1976) chose to use only the specific name *I. obesulus* when referring to the only other known population of this insular form on St Francis Island. More recently, Lyne and Mort (1981) published a taxonomic review of the genus *Isoodon* based on skull morphology and reiterated the conclusions of Thomas (1922) and Jones (1924), retaining the name *I. nauticus.* However, Strahan (1982) has since listed it at subspecific status in the Mammal Index published by the Australian Museum. As this is the most recent taxonomic treatment, we have adopted that classification for this paper. Clearly, more taxonomic studies are needed.

In this paper we present preliminary data collected during an ongoing study of the Greater Stick-nest Rat *(Leporillus conditor)* on West Franklin Island. As a part of that study, many *I. o. nauticus* were caught in traps set for the rats and these were measured, marked and released. Preliminary results of the study of *L. conditor* were presented by Read (1984).

STUDY AREA

Location and description

The Franklin Islands (32°27′S, 133°27′E) are situated 40 km south of the South Australian mainland township of Ceduna (Fig. 1). They consist of two small rocky islets and two larger islands each approximately 2.5 km long, 1.8 km

[1]South Australian National Parks and Wildlife Service, P.O. Box 902, Norwood, South Australia, Australia 5067.
[2]South Australian Museum, North Terrace, Adelaide, South Australia, Australia 5000.
[3]Present address: 35 Woodville Street, North Perth, Western Australia, Australia 6006.
Pages 345–56 *in* BANDICOOTS AND BILBIES, ed by J. H. Seebeck, P. R. Brown, R. L. Wallis and C. M. Kemper. Surrey Beatty & Sons, Sydney, 1990.

wide and 240 ha in area. The two main islands are joined by a 400 m long sand-bar which is exposed at low tide.

Land-use history

These islands constitute the easternmost outliers of Nuyts Archipelago and are part of the Nuyts Archipelago Conservation Park which was dedicated in 1967 under the control of the South Australian National Parks and Wildlife Service. Prior to dedication the islands were used intermittently for sheep grazing. However, due to the lack of potable water and to the logistical problems of transporting the stock, no long-term grazing option was ever taken up. Sheep were last grazed there in about 1925 (Read 1984).

Geomorphology

The main islands are characterized by three major features: (1) a narrow coastal fringe of granite slabs and boulders, interspersed with sandy beaches; (2) steep sandy slopes with outcropping limestone leading from the coastal fringe; and (3) a plateau of consolidated white sands over limestone, perforated by the burrows of Short-tailed Shearwaters *(Puffinus tenuirostris)* and Little Penguins *(Eudyptula minor).* Soils are mostly skeletal and support only very sparse perennial vegetation. Permanent fresh water is absent.

Climate

The climate of the islands is typically Mediterranean with relatively cool wet winters and warm to hot, dry summers. The average annual rainfall (as recorded at Ceduna) is 315 mm, with approximately half of this amount usually recorded during May to August. The harshest months are January to April when temperatures are highest, rainfalls are least, and annual and ephemeral vegetation has dried off and exposed much of the island's surface.

Vegetation

The vegetation of the Franklin Islands was originally described by Osborn (1922) and later by Robinson (1975) and Read (1984). On the plateau, the predominant perennial plant is *Nitraria billardierei* which is commonly overgrown by *Tetragonia implexicoma.* It grows in scattered, dense clumps on the more skeletal soils and occurs in assoication with smaller shrubs of *Threlkeldia diffusa, Enchylaena tomentosa, Frankenia pauciflora* and *F. sessilis.* Deeper sandy soils on the plateau are dominated by a low shrubland of *Rhagodia crassifolia.* In places where limestone outcrops, *F. pauciflora* is most common, though frequently in association with *Disphyma crassifolium* and *Carpobrotus rossii.*

Other fauna

In addition to *I. o. nauticus* and *L. conditor,* other vertebrate species are present in unusually high densities. The most significant component of the avifauna is the population of Short-tailed Shearwaters *(Puffinus tenuirostris)* which is present from October until May. During that period an estimated 100 000 birds (Read 1984) occupy the two main islands, undoubtedly providing a significant source of nutrients to an otherwise impoverished island environment. They nest in burrows which afford important shelter sites for *I. o. nauticus, L. conditor* and abundant reptiles. Similarly, Little Penguins *(Eudyptula minor)* which are present throughout the year, provide further nutrients and maintain additional burrows.

Other important fauna include two predators — the Barn Owl *(Tyto alba)* and Black Tiger Snake *(Notechis ater niger)* — known, from this study, to prey upon both *I. o. nauticus* and *L. conditor.* Barn owls breed on the Franklin Islands, and from one to five individuals are usually observed. A population of about 490 adult male and 420 adult female Black Tiger Snakes has been estimated for West Franklin Island (Schwaner 1985).

TRAPPING METHODS

Bandicoots were captured in aluminium Elliott box traps set on each of two eight cell × eight cell trapping grids.

One trapping grid, referred to as the "plateau" grid, encompassed a low open shrubland of scattered *N. billardierei* clumps. The other, referred to as the "valley" grid, encompassed a low open shrubland dominated by *Rhagodia crassifolia.* Each grid cell measured 50 m × 50 m and each grid, covering an area of 16 ha, measured 400 m × 400 m. The location and orientation of these grids is shown in Figure 1. The pattern of vegetation accross the two grids is shown in Figure 2.

During each field trip 128 traps were set on each grid for at least four nights (i.e., two traps in each of 64 cells per grid) (see Read 1984 for further details). Data presented are from field trips in December 1983, February, May, June and October 1984 and January 1985. (During these trips, data were also obtained from two linear, coastal trapping transects (50 traps each) but these are used in this paper only for estimating the islands' total population size).

Recording

Each *I. o. nauticus* captured was sexed and weighed, its pes and head lengths were measured, and its reproductive state recorded. For the latter the females' pouches were examined for

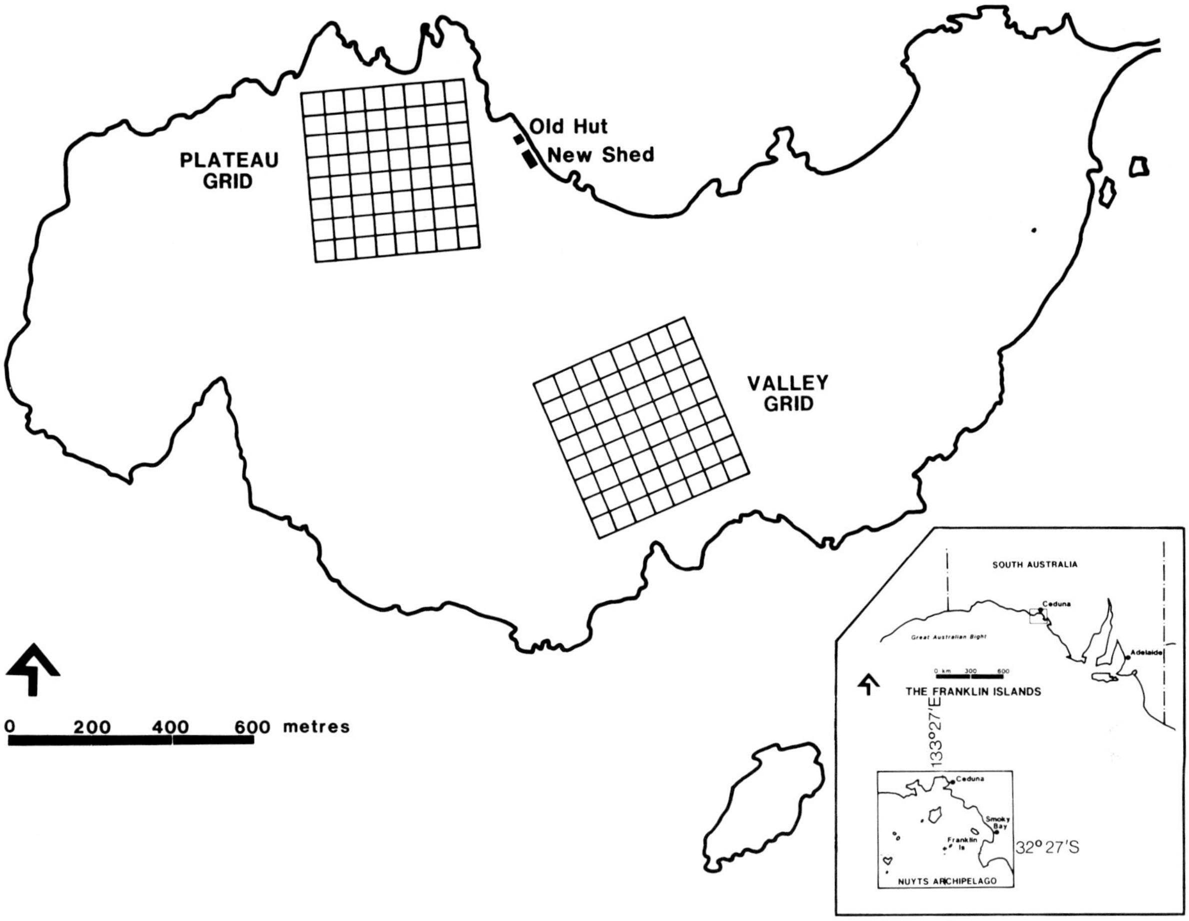

Fig. 1. The location of the Franklin Islands and position of the study grids in West Franklin Island, SA.

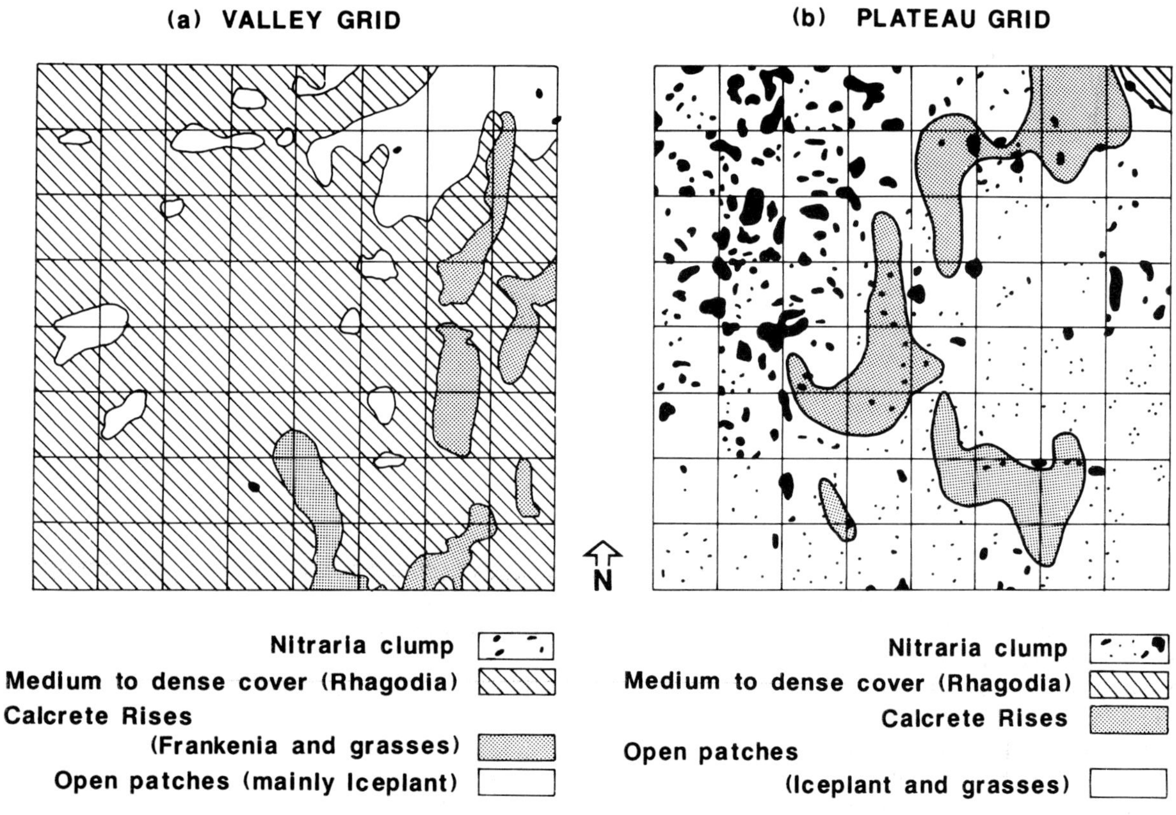

Fig. 2. Habitat maps of the "valley" and "plateau" grids on West Franklin Island.

young and/or lactating or enlarged nipples. Pouch young were counted, measured and, where possible, sexed. All independent animals captured were given a numbered monel metal ear-tag in each ear at the time of first capture, the capture location within the grid was noted and the animal was released.

Home range

Home range area was calculated for each animal by determining the number of 50 m × 50 m grid cells in which the animal was caught over all trapping periods. In addition, the maximum home range diameter was calculated as the distance across the grid between the two most distant grid cells in which an individual was captured. This was done to make allowance for those animals caught in edge cells and included in the calculation. Few individuals of either sex were captured in internal cells only.

Average range length

Within each trapping period the straight line distance between successive captures of each individual was measured and an average range length calculated for each sex for each period. Where an animal was captured on successive nights in the same trap a "movement" of 2 m was recorded (less than the smallest measureable distance using this trapping system). This ensured that animals which were essentially sedentary for some periods were included in the mean calculation. In addition the percentage of recaptures in the same trap on subsequent nights was calculated to determine if there was a seasonal pattern in this more sedentary behaviour.

Habitat utilization

The "plateau" and "valley" grids were established on areas of West Franklin Island representative of the two major vegetation types on the main platform of the island. They were quite different in vegetation structure and, to some extent, floristic composition (Fig. 2). Habitat preferences could then be examined broadly by comparing capture frequencies and densities for the two grids.

Furthermore, a vegetation density index was assigned to each grid cell on a scale of 1–4 (representing increasing density). Cover classes were ranked as follows: (1) 0–5%; (2) 6–20%; (3) 21–50%; (4) >50%. The total number of *I. o. nauticus* captures known for each grid cell throughout the study period was also ranked with the following classes recognized: (1) 0–5 captures; (2) 6–10 captures; (3) 11–15 captures; (4) 16–20 captures. It was then possible to attempt to determine whether animal distribution correlated with more open or more covered areas. Such a correlation might however, relate to soil type, soil depth, number of shearwater burrows, etc., as well as to vegetative cover.

Population estimates

The numbers of animals living on or close to the grids were estimated in two ways. Firstly, the minimum number known to be alive was calculated. Secondly, an estimate of numbers present was made using the Lincoln Index in its simplest form.

Recruitment

The size (weight) distribution for captured individuals was plotted for each sex, for each trapping period. A weight of 400 g was used as the dividing line between adult and subadult individuals. This showed the relative number of young *I. o. nauticus* entering the population at each trapping period.

RESULTS

Weight distributions

In this study *I. o. nauticus* was sexually dimorphic (Fig. 3), with males 50–100 g heavier than females. With the exception of a population at Cranbourne, Victoria (Table 1) they were considerably smaller and lighter than other populations of *I. obesulus*. This has previously been pointed out by Thomas (1922), Jones (1924) and Lyne and Mort (1981) in taxonomic discussions.

Sexual maturity

Female *I. o. nauticus* appear to reach sexual maturity at about 350–450 g in weight. An arbitrary cut-off of 400 g has therefore been used to distinguish between subadults and adults (Fig. 3). This has also been applied to males even though, in reality, males probably do not reach sexual maturity until they weigh about 450–500 g.

Seasonality of reproduction

By assigning 400 g as the demarcation between subadult and adult animals it was possible to estimate the relative proportions of adult females lactating during each trapping period. These are shown in Figure 4. Here the term lactating refers to females weighing more than 400 g with either pouch young present or with enlarged (usually) lactating nipples, even though young were absent. The data show that *I. o. nauticus* reproduced throughout the year with a peak of reproductive activity during spring. Reproductive activity was maintained at a higher level during the summer of 1984–85 compared with that of 1983–84 and examination of Figure 4 suggests that this may have been

Table 1. Size comparisons of adult wild *Isoodon obesulus* from four populations.

		Franklin Island, SA (this study)		Smithton, Tas. (Heinsohn 1966)		Cranbourne, Vic. (Stoddart and Braithwaite 1979)		Kuitpo, SA (Unpublished data)	
		Males	Females	Males	Females	Males	Females	Males	Females
Body weight (g)	x̄	595.3	527.8	1 165.9	946.6	614.2	476.1	1 031.2	614.5
	SE	±6.3	±6.1	±41.7	±33.0	±30.2	±18.0	±115.9	±23.3
	n	189	143	8	5	31	36	29	28
Head length (mm)	x̄	72.3	64.4	89.6	83.7	—	—	—	—
	SE	±0.04	±0.03	±0.6	±0.6	—	—	—	—
	n	94	94	8	5	—	—	—	—
Pes length (mm)	x̄	52.6	49.2	60.1	53.9	—	—	—	—
	SE	±0.02	±0.02	±0.4	±0.7	—	—	—	—
	n	92	95	8	5	—	—	—	—

in response to higher rainfalls during July/August/September of 1984 than during the same period in 1983. This may indicate that reproduction is correlated with increased vegetation growth and cover and the associated availability of food. Further trapping data and studies of diet may allow these observations to be tested.

Litter size

During the six trapping periods a total of 81 litters and 167 pouch young were recorded (mean litter size 2.06 ± 0.83 SD). Litters were mostly of one to three young, although two litters of four were recorded. This can be compared with a mean litter size of 3.36 recorded for *I. obesulus* in the Mt. Lofty Ranges of South Australia (Field Naturalists Society of South Australia Mammal Club, unpublished data), 2.80 recorded for *I. obesulus* in Tasmania (Heinsohn 1966) and 3.04 recorded for *I. obesulus* at Cranbourne in Victoria (Stoddard and Braithwaite 1979). These data are summarized in Table 2.

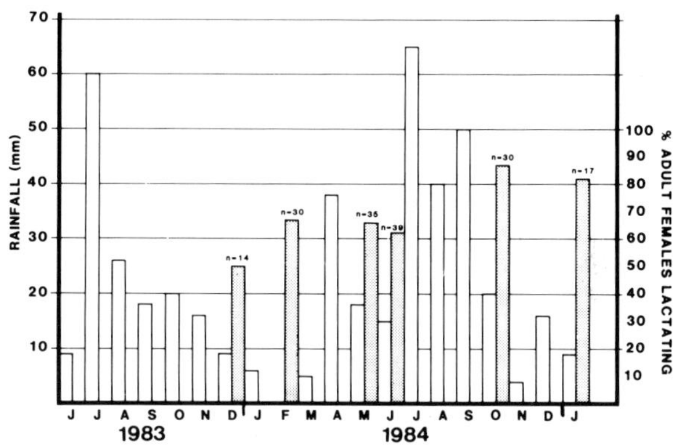

Fig. 4. Monthly rainfall at Ceduna, SA, from December 1983 to January 1985 (open columns) compared with the proportions of lactating female *Isoodon obesulus nauticus* on West Franklin Island, expressed as a percentage of total adult females (stippled columns).

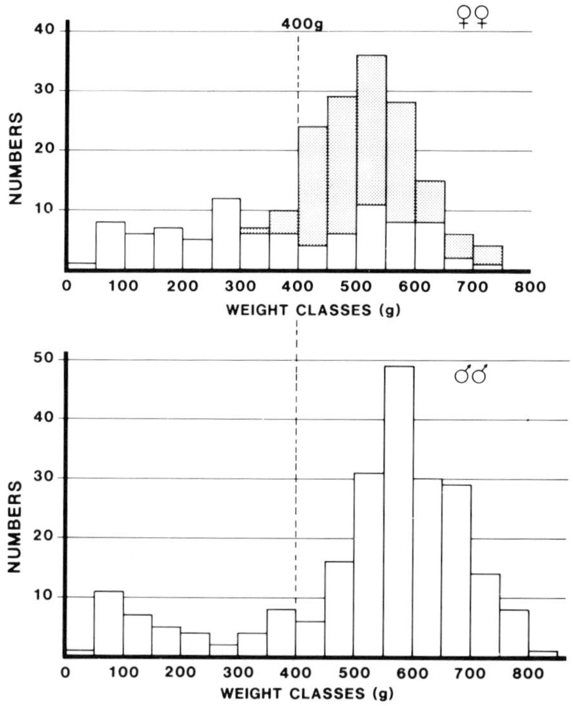

Fig. 3. Body weight distributions of male and female *Isoodon obesulus nauticus* on West Franklin Island, SA. Shaded bars indicate lactating females (including females with pouch young and females with enlarged nipples).

As with reproductive activity, the mean litter size for each trapping period on Franklin Island, shown in Figure 5, was not uniform throughout the year. Smaller litters were recorded during relatively dry periods (e.g., December 1983 and February 1984) and larger litters following good rains (e.g., May and October 1984).

Reproductive capacity

Female *I. o. nauticus* on Franklin Island can produce at least four and possibly five litters per year. One female caught in February, May, June and October had separate litters of three pouch young present on each occasion. Another caught in December, February, May and June had separate litters on three occasions and evidence (i.e., lactating nipples) of two young having just vacated the pouch on the

Table 2. Comparison of litter sizes recorded for four populations of *Isoodon obesulus*.

	Franklin Island, SA (this study)	Cranbourne, Vic. (Stoddart and Braithwaite 1979)	Smithton, Tas. (Heinsohn 1966)	Kuitpo, SA (Unpublished data)
Mean litter size	2.06	3.04	2.80	3.36
SE	±0.1	±0.2	±0.3	±0.3
No. litters	81	56	15	14
Range	1–4	1–6	1–4	2–5

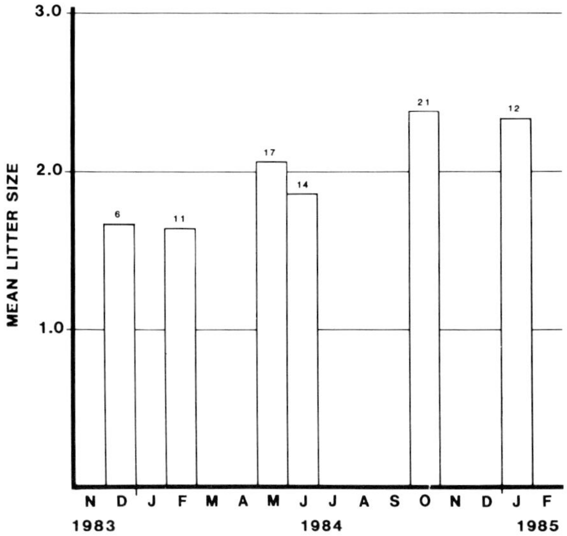

Fig. 5. Variation in mean litter size of *Isoodon obesulus nauticus* from West Franklin Island, December 1983–January 1985.

other. However, not all females are as fecund as this. Of the 24 adult females caught during three or more trapping periods only nine (38%) were lactating at each capture and, on average, lactation was recorded at only 66% of captures.

Larger females did not necessarily have larger litters. In fact, mean litter size for females heavier than 600 g ($\bar{x}\pm$SE; 1.6±0.2) was significantly less (t = 2.3; P = 0.023; d.f. = 99) than for females 400–600 g ($\bar{x}\pm$SE; 2.1±0.1). Furthermore, of the 28 females recorded with more than one litter during the study period 12 had no sequential change, seven had a sequential decrease and nine had a sequential increase in litter size with increasing age.

Mortality of pouch young

Only the May and June field trips, with from 40–49 days between captures, were in sufficient proximity to enable the success or failure of particular litters to be determined, assuming a pouch life of 55–60 days (Stoddart and Braithwaite 1979). From the data available it is possible to demonstrate that at least 22% of pouch young (i.e., at least seven of the 32 originally recorded in May) were lost from 31% of the litters (i.e., five of the 16 litters) between the two trapping periods. However, only one of the litters affected was lost entirely. One, two or three young survived to weaning in most litters. These data are summarized in Table 3.

Table 3. Changes in state of lacatation for 16 female *Isoodon obesulus nauticus* captured in May 1984 and again 6–7 weeks later in June 1984, West Franklin, South Australia. $^{+}$Number in brackets is minimum number lost which is based on approximate ages of pouch young measured in May relative to expected ages and sizes in June.

Comparative reproductive state	Incidence
May 1984	
No. with new litters	4
No. with enlarged nipples from a previous litter, but no young present	11
Total No. lactating — still or again	15
June 1984	
No. with one less nipple enlarged than in May	3(3^{+})
No. with two less nipples enlarged than in May	1(2^{+})
No. with one more nipple enlarged than in May (2→3)	1(2^{+})

Sex ratios of pouch young

The sex of each pouch young was recorded, where possible, during the May, June, October and January trapping periods. Consistently more females were recorded than males (Fig. 6) with an overall sex ratio of 1.41. This bias is also evident in capture data for subadult animals (except for May 1984) shown in Figure 9.

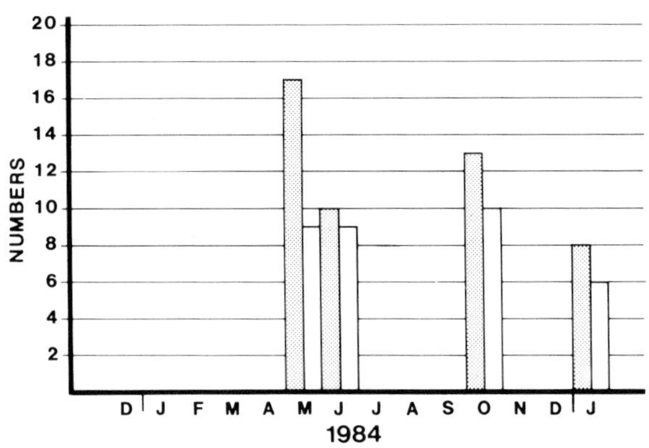

Fig. 6. Sex ratios of pouch young *Isoodon obesulus nauticus*, West Franklin Island, December 1983–January 1985. Females = stippled columns, males = open columns.

Conversely, with the exception of the January 1985 data, the sex ratios of adult animals trapped (Fig. 7) were nearer parity and biased

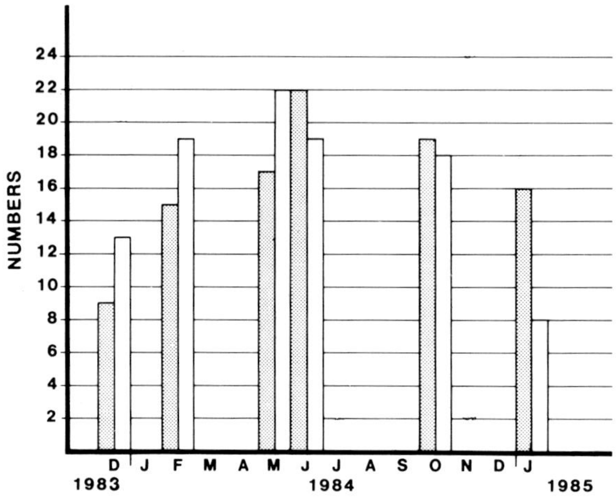

Fig. 7. Sex ratios of adult *Isoodon obesulus nauticus*, West Franklin Island, December 1983–January 1985. Females = stippled columns, males = open columns.

slightly towards males. This may be indicative of either: (a) the presence of more males in the population, due to greater longevity, as suggested by Stoddart and Braithwaite (1979) or; (b) a greater "proneness to capture" of adult males due to a variety of reasons. However, with the apparent reversal of this trend in the January 1985 trapping period, more field data are needed to help clarify these observations.

Home range

Although there are inherent problems with interpreting trap-revealed home range patterns, the trapping system designed for this project (Read 1984) goes as far as possible within the time, labour and equipment constraints used, to entice individual *I. o. nauticus* to the limits of their home ranges before encountering a trap. With the number of individuals that were present on the grids it is extremely difficult to depict home ranges graphically. Plots of animals captured more than 10 times during the study period revealed that on both grids: (a) individual animals do occupy discrete home ranges; (b) the male home ranges are larger than those of the females and (c) there is some overlap between animals of the same sex and complete overlap between animals of different sex.

Animals with 10 or more captures give by no means the whole picture, however. In an attempt to quantify these patterns, two statistics (home range area and maximum home range diameter) were calculated, for animals with five or more captures as well as for those captured on 10 or more occasions (Tables 4 and 5).

The mean home range area of males was consistently larger than for females on both grids and for all comparisons (Table 4). The estimates derived from individuals with more than 10 captures are likely to be more realistic than those from individuals with a lower number of captures.

Table 4. Mean home range area (ha) calculated for male and female *Isoodon obesulus nauticus* with more than 10 captures and more than five captures on the plateau and valley grids respectively, West Franklin Island, South Australia.

		Plateau		Valley	
		Males	Females	Males	Females
10 captures	$\bar{x}$	2.1	1.5	2.2	1.6
	SE	±0.3	±0.2	±0.2	±0.2
	n	10	11	10	7
5 captures	$\bar{x}$	1.9	1.2	1.7	1.5
	SE	±0.2	±0.1	±0.1	±0.1
	n	20	22	24	19

Table 5. Mean maximum home range diameter (m) calculated for male and female *Isoodon obesulus nauticus* with more than 10 and more than five captures on the plateau and valley grids respectively, West Franklin Island, South Australia.

		Plateau		Valley	
		Males	Females	Males	Females
10 captures	$\bar{x}$	200	168	220	157
	SE	±13.0	±16.9	±20.0	±17.0
	n	10	11	10	7
5 captures	$\bar{x}$	197	155	179	145
	SE	±11.2	±10.4	±12.0	±9.2
	n	20	22	24	19

These show that males have a mean home range of between 2.1 and 2.2 ha while females occupy between 1.5 and 1.6 ha. Maximum home range diameters (Table 5) also show consistent differences between the sexes with that for males being between 200 and 220 m and for females 157 to 168 m.

Mean range length

The changes in range length between successive captures for the six trapping periods is shown in Table 6. The data for some months, particularly January 1985, are limited, but there is a consistent trend of seasonal change discernible throughout Table 6. The greatest movement between captures was recorded in December with a general decrease through February and May to a minimum in June followed by a sharp increase in movement in the October trapping period. It should also be noted that as with the home range data outlined above, males showed a consistently greater mean range length than females throughout the study.

An alternative method of assessing the extent of this seasonal change in movement is to examine the distribution of recaptures on successive nights at the same trap site expressed as a percentage of the total captures. This is shown in Table 7. The marked increase in percentage of recaptures at the same site on subsequent nights in June indicates that the population at this time is less mobile than at other times. This is the same pattern shown by mean range lengths.

Table 6. Changes in mean range length for male and female *Isoodon obesulus nauticus* calculated for each of six trapping periods, December 1983–January 1985, West Franklin Island, South Australia.

	December			February			May			June			November			January		
	x̄	SD	n	x̄	SD	n	x̄	SD	n	x̄	SD	n	x̄	SD	n	x̄	SD	n
Plateau																		
Males																		
10 captures	77	41	10	84	68	22	64	46	23	28	22	18	55	59	7	55	21	2
5 captures	71	66	5	109	67	8	68	56	13	57	47	10	51	36	4	—	—	—
Females																		
10 captures	50	57	2	51	26	30	59	46	33	34	32	25	38	28	11	43	18	2
5 captures	—		—	44	36	15	53	25	13	26	16	13	45	50	2	30	21	2
Valley																		
Males																		
10 captures	117	107	16	72	46	30	79	48	17	66	55	18	30	20	8	—		—
5 captures	77	74	7	80	43	26	80	73	25	53	37	18	35	34	6	—		—
Females																		
10 captures	80	71	3	56	35	25	101	13	22	66	49	19	36	25	7	—		—
5 captures	87	60	4	102	43	4	42	28	10	33	21	15	93	10	2	—		—

Habitat Use

The relationships between habitat density (ranked on an increasing scale of 1–4) and numbers of *I. o. nauticus* captured (ranked on an increasing scale of 1–4) are shown in Tables 8 and 9.

On the plateau grid (Table 8), reading from the top left hand corner there were, for example, four grid cells where the lowest habitat density class (1) corresponded with the lowest number of captures (1). It can be seen that no clear association of capture classes and habitat classes is evident (see chi-square values) although there may be a tendency for the highest captures (classes 3 and 4) to be in the more open habitats (classes 1 and 2). On the valley grid (Table 9) there were more captures in the denser habitats (classes 3 and 4). This apparent contradiction between the two grids may be explained by the marked difference in vegetation structure between them (Fig. 2). The valley grid has large continuous areas of evenly distributed *Rhagodia crassifolia* shrubs which fall into the higher habitat density classes while the plateau grid is dominated by widely spaced clumps of *Nitraria billardierei*. Although individually these clumps provide habitat as dense as, or denser than, the *Rhagodia* shrubland the total grid cell was normally classed at a lower overall habitat density. Clearly further details of *I. o. nauticus* habitat preference need to be examined. For example, radio-tracking studies may be useful in indicating the relative use of individual patches of habitat. They may also give an indication of the uses of muttonbird burrows as nesting and/or shelter sites.

Table 7. Recaptures of *Isoodon obesulus nauticus* individuals at the same trap site on subsequent nights, expressed as a percentage of the total captures for each trapping period, December 1983–January 1985, West Franklin Island, South Australia.

	Dec.	Feb.	May	June	Nov.	Jan.
Plateau						
Males						
%	0	4.5	4.3	33.3	16.7	0
n	10	22	23	18	6	3
Females						
%	50	8.8	9.1	24	27.3	0
n	2	34	33	25	11	1
Valley						
Males						
%	0	0	12.5	20	0	—
n	4	8	8	10	4	—
Females						
%	—	7	16.7	23	—	—
n	—	15	18	13	—	—

Table 8. Correspondence between *Isoodon obesulus nauticus* capture classes (on an increasing scale of 1–4) and habitat density classes (on an increasing scale of 1–4) for the 64 cells of the plateau grid, West Franklin Island, South Australia.

		Capture classes 1	2	3	4	
Habitat density classes	1	4	10	5	2	
	2	4	8	7	3	$\chi^2 = 3.2$
						$P = 0.95$
	3	0	3	1	1	df = 9
	4	5	7	2	2	

Table 9. Correspondence between *Isoodon obesulus nauticus* capture classes (on an increasing scale of 1–4) and habitat density classes (on an increasing scale of 1–4) for the 64 cells of the valley grid, West Franklin Island, South Australia.

		Capture classes 1	2	3	4	
Habitat density classes	1	2	1	0	0	
	2	5	8	1	0	$\chi^2 = 3.96$
						$P = 0.88$
	3	4	8	5	0	df = 9
	4	6	15	6	3	

Population estimates

The number of *I. o. nauticus* living on or close to the plateau grid were estimated in two ways. Firstly, the minimum number known to be alive was calculated (Fig. 8). Secondly, an estimate of numbers present was made using the Lincoln Index in its simplest form (Fig. 8). Individuals caught during the first half of the trapping period were taken as the marked population and the results from the second half of each trapping period were used to calculate the estimated population.

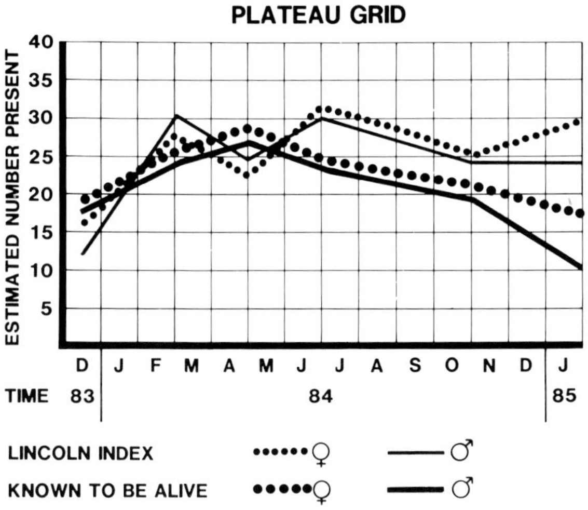

Fig. 8. Estimates of the number of *Isoodon obesulus nauticus* living on the 16 ha plateau trapping grid, West Franklin Island, December 1983–January 1985.

It should be noted that recapture rates both between and within trapping periods were high. On average, 83% of the individuals known to be alive on the grid were caught during each trapping period. Consequently the figures obtained should be reliable.

Both estimates agree closely and show an essentially stable population number of 50 individuals on the plateau grid. The drop in numbers known to be alive in January is a consequence of the method of estimation, as individuals missed in January but still living on or around the grid will be captured in future trapping periods, thus increasing future population estimates for January. Despite this there is a suggestion that numbers are lowest in summer, rise slightly during autumn and drop off slightly during winter and spring. Nevertheless, the overall pattern is one of numerical stability.

A similar pattern was found for the valley grid which had an estimated population of 46 individuals.

On the basis that the two trapping grids sampled the predominant habitats on the main platform of the island, and that data (not presented here) have been collected for two coastal transects, it is possible, with some degree of extrapolation, to calculate a minimum population estimate. Using the mean range length data presented above, an area correction can be made to the two grids to more truly represent the area being sampled by the trapping study. Using "known to be alive" estimates for the two grids, figures of 1.4 and 1.3 bandicoots ha^{-1} were calculated for the plateau and valley grids respectively. By adding estimates for the linear habitats along the coastal transects and using an average density for the other minor habitats on the main platforms of the islands (as defined by Read (1984)) an estimated minimum population size of 550 for West Franklin and 512 for East Franklin Island can be calculated. This gives a total population estimate of 1 062 for both islands.

Recruitment rates

Histograms (Fig. 9) were constructed to give a pictorial representation of the size distribution of captured individuals during each trapping period. Using 400 g as a dividing line between adult and juvenile individuals it can be seen that recruitment occurs throughout the year at a low level. Most of the animals recruited were juveniles recently independent of the pouch. A few were migrants from areas surrounding the grid and were usually caught in perimeter traps. Once again, the overall picture is one of unusual demographic stability.

A closer examination reveals that during May and June few new adults entered the population. In contrast, during October, January and February almost half the population, including adults and juveniles, were unmarked. This influx of unmarked individuals into the population is approximately equal to the disappearance of marked individuals. This results in no observable change in population number.

Once independent of the pouch, juvenile bandicoots become vulnerable to predation. On the Franklin Islands they are preyed upon by Tiger Snakes and Barn Owls.

During the 13 months of this study a total of 167 pouch young were recorded. In contrast, only 45 juveniles were trapped on the grids over the same period. This represented a loss of about 73%. With about 22% of pouch young lost before weaning (Table 3), a minimum of 50% may be lost between leaving the pouch and becoming adults.

Figure 10 shows the number of marked individuals that disappeared between trapping periods expressed as a percentage disappearence per week. A marked seasonal pattern is evident with relatively low disappearance rates in winter and relatively high disappearance rates in summer. For the spring and summer periods

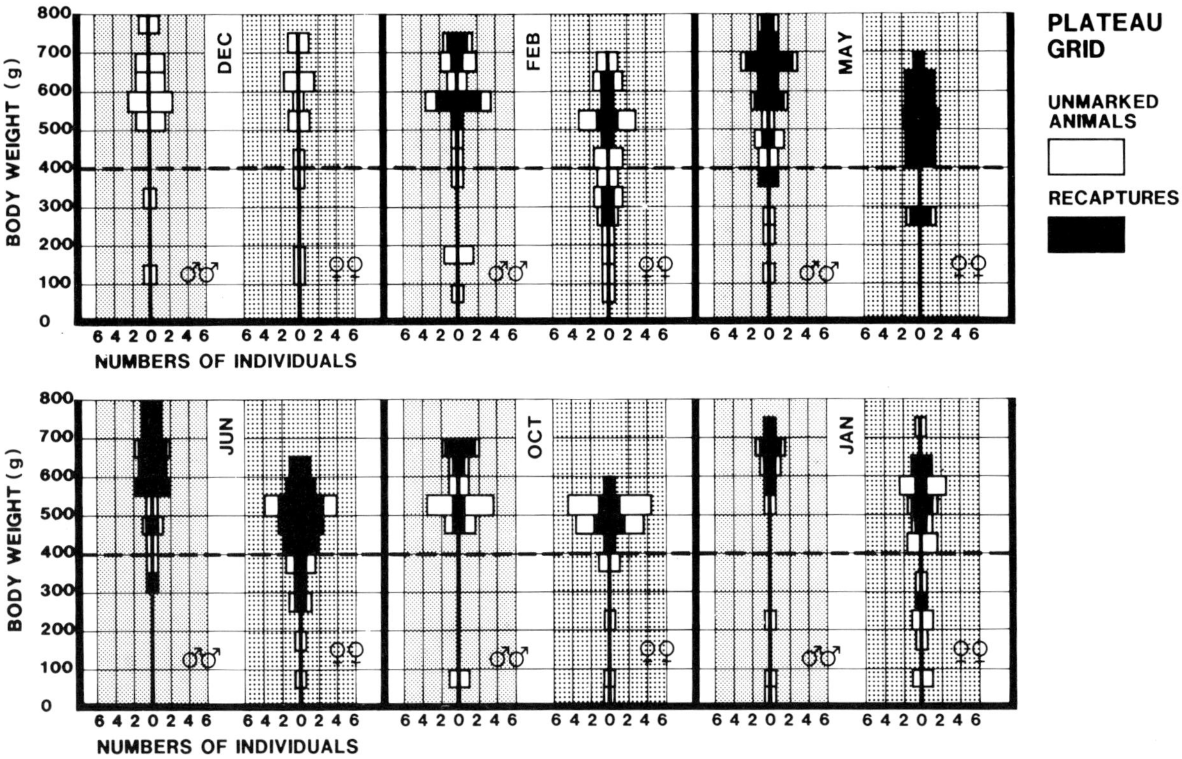

Fig. 9. Size distributions of *Isoodon obesulus nauticus* captured on the 16 ha plateau trapping grid, West Franklin Island, December 1983–January 1985. Adults >400 g, juveniles <400 g.

approximately 50% of the individuals present at one trapping period were not recaptured at later trapping periods. This disappearance pattern mirrors that shown in the histogram for appearance of new individuals (Fig. 9); that is, low recruitment rates in winter and high recruitment rates during spring and summer.

The simplest explanation for this seasonal cycle is that during winter the population is sedentary, but during spring and summer it

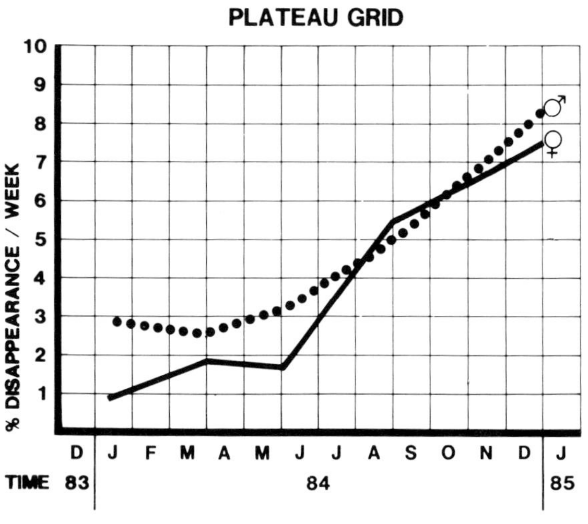

Fig. 10. Rates of disappearance of *Isoodon obesulus nauticus* from the 16 ha plateau trapping grid, West Franklin Island, December 1983–January 1985.

becomes highly mobile with individuals remaining in one area for only a short period before moving on. This is supported by the decrease in mean range length recorded in June (Table 6).

This pattern is biologically appealing. Short-tailed Shearwaters arrive on the islands in early spring and stay until late summer — the period of maximum movement of bandicoots. With an estimated density of 100 shearwater burrows ha^{-1} it is not difficult to imagine that a stable life is difficult with the shearwaters consistently "destroying" and remodelling the landscape.

An alternative explanation is simpler. Previously marked individuals have lost their tags. This tag loss could be the result of a shearwater-induced breakdown in home ranges causing increased encounters and fighting amongst individuals.

Incidental notes were kept of untagged individuals whose ears were torn in such a way as to suggest they had lost tags. At times approximately 25% of individuals lost one tag between trapping periods.

However, during October and January up to 30% may have lost both tags and thereby become unidentifiable. Loss of tags is therefore a problem in this study, especially with such infreqent trapping periods.

DISCUSSION

Isoodon obesulus nauticus of the Franklin Islands is a relatively small bandicoot with males weighing 450–800 g and females 400–750 g, breeding throughout the year, but with peaks in reproductive activity following significant falls of rain especially in winter/spring. By comparison, Heinsohn (1966) found that *I. obesulus* in northern Tasmania has a distinct breeding season which lasts for eight months (July to February) and is directly correlated with food supply and hence indirectly to rainfall and temperature. Similarly, Stoddart and Braithwaite (1979) found that the breeding season of *I. obesulus* at Cranbourne in Victoria lasts for six months (August to January). However, they suggested that this was not correlated with rainfall but rather, with changes in daylength. Braithwaite also noted that in western Victoria that species breeds throughout the year (in Stoddart and Braithwaite 1979). At Kuitpo in the Mt Lofty Ranges, South Australia, a study conducted from January 1981 to January 1983 showed that *I. obesulus* had pouch young present for only six months of the year (July to December) (Field Naturalists' Society of South Australia Mammal Club, unpublished data).

The smaller litter sizes recorded on West Franklin Island (Table 2) may be compensated for by the ability to breed throughout the year. In this way they have the capacity to produce up to five litters per year compared with only two or three produced in the northern Tasmanian, southern Victorian and mainland South Australian populations. However, most females do not reproduce continuously. Only 38% of adult females caught during three or more trapping periods were lactating at each capture and, on average, lactation was recorded at only 66% of captures.

Mean litter size appears to fluctuate considerably with changes in environmental conditions. Larger mean litter sizes were recorded following good rains. Smallest mean litter sizes were recorded at the end of a hot, dry summer. Furthermore, there does not appear to be a positive relationship between mean litter size and female body weight, unlike the population at Cranbourne where this was found to be the case (Stoddart and Braithwaite 1979).

Decreases in mean litter size appear to be due to increases in mortality of pouch young during pouch life. At least 22% of pouch young were lost before weaning during the period May to June 1984 and this affected at least 31% of litters. This coincided with a decrease in mean litter size from 2.06 to 1.86. The correlation with mortality rates of pouch young at other times is not clear. Neither is it known for certain how these mortalities occur. However, Stoddart and Braithwaite (1979) and Gemmell (1982) suggest that young are most vulnerable during weeks 2–5 of pouch life, during which time they may be expelled from the pouch by the mother. This has been observed where females have been handled during studies both in the wild and in captivity, but whether this happens under natural circumstances in not known.

The West Franklin Island population has a strong bias favouring the birth of females. This bias was evident in sex ratios recorded for both pouch young and juvenile animals captured. In the adult population trapped, sex ratios favoured males slightly, possibly reflecting the greater mobility and/or greater longevity of males, or perhaps a relative reluctance by females to enter the traps. In contrast, the sex bias (44:37) in the adult population trapped at Cranbourne favoured females (Stoddart and Braithwaite 1979).

As suggested for other populations (Heinsohn 1966; Stoddart and Braithwaite 1979) *I. o. nauticus* has a discrete home range with males occupying larger areas than females. In northern Tasmania, Heinsohn (1966) calculated that male *I. obesulus* had a mean home range area of 6.6 ha and females 2.3 ha. This compares with the mean home range areas of 2.1 to 2.2 ha for males and 1.5 to 1.6 ha for females, on both West Franklin Island grids. In a population as dense as that on West Franklin Island, there is some overlap between the home ranges of animals of the same sex but this is not as great as the overlap between sexes. More detailed studies may show that at a finer level of discrimination, males and females may in fact occupy totally discrete home ranges through temporal separation of activity or perhaps some defence of the areas around nest sites. As judged from average movement between successive captures within each trapping period (Table 6), the population appears to be at its most sedentary over winter with a major increase in movement over the spring and summer months. This may be associated with disruption due to the arrival of vast numbers of Short-tailed Shearwaters or may simply reflect a need to increase foraging range. The increased ear-tag loss in October and January may be further evidence of this increased mobility in the population with more encounters resulting in increased fighting and consequent tag loss.

Population estimates for the plateau grid for each trapping period revealed a stable population number of 50 individuals. This was maintained by a relatively low level of recruitment of about 5% per month, matched by the rate of disappearance of marked individuals. Most of the animals recruited were juveniles recently independent of the pouch. A few were migrants from areas surrounding the grid and were usually caught in perimeter traps.

Data from this study suggest that approximately 20% of pouch young die before weaning. A further 50% appear to be lost as newly-independent juveniles and of the remaining 30%, a large proportion probably do not reach sexual maturity. Stoddart and Braithwaite (1979) calculated rates of recruitment to be between 11.9 and 17.5% of all young produced, which is a similar level of recruitment to that of the Franklin Islands population.

ACKNOWLEDGEMENTS

This work was made possible by grants from World Wildlife Fund Australia, the Australian National Parks and Wildlife Service, and the National Estate Grants Programme with support from the South Australian National Parks and Wildlife Service and the South Australian Museum. Many volunteers assisted with the field work, and we are grateful for their efforts. In particular however, we are indebted to Kevin Jordan and Loraine Jansen for their tireless enthusiasm and invaluable help. Members of the Mammal Club of the Field Naturalists' Society of South Australia collected the data for the *I. obesulus* population near Kuitpo, South Australia and this was collated and supplied by Mr Graham Medlin.

REFERENCES

GEMMELL, R. T., 1982. Breeding bandicoots in Brisbane (*Isoodon macrourus;* Marsupialia, Peramelidae). *Australian Mammalogy* **5:** 187–93.

HEINSOHN, G. E., 1966. Ecology and reproduction of the Tasmanian bandicoots (*Perameles gunni* and *Isoodon obesulus*). *University of California Publications in Zoology* **80:** 1–107.

JONES, F. WOOD, 1924. *The mammals of South Australia*. Pt. 2. Government Printer: Adelaide.

LYNE, A. G. AND MORT, P. A., 1981. A comparison of skull morphology in the marsupial bandicoot genus *Isoodon:* its taxonomic implications and notes on the new species, *Isoodon arnhemensis. Australian Mammalogy* **4:** 107–33.

OSBORN, T. G. B., 1922. The flora and fauna of Nutys Archipelago and the Investigator Groups No. 3. A sketch of the ecology of the Franklin Islands. *Proceedings of the Royal Society of South Australia* **46:** 194–206.

READ, V. T., 1984. *The stick-nest rats of Australia: a preliminary report.* Unpublished report to World Wildlife Fund Australia. South Australian Department of Environment and Planning: Adelaide.

ROBINSON, A. C., 1975. The stick-nest rat *Leporillus conditor* on Franklin Island, Nuyts Archipelago, South Australia. *Austrlaian Mammalogy* **1**(4)**:** 319–27.

ROBINSON, A. C. AND SMYTH, M. E. B., 1976. The vertebrate fauna of Nutys Archipelago. South Australia. *Transactions of the Royal Society of South Australia* **100:** 171–76.

SCHWANER, T. D., 1985. Population structure of black tiger snakes, *Notechis ater niger,* on offshore islands of South Australia. Pp. 35–46 in *Biology of Australasian frogs and reptiles* ed by G. Grigg, R. Shine and H. Ehmann. Royal Zoological Society of New South Wales: Sydney.

STODDARD, D. M. AND BRAITHWAITE, R. W., 1979. A strategy for utilization of regenerating heathland habitat by the brown bandicoot (*Isoodon obesulus*; Marsupialia, Peramelidae). *Journal of Animal Ecology* **48:** 165–79.

STRAHAN, R., 1982. Mammal Index (3rd ed.). The Australian Museum: Sydney.

TATE, G. H. H., 1948. Results of the Archbold Expeditions. No. 60. Studies in the Peramelidae (Marsupialia). *Bulletin of the American Museum of Natural History* **92:** 313–46.

THOMAS, O., 1922. The bandicoots of Nutys Archipelago, South Australia, and of Cape York, North Queensland. *Annals and Magazine of Natural History* **9:** 677–79.

WATTS, C. H. S., 1974. The Nutys Islands bandicoot. *South Australian Naturalist* **49:** 20–24.

Breeding and population dynamics of *Isoodon macrourus* (Marsupialia: Peramelidae): studies from the wet-dry tropics of northern Australia

GORDON R. FRIEND[1]

Breeding, population dynamics and abundance trends of the Northern Brown Bandicoot *Isoodon macrourus* were examined for a population inhabiting tropical open forest in the Northern Territory. Forty-five males and 30 females were captured in 2 900 trapnights over 29 months. Abundance peaked in the late dry and mid wet seasons of 1981 and 1982 but declined markedly over the last 11 months of the study. During the wetter months, a large proportion of the population consisted of juveniles (particularly females) arising from seasonal breeding between August and April, but few of these animals became resident on the grids. Fecundity was high, with a mean litter size of 2.7, and up to four litters produced at about 54-day intervals by each female over a breeding season.

Rainfall, probably through its effect on food availability, seemed an important extrinsic factor influencing the onset and duration of breeding. A decline in female body weights and concomitant reduction in breeding occurred when early wet season rains failed. It is postulated, however, that the species' high rate of reproduction and dispersal of juveniles may allow a rapid increase in abundance when favourable conditions return.

It is suggested that *I. macrourus* has a variable life history strategy which allows considerable flexibility in breeding throughout its range. Such life history characteristics are likely to be well-developed in areas with less pronounced seasonality, and high levels of spatial and temporal habitat heterogeneity.

Key words: *Isoodon macrourus,* wet-dry tropics, breeding, population dynamics, abundance, seasonality, variable life history strategy.

INTRODUCTION

THE Northern Brown Bandicoot *Isoodon macrourus* (Gould) is perhaps the most common and widespread species in the family Peramelidae, extending throughout high rainfall zones from central coastal New South Wales to Cape York, and across the monsoonal (wet-dry) tropics of the Northern Territory and the Kimberley of Western Australia (Gordon 1983). Along with its close relatives *Isoodon obesulus* and *Perameles nasuta* (see Lyne 1964; Stoddart and Braithwaite 1979), this species has the shortest gestation period known for any mammal (Lyne 1974), and is capable of producing several litters at 50–60 day intervals (Mackerras and Smith 1960; Gordon 1971, 1974). These characteristics, together with early maturity, polyoestrous breeding, and minimal maternal care of young give the species a very high reproductive potential (Gordon 1971; Hall 1983).

Although breeding may occur in all months of the year, this is not a consistent finding in the studies from the south-east of the species' range. Gordon (1974) and Hall (1983) found no evidence of breeding seasonality in *I. macrourus* near Brisbane, whilst for the same region Gemmell (1982) reported that 98% of births (n=83) occurred in the period July–March. In New South Wales breeding was seasonal, occurring between July and April (Gordon 1971). Various environmental phenomena have been suggested to explain these patterns, including food availability and rainfall (Gordon 1971) and photoperiod (Gemmell 1982), but the factors involved remain unclear.

Studies of *I. macrourus* ecology in the monsoonal tropics, where temperatures are high throughout the year, daylength varies little, but rainfall is highly seasonal, could provide further insight into the factors controlling reproduction and abundance patterns. Little such work has been carried out; most information on *I. macrourus* in the region has arisen from material collected during earlier faunal surveys (e.g., Thomas 1904; Johnson 1964; Calaby and Keith 1974). More recently, however, Smith (1980) undertook a short-term study of a population in suburban Darwin, while Friend and Taylor (1985) reported the species' habitat preferences in open forest environments within Kakadu National Park.

This paper discusses aspects of *I. macrourus* population dynamics, reproduction and abundance from a live-traping study in the

[1]CSIRO Division of Wildlife and Rangelands Research, PMB 44, Winnellie, Northern Territory, Australia 0821.
Present address: Department of Conservation and Land Management, Western Australian Wildlife Research Centre, P.O. Box 51, Wanneroo, Western Australia, Australia 6065.
Pages 357–65 *in* BANDICOOTS AND BILBIES, ed by J. H. Seebeck, P. R. Brown, R. L. Wallis and C. M. Kemper. Surrey Beatty & Sons, Sydney, 1990.

wet-dry tropics of the Northern Territory, and makes some comparisons with data from southern populations.

METHODS

Study area

Studies were conducted within the CSIRO Research Station at Kapalga, 160 km east of Darwin (12°42′S, 132°22′E) in the Northern Territory. The monsoonal climate is characterized by a humid wet season between about December and March, when about 80% of the mean annual rainfall of 1 388 mm usually falls. Little rain usually falls during the dry season (May–August), while the "transition" months of April, and September to November typically experience convectional storms with considerable variability in the temporal and spatial patterning of rainfall (McCown 1981; Taylor and Tulloch 1985). Temperatures are high throughout the year, varying between a mean minimum of 18.0°C in July and a mean maximum of 37.2°C in October (meteorological data for Oenpelli 1910–84, Bureau of Meteorology, Darwin, pers. comm.). The maximum difference in daylength throughout the year is only 1 h 29 min, the shortest day occurring in June (683 min) and the longest in December (722 min) (Bureau of Meteorology, Darwin, pers. comm.).

The area studied comprised tall open-forest dominated by *Eucalyptus tetrodonta* and *E. miniata* with a relatively dense understorey of smaller trees and shrubs (e.g., *Buchanania obovata, Planchonia careya, Petalostigma pubescens* and *Acacia mimula).* The habitat is further described and illustrated in Friend and Taylor (1985).

Field sampling

The *I. macrourus* population was live-trapped on three rectangular grids (A, B and C, each with 10 columns and five rows) which were contiguous on their short axes and enclosed a total area of about 12 ha. These three grids were sequentially trapped on 13 occasions between September 1980 and January 1983, so that each was sampled every two to three months. No trapping was done during January and May 1981, and December 1982. Fifty Elliott aluminium traps (33 × 10 × 10 cm) and 25 wire-cage traps (55 × 20 × 20 cm) baited with a mixture of peanut butter, rolled oats and sardines were set on each grid for three consecutive nights and checked in the early morning. Further details of trap layout are provided in Friend (1985).

Isoodon macrourus individuals were marked on first capture with "Hauptner" numbered metal ear-tags. On their first capture in any three-night trap-session individuals were weighed to the nearest gram and their head lengths measured. In males, width of the scrotum was measured. Females were checked for general reproductive condition of the pouch, and the number, sexes (where possible) and crown-rump and head lengths (to the nearest 0.5 mm on one individual) of any pouch young were recorded.

Analysis

Data were assembled and analysed on a monthly basis. Abundance was estimated using the minimum number of animals known to be alive (KTBA) in any month (see Krebs 1966). Maximum and minimum densities (number of individuals per hectare) were calculated from KTBA estimates with the mean distance between successive captures (AvD) added as a boundary strip to the grid area (after Brant 1962). Adult male and female body weights, adult sex ratios and male scrotal widths were examined in relation to an annual seasonal cycle using data pooled by month.

Trappability was calculated for each grid as the number of animals captured in any month relative to the number known to be alive (Krebs, Keller and Tamarin 1969) and expressed as a percentage. Plots of the frequency of total periods of capture showed a marked decline after one month. Animals caught in only one month or three-night trap session were thus defined as transients, while those caught in two months or more were regarded as residents. Transiency rate was calculated from pooled monthly data as the percentage of males and females trapped that were classified as transient. The mean distance between successive captures (AvD) and the maximum distance between any two captures over the whole period of trapping were calculated for resident males and females. AvD values were also calculated on a monthly basis for males and females.

For each litter recorded, parturition date was estimated using the crown-rump and/or head length measurements in conjunction with the growth data published by Mackerras and Smith (1960). The proportion of litters born in various months and the proportion of adult females carrying pouch young each month was then calculated. For females present for 10 or more months, mean litter size and mean number of litters per year and per season were calculated and used to estimate recruitment rates (mean number of young per female per season). Mean size of and interval between successive litters were also calculated for females having two or more litters within a breeding season.

RESULTS

Abundance

Most captures (>98%) were in the wire-cage traps. Over the 2 900 wire-cage trapnights 45 male and 30 female *I. macrourus* were recorded for a total of 201 and 140 captures respectively, giving a mean trap success of 11.8%. The number of captures per individual ranged from one to 27 for males (mean 4.5, mode 1.0), and from one to 41 for females (mean 4.6, mode 1.0); 44% of the males and 60% of the females were captured only once.

The KTBA estimates fluctuated markedly (Fig. 1), but trends were fairly consistent over the first 24 months of the study. In both of these years abundance was high in October, declined within the November–January period, increased to a maximum in February, then declined sharply during the April–May period as the dry season commenced. A small increase in abundance also was recorded in June of both years (Fig. 1). Density peaked at 0.7 animals ha^{-1} in February 1981 and averaged 0.5 for the first 24 months. However, following the small June increase in 1982, the population declined considerably over the late dry, and only one male was known to be alive (giving a density of only 0.05 animals ha^{-1}) in the early wet (December–January) of 1983 when the study was terminated. A retrapping of grid A in March 1983 (not included in analysis) confirmed the presence of only one male, last captured the previous November. Subsequent retrapping of grids B and C in November 1984 and January 1985 following a late dry season wildfire failed to record any *I. macrourus*.

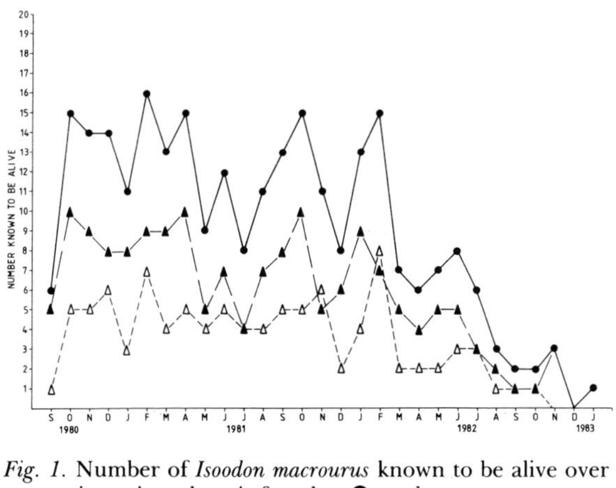

Fig. 1. Number of *Isoodon macrourus* known to be alive over time. ▲ males, △ females, ● total.

Changes in abundance over the first two years differed between sexes. Female abundance was relatively stable except for declines in the December–January period each year, followed by sharp peaks in February. The fluctuations in the total population at these times (as detailed above) are mainly due to changes in the abundance of females. Conversely, the abundance of males varied markedly and mainly contributed to the decline in the total population during the mid-dry seasons and the increases in September–October (Fig. 1). Both males and females declined consistently over the last six months of the study and no females were recorded after October 1982.

Reproduction

Twenty-seven litters were recorded during the study, but only 12 of the 30 females captured produced these. Fourteen litters were produced in the 1980/81 breeding season and 12 in the 1981/82 season, but only one had been born in the 1982/83 season before trapping ceased in late January 1983. Overall, the mean litter size was 2.7 ± 0.2 ($\bar{x}$ ± SE; n = 27) ranging from one to five. There was an average of 2.5 ± 0.5 litters per female, ranging again from one to five. A plot of body weight of females with neonatal pouch young (corrected for the estimated weight of the litter, cf. Mackerras and Smith's 1960 data) against litter size revealed no significant relationships (r = −0.027, n = 19).

Only five females were trapped for 10 months or more, and the mean litter size for these animals was 2.7 ± 0.3 (n = 21), each producing an average of 4.2 ± 0.2 (n = 5) litters per year, or 3.8 ± 0.3 (n = 4) over a complete breeding season. The chronological distribution of litter sizes for the five females was: first litter = 2.2 ± 0.6 (n = 5); second litter = 3.0 ± 0.6 (n = 5); third litter = 3.4 ± 0.4 (n = 5); fourth litter = 2.2 ± 0.7 (n = 5); fifth litter = 3.0 (n = 1). Again, however, no significant relationship between (corrected) female body weight and litter size was found for these five animals (r = −0.17, n = 15). The mean interval between successive litters was estimated at 54.1 ± 1.2 days (n = 7).

Between October and March 90–100% of adult females were carrying pouch young, and 50% were doing so in April and September. No pouch young were recorded between May and August. Estimation of the dates of parturition of litters based on pouch young crown–rump and head lengths showed that all litters were born within the period late August to mid April (Fig. 2). The highest proportion of births was in September, and about twice as many litters were born in the first half of the season (August–December, 18 litters) as in the second (January–April, nine litters), but these differences are not significant (χ^2_1 = 0.94, P >0.30). However, a plot of the mean sizes of litters against the month when each was first recorded (Fig. 3, based on data pooled for a 12-month period) indicates that although fewer litters are born in the second half of the season (January–April), their mean size is greater. The mean size of litters recorded in the September–December period was 2.4 ± 0.3 (n = 16) while that for the January–March period was 3.5 ± 0.3 (n = 10), and these means are significantly different (t = 2.6; P >0.05). Only one litter, with one young, was born in April during the study.

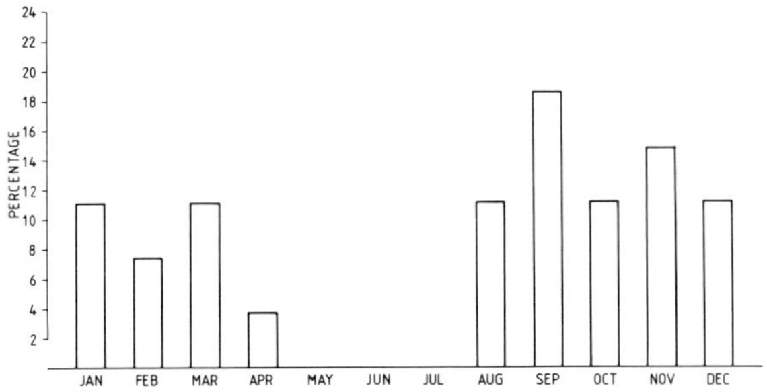

Fig. 2. Proportion (%) of total *Isoodon macrourus* litters (n = 27) born in various months (based on pooled data).

Female *I. macrourus* apparently reach sexual maturity before they are fully grown. The body weight of the smallest female with pouch young was 550 g; its age was estimated to be only about 120 days (Mackerras and Smith 1960). The youngest juveniles which were no longer attached to nipples but still accompanying the mother were estimated to be about 46 days of age, and the oldest about 55 days. Pouch life therefore apparently ceases between about 45 and 50 days, and, given that successive litters occur on average 54 days apart (see earlier), juveniles are probably weaned at about 55 days of age, and then become relatively independent. These data, in combination with a gestation period of only 12.5 days (Lyne 1974) show that females are characterized by a high reproductive output and minimal post-weaning maternal care.

For males, a regression of scrotal width against weight (using all data) revealed a significant linear relationship ($r = 0.94$ $P < 0.001$) for weights up to about 600 g. At this weight (estimated age about 100 days) there was a marked increase in scrotal size from about 16 mm to about 29 mm; scrotal widths averaged around 30 mm thereafter. These data suggest that male *I. macrourus,* like females, reach sexual maturity at about three to four months of age. The mean scrotal widths of adult males (>600 g) remained relatively constant over the seasonal cycle, except for a slight decline in the April–June (non-breeding) period.

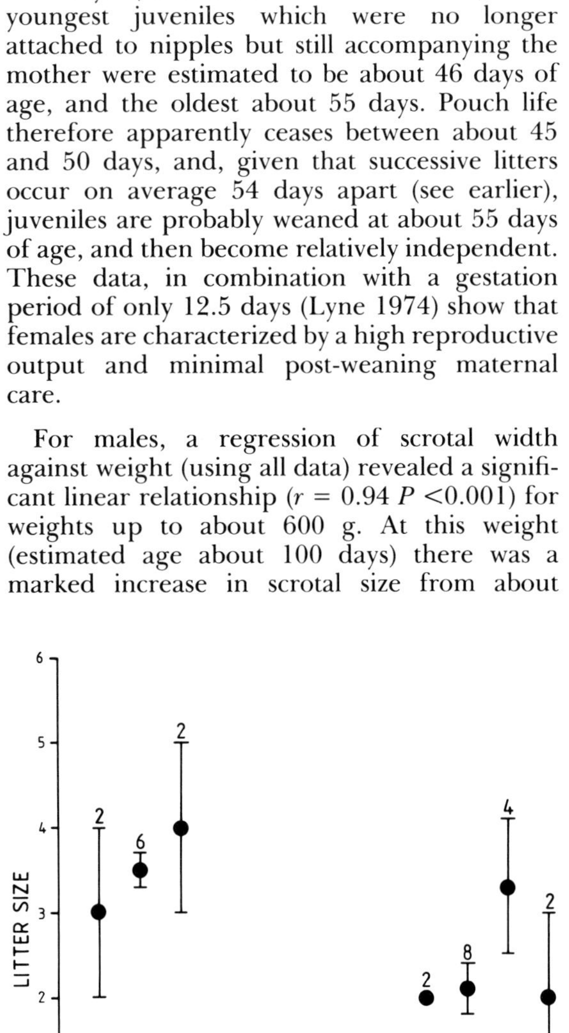

Fig. 3. Mean monthly litter size of *Isoodon macrourus* ($\bar{x} \pm$ SE; numerals are sample sizes, based on pooled data).

Population dynamics

Twenty-five males (56%) and 20 females (67%), viz. 60% of the total population, were classified as transients. Of the remaining 10 females, five were captured for 2–6 months, and five for 10–14 months. Seven of the remaining 20 males were captured for 10 or more months, and 13 for 2–8 months. On the basis of sexual maturity criteria (see earlier) immature males and females were categorized as those less than 600 g and 500 g respectively, with juveniles (arbitrarily) being those less than 200 g. Fifteen (33.3%) of the males and 18 (60%) of the females were classified as juvenile or subadult at first capture, and these animals comprised 44% and 75% of the transient males and females respectively. A distinct difference in age structure between the sexes was therefore apparent, with a high proportion of the females being immature animals, and this was reflected in the relative proportion of male and female transients in the population.

Over time however, male and female transiency rates showed similar overall trends (Fig. 4a). Transiency was relatively high during the middle part of the breeding season (October–February) particularly for females, and was presumably attributable to young animals entering the trappable population, and then dispersing from the grid area in search of new territories. For the remainder of the dry season, transiency rates in both sexes were lower but fluctuated markedly, seemingly on a two-month cycle (Fig. 4a). This phenomenon could be due to low sample size, but also could be partly attributable to cohorts of young transient animals, born in the previous season, entering the grid area in a successional pattern.

Trappability was also high during the mid-breeding season (November–February; Fig. 4b), suggesting that considerable activity and movement of animals was occurring. In both sexes trappability declined during March–May (January–May for males) as the population entered the non-breeding phase, but rose again in June–July. In the June–July period the population was probably relatively stable with generally lower abundance (Fig. 1), low transiency rates, and a high proportion of the animals known to be alive being captured (Figs 4A, 4B). As the breeding season commenced (August–September), trappability declined temporarily in

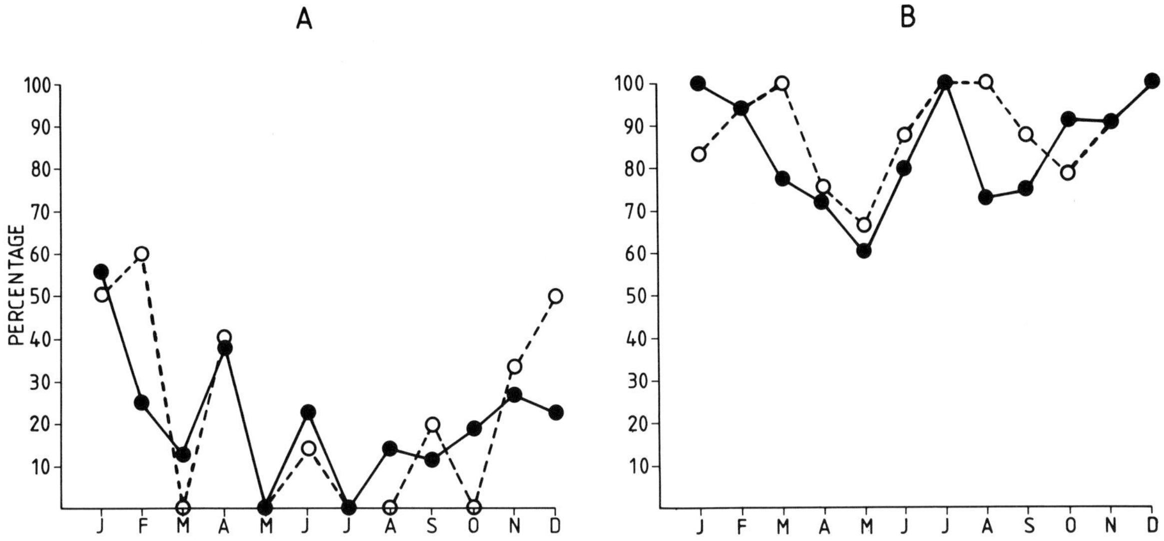

Fig. 4. Proportion of transients (A) and trappability (B) of male (●) and female (○) *Isoodon macrourus* throughout the year; based on pooled monthly data.

both sexes, perhaps because of social interactions associated with breeding.

Males moved a mean distance of 64.2 m (range 0–165 m) between successive captures, the maximum distance between any two non-successive captures being 565 m. The corresponding figures for females were 60.9 m (range 0–165 m) and 402 m. The measure of AvD (Brant 1962) added as a boundary strip around the combined grid area for calculating density was taken as 60 m. However, AvD values varied with time of year. In males, values peaked at about 90–120 m in the early and mid periods of the breeding season (August and January), but were minimal (30–40 m) in the March–April period at the end of the season. Trends for females were similar but not as marked. Such movement patterns are consistent with the trends in trappability and transiency outlined above. Home range sizes were not calculated because of the limited size of the combined grid area, the small number of animals captured multiple times and the difficulty in obtaining accurate estimates of home range areas using trap records (see Gordon 1974).

The overall sex ratio was 1.5 ($0.05 < P < 0.10$, n.s.) with adult males consistently outnumbering females, and marked peaks in the ratio occurring in December–January and in August. These peaks may have been attributable to some new males entering the grids during the early breeding season (e.g., August) and to cohorts of juveniles subsequently becoming trappable. Indeed, these fluctuations are lessened if immature animals are excluded from the calculations. Ratios were closest to unity in the mid-dry season (June–July) when the population was most stable (see above).

Adult weights ranged from 610 g to 1 780 g for males, and 530 g to 1 054 g for females. Mean weights for the three classes of animals were: adult males 1 139.4 ± 26.8 g (n = 110) and females 848.7 ± 17.3 g (n = 62); subadult males 477.3 ± 11.8 g (n = 3) and females 466.7 ± 23.5 g (n = 3); juvenile males 77.6 ± 9.2 g (n = 11) and females 69.9 ± 10.0 g (n = 13). No animals in the range 200–450 g were trapped.

Mean monthly adult weights were much more variable for males (with coefficients of variation up to 36%) than for females (with coefficients of variation up to 20%, Fig. 5). Although seasonal trends in weight were not pronounced, a general increase in mean weight occurred in the September–November period following minimums in August (females) and September (males). Such trends were presumably associated with the breeding season and, for females, may have been partly attributable to their carrying pouch young. However, female weights declined again in the December–February period when pouch young were still present and litters were significantly larger (Figs 3 and 5B). A plot of mean adult body weights for males and females over the full period of study revealed no consistent trends in the male data. For females, the early breeding season increases were apparent in both the 1980 and 1981 seasons. Over the dry season months of 1982 however, mean female weights declined markedly (concomitant with the decline in abundance of the total population, Fig. 1) and reached a minimum of about 650 g in the July–August period before females disappeared from the trap record in October.

The proportion of immature animals in the population (Fig. 6) was consistent with the

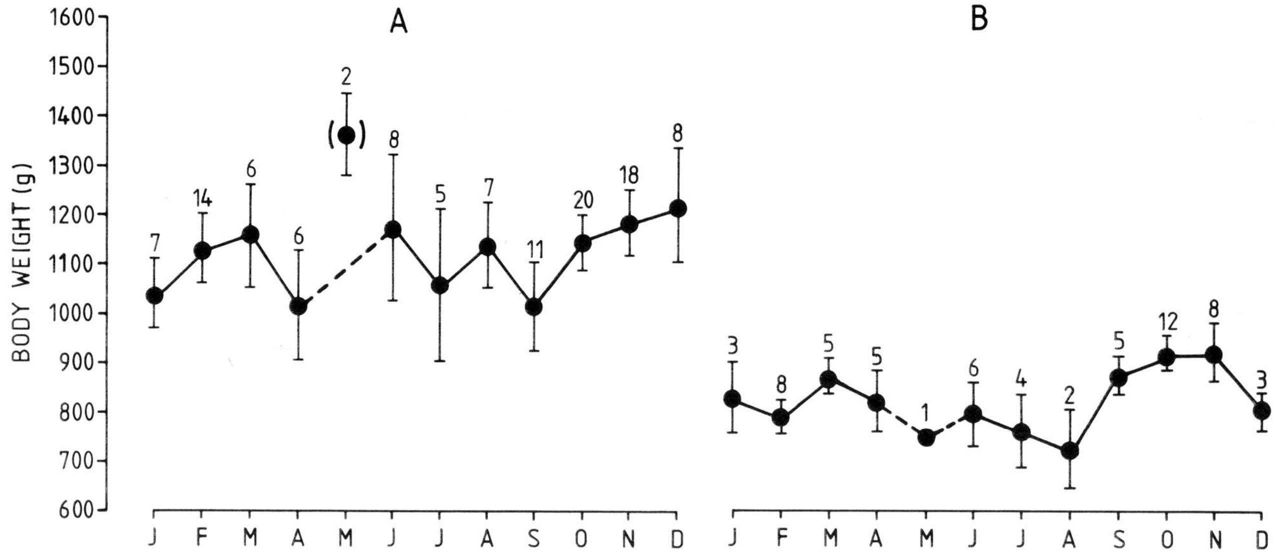

Fig. 5. Mean monthly body weights (g) of adult male (A) and female (B) *Isoodon macrourus*. x̄ ± SE; numerals are sample sizes, based on pooled monthly data. May data based on only one sample, in 1982.

estimates of the proportion of litters in various months (Fig. 2), there being about a two-month lag between births and the appearance of independent young in the trappable population. Young independent animals were first trapped in September but represented only 7–17% of the total numbers captured until the end of December. In January and February, however (particularly in 1982), immature animals comprised around 40% of the total captured (Fig. 6), consistent with the peak in births during the previous September–November (Fig. 2). Proportions declined again over March–June as breeding subsided. The high proportion recorded in May (Fig. 6) is probably due to very low sample size (five total) because no trapping was undertaken in May of 1981.

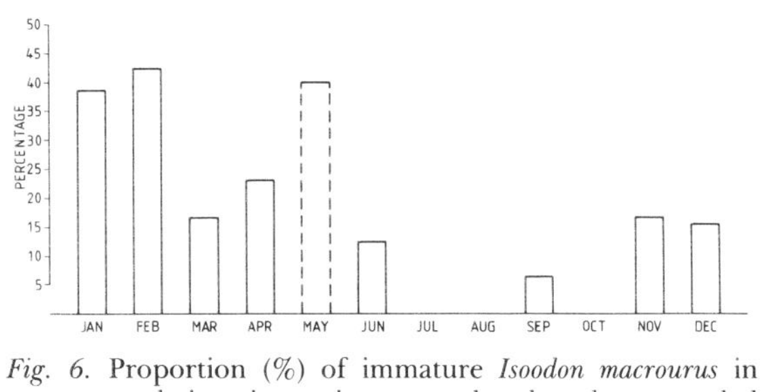

Fig. 6. Proportion (%) of immature *Isoodon macrourus* in population in various months, based on pooled monthly data. May data based on only one sample, in 1982.

From estimates of mean litter size (2.7) and mean number of litters per season (3.8) for long-term resident females (see earlier) the mean number of young per female per season is calculated as approximately 10. On this basis total (maximal) recruitment for the 1980/81 breeding season is estimated to be about 40 animals (from four resident females) and that for the 1981/82 season about 50 animals (from five resident females). However, these estimates are based on litter sizes at initial inspection and so do not take pouch young loss into account, and also do not consider that some resident females were recorded for only half of each breeding season (i.e., 12 calendar months). Totalling the number of pouch young last-recorded in each litter for all females trapped each season gives alternative estimates of 37 and 31 recruited for the respective breeding seasons. Only two juveniles were recorded in the 1982/83 season before trapping ceased. From the calendar of trap records only three male and three female immature animals became resident on the grids, and these were all subadult at first capture. The majority of young animals recruited, therefore, either died before reaching sexual maturity, dispersed from the natal area, or underwent a period of trap-shyness (e.g., animals in the 200–450 g weight range, which were never trapped).

DISCUSSION

There is little published information available on seasonal trends in abundance of wild populations of *I. macrourus*. Many studies have been laboratory orientated and/or have concentrated on aspects of reproduction, weight trends, recruitment and survival (e.g., Mackerras and Smith 1960; Lyne 1974; Gemmell 1982; Hall 1983). Studies in New South Wales (Gordon 1971) and suburban Darwin, Northern Territory (Smith 1980), recorded peak densities in April–May, which reflected recruitments from the previous breeding season. In the present study two main peaks in abundance occurred over a yearly cycle. The first, during the mid-wet (February) was mainly associated with the recruitment of juveniles (particularly females) into the trappable population, and most of these subsequently dispersed or died.

The second, in September–October, was attributable mainly to an increase in the number of males during the early stages of the breeding season, perhaps due to increased movement bringing new animals on to the grids. These data suggest that population density may be strongly influenced by a sex-related mechanism of recruitment and dispersal. However, this conclusion is tentative, and much more data, collected intensively over several years, are required for substantiation.

The factors influencing the timing, duration and intensity of breeding in *I. macrourus* are central to the question of population regulation in the species. There is apparently considerable variation in the onset and duration of breeding, even within particular regions. In the wet-dry tropics of the Northern Territory, this study and that of Smith (1980) established that breeding is restricted to the period August–April. Records from surveys of Kakadu National Park (Braithwaite, pers. comm.) also confirmed this seasonal breeding pattern. Limited records from the early Arnhem Land Expedition, however, suggested that young were present throughout the year (Johnson 1964), but more data would be required to substantiate this. In the south of the species' range (temperate–subtropical) both seasonal (Gordon 1971; Gemmell 1982) and non-seasonal (Gordon 1974; Hall 1983) breeding have been reported.

Results obtained from this study relating to the development of young and the mean interval between successive litters are similar to those from southern populations. Pouch life is about 50 days, and young are weaned at about 55–60 days, at which time a new litter may be born (Gordon 1971, 1974; Gemmell 1982; Hall 1983, this study). Sexual maturity may be attained at about four months of age. The pattern is apparently very similar in *I. obesulus* (Stoddart and Braithwaite 1979) and *P. nasuta* (Lyne 1964). Litter size in *I. macrourus*, however, appears to be lower in the tropics (2.7 this study, 2.6 Smith 1980) than at greater latitudes (3.0 in New South Wales, Gordon 1971, and 2.9 to 3.4 near Brisbane, Gordon 1974; Gemmell 1982; Hall 1983). This trend may seem to reflect the smaller body size of adult females in the tropics (e.g., 850 g this study cf. 1 046 g near Brisbane, Hall 1983). However, the lack of any positive linear relationship between female body size and litter size in some populations (Gordon 1971; Smith 1980, this study) does not support this idea, and contrasts with the findings of Hall (1983) for *I. macrourus* near Brisbane, and of Stoddart and Braithwaite (1979) for *I. obesulus* in Victoria.

There also appears to be no consistent trend in peak litter size over time for different areas. The largest litters may occur early to mid-way (Gordon 1966, 1971) or late (Gemmell 1982, this study) in the main period of breeding. Where breeding is non-seasonal there may be no obvious trend in litter size (Hall 1983).

From the limited data available for *I. macrourus* it seems that body weight/litter size relationships may be stronger in populations with non-seasonal breeding. Furthermore, in such populations, breeding intensity (as measured by the proportion of adult females carrying young per month) is generally lower but more variable than in populations with seasonal breeding (Gemmell 1982 and this study cf. Gordon 1974; Hall 1983). In the seasonal populations (particularly those in the wet-dry tropics) it appears that nearly all female bandicoots of breeding age will breed continuously throughout the season (i.e., breeding intensity is maximal). Their litter size, however, is generally smaller and/or less tied to female body size than that in the non-seasonal populations. The net result is that production of young (number per female per year) is probably similar in seasonal and non-seasonal populations. Again, more data are required to substantiate these ideas.

The factors influencing breeding in *I. macrourus* are obviously complex. Gemmell's (1982) statement that the breeding season is more definite at greater latitudes, and his speculation that its onset may therefore be coupled with the light regime, are clearly not supported by results from the Northern Territory. Daylength at Darwin latitudes varies by only 1.5 h over 12 months, yet the *I. macrourus* breeding pattern is the most seasonal yet reported.

In the wet-dry tropics food availability would appear to be a more plausible extrinsic factor influencing the onset of breeding (see also Gordon 1971). Litter and soil-dwelling invertebrates are an important component of *I. macrourus* diet (Harrison 1962; Gordon 1983), and in tropical areas the abundance of this resource is strongly influenced by moisture levels, viz. rainfall (Wolda 1978; Levings and Windsor 1982; Hertog 1983), rather than temperature or photoperiod. The occurrence of breeding in *I. macrourus* only in the months of the year where significant rainfall events usually occur, and when litter-dwelling invertebrates are most abundant (Hertog 1983), supports the idea that the availability of this resource has an important influence on breeding. Indeed, this is also the case for litter-foraging insectivores in wet-dry tropical regions elsewhere (e.g., see Nicoll 1985), and exemplifies the general importance of nutrition in mammalian reproduction (Widdowson 1981). Furthermore, perhaps because of the unreliability of the early rains in September–November (McCown 1981) female bandicoots seem to "hedge their bets" by producing many small litters during this period as they gain weight, and fewer, but larger, litters in the January–March period when the monsoonal rains usually begin. This apparently does not reflect any female body size or age structure effects (see earlier).

The decline in female body weights and lack of breeding at the time of a failure in the 1982/83 wet season rains corroborate the above hypothesis. The early months of that wet season for the Darwin-Daly District were either the driest on record (e.g., October) (Bureau of Meteorology 1982) or the driest for 10 years or more (January and February) (Bureau of Meteorology 1983). Judging from subsequent trapping this climatic event had a severe impact on the population of *I. macrourus*. Reduced invertebrate abundance at this time was also deemed responsible for similar marked reductions in abundance and breeding of the scansorial insectivore *Antechinus bellus* in this area (Friend 1985). Seasonal unpredictability, through its influence on the quantity and quality of food supply, may also explain the opportunistic or non-seasonal breeding and demographic fluctuations recorded for many herbivorous species of this region, such as *Macropus agilis* (Bolton, Newsome and Merchant 1982), *Trichosurus arnhemensis* (Kerle and Burgman 1984) and *Mesembriomys gouldii* (Friend 1987).

The variability in the onset and duration of breeding over the range of *I. macrourus* suggests, however, that there is no single or constant factor triggering breeding. The results obtained for breeding in any region may depend on particular extrinsic factors (e.g., climatic phenomena, soil type and moisture status) as well as certain intrinsic factors (e.g., social interactions, population age structure) in effect at the time of the study. Certainly the species seems able to quickly take advantage of optimal conditions with high reproductive potential, rapid development, minimal maternal care and a high rate of dispersal of juveniles (especially females). Juvenile mortality is often high (Gordon 1971; Hall 1983), particularly when conditions are sub-optimal. Population levels in a particular region may be largely determined by whether females are able to maintain weight and rear some young in patches of suitable habitat when conditions are unfavourable, and whether reproductive patterns can respond to produce many young capable of dispersion to new areas when favourable conditions return.

Breeding may thus be on a seasonal/non-seasonal continuum within any area, to enable greater flexibility in variable environments. This may entail a variable life history strategy, as found for the congeneric species *I. obesulus* (Stoddart and Braithwaite 1979). Such a breeding strategy is likely to be better developed in populations inhabiting the subtropical/temperate areas (e.g., south-east Queensland). Although fairly seasonal, these areas probably experience less severe fluctuations in soil moisture levels than those in the wet-dry tropics further north, because of better moisture-holding capacities concomitant with lower evaporation rates during the cooler winter dry season. Variation in rainfall may, however, be accentuated by fine-scale heterogeneity in habitat attributes such as soil type (e.g., sandy versus clay or loam soils), and so influence bandicoots (via the litter fauna).

In the less seasonal situations breeding intensity need not be maximal and the younger, smaller females may tend to produce the smaller litters. In the seasonal situations (e.g., the wet-dry tropics), however, it seems that breeding intensity is maximized to take advantage of the short favourable period, but that considerable flexibility in litter size is possible. The species is truly remarkable in its finely-tuned breeding adaptations to the variable environments of its range.

ACKNOWLEDGEMENTS

This study was possible only with the generous field assistance of many people, in particular M. Friend, J. Randall, R. Fletcher, K. Bardsley and A. Kerle. I wish to also thank D. Murphy for drawing the figures, and Drs R. W. Braithwaite and W. M. Lonsdale for comments on the initial draft of this paper.

REFERENCES

Bolton, B. L., Newsome, A. E. and Merchant, J. C., 1982. Reproduction in the Agile Wallaby *Macropus agilis* in the tropical lowlands of the Northern Territory: opportunism in a seasonal environment. *Australian Journal of Ecology* **7:** 262–77.

Brant, D. H., 1962. Measures of the movements and population densities of small rodents. *University of California Publications in Zoology* **62:** 105–84.

Bureau of Meteorology, 1982. Monthly weather review, Northern Territory — October. Department of Science and Technology: Darwin.

Bureau of Meteorology, 1983. Monthly weather review, Northern Territory — January, February. Department of Science and Technology: Darwin.

Calaby, J. H. and Keith, K., 1974. Mammals. Pp. 179–208 in *Fauna survey of the Port Essington District Cobourg Peninsula, Northern Territory of Australia* ed by H. J. Frith and J. H. Calaby. CSIRO Division of Wildlife Research, Technical Paper No. 28: Canberra.

Friend, G. R., 1985. Ecological studies of a population of *Antechinus bellus* (Marsupialia: Dasyuridae) in tropical northern Australia. *Australian Wildlife Research* **12:** 151–62.

Friend, G. R., 1987. Population ecology of *Mesembriomys gouldii* (Rodentia: Muridae) in the wet-dry tropics of the Northern Territory. *Australian Wildlife Research* **14:** 293–303.

FRIEND, G. R. AND TAYLOR, J. A., 1985. Habitat preferences of small mammals in tropical open forest of the Northern Territory. *Australian Journal of Ecology* **10:** 173–85.

GEMMELL, R. T., 1982. Breeding bandicoots in Brisbane (*Isoodon macrourus*; Marsupialia, Paramelidae). *Australian Mammalogy* **5:** 187–93.

GORDON, G., 1966. *A study of movements of the Short-nosed Bandicoot* Isoodon macrourus *(Gould)*. B.Sc. (Hons.) Thesis, University of Queensland: Brisbane.

GORDON, G., 1971. *A study of island populations of the Short-nosed Bandicoot* Isoodon macrourus *(Gould)*. Ph.D. Thesis, University of New South Wales: Sydney.

GORDON, G., 1974. Movements and activity of the Short-nosed Bandicoot *Isoodon macrourus* (Gould) (Marsupialia). *Mammalia* **38:** 405–31.

GORDON, G., 1983. Northern Brown Bandicoot *Isoodon macrourus*. Pp. 96–97 in *The Australian Museum complete book of Australian mammals* ed by R. Strahan. Angus & Robertson: Sydney.

HARRISON, J. L., 1962. The food of some Innisfail mammals. *Proceedings of the Royal Society of Queensland* **73:** 37–43.

HALL, L. S., 1983. Observations on body weights and breeding of the Northern Brown Bandicoot, *Isoodon macrourus*, trapped in south-east Queensland. *Australian Wildlife Research* **10:** 467–76.

HERTOG, A. L., 1983. *A survey of soil fauna and influencing factors in three habitat types in tropical Northern Territory.* Environmental Biology Certificate Thesis, Darwin Community College (unpublished).

JOHNSON, D. H., 1964. Mammals of the Arnhem Land expedition. Pp. 427–515 in *Records of the American-Australian expedition to Arnhem Land; Vol. 4 zoology* ed by R. L. Specht. Melbourne University Press: Melbourne.

KERLE, J. A. AND BURGMAN, M. A., 1984. Some aspects of the ecology of the mammal fauna of the Jabiluka area, Northern Territory. *Australian Wildlife Research* **11:** 207–22.

KREBS, C. J., 1966. Demographic changes in fluctuating populations of *Microtus californicus*. *Ecological Monographs* **36:** 239–73.

KREBS, C. J., KELLER, B. L. AND TAMARIN, R. H., 1969. *Microtus* population biology — demographic changes in fluctuating populations of *M. ochrogaster* and *M. pennsylvanicus* in southern Indiana. *Ecology* **50** 587–607.

LEVINGS, S. C. AND WINDSOR, D. M., 1982. Seasonal and annual variation in litter arthropod populations. Pp. 355–87 in *The ecology of a tropical forest — seasonal rhythms and long-term changes* ed by E. G. Leigh, A. S. Rand and D. M. Windsor. Smithsonian Institution Press: Washington, D.C.

LYNE, A. G., 1964. Observations on the breeding and growth of the marsupial *Perameles nasuta* Geoffroy, with notes on other bandicoots. *Australian Journal of Zoology* **12:** 322—39.

LYNE, A. G., 1974. Gestation period and birth in the marsupial *Isoodon macrourus*. *Australian Journal of Zoology* **22:** 303–09.

MACKERRAS, M. J. AND SMITH, R. H., 1960. Breeding the Short-nosed marsupial Bandicoot, *Isoodon macrourus* (Gould), in captivity. *Australian Journal of Zoology* **8:** 371–82.

MCCOWN, R. L., 1981. The climatic potential for beef cattle production in tropical Australia. III. Variation in the commencement, cessation and duration of the green season. *Agricultural Systems* **7:** 163–78.

NICOLL, M. E., 1985. Responses to Seychelles tropical forest seasons by a litter-foraging mammalian insectivore, *Tenrec ecaudatus*, native to Madagascar. *Journal of Animal Ecology* **54:** 71–88.

SMITH, J. R., 1980. *A study of the marsupial bandicoot* Isoodon macrourus *in Darwin.* Environmental Biology Certificate Thesis, Darwin Community College (unpublished).

STODDART, D. M. AND BRAITHWAITE, R. W., 1979. A strategy for utilization of regenerating heathland habitat by the Brown Bandicoot (*Isoodon obesulus*; Marsupialia; Peramelidae). *Journal of Animal Ecology* **48:** 165–79.

TAYLOR, J. A. AND TULLOCH, D. G., 1985. Rainfall in the wet-dry tropics: extreme events at Darwin and similarities between years during the period 1870–1983 inclusive. *Australian Journal of Ecology* **10:** 281–95.

THOMAS, O., 1904. On a collection of mammals made by Mr J. T. Tunney in Arnhem Land, Northern Territory of Australia. *Novitates Zoologicae* **11:** 222–29.

WIDDOWSON, E. M., 1981. The role of nutrition in mammalian reproduction. Pp. 145–59 in *Environmental factors in mammal reproduction* ed by D. Gilmore and B. Cook. Macmillan: London.

WOLDA, H., 1978. Seasonal fluctuations in rainfall, food and abundance of tropical insects. *Journal of Animal Ecology* **47:** 369–81.

Problems in the estimation of the effective population size of the Eastern Barred Bandicoot *Perameles gunnii* at Hamilton, Victoria

W. B. SHERWIN[1] and P. R. BROWN[2]

Since the 1930s, the population of *Perameles gunnii* on the Australian mainland has been declining. It is now restricted to the city of Hamilton and its outskirts, in western Victoria. The census size of the population is 633 or lower. Dispersal of more than 1 km is known to occur, but detailed dispersal data are not available. On the assumption that the whole population represents a single neighbourhood, available demographic data have been used to calculate the variance effective population size, which was estimated to be 67 or lower. The Hamilton population may have been at this size for 25–48 years. If current theories of conservation genetics are accepted, it is calculated that a detectable loss of genetic variability may have occurred in this period.

Key words: *Perameles gunnii*, conservation genetics, effective population size.

INTRODUCTION

THE population of *Perameles gunnii* on the Australian mainland is now restricted to the city of Hamilton and its outskirts. The Hamilton population appears to have been isolated from others, and to have remained at approximately its present size and density since at least 1960 (Seebeck 1979), or possibly since 1937 (Harper 1945). It is possible that the effective population size is so low that genetic variation is being lost, adding to the problems that we face in attempting to conserve the species. This paper documents attempts to estimate both the census size of the population and the (genetically) effective size.

From the effective size, we can estimate whether a population is likely to lose genetic variation, and how fast (Crow and Kimura 1970). Genetic variation has been considered important because, in the long term, the environment is expected to change; genetic variation may allow adaptation to these changes (Frankel 1970; Tyndale-Biscoe and Calaby 1975; Franklin 1980; Frankel and Soule 1981; Schonewald-Cox, Chambers, MacBryde and Thomas 1983). A population often contains a variety of genetically different individuals. If the environment changes, these variants may give rise to lineages which can survive in the new conditions. Therefore, a lack of genetic variation increases the chance of extinction. The maintenance of genetic variation may be influenced by population size. In large populations, genetic variation is maintained by a balance between production of variants (e.g., mutation) and their loss by chance events in each generation (the effect of selection on variation will be discussed later). Smaller populations are expected to have a higher rate of loss of genetic variation. The rate of loss is affected by the census size (N) and by the breeding system and demography. Given the appropriate information, it is possible to calculate the "variance effective population size" (Crow and Kimura 1970). This statistic is inversely proportional to the rate of loss of (additive) genetic variance (Falconer 1960). The effective size (Ne) is usually considerably lower than the census size.

Several authors have attempted to estimate the minimum effective size required for maintenance of "normal" genetic variation. Using rather limited data, Falconer (1960) and Franklin (1980) concluded that an effective size of 500 was adequate for maintenance of additive genetic variance. Lande and Barrowclough (1987) point out that this figure is only valid when heritability is 0.5, and that the effect of selection cannot be ignored (as in the above argument). They studied a number of different types and measures of genetic variation, and suggested that an effective size of a few hundred would be adequate in many cases. The measures of variation studied did not include "actual number of alleles", which may be as important as the other measures, for long-term conservation (Denniston 1978; Gale and Lawrence 1984); maintenance of actual numbers of alleles would require much larger population sizes. Despite the difficulty of giving a precise minimum size for conservation, this concept has played an important part in the recognition of conservation genetics by government authorities

[1]Department of Genetics and Human Variation, La Trobe University, Bundoora, Victoria, Australia 3083.
[2]Faculty of Applied Science, Victoria College, Rusden Campus, 662 Blackburn Road, Clayton, Victoria, Australia 3168.
Pages 367–74 *in* BANDICOOTS AND BILBIES, ed by J. H. Seebeck, P. R. Brown, R. L. Wallis and C. M. Kemper. Surrey Beatty & Sons, Sydney, 1990.

(Brown 1983; IUCN 1980; Australian Department of Home Affairs and Environment 1982). Whatever the minimum figure, it is now apparent that endangered or vulnerable populations should be characterized not only by their census size, but by their effective size as well (Bonnell and Selander 1974; Frankel and Soule 1981).

The effective size of a population is the number of individuals that contribute (approximately) equally to the next generation of parents; this number is only equal to the census size if "ideal" conditions prevail:

— dispersal high (no subdivision);

— no overlap of generations;

— random mating;

— Poisson distribution of reproductive success between individual parents;

— sex ratio 1:1;

— no fluctuations in census size.

Deviations from this pattern generally depress Ne relative to N. For example, if the adult sex ratio is very skewed (e.g., one female and 499 males), then only a small proportion of individuals mate and pass on their genes — the effective size of the population is low, and there is a high chance of loss of genetic variants each generation. Equations are available to correct for each deviation from ideality. However, schemes that include most of the corrections, such as those of Emigh and Pollak (1979) or Hill (1979), usually require complex data sets. Lande and Barrowclough (1987) offer a method that includes all corrections except assortative mating and dispersal, and requires less complex data, which may be available from ecological studies of endangered species. The present analysis uses a further simplification of this method, to estimate the effective size of *P. gunnii* at Hamilton, Victoria, using data collected in the course of an ecological investigation at the T/S/B site (Tip/Sewerage Works/Blood Paddock).

METHODS, RESULTS AND ANALYSES

Collection of data is described by Brown (1989). However, certain portions of the data were excluded, as described below, in an attempt to minimize biases.

Census size

Census size was derived from monthly known-to-be-alive totals (KTBA) at the T/S/B site. Two possible biases are that, at the beginning of the study period, KTBA is expected to rise sharply, and at the end of the study the reverse effect is expected. In the present study, a period of irregular trapping which had taken place before the main study period should prevent the first bias. The second bias is avoided by not using KTBA from the last three months of the study (over 95% of all recaptures occurred in four months or less). For the period March 1983 to May 1985, using all trapping records, the average KTBA was 16.15 ± 0.622 (se), at T/S/B. To convert this figure to a census for the whole Hamilton population, it is necessary to estimate the relationship between the area inhabited by the bandicoots at T/S/B, and the area covered by the whole Hamilton population. Brown (1989, Fig. 4) shows the T/S/B site (36 ha), and an estimate of the total area inhabited by the Hamilton population (1 412 ha, Brown 1989, Fig. 1).

Dispersal

Few data on lifetime dispersal are available. Single movements of up to 1.0 km have been observed in Hamilton (Brown, unpublished data), and 1.44 km in Tasmania (Heinsohn 1966). Initially, it is assumed that sufficient dispersal occurs for us to consider the Hamilton population to be a single panmictic unit. This assumption will be examined later.

Assortative mating

No data are available, so this factor is ignored; mates are assumed to be chosen at random with respect to genetic relationship.

Overlap of generations

Following Nei and Imaizumi (1966), the generation length (T) is defined as the mean age at breeding. One problem is that the age of animals cannot be determined after weaning, so that we can only use those few animals which have been regularly trapped from weaning. If the exact date of birth was not known, unweaned animals were aged by interpolation in Figure 16(a) of Heinsohn (1966), which gives a relationship between weight and age. Another problem is that there is no pair-bond, so age at breeding can only be determined for females, and the male generation length is assumed to be the same.

Table 1 shows age and litter size for all known litters borne by 13 females. For the births of these 73 offspring, the mean age of mothers is 13.4 ± 0.9 (se) months.

Overlap of generations

The number of individuals entering the population per generation is the overlap of generations. "Entry" implies recruitment of young, rather than immigration, in this paper.

For the analysis of a population with overlapping generations, the preferred data set includes age-specific mortality and reproductive

Table 1. Age (in months) at breeding in female *Perameles gunnii*, including size of litter.

Age	Litter size	Age	Litter size	Age	Litter size
4.0	1	5.3	2	8.4	2
7.7	2	6.7	2	4.3	2
12.0	2	5.3	1	4.2	1
14.9	2	8.4	2	5.5	3
17.0	4	11.9	3	8.3	3
5.3	1	20.6	2	10.1	2
16.4	2	22.2	2	12.1	2
22.8	2	23.6	2	4.0	1
24.5	3	29.6	1	3.9	2
27.5	2	4.5	1		
29.3	2	11.0	2		
4.7	2	19.1	1		
6.3	2	21.0	3		
8.6	2	23.7	2		

output (Emigh and Pollak 1979; Hill 1979; Lande and Barrowclough 1987). Since the difficulty of ageing most individuals precluded this approach, an alternative method is used, which involves the determination of the number of entrants to the population per generation, and correction for the variation of reproductive output between these individuals (Crow and Kimura 1972; Nei and Imaizumi 1966).

Entrants to the population were counted at weaning; new adults were considered to be immigrants, not true recruits to the population. Table 2 shows the number of entrants in a 30-month period. From these data, we may calculate the number of individuals of each sex which enter the whole Hamilton population, relying upon the assumptions that were used in the calculation of census size (above).

The number of entrants per generation is then:

$$N_m = 153.0 \text{ males}$$
$$N_f = 339.9 \text{ females}$$

These figures may be underestimates, since some of the new adults shown in Table 2 may not be immigrants, but genuine "entrants" which were not detected before weaning.

Table 2. New entrants (weanlings, i.e., independent individuals weighing less than 500 g) to the Hamilton population of *Perameles gunnii* from March 1983 to August 1985 inclusive. Test for deviation from 1:1 sex ratio for weanlings. χ^2 (1) = 4.17 $P < 0.05$.

	Female	Male	Total
Independent adults	11	39	50
Weanlings (<500 g)	20	9	29

Unequal family size

Family sizes of males are not known, and are assumed to follow the same distribution as females. Records used include all female weanlings which entered the population between March 1983 and July 1984, inclusive; therefore, these animals could be monitored for at least one generation before the end of the study in August 1985.

The breeding success of 16 females ascertained and monitored in this fashion is shown in Table 3, as production of pouch young. Not all pouch young were sexed, so the reproductive output of females is not divided into male and female offspring for this analysis. Mean number of pouch young produced in a lifetime was $\bar{y}$ = 2.6, with sample variance V_y = 28.8. However, our analysis of production of young must include mortality up to the point designated as "entry" to the population (Hill 1979). It proved impossible to follow a sufficient sample of marked pouch young of these 16 females through to weaning. However, records are available for the survival of marked pouch young of a number of mothers which were not necessarily followed for their whole lifetime (Table 4).

Table 3. Number of offspring of individual female *Perameles gunnii* at T/S/B. This table includes all females which entered the population as weanlings between March 1983 and July 1984, inclusive. Mean = 2.6, variance = 28.8.

No. of offspring	0	1	2	3	4	5	6	13	18
No. of Females	11	0	2	0	0	0	1	1	1

Table 4. Survival of marked pouch young at T/S/B: November 1980 to August 1985, inclusive.

Adult (No.)	Pouch young marked (n_i)	Weanlings surviving (l_i)	Survival (s_i)
0606	1	—	0
1212	2	—	0
1616	2	1	0.5
6262	1	1	1
3838	2	1	0.5
7575	1	—	0
8686	9	—	0
8282	9	4	0.4444
9898	3	1	0.3333
8989	3	—	0
9191	1	—	0
0402	2	—	0
8484	3	—	0
0305	1	—	0
0406	1	—	0
0105	3	—	0
0205	2	—	0
0012	1	1	1
0512	1	—	0
0712	2	—	0
0105	4	—	0

Different numbers of pouch young were marked for each mother, so calculations of survival statistics used weighting by sample size (number of young marked for each mother). The mean survival of marked young to weaning was $\bar{s}$ = 0.17, with sample variance

V_S = 0.067. Mortality of pouch young before marking has been estimated as ≤1% (Brown, unpublished data). Mean lifetime production of weanlings by individual female entrants to the population ($\bar{k}$) can be estimated as the product of mean number of pouch young produced in a lifetime ($\bar{y}$), by the mean survival of pouch young to weaning ($\bar{s}$), i.e., $\bar{k} = \bar{y}.\bar{s} = 0.43$. The sample variances of y and s may be combined in the manner shown in the appendix. The first method shown gives a sample variance $V_k = 5.0$. The ratio of V_k to $\bar{k}$ is 11.6. The second method uses different assumptions, and produces a similar value for V_k.

Since $\bar{k}$ is less than 2, we would expect the size of the population to be falling. A consistent downwards trend is not evident (Brown 1989) so the low value of $\bar{k}$ is presumably due to incompleteness of the data set. The calculations below will be performed using the assumption that $\bar{k} = 2$, but the ratio $V_k/\bar{k}$ is 11.6, as calculated above.

We can now determine the effective size of the population of each sex, using the formula of Crow and Kimura (1972) for the variance effective population size: Effective size of male population,

$$N_{em} = \frac{(2.N_m - \bar{k}/2)}{(1 + V_k)}$$

(and similarly for N_{ef}, the effective size of the female population). In this case,

$$N_{em} = 24.1 \text{ (or 24.2)}$$
$$\text{and} \quad N_{ef} = 53.7 \text{ (or 53.8)}$$

The figures in brackets are the values if $\bar{k} = 0.43$, and were not used in further calculations.

Sex ratio

Since the sex ratio in entrants was significantly different from unity (Table 3), the effective size, in 1983–85, was calculated using the formula of Crow and Kimura (1970):

$$N_e = 4/(1/N_{em} + 1/n_{ef}) = 66.6$$

Temporal fluctuations

Short-term fluctuations in census size are evident in the data of Brown (1989), but consistent trapping techniques were not used for a long enough period to provide accurate data on fluctuations in numbers between generations. Fluctuation in N_e between generations could considerably lower the apparent effective size over many generations.

DISCUSSION

Problems and further demographic research

In some cases, a suitable goal may be the correction for only one of the many factors that affect N_e (e.g., Greenwood 1976). However, for useful applications to practical problems, all relevant corrections should be made, often involving the use of many rough estimates and assumptions (Bonnell and Selander 1974). This is the approach taken in the present study; assumptions and rough estimates are shown in Table 5, with their possible effect on N_e, if better estimated. Of special interest is the assumption that male life-history (T, $\bar{k}$ and V) is identical to that of females. Two pieces of information suggest that this is not so. First, the sex-ratio shows a trend of decrease in the proportion of females with increasing age (Table 3); perhaps additional data would confirm this trend. Secondly, the home range of males is much larger than that of females in Hamilton (Brown,

Table 5. Assumptions in the estimation of N_e.

Section No. in text	Assumption/estimate	Possible effect on N_e of an improved estimate
2	Dispersal	See text
3	Assortative mating	Lower
4	Ageing of unweaned animals	?
5	Area trapped at T/S/B	?
5	Area inhabited at Hamilton	?
5	Uniform density in inhabited area	Lower
5	Equilibrium of immigration and emigration	?
5	That all entrants were detected at weaning, so all new adults recorded are migrants	Raise
6	Mean family size	Raise
6	Assumtions in the combination of variances (see appendix)	?
4,5,6	Similar life histories in both sexes	Raise
8	Fluctuations between generations	Lower

unpublished data). Heinsohn (1966) showed that the home range of Tasmanian *P. gunnii* females is 2.2 ha, while males have a range of 25.7 ha. From this information, we cannot calculate male T, $\bar{k}$, or V_k, but it is possible the values would not be the same as those for the more sedentary females.

In a stable population, mean family size, counted at birth, is expected to be two or more. The mean family size of $\bar{k} = 0.43$ suggests that the population was declining during the study period. This may be true, but an important reason for the low k value is certainly that some litters would not have been detected, when a female was not trapped for several months. A full data set may give an altered $\bar{k}$ and $V_k/\bar{k}$. When it is believed that the true value of k is below the estimate, a correction can be made to the estimated value of V_k, to give the sample variance associated with the true mean value (Crow and Morton 1955). In the present case, the true mean is supposed to be higher than the estimate, so a similar correction may be necessary, but a method is not available, so it was assumed that the ratio $V_k/\bar{k}$ is not affected by this problem.

The lack of data on dispersal is a serious shortcoming. If dispersal is low, the Hamilton population may contain a number of partially isolated subpopulations, whose effective size would be low, but would be uninfluenced by the historical decline of the population unless there have been density reductions in the remaining range. On the other hand, if dispersal is very high (e.g., average lifetime dispersal of several kilometres), the potential size of a genetic "neighbourhood" containing randomly breeding individuals may be much greater than the present population. Therefore, the range contraction would have severely reduced N_e.

Further research is needed to clarify some aspects of this study. Monitoring of the census size of the population over many generations would allow us to ascertain whether the population is declining, or showing long-term fluctuations, which would affect N_e. Information on lifetime dispersal is very important, for three reasons: (i) to determine whether differential behaviour of the sexes is causing a biased estimate of the sex ratio, and therefore an incorrect N_e; (ii) to determine whether new adults at T/S/B are immigrants or "missed" offspring of resident parents, which again will affect the estimate of N_e; and (iii) to determine the size of a "neighbourhood", or area within which random breeding occurs.

At present, we estimate the census size of the Hamilton population to be approximately 633, while the variance effective size is approximately 67. According to Franklin, this effective size is inadequate for long-term conservation, while Lande and Barrowclough (1987) suggest that it is barely adequate for maintenance of genetic variation. Both opinions are based on little data and many assumptions, and both ignore many forces that may be important, such as the possible effect of gene conversion bias on the frequency of rare allelles (Lamb 1986). This approach also ignores variation between species. Some species (e.g., those with a high mutation rate, a tendency to form isolated colonies regularly, or specialized genetic systems (James 1982) may retain their "normal" level of variation with very low census and effective sizes.

Further genetic research

In view of these misgivings about both the estimate of N_e and the critical effective size for conservation, it is advisable to determine whether the Hamilton population of *P. gunnii* is actually losing genetic variation. To do this, we must assume that before the decline, the genetic variation of the Victorian population was the same as that of the Tasmanian population. Since the Tasmanian population is still widespread and dense, the variation within it may be used as a reference for comparison with the variation in the Hamilton population at present.

The present study can be used as a pilot study, to predict the sampling necessary for detection of loss of variation at Hamilton. Variation can be measured as average heterozygosity ($\bar{H}$). For a population of effective size N_e, the average heterozygosity after t generations is expected to be:

$$\bar{H}_t = \bar{H}_o (1 - 1/2N_e)^t$$

where $\bar{H}_o$ is the initial heterozygosity (Crow and Kimura 1970). The Hamilton population has been isolated from others, and at approximately its present density and size, since at least 1960 (Seebeck 1979), or possibly since 1937 (Harper 1945). Therefore, in the 22.4 or 43.0 generations to 1985, the heterozygosity of the Hamilton population is expected to have dropped by 15.5% or 28%, respectively. Other measures of variation are expected to show as great or greater losses.

To estimate the sampling necessary to detect this loss of variation we use the method of Sokal and Rohlf (1969: 247), which requires an estimate of the coefficient of variation of H in *P. gunnii*. As insufficient data are available at present, we use simulation results of Nei (1978, Table 1). We will assume $\bar{H} = 0.57$, which is a reasonable value for mammals (Ayala and Kiger 1980; Avise and Aquadro 1982;). If 20 to 50 individuals are analysed, intra-locus variance of in $\bar{H}$ will be only 2–5% of total variance (Nei 1978). The remaining component (inter-locus variance) may be minimized by sampling a sufficiently large number of loci. Using the method of Sokal and Rohlf (1969) we can calculate the number of loci that we must analyse to detect the expected drop in heterozygosity (Table 6). The number of loci (25 or more) and the sample size (20 to 50 individuals) are quite moderate, so it should be possible to determine, with reasonable confidence, whether the *P. gunnii* population in Hamilton is suffering loss of variation. We believe that a cumulation of empirical studies such as the one suggested for *P. gunnii*, and that of Easteal (1985) will eventually establish the relationships between population size and levels of genetic variability in vertebrates.

Table 6. Number of loci necessary for detection of the difference expected between heterozygosities of the Victorian and Tasmanian populations of *Perameles gunnii* (assuming 20–50 individuals are analysed). "Power" is the chance of detecting the expected difference at the 5% significance level; i.e., the chance that the study will be successful.

		Date of isolation of Hamilton population	
		1937	1960
Desired	0.55	25	77
Power	0.80	44	140

Current management recommendations

Recommendations for management measures can be drawn from this study, and the proposed further studies. In this study, one factor that had a major effect in lowering the effective population size relative to the census size was the high variation in breeding success of different parents, shown by the value of $V_k/\bar{k}$ (11.6, compared with a value of 1 expected if death of offspring was at random with respect to parentage). The factors contributing to this were variation of lifetime production of pouch young (y), which had a coefficient of variation of 2.1, and the variation of pouch young survival (s), with a coefficient of variation of 1.6. The similarity of these coefficients suggests that each makes an important contribution to the variance of k, the lifetime production of weanlings, so management of either stage of the life cycle would be beneficial. There is probably little that can be done to alter the variance of production of pouch young, but control of predators that concentrate on younger individuals (before litters disperse) may reduce V_s, and therefore increase N_e considerably, for the same census size.

ACKNOWLEDGEMENTS

Mr P. Leppard, of the Statistics Department, Adelaide University, provided the methods shown in the Appendix. Professor H. Bennett and Dr N. Murray provided valuable comments on the analysis and manuscript. The collection of data was supported by the World Wildlife Fund, and the analysis performed while the authors were employed at the Arthur Rylah Institute for Environmental Research, Fisheries and Wildlife Division, Department of Conservation, Forests and Lands, Victoria.

REFERENCES

AUSTRALIAN DEPARTMENT OF HOME AFFAIRS AND ENVIRONMENT, 1982. Towards a national conservation strategy: a discussion paper. ADMAE: Canberrra.

AVISE, J. C. AND AQUADRO, C. F., 1982. A comparative summary of genetic differences in the vertebrates: patterns and correlations. *Evolutionary Biology* **15:** 151–86.

AYALA, F. J. AND KIGER, J. A., 1980. *Modern genetics.* Benjamin/Cummings: Menlo Park, California.

BONNELL, M. L. AND SELANDER, R. K., 1974. Elephant seals: genetic variation and near extinction. *Science* **184:** 908–09.

BROWN, A. M., 1983. Conservation Genetics in Victoria. *Resources and Planning Branch Technical Report Series No.* **1.** Fisheries and Wildlife Service, Ministry for Conservation: Victoria.

BROWN, P. R., 1989. *Management plan for the conservation of the Eastern Barred Bandicoot,* Perameles gunnii, *in Victoria.* National Parks and Wildlife Division, Arthur Rylah Institute for Environmental Research, Victorian Ministry for Conservation, Forests and Lands, Technical Report Series No. 63.

CROW, J. F. AND KIMURA, M., 1970. *An introduction to population genetics theory.* Harper and Row: New York.

CROW, J. F. AND KIMURA, M., 1972. The effective number of a population with overlapping generations: a correction and further discussion. *American Journal of Human Genetics* **24:** 1–10.

CROW, J. F. AND MORTON, N. E., 1955. Measurement of gene frequency drift in small populations. *Evolution* **9:** 202–14.

DENNISTON, C., 1978. Small population size and genetic diversity: implications for endangered species. Pp. 281–89 in *Endangered birds: management techniques for preserving threatened species* ed by S. A. Temple. University of Wisconson Press: Madison.

EASTEAL, S., 1985. The ecological genetics of introduced populations of the giant toad *Bufo marinus.* II. Effective population size. *Genetics* **110:** 107–22.

EMIGH, T. H. AND POLLACK, E., 1979. Fixation probabilities and effective population numbers in diploid populations with overlapping generations. *Theoretical Population Biology* **15:** 86–107.

FALCONER, D. S., 1960. *Introduction to quantitative genetics.* Longman: London.

FRANKEL, O. H., 1970. Variation, the essence of life. Sir William Macleay memorial lecture. *Proceedings of the Linnean Society of New South Wales* **95:** 158–69.

FRANKEL, O. H. AND SOULE, M. E., 1981. *Conservation and evolution.* Cambridge University Press: Cambridge.

FRANKLIN, I. R., 1980. Evolutionary change in small populations. Pp. 135–50 in *Conservation biology* ed by M. E. Soule and B. A. Wilcox. Sinauer: Sunderland, Massachusetts.

GALE, J. S. AND LAWRENCE, M. J., 1984. The decay of variability. Pp. 77–101 in *Crop genetic resources: conservation and evaluation* ed by J. H. W. Holden and J. T. Williams. Allen & Unwin: London.

GREENWOOD, J. J. D., 1976. Effective population numbers in the snail, *Cepaea nemoralis*. *Evolution* **28:** 513–26.

HARPER, F., 1945. *Extinct and vanishing animals of the old world.* Special publication No. 12. American Committee for Wildlife Protection, New York Zoological Park: New York.

HEINSOHN, G. E., 1966. Ecology and reproduction of the Tasmanian bandicoots (*Perameles gunni* and *Isoodon obesulus*). *University of California Publications in Zoology* **80:** 1–96.

HILL, W. G., 1979. A note on effective population size with overlapping generations. *Genetics* **92:** 317–22.

IUCN (International Union for the Conservation of Nature and Natural Resources), 1980. *World conservation strategy.* IUCN: Geneva.

JAMES, S. H., 1982. The relevance of genetic systems in *Isotoma petraea* to conservation practice. Pp. 63–71 in *Species at risk: research in Australia* ed by R. H. Groves and W. D. L. Ride. Australian Academy of Science: Canberra.

LAMB, B. C., 1986. Gene conversion disparity: factors influencing its direction and extent, with tests of assumptions and predictions in its evolutionary effects. *Genetics* **114:** 611–32.

LANDE, R. AND BARROWCLOUGH, G. F., 1987. Effective population size, genetic variation, and their use in population management. Pp. 87–123 in *Viable populations for conservation* ed by M. E. Soule. Cambridge University Press: Cambridge.

NEI, M., 1978. Genetic distance from a small number of individuals. *Genetics* **76:** 379–90.

NEI, M. AND IMAIZUMI, Y., 1966. Genetic structure of human populations. II. Differentiation of blood group gene frequencies among isolated populations. *Heredity* **21:** 183–90, 344.

SCHONEWALD-COX, C. M., CHAMBERS, S. M., MACBRYDE, B. AND THOMAS, L., 1983. *Genetics and conservation.* Benjamin/Cummings: California.

SEEBECK, J. H., 1979. Status of the Barred Bandicoot, *Perameles gunnii,* in Victoria: with a note on husbandry of a captive colony. *Australian Wildlife Research* **6:** 255–64.

SOKAL, R. R. AND ROHLF, F. J., 1969. *Biometry.* Freeman: San Francisco.

TYNDALE-BISCOE, C. H. AND CALABY, J. H., 1975. Eucalypt forests as refuges for wildlife. *Australian Forestry* **38:** 117–33.

APPENDIX

Methods for the estimation of the sample variance of k, the lifetime production of young.

Method 1

Assumptions: (i) that large sample theory can be used;
(ii) that the covariance of y and s is zero.

Method:

V_k = (second moment of y).V_s + (second moment of s).V_y
= 33.562 × 0.06686 + 0.09464 × 28.7958
= 4.9692

Calculation of the second moment of s used weighting by sample size.

Method 2

Assumption: that mean lifetime production of weanlings is equal to the product of mean production of pouch young, by mean survival of pouch young to weaning; i.e., $\bar{k} = \bar{y}, \bar{s}$, in terms of sample statistics. This may not be correct, if the covariance is non-zero; certainly the value of k obtained by this method appears to be incorrect (see main text).

Method:

V_k =(second moment of y).(second moment of s) − $\bar{y}^2.\bar{s}^2$
= 33.5625 × 0.09464 + 2.5625^2 × 0.1667^2
= 2.9938

This value would be smaller if it was assumed that $\bar{k}$ = 2.

Key to the Bandicoots and Bilbies of Australasia

This key is designed to identify adult animals. Juveniles may differ in coloration or markings.

1A Tail crested, pes with 4 digits 2

B Tail uncrested, pes with 5 digits 4

2A Manus with 5 digits 3

B Manus with 3 digits (only 2 functional)
Chaeropus ecaudatus (p. 381)

3A Tail proximally dark
Macrotis lagotis (p. 381)

B Tail wholly white
Macrotis leucura (p. 382)

4A Ears long and pointed (see drawing) .. 5

B Ears short and rounded (see drawing) .. 8

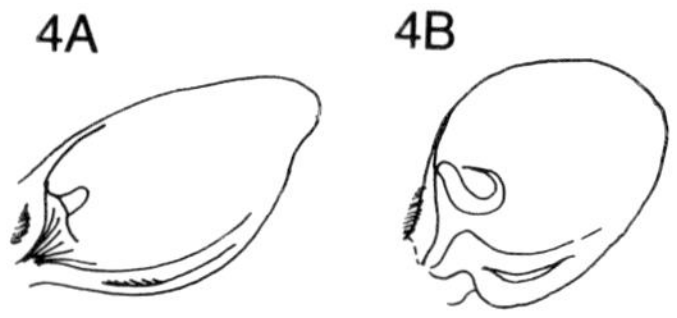

5A Fur orange, especially on shoulders
Perameles eremiana (p. 379)

B Fur not as above 6

6A Body prominently barred, especially rump 7

B Body unbarred
Perameles nasuta (p. 380)

7A Syndactylous toes arise posterior to 5th digit (see drawing); tail dorsally dark whole length
Perameles bougainville (p. 379)

B Syndactylous toes arise at same level as 5th digit (see drawing); tail dorsally dark at proximal end only
Perameles gunnii (p. 379)

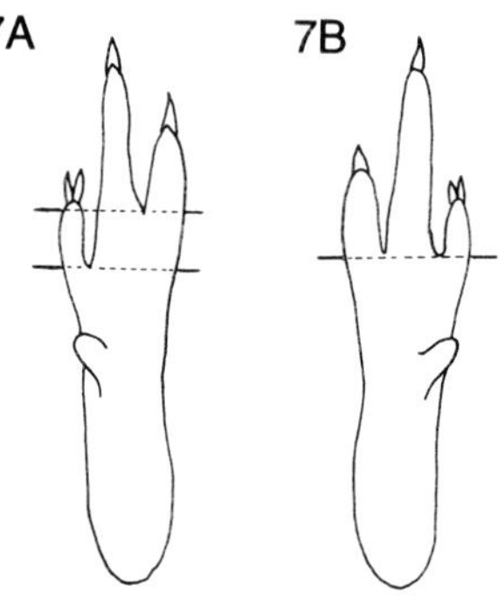

8A 4 pairs of upper incisors 9

B 5 pairs of upper incisors 10

9A Fur soft; restricted to the island of Ceram
Rhynchomeles prattorum (p. 386)

B Fur exceedingly spinous 17

10A Soles of feet wrinkled; large pear-shaped bullae 11

B Soles of feet granulated; small hemispherical bullae 13

11A Fur colour fawn/golden
Isoodon auratus (p. 380)

B Fur colour dark brown 12

12A Marked accessory vacuities in palate
Isoodon obesulus (p. 381)

B No accessory vacuities in palate
Isoodon macrourus (p. 380)

13A Dark soft fur, small size (<300 mm total length)
Microperoryctes murina (p. 383)

B Fur not soft, medium to large size (>300 mm total length) 14

14A Tail with transverse scaly tubercles on ventral surface; very large (>700 mm total length)
Peroryctes broadbenti (p. 382)

B Tail not as above (<700 mm total length) 15

15A Dorsal stripe and white tail tip present .. 16

B Dorsal stripe and white tail tip absent
Peroryctes raffrayanus (p. 382)

16A Hind foot >50 mm; head body >250 mm
Microperoryctes longicauda (p. 383)

B Hind foot <45 mm; head body <200 mm
Microperoryctes papuensis (p. 383)

17A Hallucal pad present on hind foot .. 18

B No hallucal pad present on hind foot .. 19

18A Pelage spines narrow; ♂ canine and P_3 enormous
Echymipera clara (p. 384)

B Pelage spines flattened; P_3 very narrow
Echymipera rufescens (p. 386)

19A Black crown sharply delineated from pale fur below eye; last third of guard hairs pale
Echymipera kalubu (p. 385)

B Crown not sharply delineated; last fifth of guard hairs pale
Echymipera davidi (p. 384)

The Bandicoot and Bilby Species

This list of species of living bandicoots from the Australian and New Guinean regions includes those presumed extinct (*) and those recorded only as subfossils (†).

FAMILY PERAMELIDAE

Perameles bougainville (Western Barred Bandicoot)
- *P. b. bougainville* — Shark Bay, WA
- *P. b. fasciata* — Vic*, NSW*
- *P. b. myosura* — WA (SW)*
- *P. b. notina* — WA (SE)*, SA*

Perameles eremiana (Desert Bandicoot) — WA*, NT*, SA*

Perameles gunnii (Eastern Barred Bandicoot) — Vic, Tas, Bass Strait islands†, SA*

Perameles nasuta (Long-nosed Bandicoot)
- *P. n. nasuta* — Vic, NSW, Qld (to Townsville)
- *P. n. pallescens* — Qld (Townsville to Ravenshoe)

Isoodon auratus (Golden Bandicoot)
- *I. a. auratus* — WA, NT*, SA*
- *I. a. barrowensis* — Barrow Island, WA

Isoodon macrourus (Northern Brown Bandicoot)
- *I. m. macrourus* — WA, NT
- *I. m. torosus* — NSW, Qld
- *I. m. moresbyensis* — PNG, West Irian

Isoodon obesulus (Southern Brown Bandicoot)
- *I. o. obesulus* — SA, Vic, NSW
- *I. o. affinus* — Tas
- *I. o. fusciventer* — WA (SW)
- *I. o. nauticus* — Nuyts Archipelago, SA
- *I. o. peninsulae* — Cape York, Qld

Chaeropus ecaudatus (Pig-footed Bandicoot) — WA*, NT*, SA*, Vic*, NSW*

Macrotis lagotis (Greater Bilby)
- *M. l. lagotis* — WA
- *M. l. sagitta* — WA, NT, Qld
- *M. l. cambrica* — NSW*
- *M. l. nigripes* — SA*
- *M. l. grandis* — SA*
- *M. l. interjecta* — WA (SW)*

Macrotis leucura (Lesser Bilby) — WA*, SA*, NT*, Qld*

FAMILY PERORYCTIDAE

Peroryctes broadbenti (Giant Bandicoot) — PNG

Peroryctes raffayanus (Raffray's Bandicoot)
- *P. r. raffrayanus* — PNG, West Irian
- *P. r. rothschildi* — PNG

Microperoryctes longicauda (Striped Bandicoot)
- *M. l. longicauda* — West Irian
- *M. l. ornatus* — PNG
- *M. l. dorsalis* — West Irian, PNG
- *M. l. magnus* — PNG

Microperoryctes murina (Mouse Bandicoot) — West Irian

Microperoryctes papuensis (Papuan Bandicoot) — PNG

Echymipera clara (Large-toothed Bandicoot) — West Irian

Echymipera davidi sp. nov. *(Kiriwina Bandicoot)* — *Kiriwina Island, Trobriand Group, PNG*

Echymipera kalubu (Spiny Bandicoot)
- *E. k. kalubu* — PNG, West Irian
- *E. k. oriomo* — PNG, West Irian
- *E. k. philipi* — Owi Island, Schouten Group, West Irian

Echymipera rufescens (Rufous Spiny Bandicoot)
- *E. r. rufescens* — PNG, West Irian
- *E. r. australis* — Cape York, Qld

Rhynchomeles prattorum (Ceram Island Bandicoot) — Ceram Island, West Irian

Species Accounts

(Hatched areas = former distribution; solid areas = present distribution.)

Perameles bougainville: Western Barred Bandicoot

Size:	weight 190–250 g; head-body length 200–300 mm.
Colour:	brownish-grey dorsum with darker bars on hindquarters, white venter.
Behaviour:	nocturnal; one or two animals occupy grass nest by day.
Field characteristics:	arid habitat and rump stripes good distinguishing features.
Habitat:	once widespread in arid southern and western Australia, where it inhabited dense shrubs and particularly favoured dense stands of *Casuarina*, but also occurred in open bluebush and saltbush plains; now found only on Bernier and Dorre Islands, Western Australia, where it is found in sandhills behind beaches.
Similar species:	*P. gunnii*, from which it differs in its smaller size and white venter.

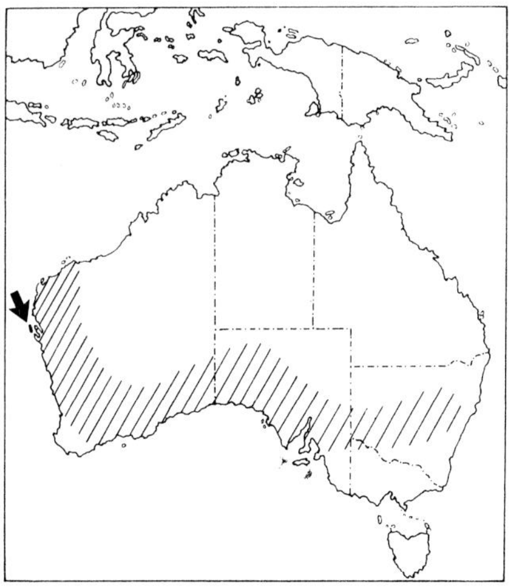

Perameles bougainville — now only found on Bernier and Dorre Islands, Shark Bay, Western Australia (arrowed).

Perameles eremiana: Desert Bandicoot

Size:	weight unknown; head-body length 235–275 mm.
Colour:	orange dorsally, paler ventrally; dark mid-dorsal stripe extending in two dark bands on either side of rump; tail bicoloured, dark dorsally.
Behaviour:	unknown, but nocturnal.
Field characteristics:	orange dorsum unique; hairy feet.
Habitat:	sandplain, sandridge desert and spinifex grassland.
Similar species:	none.

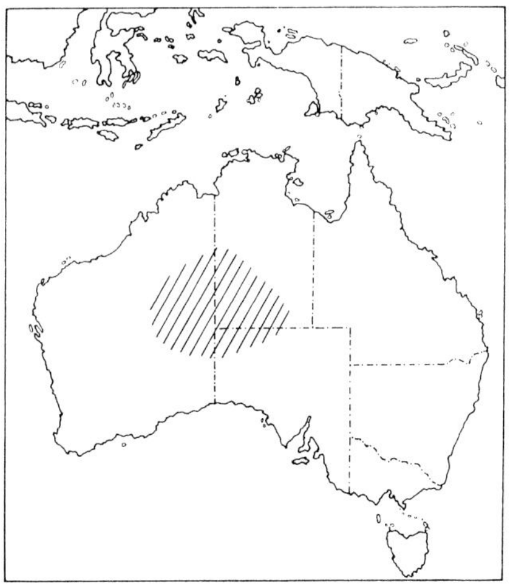

Perameles eremiana — extinct.

Perameles gunnii: Eastern Barred Bandicoot

Size:	weight 450–900 g; head-body length 270–350 mm.
Colour:	yellow-brown dorsally with dark rump stripes, slate grey venter.
Behaviour:	prefers open grasslands and flourishes near areas of pastoral development; emerges from grass nest at dusk and forages throughout night.
Field characteristics:	only extant bandicoot in eastern Australia with dark rump stripes.
Habitat:	grasslands and open woodlands in Tasmania and western Victoria.
Similar species:	*P. bougainville*, from which it differs in larger size and grey venter.

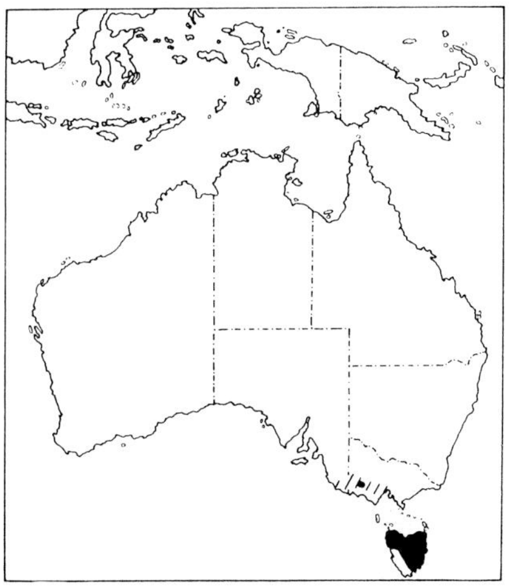

Perameles gunnii — now only found in one mainland locality (Hamilton, Victoria) and Tasmania.

***Perameles nasuta:* Long-nosed Bandicoot**

Size: weight 500–1 900 g; head-body length 310–425 mm.

Colour: drab grey-brown dorsum, white venter.

Behaviour: by day nests in a shallow hole lined with grass and leaves; by night forages, sometimes in suburban habitats, conical excavations in such environments distinctive.

Field characteristics: long, pointed ears and muzzle distinguish it from *Isoodon* species.

Habitat: wet and dry forests in eastern Australia; also suburbs of Sydney and Brisbane.

Similar species: none.

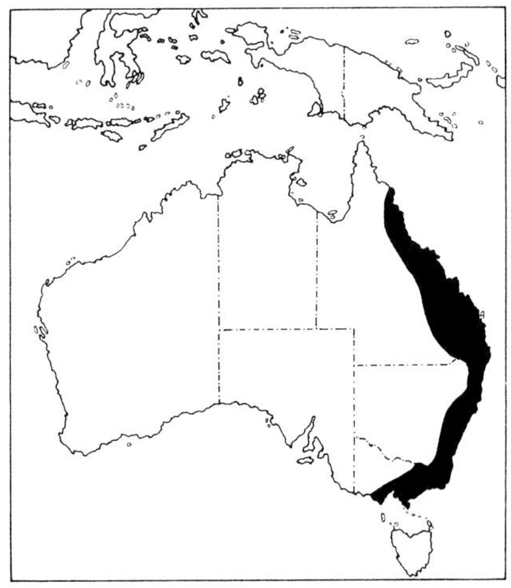

Perameles nasuta.

***Isoodon auratus:* Golden Bandicoot**

Size: weight 267–670 g; head-body length 210–295 mm.

Colour: dorsum golden-brown pencilled with black, venter white.

Behaviour: inhabits semi-arid portions of tropical Australia. Nocturnal and partially carnivorous; aggressive towards conspecifics.

Field characteristics: golden colour and very coarse, hispid fur good field characteristics

Habitat: sand dunes, spinifex, sandplains with *Acacia* and eucalypts, deciduous vine thickets.

Similar species: none.

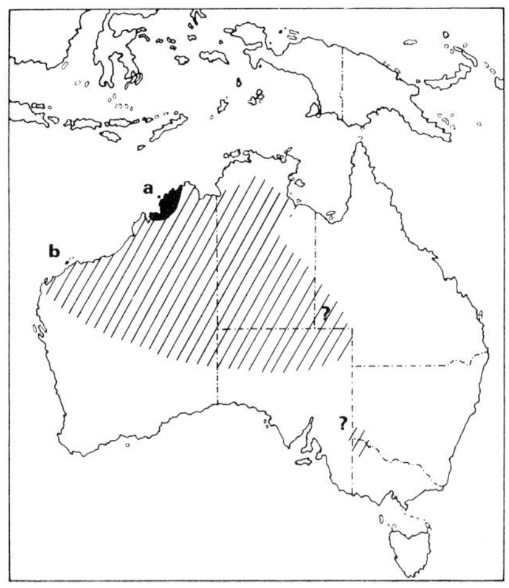

Isoodon auratus a. *I. a. auratus*
b. *I. a. barrowensis*

***Isoodon macrourus:* Northern Brown Bandicoot**

Size: weight 500–3 100 g; head-body length 300–470 mm.

Colour: brown-black above, white below.

Behaviour: day spent in a heap of ground litter over slight depression or in hollow logs; forages for invertebrates and plant matter by night.

Field characteristics: similar in appearance but larger than *I. obesulus*; grizzled appearance of fur, short, rounded ears, shortish nose and mesic environment useful distinguishing features.

Habitat: areas with low ground cover and tall grass in northern and eastern Australia; woodlands in southern New Guinea.

Similar species: *I. obesulus*, from which it differs in being larger.

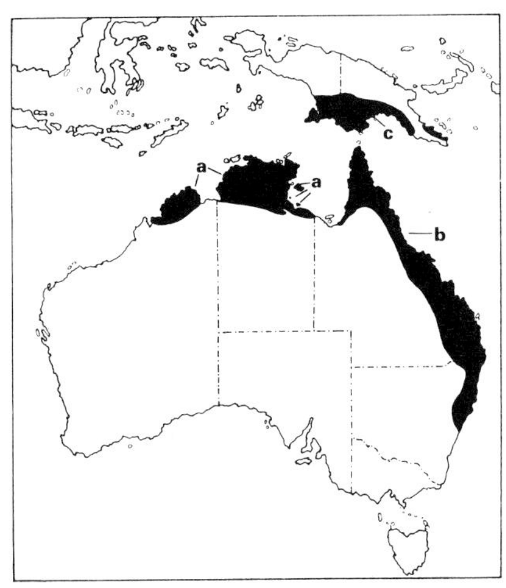

Isoodon macrourus a. *I. m. macrourus*
b. *I. m. torosus*
c. *I. m. moresbyensis*

Isoodon obesulus: Southern Brown Bandicoot

Size: weight 400–1 600 g; head-body length 280–360 mm.

Colour: black-brown dorsum, whitish venter.

Behaviour: prefers scrubby habitats with low groundcover; by day uses nest of plant material, by night forages for invertebrates.

Field characteristics: similar to *I. macrourus,* but in southern Australia and Cape York Peninsula only; and smaller.

Habitat: heath and heathy forest in southern Australia.

Similar species: *I. macrourus* from which it differs in size.

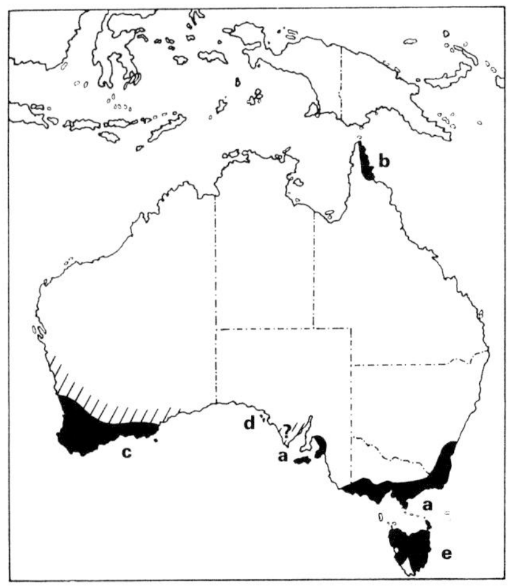

Isoodon obesulus
a. *I. o. obesulus*
b. *I. o. peninsulae*
c. *I. o. fusciventer*
d. *I. o. nauticus*
e. *I. o. affinus*

Chaeropus ecaudatus: Pig-footed Bandicoot

Size: weight unknown; head-body length 230–260 mm.

Colour: orange-brown above, fawn below, tail-tip black.

Behaviour: solitary and uncommon even before extinction; rested in shallow, covered squat.

Field characteristics: presence of only two functional digits on forefoot and single functional digit on hindfoot unique.

Habitat: now extinct; previously plains and woodland with shrub understorey.

Similar species: none.

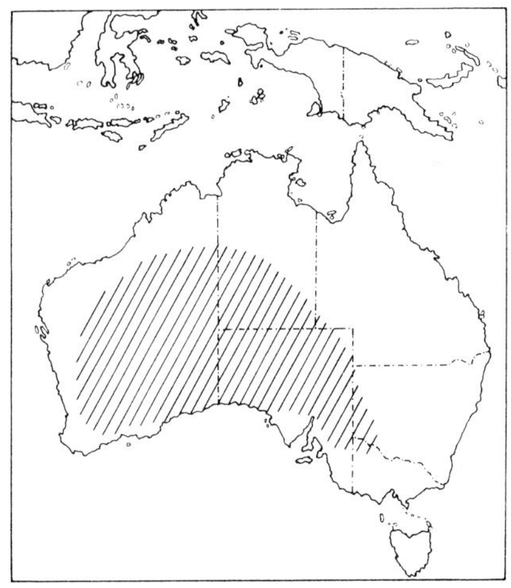

Chaeropus ecaudatus — extinct.

Macrotis lagotis: Greater Bilby

Size: weight 800–2 500 g; head-body length 290–550 mm.

Colour: bluish-grey and brownish dorsum, white venter; tail black with white tip.

Behaviour: nocturnal, burrowing species of the arid interior; burrows often found on sand dunes; cantering gait; quiet, unaggressive when handled.

Field characteristics: black, crested tail with white tip plus strong, burrowing claws a unique combination; ears extremely large, rabbit-like.

Habitat: arid zone; once very widespread, including western slopes of dividing range; former habitat savannah woodland and shrub grassland communities on sands or loams with little stone content; now only in north and centre, where inhabits hummock grasslands and *Acacia* shrublands with spinifex or tussock grass.

Similar species: *M. leucura,* from which it differs in larger size, pure white belly fur and black tail base.

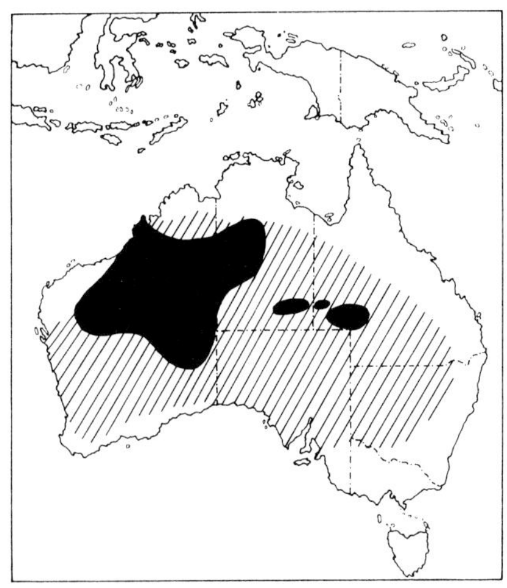

Macrotis lagotis — subspecies not distinguished on map.

Macrotis leucura: Lesser Bilby

Size: weight 360–435 g; head-body length 200–270 mm.

Colour: grey-brown above, white-tipped grey-based fur on venter; tail white dorsally.

Behaviour: burrowed in sandhills; nocturnal; aggressive when handled.

Field characteristics: crested tail with all white dorsum unique.

Habitat: now extinct, but previously sandhills in central Australia.

Similar species: *M. lagotis,* from which it differed in its smaller size, all-white tail and grey-based belly fur.

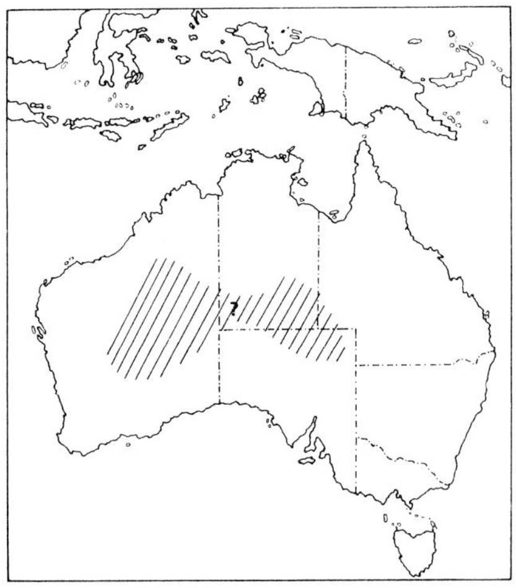

Macrotis leucura — extinct.

Peroryctes broadbenti: Giant Bandicoot

Size: weight c. 4 800 g; head-body length 595 mm.

Colour: tan, with markedly lighter sides and venter; tail uniformly dark.

Behaviour: despite its size a nervous species capable of surprising bursts of speed.

Field characteristics: largest of all bandicoots; probably only to be seen in lowlands rainforest.

Habitat: primary lowland rainforest in eastern New Guinea; possibly only below 1 200 m.

Similar species: only the smaller, darker *P. raffrayanus*.

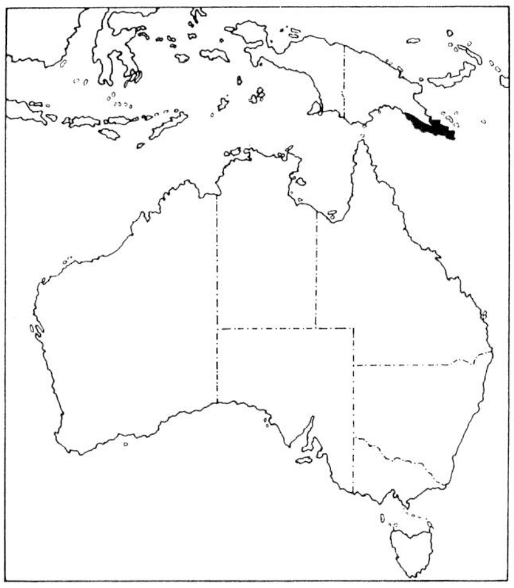

Peroryctes broadbenti.

Peroryctes raffrayanus: Raffray's Bandicoot

Size: weight 650–1 000 g; head-body length 275–372 mm.

Colour: unpatterned brown above, yellowish-white below.

Behaviour: most often encountered in late evening near fallen fruit in mid-montane rainforest; may nest communally.

Field characteristics: medium-sized bandicoot with long unicoloured tail and lacking flattened spines in fur; uncommon, only likely to be seen in primary forest well after dusk, during spotlighting.

Habitat: between 60 m and 4 000 m, but most common at c. 1 200 m in primary rainforest along the New Guinean central cordillera.

Similar species: only the much larger and more brightly coloured *P. broadbenti*.

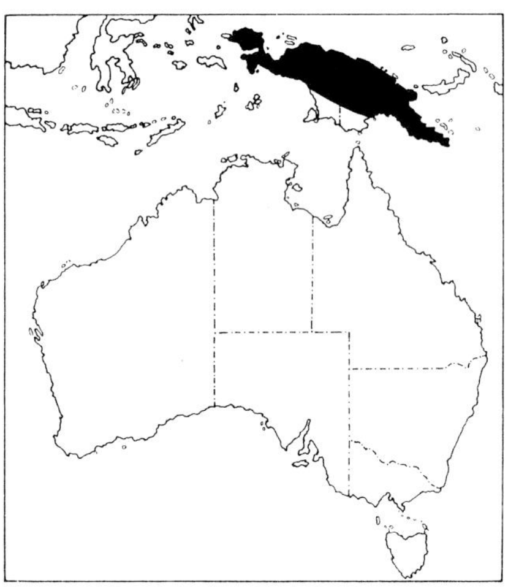

Peroryctes raffrayanus — subspecies not distinguished on map.

Microperoryctes longicauda: Striped Bandicoot

Size:	weight c. 400 g; head-body length 239–303 mm.
Colour:	tan, with stripes of varying distinctness; least distinct in western populations and most distinct in east.
Behaviour:	nests of leaves and twigs often located among roots of trees.
Field characteristics:	medium-sized; is the most common bandicoot species encountered at altitudes of over 2 000 m. Commonly encountered in upper montane rainforest (2 000–3 950 m), but can occur as low as 1 000 m in both secondary growth and primary forest. Bright coloration and white tail tip useful diagnostic features.
Habitat:	most forest types on the central cordillera above 1 000 m, also subalpine grasslands.
Similar species:	only the much smaller *M. papuensis*.

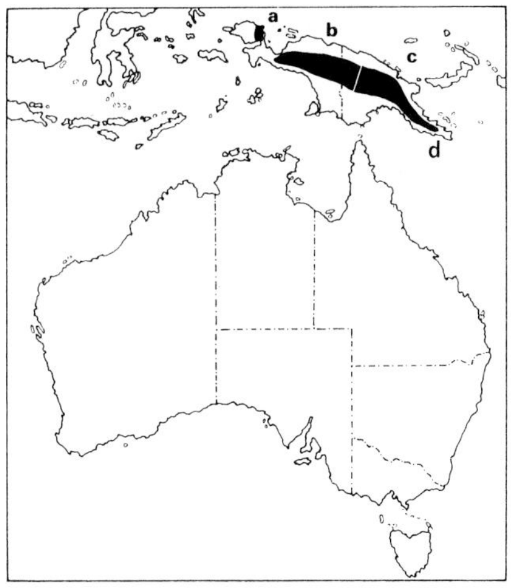

Microperoryctes longicauda
a. *M. l. longicauda*
b. *M. l. dorsalis*
c. *M. l. ornatus*
d. *M. l. magnus*

Microperoryctes murina: Mouse Bandicoot

Size:	weight unknown; head-body length 174–265 mm.
Colour:	dark brown (some with dorsal stripe), paler beneath.
Behaviour:	unknown.
Field characteristics:	very small; the very soft fur is unique among New Guinean species.
Habitat:	mid-montane rainforest in far western New Guinea.
Similar species:	none.

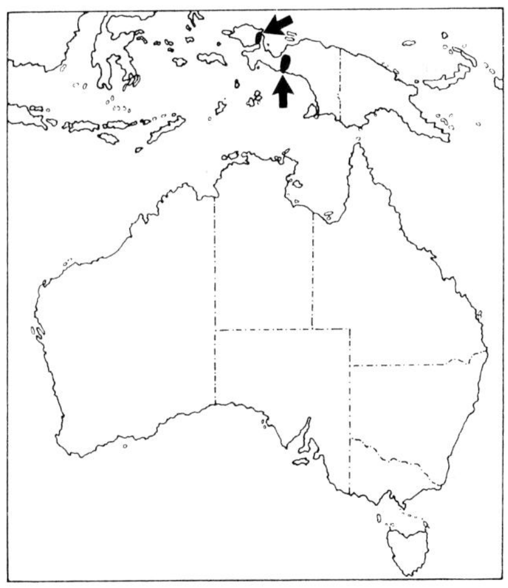

Microperoryctes murina — two separate populations known (arrowed).

Microperoryctes papuensis: Papuan Bandicoot

Size:	weight unknown; head-body length 175–200 mm.
Colour:	tan with clearly marked dorsal, flank and head stripes.
Behaviour:	unknown.
Field characteristics:	bright striping combined with small size.
Habitat:	mid-montane rainforest in far eastern New Guinea; possibly in both disturbed habitats and primary forest.
Similar species:	only the larger *M. longicauda*.

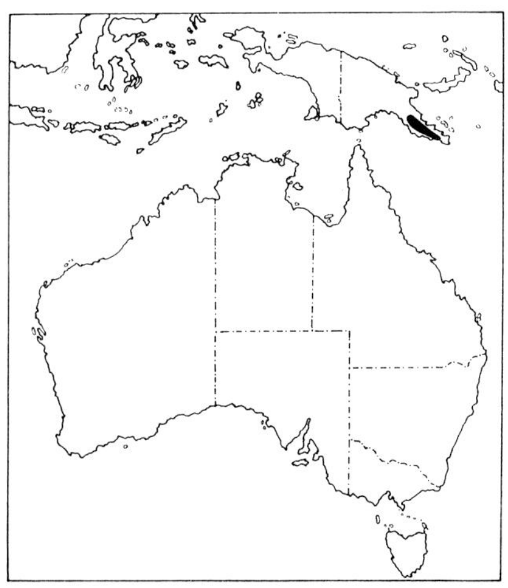

Microperoryctes papuensis.

Echymipera clara: Large-toothed Bandicoot

Size: weight 825–1 700 g; head-body length 270–410 mm.

Colour: brown dorsally; some males are paler; venter salmon or whitish.

Behaviour: strictly nocturnal and usually confined to primary forest; a shy, uncommon species.

Field characteristics: prominent black foot-pad, large head and canines of males, and narrow pelage spines are useful diagnostic features; its occurrence almost always in primary foothill forest also useful.

Habitat: primary forest between 300 and 1 200 m, north of the New Guinean central cordillera.

Similar species: all other *Echymipera* species, from which it differs in its narrow pelage spines.

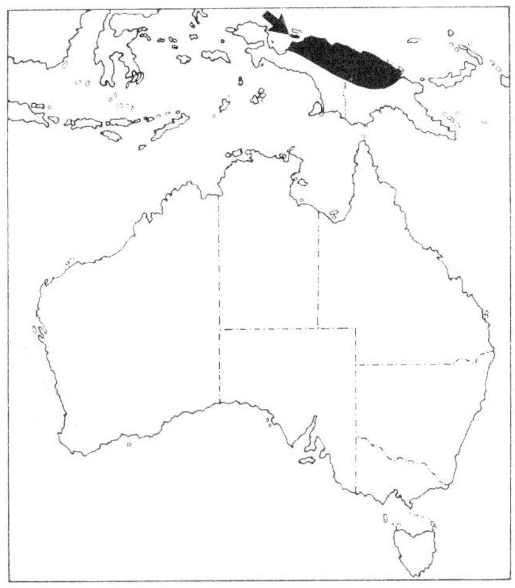

Echymipera clara — mainland West Irian, PNG and Japen Island (arrowed).

Echymipera davidi: Kiriwina Bandicoot

Size: weight 950 g; head-body length 308–397 mm.

Colour: warm brown dorsally and ventrally.

Behaviour: unknown.

Field characteristics: only bandicoot occurring on Kiriwina Island; overall brownish tones and lack of black foot-pad are good diagnostic features.

Habitat: Kiriwina Island off northeastern New Guinea. Nearly all of Kiriwina Island is gardens, grassland or regrowth, so *E. davidi* must tolerate these conditions.

Similar species: all *Echymipera* species, from which it differs in its more brownish overall colour and (except *E. kalubu*) in lacking a black foot-pad.

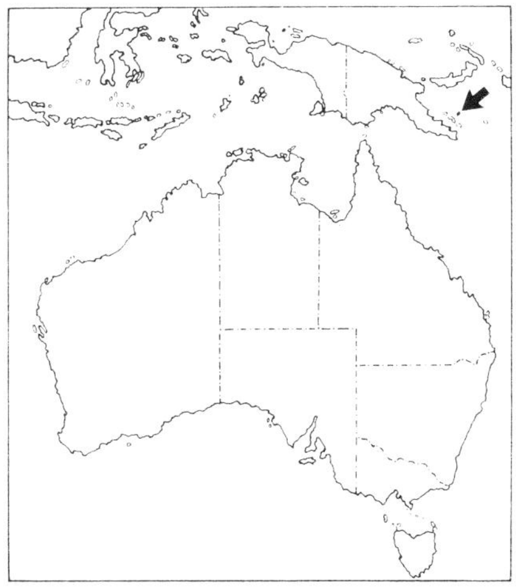

Echymipera davidi — known only from Kiriwina Island (arrowed).

Echymipera kalubu: Spiny Bandicoot

Size: weight 450–1 500 g; head-body length 225–380 mm.

Colour: brownish with black crown and white venter.

Behaviour: haunts gardens and fruiting trees by evening and night, rests by day in nest of grass or leaves. Leaves characteristic pits and scrapings in ground where it has been feeding.

Field characteristics: often seen in gardens and lowland rainforest by evening, but above 1 000 m mainly in very disturbed habitats; combination of crepuscular habits and tolerance of disturbed habitats in lowland typical only of this and *E. rufescens* on mainland New Guinea; most commonly encountered New Guinean bandicoot.

Habitat: lowland New Guinea; below 1 000 m in almost all habitats except savannah, but common in rainforest and secondary growth. Between 1 000 m and 2 000 m mainly in grossly disturbed habitats (e.g., grassland valleys).

Similar species: both *E. clara* and *E. rufescens* differ in possessing distinct black foot-pad and lacking black crown; *E. davidi* lacks black crown and is more uniform brown dorsally and ventrally. Tail shorter than in *Peroryctes* species.

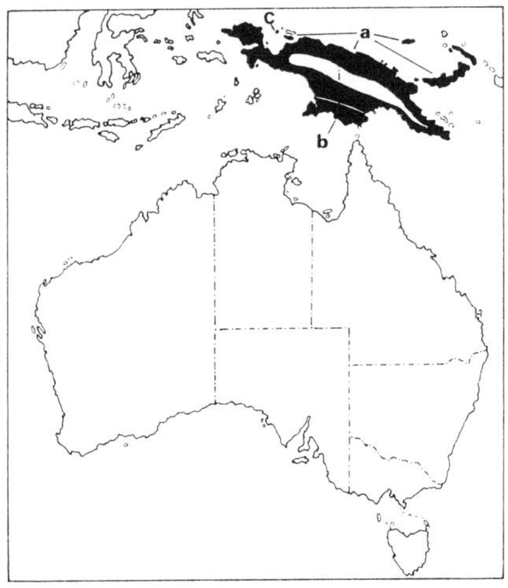

Echymipera kalubu a. *E. k. kalubu*
b. *E. k. oriomo*
c. *E. k. philipi*

Echymipera rufescens: Rufous Spiny Bandicoot

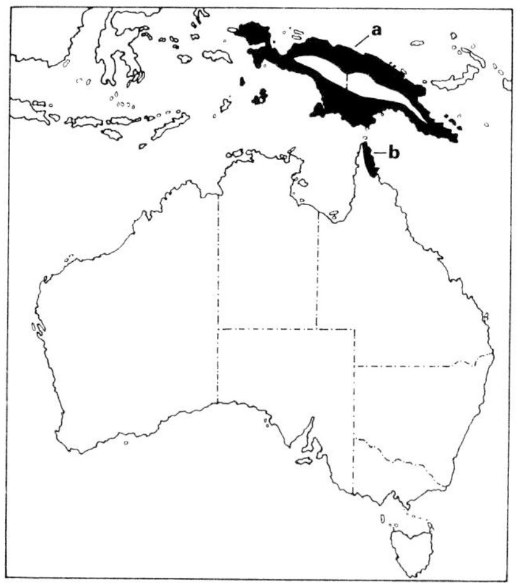

Echymipera rufescens a. *E. r. rufescens*
b. *E. r. australis*

E. r. rufescens

Size: weight unknown; head-body length 410 mm.

Colour: brownish-red dorsally, white ventrally.

Behaviour: may use burrows rather than nests to rest by day; crepuscular.

Field characteristics: only bandicoot likely to be found in very grassy habitats (e.g., airstrips).

Habitat: found only below 1 200 m, and seems to be able to tolerate more savannah-like habitats than *E. kalubu;* lowland New Guinea; disturbed areas and forest.

Similar species: can be difficult to distinguish from *E. clara,* but pelage spines are broader, and male *E. rufescens* have smaller canines; externally somewhat similar to *E. davidi* and *E. kalubu* but easily distinguished by possessing black foot-pad.

E. r. australis

Size: 500–2 000 g; head-body length 300–400 mm.

Colour: blackish above on muzzle and crown; short, almost naked, black tail.

Behaviour: nocturnal.

Field characteristics: short black tail, spiny fur, long blunt snout; most common rainforest bandicoot on Cape York.
Note: possesses only four pairs of upper incisors.

Similar species: none.

Rhynchomeles prattorum: Ceram Island Bandicoot

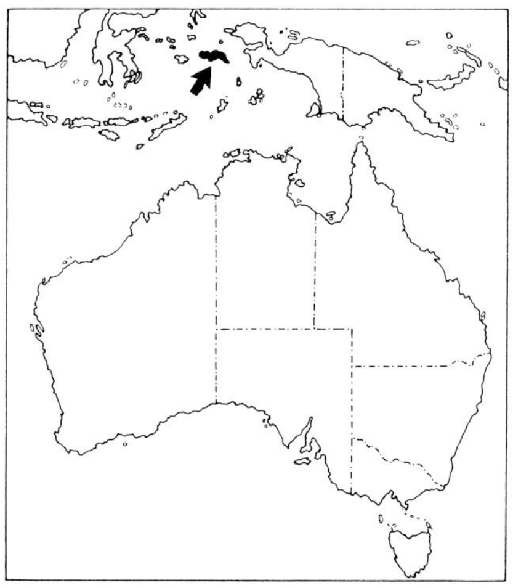

Rhynchomeles prattorum — known only from Ceram Island (arrowed).

Size: weight unknown; head-body length 320 mm.

Colour: dark chocolate brown with white patch on chest.

Behaviour: unknown.

Field characteristics: only bandicoot species known from Ceram.

Habitat: forests at 1 800 m on Mt Mansuela, Ceram.

Similar species: none.

INDEX

Bandicoots are comprehensively indexed. Other species are indexed to page number only.